Springer Monographs in Mathematics

This series publishes advanced monographs giving well-written presentations of the "state-of-the-art" in fields of mathematical research that have acquired the maturity needed for such a treatment. They are sufficiently self-contained to be accessible to more than just the intimate specialists of the subject, and sufficiently comprehensive to remain valuable references for many years. Besides the current state of knowledge in its field, an SMM volume should ideally describe its relevance to and interaction with neighbouring fields of mathematics, and give pointers to future directions of research.

More information about this series at http://www.springer.com/series/3733

Gert-Martin Greuel · Christoph Lossen
Eugenii Shustin

Singular Algebraic Curves

With an Appendix by Oleg Viro

Springer

Gert-Martin Greuel
Fachbereich Mathematik
TU Kaiserslautern
Kaiserslautern, Germany

Eugenii Shustin
School of Mathematical Sciences
Tel Aviv University
Tel Aviv, Israel

Christoph Lossen
Fachbereich Mathematik
TU Kaiserslautern
Kaiserslautern, Germany

ISSN 1439-7382 ISSN 2196-9922 (electronic)
Springer Monographs in Mathematics
ISBN 978-3-030-03349-1 ISBN 978-3-030-03350-7 (eBook)
https://doi.org/10.1007/978-3-030-03350-7

Library of Congress Control Number: 2018960191

Mathematics Subject Classification (2010): 14-02, 14Hxx, 14H10, 14H20, 14F17, 14B07, 14M25, 14J60, 14P25, 14Q05

This Springer imprint is published by the registered company Springer Nature Switzerland AG
The registered company address is: Gewerbestrasse 11, 6330 Cham, Switzerland

Für Ursula

Gert-Martin Greuel

Für Carmen, Katrin, Carolin und Benjamin

Christoph Lossen

To Mila and Boris

Eugenii Shustin

Preface

The title *Singular Algebraic Curves* designates a wide research field which has been of constant interest and importance from Descartes, Pascal, Newton to nowadays. We do not pursue the "mission impossible" to give a complete account of this topic, but concentrate mainly on the geometry of deformations and families of singular algebraic curves on algebraic surfaces defined over the complex or real field. More precisely, this book presents in detail the theory of zero-dimensional schemes related to planar curve singularities, and the cohomology vanishing theory for ideal sheaves of zero-dimensional schemes, which builds a bridge between the local and global geometry of singular algebraic curves. Moreover, the cohomological approach combines classical and modern results on the geometry of deformations and families of singular algebraic curves, on applications and computational aspects.

The problems we address include such basic and natural questions as

- What kind of singularities and how many of them can be on an algebraic curve of a given degree in the plane or in a given linear system on an algebraic surface?
- Which simultaneous deformations of singular points of a curve are possible if the curve varies in a given linear system?
- What are the geometric properties of equisingular families of curves, e.g., dimension, smoothness of "expected dimension," irreducibility?

We refer to these questions as *existence problem, deformation problem, T-smoothness problem*, and *irreducibility problem*, respectively. The T-smoothness property (i.e., smooth of expected dimension), introduced in this monograph, means that the singularities impose independent conditions in a rather strong sense on the equisingular family. It plays a central role in our approach to answer the above questions.

These questions appeared to be rather hard. After Severi's fundamental theorem, which claims that the family of plane curves of a given degree with a given number of nodes is smooth of "expected dimension," and the nodes can be independently smoothed out, not much progress, except for a very few special cases, has been done till the middle of the 1980s. However, the general development of cohomology of sheaves, deformation theory, local singularity theory, and, especially equisingularity theory, in this period (just to mention Zariski, Schlessinger, Wahl, Arnol'd and his school) played a crucial role in the further study. Harris' theorem (1985) on the irreducibility of families of plane irreducible curves of given degree and genus was a major breakthrough, though the method, based on moduli of curves and very restrictive deformation properties, applies, in fact, only in this situation.

The main idea of the approach we focus on, which can be traced back to Severi's original work, is to reformulate the above questions as H^1-vanishing conditions for the ideal sheaves of certain zero-dimensional schemes and apply H^1-vanishing criteria. The key point here is a choice and an appropriate use of the cohomology vanishing theory, which is a nontrivial and delicate task. The main advantage is that this approach provides a uniform point of view on all the problems stated, it applies to curves with arbitrary singularities and on various surfaces, and, most important, it answers the above questions in the form of *relations for numerical invariants* of curves and singularities, which cover an *asymptotically proper* range of possible values of invariants. We especially stress on the asymptotic properness of the results, because pathological properties of singular curves and their families have been observed already in the case next to nodal curves, plane curves with nodes and cusps, and this, actually, does not allow one to get complete answers for the non-nodal case. The numerical relations obtained are *asymptotically close* to the border between the regular and pathological properties of singular algebraic curves. We illustrate this just in two examples:

- The first example concerns the existence of plane curves with prescribed singularities. Given analytic or topological singularity types $S_1, \ldots, S_r$ satisfying the inequality $\mu(S_1) + \cdots + \mu(S_r) \leq \frac{1}{9}(d^2 - 2d + 3)$, we prove that then there exists an irreducible plane curve of degree d with exactly r singular points of the given types, respectively. This sufficient condition is quadratic in d and differs from the classically known necessary one, $\mu(S_1) + \cdots + \mu(S_r) \leq (d-1)^2$, only by a constant factor of d^2 in the right-hand side.
- The other example concerns the geometry of families of plane curves with nodes and cusps. Classically known is that the family of plane curves of degree d with n nodes and k cusps is smooth of expected dimension, if $k < 3d$. This is a rather restrictive condition, since asymptotically, k can be of order d^2. We prove that the inequality $4n + 9k \leq (d+3)^2$ is sufficient for the smoothness and expected dimension of the family. Moreover, the coefficients of n and k in the latter inequality are sharp for $k \geq 3$; i.e., there exist series of obstructed families with $9k = d^2 + O(d)$, or $k = 3d$ and $4n = d^2 + O(d)$.

We pay also attention to other approaches in the treatment of the problems stated, but often do not give details. Some of them are already presented in recent monographs. But the main reason for our selection is that it allows us to provide a self-contained, logically connected exposition, developing methods and explaining ideas of the main approach, and coming to concrete results and applications.

Reading of the monograph requires some familiarity with singularity theory and (local) deformation theory of complex analytic spaces as well as with basic algebraic geometry. The necessary material can be found in various sources, for example, in Hartshorne's book on algebraic geometry [HaR2] and in our book [GLS6], which originally was conceived as the first part of the present monograph. In particular, this book includes numerous references to statements and definitions introduced in [GLS6]; moreover, references to many known facts and statements used here are supplied with a double reference—one to the original source and the other to [GLS6], where one can find both a precise formulation and a proof.

Most of the statements in the text are completely proven; the others are supplied by references to monographs and articles. The main new results appear in Chap. 4, where we give explicit references to the original sources, while the statements without reference are new in this book. The main notions and theorems are illustrated by examples, often concrete numerical, elaborated with the computer algebra system SINGULAR.

The contents are divided into four chapters.

Chapter 1 is devoted to zero-dimensional schemes associated with singular points of planar curves. We especially focus on the related singularity invariants, both the classical ones and some new invariants, which appear in the last chapters of the monograph. The reader interested in the classical problems on curves on surfaces can start with Chap. 2 and look for details on zero-dimensional schemes only when needed.

Chapter 2 on global deformation theory is concerned with three specific topics which play a fundamental role in the theory of equisingular families of curves. The significance of the classical theorems of Bézout, Noether, Bertini, Plücker, Riemann–Hurwitz for the algebraic geometry of curves on the one hand and the global deformation theory of analytic spaces on the other hand go much beyond our particular goals. These topics belong to the basic knowledge of any expert in singularity theory. In addition, we introduce and discuss the notion of T-smoothness for rather general singularity types of isolated hypersurface singularities. The seminal patchworking construction, originated in works of Viro in the early 80s, arouses a growing interest thanks to various applications to real and complex algebraic geometry, singularity theory, and algebra. It plays an ultimate role in our construction of curves with prescribed singularities.

In Chap. 3, we consider several approaches to the cohomology vanishing theory for ideal sheaves of zero-dimensional schemes on algebraic surfaces, with special emphasis on numerical criteria. We expose the classical and modern contributions to the topic as well as our own results, notably intended to make general statements applicable to arbitrary singularities and zero-dimensional schemes. As an example, we mention the Castelnuovo function theory, which was brought to the study of curves with nodes and cusps by Barkats and which later allowed us to obtain the best known H^1-vanishing criteria for ideal sheaves of the most interesting zero-dimensional schemes.

The monograph culminates in Chap. 4, where we answer the questions on the non-emptiness, dimension, smoothness, and irreducibility of equisingular families of curves in the plane and on other algebraic surfaces. Some results about families of singular curves, especially those related to the non-emptiness and the smoothness problem, are supplied with natural generalizations to families of higher-dimensional projective hypersurfaces with isolated singularities. The chapter starts with an overview of the new results and methods presented in this monograph. We end up with a discussion of open problems and conjectures, related to the geometry of equisingular families of curves and hypersurfaces as well as to the tropical enumerative geometry in which the patchworking construction appears as an important ingredient, providing a link between classical and tropical algebraic-geometric objects.

Each chapter is supplied with a section devoted to historical notes and references, where we trace back the key problems and ideas related to the contents of the chapter and give a general overview of the discussed topics.

Appendix A *Patchworking Real Algebraic Varieties*, written by Oleg Viro in 1983, shows the origin of the patchworking construction and its striking applications to Hilbert's 16th problem. The ideas of this work influenced deformation theory and singularity theory and served as one of the roots of the tropical geometry. We are very grateful to Oleg Viro who kindly permitted us to include his original article into the monograph.

Acknowledgements

Our work at the monograph has been supported by the Hermann Minkowski–Minerva Center for Geometry at Tel Aviv University and by grant no. G-616-15.6/99 from the German-Israeli Foundation for Research and Development. We have significantly advanced in our project during our two "Research-in-Pairs" stays at the Mathematisches Forschungsinstitut Oberwolfach. C. Lossen was also supported by the DFG grant no. Lo 864/1-1, E. Shustin was also supported by the Bessel Research Award from the Alexander von Humboldt Foundation and by the grants no. 448/09 and 176/15 from the Israeli Science Foundation. We are indebted to our colleagues for important corrections and suggestions, notably, to Adrian Langer,

Patrick Popescu-Pampu, Joaquim Roé, Francesco Russo, Mihai Tibar, Ilya Tyomkin, and many others, including unknown referees.

Kaiserslautern, Germany — Gert-Martin Greuel
Kaiserslautern, Germany — Christoph Lossen
Tel Aviv, Israel — Eugenii Shustin
April 2018

References

[HaR2] Hartshorne, R.: Algebraic Geometry. Graduate Text in Mathematics, vol. 52. Springer, Berlin (1977)

[GLS6] Greuel, G.-M., Lossen, C., Shustin, E.: Introduction to Singularities and Deformations. Springer, Berlin (2007)

Contents

Notations and Conventions

We use mainly standard notations from complex analysis and algebraic geometry as, e.g., in the textbooks by Grauert and Remmert [GrR1, GrR2] and Hartshorne [HaR2]. In particular, every *complex space* is Hausdorff with a countable basis of the topology and hence is paracompact. For the local theory of singularities and deformations, we refer also to [GLS6]. The structure sheaf of holomorphic functions of a complex space X is denoted by $\mathcal{O}_X$ and its analytic local rings by $\mathcal{O}_{X,x}$. A smooth complex space is also called a *complex manifold.* A complex subspace X of a complex space Z, defined by an ideal sheaf $\mathcal{I}$, will be dented by $V(\mathcal{I})$ with its structure sheaf $\mathcal{O}_X = \mathcal{O}_Z/\mathcal{I}|X$. In the same way, the notation $V(I)$ is used for complex space germs defined by an ideal $I \subset \mathcal{O}_{Z,z}$, and similar for schemes.

In this book, we are mainly working in the analytic category with complex spaces and their Euclidean topology, but we use also results from algebraic geometry with algebraic schemes and their Zariski topology. We call a separated scheme of finite type over the field $\mathbb{C}$, not necessarily reduced or irreducible, simply a *(complex) algebraic scheme.* A *point in an algebraic scheme* denotes a closed point, unless we say otherwise.

By the GAGA theorems of Serre [Ser1], there is the analytification functor from the category of complex algebraic schemes to the category of complex spaces,

$$an : (\text{complex algebraic schemes}) \to (\text{complex spaces})$$

associating to every algebraic scheme Y a complex space Y_{an} consisting of the closed points of Y (hence Y_{an} and Y coincide as sets) with its Euclidean topology and the structure sheaf of holomorphic functions. Moreover, the functor *an* associates to a coherent algebraic sheaf $\mathcal{F}$ on Y a coherent analytic sheaf $\mathcal{F}_{an}$ on Y_{an} inducing natural maps of cohomology groups $H^i(Y, \mathcal{F}) \to H^i(Y_{an}, \mathcal{F}_{an})$ for all i. In the projective situation *an* is an equivalence between the categories of projective algebraic schemes and *projective complex spaces*, i.e., isomorphic to a closed complex subspace of the complex manifold $\mathbb{P}^N = \mathbb{P}^N_{\mathbb{C}}$ for some N). In particular, every closed complex subspace of $\mathbb{P}^N$ is the analytification of an algebraic subscheme of the complex projective N-space with its algebraic scheme structure (Theorem of Chow). Moreover, the functor *an* induces an equivalence between the categories of coherent algebraic sheaves on Y and coherent analytic sheaves on Y_{an}, such that the above morphism of cohomology groups is an isomorphism. For further facts about the relation between an algebraic scheme Y and its analytification Y_{an}, we refer to [Ser1] and [HaR1, Appendix B] and the references given there.

We introduce the *category of algebraic complex spaces* as the (non-full) image of the analytification functor *an*. That is, a *complex space X is algebraic* or X is an *algebraic complex space,* if there exists an algebraic scheme Y with $X \cong Y_{an}$ as complex spaces. A morphism between algebraic complex spaces $X \cong Y_{an}$ and $X' \cong Y'_{an}$ (also called *algebraic morphism*) is the analytification of a morphism between the algebraic schemes Y and Y'. In particular, an algebraic subspace of X is the analytification of a subscheme of Y. Projective complex spaces are algebraic. We use the notion *(complex) analytic variety*, respectively *(complex) algebraic variety*, as a synonym for complex space respectively algebraic complex space. A projective complex space will also be called a *projective variety.*

A complex space X is algebraic iff it has a finite covering by open sets U_i such that $(U_i, \mathcal{O}_X|U_i)$ is isomorphic to a complex model space $(V, (\mathcal{O}_{\mathbb{C}^{n_i}}/\mathcal{I})|V)$ for some n_i (c.f. [GLS6, Definition 1.35]), with $\mathcal{I}$ a coherent ideal sheaf of $\mathcal{O}_{\mathbb{C}^{n_i}}$ that can be generated by polynomials $f_1, \ldots, f_k \in \mathbb{C}[x_1, \ldots, x_{n_i}]$, such that $V \subset \mathbb{C}^{n_i}$ is the set of zeroes of $f_1, \ldots, f_k$. We call $V(\mathcal{I}) := (V, (\mathcal{O}_{\mathbb{C}^{n_i}}/\mathcal{I})|V)$ an *affine model space* of $(U_i, \mathcal{O}_X|U_i)$, which is the analytification of the affine algebraic scheme in $\mathbb{C}^{n_i}$ defined by $f_1, \ldots, f_k$. U_i is called an *affine chart* of X and we often identify $(U_i, \mathcal{O}_X|U_i)$ with its affine model space.

Note that an algebraic complex space X carries the Euclidean topology and its local rings are analytic local rings, unless we explicitly say otherwise. A *locally closed* subspace of a complex space X is the intersection of an open and a closed subspace of X. At some places, we use also the Zariski topology on an algebraic complex space X, where $U \subset X$ is called *Zariski-open* if U is the complement of a closed algebraic subspace of X. A Zariski-open subset of a projective complex space is called *quasi-projective.*

Let $X \subset \Sigma$ be a (non-empty) subspace in the category of complex spaces, algebraic varieties, or algebraic schemes, respectively. X is called a (effective) *Cartier divisor* if for each $x \in X$ the ideal of X at x is generated by a nonzero divisor in the local ring of Σ at x. If Σ is smooth, a Cartier divisor is also called a *hypersurface.* A *curve*, respectively *surface* in any of these categories denotes a space of pure dimension one respectively two. An *algebraic curve* respectively *algebraic surface* refers to an algebraic complex space of pure dimension one respectively two. By a *real algebraic variety* respectively *real analytic variety*, we mean an algebraic respectively analytic variety equipped with an anti-holomorphic involution. A *family* of complex spaces or algebraic varieties or schemes does always mean a flat family, i.e., a flat morphism in the respective category.

We speak often about *points in generic position* or a *generic element.* This will always mean points or an element of some open dense subset of the whole topological space (the topology will be the Zariski topology except we say otherwise), and this subset will either be specified immediately, or, when the precise specification is not essential, either without specification or by further listing finitely many algebraic or analytic constraints, usually within the proof. On the other hand, a *general element* or *general point* is understood scheme-theoretically, i.e., depending on indeterminate parameters or coordinates.

A *singularity* is by definition the germ (X, z) of a complex space, may be smooth. A singularity (X, z) is *isolated* if $X \backslash \{z\}$ is smooth for some representative X. Recall the definition of analytic respectively topological type of an isolated hypersurface singularity respectively a reduced plane curve singularity $(X, z) \subset (\Sigma, z)$, (Σ, z) a smooth complex germ, used throughout the book.

Two hypersurface singularities $(X, z) \subset (\Sigma, z)$ and $(X', z') \subset (\Sigma, z')$ are called *analytically equivalent* (respectively *topologically equivalent*) if there exists an analytic isomorphism (respectively a homeomorphism) of germs $(\Sigma, z) \cong (\Sigma, z')$ (respectively of small "good" neighborhoods of $z, z' \in \Sigma$) mapping (X, z) to (X', z').

The germs (X, z) and (X', z') are analytically equivalent iff their analytic local rings $\mathcal{O}_{X,z}$ and $\mathcal{O}_{X',z'}$ are isomorphic. If $(X, 0) \subset (\mathbb{C}^n, 0)$ is given by a convergent power series

$$f = \sum_{|\alpha| \geq 1} c_\alpha \boldsymbol{x}^\alpha \in \mathbb{C}\{\boldsymbol{x}\} = \mathbb{C}\{x_1, \ldots, x_n\}$$

and $(X', 0) \subset (\mathbb{C}^n, 0)$ by $g \in \mathbb{C}\{\boldsymbol{x}\}$, then $(X, 0)$ and $(X', 0)$ are analytically equivalent iff f and g are *contact equivalent* (notation $g \overset{c}{\sim} f$); i.e., there are a unit $u \in \mathbb{C}\{\boldsymbol{x}\}$ and a local $\mathbb{C}$-algebra automorphism Φ of $\mathbb{C}\{\boldsymbol{x}\}$ such that $f = u \cdot \Phi(g)$.

The transcendental definition of topological equivalence above is in fact algebraic for a *reduced plane curve singularity* (i.e., as an isolated hypersurface singularity in a smooth surface germ, given by a reduced f). Namely, two reduced plane curve singularities (C, z) and (C', z') are *topologically equivalent* iff there exists a bijection of local branches such that the Puiseux pairs of the corresponding branches coincide, as well as the pairwise intersection multiplicities of the corresponding branches. Equivalently, if they have embedded resolutions by blowing up points such that the systems of multiplicities of the reduced total transforms coincide. This definition is the preferred one since it generalizes to deformations over non-reduced base spaces (cf. [GLS6, I.34]). We shall consider topological equivalence only in the case of plane curve singularities (an equivalent algebraic definition of topological equivalence for higher-dimensional hypersurface singularities is in general unknown).

Note that the *multiplicity* or *order* of $f \in \mathcal{O}_{\Sigma, z} \cong \mathbb{C}\{\boldsymbol{x}\}$ is

$$\mathrm{mt}(f) = \mathrm{ord}(f) = \min\{|\alpha| \,|\, c_\alpha \neq 0\} = \max\{m \,|\, f \in \mathfrak{m}_z^m\},$$

where (Σ, z) is a smooth complex germ and $\mathfrak{m}_z$ denotes the maximal ideal of $\mathcal{O}_{\Sigma,z}$. Moreover, if $(D_1, z), (D_2, z) \subset (\Sigma, z)$ are two plane curve singularities in the smooth surface germ (Σ, z) defined by $f_1, f_2 \in \mathcal{O}_{\Sigma,z}$ and without common component, then

$$i_z(D_1, D_2) := i_z(f_1, f_2) := \dim_{\mathbb{C}} \mathcal{O}_{\Sigma,z}/\langle f_1, f_2 \rangle$$

is called the *intersection multiplicity* of (D_1, z) and (D_2, z) (cf. [GLS6, Proposition I.3.12] and equivalent formulas in [GLS6, Section I.3.2]).

An *analytic type* (respectively *topological type*) is the equivalence class of a hypersurface singularity (respectively a reduced plane curve singularity) with respect to analytic (respectively topological) equivalence. Deformations which fix the analytic type respectively the topological type (along some section) are called *equianalytic deformations* or *ea-deformations* respectively *equisingular deformations* or *es-deformations*. The shorthand notation ESF stands for equisingular families, i.e., families of projective hypersurfaces or projective plane curves with finitely many isolated singularities of fixed analytic or topological or, more generally, smooth singularity types.

The *simple singularities* or *ADE-singularities* (cf. [GLS6, I.2.4] for their definition and classification) play often a special role, and for them, the analytic and topological types coincide. They are denoted by A_k, $k \geq 1$, D_k, $k \geq 4$, and E_6, E_7, E_8, where, in the case of plane curves singularities, A_1, respectively A_2, denotes an *(ordinary) node* (given by $x^2 + y^2 = 0$), respectively a *cusp* (given by $x^2 + y^3 = 0$).

References

[GrR1] Grauert, H., Remmert, R.: Theory of Stein Spaces. Grundlehren der mathematischen Wissenschaften, vol. 236, Springer, Berlin (1979)

[GrR2] Grauert, H., Remmert, R.: Coherent Analytic Sheaves. Grundlehren der mathematischen Wissenschaften, vol. 265, Springer, Berlin (1984)

[HaR2] Hartshorne, R.: Algebraic Geometry. Graduate Text in Mathematics, vol. 52. Springer, Berlin (1977)

[GLS6] Greuel, G.-M., Lossen, C., Shustin, E.: Introduction to Singularities and Deformations. Springer, Berlin (2007)

[Ser1] Serre, J.-P.: Géométrie algébrique et géométrie analytique. Ann. Inst. Fourier **6**, 1–42 (1956)

[HaR1] Hartshorne, R.: Connectedness of the Hilbert scheme. Publ. Math. IHES **29**, 261–304 (1966)

Chapter 1
Zero-Dimensional Schemes for Singularities

This section is devoted to the study of zero-dimensional schemes in a smooth projective surface Σ, associated to and concentrated in the (finite) set of singular points of a reduced curve C on Σ. In this chapter, a curve (singularity) we will always mean a *reduced curve (singularity)*, unless we explicitly say the opposite. We introduce the notion of *cluster schemes* (cf. Sect. 1.1.2) which allows us to encode the topological type of the singularities of C and, in Sect. 1.1.4, a class of zero-dimensional schemes encoding the analytic type of the singularities. When dealing with zero-dimensional schemes, we use the following

Notation. Let Σ be an arbitrary smooth complex space. For $Z \subset \Sigma$ a *zero-dimensional scheme*, (i.e. a zero-dimensional complex subspace with finite support), let $\mathcal{J}_{Z/\Sigma} \subset \mathcal{O}_\Sigma$ be the corresponding ideal sheaf and $\mathcal{O}_Z = (\mathcal{O}_\Sigma/\mathcal{J}_{Z/\Sigma})|_Z$ the structure sheaf. Then we introduce

(a) the *support* of Z, $\mathrm{supp}(Z) := \{z \in \Sigma \mid \mathcal{J}_{Z/\Sigma,z} \neq \mathcal{O}_{\Sigma,z}\}$,
(b) the *degree* of Z,

$$\deg Z := \dim_{\mathbb{C}} H^0(Z, \mathcal{O}_Z) = \sum_{z \in \mathrm{supp}\, Z} \dim_{\mathbb{C}} \mathcal{O}_{\Sigma,z}/\mathcal{J}_{Z/\Sigma,z},$$

(c) and the *multiplicity* of Z at z,

$$\mathrm{mt}(Z, z) := \max \{m \in \mathbb{N} \mid \mathcal{J}_{Z/\Sigma,z} \subset \mathfrak{m}_{\Sigma,z}^m\},$$

which is the minimum order at z of the elements contained in $\mathcal{J}_{Z/\Sigma,z}$.
If the support of Z consists of only one point z, we also use the notation $\mathrm{mt}(Z)$ in place of $\mathrm{mt}(Z, z)$.

In the following we fix some further notations and prove some results about zero-dimensional schemes used later in Chap. 4 in the book. Σ denotes a smooth complex surface, while (Σ, z) stands for a smooth two-dimensional complex germ.

G.-M. Greuel et al., *Singular Algebraic Curves*, Springer Monographs in Mathematics, https://doi.org/10.1007/978-3-030-03350-7_1

1.1 Cluster and Zero-Dimensional Schemes

Let z be a singular point of the reduced curve C contained in the smooth surface Σ. Then it is well-known that we can resolve the singularity of C at z by finitely many blowing ups of points. More precisely, there exists a *good embedded resolution* of the singularity of C at z, that is, a sequence of morphisms of complex manifolds

$$\pi : \Sigma_{n+1} \xrightarrow{\pi_{n+1}} \Sigma_n \xrightarrow{\pi_n} \cdots \to \Sigma_1 \xrightarrow{\pi_1} \Sigma_0 = \Sigma$$

such that π_i is the blowing up of a point $q_{i-1} \in \Sigma_{i-1}$ infinitely near to $q_0 = z$ (see below), and such that in a neighborhood of $E = \pi^{-1}(z)$ the *reduced total transform* of C,

$$\pi^*(C)_{red} = \widetilde{C} + \sum_{i=1}^{n+1} E_i$$

is a divisor with normal crossings, that is, a hypersurface having only nodes as singularities (see [GLS6, Section I.3.3] for details). Here $\widetilde{C}$ is the *strict transform* of C and $E_i = \pi_i^{-1}(q_{i-1}) \cong \mathbb{P}^1$, $i = 1, \ldots, n+1$, is the *exceptional divisor* of π_i in Σ_i resp. in Σ_j, $j \geq i$. The Σ_i are smooth surfaces, which are projective if Σ is projective. Moreover, the corresponding resolution graph, together with the respective multiplicities of the total transform, determine the topological type of the singularity of C at z [GLS6, Theorem I.3.42].

Hence, if we want to encode topological types of plane curve singularities by zero-dimensional schemes $Z \subset \Sigma$, it is advisable to consider finite successions of blowing ups of points,

$$\pi' : \Sigma' = \Sigma_{n+1} \xrightarrow{\pi_{n+1}} \Sigma_n \longrightarrow \cdots \longrightarrow \Sigma_1 \xrightarrow{\pi_1} \Sigma_0 = \Sigma .$$

Recall that classically $E_1 \subset \Sigma_1$ and its strict transforms in Σ_i, $i > 1$, together with z, is called the *first infinitesimal neighbourhood* of $z \in \Sigma$. For $i > 1$ the ith infinitesimal neighbourhood of z consists of points in the first infinitesimal neighbourhood of some point in the $i-1$-st infinitesimal neighbourhood of z. Any point belonging to some infinitesimal neighbourhood of z is called an *infinitely near point* of $z \in \Sigma$ or a point *infinitely near to* z (see [EnC, Wae1, Cas2]). Two infinitely near points of Σ, $q' \in \Sigma'$ and $q'' \in \Sigma''$ (for another sequence of blowing ups), are identified if there exist open neighbourhoods $W' \subset \Sigma'$ of q', $W'' \subset \Sigma''$ of q'', and a Σ-isomorphism $W' \xrightarrow{\cong} W''$ mapping q' to q''.

Before we introduce the notion of constellations and clusters, which basically refer to $(n+1)$-tuples of infinitely near points and multiplicities, we should like to recall some facts from intersection theory on a blown-up surface.

For two hypersurfaces, also called *effective divisors* D_1 and D_2 on Σ, we can define the *intersection number*

$$D_1 \cdot D_2 := \sum_{x \in D_1 \cap D_2} i_x(D_1, D_2)\,,$$

which is finite iff $D_1 \cap D_2$ is a finite set. For a general definition (including self-intersection) and further properties of the *intersection pairing* of curves on surfaces we refer to [BHPV, II.10] and [HaR2, Theorem V.1.1] for algebraic schemes.

Let $\pi : \widetilde{\Sigma} \to \Sigma$ be the transformation blowing up $z \in \Sigma$ and E the exceptional divisor of π. Moreover, let $C, D \subset \Sigma$ be reduced curves, and $\widetilde{C}, \widetilde{D}$ the respective strict transforms. Then we have the following facts about *intersection numbers* or *intersection multiplicities*:

- $\widetilde{C} \cdot E = \mathrm{mt}(C, z)$,
- $\pi^*(C) \cdot E = 0$,
- $\pi^*(C) \cdot \pi^*(D) = C \cdot D$.

Here, $\pi^*(C)$ denotes the preimage $\pi^{-1}(C) \subset \widetilde{\Sigma}$ defined by the ideal sheaf $\pi^*\mathcal{J}_{C/\Sigma} \subset \mathcal{O}_{\widetilde{\Sigma}}$ where $\mathcal{J}_{C/\Sigma} \subset \mathcal{O}_{\Sigma}$ is the ideal sheaf of C (and similar for D). The first equality follows from [GLS6, Remark I.3.17.2]. The latter two follow from the projection formula [HaR2, App. A.1], [Dra], [BHPV, II.10], as π is an isomorphism outside $E = \pi^{-1}(z)$. Alternatively, one may use that any curve on Σ is linearly equivalent to the difference of two non-singular curves, meeting everywhere transversally, and not containing z (see [HaR2, V.1]).

Since $\pi^*(C) = \widetilde{C} + \mathrm{mt}(C, z) \cdot E$, the self-intersection numbers of the exceptional divisor, respectively the strict transform, can be computed as

- $E^2 = -1$,
- $\widetilde{C}^2 = C^2 - \mathrm{mt}(C, z)^2$.

1.1.1 *Constellations and Cluster*

Consider a good embedded resolution of the reduced curve singularity $(C, z) \subset (\Sigma, z)$, Σ a smooth surface. It is given by a sequence $(z = q_0, \pi_1, q_1, \ldots, \pi_m, q_m, \pi_{m+1})$ with $\pi_i : \Sigma_i \to \Sigma_{i-1}$ the blowing up of $q_{i-1} \in \Sigma_{i-1}$ (Σ_i are smooth surfaces, $\Sigma_0 = \Sigma$) and $q_i \in \Sigma_i$ infinitely near to z, such that the the strict transform of (C, z) is non-singular and the reduced total transform $\pi^{-1}(C)_{red}$, $\pi = \pi_{m+1} \circ \cdots \circ \pi_1$, has only nodes as singularities. π is called a *minimal good embedded resolution* of (C, z), if only non-nodal singularities of the reduced total transform of (C, z) are blown up in the resolution process. It is well-known that such a minimal good resolution is unique up to isomorphism over Σ. Different choices of local coordinates or a different order of blowing up points (in the case of several branches) produce different but isomorphic resolutions.

For $p \in \pi^{-1}(z)$ we denote by $C_{(p)}$ (resp. $\widehat{C}_{(p)}$) the germ at p of the strict (resp. total) transform of C and by $\mathrm{mt}\, C_{(p)}$ (resp. $\mathrm{mt}\, \widehat{C}_{(p)}$) their multiplicities (see Proposition 1.1.11 and Remark 1.1.11.1 for some relations between these multiplicities).

The concept of a constellation introduced in this section is used to describe isomorphism classes of resolutions. In particular, the minimal embedded resolution is described by what is called the "essential tree" $\mathscr{T}^*(C, z)$ (Definition 1.1.10). More generally, a constellation can be associated to a partial resolution of (C, z), respectively to further blowing ups of singular points of Σ_{n+1}. If, additionally, we assign to each point of a constellation the multiplicity of the strict transform of C through this point we get what we call a cluster (Definition 1.1.12).

An exhaustive presentation of clusters can be found in [Cas1] and in particular in [Cas2]. Nevertheless we provide all necessary definitions and illustrating examples, since zero-dimensional schemes, notably, cluster schemes will be intensively used in the sequel, allowing us to explicitly express and compute various singularity invariants responsible for a regular or pathological behavior of equisingular families of curves.

1.1.1.1 Constellations

The following definition of an (abstract) constellation does not refer to a curve singularity but is modeled along the above situation. We follow the terminology in [CGSL]. We keep the notations from above.

Definition 1.1.1 (*Constellation*) Let $\mathcal{C}on(\Sigma)$ be the set of all tuples $(q_0, \pi_1, q_1, \ldots, \pi_n, q_n)$, $n \geq -1$, where, for $i = 1, \ldots, n$

$(\mathcal{C}1) : q_0 \in \Sigma_0 := \Sigma,$
$(\mathcal{C}2) : \pi_i : \Sigma_i \to \Sigma_{i-1}$ is the blowing up of $q_{i-1} \in \Sigma_{i-1},$
$(\mathcal{C}3) : q_i \in \Sigma_i.$

We define an equivalence relation $\sim$ on $\mathcal{C}on(\Sigma)$ as follows:

$$(q_0, \pi_1, q_1, \ldots, \pi_n, q_n) \sim (q'_0, \pi'_1, q'_1, \ldots, \pi'_m, q'_m)$$

iff $m = n = -1$ or $m = n \geq 0$ and there exists a Σ-isomorphism

$$\begin{array}{ccccc}
 & \Sigma_{n+1} & \xrightarrow[\cong]{\Phi} & \Sigma'_{n+1} & \\
 \swarrow^{\pi_{n+1}} & & & & \searrow^{\pi'_{n+1}} \\
q_n \in \Sigma_n & & & & \Sigma'_n \ni q'_n \\
\downarrow \pi_n & & & & \downarrow \pi'_n \\
q_{n-1} \in \Sigma_{n-1} & & \circlearrowleft & & \Sigma'_{n-1} \ni q'_{n-1} \\
\downarrow & & & & \downarrow \\
\vdots & & & & \vdots \\
\downarrow \pi_2 & & & & \downarrow \pi'_2 \\
q_1 \in \Sigma_1 & & & & \Sigma'_1 \ni q'_1 \\
 & \searrow^{\pi_1} & & \swarrow^{\pi'_1} & \\
 & & \Sigma\,, & &
\end{array} \tag{1.1.1.1}$$

π_{n+1} (resp. π'_{n+1}) denoting the blowing up of q_n in Σ_n (resp. of q'_n in Σ'_n).

It is easy to see that a Σ-isomorphism $\Sigma_{n+1} \to \Sigma'_{n+1}$ is unique and that successively blowing up two different points in the two possible orders gives rise to Σ-isomorphic surfaces.

We call an equivalence class $\mathscr{K} \in \mathcal{C}on(\Sigma)/\sim$ a *(finite) constellation on* Σ, and the manifold Σ_{n+1} is called the *sky* of $\mathscr{K}$. Note that the latter is well-defined up to Σ-isomorphism.

If $n = -1$ then $\mathscr{K}$ is called the *empty constellation*, and we write $\mathscr{K} = \emptyset$. If $n \geq 0$ then we use the following two notations:

$$\mathscr{K} = (q_0, \pi_1, q_1, \ldots, \pi_n, q_n), \quad \text{or} \quad \mathscr{K} = (q_0, q_1, \ldots, q_n).$$

We say that a constellation $\mathscr{K}' = (q'_0, \pi'_1, q'_1, \ldots, \pi'_k, q'_k)$ is a *subconstellation* of $\mathscr{K} = (q_0, \pi_1, q_1, \ldots, \pi_n, q_n)$ (or that $\mathscr{K}$ is an *extension* of $\mathscr{K}'$) if $k \leq n$, and if for $i = k+1, \ldots, n$ there are points q'_i and blowing ups π'_i of q'_{i-1} (with $q'_k = q_k$), such that

$$(q'_0, \pi'_1, q'_1, \ldots, \pi'_k, q'_k, \pi'_{k+1}, q'_{k+1}, \ldots, \pi'_n, q'_n) = \mathcal{K}.$$

We write $\mathscr{K}' \subset \mathscr{K}$.

Notation. Let $(q_0, \pi_1, q_1, \ldots, \pi_n, q_n) \in \mathcal{C}on(\Sigma)$, then, for each $q = q_i \in \Sigma_i$ we denote by

$$E_q \subset \Sigma_{i+1} \quad \text{the exceptional divisor of} \quad \pi_{i+1} : \Sigma_{i+1} \to \Sigma_i\,,$$

respectively its strict transform $E_q \subset \Sigma_j$ on any of the surfaces $\Sigma_j, i+2 \leq j \leq n+1$.

Remark 1.1.1.1 The Σ-isomorphism (1.1.1.1) maps $E_{q_i} \subset \Sigma_{n+1}$ to some exceptional divisor $E_{q'_{\varphi(i)}} \subset \Sigma'_{n+1}$, defining in this way a permutation φ of $0, \ldots, n$. Note that

$$q_i \text{ is infinitely near to } q_j \iff q'_{\varphi(i)} \text{ is infinitely near to } q'_{\varphi(j)}. \tag{1.1.1.2}$$

Indeed, for each tuple (i, j), the intersection numbers $E_{q_i} \cdot E_{q_j}$ (on Σ_{n+1}) and $E_{q'_{\varphi(i)}} \cdot E_{q'_{\varphi(j)}}$ (on Σ'_{n+1}) coincide. In particular, a (-1)-curve $E_{q_j} \subset \Sigma_{n+1}$ is mapped to a (-1)-curve $E_{q'_{\varphi(j)}} \subset \Sigma'_{n+1}$. Since a point q_j is infinitely near to a point q_i, $i < j$, iff there exists some index $k \in \{i, i+1, \ldots, j-1\}$ such that $E_{q_j} \cdot E_{q_k} > 0$ and such that the point q_k is infinitely near to q_i, and since the analogous statement holds for $q'_{\varphi(i)}, q'_{\varphi(j)}$, a straightforward induction gives (1.1.1.2).

In view of Remark 1.1.1.1, it makes sense to refer to the *"points" of a constellation*, in particular infinitely near points, identifying q_i with $q'_{\varphi(i)}$, $i = 1, \ldots, n$. Then a constellation $\mathscr{K}'$ is a subconstellation of $\mathscr{K}$, iff each point of $\mathscr{K}'$ is also a point of $\mathscr{K}$. We write "$q \in \mathscr{K}$" to denote a point of the constellation $\mathscr{K}$.

Definition 1.1.2 (*Level, origin, graph of constellation*) Let $\mathscr{K}$ be a constellation $(q_0, \pi_1, q_1, \ldots, \pi_n, q_n)$ on Σ.

(1) We introduce a (natural) partial ordering on the points $q_0, \ldots, q_n$ by

$$q_i \leq q_j \;:\Longleftrightarrow\; q_j \text{ is infinitely near to } q_i \,.$$

(2) For each $0 \leq i \leq n$, we define the *level of* q_i as

$$\ell(q_i) := \#\{j \,|\, q_i \geq q_j\} - 1\,.$$

For instance, the level of a point q_i is 0 iff the composed map $\pi_1 \circ \ldots \circ \pi_i$ is an isomorphism locally at q_i (and π_{i+1} may be considered as blowing up the point $(\pi_1 \circ \ldots \circ \pi_i)(q_i) \in \Sigma_0 = \Sigma$). In particular, $\ell(q_0) = 0$.

(3) If no other point has level 0, then $q_i \geq q_0$ for each $0 \leq i \leq n$, and q_0 is called the *origin of the constellation* $\mathscr{K} = (q_0, \pi_1, q_1, \ldots, \pi_n, q_n)$.

(4) Finally, we define $\Gamma_{\mathscr{K}}$ to be the oriented graph whose points are in one to one correspondence with $q_0, \ldots, q_n$ and the edges with pairs (q_j, q_i) such that $\ell(q_j) = \ell(q_i) + 1$ and $q_j \geq q_i$. $\Gamma_{\mathscr{K}}$ is called the *oriented graph of the constellation* $\mathscr{K}$ or the *forest of the constellation* $\mathscr{K}$. If $\mathscr{K}$ has an origin then $\Gamma_{\mathscr{K}}$ is an oriented tree with root q_0.

By Remark 1.1.1.1, the graph $\Gamma_{\mathscr{K}}$ is independent of the chosen representative of the constellation. It is an oriented forest (i.e. a finite disjoint union of oriented trees) whose respective roots correspond to the points of level 0 in the constellation.

To shorten notation, we write $q_0, \ldots, q_n$ to denote both, the points of the constellation $\mathscr{K}$ and the corresponding vertices of the graph $\Gamma_{\mathscr{K}}$ (from the context it should be clear what the notation refers to).

Remark 1.1.2.1 Let $\mathscr{K} \neq \emptyset$ be a constellation on Σ, and let $q_{j,1}, \ldots, q_{j,k_j}$ be its points of level j, $0 \leq j \leq s := \max\{\ell(q) \mid q \in \mathscr{K}\}$. Then

$$\mathscr{K} = (q_{0,1}, \ldots, q_{0,k_0}, q_{1,1}, \ldots, q_{1,k_1}, \ldots\ldots, q_{s,k_s})\,.$$

In particular, the constellation does not depend on the numbering of the points with fixed level j.

Definition 1.1.3 (*Proximate, satellite, free*) Let $(q_0, \pi_1, q_1, \ldots, \pi_n, q_n) \in \mathcal{C}on(\Sigma)$. With the above notation we can introduce (following Enriques terminology, cf. also [Cas1, Cas2, Lip]):

(a) q_j is *proximate to* q_i, if $q_j \in E_{q_i}$, the exceptional divisor of π_{i+1}. We write

$$q_j \dashrightarrow q_i\,.$$

(b) q_j is *satellite*[1] if it is proximate to (at least) two points q_i, $0 \leq i \leq j-1$. Otherwise it is *free*.

Note that a point q_j cannot be proximate to more than two points (since the exceptional divisors have normal crossings). Moreover, all the points q_j that appear in the exceptional divisor E_{q_i} or in any of its strict transforms are proximate to q_i. On the other hand, if q_j is proximate to q_i, then $\ell(q_j) > \ell(q_i)$.

1.1.1.2 Blowing up of a Constellation

Definition 1.1.4 (*Strict and total transform of a constellation*) Let $(q_0 = z, \pi_1, q_1, \dots, \pi_n, q_n) \in \mathcal{C}on(\Sigma)$, and let $g \in \mathcal{O}_{\Sigma,z}$. We denote, for each point $q = q_i \geq z$,

- $g_{(q)}$, the *strict transform* of g under the morphism

$$\pi_{(q)} = \pi_1 \circ \cdots \circ \pi_i : \Sigma_{(q)} := (\Sigma_i, q) \to (\Sigma, z)\,,$$

- $\widehat{g}_{(q)} = \pi^{\#}_{(q)}(g)$, the *total transform* of g under $\pi_{(q)} : \Sigma_{(q)} \to (\Sigma, z)$.

Here $\pi^{\#}_{(q)} : \mathcal{O}_{\Sigma,z} \to \mathcal{O}_{\Sigma_{(q)}}$ is the algebra morphism associated to $\pi_{(q)}$.

In the same manner, we introduce $C_{(q)}$, the (germ of the) strict transform of (C, z) at q, and $\widehat{C}_{(q)}$, the (germ of the) total transform of (C, z) at q. If the germ $C_{(q)}$ is non-empty, we say that the curve C *goes through* the infinitely near point q (or that *q belongs to C*).

Remark 1.1.4.1 The proximities as well as the multiplicities of the strict, respectively total, transforms at a point q are independent of the chosen representative of a constellation.

Example 1.1.4.2 Consider $g = (x^2 - y^3)(y^2 - x^3) \in \mathbb{C}\{x, y\}$, and let $C \subset \mathbb{P}^2$ be the plane projective curve defined by the homogenization of g (consisting of two transversal cusps at the origin $z = \mathbf{0}$). A minimal good resolution of the singularity (C, z) is shown in Fig. 1.1 on p. 8.

The blowing up sequences $(q_0, \pi_1, q_1, \dots, \pi_4, q_4)$ and $(q_0, \pi_1, q_1', \dots, \pi_4', q_4')$ obtained by blowing up $q_1' = q_3$ instead of q_1 and continuing by interchanging the role of the two branches, are isomorphic by a Σ–isomorphism of the skies Σ_5 and Σ_5' and hence define the same constellation. We have, for example, $q_3 \dashrightarrow q_0$, $q_4 \dashrightarrow q_3, q_0$ but $q_4 \not\dashrightarrow q_2$. Moreover, q_0, q_1, q_3 are free while q_2 and q_4 are satellite points.

Blowing up further nodal points q_i of the intersection of the strict transform with the exceptional divisor and continuing, we get an infinite constellation, the (infinite) complete embedded resolution tree of (C, z), whose oriented graph looks as follows

---

[1]The name "satellite" point can be interpreted as follows: the position of the point on the blown-up surface is fixed by the positions of two predecessor points, as it were a kind of "geostationary satellite". Vice versa, a "free" point can move freely on some exceptional divisor.

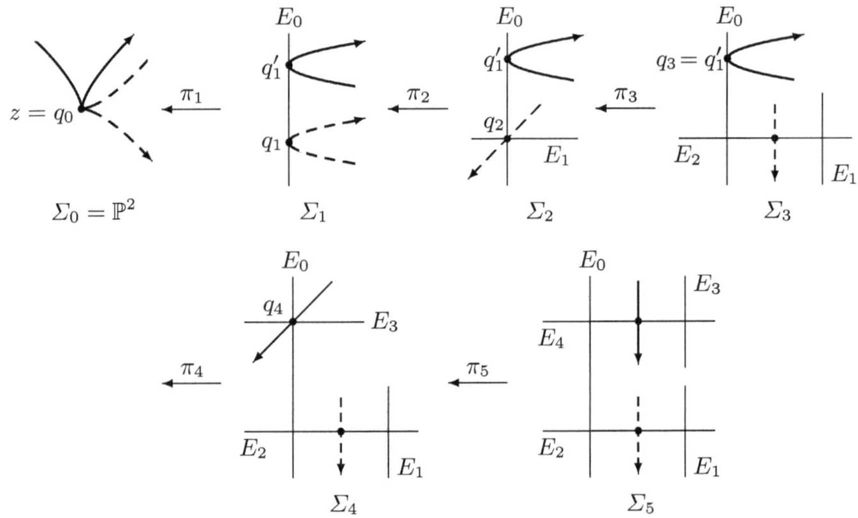

Fig. 1.1 A minimal good resolution of the germ defined by $(x^2 - y^3)(y^2 - x^3)$

(with vertices the q_i and $q_i \bullet \longleftarrow \bullet q_j$ if $q_i \leq q_j$, omitting in the picture all arrows given by reflexivity and transitivity):

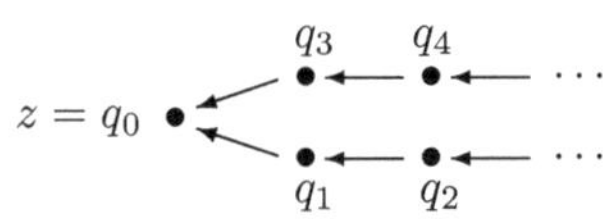

Note that $\ell(q_0) = 0$, $\ell(q_1) = \ell(q_3) = 1$, $\ell(q_2) = \ell(q_4) = 2$.

1.1.1.3 The Complete Embedded Resolution Tree

Given a reduced plane curve singularity $(C, z) \subset (\Sigma, z)$, we consider the constellations obtained from $\mathscr{K} = (z)$ by successive extension (cf. Definition 1.1.1) with points infinitely near to z. In the limit of such an extension process, we get the (infinite) complete embedded resolution tree $\mathscr{T}(C, z)$:

Definition 1.1.5 (*Infinite constellation*) By an *infinite constellation* on $\Sigma = \Sigma_0$ we denote the class of an infinite sequence $(q_0, \pi_1, q_1, \pi_2, q_2, \ldots)$ of points and blowings ups as in Definition 1.1.1 with respect to the following equivalence relation:

$$(q_0, \pi_1, q_1, \pi_2, q_2, \ldots) \sim (q_0', \pi_1', q_1', \pi_2', q_2', \ldots)$$

iff, for each $n \in \mathbb{Z}_{\geq 0}$ there exists an $m \geq n$ such that $(q_0, \pi_1, q_1, \ldots, \pi_n, q_n)$ is a subconstellation of $(q_0', \pi_1', q_1', \ldots, \pi_m', q_m')$ and $(q_0', \pi_1', q_1', \ldots, \pi_n', q_n')$ is a subconstellation of $(q_0, \pi_1, q_1, \ldots, \pi_m, q_m)$.

Definition 1.1.6 (*Complete resolution tree*) The *complete embedded resolution tree* $\mathscr{T}(C, z)$ is the unique infinite constellation $(q_0, \pi_1, q_1, \pi_2, q_2, \ldots)$ on Σ having the following properties:

(a) $q_0 = z$, and each point q_i, $i \geq 0$, belongs to C,
(b) for n sufficiently large, $\pi_1 \circ \ldots \circ \pi_n$ defines an embedded resolution of $(C, z) \subset (\Sigma, z)$, and
(c) for each q_i there are infinitely many $j \geq i$ such that q_j is infinitely near to q_i.

Note that $\mathscr{T}(C, z)$ is not a graph but a constellation.

The infinite oriented graph associated to $\mathscr{T}(C, z)$ looks as follows:

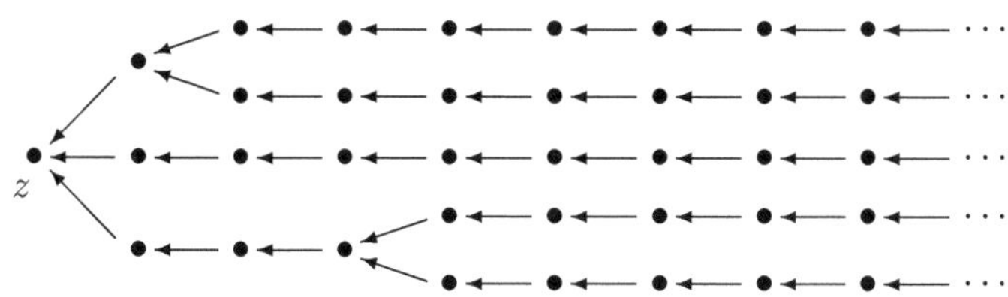

The concept of the complete embedded resolution tree is mainly used to save ink when describing the (finite) constellations which we actually work with. Indeed, all constellations with origin z can be viewed as finite subconstellations ("subtrees") of some complete embedded resolution tree $\mathscr{T}(C', z)$:

Definition 1.1.7 (*Finite subtree*) A *finite subtree of the complete embedded resolution tree* $\mathscr{T}(C, z)$ is a constellation $\mathscr{T} = (q_0, q_1, \ldots, q_n)$ on Σ which is either empty or which has origin $q_0 = z$ and the curve C goes through each q_i, $i = 1, \ldots, n$. In particular, if $\mathscr{T}$ is non-empty, the associated graph $\Gamma_{\mathscr{T}}$ is an oriented tree with root z. We use the notation $\mathscr{T} \subset \mathscr{T}(C, z)$.

If in the following we refer to points $q \in \mathscr{T}(C, z)$ we understand them as points $q = q_i$ of a finite subtree $\mathscr{T} \subset \mathscr{T}(C, z)$. Moreover, if $\mathscr{T} \subset \mathscr{T}(C, z)$ is a finite subtree, we say that a point q is a *leaf of* $\mathscr{T}$ if there no point $q' \in \mathscr{T}$ which is proximate to q.

Notation. Let (C, z), $(C', z) \subset (\Sigma, z)$ be plane curve germs *without common component*. Then we define $\mathscr{T}(C, z) \cap \mathscr{T}(C', z)$ to be the maximal finite subtree of $\mathscr{T}(C, z)$ such that all of its points belong to C'.

To see that this is a well-defined constellation on Σ, recall Noether's formula (which can be obtained by induction from [GLS6, Proposition I.3.21]):

Proposition 1.1.8 (Noether's formula) *Let* (C, z), $(C', z) \subset (\Sigma, z)$ *be curve singularities without common component, then*

$$\infty > i_z(C, C') = \sum_{q \in \mathscr{T}(C,z) \cap \mathscr{T}(C',z)} \operatorname{mt} C_{(q)} \cdot \operatorname{mt} C'_{(q)},$$

where $i_z(C, C')$ *denotes the intersection multiplicity of* C *and* C' *at* z.

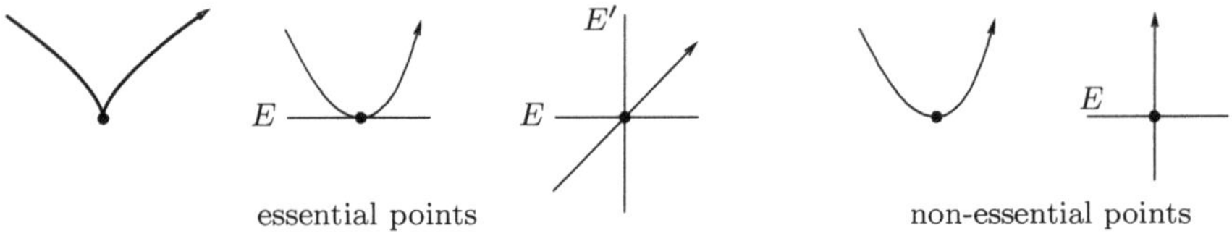

Fig. 1.2 Essential and non-essential infinitely near points

Definition 1.1.9 (*Essential point*) A point $z \neq q \in \mathscr{T}(C, z)$ is called *essential* for C iff the reduced total transform of C does not have a node at q. The origin z of $\mathscr{T}(C, z)$ is called essential for C, iff the germ (C, z) is not smooth. We call a point $q \in \mathscr{T}(C, z)$ a *singular essential point* for C if the strict transform of C at q is singular.

Notice that each satellite point in $\mathscr{T}(C, z)$ is an essential point. It may well be a non-singular essential point for C (see Fig. 1.2).

We know that, if $q \in \mathscr{T}(C, z)$ is an essential point and if $q \geq q'$, then q' is also essential. Moreover, the desingularization theorem states that there are only finitely many essential points in $\mathscr{T}(C, z)$.

Definition 1.1.10 (*Essential tree*) We introduce the *essential tree*

$$\mathscr{T}^*(C, z) \subset \mathscr{T}(C, z)$$

to be the maximal finite subtree of $\mathscr{T}(C, z)$ such that all of its points are essential.

Note that, for (C, z) a smooth germ, the essential tree $\mathscr{T}^*(C, z)$ is empty. In Example 1.1.4.2 the points $q_0, \ldots, q_4$ are essential for C while all other points of $\mathscr{T}(C, z)$ are not essential.

Proposition 1.1.11 (Proximity equality) *Let $(C, z) \subset (\Sigma, z)$ be a reduced curve singularity, and let $p \in \mathscr{T}(C, z)$. Then the multiplicity of the strict transform $C_{(p)}$ of C at p satisfies*

$$\operatorname{mt} C_{(p)} = \sum_{\substack{q \in \mathscr{T}(C,z) \\ q \dashrightarrow p}} \operatorname{mt} C_{(q)} . \tag{1.1.1.3}$$

Proof Without restriction, we may assume that $p = z$. Let $\widetilde{C}$ be the strict transform of (C, z) and E the exceptional divisor under the blowing up of $z \in \Sigma$. Then

$$\operatorname{mt}(C, z) = \widetilde{C} \cdot E = \sum_{w \in \widetilde{C} \cap E} i_w(\widetilde{C}, E) = \sum_{\substack{q \in \mathscr{T}(C,z) \\ q \dashrightarrow z}} \operatorname{mt} C_{(q)} \cdot \underbrace{\operatorname{mt} E_{(q)}}_{=1} ,$$

where the last equality holds due to Noether's formula (Proposition 1.1.8). □

Remark 1.1.11.1 (1) The difference between the multiplicities of the total and strict transform of C at a point $q \in \mathscr{T}(C, z)$ is

$$\operatorname{mt} \widehat{C}_{(q)} - \operatorname{mt} C_{(q)} = \sum_{\substack{p \in \mathscr{T}(C,z) \\ q \dashrightarrow p}} \operatorname{mt} \widehat{C}_{(p)} .$$

In particular, the multiplicities of the strict transforms of C together with the proximities ($q \dashrightarrow p$) determine the multiplicities of the total transforms.

(2) If $\mathscr{T} \subset \mathscr{T}(C, z)$ is a finite subtree and if p is a leaf of $\mathscr{T}$ then the equality (1.1.1.3) becomes a strict inequality $>$ if on the right-hand side $\mathscr{T}(C, z)$ is replaced by $\mathscr{T}$.

1.1.1.4 Cluster and Proximity Relations

The notion of a cluster is obtained by assigning virtual multiplicities to the points of a constellation (basically referring to the multiplicities of the strict transforms in a resolution process):

Definition 1.1.12 (*Cluster*) A *cluster on* Σ is a pair $(\mathscr{K}, \boldsymbol{m})$ consisting of a constellation $\mathscr{K} = (q_0, \pi_1, q_1, \ldots, \pi_n, q_n)$ on Σ and a vector $\boldsymbol{m} = (m_0, \ldots, m_n)$ of integers. The points q_i, $0 \leq i \leq n$, are called the *points of the cluster*. The integer m_i is called the *assigned (virtual) multiplicity* at the point q_i.

A *cluster on* Σ *with origin at* z is a cluster $(\mathscr{K}, \boldsymbol{m})$ where $\mathscr{K}$ is a constellation on Σ with origin z (see Definition 1.1.2).

The main results of this section are Proposition 1.1.17 and and Lemma 1.1.20. Proposition 1.1.17 gives a characterization of the clusters coming from a curve singularity (C, z) (that is, clusters such that the constellation is a subtree of the complete embedded resolution tree $\mathscr{T}(C, z)$ and the assigned multiplicities are the multiplicities of the respective strict transforms of (C, z)). Lemma 1.1.20 says that the virtual multiplicities of any cluster can be modified in a controlled way such that the newly obtained cluster comes from a curve singularity (see also [Cas1, Cas2]).

Instead of working with a cluster $(\mathscr{K}, \boldsymbol{m})$, it is frequently advisable to consider only the associated cluster graph

$$\operatorname{clg}(\mathscr{K}, \boldsymbol{m}) := (\Gamma_{\mathscr{K}}, \dashrightarrow, \boldsymbol{m}) ,$$

which encodes the essential (discrete) data of a cluster: the oriented graph (forest) $\Gamma_{\mathscr{K}}$, provided with the binary relation defined by $q \dashrightarrow p$ if q is proximate to p and with the vector of assigned multiplicities $\boldsymbol{m}$.

More generally, we introduce the notion of an (abstract) cluster graph, which a priori does not refer to a given cluster:

Definition 1.1.13 (*Cluster graph, proximity relation*) Let $\Gamma = (V, E)$ be a finite oriented graph with set of vertices V and with set of edges E and let $\geq$ be the (natural) partial ordering on V given by $p \geq q$ iff there is an oriented path in Γ from p to q.

(a) Let $\boldsymbol{m} = (m_q)_{q \in V}$ be a vector of integers, and let $(\dashrightarrow)$ be a binary relation on the set of vertices V satisfying the following properties:

 (P1) If $q \dashrightarrow p$ then $q \geq p$ and $q \neq p$.
 (P2) If $q \geq p$ and if q and p are connected by an edge then $q \dashrightarrow p$.
 (P3) If $r \geq q \geq p$ with $q \neq p$ and if $r \dashrightarrow p$ then $q \dashrightarrow p$.

 Then the triple $\mathbb{G} = (\Gamma, \dashrightarrow, \boldsymbol{m})$ is called a *cluster graph*. The vector $\boldsymbol{m}$ is also called the vector of *assigned multiplicities*.

(b) We say that the cluster graph $\mathbb{G} = (\Gamma, \dashrightarrow, \boldsymbol{m})$ *satisfies the proximity relation at p* if

$$m_p \geq \sum_{q \dashrightarrow p} m_q \,.$$

 A vertex $q \in V$ is called a *free vertex* of the cluster graph $\mathbb{G}$ iff there is at most one vertex $p \in V$ such that $q \dashrightarrow p$.

Note that if the proximity relation holds at each $p \in V$, all the assigned multiplicities m_p are non-negative (the empty sum being defined as 0).

As the assigned multiplicities in a cluster graph basically refer to multiplicities of strict transforms, the following definition of the assigned total multiplicities is modeled along the formula in Remark 1.1.11.1(1)):

Definition 1.1.14 (*Assigned total multiplicity*) Let $\mathbb{G} = (\Gamma, \dashrightarrow, \boldsymbol{m})$ be a cluster graph then we inductively define the *assigned total multiplicity* for a vertex $q \in V$ as

$$\widehat{m}_q := m_q + \sum_{q \dashrightarrow p} \widehat{m}_p \,. \tag{1.1.1.4}$$

Remark 1.1.14.1 Let $(\mathscr{K}, \boldsymbol{m})$ be a cluster on Σ. Then the oriented graph $\Gamma_{\mathscr{K}}$, provided with the binary relation defined by $q \dashrightarrow p$ if q is proximate to p and with the vector of assigned multiplicities $\boldsymbol{m}$, defines a cluster graph

$$\mathrm{clg}(\mathscr{K}, \boldsymbol{m}) = (\Gamma_{\mathscr{K}}, \dashrightarrow, \boldsymbol{m}).$$

We call it the *cluster graph defined by* $(\mathscr{K}, \boldsymbol{m})$.

The same information as encoded in the cluster graph $\mathrm{clg}(\mathscr{K}, \boldsymbol{m})$ is provided by the *bi-weighted graph of the cluster* $(\Gamma_{\mathscr{K}}, \boldsymbol{m}, \boldsymbol{\nu})$. Here, one associates to a vertex q the weights m_q, ν_q if the corresponding infinitely near point has assigned multiplicity m_q, and if ν_q is the smallest integer such that q is proximate to a point of level ν_q (see [NoV]).

Definition 1.1.15 (*Proximity relation*)

(a) A *cluster* $(\mathscr{K}, \boldsymbol{m})$ on Σ is said to *satisfy the proximity relations* if the cluster graph $\mathrm{clg}(\mathscr{K}, \boldsymbol{m})$ satisfies the proximity relation at each of its vertices.
(b) The *assigned total multiplicity* at a point $q \in \mathscr{K}$ is the total multiplicity $\widehat{m}_q$ assigned to the corresponding vertex in $\mathrm{clg}(\mathscr{K}, \boldsymbol{m})$.

Before turning to the main results of this section, we introduce a notation for the clusters coming from a plane curve singularity (C, z):

Definition 1.1.16 (*Cluster of a curve singularity*) Let $\mathscr{T} \subset \mathscr{T}(C, z)$ be any finite subtree, i.e. a constellation $\mathscr{T} = (z = q_0, q_1, \ldots, q_n)$. We define the *cluster of the curve singularity* (C, z) *w.r.t. the tree* $\mathscr{T}$ on Σ as

$$\mathcal{C}\ell(C, \mathscr{T}) := (\mathscr{T}, \boldsymbol{m})$$

by setting $m_0 := \mathrm{mt}(C, z)$ and $m_i := \mathrm{mt}\, C_{(q_i)}$, the multiplicity of the strict transform of C at q_i, $1 \leq i \leq n$. We call

$$\mathrm{clg}(C, \mathscr{T}) := \mathrm{clg}(\mathcal{C}\ell(C, \mathscr{T})) = (\Gamma_{\mathscr{T}}, \dashrightarrow, \boldsymbol{m}),$$

with proximities as in Definition 1.1.3, the *cluster graph of the curve singularity* (C, z) w.r.t. $\mathscr{T}$.

Remark 1.1.16.1 Let $f \in \mathcal{O}_{\Sigma,z}$ define (C, z) and let $\mathscr{T} = (q_0, q_1, \ldots, q_n) \subset \mathscr{T}(C, z)$ be a finite subtree. Then $\mathcal{C}\ell(C, \mathscr{T})$ is finitely determined. That is, there exists a $k \geq 0$ such that for any curve germ (C', z) defined by $f' \in \mathcal{O}_{\Sigma,z}$ having the same k–jet[2] as f, we have $\mathscr{T} \subset \mathscr{T}(C', z)$ and $\mathcal{C}\ell(C', \mathscr{T}) = \mathcal{C}\ell(C, \mathscr{T})$. This follows easily by blowing up z and induction on n.

Example 1.1.16.2 (a) Let (C, z) be an ordinary cusp and $\mathscr{T} = \mathscr{T}^*(C, z)$ the essential tree $\bullet \longleftarrow \bullet \longleftarrow \bullet$. Then the cluster graph $\mathrm{clg}(C, \mathscr{T})$ is given by the following data (the picture on the left refers to the points of the resolution process, the picture on the right to the assigned multiplicities):

Note that in the symbolic picture we omit the dashed edge (proximity) $p \dashrightarrow q$ whenever there exists an edge $p \longrightarrow q$.
(b) Let (C, z) be given by $(x^2 - y^3)(y^2 - x^3)$ and $\mathscr{T} \subset \mathscr{T}(C, z)$ some finite tree extending the essential tree (cf. Example 1.1.4.2). Then $\mathrm{clg}(C, \mathscr{T})$ is given by

[2]The *k–jet* of f is denoted by $f^{(k)}$ or $\mathrm{jet}(f, k)$ and is the image of f in the *jet-space* $J^{(k)} = \mathcal{O}_{\Sigma,z}/\mathfrak{m}_z^{k+1}$, where $\mathfrak{m}_z$ is the maximal ideal of $\mathcal{O}_{\Sigma,z}$. If we choose local coordinates of (Σ, z) we identify $f^{(k)}$ with its power series expansion up to order $\leq k$.

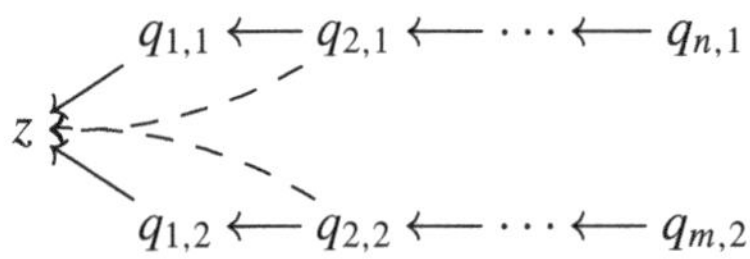
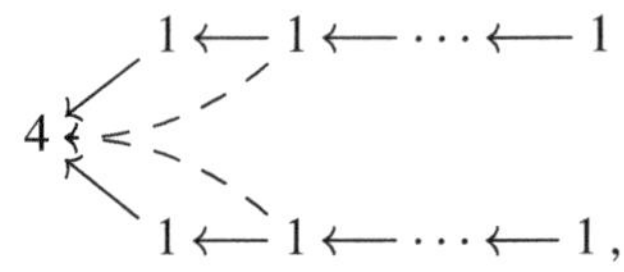

where $q_{1,1} = q_1$, $q_{2,1} = q_2$, $q_{1,2} = q_1' = q_3$, and $q_{2,2} = q_4$ in the notation of Example 1.1.4.2.
(c) Let (C, z) be given by $(x^3 - y^5)(x^2 - y^3)$ and $\mathcal{T} = \mathcal{T}^*(C, z) \subset \mathcal{T}(C, z)$ the essential tree (cf. [GLS6, Figure I.3.17, p. 192]). Then $\mathrm{clg}(C, \mathcal{T})$ is given by

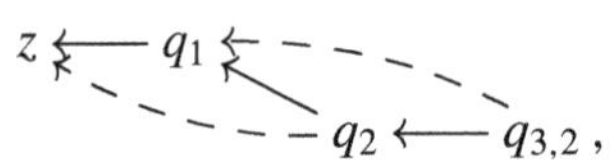
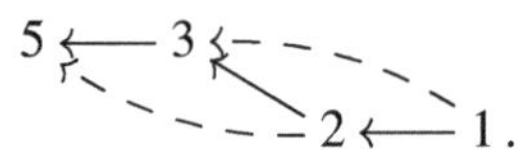

Due to Proposition 1.1.11, each cluster $\mathcal{C}\ell(C, \mathcal{T})$ satisfies the proximity relations (and the assigned total multiplicity $\widehat{m}_i := \widehat{m}_{q_i}$ as defined in Definition 1.1.13 coincides with the total multiplicity of C at the respective point q_i).

The following proposition shows that, vice versa, each cluster satisfying the proximity relations arises in the above way from some plane curve singularity:

Proposition 1.1.17 *Given a cluster $(\mathcal{K}, \boldsymbol{m})$ on Σ with origin at z. Then $(\mathcal{K}, \boldsymbol{m})$ satisfies the proximity relations iff there exist a reduced plane curve germ $(C, z) \subset (\Sigma, z)$ and a finite subtree $\mathcal{T} \subset \mathcal{T}(C, z)$ containing the essential tree $\mathcal{T}^*(C, z)$ such that*

$$(\mathcal{K}, \boldsymbol{m}) = \mathcal{C}\ell(C, \mathcal{T}). \tag{1.1.1.5}$$

Moreover, there is no point q infinitely near point to z and outside of $\mathcal{T}$ such that all curve germs (C, z) satisfying (1.1.1.5) go through q.

Proof The "if" statement follows immediately from Proposition 1.1.11, hence, it just remains to show the "only if" statement.

We proceed by induction on the highest level of a point in the constellation $\mathcal{K}$, proving the following
Claim. Let the cluster $(\mathcal{K}, \boldsymbol{m})$ satisfy the proximity relations, and let $\mathcal{K}'$ be any constellation with origin at z, containing $\mathcal{K}$ as a subconstellation. Then there exists a reduced curve germ (C, z) such that $(\mathcal{K}, \boldsymbol{m}) = \mathcal{C}\ell(C, \mathcal{T})$ for some finite subtree $\mathcal{T} \subset \mathcal{T}(C, z)$ containing the essential tree $\mathcal{T}^*(C, z)$, and such that $\mathcal{T}(C, z) \cap \mathcal{K}' = \mathcal{K}$.
If $\mathcal{K} = (z)$ then we may choose (C, z) to be an ordinary m_z-fold point with tangent directions different from those corresponding to the points of level 1 in $\mathcal{K}'$.

To apply induction we blow up z. The only problem is that we have to take care of proximities between z and points of higher level in $\mathcal{K}$.

Now, let $\pi : \Sigma_1 \to \Sigma$ be the blowing up of z, and assume that the constellation $\mathcal{K}$ has $s \geq 1$ points $q_1, \ldots, q_s$ of level 1. We consider the constellations $\mathcal{K}_1, \ldots, \mathcal{K}_s$ on Σ_1, with origin at $q_1, \ldots, q_s$, respectively, obtained from $\mathcal{K}$ by omitting the (unique) point z of level 0. Proceeding in the same manner with $\mathcal{K}'$, we obtain constellations $\mathscr{K}_j'$ on Σ_1 with origin at q_j', $j = 1, \dots, s'$, and we may assume that, for $j = 1, \dots, s \le s'$, we have $q_j' = q_j$, and that $\mathscr{K}_j$ is a subconstellation of $\mathscr{K}_j'$.

Let E denote the exceptional divisor of π, and extend the constellations $\mathscr{K}_1', \dots, \mathscr{K}_s'$ such that $\mathscr{K}_j' \cap \mathscr{T}(E, q_j) \supsetneq \mathscr{K}_j \cap \mathscr{T}(E, q_j)$, $j = 1, \dots, s$. By the induction hypothesis, we find plane curve germs $(D_j, q_j) \subset (\Sigma_1, q_j)$ and finite subtrees $\mathscr{T}_j \subset \mathscr{T}(D_j, q_j)$, $j = 1, \dots, s$, such that $\mathscr{T}_j \supset \mathscr{T}^*(D_j, q_j)$, $(\mathscr{K}_j, \boldsymbol{m}) = \mathcal{C}\ell(D_j, \mathscr{T}_j)$ and $\mathscr{T}(D_j, q_j) \cap \mathscr{K}_j' = \mathscr{K}_j$. In particular, by Noether's formula (Proposition 1.1.8),

$$i_{q_j}(D_j, E) = \sum_{q \in \mathscr{K}_j \cap \mathscr{T}(E)} \operatorname{mt} D_{j,(q)} = \sum_{q \in \mathscr{K}_j \cap \mathscr{T}(E)} m_q \,.$$

By [GLS6, Lemma I.3.20], we can blow down E, that is, there are (uniquely determined) unitangential plane curve singularities $(C_1, z), \dots, (C_s, z)$ such that (D_j, q_j) is the strict transform of (C_j, z) at q_j, $j = 1, \dots, s$. Their union $C_1 \cup \ldots \cup C_s \subset (\Sigma, z)$ is a plane germ of multiplicity $\sum_{q \dashrightarrow z} m_q$, which is at most m_z, due to the proximity relation at z. If the multiplicity equals m_z, we have found the wanted plane curve singularity (C, z). If it is strictly less, then we get this plane curve singularity by adding the appropriate number of generic smooth branches (e.g. in local coordinates we may take straight lines from a Zariski-open dense subset of all linear forms). The constellation $\mathscr{T}$ is obtained by extending (z) successively by the constellations $\mathscr{T}_j$. Note that the essential tree $\mathscr{T}^*(C, z)$ consists of z (if (C, z) is singular), the essential points of (D_j, q_j) and the points in $\mathscr{T}(D_j, q_j) \cap \mathscr{T}(E, q_j)$, $j = 1, \dots, s$. By construction, all these points are in $\mathscr{T}$. □

Remark 1.1.17.1 Let $(C, z) \subset (\Sigma, z)$ be a reduced plane curve singularity and $\mathscr{T}$ a finite (oriented) subtree of the complete embedded resolution tree $\mathscr{T}(C, z)$. We collect and reconsider the definitions made so far for this concrete situation.

Note that (by abuse of notation) $\mathscr{T}$ is not a graph but a constellation $(z = q_0, \pi_1, q_1, \dots, \pi_n, q_n)$ with $\pi_i : \Sigma_i \to \Sigma_{i-1}$ the blowing up of $q_{i-1} \in \Sigma_{i-1}$ and $q_i \in \Sigma_i$, $i = 1, \dots, n$ $(\Sigma_0 = \Sigma)$. Hence $\mathscr{T}$ corresponds to a partial resolution of (C, z) or to a non-minimal resolution by further blowing up points. If $\mathscr{T} = \mathscr{T}^*(C, z)$ is the essential tree, then the q_i are exactly the singular points that are not nodes of the reduced total transform of the minimal good embedded resolution of (C, z), with z always belonging to $\mathscr{T}$ if (C, z) is not smooth.

$\Gamma_{\mathscr{T}}$, the *graph of* $\mathscr{T}$, has $V = (q_0, \dots, q_n)$ as set of vertices and (q_i, q_j) is an oriented edge of $\Gamma_{\mathscr{T}}$ if q_j is infinitely near to q_i with level $l(q_j) = l(q_i) + 1$ (cf. Definition 1.1.2).

The *cluster of* (C, z) w.r.t. $\mathscr{T}$, is $\mathcal{C}\ell(C, \mathscr{T}) = (\mathscr{T}, \boldsymbol{m})$, $\boldsymbol{m} = (m_0, \dots, m_n)$, and has assigned multiplicities m_i, where we assign to q_0 the multiplicity m_0 of (C, z) and to q_i, $i \ge 1$, the multiplicity m_i of the strict transform $C_{(q_i)}$ at q_i of (C, z) under the morphism $\pi_{(q_i)} = \pi_1 \circ \cdots \circ \pi_i : (\Sigma_i, q_i) \to (\Sigma, z)$.

The *cluster graph* $\mathbb{G} = \operatorname{clg}(C, \mathscr{T})$ defined by $\mathcal{C}\ell(C, \mathscr{T})$ is the triple $\mathbb{G} = (\Gamma_{\mathscr{T}}, \dashrightarrow, \boldsymbol{m})$ with $q_j \dashrightarrow q_i$ if q_j is proximate to q_i, i.e. if $q_j \in E_{q_i}$, the (strict transform of the) exceptional divisor of π_{i+1}. By Proposition 1.1.11 $\mathbb{G}$ satisfies the proximity relations. Note that the multiplicities m_i of the strict transform, together with the proximities ($q_j \dashrightarrow q_i$) determine the multiplicities $\widehat{m}_i$ of the total transform $\widehat{C}_{(q_i)}$ at q_i of (C, z) by Remark 1.1.11.1.

It is well known that the multiplicities of the reduced total transform is a complete set of invariants for the topological type of a plane curve singularity (C, z) (see [Bra, Zar3] and [BrK, 8.4, Theorem 21], see also [GLS6, Theorem I.3.42]). We have hence the following proposition.

Proposition 1.1.18 *For a reduced plane curve singularity (C, z) the cluster graph $\mathbb{G} = \mathrm{clg}(C, \mathscr{T}^*(C, z))$ determines and is determined by the topological type of (C, z).*

1.1.1.5 The Unloading Principle

Throughout the following, we restrict ourselves to clusters $(\mathscr{K}, \boldsymbol{m})$ *with origin at* $z \in \Sigma$. But there is an immediate generalization to the case of clusters having more than one point of level 0.

Definition 1.1.19 (*Curve through a cluster*) We say that a plane curve C, resp. the germ (C, z), *goes through the cluster* $(\mathscr{K}, \boldsymbol{m})$ if, for each $q \in \mathscr{K}$, the total transform $\widehat{C}_{(q)}$ of C at the point q has at least the assigned total multiplicity $\widehat{m}_q = m_q + \sum_{q \dashrightarrow p} \widehat{m}_p$.

If (C, z) is defined by $g \in \mathcal{O}_{\Sigma,z}$ this means that $\mathrm{mt}(\widehat{g}_{(q)}) \geq \widehat{m}_q$, where $\widehat{g}_{(q)}$ defines $\widehat{C}_{(q)}$, and we also say that *g goes through the cluster* $(\mathscr{K}, \boldsymbol{m})$.

Consider the cluster graph $\mathrm{clg}(\mathscr{K}, \boldsymbol{m}) = (\Gamma_{\mathscr{K}}, \dashrightarrow, \boldsymbol{m})$ with set of vertices V (recall that we use the same notation for the vertex in V as for the corresponding point in $\mathscr{K}$). Assume that the proximity relation at $p_0 \in V$ is not satisfied, that is,

$$m_{p_0} < M_{p_0} := \sum_{q \dashrightarrow p_0} m_q \,.$$

Enriques introduced a rule to construct a new cluster $(\mathscr{K}', \boldsymbol{m}')$ being "close" to $(\mathscr{K}, \boldsymbol{m})$ (cf. Lemma 1.1.20(b)) and satisfying the proximity relation at p_0. He called that rule the *"unloading principle"* (principio di scaricamento [EnC, IV.17]): Let

$$r := \#\big(\{q \in \mathscr{K} \mid q \text{ is proximate to } p_0\}\big),$$

and

$$N := \min\left\{ n \in \mathbb{N} \,\middle|\, n \geq \frac{M_{p_0} - m_{p_0}}{r+1} \right\}.$$

Then we define a new cluster graph with new assigned multiplicities

$$m'_q := \begin{cases} m_q - N & \text{if } q \dashrightarrow p_0, \\ m_q + N & \text{if } q = p_0, \\ m_q & \text{otherwise,} \end{cases}$$

and new set of vertices $V' \subset V$ such that

$$q \in V' \;:\Longleftrightarrow\; \text{there exists} q \leq q' \in V \text{with } m'_{q'} \neq 0\,.$$

Furthermore, we introduce the new assigned total multiplicities $\widehat{m}'_q$ as done in (1.1.1.4). Note that there is a representative $(q_0, \pi_1, q_1, \ldots, \pi_n, q_n)$ of the constellation $\mathscr{K}$ such that $q_0, \ldots, q_{n'}$ correspond to the points in V' and $q_{n'+1}, \ldots, q_n$ correspond to the omitted points. Finally, we define the new cluster $(\mathscr{K}', \boldsymbol{m}')$ with $\mathscr{K}' := (q_0, \pi_1, q_1, \ldots, \pi_{n'}, q_{n'})$.

Lemma 1.1.20 *With the above notations, let $(\mathscr{K}', \boldsymbol{m}')$ be the cluster obtained from $(\mathscr{K}, \boldsymbol{m})$ by applying the unloading principle (at a vertex $p_0 \in V$). Then the following holds:*

(a) The cluster graph $\operatorname{clg}(\mathscr{K}', \boldsymbol{m}')$ *satisfies the proximity relation at $p_0 \in V'$.*
(b) A plane curve germ $(C, z) \subset (\Sigma, z)$ goes through the cluster $(\mathscr{K}', \boldsymbol{m}')$ iff it goes through $(\mathscr{K}, \boldsymbol{m})$.

Proof (a) We have

$$\sum_{q \dashrightarrow p_0} m'_q \;=\; \sum_{q \dashrightarrow p_0} (m_q - N) \;\leq\; M_{p_0} - (M_{p_0} - m_{p_0}) + N \;=\; m'_{p_0}\,.$$

(b) By construction, the assigned (virtual and total) multiplicities of the cluster $(\mathscr{K}, \boldsymbol{m})$ and of $(\mathscr{K}', \boldsymbol{m}')$ can only differ at p_0 and at points infinitely near to p_0. Hence, we can assume without restriction that $p_0 = z$, the origin of $\mathscr{K}$.
Step 1. We prove that $\widehat{m}_q = \widehat{m}'_q$ for each $z \neq q \in \mathscr{K}$:

Obviously, it suffices to consider those points $q \in \mathscr{K}$, with $q \dashrightarrow z$. If q is a free point, then

$$\widehat{m}'_q \;=\; m'_q + \sum_{q \dashrightarrow p} \widehat{m}'_p \;=\; m'_q + m'_z \;=\; (m_q - N) + (m_z + N) \;=\; \widehat{m}_q\,.$$

In the case of satellite points we may proceed by induction (on the level). Let $q \in \mathscr{K}$ be such that $q \dashrightarrow z$ and $q \dashrightarrow p \neq z$, in particular, $i := \ell(q) \geq 2$. Assuming that the statement holds for all points of level $j \leq i - 1$ (hence for p) we obtain

$$\widehat{m}'_q \;=\; m'_q + \widehat{m}'_z + \widehat{m}'_p \;=\; (m_q - N) + (m_z + N) + \widehat{m}_p \;=\; \widehat{m}_q\,.$$

Step 2. It remains to show that $\operatorname{mt}(C, z) \geq m'_z$ whenever $\operatorname{mt} \widehat{C}_{(q)} \geq \widehat{m}_q$ for each $q \in \mathscr{K}$.

Let $q_{1,1}, \ldots, q_{1,s}$ be the points of $\mathscr{K}$ of level 1, that is, the points in the first neighbourhood of z (which are all proximate to z). For each $j = 1, \ldots, s$, we denote by $q_{2,j}, \ldots, q_{i_j,j}$ the points of $\mathscr{K}$ of level $2, \ldots, i_j$ that are infinitely near to $q_{1,j}$ and proximate to z. By Remark 1.1.11.1 and applying induction, we obtain

$$
\begin{aligned}
\sum_{j=1}^{s} \operatorname{mt} \widehat{C}_{(q_{i_j},j)} &= \sum_{j=1}^{s} \Big(\operatorname{mt} C_{(q_{i_j},j)} + \operatorname{mt}(C,z) + \operatorname{mt} \widehat{C}_{(q_{i_j-1},j)} \Big) \\
&= \sum_{j=1}^{s} \Big(i_j \cdot \operatorname{mt}(C,z) + \sum_{i=1}^{i_j} \operatorname{mt} C_{(q_i,j)} \Big) = r \cdot \operatorname{mt}(C,z) + \sum_{q \dashrightarrow z} \operatorname{mt} C_{(q)} \\
&\leq (r+1) \cdot \operatorname{mt}(C,z)\,.
\end{aligned}
$$

On the other hand, the definition of the assigned total multiplicities implies (in the same manner) that

$$
\sum_{j=1}^{s} \widehat{m}_{q_{i_j},j} = r \cdot m_z + \sum_{q \dashrightarrow z} m_q = (r+1) \cdot m_z - m_z + M_z\,.
$$

We conclude that $\operatorname{mt}(C,z) \geq (M_z - m_z)/(r+1) + m_z$, hence (b) holds. □

Remark 1.1.20.1 In the proof of Lemma 1.1.20, we have seen that applying the unloading principle (at a vertex $p_0 \in V$) raises the sum of the assigned *total* multiplicities $\widehat{m}_q^{(i)}$ by $N > 0$. Hence, it is clear that after applying finitely many times the unloading principle at some vertex of V, the resulting cluster satisfies the proximity relations.

1.1.2 Cluster Schemes and Equisingularity

In this section, we introduce the general notion of a cluster scheme, a zero–dimensional scheme on a smooth surface Σ associated to a cluster on Σ, and relate it to equisingular deformations of plane curve singularities. We put special emphasis on the *topological singularity scheme* associated to a reduced curve singularity $(C,z) \subset (\Sigma,z)$, which is used in the construction of curves with prescribed topological singularities (see Definition 1.1.31 and Sect. 4.5). Moreover, we study the *conductor scheme* (defined by the conductor ideal $I^{cd}(C,z) \subset \mathcal{O}_{C,z}$) which is relevant to the new invariants discussed in Sect. 1.2.2. Finally, we discuss the notion of residue schemes which play an important role in Chap. 3.

1.1.2.1 Cluster Schemes

Definition 1.1.21 (*Cluster scheme*) Let $(\mathcal{K}, \boldsymbol{m})$ be a cluster, and let $z_1, \ldots, z_r \in \Sigma$ be the points of level 0 of $\mathcal{K}$. Moreover, let $\mathcal{K}_i \subset \mathcal{K}$ be the maximal subconstellation with origin z_i (that is, the constellation consisting of all points of $\mathcal{K}$ which are infinitely near to z_i) and $\boldsymbol{m}_i = (m_q)_{q \in \mathcal{K}_i}$. Then, for each $i = 1, \ldots, r$, consider the set

$$
I_i = \big\{ g \in \mathcal{O}_{\Sigma,z_i} \,\big|\, g \text{ goes through the cluster } (\mathcal{K}_i, \boldsymbol{m}_i) \big\} \subset \mathcal{O}_{\Sigma,z_i}\,.
$$

Because g goes through the cluster $(\mathscr{K}_i, \boldsymbol{m}_i)$ iff $\operatorname{mt}(\widehat{g}_{(q)}) \geq \widehat{m}_q$ for $q \in \mathscr{K}_i$, I_i is certainly an ideal. Since $\mathscr{K}_i$ consists only of finitely many points and since the multiplicity of the total transform does not drop under blowing ups, any g of sufficiently high multiplicity goes through $(\mathscr{K}_i, \boldsymbol{m}_i)$. Hence, I_i defines a zero-dimensional scheme

$$Z_i := V(I_i) \subset \Sigma$$

with support $\{z_i\}$. We call the union of these schemes,

$$Z(\mathscr{K}, \boldsymbol{m}) := Z_1 \cup \ldots \cup Z_r \subset \Sigma ,$$

the *zero-dimensional scheme defined by the cluster* $(\mathscr{K}, \boldsymbol{m})$, or, simply, a *cluster scheme* (see also Definition 1.1.30).[3] The support of $Z(\mathscr{K}, \boldsymbol{m})$ are the points of level 0 of $\mathscr{K}$.

In particular, if all points of $\mathscr{K}$ are of level 0, we call $Z(\mathscr{K}, \boldsymbol{m})$ an *(ordinary) fat point scheme*. It is defined by the ideals $I_i = \mathfrak{m}_{z_i}^{m_i} \subset \mathcal{O}_{\Sigma,z_i}$. For such a fat point scheme, we usually use the notation $Z(\boldsymbol{z}, \boldsymbol{m})$ with $\boldsymbol{z} = (z_1, \ldots, z_r)$.

Notation. The set of all cluster schemes on Σ is denoted by $\mathcal{CS}$.
Recall that Lemma 1.1.20 implies that each cluster scheme $Z(\mathscr{K}, \boldsymbol{m})$ can be defined by a cluster $(\mathscr{K}', \boldsymbol{m}')$ satisfying the proximity relations (see Remark 1.1.20.1). Thus, the following lemma (together with the unloading principle) gives a formula for computing the degree of any cluster scheme:

Lemma 1.1.22 *Let* $(\mathscr{K}, \boldsymbol{m})$ *be a cluster satisfying the proximity relations. Then*

$$\deg Z(\mathscr{K}, \boldsymbol{m}) = \sum_{q \in \mathscr{K}} \frac{m_q(m_q+1)}{2} . \tag{1.1.2.1}$$

Proof We proceed by induction on the right-hand side of (1.1.2.1). If it equals zero, that is, if there is no $q \in \mathscr{K}$ with $m_q > 0$, then the equation is trivially satisfied.

Otherwise, we can choose a point $p \in \mathscr{K}$ such that $m_p > 0$ and no point $q \in \mathscr{K}$ infinitely near to p has a positive assigned multiplicity $m_q > 0$. Consider the cluster $(\mathscr{K}', \boldsymbol{m}')$ defined by the constellation $\mathscr{K}'$ obtained from $\mathscr{K}$ when adding $m_p - 1$ free points in the first neighbourhood of p and the assigned multiplicities

$$m'_q = \begin{cases} m_p - 1 & \text{if } q = p , \\ m_q & \text{if } q \in \mathscr{K} \setminus \{p\} , \\ 1 & \text{if } q \in \mathscr{K}' \setminus \mathscr{K} . \end{cases}$$

[3] The first systematic ideal-theoretic treatment of systems of plane curves passing with assigned multiplicities through assigned infinitely near points was done by Zariski [Zar1]. He introduced the class of *complete ideals* I, which can be characterized by the fact that, for each d, the linear system of plane curves of degree d defined by polynomials in I corresponds to a *complete linear system* on some projective surface obtained from the projective plane by finitely many blowing ups of points. In these terms, the ideals I_i occurring in the definition of a cluster scheme are just $\mathfrak{m}_{z_i}$-primary complete ideals (see [Cas2, §8.3] for details).

Clearly, $(\mathscr{K}', \boldsymbol{m}')$ satisfies the proximity relations and

$$\sum_{q\in\mathscr{K}'} \frac{m'_q(m'_q+1)}{2} = \sum_{q\in\mathscr{K}} \frac{m_q(m_q+1)}{2} - 1 .$$

Hence, by the induction hypothesis, the latter equals $\deg Z(\mathscr{K}', \boldsymbol{m}')$.

Let $(\mathscr{K}'', \boldsymbol{m}'')$ be a third cluster, obtained from $(\mathscr{K}', \boldsymbol{m}')$ by adding another point $p' \notin \mathscr{K}'$ with assigned multiplicity 1. Of course, this cluster does not satisfy the proximity relation at p, and the cluster obtained from $(\mathscr{K}'', \boldsymbol{m}'')$ when applying the unloading principle at p is just our original cluster $(\mathscr{K}, \boldsymbol{m})$. In particular, $\deg Z(\mathscr{K}'', \boldsymbol{m}'') = \deg Z(\mathscr{K}, \boldsymbol{m})$ due to Lemma 1.1.20.

Moreover, by Proposition 1.1.17, there is a plane curve germ passing through $(\mathscr{K}', \boldsymbol{m}')$ but not through the extended cluster $(\mathscr{K}'', \boldsymbol{m}'')$. Hence,

$$\deg Z(\mathscr{K}'', \boldsymbol{m}'') \leq \deg Z(\mathscr{K}', \boldsymbol{m}') + 1 .$$

On the other hand, it is not difficult to see that the extra point with assigned multiplicity one imposes at most one extra condition, see Lemma 1.1.23 below (applied to a linear system of plane curves of large degree, such that the generic curve is irreducible without changing the local singularities). That is,

$$\deg Z(\mathscr{K}, \boldsymbol{m}) = \deg Z(\mathscr{K}'', \boldsymbol{m}'') \geq \deg Z(\mathscr{K}', \boldsymbol{m}') + 1 ,$$

and the above allows to conclude the equality (1.1.2.1). □

The following lemma implies that for a family of projective hypersurfaces with projective base the condition that all hypersurfaces pass through a fixed point "imposes at most one condition on the base" (under some mild conditions), i.e. reduces the dimension of the family by at most one. This applies in particular to a linear system of projective plane curves.

Lemma 1.1.23 *Let Σ and T be irreducible projective varieties and $\pi : \mathscr{X} \to T$ a family of Cartier divisors in Σ, i.e. $\mathscr{X} \subset \Sigma \times T$ is a Cartier divisor, flat over T via the projection π. Assume that the generic fibre of π is irreducible and that Σ contains at least two distinct Cartier divisors as fibres. Then, for any point $z \in \Sigma$, the set*

$$T_z := \{t \in T \ : \ z \in \mathscr{X}_t\}$$

is a non-empty subvariety of T of codimension at most one.

Proof The family $\pi : \mathscr{X} \to T$ is flat, hence open and therefore dominant since T is irreducible. By assumption there is an open dense subset of T over which the fibres of π are irreducible. It follows from Proposition 1.1.56 that $\mathscr{X}$ is irreducible. The projection $\mathrm{pr} : \mathscr{X} \to \Sigma$ (which embeds each fibre into Σ) is projective, hence the image is a closed, irreducible subvariety of Σ of dimension $\geq \dim \mathscr{X}_t = \dim \Sigma - 1$.

By assumption there are $t, t' \in T$ such that $\mathscr{X}_t \neq \mathscr{X}_{t'}$ and hence pr is onto since Σ is irreducible. We get

$$\dim \mathrm{pr}^{-1}(z) \geq \dim \mathscr{X} - \dim \Sigma = \dim T - 1.$$

Since $\{z\} \times T_z = \mathrm{pr}^{-1}(z)$, $\dim T_z \geq \dim T - 1$. □

1.1.2.2 Maximal Cluster Subschemes

There is a natural way to associate to any zero-dimensional scheme Z a cluster $\mathcal{C}\ell(Z)$. If $Z = Z(\mathscr{K}, \boldsymbol{m})$ is a cluster scheme, the associated cluster is just $(\mathscr{K}, \boldsymbol{m})$. For other schemes Z, we obtain a cluster whose associated cluster scheme is a subscheme of Z, the maximal cluster scheme contained in Z. This may, for instance, be used to estimate the degree of a zero-dimensional scheme Z (see Corollary 1.1.40 below).

Definition 1.1.24 (*Strict transform of a zero-dimensional scheme*) Let Z be a zero-dimensional scheme with support $\{z\}$, and let $I_Z \subset \mathcal{O}_{\Sigma,z}$ be the defining ideal. Further, let $\pi : \Sigma' \to \Sigma$ be the blowing up of z. Then the *strict transform* of Z is the zero-dimensional scheme in Σ', whose support is contained in the exceptional divisor E, and which (locally at a point $q \in E$) is defined by the ideal quotient $\widehat{I_Z} : I_E^{\mathrm{mt}(Z)}$. Here, $\widehat{I_Z} \subset \mathcal{O}_{\Sigma',q}$ denotes the ideal generated by the total transforms of the elements of I_Z, and I_E denotes the ideal defining E. Similarly, we define the strict transform of a zero-dimensional scheme Z if the support of Z consists of more than one point.

Definition 1.1.25 (*Cluster of a zero-dimensional scheme*) Given a zero-dimensional scheme $Z_0 := Z \subset \Sigma =: \Sigma_0$, let $\pi_i : \Sigma_i \to \Sigma_{i-1}$ be the blowing up of $\mathrm{supp}(Z_{i-1}) \subset \Sigma_{i-1}$, and let Z_i be the strict transform of Z_{i-1}, $i = 1, \ldots, r$. Note that $\mathrm{supp}(Z_i)$ consists of finitely many infinitely near points (of level i) and that $\deg Z_i$ is strictly decreasing in i (which can be seen, for instance, by using standard bases). Recall that $\deg Z = \dim_{\mathbb{C}} H^0(Z, \mathcal{O}_Z)$ and that $\mathrm{mt}(Z, z)$ is the minimum order at z of the elements contained in I_Z. We choose r being minimal with the property $\mathrm{supp}(Z_r) = \emptyset$. Then

$$\mathscr{K} := \big(\mathrm{supp}(Z_0), \mathrm{supp}(Z_1), \ldots, \mathrm{supp}(Z_{r-1})\big)$$

defines a constellation on Σ. Setting

$$m_q := \mathrm{mt}(Z_i, q) \quad \text{for each} \quad q \in \mathrm{supp}(Z_i)\,,$$

the cluster $\mathcal{C}\ell(Z) := (\mathscr{K}, \boldsymbol{m})$ is called *the cluster defined by the zero-dimensional scheme Z*. We write $\mathrm{clg}(Z) := \mathrm{clg}(\mathcal{C}\ell(Z))$ and call it the *cluster graph of the zero-dimensional scheme Z*.

Denote by Z_{cl} the zero-dimensional scheme defined by $\mathcal{C}\ell(Z)$. It is the *maximal cluster scheme* contained in Z as we shall see below in Lemma 1.1.26.

In the following, we frequently use the notation $\mathscr{T}_Z$ for the constellation $\mathscr{K}$ defined by the zero-dimensional scheme Z. Here, the letter $\mathscr{T}$ refers again to the tree structure of the graph $\Gamma_{\mathscr{K}}$.

Lemma 1.1.26 *We have $I_{Z_{cl}} \supset I_Z$. Moreover, if $I' \supset I_Z$ is any ideal of a cluster scheme then $I_{Z_{cl}} \subset I'$, hence Z_{cl} is the maximal cluster scheme contained in Z.*

Proof We sketch a proof, leaving details to the reader. It is clear that $I_{Z_{cl}} \supset I_Z$. We use now induction on $\deg Z$ with the evident base case of $\deg Z = 1$. Without loss of generality, we can assume that Z is irreducible, i.e. supported at one point z. Let $I' \supset I_Z$ be the ideal of a cluster scheme Z'. Clearly, $\mathrm{mt}(Z', z) \le \mathrm{mt}(Z_{cl}, z) = \mathrm{mt}(Z, z)$. If $\mathrm{mt}(Z', z) = \mathrm{mt}(Z, z)$, then we blow up Σ at z and apply the induction assumption. Suppose that $\mathrm{mt}(Z', z) = \mathrm{mt}(Z, z) - k$, $k > 0$. After the blow-up of z we obtain that $Z_1' \subset Z_1 \cup E^k$, E the exceptional divisor. We have to show that $Z_1' \subset (Z_{cl})_1 \cup E^k$. The scheme Z_1' can be supported in $\mathrm{supp}(Z_1)$ and at some finite set in $E \setminus \mathrm{supp}(Z_1)$. It follows that $Z_1' : E^k \subset Z_1$, which means that $\mathrm{supp}(Z_1' : E^k) \subset \mathrm{supp}(Z_1)$. Since $Z_1' : E_k$ is a cluster scheme (see Proposition 1.1.47 below), by the induction assumption $Z_1' : E^k \subset (Z_{cl})_1$, and hence $Z_1' \subset (Z_{cl})_1 \cup E^k$ implying $Z' \subset Z_{cl}$.

Example 1.1.26.1 (a) If the cluster $(\mathscr{K}, \boldsymbol{m})$ satisfies the proximity relations then Proposition 1.1.17 implies that $\mathcal{C}\ell(Z(\mathscr{K}, \boldsymbol{m})) = (\mathscr{K}, \boldsymbol{m})$.

(b) Let $Z \subset \Sigma$ be the zero-dimensional scheme with support $\{z\}$, given by the ideal $I_Z = \langle x^2, y^2 \rangle$ (x, y being local coordinates at z). Then the maximal cluster scheme $Z_{cl} \subset Z$ is defined by the ideal $\langle x^2, xy, y^2 \rangle \supsetneq I_Z$.

(c) Let $(C, z) \subset (\Sigma, z)$ be given by $f = (x^2 - y^3)(y^2 - x^3)$ (cf. Example 1.1.4.2), and let $Z \subset \Sigma$ be the zero-dimensional scheme with support $\{z\}$, given by the Tjurina ideal $\langle f, \frac{\partial f}{\partial x}, \frac{\partial f}{\partial y} \rangle$ of (C, z). Then the cluster graph of $\mathcal{C}\ell(Z)$ is given by

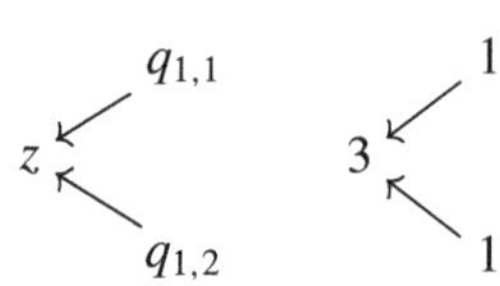

(the underlying constellation being a subconstellation of $\mathscr{T}(C, z)$, cf. Example 1.1.16.2(b)). This can be seen, for instance, by considering the following SINGULAR session (see [GrP], [DeL] for an introduction to SINGULAR):

```
ring r = 0,(x,y),ds;
option(redSB);
poly f = (x2-y3)*(y2-x3);
ideal I_Z = f, jacob(f);
I_Z = std(I_Z);            // standard basis for I_Z
I_Z;
//-> I_Z[1]=2x2y-5y4
//-> I_Z[2]=2xy2-5x4
//-> I_Z[3]=x5
```

```
//-> I_Z[4]=y5
vdim(I_Z);              // deg(Z)
//-> 10
ring r1 = 0,(u,v),dp;
map phi = r,uv,v;
map psi = r,u,uv;
ideal I1 = phi(I_Z)/v3; // strict transform of I_Z in Chart 1
vdim(std(I1));     // degree of the 0-dim scheme defined by I2
                   // in the affine plane with origin q_{1,1}
//-> 1
ideal I2 = psi(I_Z)/u3; // strict transform of I_Z in Chart 2

ring r1_local = 0,(u,v),ds;
ideal I1 = imap(r1,I1);
vdim(std(I1));     // degree of the 0-dim scheme defined by I2
                   // at the point q_{1,1}
//-> 1
ideal I2 = imap(r1,I2);
vdim(std(I2));     // degree of the 0-dim scheme defined by I2
                   // at the point q_{1,2}
//-> 1
```

In this case, we have $\deg(Z) = 10$, while $\deg(Z_{cl}) = 8$.

1.1.2.3 Equisingular Deformations

In the remaining part of this section, we restrict ourselves to cluster schemes with support $\{z\}$, that is, cluster schemes associated to clusters with origin z. By Lemma 1.1.20 they can be defined by clusters $(\mathscr{K}, \boldsymbol{m})$ satisfying the proximity relations. By Proposition 1.1.17, we know that each such cluster $(\mathscr{K}, \boldsymbol{m})$ comes from a reduced plane curve singularity $(C, z) \subset (\Sigma, z)$, that is, $(\mathscr{K}, \boldsymbol{m}) = \mathcal{C}\ell(C, \mathscr{T})$ for some finite tree $\mathscr{T} = \mathscr{T}_Z$ containing the essential tree, i.e. with $\mathscr{T}^*(C, z) \subset \mathscr{T} \subset \mathscr{T}(C, z)$. If $\mathscr{T} = (z = q_0, , q_1, \ldots, q_n)$ then $\boldsymbol{m} = (m_0, m_1, \ldots, m_n)$ with $m_i := \operatorname{mt} C_{(q_i)}$, the multiplicity of the strict transform of C at q_i.

Before we define in the next section the topological singularity scheme, recall the definition of an equisingular deformation of (C, z) from [GLS6, Definition II.2.1].

We consider an *embedded deformation* (i, Φ, σ) of $(C, z) \subset (\Sigma, z)$, *with section* over a complex germ $(T, 0)$. This is a commutative diagram

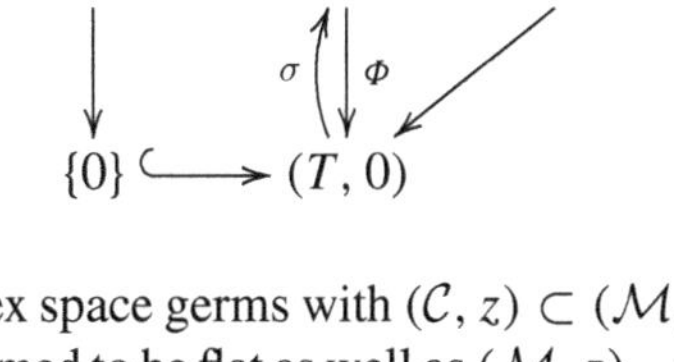

of morphisms of complex space germs with $(\mathcal{C}, z) \subset (\mathcal{M}, z)$ a hypersurface germ, σ a section of Φ. Φ is assumed to be flat as well as $(\mathcal{M}, z) \to (T, 0)$ which has (Σ, z) a special fibre. An embedded deformation (without section), denoted by (i, Φ), is given by a diagram as above but with σ deleted. We usually choose small representatives for the germs and denote them with the same letters, omitting the base points. Note that, for small representatives, we have $\mathcal{M} \simeq \Sigma \times T$ over T, and we have morphism $C \overset{i}{\hookrightarrow} \mathcal{C} \overset{\Phi}{\to} T$ with fibres $\mathcal{C}_t = \Phi^{-1}(t)$, where we identify C with $\mathcal{C}_0$ and write z instead of $(z, 0) \in \Sigma \times T$.

If $\mathcal{M} = \Sigma \times T$ with Φ the projection and if $\sigma(T) = \{z\} \times T$, then σ is called the *trivial section*. By [GLS6, Proposition II.2.2] every section can be locally trivialized by an isomorphism $\mathcal{M} \simeq \Sigma \times T$ over T.

The following (somewhat complicated) definition makes sense for arbitrary complex germs $(T, 0)$, even for Artinian ones.

Definition 1.1.27 (*Equisingular deformation*) An embedded deformation (i, Φ) of (C, z) over $(T, 0)$ is called *equisingular*, or an *es–deformation*, if the following three conditions hold:

(1) There exists a section σ of Φ, called *equimultiple section*, such that Φ is equimultiple along σ. If (C, z) is defined by $f \in \mathcal{O}_{\Sigma,z}$ with multiplicity $m = \mathrm{mt}(f)$ and if $(\mathcal{C}, z)$ is defined by $F \in \mathcal{O}_{\mathcal{M},z}$ this means that $F \in I_\sigma^m$, where I_σ denotes the ideal of $\sigma(T, 0) \subset (\mathcal{M}, z)$.
This implies that for $t \in T$ sufficiently close to 0 we have $F_t \in I^m_{\sigma(t)}$, where F_t defines the germ $(\mathcal{C}_t, \sigma(t))$ in $(\mathcal{M}_t, \sigma(t))$ and we have $I_{\sigma(t)} = \mathfrak{m}_{\mathcal{C}_t,\sigma(t)}$. Hence, the multiplicity of $(\mathcal{C}_t, \sigma(t))$ is constant for t near 0. For reduced base spaces this is equivalent to the given definition of equimultiplicity.

(2) After blowing up $\sigma(T)$ there exist equimultiple sections through each non-nodal singular point of the reduced total transform of the special fibre.
In detail: Let $\pi_1 : \mathcal{M}_{(1)} \to \mathcal{M}$ be the blowing up of $\sigma(T) \subset \mathcal{M} = \mathcal{M}_{(0)}$ and denote by $\mathcal{C}_{(1)}$ resp. $\mathcal{E}_{(1)}$ the strict transform of $\mathcal{C}$ resp. the exceptional divisor of π_1 such that $\widetilde{\mathcal{C}}_{(1)} = \mathcal{C}_{(1)} \cup \mathcal{E}_{(1)}$ is the reduced total transform of $\mathcal{C}$. Note that the restriction of π_1 over the special fibre $\Sigma = \Sigma_{(0)}$ of $\mathcal{M} \to T$ induces the blowing up $\Sigma_{(1)} \to \Sigma$ of $z \in \Sigma$, such that the reduced total transform $\widetilde{C}_{(1)} := C_{(1)} \cup E_{(1)}$ of C is the reduction of the special fibre of the composition

$$\Phi_{(1)} : \widetilde{\mathcal{C}}_{(1)} \overset{i_1}{\hookrightarrow} \mathcal{M}_{(1)} \overset{\pi_1}{\to} \mathcal{M} \to T\ .$$

Note that for $q \in \widetilde{C}_{(1)}$, the germ of $\Phi_{(1)}$ at q is an embedded deformation of $(\widetilde{C}_{(1)}, q)$ over $(T, 0)$. We then require the existence of an equimultiple section $\sigma_j^{(1)}$ of $\Phi_{(1)}$ through each non-nodal singular point of $\widetilde{C}_{(1)}$.

(3) Let $\pi_2 : \mathcal{M}_{(2)} \to \mathcal{M}_{(1)}$ be the blowing up of $\mathcal{M}_{(1)}$ along the sections $\sigma_j^{(1)}$. We require again the existence of equimultiple sections $\sigma_j^{(2)}$ through all non-nodal singularities q of the corresponding reduced total transform of (C, z). Inductively we repeat this process of blowing up equimultiple sections $\sigma_j^{(i)}$ through non-nodal singularities, until the restriction to the special fibre induces a (minimal) good embedded resolution of (C, z).

See [GLS6, Definition II.2.6] for a more detailed description. In the following we always assume that (C, z) is singular, if we talk about es-deformations of (C, z)(except we say otherwise).

Remark 1.1.27.1 We recall a few basic definitions and properties from local deformation theory (see [GLS6, Chapter II]).

A (non-embedded) *deformation with section* of (C, z) over $(T, 0)$ is a diagram as above, with $(\mathcal{M}, z)$ and the arrows to and from $(\mathcal{M}, z)$ deleted. If we delete also the section σ we call the resulting diagram just a deformation of (C, z) over $(T, 0)$. Morphisms of deformations are morphisms of diagrams and the corresponding categories are denoted by $\mathcal{D}ef_{(C,z)/(\Sigma,z)}$ (resp. $\mathcal{D}ef^{sec}_{(C,z)/(\Sigma,z)}$, resp. $\mathcal{D}ef^{fix}_{(C,z)/(\Sigma,z)}$) for embedded deformations (resp. with section, resp. with trivial section) and $\mathcal{D}ef_{(C,z)}$, resp. $\mathcal{D}ef^{sec}_{(C,z)}$, resp. $\mathcal{D}ef^{fix}_{(C,z)}$ for non-embedded deformations. By $\underline{\mathcal{D}ef}$ we denote the functor mapping a complex germ $(T, 0)$ to the set of isomorphism classes of defromations over $(T, 0)$ for the category $\mathcal{D}ef$. See also Sect. 2.2.1.1 for similar definitions from global deformation theory. By [GLS6, Proposition II.1.5] every local deformation is isomorphic to an embedded deformation and and every deformation with section is isomorphic to an embedded deformation with trivial section [GLS6, Proposition II.2.2]. This implies $\underline{\mathcal{D}ef}^{sec}_{(C,z)} \cong \underline{\mathcal{D}ef}^{fix}_{(C,z)}$.

We denote the *category of equisingular deformations* of (C, z) as in Definition 1.1.27 by $\mathcal{D}ef^{es}_{(C,z)}$. As the equimultiple section is not part of the data (we only assume the existence of an equimultiple section), we consider equisingular deformations as a subcategory of deformations without section, and $\underline{\mathcal{D}ef}^{es}_{(C,z)}$ is a subfunctor of $\underline{\mathcal{D}ef}_{(C,z)}$. Sometimes we assume the equisingular section to be given (resp. given as trivial section), i.e. we consider the *category of equisingular deformations with (trivial) section* $\mathcal{D}ef^{es,sec}_{(C,z)}$, (resp. $\mathcal{D}ef^{es,fix}_{(C,z)}$), a full subcategory of the category $\mathcal{D}ef^{sec}_{(C,z)}$ with $\underline{\mathcal{D}ef}^{es,sec}_{(C,z)} \cong \underline{\mathcal{D}ef}^{es,fix}_{(C,z)}$.

Note that, for an equisingular deformation, the equimultiple sections through all essential points are unique (cf. [GLS6, Proposition II.2.8]). That is, after blowing up an equimultiple section through a non-node of a minimal resolution, there is a unique section along which the blown up family is equimultiple.

Remark 1.1.27.2 Let (C, z) be singular and $\mathscr{T}^* = \mathscr{T}^*(C, z)$ the essential tree of (C, z). Then $z \in \mathscr{T}^*$ and the points $q \in \mathscr{T}^* \setminus \{z\}$ are exactly the non-nodal singular points of the reduced total transform of a good embedded resolution of (C, z). Hence, if we start with a good embedded resolution of (C, z), then an equisingular deformation of (C, z) over $(T, 0)$ is given by the following data:

- For each $q \in \mathscr{T}^*$ there is an embedded deformation of germs at q (or of small representatives)
$$\Phi_{(q)} : \widetilde{\mathcal{C}}_{(q)} \hookrightarrow \mathcal{M}_{(q)} \overset{\pi_{(q)}}{\to} \mathcal{M} \to T$$
of $\widetilde{C}_{(q)} \subset \Sigma_{(q)}$ together with an equimultiple section $\sigma_{(q)} : T \to \widetilde{\mathcal{C}}_{(q)}$ of $\Phi_{(q)}$, where $\widetilde{C}_{(q)}$ is the germ at q of the reduced total transform of (C, z). For $q = z$ this just means that $\Phi : \mathcal{C} \hookrightarrow \mathcal{M} \to T$ is an equimultiple deformation of (C, z) along a section σ.

- The $\mathcal{M}_{(q)}$, $q \neq z$, are obtained by blowing up sections $\sigma_{(p)}$, p of smaller level than q, with $\widetilde{\mathcal{C}}_{(q)}$ the reduced total transforms and $\Phi_{(q)}$ the composition, such that the system of sections $\{\sigma_{(q)}, q \in \mathscr{T}^*\}$ is compatible (cf. [GLS6, Definition II.2.1]).

1.1.2.4 Straight Equisingular Deformations

We introduce now a special class of equisingular deformations, which we call *straight*. They are related to the topological singularity ideal and the topological singularity scheme, a cluster scheme of exceptional importance for equisingular families of plane curves.

Assume that the (unique) equimultiple sections $\sigma_{(q)} : T \to \widetilde{\mathcal{C}}_{(q)}$ from Remark 1.1.27.2 for an es-deformation of (C, z) over a complex germ $(T, 0)$ are all trivial. We know that they can be trivialized for each q by a local isomorphism of germs $\mathcal{M}_{(q)} \simeq \Sigma_{(q)} \times T$ over T at q, but in general not simultaneously for all q by an isomorphism of $(\mathcal{M}, z)$ over T (e.g. the cross-ratio of more than three sections through one exceptional component is an invariant). Those deformations for which these sections can be simultaneously trivialized are called straight equisingular and they are closely related to the topological singularity scheme Z^s (introduced below). Straight es-deformations have been considered by J. Wahl (not under this name) and they were characterized in [GLS6, Proposition II.2.69]. We recall the main results at the end of this section.

Definition 1.1.28 (*Straight equisingular deformation*) A deformation of the (singular) reduced plane curve singularity (C, z) is called *straight equisingular* or a *straight es-deformation* if it is an equisingular deformation of (C, z) along the trivial section, such that the equimultiple sections $\sigma_{(q)}$, $q \in \mathscr{T}^*$, through the non-nodes of the reduced total transform of (C, z) from Remark 1.1.27.2 are all trivial.

We denote by $\mathcal{D}ef^s_{(C,z)}$ the category of straight es-deformations, the full subcategory of the category $\mathcal{D}ef^{sec}_{(C,z)}$ of deformations with section, and by $\underline{\mathcal{D}ef}^s_{(C,z)}$ the functor of isomorphism classes of straight es-deformations.

Let us give a concrete description of straight equisingular deformations, using the notations from Definition 1.1.27 and Remark 1.1.27.2:

Denote by $\widetilde{F}_{(q)} \in \mathcal{O}_{\mathcal{M}_{(q)}}$ a generator of the ideal of $\widetilde{\mathcal{C}}_{(q)} \subset \mathcal{M}_{(q)} = \Sigma_{(q)} \times T$ (the reduced total transform) and by $\mathfrak{m}_{(q)} \subset \mathcal{O}_{\Sigma_{(q)}}$ the maximal ideal. The condition that $\sigma_{(q)}$ is the trivial equimultiple section of $\Phi_{(q)}$ is equivalent to

$$\widetilde{F}_{(q)} \in \mathfrak{m}_{(q)}^{\widetilde{m}(q)} \cdot \mathcal{O}_{\Sigma_{(q)}} \times T \ ,$$

where $\widetilde{m}_{(q)}$ is the multiplicity of $\widetilde{C}_{(q)}$.

If $\widehat{F}_{(q)} \in \mathcal{O}_{\mathcal{M}_{(q)}}$ defines the total transform $\widehat{\mathcal{C}}_{(q)} \subset \Sigma_{(q)} \times T$ and if $\widehat{m}_{(q)}$ is the multiplicity of the total transform $\widehat{C}_{(q)}$ of the curve germ (C, z), then this is also equivalent to

$$\widehat{F}_{(q)} \in \mathfrak{m}_{(q)}^{\widehat{m}(g)} \cdot \mathcal{O}_{\Sigma_{(q)} \times T} \ .$$

This can be seen easily by induction on the number of blowing ups to resolve (C, z), using Proposition 1.1.11 and Remark 1.1.11.1(1). For $q = z$, we understand $\widetilde{\mathcal{C}}_{(q)} = \widehat{\mathcal{C}}_{(q)} = \mathcal{C}$, defined by $F \in \mathcal{O}_{\Sigma \times T}$, and both conditions mean $F \in \mathfrak{m}_z^m \cdot \mathcal{O}_{\Sigma \times T}$ with $m = \mathrm{mt}(C, z)$.

So far everything works for arbitrary complex germ $(T, 0)$. If $(T, 0)$ is reduced then the equimultiplicity condition for the trivial section $\sigma_{(q)}$, $q \in \mathscr{T}^*$, is equivalent to

$$\mathrm{mt}(C, z) = \mathrm{mt}(\mathcal{C}_t, z) \quad \text{for} \quad q = z$$

and for $q \neq z$ either to

$$\mathrm{mt}\, \widetilde{C}_{(q)} = \mathrm{mt}(\widetilde{\mathcal{C}}_{(q),t}, q)$$

or (equivalently) to

$$\mathrm{mt}\, \widehat{C}_{(q)} = \mathrm{mt}(\widehat{\mathcal{C}}_{(q),t}, q)$$

for all $t \in T$ sufficiently close to 0. $\widetilde{\mathcal{C}}_{(q),t}$ resp. $\widehat{\mathcal{C}}_{(q),t}$ denotes the reduced total, resp. the total transform of the fibre $\mathcal{C}_t$ of Φ over t.

For $(T, 0) = (\mathbb{C}, 0)$ we can describe the straight equisingularity condition even more explicitly. We do this for the total transform $\widehat{F}_{(q)} \in \mathcal{O}_{\Sigma_{(q)} \times T}$ of $F \in \mathcal{O}_{\Sigma \times T} = \mathcal{O}_{\Sigma, z}\{t\}$. Then F can be written as

$$F = f + t g_1 + t^2 g_2 + \cdots$$

with $f, g_i \in \mathcal{O}_{\Sigma, z}$ and f defining (C, z). Then $\widehat{F}_{(q)}$ reads as

$$\widehat{F}_{(q)} = \widehat{f}_{(q)} + t \widehat{g}_{1,(q)} + t^2 \widehat{g}_{2,(q)} + \cdots$$

where $\widehat{f}_{(q)}, \widehat{g}_{i,(q)} \in \mathcal{O}_{\Sigma_{(q)}}$ denote the total transforms of f, g_i.

If we fix $t = t_0 \in T$ we write $F_{t_0} = F\big|_{t=t_0} \in \mathcal{O}_{\Sigma, z}$ and $\widehat{F}_{(q),t_0} = \widehat{F}_{(q)}\big|_{t=t_0} \in \mathcal{O}_{\Sigma(q), q}$. Then the equisingularity condition is equivalent to

$$\mathrm{mt}(f) = \mathrm{mt}(F_t)\,,$$

$$\mathrm{mt}(\widehat{f}_{(q)}) = \mathrm{mt}(\widehat{F}_{(q),t})$$

for all $t \in T$ sufficiently close to 0 and for all $q \in T^*$, $q \neq z$.

Thus we get:

Lemma 1.1.29 *Let $f \in \mathbb{C}\{x, y\}$ define a reduced plane curve singularity $(C, 0)$ with essential tree $\mathscr{T}^*$ and let $F(x, y, t) = f(x, y) + \sum_{i \geq 1} t^i g_i(x, y)$ define a one-parametric deformation of $(C, 0)$. Then the following are equivalent:*

(i) *The deformation of $(C, 0)$ defined by F is equisingular and the (unique) equimultiple sections through the infinitely near points $q \in \mathscr{T}^*$ are trivial, i.e., the deformation is straight equisingular.*

(ii) For all $i \geq 1$,

$$\mathrm{mt}(f) \leq \mathrm{mt}(g_i),$$

$$\mathrm{mt}(\widehat{f}_{(q)}) \leq \mathrm{mt}(\widehat{g}_{i,(q)}) \ for \ q \in T^*, q \neq 0.$$

Let us return to cluster schemes.

Definition 1.1.30 (*Cluster ideal and scheme with underlying curve*) Let $(C, z) \subset (\Sigma, z)$ be a reduced plane curve singularity and let $\mathscr{T}$ be a finite subtree of the complete embedded resolution tree $\mathscr{T}(C, z)$. We define *the cluster ideal of* (C, z) w.r.t. $\mathscr{T}$,

$$\begin{aligned} I(C, \mathscr{T}) &:= \big\{g \in \mathcal{O}_{\Sigma,z} \,\big|\, g \text{ goes through the cluster } \mathcal{C}\ell(C, \mathscr{T})\big\} \\ &= \big\{g \in \mathcal{O}_{\Sigma,z} \,\big|\, \mathrm{mt}\,\widehat{g}_{(q)} \geq \mathrm{mt}\,\widehat{C}_{(q)} \text{ for each } q \in \mathscr{T}\big\} \subset \mathcal{O}_{\Sigma,z}\,. \end{aligned}$$

We denote by

$$Z(C, \mathscr{T}) := V(I(C, \mathscr{T}))$$

the zero-dimensional subscheme of Σ supported at $\{z\}$ and defined by the ideal $I(C, \mathscr{T}) \subset \mathcal{O}_{\Sigma,z}$ and call $Z(C, \mathscr{T})$ the *cluster scheme with underlying germ* (C, z) w.r.t. $\mathscr{T}$.

By Definitions 1.1.16 and 1.1.21 $Z(C, \mathscr{T}) = Z(\mathcal{C}\ell(C, \mathscr{T}))$ is the 0-dimensional scheme associated to the cluster $\mathcal{C}\ell(C, \mathscr{T})$. Since $\mathcal{C}\ell(C, \mathscr{T})$ satisfies the proximity relations by Proposition 1.1.11, we have $\mathcal{C}\ell(Z(C, \mathscr{T})) = \mathcal{C}\ell(C, \mathscr{T})$ by Proposition 1.1.17.

We define now the topological singularity scheme of (C, z), $Z^s(C, z)$, to be the zero-dimensional scheme associated to the essential tree $\mathscr{T}^*(C, z)$ (Definition 1.1.10), that is, $Z^s(C, z)$ is supported at $\{z\}$ and given by the cluster ideal of (C, z) w.r.t. $\mathscr{T}^*(C, z)$.

Definition 1.1.31 (*Topological singularity ideal and scheme*) We define the ideal

$$\begin{aligned} I^s(C, z) &:= I\big(C, \mathscr{T}^*(C, z)\big) \subset \mathcal{O}_{\Sigma,z} \\ &= \{g \in \mathcal{O}_{\Sigma,z} \mid \mathrm{mt}\,\widehat{g}_{(q)} \geq \mathrm{mt}\,\widehat{C}_{(q)} \text{ for each } q \in \mathscr{T}^*(C, z)\} \end{aligned}$$

and call it the*topological singularity ideal*. If (C, z) is defined by $f \in \mathcal{O}_{\Sigma,z}$ we write also $I^s(f)$. It defines the zero-dimensional scheme

$$Z^s(C, z) := V(I^s(C, z)) = Z(\mathscr{T}^*(C, z), \boldsymbol{m}),$$

the *topological singularity scheme*, where $(\mathscr{T}^*(C, z), \boldsymbol{m}) = \mathcal{C}\ell(C, \mathscr{T}^*(C, z))$.

Here $\boldsymbol{m} = \{m_q\}_{q \in \mathscr{T}^*(C,z)}$ is the vector with $m_q = \mathrm{mt}\, C_{(q)}$, the multiplicity of the strict transform of C at q.

If g defines (C, z) then $g \in I^s(C, z)$ and hence $Z^s(C, z)$ is a subscheme of C, supported at $\{z\}$.

The set of all disjoint finite unions of topological singularity schemes is denoted by $\mathcal{S}$.

Remark 1.1.31.1 This links the definition of $Z^s(C, z)$ to the definitions given in the previous sections, since $\mathcal{C}\ell(Z^s(C, z)) = \mathcal{C}\ell(C, \mathcal{T}^*(C, z))$. The topological singularity scheme $Z^s(C, z)$ is a cluster scheme, while the *topological equisingularity scheme* $Z^{es}(C, z) \subset Z^s(C, z)$, defined by the equisingularity ideal $I^{es}(C, z)$ (see Definition 1.1.63) in general not (compare their degrees in Lemma 1.1.32 and Corollary 1.1.64). We work mostly with $Z^s(C, z)$ since it is easier to handle.

By Lemma 1.1.22 we get

Lemma 1.1.32 *For $(C, z) \subset (\Sigma, z)$ a reduced plane curve singularity we have*

$$\deg Z^s(C, z) = \dim_{\mathbb{C}} \mathcal{O}_{\Sigma,z}/I^s(C, z) = \sum_{q \in \mathcal{T}^*(C,z)} \frac{m_q(m_q + 1)}{2}.$$

Example 1.1.32.1 (a) Let (C, z) be smooth. For the empty constellation $\mathcal{T}$ we obtain $I(C, \mathcal{T}) = \mathcal{O}_{\Sigma,z}$, that is, the singularity scheme $Z(C, \mathcal{T})$ is the empty scheme. If $\mathcal{T} = (z = q_0, q_1, \ldots, q_n)$ is any finite subtree of the complete embedded resolution tree $\mathcal{T}(C, z)$, and if (C, z) has the local equation $y = 0$, then $I(C, \mathcal{T}) = \langle y, x^{n+1} \rangle \subset \mathbb{C}\{x, y\}$.

(b) If (C, z) is an *ordinary* m-fold *singularity* (i.e. m smooth branches with different tangents) then $\mathcal{T}^*(C, z) = (z)$ and $I^s(C, z) = \mathfrak{m}^m_{\Sigma,z}$.

The following lemma shows the relation of cluster schemes to equisingular deformations of curve germs:

Lemma 1.1.33 *Let $Z \subset \Sigma$ be a cluster scheme with support $\{z\}$, given by the ideal $I_Z \subset \mathcal{O}_{\Sigma,z}$.*

(a) A generic element $g \in I_Z$ defines Z, in the sense that $Z = Z(C, \mathcal{T}_Z)$ for (C, z) the plane curve germ defined by g.

(b) The elements $g \in I_Z$ defining Z have no common (infinitely near) base point outside of $\mathcal{T}_Z$.

(c) Two generic elements $g, g' \in I_Z$ are topologically equivalent.

Note that "generic" in (a) means: there exists a polynomial defining Z, and if d_0 is the minimal degree of such a polynomial, then, for each $d \geq d_0$, the set of polynomials $g \in I_Z$ of degree at most d defining Z is a Zariski-open, dense subset in the vector space of all polynomials of degree at most d and contained in I_Z.

Proof (a) Let $Z = Z(C', \mathcal{T}_Z)$ for some plane curve germ $(C', z) \subset (\Sigma, z)$, and let $f \in \mathcal{O}_{\Sigma,z}$ be a defining equation. As the cluster $\mathcal{C}\ell(C', \mathcal{T}_Z)$ only depends on a sufficiently high jet of f (Remark 1.1.16.1), we may assume that f is a polynomial.

Let $d \geq \deg(f)$. Then the polynomials $g \in I_Z$ of degree $\leq d$ are parameterized by a finite dimensional vector space of positive dimension. The conditions $\operatorname{mt} \widehat{g}_{(q)} = \operatorname{mt} \widehat{C}'_{(q)}$, $q \in \mathscr{T}_Z$, define a Zariski-open subset of it. The density follows, since for almost all $t \in \mathbb{C}$ the germ $f + tg$ has precisely the assigned (total) multiplicities $\widehat{m}_q = \operatorname{mt} \widehat{C}'_{(q)}$ at each $q \in \mathscr{T}_Z$.

Statement (b) is a direct consequence of Proposition 1.1.17. (c) follows from (a) and from Proposition 1.1.18. Note that for a germ (C, z) defined by a generic element $g \in I_Z$ the essential tree $\mathscr{T}^*(C, z)$ is contained in $\mathscr{T}_Z$ as this can be assumed for (C', z) defined by g' (Proposition 1.1.17). □

We relate the topological singularity ideal $I^s(C, z)$ right now to the equisingularity ideal $I^{es}(C, z)$ and the fixed equisingularity ideal $I^{es}_{fix}(C, z)$, defined later. They satisfy (see Definition 1.1.63)

$$I^{es}(C, z) = j(f) + I^{es}_{fix}(C, z).$$

Let $j(f)$ denote the Jacobian ideal and

$$I^{ea}(C, z) = \langle f, j(f)\rangle \text{ resp. } I^{ea}_{fix}(C, z) = \langle f, \mathfrak{m}_z j(f)\rangle$$

the Tjurina resp. fixed Tjurina ideal (Definition 2.2.7). Moreover, T_ε denotes the fat point with structure sheaf $\mathbb{C}[\varepsilon]$, $\varepsilon^2 = 0$, and $\underline{\mathcal{Def}}^-_{(C,z)}(T_\varepsilon)$ is the tangent space to the deformation functor $\underline{\mathcal{Def}}^-_{(C,z)}$ (see Definition 2.2.7 and the paragraph before), i.e.,

$$\underline{\mathcal{Def}}^-_{(C,z)}(T_\varepsilon) = \{g \in \mathcal{O}_{\Sigma,z} | f + \varepsilon g \in \mathcal{Def}^-_{(C,z)}(T_\varepsilon)\}/\text{isomorphism in } \mathcal{Def}^-_{(C,z)}.$$

Corollary 1.1.34 *Let $(C, z) \subset (\Sigma, z)$ be a reduced plane curve singularity, defined by $f \in \mathcal{O}_{\Sigma,z}$.*

(1) If $g \in I^s(C, z)$, then f and $f + tg$ are topologically equivalent for all but finitely many $t \in \mathbb{C}$ and $f + tg$ defines an equisingular deformation of (C, z) over $(\mathbb{C}, 0)$ along the trivial section.

(2) We get

$$I^s(C, z) = \{g \in \mathcal{O}_{\Sigma,z} | \; f + \varepsilon g \textit{ is a straight es-deformation of } (C, z) \textit{ over } T_\varepsilon\}$$

and $I^s(C, z)/I^{ea}_{fix}(C, z) = \underline{\mathcal{Def}}^s_{(C,z)}(T_\varepsilon)$ is the tangent space to the functor of (isomorphism classes of) straight es-deformations.

(3) We have the inclusions $I^{ea}_{fix}(C, z) \subset I^{ea}(C, z) \subset I^{es}(C, z)$ and

$$I^{ea}_{fix}(C, z) \subset I^s(C, z) \subset I^{es}_{fix}(C, z) \subset I^{es}(C, z).$$

$I^{es}(C, z)/I^{ea}(C, z)$ resp. $I^{es}_{fix}(C, z)/I^{ea}_{fix}(C, z)$ is the tangent space to the functor of isomorphism classes of es-deformations resp. of es-deformations with section. We get

$$\dim_{\mathbb{C}} I^{es}(C, z)/I^{es}_{fix}(C, z) = \dim_{\mathbb{C}} I^{ea}(C, z)/I^{ea}_{fix}(C, z) = 2,$$

$$\dim_{\mathbb{C}} I^{es}_{fix}(C,z)/I^{ea}_{fix}(C,z) = \dim_{\mathbb{C}} I^{\,es}(C,z)/I^{ea}(C,z).$$

Proof (1) follows from the proof of Lemma 1.1.33, (2) from the proof of Lemma 1.1.29 applied to $f + tg, t^2 = 0$.

(3) The inclusion $I^{ea}_{fix}(C,z) \subset I^s(C,z)$ follows form the fact that for $g \in I^{ea}_{fix}(C,z)$ the deformation $f + \varepsilon g$ is trivial along the trivial section, hence straight, over T_ε. The other inclusions follow from the definitions. For the first dimension statements see Lemma 2.2.13. The second follows from the first and the following exact sequences,

$$0 \to I^{es}_{fix}(C,z)/I^{ea}_{fix}(C,z) \to I^{\,es}(C,z)/I^{ea}_{fix}(C,z) \to I^{\,es}(C,z)/I^{es}_{fix}(C,z) \to 0,$$

$$0 \to I^{ea}(C,z)/I^{ea}_{fix}(C,z) \to I^{\,es}(C,z)/I^{ea}_{fix}(C,z) \to I^{\,es}(C,z)/I^{ea}(C,z) \to 0.$$

Even more, the forgetful morphism

$$I^{es}_{fix}(C,z)/I^{ea}_{fix}(C,z) \to I^{\,es}(C,z)/I^{ea}(C,z)$$

is an isomorphism, since both spaces have the same dimension and the cokernel is $I^{\,es}(C,z)/j(f) + I^{es}_{fix}(C,z) = 0$. □

Remark 1.1.34.1 Note that straight equisingular deformations are deformations with (trivial) section while equisingular deformations are deformations without section. Forgetting the section we get a morphism from the category $\mathcal{D}ef^{s}_{(C,z)}$ of straight es-deformations to the category $\mathcal{D}ef^{es}_{(C,z)}$ of es-deformations. Saying that an equisingular deformation is straight equisingular means that it is isomorphic (as deformation without section) to the image in $\mathcal{D}ef^{es}_{(C,z)}$ of a straight equisingular deformation. The image of $\mathcal{D}ef^{s}_{(C,z)}$ in $\mathcal{D}ef^{es}_{(C,z)}$ is called the category of *straight es-deformation without section*. It follows that

$$\langle j(f), I^s(C,z)\rangle \,/\, \langle f, j(f)\rangle$$

is the tangent space to the functor of isomorphism classes of straight equisingular deformation of (C,z) without section.

Let us recall the main result about straight es-deformations from [GLS6, Proposition II.2.69] (basically due to Wahl [Wah1]) with $\tau(C,z) = \dim_{\mathbb{C}} \mathcal{O}_{\Sigma,z}/I^{ea}(C,z)$ and $\tau^{\,es}(C,z) = \dim_{\mathbb{C}} \mathcal{O}_{\Sigma,z}/I^{\,es}(C,z)$.

Proposition 1.1.35 *Let $(C,z) \subset (\Sigma,z)$ be a reduced plane curve singularity defined by $f \in \mathcal{O}_{\Sigma,z}$. Then the following are equivalent:*

(a) There are $\tau' = \tau(C,z) - \tau^{\,es}(C,z)$ elements $g_1,\dots,g_{\tau'} \in I^{\,es}(C,z)$ such that

$$\varphi^{\,es} : V\left(f + \sum\nolimits_i t_i g_i\right) \subset (\Sigma \times \mathbb{C}^{\tau'}, (z,\mathbf{0})) \xrightarrow{\text{pr}} (\mathbb{C}^{\tau'}, \mathbf{0})$$

is a semiuniversal equisingular deformation of (C,z).

(b) *There exist* $g_1, \ldots, g_{\tau'} \in I^{es}(C, z)$ *inducing a basis of* $I^{es}(C, z)/\langle f, j(f)\rangle$ *such that*

$$\varphi^{es} : V\left(f + \sum\nolimits_i t_i g_i\right) \subset (\Sigma \times \mathbb{C}^{\tau'}, (z, \mathbf{0})) \xrightarrow{\text{pr}} (\mathbb{C}^{\tau'}, \mathbf{0})$$

is a semiuniversal equisingular deformation of (C, z).

(c) *Each locally trivial deformation of the reduced exceptional divisor* E *of a minimal embedded resolution of* $(C, z) \subset (\Sigma, z)$ *is trivial.*

(d) $I^{es}(C, z) = \langle f, j(f), I^s(C, z)\rangle$.

(e) *Each equisingular deformation of* (C, z) *is straight equisingular.*

We mention also the following consequence for semiquasi-homogeneous and Newton nondegenerate singularities from [GLS6, Proposition II.2.17 and Corollary II.2.71].

Corollary 1.1.36 *Let* $(C, z) \subset (\Sigma, z)$ *be a reduced plane curve singularity defined by* $f \in \mathcal{O}_{\Sigma,z}$, *and let* $\tau' = \tau(C, z) - \tau^{es}(C, z)$.

(a) *If* $f = f_0 + f'$ *is semiquasi-homogeneous with principal part* f_0 *being quasi-homogeneous of type* $(w_1, w_2; d)$, *then*

$$I^{es}(C, z) = \langle j(f), I^s(C, z)\rangle = \langle j(f),\ x^\alpha y^\beta \mid w_1\alpha + w_2\beta \geq d\rangle$$

and a semiuniversal equisingular deformation for (C, z) *is given by*

$$\varphi^{es} : V\left(f + \sum\nolimits_{i=1}^{\tau'} t_i g_i\right) \subset (\Sigma \times \mathbb{C}^{\tau'}, (z, \mathbf{0})) \xrightarrow{\text{pr}} (\mathbb{C}^{\tau'}, \mathbf{0}),$$

for suitable $g_1, \ldots, g_{\tau'}$ *representing a* $\mathbb{C}$*-basis for the quotient*

$$\langle j(f),\ x^\alpha y^\beta \mid w_1\alpha + w_2\beta \geq d\rangle / \langle f, j(f)\rangle.$$

(b) *If* f *is Newton non-degenerate with Newton diagram* $\Gamma(f)$ *at the origin, then*

$$I^{es}(C, z) = \langle j(f), I^s(C, z)\rangle = \langle j(f),\ x^\alpha y^\beta \mid x^\alpha y^\beta \text{ has Newton order } \geq 1\rangle$$

and a semiuniversal equisingular deformation for (C, z) *is given by*

$$\varphi^{es} : V\left(f + \sum\nolimits_{i=1}^{\tau'} t_i g_i\right) \subset (\Sigma \times \mathbb{C}^{\tau'}, (z, \mathbf{0})) \xrightarrow{\text{pr}} (\mathbb{C}^{\tau'}, \mathbf{0}),$$

for suitable $g_1, \ldots, g_{\tau'}$ *representing a monomial* $\mathbb{C}$*-basis for the quotient*

$$\langle j(f),\ x^\alpha y^\beta \mid x^\alpha y^\beta \text{ has Newton order } \geq 1\rangle / \langle f, j(f)\rangle.$$

Moreover, in both cases each equisingular deformation of $(C, 0)$ *is straight equisingular.*

Remark 1.1.36.1 Note that in Proposition 1.1.35(b) and Corollary 1.1.36 not every basis has the claimed property (this was not clearly formulated in [GLS6]).

Moreover, we can extend the chosen basis of $I^{es}(C,z)/\langle f, j(f)\rangle$ to a basis $g_1,\dots,g_\tau$ of $\mathcal{O}_{\Sigma,z}/\langle f, j(f)\rangle$, showing that the base space of the semiuniversal es-deformation is the linear subspace $\{t_{\tau'+1}=\dots=t_\tau=0\}$ of the usual semiuniversal deformation of (C,z) given by $f+\sum_{i=1}^{\tau} t_i g_i$. This is a general fact of straight es-deformations, as we show now.

Proposition 1.1.37 *Let $(C,z)\subset(\Sigma,z)$ be a reduced plane curve singularity defined by $f\in\mathcal{O}_{\Sigma,z}$.*

(1) There exist $g_1,\dots,g_{\tau'}\in I^s(C,z)$ inducing a basis of the quotient $I^s(C,z)\,/\,\langle f, \mathfrak{m}_z j(f)\rangle$ such that

$$\varphi^s : V\left(f+\sum\nolimits_i t_i g_i\right)\subset(\Sigma\times\mathbb{C}^{\tau'},(z,\mathbf{0}))\xrightarrow{\text{pr}}(\mathbb{C}^{\tau'},\mathbf{0}),$$

is a semiuniversal straight equisingular deformation of (C,z).

(2) There exist $g_1,\dots,g_{\tau''}\in\langle j(f), I^s(C,z)\rangle$ inducing a basis of of the quotient $\langle j(f), I^s(C,z)\rangle\,/\,\langle f, j(f)\rangle$ such that

$$\varphi^s : V\left(f+\sum\nolimits_i t_i g_i\right)\subset(\Sigma\times\mathbb{C}^{\tau''},(z,\mathbf{0}))\xrightarrow{\text{pr}}(\mathbb{C}^{\tau''},\mathbf{0}),$$

is a semiuniversal straight equisingular deformation of (C,z) without section.

Proof (1) That the deformation φ^s is straight es follows from Lemma 1.1.29 and its proof. Extending the chosen basis of $I^s(C,z)/\langle f,\mathfrak{m}_z j(f)\rangle$ to a basis of $\mathfrak{m}_z/\langle f,\mathfrak{m}_z j(f)\rangle$ shows that the base space of φ^s is a linear subspace $B^s_{C,z}$ of the base space $B^{sec}_{C,z}\cong B^{fix}_{C,z}$ of the semiuniversal deformation of (C,z) with (trivial) section. If a given deformation over an arbitrary base germ $(T,0)$ is straight es, it can be induced by a map $\varphi:(T,0)\to B^{fix}_{C,z}$. φ must factor through $B^s_{C,z}$ by the definition of I^s because otherwise the deformation is not equimultiple along the trivial sections through all infinitely near points $q\in\mathcal{T}^*(C,z)$. This shows that φ^s is complete. Moreover, for extensions of Artinian base spaces $(T,0)\subset(T',0)$, the equimultiple deformations along the trivial sections over $(T,0)$ can can be clearly extended to equimultiple deformations along the trivial sections over $(T',0)$. Hence the versality follows. (For "complete" and "versal" see [GLS6, Definition II.1.8]).

(2) follows from (1), noting that $\langle j(f), I^s(C,z)\rangle\,/\,\langle f, j(f)\rangle$ is the tangent space to the functor of isomorphism classes of straight equisingular deformation of (C,z) without section. As in Remark 1.1.36.1 the base space of the semiuniversal straight equisingular deformation of (C,z) without section can be realized as a linear subspace of the usual semiuniversal deformation of (C,z). □

1.1.2.5 Milnor Number, δ-Invariant, 2nd-Moment

Lemma 1.1.33 shows, in particular, that although the curve singularity $(C, z) \subset (\Sigma, z)$ defined by a generic element of I_Z is not uniquely determined by Z, all topological invariants of (C, z) can be associated uniquely to Z. Therefore we define the *Milnor number* $\mu(Z)$ and the *δ-invariant* (*delta-invariant*) $\delta(Z)$ of a cluster scheme Z as those of (C, z), where (C, z) is defined by a generic element of I_Z,

$$\mu(Z) := \mu(C, z), \quad \delta(Z) := \delta(C, z) .$$

Here

$$\mu(C, z) := \mu(f) := \dim_{\mathbb{C}} \mathcal{O}_{\Sigma,z} \Big/ \Big\langle \frac{\partial f}{\partial x}, \frac{\partial f}{\partial y} \Big\rangle$$

is the *Milnor number* of (C, z) or of f ($f \in \mathcal{O}_{\Sigma,z}$ a defining power series of (C, z) and x, y local coordinates of (Σ, z)), and

$$\delta(C, z) := \dim_{\mathbb{C}} (n_* \mathcal{O}_{\overline{C}} / \mathcal{O}_C)_z$$

is the *δ-invariant* (*delta-invariant*) of (C, z) ($n : \overline{C} \to C$ the normalization of a representative C of (C, z)).

Moreover, we introduce the multiplicity of Z at an infinitely near point $q \in \mathcal{T}_Z$ to be the multiplicity of the corresponding strict transform of Z (cf. Definition 1.1.25). Lemma 1.1.33 shows that $\mathrm{mt}(Z, q) = \mathrm{mt}\, C_{(q)}$.

As a consequence, we may also express the degree of a cluster scheme (see Lemma 1.1.22) using the δ-invariant:

Lemma 1.1.38 *Let $Z \subset \Sigma$ be a cluster scheme, and let $m_q = \mathrm{mt}(Z, q)$ for $q \in \mathcal{T}_Z$. Then*

$$\deg Z \;=\; \delta(Z) + \sum_{q \in \mathcal{T}_Z} m_q .$$

Proof This follows from Lemma 1.1.22 and the well-known formula

$$\delta(Z) \;=\; \delta(C, z) \;=\; \sum_{q \in \mathcal{T}_Z} \frac{m_q(m_q - 1)}{2}\,, \tag{1.1.2.2}$$

$(C, z) \subset \Sigma$ being a generic curve germ, i.e. defined by a generic element from I_Z, passing through Z (see [Mil, Remark 10.10] and [GLS6, Proposition I.3.34]). □

Definition 1.1.39 (*δ-invariant, second moment*) We introduce the *δ-invariant* of an arbitrary zero-dimensional scheme Z as the δ-invariant of the maximal cluster scheme Z_{cl} (Definition 1.1.25) contained in Z, that is,

$$\delta(Z) := \delta(Z_{cl}) = \sum_{q \in \mathcal{T}(Z_{cl})} \frac{\mathrm{mt}(Z, q)\big(\mathrm{mt}(Z, q) - 1\big)}{2} .$$

Similarly, we introduce the *second moment* of a zero-dimensional scheme Z,

$$M_2(Z) := M_2(Z_{cl}) := \sum_{q\in\mathcal{T}(Z_{cl})} \mathrm{mt}(Z,q)^2 .$$

Note that, due to Noether's formula (Proposition 1.1.8) and Lemma 1.1.33(a), $M_2(Z)$ is the intersection number of two different generic elements g, f in I_Z, implying the second inequality resp. equality of the following corollary. Since $Z \subset \{f = 0\}\cap\{g = 0\}$, we get the first inequality for the second moment:

Corollary 1.1.40 *Let Z be a zero-dimensional scheme on Σ. Then*

$$\deg Z \le M_2(Z) \le \deg Z + \delta(Z) .$$

Moreover, if Z is a cluster scheme then $M_2(Z) = \deg Z + \delta(Z)$. If Z is (locally) a complete intersection scheme then $M_2(Z) = \deg Z$.

Example 1.1.40.1 Let $(C, z) \subset (\Sigma, z)$ be given by $f = (x^2 - y^3)(y^2 - x^3)$, and let $Z \subset \Sigma$ be the zero-dimensional scheme with support $\{z\}$ given by the Tjurina ideal of (C, z) (see Example 1.1.26.1(c)). Then

$$\deg Z = 10 < M_2(Z) = 11 < 13 = \deg Z + \delta(Z) .$$

We close this part by giving an estimate for the degree of a topological singularity scheme $Z^s(C, z)$ in terms of the δ-invariant:

Corollary 1.1.41 *Let $(C, z) \subset (\Sigma, z)$ be a reduced plane curve singularity. Then*

$$\deg Z^s(C, z) \le 3\,\delta(C, z) ,$$

unless (C, z) is an A_{2k}-singularity, $k \ge 1$. In the latter case,

$$\deg Z^s(C, z) = 3k + 2 = 3\,\delta(C, z) + 2 .$$

Proof The number of essential points $q \in \mathcal{T}^*(C, z)$ with $m_q := \mathrm{mt}\, C_{(q)} = 1$ does not exceed $\mathrm{mt}(C, z)$. Hence, by Lemma 1.1.22, we obtain

$$\begin{aligned}\deg Z^s(C, z) &\le \sum_{m_q>1} \frac{m_q(m_q+1)}{2} + \mathrm{mt}(C, z)\\ &= 3 \sum_{q\in\mathcal{T}^*(C,z)} \frac{m_q(m_q-1)}{2} - \sum_{m_q\ge 3} m_q(m_q-2) + \mathrm{mt}(C, z).\end{aligned}$$

In particular, (1.1.2.2) implies the above inequality if $\mathrm{mt}(C, z) \ge 3$. For A_μ-singularities see the following example. □

Example 1.1.41.1 (1) If $\mathrm{mt}(C, z) = 2$, that is, if (C, z) is an A_μ-singularity, then one computes easily

$$\deg Z^s(A_\mu) = \begin{cases} 3k = 3\delta(A_\mu) & \text{if } \mu = 2k-1\,, \\ 3k+2 = 3\delta(A_\mu)+2 & \text{if } \mu = 2k\,, \end{cases}$$

(2) For the remaining simple singularities we compute

$$\deg Z^s(D_\mu) = \begin{cases} 3k-1 & \text{if } \mu = 2k-1\,, \\ 3k & \text{if } \mu = 2k\,, \end{cases}$$

and $\deg Z^s(E_\mu) = \mu + 3$ for $\mu = 6, 7, 8$.

(3) If (C, z) is an ordinary m-fold singularity we get from Example 1.1.32.1(b)

$$\deg Z^s(C, z) = \frac{m(m+1)}{2}\,.$$

1.1.2.6 Conductor Scheme and Residue Scheme

We introduce two further important zero-dimensional schemes, the conductor scheme and the residue scheme, which turn out to be cluster schemes.

Definition 1.1.42 (*Conductor Scheme*) Let $(C, z) \subset (\Sigma, z)$ be a reduced plane curve germ. The ideal

$$I^{cd}(C, z) = I^{cd}(\mathcal{O}_{C,z}) = \mathrm{Ann}_{\mathcal{O}_{C,z}}(\overline{\mathcal{O}}/\mathcal{O}_{C,z}) \subset \mathcal{O}_{C,z}$$

is called the *conductor ideal* (where $\mathcal{O}_{C,z} \hookrightarrow \overline{\mathcal{O}}$ is the normalization). More generally, for any ideal $I \subset \mathcal{O}_{C,z}$ we denote by

$$I^{cd}(I) = \mathrm{Ann}_{\mathcal{O}_{C,z}}(\overline{\mathcal{O}}/I) \subset I$$

the *conductor ideal of I*, which is an $\overline{\mathcal{O}}$-ideal of $\mathcal{O}_{C,z}$.

The ideal $I^{cd}(C, z)$ defines a zero-dimensional scheme

$$Z^{cd}(C, z) := V(I^{cd}(C, z)) \subset (C, z),$$

called the *conductor scheme* of (C, z).

Remark 1.1.42.1 (1) The *degree of the conductor scheme*, also called the *degree of the conductor* is

$$\deg Z^{cd}(C, z) = \dim_{\mathbb{C}} \mathcal{O}_{C,z}/I^{cd}(C, z) = \delta(C, z)$$

by [GLS6, Formula (I.3.4.11)].

(2) Let (C_i, z), $i = 1, \cdots, r$, be the branches of (C, z), then $\overline{\mathcal{O}} \cong \prod_{i=1}^r \mathbb{C}\{t_i\}$ and for $\varphi \in \mathcal{O}_{C,z} \subset \overline{\mathcal{O}}$ we set

$$\operatorname{ord} \varphi\big|_{C_i} := \text{ the } t_i\text{-order of } \varphi.$$

This is also the intersection multiplicity of $i_z(C_i, V(\varphi))$ of the germs (C_i, z) and $(V(\varphi), z)$ (cf. [GLS6, Definition (I.3.9)]).

(3) $I^{cd}(C, z)$ is a principal $\overline{\mathcal{O}}$-ideal and hence generated by one element of the form $(t_1^{c_1}, \cdots, t_r^{c_r}) \in \overline{\mathcal{O}}$. We get

$$\varphi \in I^{cd}(C, z) \Leftrightarrow \operatorname{ord} \varphi\big|_{C_i} \geq c_i, \ i = 1, \cdots, r.$$

(4) Moreover, by [GLS6, Formula (I.3.4.10)] we have

$$c_i = 2\delta(C_i, z) + \sum_{j \neq i} i_z(C_i, C_j), \ i = 1, \ldots, r, \tag{1.1.2.3}$$

and by [GLS6, Formula (I.3.4.12)]

$$\dim_{\mathbb{C}} \overline{\mathcal{O}}/I^{cd}(C, z) = c_1 + \cdots + c_r = 2\delta(C, z).$$

(5) The preceding local properties yield the following global one. Let C be a reduced, irreducible projective curve of a smooth surface Σ, $\nu : C^\vee \to C$ its normalization, $D^\vee$ an effective divisor on $C^\vee$ supported at a finite set F. Then $\nu_* \mathcal{O}_{C^\vee}(D^\vee) = \mathcal{O}_C(D)$, where D is a Cartier divisor on C supported at $\operatorname{Sing}(C) \cup \nu(F)$ and defined as follows:

- if $z \in \nu(F) \setminus \operatorname{Sing}(C)$ and $mz^\vee$ is a component of $D^\vee$ supported at $z^\vee = \nu^{-1}(z)$, $m \geq 1$, then mz is the component of D supported at z;
- if $z \in \operatorname{Sing}(C)$, $C_1, \ldots, C_r$ all the components of the germ (C, z), and $\{z_1, \ldots, z_r\} = \nu^{-1}(z) \subset C^\vee$ so that the germ (C_i, z) lifts to the germ $(C^\vee, z_i)$, $i = 1, \ldots, r$, then the component of D supported at z is defined by the ideal

$$I = \{\varphi \in \mathcal{O}_{C,z} \ : \ \operatorname{ord} \varphi\big|_{C_i} \geq c_i + k_i, \ i = 1, \ldots, r\},$$

where $c_1, \ldots, c_r$ are given by (1.1.2.3), and $k_1 z_1 + \cdots + k_r z_r$ is the part of $D^\vee$ supported at $\{z_1, \ldots, z_r\}$.

The following lemma shows that the conductor scheme is a cluster scheme and that each cluster scheme $Z(C, \mathcal{T})$ with $\mathcal{T} \supset \mathcal{T}^*(C, z)$ contains the conductor scheme.

Lemma 1.1.43 *$Z^{cd}(C, z)$ is a cluster scheme given by the constellation consisting of the singular essential points $\mathcal{T}^{cd}(C, z) \subset \mathcal{T}^*(C, z)$, with assigned multiplicities $m_q^{cd} = \operatorname{mt} C_{(q)} - 1$ for each $q \in \mathcal{T}^{cd}(C, z)$.*

We prove this lemma as a special case (a) of the following more general statement:

Lemma 1.1.44 *Let $(C, z) \subset (\Sigma, z)$ be a reduced plane curve singularity with irreducible components $Q_1, \ldots, Q_s$. Let $\mathscr{T} \subset \mathscr{T}(C, z)$ be a finite subtree containing $\mathscr{T}^{\,\mathrm{cd}}(C, z)$, and let, for each $q \in \mathscr{T}$,*

$$m_q \in \{\mathrm{mt}\, C_{(q)} - 1\,,\ \mathrm{mt}\, C_{(q)}\} \tag{1.1.2.4}$$

be such that the proximity relations are satisfied. Moreover, let Z be the cluster scheme $Z(\mathscr{T}, \boldsymbol{m})$. Then we have

$$I_Z = \left\{ f \in \mathcal{O}_{C,z} \,\middle|\, i_z(f, Q_j) \geq k_j(Z) \text{ for each } j = 1, \ldots, s \right\}$$

with

$$k_j(Z) := \sum_{q \in \mathscr{T}} m_q \cdot \mathrm{mt}\, Q_{j,(q)}\,, \quad j = 1, \ldots, s\,.$$

In particular, $k_1(Z) + \ldots + k_s(Z) = \deg Z + \delta(C, z)$, and the following holds:

(a) If $\mathscr{T} = \mathscr{T}^{\,\mathrm{cd}}$ and $m_q = \mathrm{mt}\, C_{(q)} - 1$ for each $q \in \mathscr{T}$, then

$$k_j(Z) = 2\delta(Q_j) + \sum_{i \neq j} i_z(Q_i, Q_j) = \mathrm{cd}\big(\mathcal{O}_{C,z}\big)_j\,,$$

for each $j = 1, \ldots, s$. Here, $\mathrm{cd}\big(\mathcal{O}_{C,z}\big)_j$ denotes the jth component of the conductor of the semigroup $\Gamma(\mathcal{O}_{C,z})$ (cf. *[GLS6, Section I.3.4, page 214]).*

(b) If $\mathscr{T} \supset \mathscr{T}^{\,\mathrm{cd}}$ and $Z = Z(C, \mathscr{T})$, then, for each $j = 1, \ldots, s$,

$$k_j(Z) = 2\delta(Q_j) + \sum_{i \neq j} i_z(Q_i, Q_j) + \sum_{q \in \mathscr{T} \cap \mathscr{T}(Q_j)} \mathrm{mt}\, Q_{j,(q)}\,.$$

Proof As shown before, the multiplicity of the strict transform at $q \in \mathscr{T}$ of a generic element $g \in I_Z$ satisfies $\mathrm{mt}\, g_{(q)} = \mathrm{mt}(Z, q) = m_q$. In particular, for each branch Q_j of (C, z) and each $f \in I_Z$,

$$i_z(f, Q_j) \geq i_z(g, Q_j) \geq \sum_{q \in \mathscr{T}} m_q \cdot \mathrm{mt}\, Q_{j,(q)} = k_j(Z) =: k_j\,.$$

Hence, $I_Z \subset J := \bigcap_{j=1}^{s} \left\{ f \in \mathcal{O}_{\Sigma,z} \mid i_z(f, Q_j) \geq k_j \right\}$. Considering both as ideals in $\mathcal{O}_{C,z}$, we show that $\dim_{\mathbb{C}}(\mathcal{O}_{C,z}/J) \geq \deg Z$ which implies $I_Z = J$. To do so, let n denote the normalization

$$n : \mathcal{O}_{C,z} \hookrightarrow \prod_{j=1}^{s} \mathbb{C}\{t_j\} =: \overline{\mathcal{O}}\,,$$

and consider the image $n(J) \subset \overline{\mathcal{O}}$. For an element $f \in \overline{\mathcal{O}}$ the conditions on the intersection multiplicities $i_z(f, Q_j)$ read as $f \in \prod_{j=1}^{s} t_j^{k_j} \cdot \mathbb{C}\{t_j\}$. Hence,

$$\begin{aligned}
\dim_{\mathbb{C}}\left(\mathcal{O}_{C,z}/J\right) &\geq \dim_{\mathbb{C}}\Big(\overline{\mathcal{O}}\Big/\prod_{j=1}^{s} t_j^{k_j}\cdot\mathbb{C}\{t_j\}\Big) - \dim_{\mathbb{C}}\left(\overline{\mathcal{O}}/\mathcal{O}_{C,z}\right)\\
&= \sum_{j=1}^{s} k_j - \delta(C,z) = \sum_{q\in\mathscr{T}} m_q\cdot \operatorname{mt} C_{(q)} - \sum_{q\in\mathscr{T}} \frac{\operatorname{mt} C_{(q)}(\operatorname{mt} C_{(q)}-1)}{2}\\
&= \sum_{q\in\mathscr{T}} \frac{m_q(m_q+1)}{2} = \deg Z
\end{aligned}$$

by the assumption (1.1.2.4) and Lemma 1.1.22. The statements (a) and (b) are now obtained by a straightforward computation, using that

$$\delta(C,z) = \sum_{i=1}^{s} \delta(Q_i) + \sum_{i<k} i_z(Q_i, Q_k)$$

(see [HiH], [Mil, Remark 10.10], or [GLS6, Lemma I.3.32]), and applying [GLS6, formula (I.3.4.10)]. □

Corollary 1.1.45 *Let $(C,z) \subset (\Sigma,z)$ be a reduced plane curve singularity, let $\mathcal{O}_{C,z} \hookrightarrow \overline{\mathcal{O}}$ be the normalization, and let Z be a zero-dimensional scheme. Then the following are equivalent:*

(a) Z is defined by an $\overline{\mathcal{O}}$-ideal.
(b) Z is a cluster scheme $Z(\mathscr{T},\mathfrak{m})$ with $\mathscr{T} \subset \mathscr{T}(C,z)$ a finite subtree containing $\mathscr{T}^{cd}(C,z)$, and with $m_q \in \{\operatorname{mt} C_{(q)}-1\,,\ \operatorname{mt} C_{(q)}\}$ for each $q \in \mathscr{T}$.

Proof Let $Q_1,\dots,Q_s$ be the irreducible components of (C,z).

The implication $(b) \Rightarrow (a)$ follows from Lemma 1.1.43, respectively its proof. Indeed, let $Z = Z(\mathscr{T},\mathfrak{m})$ with $\mathscr{T}^{cd}(C,z) \subset \mathscr{T} \subset \mathscr{T}(C,z)$, such that (1.1.2.4) holds. Then we have shown that (via the normalization map)

$$I_Z = \prod_{j=1}^{s} t_j^{k_j(Z)}\mathbb{C}\{t_j\}\,.$$

For $(a) \Rightarrow (b)$, note that each zero-dimensional $\overline{\mathcal{O}}$-ideal $I \subset \mathcal{O}_{C,z}$ is contained in the conductor ideal $I^{cd}(\mathcal{O}_{C,z})$ and corresponds to a zero-dimensional ideal in $\overline{\mathcal{O}} = \prod_{j=1}^{s}\mathbb{C}\{t_j\}$. Hence, $I = \prod_{j=1}^{s} t_j^{k_j}\cdot\mathbb{C}\{t_j\}$ for some $k_j \geq \operatorname{cd}(\mathcal{O}_{C,z})_j$. By the above, it only remains to show that, for each choice of $(k_1,\dots,k_s)$, we find a finite subtree $\mathscr{T} \subset \mathscr{T}(C,z)$ extending $\mathscr{T}^{cd}$, and assigned multiplicities m_q, $q \in \mathscr{T}$, such that

$$k_j = \sum_{q\in\mathscr{T}} m_q\cdot \operatorname{mt} Q_{j,(q)}\,.$$

But this is clear: let $\mathscr{T}$ be the extension of $\mathscr{T}^{cd}$ by $\ell_j := k_j - \operatorname{cd}(\mathcal{O}_{C,z})_j \geq 0$ points of $\mathscr{T}(Q_j)$, $j = 1,\dots,s$, and set

$$m_q := \begin{cases} \operatorname{mt} C_{(q)} - 1 & \text{for } q \in \mathscr{T}^{cd}, \\ 1 & \text{for } q \in \mathscr{T} \setminus \mathscr{T}^{cd}. \end{cases}$$

Note that the cluster $(\mathscr{T}, \boldsymbol{m})$ does not necessarily satisfy the proximity relations. However, successively applying the unloading principle preserves the property (1.1.2.4). □

Definition 1.1.46 (*Residue of a Cluster Scheme*) Let $Z \subset \Sigma$ be a zero-dimensional scheme and $F \subset \Sigma$ a curve. Then we define the *residue of Z by F* to be the zero-dimensional scheme $Z : F$ defined at each point z in the support of Z by the ideal quotient

$$I_{Z,z} : \langle f \rangle := \left\{ g \in \mathcal{O}_{\Sigma,z} \;\middle|\; gf \in I_{Z,z} \right\} \subset \mathcal{O}_{\Sigma,z} ,$$

where $I_{Z,z} \subset \mathcal{O}_{\Sigma,z}$ is the ideal of the component of Z concentrated at z, and $f = 0$ is a local equation of the germ $(F, z) \subset (\Sigma, z)$ (if non-empty).

Proposition 1.1.47 *Let $Z \subset \Sigma$ be a cluster scheme with support $\{z\}$, given by the ideal $I_Z \subset \mathcal{O}_{\Sigma,z}$, and let $(L, z) \subset (\Sigma, z)$ be a smooth curve germ. Then the following hold:*

(a) The residue $Z : L$ is a cluster scheme, and its tree $\mathscr{T}_{Z:L}$ is a subtree of $\mathscr{T}_Z$.
(b) $\operatorname{mt}(Z, q) - 1 \le \operatorname{mt}(Z{:}L,\, q) \le \operatorname{mt}(Z, q)$ for all $q \in \mathscr{T}_Z$.

Proof (a) Let $Z = Z(C, \mathscr{T}_Z)$ and $(\mathscr{K}, \boldsymbol{m}) := \mathcal{C}\ell(C, \mathscr{T}_Z)$ be the cluster defined by the points $q \in \mathscr{T}_Z$ with assigned multiplicities $m_q := \operatorname{mt} C_{(q)} = \operatorname{mt}(Z, q)$. Moreover, let

$$m_q^{(0)} := \begin{cases} m_q - 1 & \text{if } q \in \mathscr{T}_Z \cap \mathscr{T}(L, z), \\ m_q & \text{otherwise}. \end{cases}$$

By definition, a curve D goes through the cluster $(\mathscr{K}, \boldsymbol{m}^{(0)})$ iff $D \cup L$ goes through $\mathcal{C}\ell(C, \mathscr{T}_Z)$.
Case 1: $(\mathscr{K}, \boldsymbol{m}^{(0)})$ satisfies the proximity relations.
Then Proposition 1.1.17 gives the existence of a reduced plane curve germ (C', z) such that $(\mathscr{K}, \boldsymbol{m}^{(0)}) = \mathcal{C}\ell(C', \mathscr{T}')$ and $\mathscr{T}'$ contains the essential tree $\mathscr{T}^*(C', z)$. Here $\mathscr{T}' \subset \mathscr{T}_Z$ is the subconstellation obtained by removing those points with (new) assigned multiplicities $m_q^{(0)} = 0$.
Case 2: $(\mathscr{K}, \boldsymbol{m}^{(0)})$ does not satisfy all proximity relations.
Then, by k times applying the unloading principle at some vertex, we get a cluster $(\mathscr{K}^{(k)}, \boldsymbol{m}^{(k)})$ such that $(\mathscr{K}^{(k)}, \boldsymbol{m}^{(k)})$ satisfies the proximity relations and a curve D goes through the cluster $(\mathscr{K}^{(k)}, \boldsymbol{m}^{(k)})$ iff $D \cup L$ goes through $\mathcal{C}\ell(C, \mathscr{T}_Z)$ (see Lemma 1.1.20 and Remark 1.1.20.1). Now, we may use the same arguments as in Case 1.
(b) We prove by induction that in each step we have

$$m_q^{(i)} \in \left\{ m_q - 1,\, m_q \right\} \text{ for each } q \in \mathscr{T}_Z .$$

For $i = 0$ the statement is obvious. If the cluster $(\mathscr{K}^{(i)}, \boldsymbol{m}^{(i)})$ does not satisfy the proximity relation at a point $p = p^{(i)}$ then the induction hypothesis implies

$$m_p^{(i)} = m_p - 1\,, \quad m_q^{(i)} = m_q \ \text{ for all } q \dashrightarrow p\,.$$

In particular, the unloading principle leads to virtual multiplicities

$$m_p^{(i+1)} = m_p\,, \quad m_q^{(i+1)} = m_q - 1 \ \text{ for all } q \dashrightarrow p\,,$$

while the other assigned multiplicities remain unchanged. Hence the statement. □

1.1.3 The Hilbert Scheme of Cluster Schemes

Throughout this section let $\Sigma \hookrightarrow \mathbb{P}^N$ be a smooth projective surface. In the following we recall the notion of families (of zero-dimensional schemes, respectively curves) and some well-known results on such families. We consider the *Hilbert scheme of points* on Σ and the subscheme of families of simultaneous resolvable points with fixed cluster graph. Recall that a "family" does always mean "flat family".

1.1.3.1 The Hilbert Scheme of Points

We recall the *Hilbert scheme of points* on Σ, which is a projective variety that parameterizes all zero-dimensional subschemes of the projective surface Σ. We start with the notion of a family of zero-dimensional schemes. We have to distinguish between algebraic families and analytic families. Except we say otherwise, we work in the analytic category.

Definition 1.1.48 (*Family of zero-dimensional schemes*)

1. Let T be an arbitrary complex space. A *family of zero-dimensional schemes (of degree n) on Σ over T* is a commutative diagram of complex spaces

$$\begin{array}{ccc} \mathcal{Z} & \overset{j}{\hookrightarrow} & \Sigma \times T \\ & {\scriptstyle\varphi}\searrow \quad \swarrow{\scriptstyle\mathrm{pr}} & \\ & T & \end{array}$$

   with $j : \mathcal{Z} \hookrightarrow \Sigma \times T$ a closed embedding, pr the projection and $\varphi = \mathrm{pr} \circ j$ finite and flat. The fibre $\mathcal{Z}_t := \varphi^{-1}(t)$, $t \in T$, is mapped by j to a zero-dimensional subscheme of Σ and if its degree is n we say that $j : \mathcal{Z} \hookrightarrow \Sigma \times T$ is a *family of zero-dimensional schemes of degree n*. Sometimes we call such a family an *analytic family* in order to distinguish it form an algebraic family.
2. If $j : \mathcal{Z} \cong Z \times T$ over T for some zero–dimensional scheme $Z \subset \Sigma$, we say that $\mathcal{Z} \to T$ is a *trivial family*.

3. A *family with sections* is a commutative diagram as above, together with disjoint sections $\sigma_1, \dots, \sigma_r : T \to \mathcal{Z}$ of φ. The sections are called *trivial* if σ_i is an isomorphism $T \to \{z_i\} \times T$ for some $z_i \in \Sigma$.
4. Note that φ is a projective morphism[4] and the fibres $\mathcal{Z}_t$ are projective algebraic varieties. If T, $\mathcal{Z}$ and j are algebraic then we get a diagram in the category of algebraic complex spaces and we call φ an *algebraic family of zero-dimensional schemes* on Σ over T.

We usually identify two families $j : \mathcal{Z} \hookrightarrow \Sigma \times T$ and $j' : \mathcal{Z}' \hookrightarrow \Sigma \times T$ if the images $j(\mathcal{Z})$ and $j'(\mathcal{Z}')$ coincide, and then we get subspaces $\mathcal{Z} \subset \Sigma \times T$ (of degree n) which are finite and flat over T. For algebraic families this is the Hilbert functor:

The *Hilbert functor* $\mathcal{H}ilb^n_\Sigma$ associates to an algebraic variety T the set

$$\mathcal{H}ilb^n_\Sigma(T) := \left\{ \begin{array}{l} \text{closed algebraic subvarieties } \mathcal{Z} \subset \Sigma \times T \\ \text{which are finite of degree } n \text{ and flat over } T \end{array} \right\}.$$

This functor is known to be representable by a smooth projective algebraic variety of dimension $2n$, denoted by Hilb^n_Σ, the *Hilbert schemeof points in* Σ, which is connected if Σ is connected (cf. [HaR1, Fog], or the overview article [Iar2]). Representability means that there is a closed algebraic subvariety $\mathcal{U}^n_\Sigma \subset \Sigma \times \mathrm{Hilb}^n_\Sigma$, finite of degree n and flat over Hilb^n_Σ, such that the *universal family*

$$\begin{array}{ccc} \mathcal{U}^n_\Sigma & \hookrightarrow & \Sigma \times \mathrm{Hilb}^n_\Sigma \\ & {\scriptstyle\varphi}\searrow \quad \swarrow{\scriptstyle \mathrm{pr}} & \\ & \mathrm{Hilb}^n_\Sigma & \end{array}$$

has the following *universal property*: each element of $\mathcal{H}ilb^n_\Sigma(T)$, T an algebraic variety, can be induced from φ via base change by a *unique* (algebraic) morphism $T \to \mathrm{Hilb}^n_\Sigma$. In particular, the points of Hilb^n_Σ correspond bijectively to the zero-dimensional schemes of degree n on Σ.

Remark 1.1.48.1 1. In the given references the properties of Hilb^n_Σ are proven for algebraic schemes. Our formulation for algebraic varieties (i.e. algebraic complex spaces) follows immediately by analytification (cf. Notations and Conventions).
2. An analogous functor can be defined for arbitrary complex spaces T, associating to T the set of closed complex subspaces $\mathcal{Z} \subset \Sigma \times T$, finite of degree n and flat over T. This is the *Douady functor*, denoted by $\mathcal{D}ou^n_\Sigma$, which is representable by a complex space Dou^n_Σ, the *Douady space* (c.f. [Dou2, Dou1, Bin]). Hence, there exists a closed subspace $\mathcal{V}^n_\Sigma \subset \Sigma \times \mathrm{Dou}^n_\Sigma$, finite of degree n and flat over Dou^n_Σ, such that each element of $\mathcal{D}ou^n_\Sigma(T)$, T a complex space, can be induced from $\mathcal{V}^n_\Sigma \to \mathrm{Dou}^n_\Sigma$ via base change by a unique analytic morphism $T \to \mathrm{Dou}^n_\Sigma$.
3. The Hilbert scheme and the Douady space exist in much grater generality. In our situation of zero-dimensional schemes on the smooth projective surface Σ,

[4] We say that a morphism $X \to T$ is *projective*, or that X is *projective over* T, if φ factors through a closed immersion $X \hookrightarrow \mathbb{P}^N \times T$ for some N, followed by the projection $\mathbb{P}^N \times T \to T$.

it can be shown that Hilb^n_Σ and Dou^n_Σ are isomorphic and that $\mathscr{U}^n_\Sigma \subset \Sigma \times \mathrm{Hilb}^n_\Sigma$ is isomorhic to $\mathscr{V}^n_\Sigma \subset \Sigma \times \mathrm{Dou}^n_\Sigma$ over this isomorphism. In particular, $\mathscr{U}^n_\Sigma \subset \Sigma \times \mathrm{Hilb}^n_\Sigma$ has the universal property for complex spaces T and analytic morphisms $T \to \mathrm{Hilb}^n_\Sigma$. We prove this fact in greater generality in Sect. 2.2.1.1, Proposition 2.2.28.

In the following we basically work in the complex analytic category and use the universal property of Hilb^n_Σ for families of zero-dimensional schemes over complex spaces. We use also the projectivity of Hilb^n_Σ.

There exists a birational morphism, the *Hilbert–Chow morphism* (cf. [Fog, Mum2])

$$\Phi : \mathrm{Hilb}^n_\Sigma \longrightarrow \mathrm{Sym}^n \Sigma\,,$$

$\mathrm{Sym}^n \Sigma$ denoting the nth symmetric power of Σ. It can be thought of as assigning to a closed subscheme $Z \subset \Sigma$ of degree n the 0-cycle consisting of the points of Z with multiplicities given by the length of their local rings on Z (cf. [Fog, Cor. 2.6]). If Z consists of n distinct simple points $z_1, \dots, z_n$ then Φ maps Z to the image of $(z_1, \dots, z_n) \in \Sigma^n$ in $\mathrm{Sym}^n\, \Sigma = \Sigma^n / S_n$ where S_n is the symmetric group permuting the coordinates. By loc. cit. Φ can be extended to all of Hilb^n_Σ (for a generalization see (2.2.2.1) in Sect. 2.2.2.2).

Definition 1.1.49 (*Punctual Hilbert functor and scheme*) Let $z_0 \in \Sigma$ and x, y be local coordinates at z_0. Then we have the *punctual Hilbert functor* $\mathcal{H}ilb^n_{\mathbb{C}\{x,y\}}$, which associates to a complex space T the set

$$\mathcal{H}ilb^n_{\mathbb{C}\{x,y\}}(T) := \big\{ (\mathscr{Z} \subset \Sigma \times T) \in \mathcal{H}ilb^n_\Sigma(T) \,\big|\, \mathrm{supp}(\mathscr{Z}) \subset \{z_0\} \times T \big\}.$$

Fogarty [Fog] and Hartshorne [HaR1] showed that the punctual Hilbert functor is representable (in the algebraic category) by a connected projective variety $\mathrm{Hilb}^n_{\mathbb{C}\{x,y\}}$, the *punctual Hilbert scheme*, which can be identified with the closed subvariety $\Phi^{-1}(nz_0) \subset \mathrm{Hilb}^n_\Sigma$. By Remark 1.1.48.1 this holds also in the analytic category. Briançon [Bri] proved (in the analytic categroy) that $\mathrm{Hilb}^n_{\mathbb{C}\{x,y\}}$ is irreducible (in general not reduced) of dimension $n - 1$.

1.1.3.2 The Hilbert Scheme of a Cluster Graph

Even if we are mainly interested in singularity schemes, we consider in the sequel the more general situation of families of zero-dimensional schemes with prescribed cluster graph (satisfying the proximity relations).

A family of zero-dimensional schemes $\mathscr{Z} \hookrightarrow \Sigma \times T \to T$ with section $\sigma : T \to \mathscr{Z}$ is called *equimultiple (of multiplicity m) along* σ if for any $t \in T$ we have $m = \mathrm{mt}(\mathscr{Z}_t, \sigma(t))$ and

$$\mathcal{I}_{(\mathscr{Z},t)} \subset \mathcal{I}^m_{(\sigma(T),\sigma(t))},$$

where $\mathcal{I}_{\mathcal{Z}} \subset \mathcal{O}_{\Sigma\times T}$ denotes the ideal sheaf of $\mathcal{Z}$ and $\mathcal{I}_{\sigma(T)} \subset \mathcal{O}_{\Sigma\times T}$ the ideal sheaf of the section σ. This imposes a nontrivial condition also over an Artinian base space.

Definition 1.1.50 (*Resolvable family*) Let $\mathcal{Z} \hookrightarrow \Sigma \times T \to T$ be a family of zero-dimensional schemes over a complex space T. We say that the family is *resolvable by blowing up sections* if there exist an N and for i= $0, \dots, N$ pairwise disjoint sections

$$\sigma_1^{(i)}, \dots, \sigma_{k_i}^{(i)} : T \longrightarrow \mathscr{X}^{(i)},$$

and morphisms $\pi_i : \mathscr{X}^{(i+1)} \to \mathscr{X}^{(i)}$ such that the following holds:

- $\mathscr{X}^{(0)} = \Sigma \times T$, $\mathcal{Z}^{(0)} = \mathcal{Z}$,
- $\pi_i : \mathscr{X}^{(i+1)} \to \mathscr{X}^{(i)}$ is the blowing up of $\mathscr{X}^{(i)}$ along $\sigma_1^{(i)}, \dots, \sigma_{k_i}^{(i)}$ and $\mathcal{Z}^{(i+1)}$ is the strict transform of $\mathcal{Z}^{(i)}$,
- the family $(\mathcal{Z}^{(i)} \hookrightarrow \mathscr{X}^{(i)} \to T)$ is *equimultiple along the sections* $\sigma_j^{(i)}$, $j = 1, \dots, k_i$,
- $\operatorname{supp}(\mathcal{Z}^{(i)}) = \bigcup_{j=1}^{k_i} \sigma_j^{(i)}(T)$,
- $\operatorname{supp}(\mathcal{Z}^{(N+1)}) = \emptyset$.

Example 1.1.50.1 (a) Any trivial family $Z \times T \hookrightarrow \Sigma \times T \to T$, Z a zero-dimensional scheme on Σ, is resolvable by blowing up (trivial) sections.
(b) Consider the 1-parameter family $\mathcal{Z} \hookrightarrow \Sigma \times \mathbb{C} \to \mathbb{C}$, given by

$$\mathcal{Z}_t := Z(C_t, \mathscr{T}^*(C_t, z)),$$

(C_t, z) the reduced plane curve germ with local equation $(x+ty)^2 - y^3$ (that is, $I_{\mathcal{Z}_t} = \langle (x+ty)^2, xy^2, y^3 \rangle$) for all $t \in \mathbb{C}$. Then we have

- $\operatorname{supp}(\mathcal{Z}^{(0)}) = \operatorname{supp}(\mathcal{Z}) = \{z\} \times \mathbb{C}$,
- $\mathcal{Z}^{(0)}$ is equimultiple along the (trivial) section $\sigma^{(0)} : t \mapsto (x, y; t) = (0, 0; t)$,
- $\mathscr{X}^{(1)} = \Sigma' \times \mathbb{C}$, where $\Sigma' \to \Sigma$ is the blowing up of $z \in \Sigma$.
- In the affine chart $\mathscr{X}^{(1)} \supset U \times \mathbb{C} \to \Sigma \times \mathbb{C}$, $(u, v; t) \mapsto (uv, v; t)$, the strict transform $\mathcal{Z}^{(1)}$ is supported along $\sigma^{(1)} : t \mapsto (u, v; t) = (-t, 0; t)$ and given by the ideal $\langle (u+t)^2, v \rangle$. In particular, $\mathcal{Z}^{(1)}$ is equimultiple along $\sigma^{(1)}$.
- Let $\mathscr{X}^{(2)} \to \mathscr{X}^{(1)}$ be the blowing up of $\mathscr{X}^{(1)}$ along $\sigma^{(1)}$. Then, in the affine chart $\mathscr{X}^{(2)} \supset U' \times \mathbb{C} \to \Sigma' \times \mathbb{C}$, $(u', v'; t) \mapsto (u, v; t) = (u'-t, (u'-t)v'; t)$, the strict transform $\mathcal{Z}^{(2)}$ is supported along $\sigma^{(1)} : t \mapsto (u', v'; t) = (0, 0; t)$, and given by the ideal $\langle u', v' \rangle$. In particular, $\mathcal{Z}^{(2)}$ is equimultiple along $\sigma^{(2)}$.
- Let $\mathscr{X}^{(3)} \to \mathscr{X}^{(2)}$ be the blowing up of $\mathscr{X}^{(2)}$ along $\sigma^{(2)}$. Then the strict transform $\mathcal{Z}^{(3)}$ is empty.

Hence, $\mathcal{Z}$ is resolvable by blowing up sections.

Let $\mathbb{G} = (\Gamma, \dashrightarrow, \boldsymbol{m})$ be a cluster graph (Definition 1.1.13) with set of vertices V, satisfying the proximity relations, and set (cf. Lemma 1.1.22)

$$n := \sum_{q \in V} \frac{m_q(m_q+1)}{2}.$$

Definition 1.1.51 (*Hilbert functor fixing a cluster graph*) Let $\mathbb{G}$ be the cluster graph of a zero-dimensional scheme on Σ (cf. Definition 1.1.25). We define the*Hilbert functor with fixed cluster graph* $\mathbb{G}$, $\mathcal{H}ilb_\Sigma^{\mathbb{G}}$, on the category of reduced complex spaces T by associating to T the set of all families of zero-dimensional schemes $(\mathcal{Z} \subset \Sigma \times T) \in \mathcal{H}ilb_\Sigma^n(T)$ having the following properties:

($\mathbb{G}$1) There exists a finite disjoint union T' of irreducible reduced complex spaces and a finite surjective morphism $\alpha : T' \to T$ such that the induced family

$$\begin{array}{ccc} \alpha^*\mathcal{Z} & \hookrightarrow & \Sigma \times T' \\ & \searrow \quad \swarrow_{\text{pr}} & \\ & T' & \end{array}$$

is resolvable by blowing up sections $\sigma_j^{(i)} : T' \to \mathscr{X}^{(i)}$ (cf. Definition 1.1.50 for the notations) and, additionally, if F is any component of the exceptional divisor of $\mathscr{X}^{(i)} \to \Sigma \times T'$, then the image of $\sigma_j^{(i)}$ is either contained in F or it has empty intersection with F, $1 \leq j \leq k_i$, $1 \leq i \leq N$.

($\mathbb{G}$2) The cluster graph defined by each fibre $\mathcal{Z}_t \subset \Sigma$, $t \in T$, is constant and satisfies $\text{clg}(\mathcal{Z}_t) = \mathbb{G}$.

Note that if ($\mathbb{G}$1) is satisfied then ($\mathbb{G}$2) is equivalent to

($\mathbb{G}2'$) in each connected component of T there exists a point t such that the fibre $\mathcal{Z}_t \subset \Sigma$ satisfies $\text{clg}(\mathcal{Z}_t) = \mathbb{G}$.

Note that the cluster graph being constant imposes no condition over fat points. Therefore we define the functor $\mathcal{H}ilb_\Sigma^{\mathbb{G}}$ only for reduced base spaces.

The next proposition says that ($\mathbb{G}$1) follows already from ($\mathbb{G}$2).

Proposition 1.1.52 *Let T be a reduced complex space, and $\mathbb{G} = (\Gamma, \dashrightarrow, \boldsymbol{m})$ be a cluster graph with set of vertices V. Then*

$$\mathcal{H}ilb_\Sigma^{\mathbb{G}}(T) = \big\{(\mathcal{Z} \subset \Sigma \times T) \in \mathcal{H}ilb_\Sigma^n(T) \,\big|\, \text{clg}(\mathcal{Z}_t) = \mathbb{G} \text{ for all } t \in T\big\}.$$

For the proof (given in [NoV] for the algebraic category, respectively in [Ris1] for $T \subset \mathbb{C}$ an open subset), we use the following semicontinuity statement for families of zero-dimensional schemes which is interesting in itself:

Lemma 1.1.53 *Let T be reduced and $\mathcal{J} \subset \mathcal{O}_{\Sigma \times T}$ a coherent sheaf of ideals such that the induced morphism $V(\mathcal{J}) \to T$ is finite. Then, for any $t \in T$, there exists a polynomial*

$$P_t(w) = \frac{e_{\mathcal{J}}(t)}{2} w^2 + (\textit{lower terms in} w) \in \mathbb{Q}[w], \quad e_{\mathcal{J}}(t) \in \mathbb{N},$$

(the Hilbert–Samuel polynomial) such that, for m sufficiently large,

$$P_t(m) = \dim_{\mathbb{C}}\big(\mathcal{O}_\Sigma/(\mathcal{J}_t)^m\big),$$

where $\mathcal{J}_t \subset \mathcal{O}_\Sigma = \mathcal{O}_{\Sigma\times\{t\}}$ denotes the sheaf of ideals induced by $\mathcal{J}$ on the fibre $\Sigma \times \{t\}$. Moreover,

(a) if $(\mathcal{K}, \boldsymbol{m}) = \mathcal{C}\ell(V(\mathcal{J}_t))$ is the cluster defined by the zero-dimensional scheme $V(\mathcal{J}_t) \subset \Sigma$, then

$$e_{\mathcal{J}}(t) = \sum_{q\in\mathcal{K}} m_q^2 = M_2\big(V(\mathcal{J}_t)\big),$$

with M_2 the second moment;

(b) the function $e_{\mathcal{J}} : T \to \mathbb{Z}$, $t \mapsto e_{\mathcal{J}}(t)$ is upper semicontinuous in the Zariski topology.

Proof The existence of the Hilbert–Samuel polynomial is proven, e.g., in [Ser2, II–25] or [GrP, Section 5.5]. For (a) we refer to [Ris1, Cor. 1.2] and (b) follows since the fibre dimension of the coherent sheaves $\mathcal{O}_{\Sigma\times T}/\mathcal{J}^m$ is upper semicontinuous. □

Remark 1.1.53.1 If in the situation of Lemma 1.1.53 each $V(\mathcal{J}_t) \subset \Sigma$, $t \in T$, is a cluster scheme, then $\mathcal{O}_{\Sigma\times T}/\mathcal{J}$ is $\mathcal{O}_T$-flat iff $e_{\mathcal{J}}(t)$ is locally constant on T. Indeed, $\mathcal{O}_{\Sigma\times T}/\mathcal{J}$ is $\mathcal{O}_T$-flat iff $\dim_{\mathbb{C}} \mathcal{O}_\Sigma/\mathcal{J}_t = \sum_{q\in\mathcal{K}} m_q(m_q+1)/2$ (Lemma 1.1.22) is locally constant [GLS6, Theorem I.1.78 and Theorem I.1.80 (2)] which is equivalent to $e_{\mathcal{J}}(t)$ being locally constant by (a).

In the general situation, however, this does not hold. Consider, for instance, the coherent sheaf of ideals $\mathcal{J} \subset \mathcal{O}_{\Sigma\times\mathbb{C}}$ with $\mathcal{J}_{t,z} = \langle x^2, y^2, txy\rangle \subset \mathcal{O}_{\Sigma,z}$, $\mathcal{J}_{t,q} = \mathcal{O}_{\Sigma,q}$ for $q \neq z$. Then $P_0(w) = 2w^2 + 2w$, $P_t(w) = 2w^2 + w$ for $t \neq 0$, hence $e_{\mathcal{J}}(t) = 4$ for all $t \in \mathbb{C}$, but $\mathcal{O}_{\Sigma\times T}/\mathcal{J}$ is not $\mathcal{O}_T$-flat, as $\deg V(\mathcal{J}_0) = 4 > 3 = \deg V(\mathcal{J}_t)$ for $t \neq 0$.

Proof of Proposition 1.1.52. We have to show that any family of zero-dimensional schemes on Σ over T with $\operatorname{clg}(\mathcal{Z}_t) = \mathbb{G}$ for all $t \in T$ satisfies the condition ($\mathbb{G}$1). To be able to proceed by induction, we actually prove the analogous statement in a slightly more general situation given by the following diagram

$$\begin{array}{ccc} \mathcal{Z} & \hookrightarrow & \mathcal{X} \\ & {\scriptstyle\varphi}\searrow \quad \swarrow{\scriptstyle\pi} & \\ & T\,, & \end{array}$$

where $\pi : \mathcal{X} \to T$ is a flat family of smooth projective surfaces, $\mathcal{Z} \hookrightarrow \mathcal{X}$ is a closed embedding such that $\varphi := \pi|_{\mathcal{Z}}$ is a finite morphism, and for any fibre $\mathcal{Z}_t := \varphi^{-1}(t)$, $t \in T$, we have $\operatorname{clg}(\mathcal{Z}_t) = \mathbb{G}$. Obviously, it suffices to consider the case that T is irreducible.

Let $\mathcal{Z}^i$, $i = 1, \ldots, s$, be those irreducible components of the reduction $\mathcal{Z}_{\text{red}}$ with $\varphi(\mathcal{Z}^i) = T$ and let $\mathcal{Z}' \subset \mathcal{Z}$ be the subscheme given by

$$\mathcal{O}_{\mathcal{Z}'} = \mathcal{O}_{\mathcal{Z}}\big|_{\mathcal{Z}^1\cup\ldots\cup\mathcal{Z}^s}\,.$$

Denote by $\mathfrak{Z} := \overline{\mathcal{Z}_{\mathrm{red}} \setminus \mathcal{Z}'_{\mathrm{red}}}$ the union of the remaining components.
Step 1. We show that $\mathcal{Z}' = \mathcal{Z}$. In fact, this follows since flat maps are open [GLS6, Theorem I.1.84]. However, we give a direct argument here by using the semicontinuity of $e_{\mathcal{J}}(t)$.
Assume $\mathfrak{Z} \neq \emptyset$. Note that φ is a finite morphism, whence $\varphi(\mathfrak{Z}) \subsetneqq T$ is a (proper) analytic subset. Let $\mathcal{J} \subset \mathcal{O}_{\mathscr{X}}$ (resp. $\mathcal{J}' \subset \mathcal{O}_{\mathscr{X}}$) denote the (coherent) ideal sheaf defining $\mathcal{Z}$ (resp. $\mathcal{Z}'$). Then, by Lemma 1.1.53(a) and the assumption (constant cluster graph), we obtain for any $t \in \varphi\big(\mathfrak{Z} \setminus (\mathcal{Z}'_{\mathrm{red}} \cap \mathfrak{Z})\big)$ and any $\tau \in T \setminus \varphi(\mathfrak{Z})$ that

$$e_{\mathcal{J}'}(\tau) = e_{\mathcal{J}}(\tau) = \sum_{q\in V} m_q^2 = e_{\mathcal{J}}(t) = \sum_{q\in V_t} m_q^2 > \sum_{q\in V'_t} m_q^2 = e_{\mathcal{J}'}(t),$$

where $V'_t \subset V_t$ denotes the set of vertices of $\mathrm{clg}(\mathcal{Z}'_t) \subset \mathrm{clg}(\mathcal{Z}_t)$. But this contradicts the semicontinuity of $e_{\mathcal{J}'}$ (Lemma 1.1.53(b)).
Step 2. We show that after a base change as allowed in condition ($\mathbb{G}$1), we can assume that φ induces isomorphisms of reduced complex spaces $\mathcal{Z}^i \xrightarrow{\cong} T$, $i = 1, \ldots, s$. In particular, it follows that $s =: k_0$ is the number of vertices of level 0 in $\mathbb{G}$.

We proceed by induction. Assume $\mathcal{Z}^i \xrightarrow{\cong} T$, $i = 1, \ldots, \nu-1$, and consider the family

$$\begin{array}{ccc} \alpha_\nu^*\mathcal{Z} & \hookrightarrow & \alpha_\nu^*\mathscr{X} \\ & {\scriptstyle \alpha_\nu^*\varphi}\searrow \quad \swarrow{\scriptstyle \mathrm{pr}} & \\ & \mathcal{Z}^\nu & \end{array}$$

induced by the finite surjective morphism $\alpha_\nu := \varphi|_{\mathcal{Z}^\nu} : \mathcal{Z}^\nu \to T$. As before, we obtain a decomposition into irreducible components

$$(\alpha_\nu^*\mathcal{Z})_{\mathrm{red}} = \alpha_\nu^*\mathcal{Z}^1 \cup \ldots \cup \alpha_\nu^*\mathcal{Z}^{\nu-1} \cup \widetilde{\mathcal{Z}}^\nu \cup \ldots \cup \widetilde{\mathcal{Z}}^{s'}, \quad s' \geq s\,,$$

where $\alpha_\nu^*\varphi : \alpha_\nu^*\mathcal{Z}^i \xrightarrow{\cong} \mathcal{Z}^\nu$, $i = 1, \ldots, \nu-1$, and $\alpha_\nu^*\varphi(\widetilde{\mathcal{Z}}^i) = \mathcal{Z}^\nu$, $i = \nu, \ldots, s'$. Moreover, we can assume that $\widetilde{\mathcal{Z}}^\nu$ is the diagonal in $\alpha_\nu^*\mathcal{Z}^\nu = \mathcal{Z}^\nu \times_T \mathcal{Z}^\nu$, in particular,

$$\alpha_\nu^*\varphi : \widetilde{\mathcal{Z}}^\nu \xrightarrow{\cong} \mathcal{Z}^\nu\,.$$

Hence, after the base change α_ν we have ν components of the zero-dimensional scheme $(\alpha_\nu^*\mathcal{Z})_{\mathrm{red}}$ being isomorphic to the base $\mathcal{Z}^\nu$. This proves *Step 2.*
Step 3. We prove that $\mathcal{Z}^i \cap \mathcal{Z}^j = \emptyset$ for any $i \neq j$, $1 \leq i, j \leq s$.

Assume that there exists a $w \in \mathcal{Z}^i \cap \mathcal{Z}^j$, $t := \varphi(w) \in T$. Then $\mathrm{clg}(\mathcal{Z}_t)$ has less than k_0 vertices of level 0 contradicting the assumption, since $s = k_0$ by *Step 2.*

Since then $(\mathcal{Z}^i)_{red} \cong T$ for $i = 1, \cdots, k_0$ we conclude that after a base change as in ($\mathbb{G}$1) there exist disjoint (smooth) sections

$$\sigma_1^{(0)},\ldots,\sigma_{k_0}^{(0)} : T \longrightarrow \mathcal{Z}_{red} \subset \mathscr{X}\,, \quad \operatorname{supp}(\mathcal{Z}) = \sigma_1^{(0)}(T) \cup \ldots \cup \sigma_{k_0}^{(0)}(T)\,.$$

In addition, since the functions $t \mapsto \operatorname{mt}(\mathcal{Z}_t, \sigma_j^{(0)}(t))$, $j = 1,\ldots,k_0$, are upper semicontinuous and since $t \mapsto \operatorname{clg}(\mathcal{Z}_t)$ is constant (in particular, the occurring multiplicities at vertices of level 0 are fixed), it follows that the family $\mathcal{Z} \hookrightarrow \mathscr{X} \to T$ is equimultiple along the sections $\sigma_j^{(0)}$.

Step 4. We show (by induction on n) that after a base change as allowed in ($\mathbb{G}$1) we may assume that the family $\mathcal{Z} \hookrightarrow \mathscr{X} \to T$ is resolvable by blowing up sections $\sigma_j^{(i)}$, $j = 1,\ldots k_i$, $i = 0,\ldots,N$.

Let $\mathscr{X}^{(1)} \to \mathscr{X}^{(0)} := \mathscr{X}$ be the blowing up of $\mathscr{X}$ along the sections $\sigma_j^{(0)}$, $j = 1,\ldots,k_0$, and let $\mathcal{Z}^{(1)} \subset \mathscr{X}^{(1)}$ be the strict transform of $\mathcal{Z} \subset \mathscr{X}$. Then, clearly, $\mathcal{Z}^{(1)} \hookrightarrow \mathscr{X}^{(1)} \to T$ is again a family of the above type (with fixed cluster graph $\mathbb{G}^{(1)}$ obtained from $\mathbb{G}$ by omitting all vertices $q_{0,j}$ of level 0 and the corresponding proximities $p \dashrightarrow q_{0,j}$). Hence, by the induction hypothesis, there exists a base change $\alpha : T' \to T$ (as in ($\mathbb{G}$1)) such that the induced family

$$\begin{array}{ccc} \alpha^*\mathcal{Z}^{(1)} & \hookrightarrow & \alpha^*\mathscr{X}^{(1)} = \mathscr{X}^{(1)} \times_T T' \\ & \searrow \quad \swarrow & \\ & T' & \end{array}$$

is resolvable by blowing up $\sigma_1^{(i)},\ldots,\sigma_{k_i}^{(i)} : T' \to \alpha^*\mathscr{X}^{(i)}$, $i = 1,\ldots,N$.

On the other hand, $\alpha^*\mathcal{Z}^{(1)} = \mathcal{Z}^{(1)} \times_T T'$ is naturally isomorphic to the blowing up of $\mathscr{X} \times_T T'$ along the sections $\alpha^*\sigma_j^{(0)}$, $j = 1,\ldots,k_0$ (cf. [NoV, Prop. 4.3]), whence the statement.

Step 5. We show that for any irreducible component F of the exceptional divisor of $\mathscr{X}^{(i)} \to \mathscr{X}$, $1 \le i \le N+1$, and any $1 \le j \le k_i$ either $\sigma_j^{(i)}(T)$ is contained in F or it has empty intersection with F.

Let $E_\nu^{(\ell)} \subset \mathscr{X}^{(i)}$ be the strict transform of the exceptional divisor obtained by blowing up $\mathscr{X}^{(\ell)}$, $0 \le \ell < i$, along the section $\sigma_\nu^{(\ell)}$, $\nu = 1,\ldots,k_\ell$. Then the points in

$$\bigcup_{j=1}^{k_i} \bigcup_{\nu=1}^{k_\ell} \sigma_j^{(i)}(t) \cap E_\nu^{(\ell)}$$

correspond precisely to those vertices of level i in the cluster graph $\operatorname{clg}(\mathcal{Z}_t) = \mathbb{G}$ being proximate to a vertex of level ℓ. Since the number of such vertices is constant (in t), and since the sets $\{t \in T \mid \sigma_j^{(i)}(t) \in E_\nu^{(\ell)}\}$ are analytic subsets of T, the irreducibility of T implies for any ν, j that either $\sigma_j^{(i)}(T) \subset E_\nu^{(\ell)}$, or $\sigma_j^{(i)}(T) \cap E_\nu^{(\ell)} = \emptyset$. □

1.1.3.3 The Rooted and the Punctual Hilbert Scheme

Fix now a cluster graph $\mathbb{G}$ that satisfies the proximity relations. We introduce the rooted Hilbert functor $\mathcal{H}ilb^{\mathbb{G}}_{\mathbb{C}\{x,y\}}$, a subfunctor of the punctual Hilbert functor

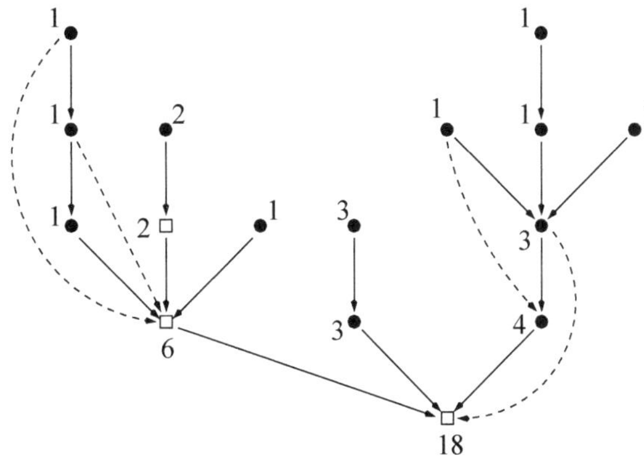

Fig. 1.3 A cluster graph $\mathbb{G}$ with a subset V_0 (marked by "□") of the set of vertices V

$\mathcal{H}ilb^n_{\mathbb{C}\{x,y\}}$ (Definition 1.1.49), consisting of families of zero-dimensional schemes with constant cluster graph $\mathbb{G}$ and with trivial section through the root of $\mathbb{G}$.

For the proof we need a generalization, the rooted Hilbert functor $\mathcal{H}ilb^{\mathbb{G},V_0}_{\mathscr{K}_0}$, a subfunctor of $\mathcal{H}ilb^n_\Sigma$ consisting of families of zero-dimensional schemes on Σ with constant cluster graph $\mathbb{G}$ and with trivial sections through a given set of vertices V_0 of $\mathbb{G}$ containing the root of $\mathbb{G}$ (Fig. 1.3). If V_0 is precisely the root of $\mathbb{G}$, then $\mathcal{H}ilb^{\mathbb{G},V_0}_{\mathscr{K}_0} = \mathcal{H}ilb^{\mathbb{G}}_{\mathbb{C}\{x,y\}}$. We show that the rooted Hilbert functor $\mathcal{H}ilb^{\mathbb{G},V_0}_{\mathscr{K}_0}$ is representable by a locally closed algebraic subvariety of Hilb^n_Σ, resp. of the punctual Hilbert scheme $\mathrm{Hilb}^n_{\mathbb{C}\{x,y\}}$ if V_0 is the root of $\mathbb{G}$. Moreover, we give a formula for its dimension.

Definition 1.1.54 (*Rooted Hilbert functor*) Let $\mathbb{G} = (\Gamma, \dashrightarrow, \boldsymbol{m})$ be a cluster graph satisfying the proximity relations, such that $\Gamma = (V, E)$ is an (oriented) *tree*. Let $V_0 \subset V$ be any subset containing the root of Γ and satisfying

$$(q \in V_0\,,\;\; p \leq q \implies p \in V_0). \tag{1.1.3.1}$$

Moreover, consider the subgraph $\Gamma_0 = (V_0, E_0)$ of Γ with $E_0 \subset E$ the set of edges with both endpoints in V_0. Let $(\mathscr{K}_0, \boldsymbol{m})$ be any cluster with origin at z defining the cluster graph $\mathbb{G}_0 = (\Gamma_0, \dashrightarrow, \boldsymbol{m})$ (obtained by restricting the binary relation $\dashrightarrow$ on V to V_0, cf. Remark 1.1.14.1).

Then we define the *rooted Hilbert functor* $\mathcal{H}ilb^{\mathbb{G},V_0}_{\mathscr{K}_0}$ by associating to a reduced complex space T the set of all families $(\mathcal{Z} \subset \Sigma \times T) \in \mathcal{H}ilb^n_\Sigma(T)$ satisfying ($\mathbb{G}$1), ($\mathbb{G}$2) of Definition 1.1.51 and, additionally,

($\mathbb{G}V_0$) the sections $\sigma^{(i)}_j$ from Definition 1.1.51 ($\mathbb{G}$1) passing through (the infinitely near points in $\mathcal{C}\ell(\mathcal{Z}_t)$ corresponding to) the vertices $p_{i,j} \in V_0$ are trivial sections.

If V_0 consists precisely of the root of Γ, that is, $\mathscr{K}_0 = (z)$, then we require that the initial section $\sigma = \sigma^{(0)}_1$ through z is trivial and we write $\mathcal{H}ilb^{\mathbb{G}}_{\mathbb{C}\{x,y\}}$ instead of $\mathcal{H}ilb^{\mathbb{G},V_0}_{\mathscr{K}_0}$ and call it the *punctual Hilbert functor fixing* $\mathbb{G}$.

Remark 1.1.54.1 (1) Note that a family in $\mathcal{H}ilb^{\mathbb{G},V_0}_{\mathscr{K}_0}(T)$ is trivial, given by a subspace $\mathcal{Z}_t \times T \subset \Sigma \times T$, iff the sections through the infinitely near points corresponding to all vertices in V are trivial.

(2) Let $(C, z) \subset (\Sigma, z)$ be a reduced plane curve singularity and $\mathscr{T} \subset \mathscr{T}(C, z)$ a finite subtree of the complete embedded resolution tree of (C, z). The cluster $\mathcal{C}\ell(C, \mathscr{T})$ consists of $\mathscr{T}$ and the multiplicities of the strict transform of (C, z) at the points of $\mathscr{T}$ (Definition 1.1.16). Let $\mathbb{G} = \mathrm{clg}(\mathcal{C}\ell(C, \mathscr{T}))$ be its cluster graph (Definition 1.1.16). Since the cluster scheme of $\mathcal{C}\ell(C, \mathscr{T})$ (Definition 1.1.21) coincides with the 0-dimensional scheme $Z(C, \mathscr{T})$ defined by the cluster ideal

$$I(C, \mathscr{T}) = \{g \in \mathcal{O}_{\Sigma,z} \mid g \text{ goes through } \mathcal{C}\ell(C, \mathscr{T})\}$$

(cf. Definition 1.1.30), we get $Z(C, \mathscr{T}) \in \mathcal{H}ilb^{\mathbb{G}}_{\mathbb{C}\{x,y\}}$. In particular, if $\mathscr{T} = \mathscr{T}^*$ is the essential tree, then $Z(C, \mathscr{T}^*)$ is the topological singularity scheme $Z^s(C, z)$ (cf. Definition 1.1.31) and we get

$$Z^s(C, z) \in \mathcal{H}ilb^{\mathbb{G}}_{\mathbb{C}\{x,y\}} .$$

Theorem 1.1.55 *Let $\mathbb{G}, V, V_0, \mathscr{K}_0$ be as in Definition 1.1.54, and*

$$n = \sum_{q \in V} \frac{m_q(m_q + 1)}{2} .$$

Then the rooted functor $\mathcal{H}ilb^{\mathbb{G},V_0}_{\mathscr{K}_0}$ (on the category of reduced complex spaces) is representable by a reduced, quasi-projective subvariety $\mathrm{Hilb}^{\mathbb{G},V_0}_{\mathscr{K}_0}$ of the projective variety Hilb^n_Σ. It is called the rooted Hilbert scheme and is of dimension equal to the number of free vertices in $V \setminus V_0$ (cf. Definition 1.1.13).

In particular, the punctual functor $\mathcal{H}ilb^{\mathbb{G}}_{\mathbb{C}\{x,y\}}$ is representable by a quasi-projective subvariety $\mathrm{Hilb}^{\mathbb{G}}_{\mathbb{C}\{x,y\}}$ of the projective variety $\mathrm{Hilb}^n_{\mathbb{C}\{x,y\}}$ of dimension equal to the number of free vertices in $V \setminus \{z\}$.

Proof We proceed by induction on n. For $n = 1$ there is nothing to show ($\mathrm{Hilb}^1_{\mathbb{C}\{x,y\}}$ is just one point). Let $n \geq 2$ and $m := m_z$, z denoting the root of Γ (according to our notation, cf. Definition 1.1.2).

Step 1. The subfunctor $\mathcal{H}^{n,m}_{\mathbb{C}\{x,y\}}$ of $\mathcal{H}ilb^n_{\mathbb{C}\{x,y\}}$ given by

$$\mathcal{H}^{n,m}_{\mathbb{C}\{x,y\}}(T) = \left\{(\mathcal{Z} \subset \Sigma \times T) \in \mathcal{H}ilb^n_{\mathbb{C}\{x,y\}}(T) \,\middle|\, \mathrm{mt}(\mathcal{Z}_t) = m \;\; \forall t \in T\right\}$$

is representable by a (Zariski-) locally closed subvariety $\mathrm{H}^{n,m}_{\mathbb{C}\{x,y\}} \subset \mathrm{Hilb}^n_{\mathbb{C}\{x,y\}}$.

This can be seen as follows: consider the description of $\mathrm{Hilb}^n_{\mathbb{C}\{x,y\}}$ as an algebraic subset of the Grassmannian of codimension n vector spaces of $\mathbb{C}\{x, y\}/\mathfrak{m}^n$ given by Briançon. In the local chart U associated to given stairs (cf. [Bri, II 2.1]) the subspace

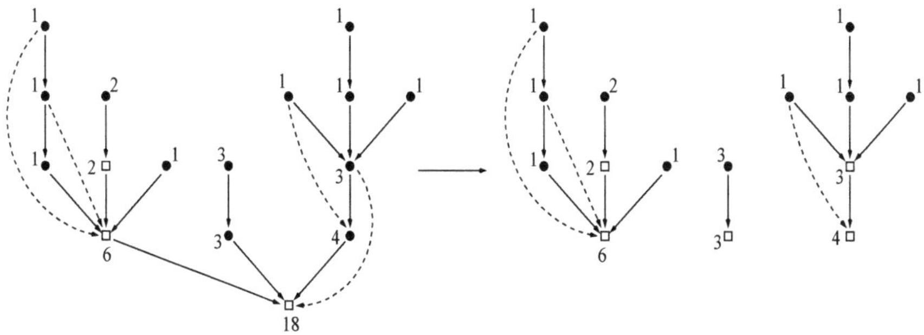

Fig. 1.4 A cluster graph $\mathbb{G}$ (with subset V_0, marked by □) and the cluster graphs $\mathbb{G}^{(i)}$ (with subsets $V_0^{(i)}$), $i = 1, 2, 3$

$\mathrm{H}^{n,m}_{\mathbb{C}\{x,y\}} \subset \mathrm{Hilb}^n_{\mathbb{C}\{x,y\}}$ is defined by the vanishing of all $\lambda_{\alpha,\beta,i,j}$ with $i + j < m$ and the condition that not all $\lambda_{\alpha,\beta,i,j}$, $i + j = m$ vanish.

Step 2. We introduce the following notations:

(a) $\mathscr{U} \hookrightarrow \Sigma \times \mathrm{H}^{n,m}_{\mathbb{C}\{x,y\}} \to \mathrm{H}^{n,m}_{\mathbb{C}\{x,y\}}$ denotes the universal family, $\pi : \Sigma' \to \Sigma$ (respectively $\pi : \Sigma' \times \mathrm{H}^{n,m}_{\mathbb{C}\{x,y\}} \to \Sigma \times \mathrm{H}^{n,m}_{\mathbb{C}\{x,y\}}$) the blowing up of $z \in \Sigma$ (respectively of the trivial section $t \mapsto (z, t)$ in $\Sigma \times \mathrm{H}^{n,m}_{\mathbb{C}\{x,y\}}$), and $E' \subset \Sigma'$ the exceptional divisor of π.

(b) $\mathbb{G}^{(i)} := (\Gamma^{(i)}, \dashrightarrow, \boldsymbol{m})$ ($\Gamma^{(i)}$ an oriented tree with set of vertices $V^{(i)}$), $i = 1, \ldots, k_1$, denote the cluster graphs obtained from $\mathbb{G}$ by removing the root z (cf. Fig. 1.4).

Set

$$n_i := \sum_{q\in V^{(i)}} \frac{m_q(m_q+1)}{2}, \quad \widetilde{n} := n_1 + \cdots + n_{k_1} = n - \frac{m(m+1)}{2}.$$

Without restriction, assume that the roots of $\Gamma^{(1)}, \ldots, \Gamma^{(s)}$, $0 \le s \le k_1$, are vertices in V_0 (corresponding precisely to the infinitely near points $q_{1,1}, \ldots, q_{1,s}$ of level 1 in the constellation $\mathscr{K}_0$), while the roots of $\Gamma^{(i)}$, $i > s$, are not in V_0. We introduce the subsets

$$V_0^{(i)} := \big(V_0 \cap V^{(i)}\big) \cup \underbrace{\big(\{p \mid p \dashrightarrow z\} \cap V^{(i)}\big)}_{\substack{\| \\ \{p_{\ell_i,i} \to \cdots \to p_{1,i}\}}} \subset V^{(i)}, \quad i = 1, \ldots, k_1,$$

which (clearly) satisfy (1.1.3.1), and which correspond to constellations $\mathscr{K}_0^{(i)}$ with origin $p_{1,i} \in E' \subset \Sigma'$, $i = 1, \ldots, k_1$, given by

- those points of the constellation $\mathscr{K}_0$ which are infinitely near to $q_{1,i}$ (in particular, $p_{1,i} = q_{1,i}$ for $i = 1, \ldots, s$),

- all the intersection points $p_{j,i}$, $j = 2, \ldots, \ell_i$, of the strict transform of E' with the exceptional divisor of $\pi_{j,i} : \Sigma_{j,i} \to \Sigma_{j-1,i}$, the blowing up of $p_{j-1,i} \in \Sigma_{j-1,i}$. (Here $\Sigma_{1,i} = \Sigma'$.)

Note that the points $p_{1,1}, \ldots, p_{1,s}$ are fixed by $\mathscr{K}_0$, while $p_{1,s+1}, \ldots, p_{1,k_1}$ can be chosen arbitrarily in E', such that all the $p_{1,i}$ are pairwise distinct.
Step 3. Let $t \in \mathrm{H}^{n,m}_{\mathbb{C}\{x,y\}}$ be such that $\mathcal{C}\ell(\mathscr{U}_t)$ defines the cluster graph $\mathbb{G}$ and the infinitely near points corresponding to the vertices in V_0 are in the prescribed position given by the constellation $\mathscr{K}_0$. We show that there exists a Cartesian diagram of germs

$$
\begin{array}{ccccc}
\big(\mathrm{Hilb}^{n}_{\mathbb{C}\{x,y\}}, t\big) & & & & \\
\big\uparrow{\scriptstyle \text{closed}} & & & & \\
\big(\mathrm{H}^{n,m}_{\mathbb{C}\{x,y\}}, t\big) & & & & \\
\big\uparrow{\scriptstyle \text{closed}} & & & & \\
(H,t) & \xrightarrow[(i)]{\psi} & \big(\mathrm{Hilb}^{\widetilde{n}}_{\Sigma'}, \psi(t)\big) & \xrightarrow[\cong]{\zeta} & \prod_{i=1}^{k_1}\big(\mathrm{Hilb}^{n_i}_{\Sigma'}, \psi(t)_i\big) \\
\big\uparrow & & & & {\scriptstyle (ii)}\big\uparrow{\scriptstyle \text{closed}} \\
 & & & & \prod_{i=1}^{s}\big(\mathrm{Hilb}^{n_i}_{\mathbb{C}\{x,y\}}, \psi(t)_i\big) \times \prod_{i=s+1}^{k_1}\big(E' \times \mathrm{Hilb}^{n_i}_{\mathbb{C}\{x,y\}}, (p_{1,i}, \psi(t)_i)\big) \\
\big| & & & & {\scriptstyle (iii)}\big\uparrow{\scriptstyle \text{closed}} \\
\big(\mathrm{Hilb}^{\mathbb{G},V_0}_{\mathscr{K}_0}, t\big) & \xrightarrow[\zeta\circ\psi]{} & & & \prod_{i=1}^{s}\big(\mathrm{Hilb}^{\mathbb{G}^{(i)},V_0^{(i)}}_{\mathscr{K}_0^{(i)}}, \psi(t)_i\big) \times \prod_{i=s+1}^{k_1}\big(E' \times \mathrm{Hilb}^{\mathbb{G}^{(i)},V_0^{(i)}}_{\mathscr{K}_0^{(i)}}, (p_{1,i}, \psi(t)_i)\big)
\end{array}
$$

actually defined in a Zariski-open neighbourhood of of the origin of the germs. This obviously implies the statement of Theorem 1.1.55. We explain the morphisms in the diagram:
(i) Consider the strict transform of the universal family $\mathscr{U} \hookrightarrow \Sigma \times \mathrm{H}^{n,m}_{\mathbb{C}\{x,y\}}$ over $\mathrm{H}^{n,m}_{\mathbb{C}\{x,y\}}$,

$$
\begin{array}{ccc}
\widetilde{\mathscr{U}} & \hookrightarrow & \Sigma' \times \mathrm{H}^{n,m}_{\mathbb{C}\{x,y\}}\,, \\
 & {\scriptstyle \varphi}\searrow \quad \swarrow{\scriptstyle \mathrm{pr}} & \\
 & \mathrm{H}^{n,m}_{\mathbb{C}\{x,y\}} &
\end{array}
$$

where the closed embedding $\widetilde{\mathscr{U}} \hookrightarrow \Sigma' \times \mathrm{H}^{n,m}_{\mathbb{C}\{x,y\}}$ is given by the ideal (sheaf) $\mathcal{J}_{\widetilde{\mathscr{U}}}$,

$$
\mathcal{J}_{\widetilde{\mathscr{U}}}(U) := \big\{\widehat{g} \,\big|\, g \in \mathcal{J}_{\mathscr{U}}(U)\big\} : (\mathcal{I}(U))^m,
$$

$\widehat{g}$ denoting the total transform of g under π and $\mathcal{I}$ the ideal of the exceptional divisor in $\Sigma' \times \mathrm{H}^{n,m}_{\mathbb{C}\{x,y\}}$.

By semicontinuity of the fibre dimension of the finite morphism φ, it follows that there is a (Zariski-) locally closed subvariety $H \subset \mathrm{H}^{n,m}_{\mathbb{C}\{x,y\}}$ such that $t \in H$, and $\dim_{\mathbb{C}}(\widetilde{\mathscr{U}_h}) = \widetilde{n}$ for each $h \in H$.

In particular, the restriction of φ to the preimage of H defines a flat morphism. Hence, the universal property of $\mathrm{Hilb}^{\widetilde{n}}_{\Sigma'}$ implies the existence of a morphism $\psi : H \to \mathrm{Hilb}^{\widetilde{n}}_{\Sigma'}$.

(ii) There is an isomorphism

$$\zeta : \left(\mathrm{Hilb}^{\widetilde{n}}_{\Sigma'}, \psi(t)\right) \xrightarrow{\cong} \prod_{i=1}^{k_1} \left(\mathrm{Hilb}^{n_i}_{\Sigma'}, \psi(t)_i\right)$$

and we can consider the (Hilbert–Chow) morphism

$$\phi = (\phi_1, \ldots, \phi_{k_1}) : \prod_{i=1}^{k_1} \left(\mathrm{Hilb}^{n_i}_{\Sigma'}, \psi(t)_i\right) \longrightarrow \prod_{i=1}^{k_1} \left(\mathrm{Sym}^{n_i} \Sigma', n_i \cdot p_{1,i}\right),$$

defined in a Zariski-open neighbourhood of the origin of the germs. The preimages under ϕ_i of the (germs at $n_i \cdot p_{1,i}$ of the) locally closed subsets

$$\Delta^{(i)} := \begin{cases} \{n_i \cdot p_{1,i}\} & \text{if } 1 \leq i \leq s \\ \{n_i \cdot w \mid w \in E'\} & \text{if } s < i \leq k_1 \end{cases}$$

are (locally) isomorphic to $\mathrm{Hilb}^{n_i}_{\mathbb{C}\{x,y\}}$ (if $i \leq s$), and $E' \times \mathrm{Hilb}^{n_i}_{\mathbb{C}\{x,y\}}$ (if $i > s$), respectively.

(iii) Finally, locally at t, $\mathrm{Hilb}^{\mathbb{G},V_0}_{\mathscr{K}_0}$ is the preimage under $\zeta \circ \psi$ of

$$\prod_{i=1}^{s} \mathrm{Hilb}^{\mathbb{G}^{(i)},V_0^{(i)}}_{\mathscr{K}_0^{(i)}} \times \prod_{i=s+1}^{k_1} \left(E' \times \mathrm{Hilb}^{\mathbb{G}^{(i)},V_0^{(i)}}_{\mathscr{K}_0^{(i)}}\right)$$
$$\subset \prod_{i=1}^{s} \mathrm{Hilb}^{n_i}_{\mathbb{C}\{x,y\}} \times \prod_{i=s+1}^{k_1} \left(E' \times \mathrm{Hilb}^{n_i}_{\mathbb{C}\{x,y\}}\right)$$

which, by the induction hypothesis, is a locally closed subvariety. The dimension statement follows from Step 2 and by induction. □

1.1.3.4 Irreducibility of the Rooted Hilbert Scheme

The aim of this section is to prove the irreducibility of the rooted Hilbert scheme $\mathrm{Hilb}^{\mathbb{G},V_0}_{\mathscr{K}_0}$ and the punctual Hilbert scheme $\mathrm{Hilb}^{\mathbb{G}}_{\mathbb{C}\{x,y\}}$ (Definition 1.1.54).

The following proposition is used several times in this book.

Proposition 1.1.56 *Let $f : X \to Y$ be a dominant morphism of algebraic varieties with Y irreducible and $f^{-1}(y)$ irreducible of dimension r for all non-empty fibres. If* $\dim(X, x) \geq \dim Y + r$ *for all $x \in X$, then X is irreducible with* $\dim X = \dim Y + r$.

Proof We show first that each irreducible component X' of X dominates Y. Assume this is not the case. The image $Y' := \overline{f(X')}$ is irreducible of dimension $\dim Y' < \dim Y$. Applying [GLS6, Theorem I.2.35] to $f : X' \to Y'$, we get $\dim X' = \dim Y' + r < \dim Y + r$, contradicting the assumption $\dim(X, x) \geq \dim Y + r$ for all $x \in X$.

Hence, if X is reducible with $X = X' \cup X''$ and X', X'' proper closed subvarieties of X, it follows from above that there is a point $y \in f(X') \cap f(X'')$. Hence $f^{-1}(y) \subset (f^{-1}(y) \cap X') \cup (f^{-1}(y) \cap X'')$ is reducible, a contradiction. Thus, X is irreducible. □

Theorem 1.1.57 *Let* $\mathbb{G}, V, V_0, \mathscr{K}_0$ *be as in Definition 1.1.54. Then the rooted Hilbert scheme* $\mathrm{Hilb}_{\mathscr{K}_0}^{\mathbb{G},V_0}$ *is irreducible.*

In particular, $\mathrm{Hilb}_{\mathbb{C}\{x,y\}}^{\mathbb{G}}$ *is an integral quasi-projective subvariety of* $\mathrm{Hilb}_{\mathbb{C}\{x,y\}}^{n}$, *$n$ as in Theorem 1.1.55, of dimension equal to the number of free vertices in* $V \setminus \{z\}$ *(cf. Definition 1.1.13).*

By Theorem 1.1.55 $\mathrm{Hilb}_{\mathscr{K}_0}^{\mathbb{G},V_0}$ represents the rooted Hilbert functor $\mathcal{H}ilb_{\mathscr{K}_0}^{\mathbb{G},V_0}$ on the category of reduced complex spaces. Let us formulate what this means for $\mathrm{Hilb}_{\mathbb{C}\{x,y\}}^{\mathbb{G}}$:

There exists a universal family $\mathscr{Z}^{\mathbb{G}} \subset \Sigma \times \mathrm{Hilb}_{\mathbb{C}\{x,y\}}^{\mathbb{G}} \to \mathrm{Hilb}_{\mathbb{C}\{x,y\}}^{\mathbb{G}}$, such that every family of zero-dimensional schemes $\mathscr{Z} \subset \Sigma \times T$ over a reduced complex space T is induced from the universal family by a unique morphism $T \to \mathrm{Hilb}_{\mathbb{C}\{x,y\}}^{\mathbb{G}}$ if the following holds: for each $t \in T$ the zero-dimensional scheme $\mathscr{Z}_t$ has support $\{z\}$ and satisfies $\mathrm{clg}(\mathscr{Z}_t) = \mathbb{G}$.

We shall apply this in the next section to the case when $\mathbb{G} = \mathrm{clg}(Z^s(C, z))$ is the cluster graph of the singularity scheme $Z^s(C, z)$, i.e., the graph of the cluster $\mathcal{C}\ell(C, \mathscr{T}^*(C, z)) = \mathcal{C}\ell(Z^s(C, z))$ where (C, z) is a reduced plane curve singularity (cf. Definition 1.1.31 and Remark 1.1.31.1).

Proof To show irreducibility we proceed again by induction on n. With the notations introduced in the proof of Theorem 1.1.55, we can assume that the first ℓ triples

$$\left(\mathbb{G}^{(i)}, V_0 \cap V^{(i)}, V_0^{(i)}\right), \quad i = 1, \ldots, \ell,$$

are pairwise different and occur precisely ν_i-times among all such triples (in particular, $\nu_1 + \cdots + \nu_\ell = k_1$). Recall that we assumed $V_0 \cap V^{(i)} \neq \emptyset$ precisely for $i = 1, \ldots, s \leq \ell$. (Note that $\nu_i = 1$ if $V_0 \cap V^{(i)} \neq \emptyset$).

For any $i = 1, \ldots, \ell$, let $\widetilde{\mathscr{Z}}^{(i)}$ be the union of those connected components of the strict transform

$$\begin{array}{ccc} \widetilde{\mathscr{Z}} & \hookrightarrow & \Sigma' \times \mathrm{Hilb}_{\mathscr{K}_0}^{\mathbb{G},V_0} \\ & {\scriptstyle\varphi}\searrow \quad \swarrow{\scriptstyle \mathrm{pr}} & \\ & \mathrm{Hilb}_{\mathscr{K}_0}^{\mathbb{G},V_0} & \end{array}$$

of the universal family $\mathscr{Z} \hookrightarrow \Sigma \times \mathrm{Hilb}_{\mathscr{K}_0}^{\mathbb{G},V_0} \to \mathrm{Hilb}_{\mathscr{K}_0}^{\mathbb{G},V_0}$, which satisfy

- $\operatorname{clg}(\widetilde{\mathscr{Z}}_t^{(i)}, x) = \mathbb{G}^{(i)}$,
- the infinitely near points of $\mathcal{C}\ell(\widetilde{\mathscr{Z}}_t^{(i)}, x)$ corresponding to the vertices in $V_0 \cap V^{(i)}$ are in the prescribed position given by the constellation $\mathscr{K}_0$,
- the infinitely near points of $\mathcal{C}\ell(\widetilde{\mathscr{Z}}_t^{(i)}, x)$ corresponding to the vertices in $V_0^{(i)}$ are on E' (respectively on its strict transform).

for all $x \in \operatorname{supp}(\widetilde{\mathscr{Z}}_t^{(i)})$, $t \in \operatorname{Hilb}_{\mathscr{K}_0}^{\mathbb{G}, V_0}$. In particular,

$$\widetilde{\mathscr{Z}} = \widetilde{\mathscr{Z}}^{(1)} \cup \ldots \cup \widetilde{\mathscr{Z}}^{(\ell)},$$

and the fibres of the restriction of φ, $\varphi_i : \widetilde{\mathscr{Z}}^{(i)} \to \operatorname{Hilb}_{\mathscr{K}_0}^{\mathbb{G}, V_0}$ have constant (vector space) dimension ($= \nu_i n_i$). Hence, the φ_i are flat and, by the universal property of $\operatorname{Hilb}_{\Sigma'}^{\nu_i n_i}$, we obtain morphisms

$$\operatorname{Hilb}_{\mathscr{K}_0}^{\mathbb{G}, V_0} \xrightarrow{\rho_i} \operatorname{Hilb}_{\Sigma'}^{\nu_i n_i} \xrightarrow{\varphi_i} \operatorname{Sym}^{\nu_i n_i} \Sigma', \quad i = 1, \ldots, \ell.$$

We complete the proof by showing that the composed morphism

$$\varphi \circ \rho := (\varphi_1 \circ \rho_1, \ldots, \varphi_\ell \circ \rho_\ell) : \operatorname{Hilb}_{\mathscr{K}_0}^{\mathbb{G}, V_0} \longrightarrow \operatorname{Sym}^{\nu_1 n_1} \Sigma' \times \ldots \times \operatorname{Sym}^{\nu_\ell n_\ell} \Sigma'$$

is dominant on the irreducible set $\Delta_1 \times \ldots \times \Delta_\ell$ with irreducible and equidimensional fibres and apply Theorem 1.1.55 and Proposition 1.1.56.

Here, $\Delta_i := \{n_i \cdot q_{1,i}\}$ if $1 \le i \le s$ ($q_{1,i}$ being the infinitely near point in $\mathscr{K}_0$ corresponding to the root of $\Gamma^{(i)}$), and $\Delta_i := \left\{ \sum_{j=1}^{\nu_i} n_i w_{i,j} \,\middle|\, w_{i,j} \in E' \right\}$ if $s < i \le \ell$.

Let $(w_{i,j})_{i,j}$ be any k_1-tuple of *pairwise different* points $w_{i,j} \in E'$, $w_{i,1} = q_{1,i}$ if $1 \le i \le s$ ($j = 1, \ldots, \nu_i$, $i = 1, \ldots, \ell$). By Proposition 1.1.17, there exists a reduced plane curve germ $(C, z) \subset (\Sigma, z)$ and a subconstellation $\mathscr{T} \subset \mathscr{T}^*(C, z)$ such that $(\mathscr{K}_0, \boldsymbol{m}) = \mathcal{C}\ell(C, \mathscr{T})$. Adding branches with tangent directions $w_{i,j}$, $j = 1, \ldots, \nu_i$, $i = s+1, \ldots, \ell$, we can easily construct a plane curve germ $(C_{\boldsymbol{w}}, z)$ and a subconstellation $\mathscr{T}_{\boldsymbol{w}} \subset \mathscr{T}(C_{\boldsymbol{w}}, z)$ such that $\mathbb{G} = \operatorname{clg}(C_{\boldsymbol{w}}, \mathscr{T}_{\boldsymbol{w}})$. By construction, $Z_{\boldsymbol{w}} := Z(C_{\boldsymbol{w}}, \mathscr{T}_{\boldsymbol{w}})$ corresponds to a point in the fibre $(\phi \circ \rho)^{-1}\big(\sum_{j=1}^{\nu_1} n_1 w_{1,j}, \ldots, \sum_{j=1}^{\nu_\ell} n_\ell w_{\ell,j}\big)$. On the other hand, any point in the image is of this form and the fibre is isomorphic to

$$\prod_{j=1}^{\nu_1} \operatorname{Hilb}_{\mathscr{K}_0^{(1)}}^{\mathbb{G}^{(1)}, V_0^{(1)}} \times \cdots \times \prod_{j=1}^{\nu_\ell} \operatorname{Hilb}_{\mathscr{K}_0^{(\ell)}}^{\mathbb{G}^{(\ell)}, V_0^{(\ell)}}.$$

Hence, by the induction hypothesis, the fibres are irreducible and equidimensional.

In the same manner, the dimension statement follows from the induction hypothesis, since the dimension of the image of $\varphi \circ \rho$ equals the number of free points of level 1 in $\mathbb{G} \setminus V_0$. □

1.1.3.5 Plane Curves with One Singular Point

Using the results of the previous section we study families of reduced curves in $\mathbb{P}^2$ of fixed degree having exactly one singularity of given topological type S, encoded by the topological singularity scheme Z^s (in the next section we consider families of hypersurfaces in $\mathbb{P}^n$ with exactly one isolated singularity of fixed analytic type Z^a). We derive in particular a formula for the dimension of this family.

In the following we consider first the case where the singular point is fixed and afterwards we let the singular point move in $\mathbb{P}^2$. Let Σ denote a smooth complex surface.

Definition 1.1.58 (*Punctual Hilbert Scheme of a Topological Type*) Let $(C, z) \subset (\Sigma, z)$ be a reduced plane curve singularity, $\mathscr{T}^* = \mathscr{T}^*(C, z)$ the essential tree and $\mathbb{G} = (\Gamma_{\mathscr{T}^*}, \dashrightarrow, \boldsymbol{m})$ the cluster graph defined by $\mathcal{C}\ell(C, \mathscr{T}^*)$. Since $\mathbb{G} = \mathrm{clg}(Z^s(C, z))$ is a complete invariant of the topological type S of (C, z) by Proposition 1.1.18, it depends only on S and we can introduce the *punctual Hilbert scheme associated to a topological type S*,

$$\mathcal{H}_0^s(S) := \mathrm{Hilb}^{\mathbb{G}}_{\mathbb{C}\{x,y\}} .$$

Note that, after choosing for $w \in \Sigma$ an isomorphism $\mathcal{O}_{\Sigma,w} \cong \mathbb{C}\{x, y\}$, we have for any plane curve singularity (C, w),

$$Z^s(C, w) \in \mathcal{H}_0^s(S)$$

iff (C, w) has topological type S, where the 0-dimensional scheme $Z^s(C, w)$ is the topological singularity scheme defined by $I^s(C, w)$ (Definition 1.1.31).

The Hilbert scheme $\mathcal{H}_0^s(S)$ is by Theorem 1.1.57 an irreducible quasi-projective algebraic subvariety of $\mathrm{Hilb}^n_{\mathbb{C}\{x,y\}}$. Its points are (by Proposition 1.1.17) zero-dimensional schemes $Z^s(C, w) \in \mathcal{H}_0^s(S)$ with constant cluster graph $\mathbb{G} = \mathrm{clg}(Z^s(C, w))$ of degree $\dim_{\mathbb{C}} \mathcal{O}_{\Sigma,z}/I^s(C, w) = n = \sum_{q\in\mathscr{T}^*(C,w)} \frac{m_q(m_q+1)}{2}$, m_q the multiplicity of the strict transform of (C, w) at q (Lemma 1.1.32).

Definition 1.1.59 We denote by

$$V_d(S) \subset V_d := \mathbb{P}\big(H^0(\mathcal{O}_{\mathbb{P}^2}(d))\big) = \mathbb{P}^N,\ N = \frac{d(d+3)}{2},$$

the *variety of reduced curves $C \subset \mathbb{P}^2$ of degree d having exactly one singular point of topological type S.*

More precisely, we require that the universal family $\mathscr{U}_d \to V_d$ admits an *analytic* section over $V_d(S)$, picking the singular point, along which the family is equisingular (by Proposition 2.2.6 the section is unique).

A priori $V_d(S)$ is a set, but we show in Proposition 1.1.61 that it is in fact a quasi-projective subvariety (with its reduced structure). For a generalization see

Sect. 2.2.2, in particular Sect. 2.2.2.2, where we show that $V_d(S)$ can be endowed with a not necessarily reduced *analytic* structure such that the universal family $\mathscr{U}_d(S) = \{(z, C) \mid z \in C\} \subset \mathbb{P}^2 \times V_d(S) \to V_d(S)$ represents equisingular families over arbitrary (not necessarily reduced) base spaces with one singular point of type S (Theorem 2.2.32).

Recall a few general facts about plane curves of degree d. Let

$$F(x, a) = \sum_{i+j+k=d} a_{i,j,k} x_0^i x_1^j x_2^k$$

be a homogeneous polynomial in (x_0^i, x_1^j, x_2^k) of degree d with coefficients $a = (a_{i,j,k})$ considered as variables. For fixed $a \in \mathbb{P}^N$, N = d(d+3)/2, the polynomial F_a, $F_a(x) = F(x, a)$, defines a curve of degree d in $\mathbb{P}^2$.

Lemma 1.1.60 *Consider the projective variety*

$$\mathscr{U}_d = \{(p, C) \mid p \in C\} = \{(p, a) \in \mathbb{P}^2 \times \mathbb{P}^N \mid F(p, a) = 0\} \subset \mathbb{P}^2 \times V_d$$

and the projection $pr : \mathscr{U}_d \to V_d$, the universal family of plane curves of degree d. In the following we use the Zariski topology.

1. *The set of reduced curves $V_d^{red} \subset V_d$ is an open, dense subvariety of V_d.*
2. *The set of singular curves $V_d^{sing} \subset V_d$ is a closed subvariety (the "discriminant"). If $d \geq 2$ then V_d^{sing} is of codimension one in V_d and the set of curves having a node as only singular point is a smooth, open and dense subvariety of V_d^{sing}. Moreover, if $d \geq 3$ the variety of curves having a cusp as only singular point is smooth and has codimension one in V_d^{sing}.*
3. *Denote by $V_d^{(1)} \subset V_d$ the set of curves of degree d having exactly one singular point such that the map $\Phi_d : V_d^{(1)} \to \mathscr{U}_d$ mapping C to (p, C), with p the singular point of C, is an analytic section of pr. Then $V_d^{(1)}$ is an open and dense subset of V_d^{sing}, which is non-empty if $d \geq 2$.*

Proof (1) The set $V_d^{red} = \{a \in \mathbb{P}^N \mid F_a \text{ is reduced}\}$ is an open subvariety of $\mathbb{P}^N = V_d$ since the condition that F is not reduced, i.e factors with a multiple factor, is a closed condition for the variables a. V_d^{red} is dense, since it contains all smooth curves.

(2) V_d^{sing} is closed since it is the image under $pr : \mathscr{U}_d \to V_d$ of

$$\mathscr{S}_d := \{(p, a) \in \mathscr{U}_d \mid F_{x_0}(p, a) = F_{x_1}(p, a) = F_{x_2}(p, a) = 0\},$$

which is a closed subvariety of $\mathscr{U}_d$, the singular locus of pr. Since there are smooth curves in V_d for any d, V_d^{sing} has codimension ≥ 1.

For $d = 2$ the union of 2 lines define 1-nodal curves being dense and open in V_2^{sing}. Now let p be a singular point of a curve C of degree $d \geq 3$ defined by F. Take two non-tangent lines through p defined by G_1 and $d - 2$ generic lines, defined by G_2, not passing through p. Choose other copies of 2 lines H_1 through p and

$d-2$ generic lines H_2 not through p such that p is the only intersection point of the three curves of degree d defined by $G := G_1 \cdot G_2$, $H := H_1 \cdot H_2$ and F. Consider the linear system $\lambda_0 F + \lambda_1 G + \lambda_2 H$. By Bertini's theorem (Corollary 2.1.22), the generic element of this system is a curve of degree d, which is non-singular outside p. A local analysis at p shows that it has a node at p. Hence the set of 1-nodal curves is dense in V_d^{sing}. It is smooth of codimension 1 in V_d by Corollary 4.3.6. It is also open in V_d^{sing}, since the the 1-nodal curves have total Milnor number 1 and the Milnor number is semicontinuous.

If $d \geq 3$ let $F_1 = zx^2 - y^3$ define the cuspidal cubic in $\mathbb{P}^2$ with singular point $p = 0$. For $d > 3$ let F_2 define $d-3$ generic lines not passing through 0 and set $F = F_1 \cdot F_2$. Moreover, choose 2 different copies of 3 lines passing through 0 defined by G_1 and H_1 and choose 2 different copies $d-3$ generic lines not passing through 0 defined by G_2 and H_2 such that 0 is the only intersection point of $G := G_1 \cdot G_2 = 0$, $H := H_1 \cdot H_2 = 0$, $F = 0$. As before we get by Bertini that the generic member of the linear system $\lambda_0 F + \lambda_1 G + \lambda_2 H$ has a cusp at 0 and no other singularity. By Corollary 4.3.6 the set of 1-cuspidal curves is smooth of codimension 2 in V_d.

(3) Set $\mathscr{U}_d^{red} = pr^{-1}(V_d^{red})$ and endow $\mathscr{U}_d^{red} \cap \mathscr{S}_d$ with its reduced structure. We consider the restriction of pr,

$$\pi : \mathscr{U}_d^{red} \cap \mathscr{S}_d \to V_d^{red} \cap V_d^{sing},$$

which is a finite morphism since reduced curves have only finitely many singularities. The set of reduced singular curves is hence an open (and dense by (2)) subvariety of V_d^{sing}.

Set $V' = \{C \in V_d^{red} \cap V_d^{sing} \mid \dim_{\mathbb{C}} H^0(\mathcal{O}_{\pi^{-1}(C)}) = 1\}$. By the semicontinuity of the fibre dimension of a finite morphism, V' is an open dense subset of $V_d^{red} \cap V_d^{sing}$ (since each curve has at least one singularity). $\dim_{\mathbb{C}} H^0(\mathcal{O}_{\pi^{-1}(C)}) = 1$ is equivalent to C having only 1 singular point, say p, and $\mathfrak{m}_p = \mathfrak{m}_C \cdot \mathcal{O}_{\mathscr{S}_d,p}$, where $\mathfrak{m}_p$ is the maximal ideal of $\mathcal{O}_{\mathscr{S}_d,p}$ and $\mathfrak{m}_C$ the maximal ideal of $\mathcal{O}_{V_d^{sing},C}$, i.e. π is unramified at p. The non-flat locus of π is closed and nowhere dense ($\mathscr{U}_d^{red} \cap \mathscr{S}_d$ is reduced) and if V'' denotes the complement of its image, then V'' is open and dense in $V_d^{red} \cap V_d^{sing}$. Hence $V^{(1)} := V' \cap V''$ is open and dense in $V_d^{red} \cap V_d^{sing}$ and hence in V_d^{sing}. It is the set of curves C with 1 singular point p such that π is unramified and flat, hence étale at p (cf. [AK, VI 3.1 and 4.2]). This is equivalent to π being locally (in the Euclidean topology) an isomorphism and that pr admits an analytic section given by Φ_d. If $d \geq 2$, $V^{(1)}$ contains the 1-nodal curves and is hence not empty. □

Note that picking the singular point in a family of curves having only one singular point is in general not an analytic section. Consider e.g. the deformation of a cuspidal cubic into a nodal cubic in V_3, which is a family of curves having only one singular point. The singular locus of pr is smooth while the discriminant is singular at the cuspidal cubic, hence picking the singular point can not be a section (see also Example 2.10.2).

Remark 1.1.60.1 Let

$$\Phi_d : V_d(S) \to \mathbb{P}^2$$

be the morphism sending C to its singular point. Consider the universal family $\mathcal{U}_d(S) \hookrightarrow \mathbb{P}^2 \times V_d(S) \to V_d(S)$ of reduced plane curves C of degree d having a singularity of type S along the section $\Phi_d \times \mathrm{id} : V_d(S) \to \mathbb{P}^2 \times V_d(S)$ as only singularity. Associating to (C, w) the zero dimensional scheme $Z^s(C, w)$, $w = \Phi_d(C)$, defines a family $\mathcal{Z}^s \hookrightarrow \mathbb{P}^2 \times V_d(S) \to V_d(S)$ of singularity schemes with fibres $Z^s(C, w)$, supported along the image of $\Phi_d \times \mathrm{id}$. Using the projective transformations of $\mathbb{P}^2$, we see that Φ_d is surjective.

We need to trivialize the section (over some open set). There exists an affine subset $\mathbb{A}^2 \subset \mathbb{P}^2$ such that the complementary line L_∞ satisfies

$$V := V_d(S) \setminus \Phi_d^{-1}(L_\infty) \underset{\text{dense}}{\hookrightarrow} V_d(S)\,.$$

Note that V is Zariski-open in $V_d(S)$ as it is the intersection with $V_d(S)$ of the complement in V_d of $pr(\mathscr{S}_d \cap L_\infty)$ (in the notations of the proof of Lemma 1.1.60).

We consider the induced family $\mathcal{Z}^s \hookrightarrow \mathbb{A}^2 \times V \to V$. Applying the translation

$$\mathbb{A}^2 \times V \longrightarrow \mathbb{A}^2 \times V\,,\ (\boldsymbol{x}\,;\ C) \longmapsto \big(\boldsymbol{x} - \Phi_d(C);\ C\big)$$

leads to a family over V of zero-dimensional schemes in $\mathbb{A}^2$, supported along the trivial section.

It follows that there exists a morphism

$$\Psi_d : V \longrightarrow \mathcal{H}^s(S) := \mathbb{P}^2 \times \mathcal{H}^s_0(S)\,, \tag{1.1.3.2}$$

mapping a curve $C \in V$ with singularity at w to the tuple $\big(w;\ \tau_{0w}(Z^s(C, w))\big)$, where $\tau_{0w}(Z^s(C, w))$ denotes the scheme with structure sheaf the pull back of the structure sheaf of $Z^s(C, w)$ via the translation τ_{0w} mapping 0 to w.

Proposition 1.1.61 *The set $V_d(S)$ of reduced curves $C \subset \mathbb{P}^2$ of degree d having exactly one singular point of topological type S is a quasi-projective subvariety of $\mathbb{P}\big(H^0(\mathcal{O}_{\mathbb{P}^2}(d))\big)$.*

Proof We use the notations of Remark 1.1.60.1 and (the proof of) Lemma 1.1.60. We need only to show that V from Remark 1.1.60.1 is quasi-projective. Set $\widetilde{V}^{(1)} = pr^{-1}(V^{(1)})$ and consider

$$\widetilde{V} := \{(w, C, Z) \mid (w, C) \in \mathscr{S}_d,\ C \in V^{(1)},\ Z = \tau_{0w}(Z^s(C, w))\},$$

which is Zariski-closed (gven by equations derived from $Z = \tau_{0w}(Z^s(C, w))$) in the quasi-projective variety $(\widetilde{V}^{(1)} \cap \mathscr{S}_d) \times \mathcal{H}^s_0(S)$. Therefore $V = \pi_2(\widetilde{V})$, π_2 the 2nd projection, is a closed subvariety of $V^{(1)}$ and hence quasi-projective. □

Theorem 1.1.57 gives a nice formula for the dimension of $\mathcal{H}_0^s(S)$ as the number of free points in the essential tree of the curve singularity (C, z) having topological type S. We use the above family $\Psi_d : V \to \mathcal{H}^s(S)$ from (1.1.3.2) and global deformation theory (cf. Chap. 2) for a different approach to determine an alternative formula for the dimension of $\mathcal{H}_0^s(S)$:

Proposition 1.1.62 *Let S be a topological type of a plane curve singularity (C, z) given by $f \in \mathbb{C}\{x, y\}$.*

(1) There exists a d_0 such that for $d \geq d_0$ the morphism

$$\Psi_d : V \to \mathbb{P}^2 \times \mathcal{H}_0^s(S)$$

is dominant with fibres $\Psi_d^{-1}(\Psi_d(C))$ being Zariski-open dense subsets of the projective space $\mathbb{P}(H^0(\mathcal{J}_{Z^s(C,z)/\mathbb{P}^2}(d)))$.

(2) $\mathcal{H}_0^s(S)$ is a quasi-projective, irreducible variety of dimension

$$\dim \mathcal{H}_0^s(S) = \deg Z^s(C, z) - \tau^{es}(S) - 2 = \deg Z^s(C, z) - \tau^{es}_{fix}(S),$$

with $\tau^{es}(S)$ and $\tau^{es}_{fix}(S)$ as in the following definition (for $\deg Z^s(C, z)$ *see Lemma 1.1.32).*

Definition 1.1.63 (*Equisingularity ideal and scheme, and τ^{es}*)
Let $f \in \mathcal{O}_{\Sigma,z}$ define (C, z). We define the *equisingularity ideal* $I^{es}(S)$ (cf. [Wah1], [GLS6, Section II.2.2]) as

$$I^{es}(C, z) = I^{es}(f) = \left\{ g \in \mathcal{O}_{\Sigma,z} \;\middle|\; \begin{array}{c} \text{There is a section } \sigma \text{ such that} \\ f + \varepsilon g \text{ is an } es-\text{deformation} \\ \text{of } (C, z) \text{ over } T_\varepsilon \text{ along } \sigma \end{array} \right\}.$$

It defines the zero-dimensional scheme $Z^{es}(S) = Z^{es}(C, z)$, called the *(topological) equisingularity scheme* of degree $\tau^{es}(S)$,[5] with

$$\tau^{es}(S) := \tau^{es}(C, z) = \tau^{es}(f) = \dim_{\mathbb{C}} \mathbb{C}\{x, y\}/I^{es}(f).$$

Moreover, we define the *fixed equisingularity ideal* contained in $I^{es}(f)$ by

$$I^{es}_{fix}(C, z) = I^{es}_{fix}(f) = \left\{ g \in \mathcal{O}_{\Sigma,z} \;\middle|\; \begin{array}{c} f + \varepsilon g \text{ is an } es\text{-deformation of } (C, z) \\ \text{over } T_\varepsilon \text{ along the trivial section} \end{array} \right\}$$

satisfying (by [GLS6, Proposition II.2.14])

---

[5] τ^{es} is equal to the codimension of the μ–constant stratum in the ($\tau(C, z)$–dimensional) base space of the semiuniversal deformation of (cf. [GLS6, Corollary II.2.68]), which is also equal to $\mu(C, z) - \text{mod}\,(C, z)$ where $\text{mod}\,(C, z)$ is the right–modality of (C, z) in the sense of Arnol'd [GLS6, Section II.2.2, Remarks and Exercises, p. 373].

$$I^{es}(C,z) = I^{es}_{fix}(C,z) + j(f) \text{ and}$$

$$\tau^{es}_{fix}(S) := \tau^{es}_{fix}(f) = \dim_{\mathbb{C}} \mathcal{O}_{\Sigma,z}/I^{es}_{\text{fix}}(f) = \tau^{es}(f) + 2.$$

$I^{es}_{fix}(C,z)$ defines the *fixed (topological) equisingularity scheme* $Z^{es}_{fix}(S) = Z^{es}_{fix}(C,z)$ of degree $\tau^{es}_{fix}(S)$

Proof of Proposition 1.1.62. We have seen in Remark 1.1.16.1 that there exists an integer m such that the scheme $Z^s(C,z)$ depends only on the $(m-1)$–jet of the equation of (C,z). Hence, for $d_0 \geq m$ the above morphism $\Psi_{d_0} : V \to \mathbb{P}^2 \times \mathcal{H}^s_0(S)$, defined on an open dense subset $V \subset V_{d_0}(S)$, is dominant.

Note that each fibre $\Psi^{-1}_{d_0}(\Psi_{d_0}(C))$, $C \in V$, is a non-empty sub set of $\mathbb{P}\big(H^0(\mathcal{J}_{Z^s(C,z)/\mathbb{P}^2}(d_0))\big)$ consisting of curves $C' \in V$ with $Z^s(C',z') = Z^s(C,z)$ (for $z' = \Phi_d(C')$). This means that C', given by $g \in H^0(\mathcal{O}_{\mathbb{P}^2}(d))$, belongs to the fibre of Ψ_d iff $I^s(g) = I^s(f)$, that is, iff $g \in H^0(\mathcal{J}_{Z^s(C.z)/\mathbb{P}^2}(d_0))$. Hence

$$\Psi^{-1}_{d_0}(\Psi_{d_0}(C)) = V \cap \mathbb{P}(H^0(\mathcal{J}_{Z^s(C,z)/\mathbb{P}^2}(d_0)))\,.$$

Now, by Lemma 1.1.33 the subset of elements $g \in H^0(\mathcal{J}_{Z^s(C,z)/\mathbb{P}^2}(d_0))$ defining $Z^s(C,z)$ is a Zariski-open and dense. This shows the openness of the fibre and hence proves (1).

To see the dimension statement in (2), note that by [GLS6, Theorem I.2.35],

$$\begin{aligned}\dim \mathcal{H}^s_0(S) + 2 &= \dim \Psi_{d_0}(V)\\ &= \dim V_{d_0}(S) - h^0\big(\mathcal{J}_{Z^s(C',z')/\mathbb{P}^2}(d_0)\big) + 1\end{aligned} \tag{1.1.3.3}$$

Since $\deg Z^s(C',z')$ is constant for $C' \in V$, we can assume that $d \geq d_0$ is chosen sufficiently large such that $H^1\big(\mathcal{J}_{Z^s(C',z')/\mathbb{P}^2}(d)\big)$ vanishes for all $C' \in V$ (using Serre vanishing). Then the long exact cohomology sequence associated to

$$0 \longrightarrow \mathcal{J}_{Z^s(C',z')/\mathbb{P}^2}(d) \longrightarrow \mathcal{O}_{\mathbb{P}^2}(d) \longrightarrow \mathcal{O}_{Z^s(C',z')} \longrightarrow 0$$

implies

$$h^0\big(\mathcal{J}_{Z^s(C',z')/\mathbb{P}^2}(d)\big) = \frac{d(d+3)}{2} + 1 - \deg Z^s(C,z)\,. \tag{1.1.3.4}$$

Moreover, $Z^s(C',z') \supset Z^{es}(C',z')$ (Corollary 1.1.34), whence the above H^1-vanishing implies that, for each $C' \in V$, $H^1\big(\mathcal{J}_{Z^{es}(C',z')/\mathbb{P}^2}(d)\big)$ vanishes, too. Finally, we can apply global deformation theory (Theorem 2.40(c), below) to deduce that $V_d(S)$ is smooth of the "expected" dimension

$$\dim V_d(S) = \frac{d(d+3)}{2} - \tau^{es}(S)\,.$$

Combining this equality with (1.1.3.3) and (1.1.3.4), we obtain the above formula for $\dim \mathcal{H}_0^s(S)$. The quasi-projectivity and irreducibility were shown in Theorem 1.1.57. □

Remark 1.1.63.1 The topological singularity scheme $Z^s(C, z)$ is defined by the ideal

$$I^s(f) = \{g \in \mathcal{O}_{\Sigma,z} \mid \operatorname{mt} \widehat{g}_{(q)} \geq \operatorname{mt} \widehat{C}_{(q)} \text{ for all } q \in \mathscr{T}^*\}$$

(cf. Definition 1.1.31), where the points q of the essential tree $\mathscr{T}^* = \mathscr{T}^*(C, z)$ are the non-nodes of the minimal good embedded resolution of (C, z) (including z if (C, z) is singular). By Corollary 1.1.34 we have $I^s(f) \subset I_{\text{fix}}^{es}(f)$ and from Proposition 1.1.62 we get

$$\dim \mathcal{H}_0^s(S) = \dim_{\mathbb{C}} I_{\text{fix}}^{es}(f)/I^s(f)\,,$$

where $\dim \mathcal{H}_0^s(S)$ is the number of free vertices $q \in \mathscr{T}^* \setminus \{z\}$ (Theorem 1.1.57).

Note that for *es*-deformations there exists (unique) sections $\sigma_{(q)}$ through $q \in \mathscr{T}^*$ along which the deformation of the reduced total transform is equimultiple. For *es*-deformations $f + tg$ over $(\mathbb{C}, 0)$ with $g \in I^s(f)$ these sections are all trivial. For arbitrary *es*-deformations with tangent directions in $I_{\text{fix}}^{es}(f)$ the sections through satellite points $q \in \mathscr{T}^*$ have to stay at satellite points. But the sections through free points in $\mathscr{T}^* \setminus \{z\}$ may move along the exceptional divisor, giving one degree of freedom for every free point. Since $\mathcal{H}_0^s(S)$ represents es-deformations with trivial initial section, we get a geometric interpretation of the formula $\dim \mathcal{H}_0^s(S) = \#$(free vertices $q \in \mathscr{T}^* \setminus \{z\}$).

As an immediate consequence of Lemma 1.1.22, Proposition 1.1.62 and Theorem 1.1.57, we get the following handy formula for $\deg Z^{es}(C, z) = \tau^{es}(C, z)$, which equals the codimension of the μ-constant stratum in the semiuniversal deformation base:

Corollary 1.1.64 *Let $(C, z) \subset (\Sigma, z)$ be a reduced plane curve singularity, let $\mathscr{T}^*(C, z)$ denote the essential tree, and let, for each $p \in \mathscr{T}^*(C, z)$, m_p denote the multiplicity of the strict transform of C at p. Then*

$$\tau^{es}(C, z) = \sum_{p \in \mathscr{T}^*(C,z)} \frac{m_p(m_p + 1)}{2} - \#(\textit{free points in } \mathscr{T}^*(C, z)) - 1\,.$$

This formula has been generalized to algebroid curves in arbitrary characteristic in [CGL1, Theorem 6.2(5)].

At the end of this section let us relate $\mathcal{H}_0^s(S)$ to straight equisingular deformations.

Remark 1.1.64.1 For a reduced plane curve singularity $(C, z) \subset (\Sigma, z)$ we consider the following base spaces of semiuniversal deformations of (C, z) (all of them are smooth):

- $B_{C,z}$ the base space of the (usual) semiuniversal deformation of dimension $\tau^{ea}(C, z) = \tau(C, z)$,

- $B^{fix}_{C,z}$ the base space of the semiuniversal deformation with (trivial) section of dimension $\tau^{ea}_{fix}(C, z) = \tau_{fix}(C, z)$,
- $B^{es}_{C,z}$ the base space of the seminuniversal equisingular deformation of dimension $\tau^{ea}(C, z) - \tau^{es}(C, z)$,
- $B^{es,fix}_{C,z}$ the base space of the seminuniversal es-deformation with section of dimension $\tau^{ea}_{fix}(C, z) - \tau^{es}_{fix}(C, z) = \tau^{ea}(C, z) - \tau^{es}(C, z)$,
- $B^{s}_{C,z}$ the base space of the seminuniversal straight es-deformation of dimension $\tau^{ea}_{fix}(C, z) - \tau^{s}(C, z)$, with $\tau^{s}(C, z) := \dim_{\mathbb{C}} \mathcal{O}_{\Sigma,z}/I^{s}(C, z) = \deg Z^{s}(C, z)$.

For the definition of $\tau, \tau^{ea}, \tau_{fix}, \tau^{ea}_{fix}$ see Sect. 1.1.4.1, for $\tau^{es}, \tau^{es}_{fix}$ Definition 1.1.63 and for the relation Lemma 2.2.13. By [GLS6, Theorem II.2.61, Corollary II.2.67 and Proposition II.2.63] we know that $B^{es}_{C,z}$ is isomorphic to the μ-constant stratum in $B_{C,z}$, while $B^{es,fix}_{C,z}$ is isomorphic to the μ-constant stratum in $B^{fix}_{C,z}$, both are smooth.

In general (cf. Remark 2.2.2.1), deformations with section are isomorphic to deformations with trivial section and we identify the corresponding semiuniversal base spaces $B^{sec}_{(C,z)} = B^{fix}_{(C,z)}$. Moreover, the forgetful morphism of functors from deformations with section to deformations without section is smooth. It follows that the induced morphism of base spaces $B^{es,fix}_{C,z} \to B^{es}_{C,z}$ is smooth (flat with smooth fibre). Hence both are smooth and of the same dimension by Corollary 1.1.34 and we get that the morphism $B^{fix,es}_{C,z} \to B^{es}_{C,z}$ is an isomorphism (reflecting the fact that the equisingular section is unique).

Remark 1.1.64.2 Equisingular deformations of (C, z) along the trivial section over a reduced germ T are exactly those, which are equimultiple along the trivial section and equimultiple along the (not necessarily trivial) sections $\sigma_{(q)}, q \in \mathscr{T}^*$, through the non-nodes of the reduced total transform of (C, z) in a resolution of (C, z). They have a fixed cluster graph and hence induce a family of zero-dimensional schemes (a deformation of $Z^{s}(C, z)$) in $\mathcal{H}ilb^{\mathbb{G}}_{\mathbb{C}\{x,y\}}(T)$. Since this association is functorial in T and respects isomorphism classes, we get a morphism of functors

$$\underline{\mathcal{D}ef}^{es,fix}_{C,z} \to \mathcal{H}ilb^{\mathbb{G}}_{\mathbb{C}\{x,y\}},$$

$\mathbb{G} = \mathrm{clg}(Z^{s}(C, z))$, where $\underline{\mathcal{D}ef}^{es,fix}_{C,z}$ is the functor of isomorphism classes of equisingular deformations of (C, z) with (trivial) section (cf. Definitions 2.2.2 and 2.2.3). By Remark 1.1.54.1 the fibres are exactly the straight equisingular deformations (Definition 1.1.28).

The morphism of functors induces a morphism (of germs)

$$\psi : B^{es,fix}_{C,z} \to \mathcal{H}^{s}_{0}(S),$$

which is surjective (by Proposition 1.1.17) with fibre over $Z^{s}(C, z)$ being isomorphic to $B^{s}_{C,z}$. It follows from the theorems of Frisch and Bertini (cf. [BiF, Corollary 2.1]) that there are points $Z^{s}(C', z')$ in $\mathcal{H}^{s}_{0}(S)$ arbitrary close to $Z^{s}(C, z)$, over which ψ is flat and hence satisfies

$$\dim \mathcal{H}_0^s(S) = \tau(C', z') - \tau^{es}(C', z') - \dim_{\mathbb{C}} I^s(C', z')/I^{ea}(C', z')$$
$$= \deg Z^s(C', z') - \tau^{es}_{fix}(C', z').$$

Since $\deg Z^s(C', z') = \deg Z^s(C, z)$ and since $\mathcal{H}_0^s(S)$ is irreducible (by Proposition 1.1.62) its dimension is constant and thus we get $\tau^{es}_{fix}(C', z') = \tau^{es}_{fix}(C, z)$. Hence we obtain, by local deformation theory, another proof of the formula

$$\dim \mathcal{H}_0^s(S) = \deg Z^s(C, z) - \tau^{es}_{fix}(C, z).$$

1.1.4 Zero-Dimensional Schemes for Analytic Types

In the previous sections we considered zero-dimensional schemes associated to topological types of plane curve singularities, in particular, the schemes Z^s. Here we introduce the analogue of the schemes Z^s, namely the schemes Z^a for analytic types in the more general context of germs of $(n-1)$–dimensional hypersurfaces $F \subset X$ with isolated singularities, X an n-dimensional smooth projective variety.

1.1.4.1 Equianalytic Ideals

Two germs of hypersurfaces $(F, z) \subset (X, z)$, resp. $(G, w) \subset (X, w)$, given by $f \in \mathcal{O}_{X,z}$, resp. $g \in \mathcal{O}_{X,w}$, are called *analytically equivalent* if there exists an analytic isomorphism $\psi : (X, z) \xrightarrow{\cong} (X, w)$ mapping (F, z) onto (G, w). This is equivalent to $f = \psi^\sharp(u \cdot g)$ for some unit $u \in \mathcal{O}_{X,w}$, where $\psi^\sharp : \mathcal{O}_{X,w} \to \mathcal{O}_{X,z}$ denotes the isomorphism of local rings induced by ψ.

We then say also that f and g are *analytically equivalent* or *contact equivalent* (notation: $f \overset{c}{\sim} g$) (cf. [GLS6, Definitions I.2.9, I.3.30]). The equivalence class of (F, z) (or of f) is called the *analytic type* of (F, z) (or of f).

Later we shall also consider right equivalence, where f and g are called *right equivalent* (notation $f \overset{r}{\sim} g$) if $f = \psi^\sharp(g)$ for ψ as above.

We define below ideals $I^a(F, z)$ and zero-dimensional schemes $Z^a(F, z)$ codifying the analytic type of (F, z). The ideals $I^a(F, z)$ will be contained in the Tjurina ideal of (F, z) and satisfy some additional conditions which are not satisfied by the Tjurina ideal itself (they reflect linear pieces inside the contact orbit).

Let us recall, first, the definition of the *Tjurina ideal* or *equianalytic ideal* of a power series $f \in \mathbb{C}\{\boldsymbol{x}\}$, $\boldsymbol{x} = (x_1, \ldots, x_n)$,

$$I^{ea}(f) := \langle f, j(f) \rangle \subset \mathbb{C}\{\boldsymbol{x}\},$$

where $j(f) := \langle \frac{\partial f}{\partial x_1}, \ldots, \frac{\partial f}{\partial x_n} \rangle$ is the *Jacobian ideal* or *Milnor ideal* of f. The numbers

$$\tau^{ea}(f) := \tau(f) := \dim \mathbb{C}\{\boldsymbol{x}\}/\langle f, j(f)\rangle, \quad \text{respectively}$$
$$\mu(f) := \dim_{\mathbb{C}} \mathbb{C}\{\boldsymbol{x}\}/j(f)$$

are called the *Tjurina number* respectively the *Milnor number* of f.

We shall also consider the *fixed equianalytic ideal*

$$I^{ea}_{fix}(f) := \langle f\rangle + \mathfrak{m} \cdot j(f)\,,$$

where $\mathfrak{m}$ is the maximal ideal of $\mathbb{C}\{\boldsymbol{x}\}$, and the *fixed Tjurina number*

$$\tau^{ea}_{fix}(f) := \tau_{fix}(f) := \dim_{\mathbb{C}} \mathbb{C}\{\boldsymbol{x}\}/\langle f, \mathfrak{m} \cdot j(f)\rangle\,.$$

The *fixed Milnor number* is

$$\mu_{fix}(f) := \dim_{\mathbb{C}} \mathbb{C}\{\boldsymbol{x}\}/\mathfrak{m} \cdot j(f)\,.$$

It is easy to see that $\mu_{\text{fix}} = \mu + n$. We also have $\tau_{\text{fix}} = \tau + n$, which is more difficult and proved in Lemma 2.2.13.

If $z \in X$ is a point, choose local coordinates $\boldsymbol{x} = (x_1, \ldots, x_n)$ of X at z and let $\psi : \mathcal{O}_{X,z} \to \mathbb{C}\{\boldsymbol{x}\}$ be the corresponding isomorphism of analytic algebras. Then we define for $f \in \mathcal{O}_{X,z}$

$$j(f) := \psi^{-1}(j(\psi(f)))\,,$$

$$I^{ea}(f) := \psi^{-1}(I^{ea}(\psi(f)))\,,$$

$$I^{ea}_{fix}(f) := \psi^{-1}(I^{ea}_{fix}(\psi(f)))\,.$$

These definitions are independent of the chosen local coordinates. In fact, for any analytic automorphism $\varphi \in \operatorname{Aut} \mathbb{C}\{\boldsymbol{x}\}$ and any unit $u \in \mathbb{C}\{\boldsymbol{x}\}^*$, we have for $f \in \mathbb{C}\{\boldsymbol{x}\}$

$$j(\varphi(f)) = \varphi(j(f))\,,$$

$$I^{ea}(\varphi(u \cdot f)) = \varphi(I^{ea}(f))\,,$$

$$I^{ea}_{fix}(\varphi(u \cdot f)) = \varphi(I^{ea}_{fix}(f))\,,$$

which follows from the chain rule and the product rule. Moreover, $I^{ea}(f)$ and $I^{ea}_{fix}(f)$ depend only on the ideal $\langle f\rangle$ but not on its generator. Therefore we can set for a hypersurface germ $(F, z) \subset (X, z)$

$$I^{ea}(F, z) := I^{ea}(f), \quad I^{ea}_{fix}(F, z) := I^{ea}_{fix}(f)\,,$$

where $f \in \mathcal{O}_{X,z}$ is any generator of the ideal of $(F, z) \subset (X, z)$.

Note that $I^{ea}_{fix}(f)$ may be considered as the tangent space at f to the orbit of f under the contact group (cf. Remark 1.1.66.1).

We denote by

$$Z^{ea}(F, z) \quad \text{resp.} \quad Z^{ea}_{fix}(F, z)$$

the zero-dimensional schemes of degree $\tau(f)$ resp. $\tau_{fix}(f)$ defined by $I^{ea}(f)$ resp. $I^{ea}_{fix}(f)$ and call it the *equianalytic (singularity) scheme* resp. the *fixed equianalytic (singularity) scheme.*

1.1.4.2 Analytic Singularity Ideal and Scheme

In this section we introduce the analytic singularity ideal and the corresponding scheme for arbitrary isolated hypersurface singularities, a kind of analogue of the topological singularity ideal and scheme (cf. Definition 1.1.31) for plane curve singularities.

Let X be a smooth n-dimensional complex space, $z \in X$ and $\mathfrak{m}_z$ the maximal ideal of $\mathcal{O}_{X,z}$. Recall that $f, g \in \mathcal{O}_{X,z}$ are called contact or analytically equivalent, $g \overset{c}{\sim} f$, if the local rings $\mathcal{O}_{X,z}/\langle f \rangle$ and $\mathcal{O}_{X,z}/\langle g \rangle$ are isomorphic, i.e., there exist an automorphism Φ of $\mathcal{O}_{X,z}$ and a unit $u \in \mathcal{O}^*_{X,z}$ unit such that $g = u \cdot \Phi(f)$.

Definition 1.1.65 A mapping $I : g \mapsto I(g)$, which associates to every $g \in \mathcal{O}_{X,z}$ with isolated singularity a zero-dimensional ideal $I(g) \subset \mathcal{O}_{X,z}$ is said to satisfy the *analytic singularity conditions* if the following holds:

(a) $g \in I(g)$.
(b) A generic[6] $h \in I(g)$ is contact equivalent to g and satisfies $I(h) = I(g)$.
(c) If $h - g \in \mathfrak{m}_z \cdot I(g)$ then $h \overset{c}{\sim} g$.
(d) For $\Phi \in Aut(\mathcal{O}_{X,z})$ and $u \in \mathcal{O}^*_{X,z}$ we have $I(\Phi(g)) = \Phi(I(g))$ and $I(u \cdot g) = I(g)$. In particular, if $f \overset{c}{\sim} g$ iff $I(f)$ and $I(g)$ are isomorphic.
(e) There exists an $m > 1$ such that $I(g)$ is *m-determined*, i.e. $I(g) = I(h)$ for all h s.t. $g - h \in \mathfrak{m}_z^{m+1}$.

Below, we give examples of such a mapping determined by the ideal $I^a(f)$ resp. $\widetilde{I}^a(f)$ (cf. Definition 1.1.66 resp. Definition 1.1.73). The ideals $I^{ea}(f)$ and $I^{ea}_{fix}(f)$ from the previous section do, however, not satisfy all conditions from Definition 1.1.65 (cf. Remark 1.1.67.2).

Property (d) implies that the isomorphism class of $I(g)$ is an invariant of the analytic type of g. Moreover, $I(g)$ is independent of the choice of the generator g of the ideal $\langle g \rangle$ and we can define $I(F, z) := I(f)$ for a hypersurface germ

[6]Condition (b) for a"generic" h means the following: let $x_1, \ldots, x_n$ be local coordinates for X at w and $d \geq m$ (as in (e)), then the set of polynomials $h \in I(g) \cap \mathbb{C}\{x\}$ of degree $\leq d$ satisfying $h \overset{c}{\sim} g$ and $I(h) = I(g)$ is a Zariski-open, dense subset of $I(g) \cap \mathbb{C}[x]_{\leq d}$.

$(F, z) \subset (X, z)$ whose ideal is generated by f. By (e) and (a) $\mathfrak{m}_z^{m+1} \subset I(f)$ and we denote by

$$Z^I(F, z) := V(I(F, z)) = V(I(f)) \subset X$$

the zero-dimensional scheme defined by $I(f)$ of degree $\dim_{\mathbb{C}} \mathcal{O}_{X,z}/I(f)$.

Definition 1.1.66 (1) Among all mappings $I : g \mapsto I(g)$ that satisfy the analytic singularity conditions we may choose one with the property that $I(g)$ has maximal size for all g, i.e., minimal possible colength in $\mathcal{O}_{X,z}$. Denote this choice by I^a and call it a *(maximal) analytic singularity mapping*. For $f \in \mathcal{O}_{X,z}$ with isolated singularity, the ideal $I^a(f)$ is called an *analytic singularity ideal* of f.

(2) If $(F, z) \subset (X, z)$ is a hypersurface germ with isolated singularity given by $f \in \mathcal{O}_{X,z}$ then we set

$$I^a(F, z) := I^a(f), \quad Z^a(F, z) := Z^a(f) := V(I^a(F, z)) \subset X,$$

and call $Z^a(F, z)$ an *analytic singularity scheme* of F at z.

(3) If S is the analytic type of (F, z), we introduce the degree of $Z^a(S)$,

$$\deg Z^a(S) := \deg Z^a(F, z) := \deg Z^a(f) = \dim_{\mathbb{C}} \mathcal{O}_{X,x}/I^a(f),$$

which is independent of the choice of f representing S. By Lemmas 1.1.67 and 2.2.13 we have

$$\deg Z^a(f) \geq \dim_{\mathbb{C}} \mathcal{O}_{X,z}/I^{ea}_{\text{fix}}(f) = \tau(f) + n .$$

(4) Moreover, since $Z^a(F, z) = Z^a(f)$ is zero-dimensional, we can define

$$\nu^a(F, z) := \nu^a(f) := \min \left\{ \nu \in \mathbb{Z} \,\middle|\, \mathfrak{m}_z^{\nu+1} \subset I^a(f) \right\}.$$

and call it the *analytic singularity level* of f (see also Definition 1.2.2).

Note that the map I^a is not uniquely determined by the maximality condition and hence $I^a(f)$ may not be uniquely determined by f. On the other hand, the codimension of $I^a(f)$ is an invariant of the contact class of f. Since we are not interested in the analytic structure of $I^a(f)$ but only in its codimension, we just choose one map I^a and we sometimes speak about "the" analytic singularity ideal resp. scheme of f.

1.1.4.3 Hypersurfaces with One Isolated Singularity

Let X be an n-dimensional complex manifold. Analytic equivalence between isolated hypersurface singularities in (X, z) can be described by an algebraic group action on a sufficiently high jet space of the local ring $\mathcal{O}_{X,z}$. We exploit this fact in this section.

Remark 1.1.66.1 We have to consider the *contact group*

$$\mathcal{K} := \mathcal{O}^*_{X,z} \ltimes \mathcal{R},$$

acting on $\mathcal{O}_{X,z}$ via $(u, \Phi) \circ f = u \cdot \Phi(f)$ for $f \in \mathcal{O}_{X,z}$, $u \in \mathcal{O}^*_{X,z}$ and $\Phi \in \mathcal{R}$, where $\mathcal{R} = \operatorname{Aut} \mathcal{O}_{X,z}$ is the the *right group* of automorphisms of the local analytic $\mathbb{C}$-algebra $\mathcal{O}_{X,z}$. Then $g \in \mathcal{O}_{X,z}$ is contact equivalent to f, iff g is contained in the $\mathcal{K}$-orbit $\mathcal{K} \cdot f$ of f.

Passing to k-jets of u (Remark 1.1.16.1) and Φ we get the contact group $\mathcal{K}^{(k)}$ which acts on the jet-space $J^{(k)} := \mathcal{O}_{X,z}/\mathfrak{m}_z^{k+1}$ (cf. [GLS6, I.2.3]). We tacitly identify elements of $J^{(k)}$ with power series expansions of elements in $\mathcal{O}_{X,z}$ up to order k. The group $\mathcal{K}^{(k)}$ is an algebraic group and the orbits are smooth and irreducible algebraic subvarieties of $J^{(k)}$, parameterizing contact classes of isolated hypersurface singularities for sufficiently high k.

Namely, if f is contact k-determined then g is contact equivalent to f iff $g^{(k)}$ is contained in the $\mathcal{K}^{(k)}$-orbit of $f^{(k)}$. Moreover, by [GLS6, Proposition I.2.38] the tangent space of $\mathcal{K}^{(k)} \cdot f^{(k)}$ at $f^{(k)}$ is

$$T_{f^{(k)}}(\mathcal{K}^{(k)} \cdot f^{(k)}) = (I^{ea}_{fix}(f) + \mathfrak{m}_z^{k+1})/\mathfrak{m}^{k+1} .$$

In this sense we say that I^{ea}_{fix} is the tangent space at f to the contact orbit $\mathcal{K} \cdot f$.

Remark 1.1.66.2 The group $\mathcal{K}$ acts also on the set of ideals $I \subset \mathcal{O}_{X,z}$. If I has finite codimension, say c, in $\mathcal{O}_{X,z}$ then $\mathfrak{m}_z^{k+1} \subset I$ for some k and we have an exact sequence

$$0 \to I/\mathfrak{m}_z^{k+1} \to J^{(k)} = \mathcal{O}_{X,z}/\mathfrak{m}_z^{k+1} \to \mathcal{O}_{X,z}/I \to 0.$$

We can identify the set of ideals of codimension c in $\mathcal{O}_{X,z}$ with the set of ideals of codimension c in $J^{(k)}$, which is an algebraic subvariety of the Grassmannian of vector spaces of codimension c in $J^{(k)}$. Choosing local coordinates $\boldsymbol{x} = (x_1, \cdots, x_n)$ of (X, z), this is in fact the *punctual Hilbert scheme* $\operatorname{Hilb}^c_{\mathbb{C}\{\boldsymbol{x}\}}$, parameterizing zero-dimensional schemes of degree c in X with support in z (cf. [Bri, II.2.1.], see also Definition 1.1.49).

We identify the k-jet $I^{(k)} \subset J^{(k)}$ of an ideal I of codimension c with the corresponding point in the punctual Hilbert scheme $\operatorname{Hilb}^c_{\mathbb{C}\{\boldsymbol{x}\}}$. The algebraic group $\mathcal{K}^{(k)}$ acts algebraically on $\operatorname{Hilb}^c_{\mathbb{C}\{\boldsymbol{x}\}}$, which coincides with the action of $\mathcal{R}^{(k)}$, since multiplication with a unit does not change an ideal. We denote the orbit of $I^{(k)}$ by

$$\mathcal{K}^{(k)} \cdot I^{(k)} = \mathcal{R}^{(k)} \cdot I^{(k)} = \{\Phi^{(k)}(I^{(k)}) = \Phi(I)^{(k)} \mid \Phi \in \mathcal{R}\},$$

which is an irreducible, smooth algebraic subvariety of the projective variety $\operatorname{Hilb}^c_{\mathbb{C}\{\boldsymbol{x}\}}$.

Lemma 1.1.67 *Let $f \in \mathcal{O}_{X,z}$ define an isolated singularity. For any I satisfying the analytic singularity conditions of Definition 1.1.65 we have $I(f) \subset I^{ea}_{fix}(f)$; in particular,*

$$I^a(f) \subset I^{ea}_{fix}(f)\,.$$

Proof Choose m for I as in (e) such that $I(f) = I(f^{(k)})$ for $k \geq m$. Let $k \geq m$ satisfy also $f^{(k)} \overset{c}{\sim} f$ (f is finitely determined as it has an isolated singularity) and consider the action of $\mathcal{K}^{(k)}$ on $J^{(k)}$. By conditions (a) and (b) the orbit of $f^{(k)}$ is dense in $J^{(k)} \cap I(f^{(k)})$. Since orbits are Zariski-open in their closure (cf. [GLS6, Theorem 2.34]), $\mathcal{K}^{(k)} \cdot f^{(k)} \cap I^{(k)}(f)$ is open in $I^{(k)}(f)$, the k-jet of $I(f)$.

Hence, for any $h \in I(f)$ and sufficiently small $t \in \mathbb{C}$, $f^{(k)} + th^{(k)}$ is contained in the orbit of $f^{(k)}$. This implies that the line $f^{(k)} + th^{(k)}$, $t \in \mathbb{C}$, is contained in the tangent space $T_{f^{(k)}}(\mathcal{K}^{(k)} \cdot f^{(k)})$. Since this holds for arbitrary big k we get that $f + h$ and hence h is contained in $I^{ea}_{fix}(f)$. □

Remark 1.1.67.1 The proof of Lemma 1.1.67 shows that for any element $h \in I(f)$ we have for all $t \in \mathbb{C}$ which are sufficiently close to 0

$$f \overset{c}{\sim} f + th \quad \text{and} \quad I(f) = I(f + th)\,.$$

In fact, since I is an ideal, the line $L = \{f + th \mid t \in \mathbb{C}\}$ is contained in $I(f)$. By condition (b) there exists for each $k \geq m$ a Zariski-open, dense neighborhood $U^{(k)} \subset I^{(k)}(f)$ of $f^{(k)}$ such that $g \overset{c}{\sim} f^{(k)}$ and $I^{(k)}(g) = I^{(k)}(f)$ for each $g \in U^{(k)}$. $L^{(k)}$ is contained in $U^{(k)}$ except at finitely many points, which means $f^{(k)} \overset{c}{\sim} f^{(k)} + th^{(k)}$, $I^{(k)}(f) = I^{(k)}(f + th)$ except for finitely many t and for all $k \geq m$. This implies the claim.

Remark 1.1.67.2 (1) If f defines an isolated singularity then $I^{ea}(f) = \langle f, j(f)\rangle$ and $I^{ea}_{fix}(f) = \langle f, \mathfrak{m}j(f)\rangle$ satisfy conditions (a), (d), and (e) but in general not (b) and (c) of Definition 1.1.65. While (a) and (d) are clear, (e) follows from the following Lemma 1.1.69.

(2) To see that (b) and (c) do not hold for $I^{ea}_{fix}(f)$, consider $f = x^3 + y^7$ and $g = f + x^2y^2$. Then $f - g \in \mathfrak{m}I^{ea}_{fix}(f)$ but f is not contact equivalent to g ($\mu(f) = 12$, $\mu(g) = 11$), hence(c) fails. (b) does also fail since $f + tx^2y^2$ is not contact equivalent to f for all $t \neq 0$.

Definition 1.1.68 Let $f \in \mathbb{C}\{\boldsymbol{x}\}$ define an isolated singularity. We set

$$\nu^{ea}(f) := \min\{\nu \mid \mathfrak{m}^{\nu+1} \subset I^{ea}(f)\}\,,$$

$$\nu^{ea}_{fix}(f) := \min\{\nu \mid \mathfrak{m}^{\nu+1} \subset I^{ea}_{fix}(f)\}\,,$$

and call it the *equianalytic* (resp. *fixed equianalytic*) *level* of f.

Lemma 1.1.69 *For $f \in \mathbb{C}\{\boldsymbol{x}\}$ an isolated singularity we have $\nu^{ea}(f) \leq \nu^{ea}_{fix}(f) \leq \nu^{ea}(f) + 1$. Moreover, the following holds:*

(a) f and $I^{ea}_{fix}(f)$ are $(\nu^{ea}_{fix}(f) + 1)$-determined,

(b) $I^{ea}(f)$ *is* $(\nu^{ea}(f)+1)$*-determined.*

Moreover, $\nu^{ea}_{fix}(f) \leq \nu^a(f)$*, with* $\nu^a(f)$ *the analytic singularity level (Definition 1.1.66).*

Proof Since $\mathfrak{m}I^{ea}(f) \subset I^{ea}_{fix}(f) \subset I^{ea}(f)$ the inequalities hold.

To see (a) let $m = \nu^{ea}_{fix}(f)+1$ and $f-g \in \mathfrak{m}^{m+1}$. Then $\frac{\partial f}{\partial x_i} - \frac{\partial g}{\partial x_i} \in \mathfrak{m}^m$ and hence, since $\mathfrak{m}^{m+1} \subset I^{ea}_{fix}(f)$,

$$I^{ea}_{fix}(f) = \mathfrak{m}\cdot j(f) + \langle f\rangle + \mathfrak{m}^{m+1} = \mathfrak{m}\cdot j(g) + \langle g\rangle + \mathfrak{m}^{m+1} = I^{ea}_{fix}(g) + \mathfrak{m}^{m+1}\,.$$

Moreover, $\mathfrak{m}^{m+1} \subset \mathfrak{m}I^{ea}_{fix}(f)$ and hence f is m-determined by [GLS6, Theorem I.2.23]. Therefore f and g are contact equivalent, which implies that $I^{ea}_{fix}(f)$ is isomorphic to $I^{ea}_{fix}(g)$. Hence $\mathfrak{m}^m \subset I^{ea}_{fix}(g)$, showing that $I^{ea}_{fix}(f) = I^{ea}_{fix}(g)$.

A similar argument works for $I^{ea}(f)$ in (b). The last statement follows since $I^a(f) \subset I^{ea}_{fix}(f)$ by Lemma 1.1.67. □

Remark 1.1.69.1 Let $f \in \mathcal{O}_{X,z}$ define an isolated singularity of analytic type S and let I^a be a maximal analytic singularity mapping on $\mathcal{O}_{X,z}$ (Definition 1.1.65). Let $I^a(f)$ be k-determined, i.e. $k \geq \nu^a(f)$, then $\mathfrak{m}_z^{k+1}$ is contained in $I^a(f)$ and $c := \dim_{\mathbb{C}} \mathcal{O}_{X,z}/I^a(f)$ is finite. We consider the k-jet $I^a(f)^{(k)} \subset J^{(k)}$ of $I^a(f)$ and define

$$\mathcal{H}^a_0(S) := \mathcal{K}^{(k)}\cdot I^a(f)^{(k)} \subset \mathrm{Hilb}^c_{\mathbb{C}\{x\}}$$

the orbit of $I^a(f)^{(k)}$ under the action of $\mathcal{K}^{(k)}$ (see Remark 1.1.66.2). We call it the *punctual Hilbert scheme associated to an analytic type S.*

If (F,z) is a hypersurface singularity defined by $f \in \mathcal{O}_{X,z}$ and $Z^a(f) = Z^a(F,z)$ the zero dimensional scheme defined by $I^a(f)$ we write also $Z^a(F,z) \in \mathcal{H}^a_0(S)$. Since $\Phi^{(k)}\cdot I^a(f)^{(k)} = I^a(\Phi(f))^{(k)}$ by (d) of Definition 1.1.65, $\mathcal{H}^a_0(S)$ parameterizes the analytic singularity schemes of hypersurface singularities in $\mathcal{O}_{(X,z)}$ of analytic singularity type S.

As orbit of an irreducible algebraic group $\mathcal{H}^a_0(S)$ is a smooth, irreducible algebraic variety. Since the orbit of an algebraic group is open in its closure, $\mathcal{H}^a_0(S)$ is a quasi-projective subvariety of the projective variety $\mathrm{Hilb}^c_{\mathbb{C}\{x\}}$, which carries the reduced structure.

Let $w \in X$ and $\phi : \mathcal{O}_{X,w} \cong \mathcal{O}_{X,z}$ be an isomorphism. Then we have, by Definition 1.1.65(d), for any isolated hypersurface singularity $(F,w) \subset (X,w)$ defined by $g \in \mathcal{O}_{X,w}$,

$$Z^a(\phi(g)) \in \mathcal{H}^a_0(S)$$

iff g is of analytic type S.

Definition 1.1.70 We denote by

$$V^a_d(S) \subset V_d := \mathbb{P}\big(H^0(\mathcal{O}_{\mathbb{P}^n}(d))\big) = \mathbb{P}^{\binom{n+d}{n}-1}$$

the *variety of reduced hypersurfaces F in $\mathbb{P}^n$ of degree d having exactly one singular point, which is isolated, of analytic type S.*

More precisely, we require that the universal family $\mathscr{U}_d \to V_d$ of hypersurfaces of degree d admits an analytic section over $V_d^a(S)$, picking the singular point, along which the family is equianalytic (by Proposition 2.2.6 the section is unique if S is not a smooth type).

$V_d^a(S)$ is, first of all, a set. But we show in Proposition 1.1.72 below that it is a quasi-projective variety. For a generalization see Sect. 2.2.2.2, where we show that we can impose an analytic structure on $V_d^a(S)$ (not necessarily reduced), such that the universal family $\mathscr{U}_d^a(S) := \{(z, F) \mid z \in F\} \subset \mathbb{P}^n \times V_d^a(S) \to V_d^a(S)$ represents equianalytic families with one singular point of analytic type S (Theorem 2.2.32).

Remark 1.1.70.1 The following items 1. and 2. follow similar as in Lemma 1.1.60 for plane curves.

1. The set V_d^{iso} of hypersurfaces of degree d with isolated singularities is a codimension 1 closed subvariety of V_d (if $d > 1$), with hypersurfaces having exactly one singularity of type A_1 being open and dense.
2. Let $V_d^{(1)}$ be the set of hypersurfaces in V_d having exactly one singularity and the property that $\Phi_d : V_d^{(1)} \to \mathscr{U}_d$, $F \mapsto (p, F)$, with p the singular point of F, is an analytic section of $\mathscr{U}_d^a(S) \to V_d^a(S)$. Then $V_d^{(1)}$ is an open and dense subset of V_d^{iso}.
3. As in Remark 1.1.60.1, there exists an affine chart $\mathbb{A}^n \subset \mathbb{P}^n$ containing z and a Zariski-open dense subset $V \subset V_d^a(S)$ such that $w = \Phi_d(F) \in \mathbb{A}^n$ for each $F \in V$. In particular, we can define a morphism

$$V_d^a(S) \underset{\text{open,dense}}{\supset} V \xrightarrow{\Psi_d} \mathcal{H}^a(S) := \mathbb{P}^n \times \mathcal{H}_0^a(S)\,, \quad F \longmapsto \big(w, \tau_{zw}(Z^a(F, w))\big)\,, \tag{1.1.4.1}$$

where τ_{zw} denotes the translation mapping z to w. Note that $\mathcal{H}^a(S)$ is irreducible because the orbit $\mathcal{H}_0^a(S)$ is irreducible.

As in the case of topological types, we can apply global deformation theory to determine the dimension of $\mathcal{H}_0^a(S)$:

Lemma 1.1.71 *Let S be the analytic type of an isolated hypersurface singularity $(F, z) \subset (X, z)$. Then $\mathcal{H}_0^a(S)$ is a smooth, irreducible quasi-projective variety of dimension*

$$\dim \mathcal{H}_0^a(S) = \deg Z^a(S) - \tau(S) - n = \deg Z^a(S) - \tau_{fix}(S)\,,$$

where $\tau(S)$ is the Tjurina number *of f and* $\deg Z^a(S) = \dim_{\mathbb{C}} \mathcal{O}_{X,z}/I^a(f)$, *where f defines (F, z).*

The proof of the lemma (that the last equality follows from Lemma 2.2.13) resp. of the following proposition is analogous to the proof of Proposition 1.1.62, resp. of Proposition 1.1.61, hence left as an exercise.

Proposition 1.1.72 *The variety $V_d^a(S)$ of reduced hypersurfaces $F \subset \mathbb{P}^n$ of degree d having exactly one singular point of analytic type S is a quasi-projective subvariety of $\mathbb{P}\big(H^0(\mathcal{O}_{\mathbb{P}^n}(d))\big)$.*

In general, the schemes $Z^a(F, z)$ are difficult to handle since there is no concrete description of $I^a(F, z)$, which would be needed, for instance, to determine the degree of $Z^a(F, z)$. Therefore we introduce in the next section the ideal $\widetilde{I}^a$ and the scheme $\widetilde{Z}^a$. Of course, there are special cases, where we can describe $I^a(F, z)$ explicitly; for instance, for a simple plane curve singularity (C, z) we just have $Z^a(C, z) = Z^s(C, z)$ (cf. [GLS6, Section I.2.4]).

1.1.4.4 The Singularity Ideal $\widetilde{I}^a$ and the Scheme $\widetilde{Z}^a$

To be able to estimate $\deg Z^a(S)$ for arbitrary singularities, we consider a mapping $g \mapsto I(g)$ on $\mathcal{O}_{X,z}$ satisfying the analytic singularity properties (a)–(e) of Definition 1.1.65, but where $I(g)$ is not necessarily of maximal size.

Definition 1.1.73 Let $f \in \mathcal{O}_{X,z}$ define an isolated hypersurface singularity and let $I^{ea}(f) = \langle f, j(f)\rangle$ denote the equianalytic ideal. We introduce the ideal

$$\widetilde{I}^a(f) := \big\{ h \in \mathcal{O}_{X,z} \bigm| \langle h, j(h)\rangle \subset \langle f, j(f)\rangle \big\} .$$

Note that, if $\boldsymbol{x} = (x_1, \dots, x_n)$ are local coordinates at z and if $f \in \mathbb{C}\{\boldsymbol{x}\}$ then

$$\widetilde{I}^a(f) = \left\{ \alpha_0 f + \sum_{i=1}^{n} \alpha_i \frac{\partial f}{\partial x_i} \;\middle|\; \begin{array}{l} \alpha_0, \alpha_1, \dots, \alpha_n \in \mathbb{C}\{\boldsymbol{x}\}, \\ (\alpha_1, \dots, \alpha_n) \cdot H(f)(\boldsymbol{x}) \equiv 0 \bmod \langle f, j(f)\rangle \end{array} \right\} \tag{1.1.4.2}$$

where $H(f)(\boldsymbol{x})$ denotes the Hessian matrix of f.

Clearly $\widetilde{I}^a(f) \subset I^{ea}(f)$ is an ideal containing f. Since $\mathfrak{m}^{\tau+1} \subset \langle f, j(f)\rangle, \tau = \tau(f)$, each monomial $h \in \mathfrak{m}^{\tau+2}$ is in $\widetilde{I}^a(f)$. Therefore $\mathfrak{m}^{\tau+2} \subset \widetilde{I}^a(f)$, showing that $\widetilde{I}^a(f)$ is already determined by the $(\tau{+}1)$–jet of f. Hence $\widetilde{I}^a(f)$ satisfies condition (a) and (e) of Definition 1.1.65 and we show below that the mapping $\widetilde{I}^a$ on $\mathcal{O}_{X,z}$ satisfies also the conditions (b), (c) and (d) (Lemmas 1.1.74, 1.1.75). Therefore $\widetilde{I}^a(f)$ satisfies the analytic singularity conditions of Definition 1.1.65 and

$$\widetilde{I}^a(f) \subset I^a(f) \subset I^{ea}_{fix}(f) .$$

Moreover, the description (1.1.4.2) of $\widetilde{I}^a(f)$ provides an algorithm, using standard bases, to compute $\widetilde{I}^a(f)$.

Example 1.1.73.1 Let $f(x, y) := x^6 + y^6 - x^3y^3$. Then the following procedure, written in the syntax of SINGULAR [GrP], computes the ideal $\widetilde{I}^a(f)$, respectively its codimension:

```
ring r = 0,(x,y),ds;
poly f = x6+y6-x3*y3;
ideal I1 = f,jacob(f);
qring q = std(I1);             // quotient ring C{x,y}/<f,j(f)>
poly f = imap(r,f);
matrix H = jacob(jacob(f));    // the Hessian matrix
module A = syz(H);             // H*A=0
A = transpose(A);
setring r;
module A = imap(q,A);              // module generated by rows of A
vector v = [diff(f,x),diff(f,y)];
ideal I = f,A*v,jacob(f)*jacob(f);
I = std(I);
I;
//-> I[1]=x6-x3y3+y6
//-> I[2]=x5y3-5x2y6
//-> I[3]=4x3y5-5y8
//-> I[4]=y9
//-> I[5]=x2y8
vdim(I);
//-> 39
```

Thus, $\widetilde{I}^a(f) = \langle f, x^5y^3 - 5x^2y^6, 4x^3y^5 - 5y^8, y^9, x^2y^8\rangle \subset \mathbb{C}\{x, y\}$ defines a zero-dimensional scheme of degree 39.

Lemma 1.1.74 *Let $f \in \mathcal{O}_{X,z}$ define an isolated singularity, let $\Psi : (X, w) \to (X, z)$ be the germ of an analytic isomorphism and $u \in \mathcal{O}_{X,z}$ a unit. Then*

$$\Psi^\sharp \widetilde{I}^a(u \cdot f) = \widetilde{I}^a(\Psi^\sharp f).$$

Proof Apply the chain rule and the product rule. □

In particular, for a hypersurface germ $(F, z) \subset (X, z)$ with isolated singularity defined by f we can introduce

$$\widetilde{I}^a(F, z) := \widetilde{I}^a(f) \quad \text{and} \quad \widetilde{Z}^a(F, z) := V(\widetilde{I}^a(F, z))$$

with $\deg \widetilde{Z}^a(F, z) = \dim_{\mathbb{C}} \mathcal{O}_{X,x}/\widetilde{I}^a(f)$.

Lemma 1.1.75 *Let $f \in \mathbb{C}\{\boldsymbol{x}\}$ define an isolated singularity. Then*

(a) a generic element $g \in \widetilde{I}^a(f)$ satisfies $g \overset{c}{\sim} f$ and $\widetilde{I}^a(g) = \widetilde{I}^a(f)$ (then we also say that g defines $\widetilde{I}^a(f)$).
(b) If, for $g \in \mathbb{C}\{\boldsymbol{x}\}$, $g - f \in \mathfrak{m} \cdot \widetilde{I}^a(f)$ then $g \overset{c}{\sim} f$.

Proof (a) Let $d \geq \tau + 1$. Then every element $g \in \widetilde{I}^a(f)$ is contact equivalent to the polynomial $g^{(d)}$ of degree $\leq d$ and satisfies $\widetilde{I}^a(g) = \widetilde{I}^a(g^{(d)})$, by the remarks above. The polynomials $g \in \widetilde{I}^a(f)$ of degree $\leq d$ are parameterized by a finite dimensional vector space of positive dimension. Since $\langle g, j(g)\rangle \subset \langle f, j(f)\rangle$, we have $\tau(g) \geq \tau(f)$ and equality holds exactly if $\langle g, j(g)\rangle = \langle f, j(f)\rangle$, that is,

exactly if $\widetilde{I}^a(g) = \widetilde{I}^a(f)$. The theorem of Mather-Yau [MaY] implies that f is contact equivalent to g iff $\langle g, j(g)\rangle = \langle f, j(f)\rangle$ (see also [GLS6, Theorem I.2.26]). Since the set of all g with minimal possible Tjurina number $\tau(g) = \tau(f)$ is a non-empty Zariski-open set, the statement follows for such "generic" g.
(b) Let $h := g - f$. Then, by assumption, $h = ah'$ with $a \in \mathfrak{m}$, $h' \in \mathbb{C}\{\mathbf{x}\}$ satisfying $\langle h', j(h')\rangle \subset \langle f, j(f)\rangle$. By [GLS6, Corollary I.2.27] it follows that $f + th' \overset{c}{\sim} f$ for almost all $t \in \mathbb{C}$. In particular, h' is in the tangent space to the contact orbit, which is $I^{ea}_{fix}(f) = \langle f\rangle + \mathfrak{m} \cdot j(f)$ [GLS6, Proposition I.2.38]. We conclude that h is of the form

$$h = \alpha_0 f + \sum_{i=1}^{n} \alpha_i \frac{\partial f}{\partial x_i}, \quad \alpha_0 \in \mathfrak{m}, \quad \alpha_1, \ldots, \alpha_n \in \mathfrak{m}^2.$$

Since we have $(\alpha_1, \ldots, \alpha_n) \cdot H(f)(\boldsymbol{x}) \equiv 0 \bmod \langle f, j(f)\rangle$ (see the description (1.1.4.2) of $\widetilde{I}^a(f)$), it is not difficult to derive

$$j(h - \alpha_0 f) \subset \mathfrak{m} \cdot \langle f, j(f)\rangle = \mathfrak{m} \cdot \langle uf, j(uf)\rangle,$$

where $u = 1 + \alpha_0 \in \mathbb{C}\{\boldsymbol{x}\}^*$. Hence, by [GLS6, Cor. I.2.27 (2)], we have $uf + t(h - \alpha_0 f) \overset{c}{\sim} uf$ for all $t \in \mathbb{C}$. In particular,

$$f \overset{c}{\sim} uf \overset{c}{\sim} uf + h - \alpha_0 f = f + h = g.$$

□

Proposition 1.1.76 *Let (X, z) be smooth of dimension n and $(F, z) \subset (X, z)$ a hypersurface germ with isolated singularity. Then*

$$\deg Z^a(F, z) \le \deg \widetilde{Z}^a(F, z) \le (n + 1) \cdot \tau(F, z).$$

Proof The second inequality holds since, by (1.1.4.2), the map

$$(\mathbb{C}\{\boldsymbol{x}\}/I^{ea}(f))^n \to I^{ea}(f)/\widetilde{I}^a(f), \quad (\alpha_1, \ldots, \alpha_n) \mapsto \sum_{i=1}^{n} \alpha_i \frac{\partial f}{\partial x_i},$$

is well defined and a surjection and since we have an exact sequence

$$0 \to I^{ea}(f)/\widetilde{I}^a(f) \to \mathbb{C}\{\boldsymbol{x}\}/\widetilde{I}^a(f) \to \mathbb{C}\{\boldsymbol{x}\}/I^{ea}(f) \to 0.$$

The first inequality holds since the mapping $g \mapsto \widetilde{I}^a(g)$ satisfies the conditions (a)–(e) and since $I^a(g)$ is characterized by having the minimal colength among all ideals satisfying (a)–(e). □

A more thorough reasoning shows that in the case of a plane curve singularity, one can actually give a better bound for $\deg Z^a(C, z)$. In the proof (and later) we will use the notion of a generic polar.

Definition 1.1.77 (*Generic polar*) For $f \in \mathbb{C}\{x_1, \cdots, x_n\}$ and $P = (a_1, \cdots, a_n) \in \mathbb{C}^n \setminus \{0\}$ the linear combination $\Pi_P = a_1 f_{x_1} + \cdots + a_n f_{x_n}$, resp. if $\Pi_P \in \mathfrak{m}$ the hypersurface germ $V(\Pi_P) \subset (\mathbb{C}^n, 0)$, is called the *polar*, resp. the *polar hypersurface germ*, of f w.r.t. P. Π_P resp. $V(\Pi_P)$ is called a *generic polar* if there exists a Zariski open dense set $U \subset \mathbb{P}^{n-1}$, depending on the context, with $(a_1 : \cdots : a_n) \in U$.

For $f \in \mathcal{O}_{\Sigma,z}$, (Σ, z) a smooth n-dimensional germ, a polar is defined w.r.t. some local coordinates of (Σ, z). This is independent of the coordinates in the sense, that if ψ is an automorphism of $\mathbb{C}\{x_1, \cdots, x_n\}$ with linear part ψ', then the polar of $\psi(f)$ is the polar of f w.r.t. $\psi'(P)$, as follows from the chain rule.

Lemma 1.1.78 *Let $(C, z) \subset (\Sigma, z)$ be a plane curve singularity. Then*

$$\deg Z^a(C, z) \le 2\mu(C, z)\,, \qquad (1.1.4.3)$$

unless (C, z) is a node or a cusp. If (C, z) is an A_μ-singularity, $\mu \ge 1$, then

$$\deg Z^a(C, z) = \deg Z^s(C, z) = \begin{cases} 3k = \frac{3}{2}\mu(C, z) + \frac{3}{2} & \text{if } \mu = 2k - 1\,, \\ 3k + 2 = \frac{3}{2}\mu(C, z) + 2 & \text{if } \mu = 2k\,. \end{cases}$$

Proof We establish the inequality (1.1.4.3) first for simple singularities. In this case $\deg Z^a(C, z) = \deg Z^s(C, z)$ and it is not difficult to compute

$$\deg Z^a(A_\mu) = \left\lfloor \frac{3\mu + 4}{2} \right\rfloor, \quad \deg Z^a(D_\mu) = \left\lfloor \frac{3\mu + 1}{2} \right\rfloor, \quad \mu \ge 1$$

and $\deg Z^a(E_\mu) = \mu + 3$, $\mu = 6, 7, 8$. In particular, (1.1.4.3) holds true, unless (C, z) is a node (A_1-singularity), or a cusp (A_2-singularity).

Now, let (C, z) be a non-simple singularity and given by a local equation $f = 0$. Moreover, let $\Pi_1 = af_x + bf_y$, $\Pi_2 = a'f_x + b'f_y$ be two generic polars of f and let Π_{11}, Π_{12} be two generic polars of Π_1. Then (1.1.4.2) implies

$$\widetilde{I}^a(f) \supset \left\{\alpha f + \beta \Pi_1 \,\middle|\, \alpha \in \mathbb{C}\{x, y\}\,,\ \beta \in J\right\},$$

where

$$J := \left\{h \in \mathbb{C}\{x, y\} \,\middle|\, h\Pi_{11}, h\Pi_{12} \in \langle \Pi_1, \Pi_2 \rangle\right\}.$$

By the local version of Noether's fundamental ("$AF + BG$") theorem (Theorem 2.1.26), the above ideal J contains the ideal

$$I := \left\{h \in \mathbb{C}\{x, y\} \,\middle|\, i_0(h, g_j) \ge i_0(\Pi_2, g_j) - \operatorname{mt} g_j + 1 \text{ for all } j\right\}\,,$$

where $\Pi_1 = g_1 \cdot \ldots \cdot g_s$ is the irreducible decomposition of $\Pi_1 \in \mathbb{C}\{x, y\}$. The number of conditions imposed to elements of I can be easily computed, leading to the estimate

$$\begin{aligned}\dim \mathbb{C}\{x,y\}/J \le \dim \mathbb{C}\{x,y\}/I &= \sum_{j=1}^{s}\big(i_0(\Pi_2,g_j) - \operatorname{mt} g_j + 1\big) - \delta(\Pi_1)\\ &\le \sum_{j=1}^{s} i_0(\Pi_2,g_j) - \delta(\Pi_1) = \mu(f) - \delta(\Pi_1)\,.\end{aligned}$$

Since f is not simple, $\delta(\Pi_1) \ge \operatorname{mt}(f) - 1$. Finally, the exact sequence

$$\mathbb{C}\{x,y\}/J \xrightarrow{\cdot\Pi_1} \dim \mathbb{C}\{x,y\}/\widetilde{I}^a(f) \longrightarrow \dim \mathbb{C}\{x,y\}/\langle f,\Pi_1\rangle \longrightarrow 0$$

allows us to conclude

$$\begin{aligned}\deg Z^a(C,z) &\le \dim \mathbb{C}\{x,y\}/\widetilde{I}^a(f) \le \dim \mathbb{C}\{x,y\}/J + \dim \mathbb{C}\{x,y\}/\langle f,\Pi_1\rangle\\ &\le (\mu(f) - \operatorname{mt}(f) + 1) + (\mu(f) + \operatorname{mt}(f) - 1) = 2\mu(C,z)\,,\end{aligned}$$

the latter inequality being due to [GLS6, Proposition I.3.38]. □

Example 1.1.78.1 (a) Let $(C,z) \subset (\Sigma,z)$ be the reduced plane curve germ given by the local equation $y^q + x^p = 0$. Then

$$\widetilde{I}^a(C,z) = \langle y^q,\ x^{p-1}y^{q-1},\ x^p\rangle\,,$$

and the zero-dimensional scheme $\widetilde{Z}^a(C,z)$ is of degree $\tau(C,z) + p + q - 1$.
(b) Let $(C',z) \subset (\Sigma,z)$ be the ordinary four-fold point with local equation $x^4 - x^2y^2 + y^4 = 0$. Then

$$\widetilde{I}^a(C',z) = \langle x^4 - x^2y^2 + y^4,\ x^3y^2 - 6xy^4,\ 5x^2y^3 + 6y^5,\ xy^5,\ y^6\rangle\,,$$

whence $\deg \widetilde{Z}^a(C',z) = 16$. Note that (C',z) is topologically equivalent to the germ (C,z) with local equation $x^4 + y^4 = 0$, having $\deg \widetilde{Z}^a(C,z) = 15$. In particular, this shows that $\deg \widetilde{Z}^a(C,z)$ is not an invariant of the topological type.
(c) For ordinary four-fold points $(C,z) \subset (\Sigma,z)$ we can easily define a better collection of ideals than $\widetilde{I}^a(C,z)$. Consider the (cluster scheme) ideal

$$I(C,\mathscr{T}) := \big\{g \in \mathcal{O}_{\Sigma,z} \,\big|\, \operatorname{mt} g \ge 4\,,\ \operatorname{mt} \hat{g}_{(q)} \ge 5,\ \ z \ne q \in \mathscr{T}\big\} \subset \mathcal{O}_{\Sigma,z},$$

with $\mathscr{T}$ the constellation consisting of z and the first non-essential infinitely near point of each of the four branches of (C,z). Note that $I(C,\mathscr{T})$ is the tangent space to deformations of (C,z) fixing the position of the singularity at z and the tangents to the four branches. Since the analytic type of an ordinary 4-fold point depends precisely on the cross-ratios of the tangents, the ideals $I(C,\mathscr{T})$ satisfy (a)–(e). In particular, we obtain $\deg Z^a(C,z) \le \deg Z(C,\mathscr{T}) = \delta(C,z) + \sum_{q\in\mathscr{T}} \operatorname{mt} C_{(q)} = 14$.

1.2 Non-classical Singularity Invariants

The most important classical numerical invariants of reduced plane curve singularities are the Milnor number μ, the Tjurina number τ, the δ-, and the κ-invariant (definitions and a complete account can be found in [GLS6, Sections I.2 and I.3.4]). In the following we shall introduce some further invariants which will play a special rôle when studying the geometric properties of equisingular families of curves on surfaces in Chapter IV: the analytic, respectively topological, deformation–determinacy, the β-invariant (and the local isomorphism defect), and the γ-invariant of a reduced plane curve singularity. This section is completely local, i.e. we consider power series $f \in \mathbb{C}\{\boldsymbol{x}\} = \mathbb{C}\{x_1, \dots, x_n\}$ or germs of curves (C, z) in a smooth surface germ (Σ, z).

1.2.1 Determinacy Bounds

It is well known that a power series $f \in \mathbb{C}\{\boldsymbol{x}\}$ with isolated singularity is determined, up to analytic equivalence, by a sufficiently high jet, that is, by its Taylor series expansion up to a sufficiently high order. For details see [GLS6, Section I.2.2], where we gave determinacy bounds for contact equivalence as wells as for right equivalence.

1.2.1.1 Deformation Determinacy

In this section, we introduce the concept of *deformation-determinacy* for analytic as well as for topological equivalence (for plane curve singularities). For our purposes deformation determinacy is even more important than (absolute) determinacy.

Definition 1.2.1 Let $f \in \mathbb{C}\{\boldsymbol{x}\} = \mathbb{C}\{x_1, \dots, x_n\}$ be any power series, $\mathfrak{m}$ the maximal ideal of $\mathbb{C}\{\boldsymbol{x}\}$, and $k \in \mathbb{Z}$.

(1) f is *k-analytic deformation-determined* if for any $\nu \geq k$ and any $g \in \mathfrak{m}^{\nu+1}$

$$f + tg \quad \text{is contact equivalent to} \quad f$$

for all $t \in \mathbb{C}$ sufficiently close to 0. The minimal such k is called the *analytic deformation-determinacy* of f and denoted by $\mathrm{ddt}^a(f)$.

(2) If $n = 2$, then f is *k-topological deformation-determined* if for any $\nu \geq k$ and any $g \in \mathfrak{m}^{\nu+1}$

$$f + tg \quad \text{is topologically equivalent to} \quad f$$

for all $t \in \mathbb{C}$ sufficiently close to 0. The minimal such k is called to *topological deformation-determinacy* of f and denoted by $\mathrm{ddt}^s(f)$.

Condition (1) (resp. (2)) can be equivalently formulated by saying that the power series $f + tg \in \mathbb{C}\{\boldsymbol{x}, t\}$ defines an analytically (resp. topologically) trivial deformation of the curve singularity defined by f.

Note that the (abstract) determinacy requires $f + tg$ for *all* $t \in \mathbb{C}$ to be contact, respectively topologically equivalent to f. Hence we have

$$\mathrm{ddt}^a(f) \leq \text{ analytic determinacy of } f\ ,$$

$$\mathrm{ddt}^s(f) \leq \text{ topological determinacy of } f\ .$$

Already the example of a node $f = x^2 + y^2$ shows that we usually have strict inequalities: f is 2-determined but $\mathrm{ddt}^a(f) = \mathrm{ddt}^s(f) = 1$, e.g. $f - x^2 \overset{c}{\not\sim} f$ but $f - tx^2 \overset{c}{\sim} f$ for all $t \neq 1$. For deformation-determinacy no cancellation of terms can appear since we consider arbitrary sufficiently small t.

Definition 1.2.2 For $f \in \mathbb{C}\{x_1, \ldots, x_n\}$ we call

$$\nu^a(f) := \min\{\nu \mid \mathfrak{m}^{\nu+1} \subset I^a(f)\}$$

the *analytic singularity level* of f. For $n = 2$ we call

$$\nu^s(f) := \min\{\nu \mid \mathfrak{m}^{\nu+1} \subset I^s(f)\} \tag{1.2.1.1}$$

the *topological singularity level* of f. Here $I^a(f)$ and $I^s(f)$ are the analytic singularity ideal (cf. Definition 1.1.66) and the topological singularity ideal (cf. Definition 1.1.31).

We want to relate $\mathrm{ddt}^a(f)$ to $\nu^a(f)$ and $\mathrm{ddt}^s(f)$ to $\nu^s(f)$.

Let us first show how the ideal $I^{ea}_{fix}(f) = \langle f, \mathfrak{m}j(f)\rangle$ is related to deformations of f.

Lemma 1.2.3 *Let $f \in \mathbb{C}\{\boldsymbol{x}\}$ define an isolated singularity and $g \in \mathbb{C}\{\boldsymbol{x}\}$.*

(1) If $f + tg \overset{c}{\sim} f$ for all $t \in \mathbb{C}$ close to 0 then $g \in I^{ea}_{fix}(f)$.
(2) If $g \in I^{ea}_{fix}(f)$ then there exists $F \in \mathbb{C}\{\boldsymbol{x}, t\}$ of the form

$$f_t(\boldsymbol{x}) := F(\boldsymbol{x}, t) = f(\boldsymbol{x}) + tg(\boldsymbol{x}) + t^2h(\boldsymbol{x}, t)$$

for some $h \in \mathbb{C}\{\boldsymbol{x}, t\}$ such that $f_t \overset{c}{\sim} f$ for all $t \in \mathbb{C}$ close to 0.

Proof Recall that $I^{ea}_{fix}(f)$ is the tangent space at f to the orbit of f under the contact group $\mathcal{K}$ (cf. Remark 1.1.66.1).

(1) If $f + tg \overset{c}{\sim} f$ for t close to 0, then the germ of the curve $\gamma(t) = f + tg$ at 0 is contained in $\mathcal{K} \cdot f$, hence g is tangent to $\mathcal{K} \cdot f$.

(2) Let g be tangent at f to the orbit of f under the action of $\mathcal{K}$ (or rather its k-jet for arbitrary big k). Since orbits are smooth, any tangent vector is the tangent

to some germ of a complex curve contained in the orbit. Such a germ of a curve can be parameterized with $t \mapsto f_t$ as above. □

We show now that analytic determinacy and deformation-determinacy differ at most by 1. Recall that $\nu^{ea}_{fix}(f)$ is the minimal ν such that $\mathfrak{m}^{\nu+1} \subset I^{ea}_{fix}(f)$.

Lemma 1.2.4 *If $f \in \mathbb{C}\{\boldsymbol{x}\}$ defines an isolated singularity, then*

$$\nu^{ea}_{fix}(f) \le ddt^a(f) \le analytic\ determinacy\,(f) \le \nu^{ea}_{fix}(f) + 1\,.$$

Proof If $\mathrm{ddt}^a(f) = \nu$ then $f + tg \overset{c}{\sim} f$ for all t close to 0 and all $g \in \mathfrak{m}^{\nu+1}$. By Lemma 1.2.3(1) $\mathfrak{m}^{\nu+1} \subset I^{ea}_{fix}(f)$, showing the first inequality. Since the second inequality is trivial, we consider the third one and assume $\mathfrak{m}^{\nu+1} \subset I^{ea}_{fix}(f)$. Then $\mathfrak{m}^{\nu+2} \subset \mathfrak{m} I^{ea}_{fix}(f)$ and, by [GLS6, Theorem I.2.23], f is $(\nu + 1)$-determined. □

The following lemma provides better bounds for the deformation-determinacy and relates the different 0-dimensional ideals

$$\widetilde{I}^a(f) \subset I^a(f) \subset I^{ea}_{fix}(f) \subset I^{ea}(f)$$

introduced in the preceding section.

Lemma 1.2.5 *Let $f \in \mathbb{C}\{\boldsymbol{x}\}$ define an isolated singularity. Then the following inequalities hold:*

$$\nu^{ea}(f) \le \nu^{ea}_{fix}(f) \le ddt^a(f) \le \nu^a(f) \le \widetilde{\nu}^a(f) = \nu^{ea}(f) + 1 \le \tau(f).$$

Proof The first and the fourth inequalities are clear because of the inclusion of the corresponding ideals. The last inequality follows from $\mathfrak{m}^{\tau+1} \subset I^{ea}(f)$, because $I^{ea}(f)$ has codimension $\tau = \tau(f)$ in $\mathbb{C}\{\mathbf{x}\}$. To see $\widetilde{\nu}^a(f) = \nu^a(f)+1$ let $\nu = \nu^a(f)$ and $\mathfrak{m}^{\nu+1} \subset I^{ea}(f) = \langle f, j(f)\rangle$. Hence, for $h \in \mathfrak{m}^{\nu+2}$ we have $\langle h, j(h)\rangle \subset \mathfrak{m}^{\nu+1}$. That is, $h \in \widetilde{I}^{ea}(f)$ and $\widetilde{\nu}^a(f) \le \nu^a(f)+1$. Now let $\nu = \widetilde{\nu}^a(f)$ and $\mathfrak{m}^{\nu+1} \subset \widetilde{I}^{ea}(f)$, i.e. $\langle h, j(h)\rangle \subset \langle f, j(f)\rangle$ for all $h \in \mathfrak{m}^{\nu+1}$. Since the partials of all $h \in \mathfrak{m}^{\nu+1}$ generate $\mathfrak{m}^{\nu}$, we have $\mathfrak{m}^{\nu} \subset \langle h, j(h)\rangle$ and therefore $\nu^a(f) \le \widetilde{\nu}^a(f) - 1$. □

The invariants ν^{ea}, ν^{ea}_{fix}, $\widetilde{\nu}^a$ and $\ll$ can effectively be computed by Gröbner bases methods (cf. e.g. Example 1.1.73.1); this holds also for ddt^a, using Lemma 1.2.3(1). For example, we have for $A_k(x^2 + y^{k+1}) : \nu^{ea} = k - 1$, $\nu^{ea}_{\mathrm{fix}} = k$; for $D_k(yx^2 + y^k)$: $\nu^{ea}_{\mathrm{fix}} = k - 2$, $\mathrm{ddt}^a = k - 1$; for $(x^5 + y^5)$: $\mathrm{ddt}^a = 6$, $\widetilde{\nu}^a = 7$. So far we have no counterexample to $\mathrm{ddt}^a = \nu^a$.

Problem 1.2.6 Does $\mathrm{ddt}^a = \nu^a(f)$ always hold for $f \in \mathfrak{m} \subset \mathbb{C}\{x_1, \ldots, x_n\}$?

For topological types we can, however, prove:

Lemma 1.2.7 *Let $f \in \mathbb{C}\{\mathbf{x}\}$ define an isolated singularity. Then* $\mathrm{ddt}^s(f) = \nu^s(f)$.

Proof Let $\nu = \mathrm{ddt}^s(f)$ and $g \in \mathfrak{m}^{\nu+1}$. Then $f + tg$ is topologically equivalent to f for $t \in \mathbb{C}$ sufficiently close to 0. By Lemma 1.1.33 and the definition of I^s this is equivalent to $g \in I^s$. □

The last lemma shows that the topological singularity level ν^s is a topological invariant for reduced plane curve singularities. Since ν^a is an analytic invariant for isolated hypersurface singularities, we can write

$$\nu^a(S) := \nu^a(F, z) := \nu^a(f) \quad \big(\text{resp. } \nu^s(S) := \nu^s(C, z) := \nu^s(f)\big)$$

if f is a local equation for (F, z), resp. (C, z), and S the corresponding analytic (resp. topological) type.

1.2.1.2 Bounds for the Deformation Determinacy

We shall now give for a reduced plane curve singularity (C, z) bounds for the topological deformation-determinacy $\mathrm{ddt}^s(C, z) = \nu^s(C, z)$ in terms of classical invariants.

In view of (1.2.1.1), Lemma 1.1.44 allows to compute the topological deformation determinacy of a plane curve germ as soon as one knows the respective system of multiplicity sequences, or, equivalently, the cluster graph:

Proposition 1.2.8 *Let $(C, z) \subset (\Sigma, z)$ be a reduced plane curve singularity, and let $Q_1, \ldots, Q_s$ be its local branches. Then the topological deformation determinacy $\nu^s(C, z)$ is given by*

$$\min\left\{\nu \in \mathbb{N} \;\middle|\; \nu + 1 \geq \max_j \frac{2\delta(Q_j) + \sum_{i \neq j} i_z(Q_i, Q_j) + \sum_{q \in \mathscr{T}^* \cap Q_j} \mathrm{mt}\, Q_{j,(q)}}{\mathrm{mt}\, Q_j}\right\}$$

where $Q_{j,(q)}$ denotes the strict transform of Q_j at $q \in \mathscr{T}^* := \mathscr{T}^*(C, z)$.

Note that the formula for ν^s given in [Lic] is not correct, at least in the case of several branches, as can be seen for A_{2k+1}-singularities.

Example 1.2.8.1 (a) Let (C, z) be an A_{2k}-singularity. Then the cluster graph $\mathrm{clg}(C, \mathscr{T}^*)$, $\mathscr{T}^* = \mathscr{T}^*(C, z)$, is given by

$$z \leftarrow q_1 \leftarrow \cdots \leftarrow q_{k-1} \leftarrow q_k \nwarrow q_{k+1}\ (\text{dashed arrow } q_{k+1} \dashrightarrow q_{k-1}), \qquad 2 \leftarrow 2 \leftarrow \cdots \leftarrow 2 \leftarrow 1 \nwarrow 1\ (\text{dashed arrow } 1 \dashrightarrow 2),$$

and we obtain $\nu^s(C, z) = 2k$.

(b) Let (C, z) be an A_{2k-1}-singularity. Then the cluster graph $\mathrm{clg}(C, \mathscr{T}^*)$, $\mathscr{T}^* = \mathscr{T}^*(C, z)$, is given by

$$z \longleftarrow q_1 \longleftarrow \cdots \longleftarrow q_{k-1}, \qquad 2 \longleftarrow 2 \longleftarrow \cdots \longleftarrow 2,$$

and we obtain $\nu^s(C, z) = 2k - 1$.
(c) Let (C, z) be given by the local equation $(x^3 - y^5)(x^2 - y^3) = 0$. Then $\mathrm{clg}(C, \mathscr{T}^*)$ is given by

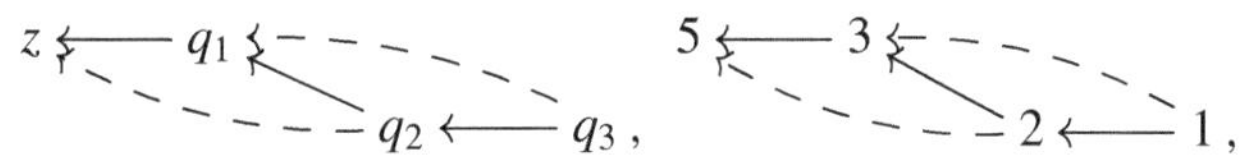

and we obtain

$$\nu^s(C, z) = \min\left\{\nu \in \mathbb{Z} \,\middle|\, \nu + 1 \geq \max\left(\frac{2+9+4}{2}, \frac{8+9+7}{3}\right)\right\} = 7\,.$$

We can estimate $\nu^s(C, z)$ in terms of $\tau^{es}(C, z)$, the codimension of the μ-const stratum in the semiuniversal deformation of (C, z), respectively in terms of $\delta(C, z)$. Recall that $\delta(C, z)$ is the codimension of the *equiclassical stratum* (cf. [DiH]) in the semiuniversal deformation of (C, z), whence $\delta(C, z) \leq \tau^{es}(C, z)$.

Lemma 1.2.9 $\nu^s(C, z) \leq \tau^{es}(C, z)$ *for any reduced plane curve singularity* (C, z). *If all branches of* (C, z) *have at least multiplicity* 3 *then we have even* $\nu^s(C, z) \leq \delta(C, z)$.

Proof If (C, z) is an A_k-singularity, we have $\tau^{es}(C, z) = \tau(C, z) = k$ and the statement follows from the above. Let $\mathrm{mt}(C, z) \geq 3$ and $Q_1, \ldots, Q_s$ be the local branches of (C, z).
Case 1: (C, z) is irreducible. Then, by Proposition 1.2.8, we have

$$\nu^s(C, z) = \min\left\{\nu \in \mathbb{N} \,\middle|\, \nu + 1 \geq \frac{2\delta(C, z) + \sum_{q\in\mathscr{T}^*} m_q}{\mathrm{mt}(C, z)}\right\},$$

$m_q := \mathrm{mt}\, C_{(q)}$, $\mathscr{T}^* = \mathscr{T}^*(C, z)$. If $\mathrm{mt}(C, z) = 3$ then $\#\{q \in \mathscr{T}^* | m_q \leq 2\} \leq 3$, whence

$$\sum_{q\in\mathscr{T}^*} m_q \leq \sum_{q\in\mathscr{T}^*} \frac{m_q(m_q - 1)}{2} + 3 = \delta(C, z) + \mathrm{mt}(C, z)\,. \tag{1.2.1.2}$$

If $\mathrm{mt}(C, z) \geq 4$, we know at least that $\#\{q \in \mathscr{T}^* | m_q = 1\} \leq \mathrm{mt}(C, z)$. Thus,

$$\sum_{q\in\mathscr{T}^*} m_q \leq 2 \sum_{q\in\mathscr{T}^*} \frac{m_q(m_q - 1)}{2} + \mathrm{mt}(C, z) = 2\delta(C, z) + \mathrm{mt}(C, z)\,.$$

Case 2: (C, z) is reducible. For any j we have to estimate

$$\frac{2\delta(Q_j) + \sum_{i\neq j} i_z(Q_i, Q_j) + \sum_{q\in\mathscr{T}^*\cap Q_j} \mathrm{mt}\, Q_{j,(q)}}{\mathrm{mt}\, Q_j} - 1\,. \tag{1.2.1.3}$$

Each point $q \in \mathscr{T}^* \cap Q_j$ is either essential for the branch Q_j itself, or it lies in the intersection $\mathscr{T}(Q_j) \cap \mathscr{T}(Q_i)$ for some $i \neq j$. Hence, if $\mathrm{mt}\, Q_j \geq 3$ then (1.2.1.3) does not exceed

$$\frac{2\delta(Q_j) + \sum_{q\in\mathscr{T}^*(Q_j)} \mathrm{mt}\, Q_{j,(q)} + 2\sum_{i\neq j} i_z(Q_i, Q_j)}{\mathrm{mt}\, Q_j} - 1$$
$$\overset{(1.2.1.2)}{\leq} \frac{3\delta(Q_j) + 2\sum_{i\neq j} i_z(Q_i, Q_j)}{3}$$
$$< \sum_{i=1}^{s} \delta(Q_i) + \sum_{i<k} i_z(Q_i, Q_k) \;=\; \delta(C, z)\,.$$

If $\mathrm{mt}\, Q_j = 2$ we can estimate (1.2.1.3) as

$$\frac{\sum\limits_{q\in\mathscr{T}^*(Q_j)} \mathrm{mt}\, Q_{j,(q)} \left(\mathrm{mt}\, Q_{j,(q)} - 1\right) + \sum\limits_{q\in\mathscr{T}^*\cap Q_j} \mathrm{mt}\, Q_{j,(q)} \left(m_q - \mathrm{mt}\, Q_{j,(q)} + 1\right)}{2} - 1$$
$$\leq \sum_{q\in\mathscr{T}^*\cap Q_j} \frac{m_q \cdot \mathrm{mt}\, Q_{j,(q)}}{2} - 1 \leq \sum_{q\in\mathscr{T}^*} \frac{m_q(m_q+1)}{2} - \#\{q\in\mathscr{T}^*\} - 1$$
$$\leq \tau^{es}(C, z)\,.$$

(Recall that $m_z \geq 3$ and that there are at most two points in $\mathscr{T}^* \cap Q_j$ with $m_q = 1$.) Finally, for a smooth branch Q_j, that is, $\mathrm{mt}\, Q_j = 1$, (1.2.1.3) can be estimated as

$$\sum_{q\in\mathscr{T}^*\cap Q_j} m_q - 1 \leq \sum_{q\in\mathscr{T}^*} \frac{m_q(m_q+1)}{2} - \#\{q\in\mathscr{T}^*\} - 1 \leq \tau^{es}(C, z)\,,$$

since there is no point in $\mathscr{T}^* \cap Q_j$ with $m_q < 2$ (for the latter inequality see Corollary 1.1.64). □

1.2.2 New Topological Invariants

The singularity invariants discussed in this section (as well as the γ-invariant treated in the next section) are related to the H^1-vanishing criteria for the ideal sheaves of zero-dimensional schemes on curves, the key tool in our study of families of singular algebraic curves on surfaces. Particularly, the local isomorphism defect naturally appears in the Riemann–Roch type H^1-vanishing criteria (cf. Proposition 3.1.2 and Corollaries 3.1.3, 3.1.4). It applies to analytical as well as to topological singularity types, however, it is hard to compute. In turn, the β-invariant is easy to compute and it provides an efficient bound for the local isomorphism defect related to the topological equisingular deformation of curves. Namely, the β-invariant characterizes the minimal cluster schemes which contain the topological singularity schemes whose ideals are tangent to the topological equisingular families. More precisely, the ideals of the above cluster schemes are tangent spaces to the topological equisingular families with additional restrictions to sections (like fixing the singular point and certain infinitely near points).

Throughout this section, let $(C, z) \subset (\Sigma, z)$ be a reduced plane curve singularity in the smooth surface Σ, given by a local equation $f \in \mathbb{C}\{x, y\}$.

1.2.2.1 The Topological Cluster Scheme

Definition 1.2.10 We introduce the *topological cluster scheme*

$$Z^{\,top}(C, z) := Z(\mathcal{T}^*(C, z), \boldsymbol{m}^{\,top})$$

where the assigned multiplicities m_q^{top}, $q \in \mathcal{T}^*(C, z)$, are given as follows:

(a) If (C, z) is an A_k-singularity, $k \geq 1$, then we assign to an essential point q (cf. Example 1.2.10.1, below)

- $m_q^{top} = 0$, if $\mathrm{mt}\, C_{(q)} = 1$ and $C_{(q)}$ intersects (each component of) the germ of the exceptional divisor at q transversally.
- $m_q^{top} = 1$, if q is the last point with $\mathrm{mt}\, C_{(q)} = 2$, or if q is a non-singular point of $C_{(q)}$ and $C_{(q)}$ is tangent to the exceptional divisor at q,
- $m_q^{top} = 2$ otherwise.

(b) If $\mathrm{mt}(C, z) \geq 3$, then we assign

- $m_q^{top} = 0$, if q satisfies the following conditions:
  $\mathrm{mt}\, C_{(q)} = 1$ and $C_{(q)}$ intersects (each component of) the germ of the exceptional divisor at q transversally, and
  if p denotes the unique point such that q is proximate to p and $\mathrm{mt}\, C_{(p)} \geq 2$, then either $p = z$ or $C_{(p)}$ is not tangential to the germ of the exceptional divisor at p.
- $m_q^{top} = \mathrm{mt}\, C_{(q)}$ otherwise.

We also introduce $\mathcal{T}^{\,top} \subset \mathcal{T}^*(C, z)$ as the constellation obtained by omitting those essential points with assigned multiplicity $m_q^{top} = 0$.

By Proposition 1.2.11 and Corollary 1.1.34 we have $Z^{es} \subset Z^{top} \subset Z^s$. We illustrate this with a few examples.

Example 1.2.10.1 (a) Let (C, z) be an A_{2k}-singularity, $k \geq 2$. Then the cluster graph $\mathrm{clg}(\mathcal{T}^{\,top}, \boldsymbol{m}^{\,top})$, is given by (cf. Example 1.2.8.1)

$$z \longleftarrow q_1 \longleftarrow \cdots \longleftarrow q_{k-2} \longleftarrow q_{k-1} \longleftarrow q_k\,, \qquad 2 \longleftarrow 2 \longleftarrow \cdots \longleftarrow 2 \longleftarrow 1 \longleftarrow 1\,,$$

and we obtain $\deg Z^{\,top}(C, z) = 3k - 1 = \deg Z^s(C, z) - 3$.

(b) Let (C, z) be an A_{2k-1}-singularity. Then $\mathrm{clg}(\mathscr{T}^{top}, \boldsymbol{m}^{top})$ is given by

$$z \longleftarrow q_1 \longleftarrow \cdots \longleftarrow q_{k-2} \longleftarrow q_{k-1}\,, \qquad 2 \longleftarrow 2 \longleftarrow \cdots \longleftarrow 2 \longleftarrow 1\,,$$

and we obtain $\deg Z^{top}(C, z) = 3k - 2 = \deg Z^s(C, z) - 2$.

(c) Let (C, z) be given by $f = x^3 - y^4$. Then $\mathrm{clg}(\mathscr{T}^{top}, \boldsymbol{m}^{top})$, is given by

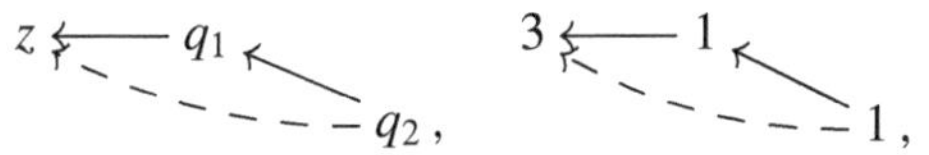

and we obtain $\deg Z^{top}(C, z) = 5 = \deg Z^s(C, z) - 1$.

(d) Let (C, z) be given by $f = x^3 - y^5$. Then $\mathrm{clg}(\mathscr{T}^{top}, \boldsymbol{m}^{top})$ is given by

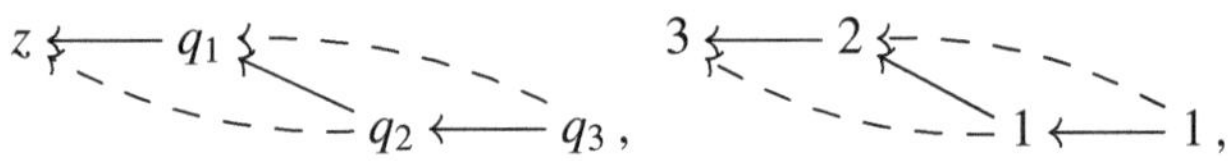

that is, $(\mathscr{T}^{top}, \boldsymbol{m}^{top}) = \mathcal{C}\ell\big(C, \mathscr{T}^*(C, z)\big)$, and we obtain $\deg Z^{top}(C, z) = \deg Z^s(C, z) = 11$.

The characterizing property of the topological cluster scheme is given by the following proposition (see also the proof for further characterizations):

Proposition 1.2.11 *The scheme $Z^{top}(C, z)$ is the minimal cluster scheme containing $Z^{es}(C, z)$. In other words, the defining ideal $I^{top}(C, z)$ of $Z^{top}(C, z)$ is the maximal $\overline{\mathcal{O}}$-ideal contained in $I^{es}(C, z)$ ($\mathcal{O}_{C,z} \hookrightarrow \overline{\mathcal{O}}$ the normalization).*

Proof Step 1. We show that $Z^{top}(C, z) \supset Z^{es}(C, z)$.
Case A. Let (C, z) be an A_μ-singularity, $\mu \geq 1$, given by the local equation $f = y^2 - x^{\mu+1}$. If $\mu = 2k - 1$, then (C, z) decomposes into two smooth branches, $Q_1 = y - x^k$ and $Q_2 = y + x^k$. Lemma 1.1.44 gives

$$f \in I^{top}(C, z) \iff i_z(f, Q_i) \geq 2k - 1 \text{ for } i = 1, 2\,.$$

If $\mu = 2k$, then the germ (C, z) is irreducible, and Lemma 1.1.44 gives

$$f \in I^{top}(C, z) \iff i_z(f, C) \geq 4k - 1\,.$$

In both cases, we obtain

$$I^{top}(C, z) = \langle y^2, x^\mu, yx^\ell \rangle, \quad \ell = \left[\frac{\mu - 1}{2}\right],$$

thus, $I^{top}(C, z)$ is contained in $I^{es}(C, z) = \langle y, x^\mu \rangle$.
Case B. Let $\mathrm{mt}(C, z) \geq 3$. Then

$$I^{top}(f) = \left\{ g \in I^{es}(f) \;\middle|\; \begin{array}{c}\text{The equimultiple sections through the}\\ \text{singular points of the strict transform}\\ \text{of } (C,z) \text{ are trivial sections}\end{array} \right\}. \qquad (1.2.2.1)$$

To see this, recall that $f+\varepsilon g$ defines an equisingular deformation iff, for each $q \in \mathscr{T}^*(C,z)$, the induced deformation of the reduced total transform of (C,z) at q is equimultiple along a section σ_q (and these sections have to be compatible). Let $\mathscr{T} \subset \mathscr{T}^*(C,z)$ be a subconstellation such that the sections σ_q through all points $q \in \mathscr{T}$ are trivial. Then the equimultiplicity of the reduced total transform along σ_q, $q \in \mathscr{T}$, is equivalent to the equimultiplicity of the total transform along σ_q, hence, to $\operatorname{mt} \widehat{g}_{(q)} \geq \operatorname{mt} \widehat{f}_{(q)}$.

Thus, to get (1.2.2.1), it suffices to consider the points $q \in \mathscr{T}^*(C,z)$, where the strict transform of (C,z) is smooth. For each branch Q of (C,z), let q_Q be the point of lowest level in $\mathscr{T}^*(C,z) \cap \mathscr{T}(Q)$ where the strict transform of (C,z) is smooth (if such a point exists). Locally at q_Q, there are two possible situations: either q_Q is a free point, or it is a satellite point.

Let q_Q be a satellite point. If we assume then that the equimultiple sections through the singular essential points are trivial sections, the equimultiple sections through all the points $q \in \mathscr{T}^*(C,z)$ that are infinitely near to q_Q (in particular through q_Q itself) are trivial sections, too. On the other hand, q_Q being a satellite point implies that for the predecessor p of q_Q we have $\operatorname{mt} C_{(p)} \geq 2$, and $C_{(p)}$ is tangential to the germ of the exceptional divisor at p. Hence, on both sides of (1.2.2.1), the condition imposed to g at a point $q \in \mathscr{T}^*(C,z)$ that is infinitely near to q_Q reads as $\operatorname{mt} \widehat{g}_{(q)} \geq \operatorname{mt} \widehat{f}_{(q)}$.

Now, assume that q_Q is a free point, and denote by $E = \{e = 0\}$ the germ of the exceptional divisor at q. If $f+\varepsilon g$ defines an equisingular deformation of (C,z) such that the equimultiple sections through the singular essential points are trivial sections, then the proof of [GLS6, Proposition II.2.14] shows that $\widehat{g}_{(q_Q)} \in e^m \cdot I_E^{es}(f_{(q_Q)})$ with $m := \operatorname{mt} \widehat{f}_{(q_Q)} - \operatorname{mt} f_{(q_Q)}$. But, as $f_{(q_Q)}$ is smooth,

$$I_E^{es}(f_{(q_Q)}) = \left\{ h \in \mathcal{O}_{\Sigma_{(q_Q)},q_Q} \;\middle|\; i_{q_Q}(h,E) \geq i_{q_Q}(f_{(q_Q)},E) - 1 \right\}$$

(see Remark A.4.0.1). As E is smooth, Noether's formula (Proposition 1.1.8) shows that the condition imposed to g at a point $q \in \mathscr{T}^*(C,z)$ infinitely near to q_Q reads as $\operatorname{mt} \widehat{g}_{(q)} \geq \operatorname{mt} \widehat{f}_{(q)}$, unless q is a leaf of $\mathscr{T}^*(C,z)$. For a leaf of $\mathscr{T}^*(C,z)$ infinitely to q_Q, there is no imposed condition. On the other hand, these conditions are precisely the conditions imposed at q for g to define a curve germ going through the cluster $(\mathscr{T}^{top}, \boldsymbol{m}^{top})$.

If (C,z) is a non-singular curve germ, L is a non-singular curve through z, and $i_z(C,L) \geq 2$, then the tangent space at (C,z) to the family of non-singular germs (C',z') such that $z' \in L$, $i_{z'}(C',L) = i_z(C,L)$, is the ideal

$$\left\{ f \in \mathcal{O}_{C,z} \;\middle|\; i_z(f,C) \geq i_z(C,L) - 1 \right\},$$

that is, the ideal of the cluster scheme in (C,z), consisting of $i_z(C,L) - 1$ successive infinitely near points of multiplicity 1 (cf. Lemma 1.2.12, below).

Step 2. We show the minimality of $Z^{top}(C, z)$.
Case A. Let (C, z) be an A_μ-singularity, $\mu \geq 1$. If, in the case $\mu = 2k - 1$, a cluster scheme $Z \subset (C, z)$ is defined by intersection multiplicities $k_1 < 2k - 1$, $k_2 \geq k_1$, then $I(Z)$ contains an element $(y - x^{k_1})(y + x^{k_2} + x^k)$, which does not belong to $I^{es}(C, z)$. Similarly, if, in the case $\mu = 2k$, a cluster scheme $Z \subset (C, z)$ is defined by an intersection multiplicity $k_1 < 4k - 1$, then $I(Z)$ contains $f = x^{2k-1}$, for which $i_z(f, C) = 4k - 2$, and which is not in $I^{es}(C, z)$. Thus, we establish the minimality of $Z^{top}(C, z)$.
Case B. Let $m = \mathrm{mt}(C, z) \geq 3$, and assume there exists a cluster scheme Z such that $Z^{top}(C, z) \supsetneq Z \supset Z^{es}(C, z)$. It is not difficult to see that such a scheme Z would be contained in a cluster scheme Z' given by the constellation $\mathscr{T}^{top}$ and assigned multiplicities

- $m'_p = m_p^{top} - 1$ for some "leaf" $p \in \mathscr{T}^{top}$, that is, a point p such that no point $q \in \mathscr{T}^{top}$ is proximate to p, and
- $m'_q = m_q^{top}$ for all points $q \in \mathscr{T}^{top} \setminus \{p\}$.

Now there are exactly three possibilities: either $p = z$, or p lies on some component $E_{(p)}$ of the exceptional divisor and satisfies $m_p^{top} = \mathrm{mt}\, C_{(p)} \geq 2$, or $m_p^{top} = \mathrm{mt}\, C_{(p)} = 1$ and $i_p(C_{(p)}, E_{(p)}) \geq 2$.

In the first case, (C, z) would be an ordinary m-tuple point, in particular, $I'_Z = \mathfrak{m}_z^{m-1} \not\subset I^{es}(C, z) = \langle j(f), \mathfrak{m}_z^m \rangle$, contradicting our assumption. In the other two cases, the first order deformations $f + \varepsilon g$, $g \in I_{Z'}$, respectively its strict transforms, are equimultiple along trivial sections through all points $q \in \mathscr{T}^{top} \setminus \{p\}$. To obtain $g \in I^{es}(C, z)$, the strict transform $f_{(p)} + \varepsilon g_{(p)}$ has to be equimultiple along some section $T_\varepsilon \to E_{(p)} \times T_\varepsilon$. But this is certainly not true for every $g \in I_{Z'}$ (cf. Lemma 1.2.12, below). □

In the above proof of Proposition 1.2.11 we made several times use of the following easy fact:

Lemma 1.2.12 *Let $(C, z) \subset (\Sigma, z)$ be a reduced plane curve germ, let $L \subset \Sigma$ be a curve such that the germ $(L, z) \subset (\Sigma, z)$ is non-singular and intersects (C, z) with multiplicity at least 2. Then the following holds:*

(a) If (C, z) is non-singular, then the tangent space at (C, z) to the family of non-singular germs (C', z') with $z' \in L$ and $i_{z'}(C', L) = i_z(C, L)$, is the ideal

$$\left\{ f \in \mathcal{O}_{C,z} \,\middle|\, i_z(C, f) \geq i_z(C, L) - 1 \right\}.$$

(b) If $\mathrm{mt}(C, z) = m \geq 2$, then the tangent space at (C, z) to the family of germs (C', z') with $z' \in L$ and $\mathrm{mt}(C', z') = m$ is an ideal J satisfying

$$\mathfrak{m}_z^m \subsetneq J \subsetneq \mathfrak{m}_z^{m-1},$$

$\mathfrak{m}_z \subset \mathcal{O}_{\Sigma,z}$ denoting the maximal ideal at z.

Proof We can use local coordinates x, y in a neighbourhood of $z = (0, 0)$ such that $L = \{x = 0\}$.

(a) Assume that $i_z(C, L) = m$, that is,

$$f(0, y) = c_m y^m + \text{higher terms in } y\,, \quad c_m \neq 0\,.$$

Then g belongs to the above tangent space iff the first order deformation $f + \varepsilon g$ satisfies (for some $\alpha \in \mathbb{C}$)

$$\begin{aligned} f(0, y) + \varepsilon g(0, y) &= (y + \varepsilon\alpha)^m + \text{higher terms in } (y + \varepsilon\alpha) \\ &= \varepsilon\alpha y^{m-1} + \text{higher terms in y}\,, \end{aligned}$$

that is, iff $i_z(g, L) \geq m - 1$.

(b) A first order deformation $f + \varepsilon g$ defines an equimultiple deformation along a section $T_\varepsilon \to L \times T_\varepsilon$ iff, for some $\alpha \in \mathbb{C}$,

$$f + \varepsilon g \in \langle x, y + \varepsilon\alpha \rangle^m\,.$$

Hence, $J = \langle x^{m-1}, \mathfrak{m}^m \rangle$. □

1.2.2.2 The β-Invariant

In the following we introduce the β-invariant of a cluster scheme containing the conductor scheme.

Definition 1.2.13 Let (C, z) be a reduced plane curve germ, $(D, z) \subset (C, z)$ a subgerm, and (C', z) the union of those components of (C, z) which are not in (D, z). Then we introduce the β-invariants

$$\begin{aligned} \beta(C, z) &:= \deg Z^{\,top}(C, z) - \deg Z^{\,cd}(C, z)\,, \\ \beta_D(C, z) &:= \deg(Z^{\,top}(C, z) \cap D) - \deg(Z^{\,cd}(C, z) \cap D)\,, \end{aligned}$$

Due to Lemma 1.1.43, the topological cluster scheme $Z^{\,top}(C, z)$ (Definition 1.2.10) contains the conductor scheme $Z^{\,cd}(C, z)$. Hence, the β-invariants are non-negative numbers. Moreover, the equality

$$\deg Z^{\,cd}(C, z) \;=\; \delta(C, z) \tag{1.2.2.2}$$

(cf. [GLS6, formula (I.3.4.11)]), respectively the following lemma, give

$$\begin{aligned} \beta(C, z) &= \deg Z^{\,top}(C, z) - \delta(C, z)\,, \\ \beta_D(C, z) &= \deg(Z^{\,top}(C, z) \cap D) - \delta(D, z) - i_z(D, C')\,. \end{aligned}$$

Lemma 1.2.14 *If* (C,z) *decomposes into the union* $(C,z)=(D,z)\cup(C',z)$ *then* $Z^{\,cd}(C,z):D=Z^{\,cd}(C',z)$. *In particular,*

$$\deg(Z^{\,cd}(C,z)\cap D)=\delta(D,z)+i_z(D,C')\,.$$

Proof By Lemma 1.1.43, the conductor scheme $Z^{\,cd}(C,z)$ is a cluster scheme, given by the constellation of singular essential points $\mathscr{T}^{\,cd}\subset\mathscr{T}^*(C,z)$, and assigned multiplicities $m_q^{cd}=\mathrm{mt}\,C_{(q)}-1$. Since the analogous statement holds true for $Z^{\,cd}(C',z)$, and since $\mathrm{mt}\,C'_{(q)}-1=m_q^{cd}-\mathrm{mt}\,D_{(q)}$, the first statement is obvious. Moreover, we obtain

$$\begin{aligned}\deg\big(Z^{\,cd}(C,z)\cap D\big)&=\deg Z^{\,cd}(C,z)-\deg\big(Z^{\,cd}(C,z):D\big)\\&=\deg Z^{\,cd}(C,z)-\deg Z^{\,cd}(C',z)\;=\;\delta(C,z)-\delta(C',z)\\&=\delta(D,z)+i_z(D,C')\,,\end{aligned}$$

the latter two equalities due to (1.2.2.2), respectively Hironaka's lemma [HiH] (see also [GLS6, Lemma I.3.32]). □

Lemma 1.2.15 *Let* (C,z) *be a reduced plane curve germ,* $(D,z)\subsetneq(C,z)$ *a proper subgerm and* (C',z) *the union of those components of* (C,z) *which are not in* (D,z). *Then*

$$\beta(C,z)=\begin{cases}\mu(C,z)-1 & \text{if } (C,z) \text{ is an } A_\mu\text{-singularity}\,,\\ \displaystyle\sum_{q\in\mathscr{T}^{\,top}(C,z)}\mathrm{mt}\,C_{(q)} & \text{otherwise}\,;\end{cases}$$

$$\beta_D(C,z)=\begin{cases}i_z(D,C')-1 & \text{if } (C,z) \text{ is an ADE-singularity}\,,\\ \displaystyle\sum_{q\in\mathscr{T}^{\,top}(C,z)}\mathrm{mt}\,D_{(q)} & \text{otherwise}\,.\end{cases}$$

In particular, if $Q_1,\dots,Q_s$ *are the local branches of* (C,z) *then*

$$\beta(C,z)\;=\;\sum_{i=1}^{s}\beta_{Q_i}(C,z)\,.$$

Exercise 1.2.16 (1) Prove Lemma 1.2.15.

(2) Let (C,z) be a plane curve singularity, which is topologically equivalent to $\{y^k-x^\ell=0\}$, $3\le k\le\ell$. Prove that

$$\beta(C,z)=\begin{cases}k+\ell-\gcd(k,\ell), & \text{if } \ell\not\equiv 1 \bmod k,\\ k+\ell-2, & \text{if } \ell\equiv 1 \bmod k.\end{cases}$$

More generally, we define the *β-invariant of a cluster scheme*:

Definition 1.2.17 Let (C, z) be a reduced plane curve germ, and Z a cluster scheme containing the conductor scheme $Z^{cd}(C, z)$. Then we introduce

$$\begin{aligned}\beta(Z; C, z) &:= \deg Z - \delta(C, z),\\ \beta_D(Z; C, z) &:= \deg(Z \cap D) - \delta(D, z) - i_z(D, C'),\end{aligned}$$

where $(D, z) \subset (C, z)$ is a subgerm, and (C', z) the union of the remaining components of (C, z).

Note that all the formulae of Lemma 1.2.15 hold when replacing $\mathscr{T}^{top}(C, z)$ by any finite subtree $\mathscr{T} \subset \mathscr{T}(C, z)$ containing $\mathscr{T}^{cd}(C, z)$.

Example 1.2.17.1 Let $z = (0, 0) \in \mathbb{C}^2$, and (C, z) be a curve germ defined by a polynomial $F(x, y)$ with a convenient Newton diagram $\Gamma(F, 0)$, which meets the coordinate axes at integral points $(m_1, 0)$ and $(0, m_2)$.

Assume that the truncations F^σ of F on edges σ of $\Gamma(F, 0)$ have no multiple factors different from x and y. Then the ideal, generated by the monomials lying on $\Gamma(F, 0)$ or above, defines a cluster scheme $Z \subset (C, z)$. This scheme contains the conductor scheme $Z^{cd}(C, z)$, and

$$\beta(Z; C, z) = m_1 + m_2 - \#(\Gamma(F, 0) \cap \mathbb{Z}^2) + 1\,.$$

This immediately follows from the fact that $\deg Z$ is the number of integral points with non-negative coordinates lying strictly below $\Gamma(F, 0)$, and $\delta(C, z)$ is the number of integral points with positive coordinates lying on $\Gamma(F, 0)$ or below.

1.2.2.3 The Isomorphism Defect

We introduce now the isomorphism defect for any zero-dimensional scheme Z, which is closely related to the β-invariant if Z is a cluster scheme, containing the conductor scheme (cf. Proposition 1.2.21, Corollary 1.2.22, and Conjecture 1.2.23).

Definition 1.2.18 Let $Z \subset (C, z)$ be a zero-dimensional scheme with ideal $I_Z \subset \mathcal{O}_{C,z}$. I_Z has finite codimension in $\mathcal{O}_{C,z}$ and we define the *local isomorphism defect* of Z in C at z as

$$\mathrm{isod}_z(Z, C) := \min\left(\dim_{\mathbb{C}} \mathrm{Coker}(\varphi : I_Z \hookrightarrow \mathcal{O}_{C,z})\right),$$

the minimum being taken over all injective $\mathcal{O}_{C,z}$-linear maps $\varphi : I_Z \to \mathcal{O}_{C,z}$. Moreover, for any subgerm $(D, z) \subset (C, z)$, we define

$$\mathrm{isod}_{D,z}(Z, C) := \min\left(\dim_{\mathbb{C}} \mathrm{Coker}(\varphi_D : I_{Z\cap D} \hookrightarrow \mathcal{O}_{D,z})\right),$$

the minimum being taken over all injective maps φ_D, which are induced by an $\mathcal{O}_{C,z}$-linear map $\varphi : I_Z \to \mathcal{O}_{C,z}$.

Lemma 1.2.19 *Let (C, z) be a singular reduced plane curve germ. Then*

$$\mathrm{isod}_z(Z^{ea}(C, z), C) \geq 1\,.$$

Moreover, $\mathrm{isod}_z(Z^{ea}(C, z), C) = 1$ *iff (C, z) is quasi-homogeneous.*

Proof Let x, y be local coordinates at z, and $f \in \mathbb{C}\{x, y\}$ a local equation for (C, z). Due to the finite determinacy, we can even assume that f is a polynomial, that is, $f \in \mathbb{C}[x, y]$. We have to consider injective, $\mathbb{C}\{x, y\}/\langle f\rangle$-linear maps $\varphi : \langle f_x, f_y\rangle \longrightarrow \mathbb{C}\{x, y\}/\langle f\rangle$.

Assume that φ is an isomorphism, that is, $\langle f_x, f_y\rangle$ is generated by a single element $\alpha f_x+\beta f_y$. But then $i(f, \alpha f_x+\beta f_y) = \dim \mathbb{C}\{x, y\}/\langle f, f_x, f_y\rangle$, which contradicts [Shu10, Lemma 4.1] (cf. [GLS6, Lemma I.3.40]).

Moreover, if φ surjects onto the maximal ideal $\langle x, y\rangle$, then the Jacobian ideal can be generated by two polars g_1, g_2, satisfying $y \cdot g_1 - x \cdot g_2 \equiv 0 \bmod f$. For degree reasons, this implies that $y \cdot g_1 - x \cdot g_2 = cf$ for some $c \in \mathbb{C}$. Since $i(g_1, g_2) = \mu(f) < \infty$, $c = 0$ implies $g_1 = y, g_2 = x$. In any case, $f \in \langle g_1, g_2\rangle = \langle f_x, f_y\rangle$, that is, $\mu(C, z) = \tau(C, z)$, which yields (C, z) quasihomogeneous by K. Saito theorem [Sai].

Finally, if f is a quasi-homogeneous polynomial, then $f_x \mapsto y$, $f_y \mapsto -x$ defines a $\mathbb{C}\{x, y\}/\langle f\rangle$-linear map φ as above, surjecting onto the maximal ideal. □

Exercise 1.2.20 Prove that $\mathrm{isod}_z(Z^{es}(C, z), C) \geq 1$ for any singular reduced plane curve germ (C, z).

Let us turn to the question how to compute (or at least, find good estimates for) the isomorphism defects $\mathrm{isod}_z(Z, C)$, $\mathrm{isod}_{D,z}(Z, C)$, with $Z \subset (C, z)$ any zero-dimensional scheme.

A computationally accessible (and quite effective) method is to study the semigroups $\Gamma(\mathcal{O}) = \boldsymbol{v}(\mathcal{O})$ and $\Gamma(I_Z) = \boldsymbol{v}(I_Z)$ ($\boldsymbol{v}$ the valuation map, cf. [GLS6, Section I.3.4, page 214]): let $\varphi : I_Z \hookrightarrow \mathcal{O}_{C,z}$ be any $\mathcal{O}_{C,z}$-linear map, and let $\Gamma(I_Z) \to \Gamma(\mathcal{O})$ be the induced morphism of semigroups. It is clear that φ maps the conductor ideal of I_Z to the conductor ideal of $\mathcal{O}$. In particular, we can estimate the dimension of the cokernel of φ by computing the difference of the number of elements in $\Gamma(\mathcal{O})$ and $\Gamma(I_Z)$ below the respective conductor. In other words, we obtain

$$\mathrm{isod}_z(Z, C) \geq \underbrace{\dim_{\mathbb{C}}\big(\mathcal{O}/I^{cd}(\mathcal{O})\big)}_{=\,\delta(C,z)} - \dim_{\mathbb{C}}\big(I_Z/I^{cd}(I_Z)\big)\,, \tag{1.2.2.3}$$

due to (1.2.2.2), respectively, if C decomposes as $D \cup C'$ then, due to [GLS6, formula (I.3.4.10)],

$$\mathrm{isod}_{D,z}(Z, C) \geq \delta(D, z) + i_z(D, C') - \dim_{\mathbb{C}}\big(\big(I_Z/I^{cd}(I_Z)\big) \otimes \mathcal{O}_{D,z}\big)\,. \tag{1.2.2.4}$$

Note that these bounds are optimal for this approach, since the symmetry of the semigroup (see [GLS6, formula (I.3.4.9)]) implies the existence of a morphism $\Gamma(I_Z) \to \Gamma(\mathcal{O})$ mapping the conductor of $\Gamma(I_Z)$, $\mathrm{cd}(I_Z)$, to the conductor of $\Gamma(\mathcal{O})$. This can be seen as follows: if $\boldsymbol{b} \in \Gamma(I_Z)$ then it is clear that $\mathrm{cd}(I_Z) - \mathbf{1} - \boldsymbol{b}$ is a maximal in $\Gamma(\mathcal{O})$, hence, by the above symmetry, $\mathrm{cd}(\mathcal{O}) - \mathrm{cd}(I_Z) + \boldsymbol{b} \in \Gamma(\mathcal{O})$, and the latter is the image of $\boldsymbol{b}$.

Example 1.2.20.1 (a) For (C, z) a simple (ADE-)singularity, we obtain

$$\mathrm{isod}_z(Z^{es}(C, z), C) = 1\,.$$

(b) If (C, z) is an ordinary singularity of multiplicity at least 3, then

$$\mathrm{isod}_z(Z^{es}(C, z), C) = \tau^{es}(C, z) - \mathrm{mt}(C, z)\,.$$

Moreover, if Q is a (smooth) branch of (C, z), then

$$\mathrm{isod}_{Q,z}(Z^{es}(C, z), C) = \mathrm{mt}(C, z) - 2\,.$$

Proposition 1.2.21 *Let $(C, z) \subset (\Sigma, z)$ be a reduced plane curve germ, let $(D, z) \subset (C, z)$ be a subgerm, and let (C', z) denote the union of those components of (C, z) which are not in (D, z).*

(1) Let $Z_1 \subset Z_2 \subset (C, z)$ be zero-dimensional schemes. Then

$$\deg Z_1 - \mathrm{isod}_z(Z_1, C) \le \deg Z_2 - \mathrm{isod}_z(Z_2, C)\,,$$
$$\deg(Z_1 \cap D) - \mathrm{isod}_{D,z}(Z_1, C) \le \deg(Z_2 \cap D) - \mathrm{isod}_{D,z}(Z_2, C)\,.$$

(2) Let $Z = Z(\mathcal{T}, \boldsymbol{m})$ be a cluster scheme with $\mathcal{T}$ a finite subtree of $\mathcal{T}(C, z)$ containing $\mathcal{T}^{cd}(C, z)$ and with assigned multiplicities

$$m_q \in \{\mathrm{mt}\, C_{(q)} - 1\,,\ \mathrm{mt}\, C_{(q)}\}\,, \quad q \in \mathcal{T}\,,$$

such that the proximity relations are satisfied. Then

$$\mathrm{isod}_z(Z, C) = \delta(C, z)\,, \quad \mathrm{isod}_{D,z}(Z, C) = \delta(D, z) + i_z(D, C')\,.$$

In particular, we obtain in this case

$$\beta(Z; C, z) = \deg Z - \mathrm{isod}_z(Z, C)\,,$$
$$\beta_D(Z; C, z) = \deg(Z \cap D) - \mathrm{isod}_{D,z}(Z, C)\,.$$

Moreover, $Z : C'$ is a cluster scheme contained in (D, z), with assigned multiplicities $m'_q \in \{\mathrm{mt}\, D_{(q)} - 1\,,\ \mathrm{mt}\, D_{(q)}\}$ for $q \in \mathcal{T} \cap \mathcal{T}(D, z)$, satisfying

$$\deg(Z:C') = \deg(Z\cap D) - i_z(D, C')$$

and, for each irreducible component Q_i *of* (D, z),

$$\deg\big((Z:C')\cap Q_i\big) = \deg(Z\cap Q_i) - i_z(Q_i, C')\,.$$

Hence, we get in this case

$$\deg(Z\cap D) - \operatorname{isod}_{D,z}(Z, C) = \deg(Z:C') - \operatorname{isod}_z(Z:C', D)$$

and, for each irreducible component Q_i *of* (D, z),

$$\begin{aligned}&\deg(Z\cap Q_i) - \operatorname{isod}_{Q_i,z}(Z, C)\\ &\qquad = \deg(Z:C')\cap Q_i - \operatorname{isod}_z\big((Z:C')\cap Q_i, D\big)\,.\end{aligned}$$

Example 1.2.21.1 For an m-fold ordinary singularity (C, z) formula (1.2.21.1) yields that

$$\operatorname{isod}_z(Z^s(C, z), C) = \delta(C, z) = \frac{m(m-1)}{2}\,.$$

Proof (1) follows from the definition of the isomorphism defect, when considering restrictions of injective morphisms $I_{Z_1} \to \mathcal{O}_{C,z}$, $I_{Z_1\cap D} \to \mathcal{O}_{D,z}$ to I_{Z_2}, $I_{Z_2\cap D}$, respectively. Indeed, let $\varphi : I_{Z_1} \hookrightarrow \mathcal{O}_{C,z}$ be an injective morphism and consider the induced commutative diagram

$$\begin{array}{ccccccccc} 0 & \longrightarrow & I_{Z_1} & \overset{\varphi}{\hookrightarrow} & \mathcal{O}_{C,z} & \longrightarrow & \operatorname{Coker}(\varphi) & \longrightarrow & 0\\ & & {\scriptstyle incl}\uparrow & & \| & & \uparrow & & \\ 0 & \longrightarrow & I_{Z_2} & \overset{\psi}{\hookrightarrow} & \mathcal{O}_{C,z} & \longrightarrow & \operatorname{Coker}(\psi) & \longrightarrow & 0\,.\end{array}$$

The snake lemma implies $\dim_{\mathbb{C}}\operatorname{Coker}(\psi) - \dim_{\mathbb{C}}\operatorname{Coker}(\varphi) = \dim_{\mathbb{C}} I_{Z_1}/I_{Z_2}$. By definition of the isomorphism defect, we get

$$\operatorname{isod}_z(Z_2, C) \le \operatorname{isod}_z(Z_1, C) + \deg Z_2 - \deg Z_1\,.$$

The proof of the second statement is analogous.
(2) Lemma 1.1.44 shows that I_Z is, indeed, an $\overline{\mathcal{O}}$-ideal ($\mathcal{O}_{C,z} \hookrightarrow \overline{\mathcal{O}}$ the normalization). Hence, the minimal dimension of the cokernel is realized by mapping I_Z onto $I^{cd}(C, z)$. In particular, $\operatorname{isod}_z(Z, C) = \dim_{\mathbb{C}}\big(\mathcal{O}_{C,z}/I^{cd}(C, z)\big)$ and $\operatorname{isod}_{D,z}(Z, C) = \deg\big(Z^{cd}(C, z)\cap D\big)$, hence, the first statement follows from (1.2.2.2), respectively Lemma 1.2.14.

The reduction $Z : C'$ is a cluster scheme due to Proposition 1.1.47(1). Moreover, all intermediate multiplicity vectors $\boldsymbol{m}^{(0)}, \ldots, \boldsymbol{m}^{(\ell)} = \boldsymbol{m}'$ appearing during the successive unloading procedure in the proof of Proposition 1.1.47 satisfy

$m_q^{(i)} \in \{\mathrm{mt}\, D_{(q)} - 1\,,\, \mathrm{mt}\, D_{(q)}\}$. Since unloading is only applied at points p with $m_p^{(i)} = M_p^{(i)} - 1$, we get

$$\sum_{q\in\mathcal{T}} \frac{m_q^{(i)}(m_q^{(i)}+1)}{2} = \sum_{q\in\mathcal{T}} \frac{m_q^{(i-1)}(m_q^{(i-1)}+1)}{2},$$

and, hence,

$$\deg(Z:C') = \sum_{q\in\mathcal{T}} \frac{m'_q(m'_q+1)}{2} = \sum_{q\in\mathcal{T}} \frac{m_q^{(0)}(m_q^{(0)}+1)}{2}.$$

Since either $m_q^{(0)} = m_q - \mathrm{mt}\, C'_{(q)} \in \{\mathrm{mt}\, D_{(q)} - 1\,,\, \mathrm{mt}\, D_{(q)}\}$, or $m_q^{(0)} = 0$, each non-zero summand equals

$$\mathrm{mt}\, D_{(q)} \cdot \left(m_q - \frac{\mathrm{mt}\, D_{(q)} - 1}{2}\right) - \mathrm{mt}\, D_{(q)} \cdot \mathrm{mt}\, C'_{(q)},$$

and, by adding up, we get $\deg(Z \cap D) - i_z(C', D)$. □

As an immediate corollary we obtain

Corollary 1.2.22 *Let C, D, Z be as in Proposition 1.2.21(2), and $Z' \subset (C, z)$ any zero-dimensional scheme such that $Z^{cd}(C, z) \subset Z' \subset Z$. Then*

$$\begin{aligned}\beta(Z; C, z) &\geq \deg Z' - \mathrm{isod}_z(Z', C),\\ \beta_D(Z; C, z) &\geq \deg(Z' \cap D) - \mathrm{isod}_{D,z}(Z', C).\end{aligned}$$

In particular,

$$\begin{aligned}\beta(C, z) &\geq \tau^{es}(C, z) - \mathrm{isod}_z(Z^{es}(C, z), C), \qquad (1.2.2.5)\\ \beta_D(C, z) &\geq \deg(Z^{es}(C, z) \cap D) - \mathrm{isod}_{D,z}(Z^{es}(C, z), C).\end{aligned}$$

Conjecture 1.2.23 Let (C, z) be a reduced plane curve singularity, let $(D, z) \subset (C, z)$ be a subgerm, and let $Z' \subset (C, z)$ be a zero-dimensional scheme such that $Z^{cd}(C, z) \subset Z'$. Moreover, let $Z \subset (C, z)$ be the minimal cluster scheme containing Z'. Then

$$\begin{aligned}\beta(Z; C, z) &= \deg Z' - \mathrm{isod}_z(Z', C),\\ \beta_D(Z; C, z) &= \deg(Z' \cap D) - \mathrm{isod}_{D,z}(Z', C).\end{aligned}$$

In particular,

$$\beta(C, z) = \tau^{es}(C, z) - \mathrm{isod}_z(Z^{es}(C, z), C), \qquad (1.2.2.6)$$

$$\beta_D(C, z) = \deg(Z^{es}(C, z) \cap D) - \mathrm{isod}_{D,z}(Z^{es}(C, z), C). \qquad (1.2.2.7)$$

Conjecture 1.2.23 holds if the bounds (1.2.2.3), respectively (1.2.2.4) are sharp for the scheme Z', that is, if

$$\mathrm{isod}_z(Z', C) = \delta(C, z) - \dim_{\mathbb{C}}\big(I_{Z'}/I^{cd}(I_{Z'})\big),$$
$$\mathrm{isod}_{D,z}(Z', C) = \delta(D, z) + i_z(D, C') - \dim_{\mathbb{C}}\big(\big(I_{Z'}/I^{cd}(I_{Z'})\big) \otimes \mathcal{O}_{D,z}\big).$$

Indeed, Lemma 1.1.44 shows that an ideal $I \subset I^{cd}(C, z)$ defines a cluster scheme iff I is an $\overline{\mathcal{O}}$-ideal ($\mathcal{O}_{C,z} \hookrightarrow \overline{\mathcal{O}}$ the normalization). Hence, $I^{cd}(I_{Z'})$ defines the minimal cluster scheme containing Z', and the statement follows from Proposition 1.2.21(2).

In particular, the equalities (1.2.2.6) and (1.2.2.7) hold for simple singularities (see Lemma 1.2.15 and Example 1.2.20.1). More generally, the following exercise treats the case of quasi-homogeneous singularities of type $\{y^k - x^\ell = 0\}$.

Exercise 1.2.24 Let (C, z) be a plane curve singularity with local equation $f = y^k - x^\ell$, $3 \le k \le \ell$. Prove that

$$\mathrm{isod}_z(Z^{es}(C, z), C) = \frac{k\ell - k - \ell + \gcd(k, \ell)}{2} - \left\lfloor \frac{\ell}{k} \right\rfloor - \varepsilon,$$

where $\varepsilon = 1$ if $\ell \equiv 0, 1 \bmod k$, and $\varepsilon = 2$ otherwise. Moreover, prove that

$$\tau^{es}(C, z) = \frac{k\ell + k + \ell - \gcd(k, \ell)}{2} - \left\lfloor \frac{\ell}{k} \right\rfloor - \varepsilon',$$

where $\varepsilon' = 1$ if k divides ℓ and $\varepsilon' = 2$ otherwise. Use Exercise 1.2.16 to conclude that $\tau^{es}(C, z) - \mathrm{isod}_z(Z^{es}(C, z), C) = \beta(C, z)$.

1.2.3 New Analytic Invariants

Here we introduce new (analytic) invariants of plane curve singular points, which will appear in the H^1-vanishing criteria for the ideal sheaves of zero-dimensional subschemes of the plane, in the asymptotically proper sufficient criteria for the smoothness of equisingular families of curves and for the versality of deformations induced by curves of a given degree (see Sects. 3.5, 4.3.2, and 4.4.1).

1.2.3.1 The γ-Invariant

Let $f \in \langle x, y\rangle^2 \subset \mathbb{C}\{x, y\}$ be a reduced power series and $I_Z \subset \langle x, y\rangle \subset \mathbb{C}\{x, y\}$ be any ideal containing the Tjurina ideal $I^{ea}(f) = \langle f, j(f)\rangle$, defining a zero-dimensional scheme $\emptyset \ne Z \subset Z^{ea}(f) \subset (\mathbb{C}^2, 0)$. Then we introduce several γ-invariants.

Definition 1.2.25 With the notations above call

$$\gamma\big(f;Z\big) := \max\left\{4\deg Z\,,\ \max_g\left\{\frac{i(f,g)^2}{i(f,g)-\deg Z}\right\}\right\},$$

the *γ-invariant of f with respect to Z*, where the inner maximum is taken over all power series $g \in I_Z$ satisfying $1 \le i(f,g) \le 2\deg Z$. Note that $\gamma\big(f;Z\big)$ is a finite rational number due to [Shu10, Lemma 4.1] (cf. [GLS6, Lemma I.3.40]). We set

$$\gamma^{ea}(f) := \max\big\{\gamma\big(f;Z\big)\,\big|\,Z \subset Z^{ea}(f) \text{ a (local) complete intersection}\big\}\,,$$
$$\gamma^{es}(f) := \max\big\{\gamma\big(f;Z\big)\,\big|\,Z \subset Z^{es}(f) \text{ a (local) complete intersection}\big\}\,.$$

Moreover, we set $\gamma^{ea}(f) := \gamma^{es}(f) := 0$ for $f \in \mathbb{C}\{x,y\} \setminus \langle x,y\rangle^2$. We call $\gamma^{ea}(f)$, resp. $\gamma^{es}(f)$ the *equianalytic*, resp. *equisingular γ-invariant of f.*

Remark 1.2.25.1 $I^{ea}(f)$ and $I^{es}(f)$ are independent of the chosen local coordinates, that is, for any unit $u \in \mathbb{C}\{x,y\}$ and any local analytic isomorphism $\Phi : \mathbb{C}\{x,y\} \to \mathbb{C}\{x,y\}$ we have $I^{ea}(uf\circ\Phi) = I^{ea}(f)\circ\Phi$ and similar for es. Hence, for any reduced curve $C \subset \Sigma$, Σ a smooth surface, and any $z \in \Sigma$ we can introduce

$$\gamma(C;Z) := \gamma(f;Z)\,,\quad \gamma^{es}(C,z) := \gamma^{es}(f)\,,\quad \gamma^{ea}(C,z) := \gamma^{ea}(f)\,,$$

where $f \in \mathbb{C}\{x,y\}$ is a local equation for $(C,z) \subset (\Sigma,z)$.

A first, general (but rough), estimate for these invariants is the following:

Proposition 1.2.26 *Let $f \in \mathbb{C}\{x,y\}$ be reduced, and let $I_Z \subset \langle x,y\rangle \subset \mathbb{C}\{x,y\}$ be an ideal containing the Tjurina ideal $I^{ea}(f)$. Then we can estimate*

$$4\cdot\deg Z \le \gamma(f;Z) \le \big(\deg Z+1\big)^2.$$

If, additionally, $I_Z \supset I^{es}(f)$ then

$$\frac{\kappa(f)^2}{\kappa(f)-\deg Z} \le \gamma(f;Z) \le \big(\deg Z+1\big)^2,$$

where $\kappa(f)$ denotes the intersection number of f with a generic polar of f. In particular, we obtain for any reduced plane curve singularity $(C,0) \subset (\mathbb{C}^2,0)$

$$4\cdot\tau_{ci}(C,0) \le \gamma^{ea}(C,0) \le \big(\tau_{ci}(C,0)+1\big)^2,$$
$$4\cdot\tau^{es}_{ci}(C,0) \le \gamma^{es}(C,0) \le \big(\tau^{es}_{ci}(C,0)+1\big)^2,$$

the invariants τ_{ci}, respectively τ^{es}_{ci}, being defined as follows:

$$\tau_{ci}(C,0) := \max\big\{\deg(Z) \,\big|\, Z \subset Z^{ea}(C,0) \text{ a (local) complete intersection}\big\},$$
$$\tau_{ci}^{es}(C,0) := \max\big\{\deg(Z) \,\big|\, Z \subset Z^{es}(C,0) \text{ a (local) complete intersection}\big\}.$$

For the κ-invariant, $\kappa(f) := i(f, \alpha\partial f/\partial x + \beta\partial f/\partial y)$ with $(\alpha : \beta) \in \mathbb{P}^1_{\mathbb{C}}$ generic, see also [GLS6, Proposition I.3.38].

Proof Step 1. We prove the upper bound $\gamma(f, Z) \leq (\deg Z + 1)^2$.
Since $4 \deg Z \leq (\deg Z + 1)^2$, it remains to consider power series $g \in \langle x, y\rangle$ such that $i(f, g) \leq 2 \deg Z$. The quotient

$$\frac{i(f,g)^2}{i(f,g) - \deg(Z \cap \{g = 0\})}$$

takes its maximum for the minimal possible intersection number. Hence,

$$\frac{i(f,g)^2}{i(f,g) - \deg Z} \leq (\deg Z + 1)^2 .$$

Step 2. If we assume additionally that $Z^{es}(f)$ is contained in Z then a generic polar g of f is contained in I_Z, satisfies $i(f, g) = \kappa(f)$, and the latter value is bounded by

$$\deg Z^{es}(f) = \tau^{es}(f) \leq \tau(f) < \mu(f) + \mathrm{mt}(f) - 1 = \kappa(f) \leq 2 \deg Z^{es}(f) .$$

Here the second inequality comes from

$$\deg Z^{es}(f) \geq \kappa(f) - \delta(f) \quad \text{and} \quad Z^{es}(f) \geq \delta(f)$$

which follows from the Gudkov–Shustin theorem [GuS] (cf. [GLS6, Lemma II.2.18]). Hence, $\gamma(f; Z) \geq \kappa(f)^2 / \big(\kappa(f) - \deg Z\big)$. □

Corollary 1.2.27

$$\gamma^{ea}(f) \leq (\tau(f) + 1)^2, \quad \gamma^{es}(f) \leq (\tau^{es}(f) + 1)^2 .$$

1.2.3.2 Estimates for the γ-Invariant

The last estimates for γ are general, but not very good for specific cases. The following proposition computes, respectively gives better estimates, for the γ-invariants in the case of semiquasi-homogeneous singularities.

Proposition 1.2.28 *Let $f \in \mathbb{C}\{x, y\}$ be a semiquasi-homogeneous power series.*

(a) If f defines a simple singularity then

$$\gamma^{es}(f) = \gamma^{ea}(f) = \begin{cases} (k+1)^2 & \textit{if } f \textit{ defines an } A_k - \textit{singularity}, \\ (k-1)^2 & \textit{if } f \textit{ defines a } D_k - \textit{singularity}, k \geq 9, \\ \frac{1}{2}(k+2)^2 & \textit{if } f \textit{ defines a } D_k - \textit{or } E_k - \textit{sing.}, k \leq 8. \end{cases} \tag{1.2.3.1}$$

(b) If f is homogeneous of degree $m \geq 2$, or, in other words, if f defines an ordinary singularity of multiplicity m, then

$$\gamma^{es}(C, z) = \begin{cases} 4 & \textit{if } m = 2, \\ 2m^2 & \textit{if } m \geq 3. \end{cases}$$

(c) If f is convenient *with principal part $f_0 = \sum_{\alpha p+\beta q=d} c_{\alpha\beta} x^\alpha y^\beta$, where $q > p \geq 3$, then*

$$q^2 \leq \gamma^{es}(f) \leq \begin{cases} 3q^2 & \textit{if } \frac{q}{p} \in (1, 2) \\ 2q^2 & \textit{if } \frac{q}{p} \in [2, 4) \\ q^2 & \textit{if } \frac{q}{p} \geq 4. \end{cases}$$

Remark 1.2.28.1 Note that the invariant γ^{es} does not behave upper semicontinuously. Consider, for instance, the deformation $f_t := y^q + x^q + ty^{q-1}$ of the ordinary singularity f_0 with multiplicity $q \geq 6$. As computed above, $\gamma^{es}(f_0) = 2q^2$, while for $t \neq 0$ we have

$$\gamma^{es}(f_t) \geq \frac{i(f_t, x^3)^2}{i(f_t, x^3) - \dim_{\mathbb{C}} \mathbb{C}\{x, y\}/\langle y^{q-2}, x^3\rangle} = 3(q-1)^2 > \gamma^{es}(f_0).$$

Proof of Proposition 1.2.28 *(a).* It suffices to consider the normal forms (see page 145 (Vol. I)):

(a)A_k – singularity: $f = y^{k+1} - x^2$, (b)D_k – singularity: $f = y^{k-1} - yx^2$,
(c)E_6 – singularity: $f = y^4 - x^3$, (d)E_7 – singularity: $f = xy^3 - x^3$,
(e) E_8 – singularity: $f = y^5 - x^3$.

In any case, the equisingularity and Tjurina ideal coincide (see also Fig. 1.5). Let $I_Z \supset I^{ea}(f)$ be a complete intersection ideal (defining the zero-dimensional scheme Z). Since $4\deg(Z) \leq 4\deg(Z^{ea}(f)) = 4k$, which is obviously below the bounds on the right-hand side of (1.2.3.1), it remains to consider power series $g \in I_Z$ with $i(f, g) < 2\deg Z$ and to determine the maximum of

$$\frac{i(f, g)^2}{i(f, g) - \deg Z}. \tag{1.2.3.2}$$

For fixed $\deg Z$, this maximum is attained at the minimal intersection number $i(f, g) > \deg Z$. In particular, (1.2.3.2) does not exceed $(\deg Z + 1)^2 \leq (k+1)^2$. Let's consider the above normal forms:

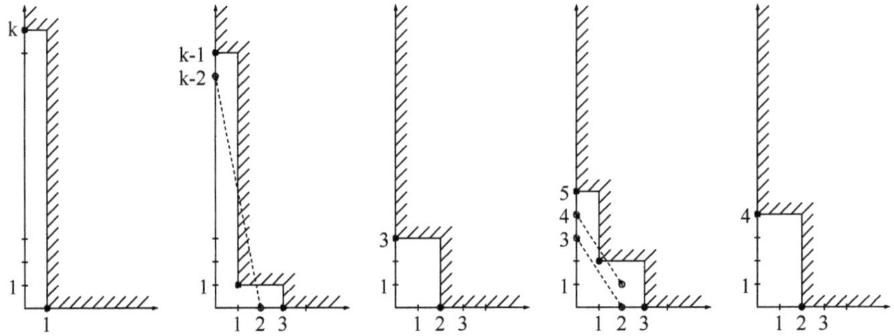

Fig. 1.5 The equisingularity ideals of an A_k-, D_k-, E_6-, E_7-, E_8-singularity, respectively (the filled circles symbolize monomials in $I^{ea}(f)$, the white circles and dashed lines correspond to linear combinations of monomials in $I^{ea}(f)$)

(a) the maximum $(k+1)^2$ is attained for a generic polar, e.g. for $g := x$.

(b) for $\deg Z \le k-2$, the maximum $(k-1)^2$ is attained at $g := x$.
If $\deg Z \ge k-1$, then, necessarily, $g \in I_Z \subset \langle x^2, y^{k-2}, xy \rangle$. In any case, $i = i(f,g)$ exceeds $k+2$, whence $i^2/(i - \deg Z) \le (k+2)^2/2$. On the other hand, the maximum $(k+2)^2/2$ is attained for a generic polar, e.g. for $g := f_x + f_y = (k-1)y^{k-2} - x^2 - 2xy$.

(c)–(e) proceeding as in the case of D_k-singularities, we easily obtain that the maxima are attained for generic polars.
The proof of properties (b) and (c) is given in the next section. □

1.2.3.3 The Slope of the Hilbert–Samuel Function

The proof of Proposition 1.2.28(b), (c) is based on a closer study of the Hilbert–Samuel function of zero-dimensional local rings (see also [GLS6, page 207]).

Definition 1.2.29 Let $A = \mathbb{C}\{x, y\}$ and $I \subset A$ an ideal defining the zero-dimensional scheme $Z \subset (\mathbb{C}^2, 0)$. Then the *Hilbert Samuel function* of A/I is defined to be

$$H^1_{A/I} : \mathbb{Z}_{\ge 0} \longrightarrow \mathbb{Z}_{\ge 0}, \quad d \longmapsto \dim_{\mathbb{C}} \left(\mathbb{C}\{x, y\} / (I + \mathfrak{m}^{d+1}) \right),$$

with $\mathfrak{m} = \langle x, y \rangle$, the maximal ideal of A. Moreover, the *slope of the Hilbert Samuel function* at $d \in \mathbb{Z}_{\ge 0}$ is

$$H^0_{A/I}(d) := H^1_{A/I}(d) - H^1_{A/I}(d-1) = \dim_{\mathbb{C}} \left(\mathfrak{m}^d / ((I \cap \mathfrak{m}^d) + \mathfrak{m}^{d+1}) \right).$$

Setting $H^1_{A/I}(-1) := 0$, the slope has the following, obvious, properties:[7]

[7] See Notations in this chapter and Sect. 3.5 for the global analogue.

(a) if $I \subset J$ then $H^0_{A/I}(d) \geq H^0_{A/J}(d)$ for any $d \in \mathbb{Z}_{\geq 0}$,
(b) $H^0_{A/I}(0) + H^0_{A/I}(1) + H^0_{A/I}(2) + \ldots = \deg Z$,
(c) $H^0_{A/I}(d) = d+1$ for any $0 \leq d < \mathrm{mt}(Z)$,
(d) $H^0_{A/I}(d) \leq H^0_{A/I}(d-1)$ for any $d \geq \mathrm{mt}(Z)$.

Moreover, it has turned out that the Hilbert–Samuel function (and, hence, its slope) depends only on the leading ideal w.r.t. any local degree ordering[8] (see also [GrP, Proposition 5.5.7]):

Lemma 1.2.30 *Let $I \subset A = \mathbb{C}\{x, y\}$ be a zero-dimensional ideal, and let $L(I) \subset A$ denote the leading ideal of I w.r.t. a local degree ordering. Then*

$$H^1_{A/I} = H^1_{A/L(I)}, \qquad H^0_{A/I} = H^0_{A/L(I)}.$$

Proof We have to prove that $\dim_{\mathbb{C}} A/(I + \mathfrak{m}^d) = \dim_{\mathbb{C}} A/(L(I) + \mathfrak{m}^d)$ for each $d \geq 0$. Of course, the set $S := \{\boldsymbol{x}^\alpha \notin L(I) \mid |\alpha| < d\}$ represents a $\mathbb{C}$-basis of $A/(L(I) + \mathfrak{m}^d) \cong \mathbb{C}[x, y]/(L(I) + \mathfrak{m}^d)$. On the other hand, using reduction by a standard basis (w.r.t. the chosen local degree ordering) for I, we can write each $f \in \mathbb{C}[x, y]$ as $f \equiv g + \sum_{\boldsymbol{x}^\alpha \in S} c_\alpha \boldsymbol{x}^\alpha \mod \mathfrak{m}^d$ for some $g \in I$ and uniquely determined complex coefficients c_α. Hence, S also represents a $\mathbb{C}$-basis of $\mathbb{C}[x, y]/(I + \mathfrak{m}^d) \cong A/(I + \mathfrak{m}^d)$. □

Finally, the following lemma provides a lower bound for the minimal number of generators of a zero-dimensional ideal I in terms of the slope of the Hilbert–Samuel function:

Lemma 1.2.31 *Let $I \subset A = \mathbb{C}\{x, y\}$ be an ideal, defining the zero-dimensional scheme $Z \subset (\mathbb{C}^2, 0)$. The minimal number of generators of I, $mng(I)$, satisfies*

$$mng(I) \geq 1 + \sup_{d \geq \mathrm{mt}(Z)} \left(H^0_{A/I}(d-1) - H^0_{A/I}(d)\right).$$

Proof See [Iar1, Theorem 4.3]. □

Together with property (a), the latter can be used to give a bound for the degree of a (local) complete intersection $Z \subset Z(m)$, where $Z(m) \subset (\mathbb{C}^2, 0)$ denotes the ordinary fat point scheme defined by $\mathfrak{m}^m \subset A = \mathbb{C}\{x, y\}$.

Corollary 1.2.32 *Let $Z \subset Z(m)$ be a (local) complete intersection. Then*

$$\deg Z \leq \sum_{i=1}^{\lceil m/2 \rceil} (m - 2i + 2) = \left\lceil \frac{m}{2} \right\rceil \cdot \left(m - \left\lceil \frac{m}{2} \right\rceil + 1\right) \leq \frac{m^2 + 2m + 1}{4}.$$

---

[8]A semigroup ordering on the monomials in $\boldsymbol{x} = (x_1, \ldots, x_n)$ is called a *local degree ordering* if it satisfies $x^\alpha > x^\beta$ for $|\alpha| < |\beta|$.

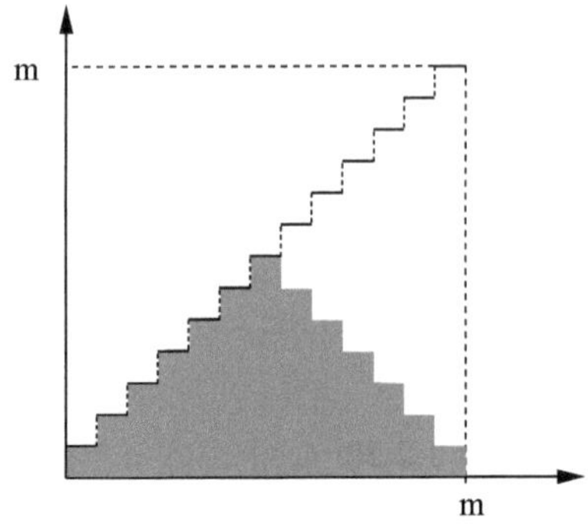

Fig. 1.6 The graph of $H^0_{A/\mathfrak{m}^m}$. The area of the shaded region is the maximal possible value of $\dim_{\mathbb{C}}(A/I_Z)$ for a (local) complete intersection $Z \subset Z(m)$

Proof This is an immediate consequence of Lemma 1.2.31 and the above properties of H^0_{A/I_Z}, since

$$H^0_{A/\mathfrak{m}^m}(d) = \begin{cases} d+1 & \text{if } 0 \le d \le m-1\,, \\ 0 & \text{otherwise} \end{cases}$$

(see also Fig. 1.6). □

Proof of Proposition 1.2.28 *(b).* The cases $m = 2, 3$ being covered by Proposition 1.2.28(a), we restrict ourselves to the case $m \ge 4$. Let $f \in A = \mathbb{C}\{x, y\}$ be a local equation for $(C, z) \subset (\mathbb{P}^2, z)$ and $I_Z \supset I^{es}(f)$ a complete intersection ideal (defining the zero-dimensional scheme $Z \subset Z^{es}(f)$. Moreover, let $g \in I_Z$ be any power series. Recall that $I^{es}(f) = I^{ea}(f) + \mathfrak{m}^m \subset A$ (see [GLS6, Proposition II.2.14]). In particular, Corollary 1.2.32 gives

$$4 \deg Z \le \begin{cases} m^2 + 2m & \text{if } m \text{ is even} \\ m^2 + 2m + 1 & \text{if } m \text{ is odd.} \end{cases} \tag{1.2.3.3}$$

Moreover, the slope of the Hilbert Samuel function satisfies

$$H^0_{A/I_Z}(d) \le H^0_{A/\mathfrak{m}^m}(d) = \begin{cases} d+1 & \text{if } 0 \le d \le m-1\,, \\ 0 & \text{otherwise} \end{cases}$$

and, since $g \in I_Z$, $H^0_{A/I_Z}(d) \le \mathrm{mt}(g)$. On the other hand, I_Z being a complete intersection ideal, Lemma 1.2.31 implies that

$$H^0_{A/I_Z}(d) \le H^0_{A/I_Z}(d+1) + 1 \le m - d$$

(cf. Fig. 1.6). Assume now that $g \in I_Z$ satisfies $i(f, g) < 2 \deg Z$, in particular, $\mathrm{mt}(g) \le (m+1)/2$, due to (1.2.3.3). Then, by the above, we can conclude

$$\deg Z = \sum_{d=0}^{m-1} H^0_{A/I_Z}(d) \le \mathrm{mt}(g) \cdot m - \sum_{i=1}^{\mathrm{mt}(g)-1} 2i = \mathrm{mt}(g) \cdot m - \mathrm{mt}(g)^2 + \mathrm{mt}(g)\,.$$

Again, for fixed $\deg Z$, the maximal value of (1.2.3.2) is attained at the minimal possible intersection number $i(f,g)$. Since $i(f,g) \geq m \cdot \mathrm{mt}(g)$ (see [GLS6, Exercise I.3.2.1]), we obtain

$$\frac{i(f,g)^2}{i(f,g) - \deg Z} \leq \frac{\mathrm{mt}(g)^2 m^2}{\mathrm{mt}(g)^2 - \mathrm{mt}(g)} = \left(1 + \frac{1}{\mathrm{mt}(g)-1}\right) m^2,$$

for $\mathrm{mt}(g) \geq 2$. If $\mathrm{mt}(g) = 1$ then, as both partial derivatives are in I_Z, we get $\deg Z \leq m - 1$, and the left-hand side cannot exceed m^2.

On the other hand, we show that $\gamma^{es}(f) \geq 2m^2$: let

$$f_x := \frac{\partial f}{\partial x} = \prod_{i=1}^{m-1} (\alpha_i x - \beta_i y + \text{higher terms in } x, y) =: \prod_{i=1}^{m-1} g_i$$

(having no common tangent with f, since f is reduced). The complete intersection ideal $I_Z := \langle f_y, g_1 g_2 \rangle$ contains $I^{es}(f)$. Choosing $g := g_1 g_2$, we obtain $i(f,g) = 2m$ and $\deg Zg = 2m-2$, whence $\gamma^{es}(f) \geq 2m^2$. □

For the proof of Proposition 1.2.28(c) we need the following lemma, giving a bound for the intersection multiplicity of two plane curve germs in the case that one of them is given by a convenient semiquasi-homogeneous power series.

Lemma 1.2.33 *Let $f, g \in \mathbb{C}\{x, y\}$ with f a convenient semiquasi-homogeneous power series of $\boldsymbol{w}$-degree d, and $x^\alpha y^\beta$ the leading monomial of g w.r.t. a local $\boldsymbol{w}$-weighted degree ordering*[9] *$<_{\boldsymbol{w}}$. Then the intersection multiplicity $i(f,g)$ of f and g at 0 satisfies $i(f,g) \geq \alpha p + \beta q$.*

Proof Since f and $x^q - y^p$ are topologically equivalent, the Puiseux pairs of their r factors coincide ($r = \gcd(p,q)$). Hence, the branches of the plane curve singularities defined by f have parameterizations

$$x(t) = t^{\tilde{p}}, \quad y(t) = c_j t^{\tilde{q}} + (\text{higher terms in } t),$$

$c_j \neq 0$, $\tilde{p} = p/r$, $\tilde{q} = q/r$. It follows that

$$i(f,g) = \mathrm{ord}\big(g(x(t), y(t))\big) \geq \sum_{j=1}^{r} (\alpha\tilde{p} + \beta\tilde{q}) = \alpha p + \beta q.$$

Proof of Proposition 1.2.28*(c)*. Let $I_Z \subsetneq \mathbb{C}\{x,y\}$ be a complete intersection ideal containing $I^{es}(f)$, and let $0 \neq Z \subset Z^{es}(f)$ be the corresponding zero-dimensional scheme. Then we have the following estimates for $4\deg Z$:

- $4\deg Z \leq q^2 + 2q + 1 \leq 2q^2$, by Corollary 1.2.32.

[9] A semigroup ordering on the monomials in x, y is called a *local $\boldsymbol{w}$-weighted degree ordering* if it satisfies $x^\alpha y^\beta > x^a y^b$ for $w_1\alpha + w_2\beta < w_1 a + w_2 b$, where $\boldsymbol{w} = (w_1, w_2)$.

- $4\deg Z \leq 4\deg Z^{es}(f) \leq 2(p+1)(q+1) < q^2$, as $3 \leq p < \frac{1}{3}q$.

It remains to consider power series $g \in I_Z$ with $i(f,g) < 2\deg Z$.

Let $s := q/p$. Then we have to show that

$$\frac{i(f,g)^2}{i(f,g) - \deg Z} \leq \begin{cases} 3q^2 & \text{if } s \in \big(1,2\big) \\ 2q^2 & \text{if } s \in \big[2,4\big) \\ q^2 & \text{if } s \geq 4. \end{cases} \tag{1.2.3.4}$$

Again, the maximum of the left-hand side is taken at the minimal possible value for $i(f,g)$. Let

$$g = x^\alpha y^\beta + (\text{lower terms w.r.t. } <_{(p,q)}) \in I_Z\,,$$

with $<_{(p,q)}$ a local (p,q)-weighted degree ordering. Then, by Lemma 1.2.33 (and since the denominator is at least 1), the left-hand side of (1.2.3.4) does not exceed

$$\min\left\{(\alpha p + \beta q)^2,\ \frac{(\alpha p + \beta q)^2}{(\alpha p + \beta q) - \deg Z}\right\}. \tag{1.2.3.5}$$

Let's consider more closely the possible situations. Since the intersection multiplicity $i(f,g)$ is chosen minimally for $g \in I_Z \supset I^{es}(f)$, it is clear that $\alpha p + \beta q \leq pq$. Moreover, $\beta \leq \frac{q+1}{2}$, since otherwise

$$i(f,g) \geq \beta q \geq \left(\left\lfloor\frac{(q+1)}{2}\right\rfloor + 1\right)\cdot q \geq \left\lfloor\frac{q^2+2q+1}{2}\right\rfloor \geq 2\deg Z\,,$$

the latest inequality again being implied by Corollary 1.2.32.

Case 1. $\alpha = 0,\ \beta \geq 1$. If $\beta = 1$ then (1.2.3.5) is obviously bounded by q^2. Hence, we can assume $\beta \geq 2$. Since I_Z is a complete intersection ideal with $I_Z \supset I^{es}(f) \supset \mathfrak{m}^q$ and since the multiplicity of I_Z is at most $\mathrm{mt}(g) = \beta$, the above properties of the slope of the Hilbert–Samuel function imply immediately (cf. the proof of Corollary 1.2.32)

$$\deg Z \;=\; \dim_{\mathbb{C}} \mathbb{C}\{x,y\}\big/I_Z \;\leq\; \beta q - \beta^2 + \beta\,,$$

and it follows that (1.2.3.5) does not exceed $2q^2$.

If $s \geq 3$ then we can improve this bound (by counting the monomials in the shaded region of Fig. 1.7): in any case

$$\deg Z \leq \beta q - 1 - \lfloor s \rfloor \cdot \frac{\beta(\beta-1)}{2}\,.$$

In particular, (1.2.3.5) is bounded by q^2 if $s \geq 3$.

Case 2. $\alpha = 1,\ \beta \geq 1$. In this case, counting the monomials in the shaded region of Fig. 1.8 gives the estimate

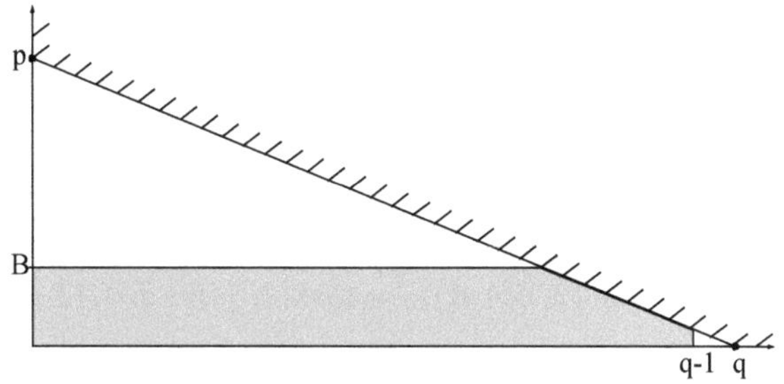

Fig. 1.7 The ideal $I := I^{es}_{fix}(f) + \langle \frac{\partial f}{\partial x} \rangle + \langle y^\beta \rangle$. The monomials inside the shaded area generate $\mathbb{C}\{x,y\}/I$

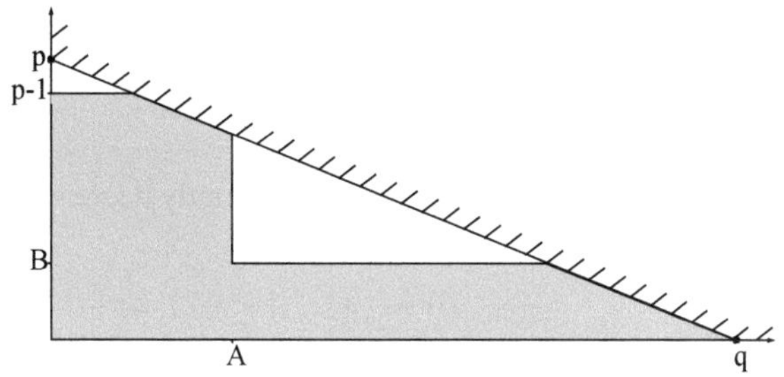

Fig. 1.8 The ideal $I := I^{es}_{fix}(f) + \langle \frac{\partial f}{\partial y} \rangle + \langle x^\alpha y^\beta \rangle$. The monomials inside the shaded area generate $\mathbb{C}\{x,y\}/I$

$$\begin{aligned}\deg Z &\le \dim_{\mathbb{C}} \mathbb{C}\{x,y\}\big/\big(I^{es}_{fix}(f) + \langle y^{p-1}, xy^\beta\rangle\big)\\ &\le p - 1 + \beta(q-1) - \lfloor s \rfloor \frac{\beta(\beta-1)}{2}\,.\end{aligned}$$

It follows that (1.2.3.5) is bounded by

$$\frac{\left(\frac{1}{s}+\beta\right)^2}{\beta + \lfloor s \rfloor \frac{\beta^2-\beta}{2} + 1} \cdot q^2 \le \begin{cases} \dfrac{2(\beta+1)^2}{\beta^2+\beta+2} q^2 \le \dfrac{16}{7} q^2 & \text{if } 1 \le s < 2 \\ \dfrac{\left(\beta+\frac{1}{2}\right)^2}{\beta^2+1} q^2 \le \dfrac{5}{4} q^2 & \text{if } 2 \le s < 3 \\ \dfrac{2\left(\beta+\frac{1}{3}\right)^2}{3\beta^2-\beta+2} q^2 < q^2 & \text{if } 3 \le s \end{cases}$$

Case 3. $\alpha \ge 2,\ \beta \ge 1$. We can estimate (cf. Fig. 1.8)

$$\begin{aligned}\deg Z &\le \dim_{\mathbb{C}} \mathbb{C}\{x,y\}\big/\big(I^{es}_{fix}(f) + \langle y^{p-1}, x^\alpha y^\beta\rangle\big)\\ &\le \left(\alpha p - \sum_{i=1}^{\alpha-1} \left\lfloor \frac{i}{s} \right\rfloor - \min\left\{\lceil s \rceil, \alpha\right\}\right) + \left(\beta(q-\alpha) - \sum_{j=1}^{\beta-1} \lfloor js \rfloor\right)\\ &\le \alpha p + \beta q - \left(\alpha\beta + \sum_{i=1}^{\alpha-1}\left(\frac{i}{s}-1\right) + \sum_{j=1}^{\beta-1}(js-1) + \min\left\{\lceil s \rceil, \alpha\right\}\right)\end{aligned}$$

In any case, the left-hand side of (1.2.3.5) does not exceed $\varphi(\alpha,\beta)\cdot q^2$, where

$$\varphi(\alpha,\beta) := \frac{(\frac{\alpha}{s}+\beta)^2}{(\alpha-1)(\beta-1)+\frac{\alpha(\alpha-1)}{2s}+s\cdot\frac{\beta(\beta-1)}{2}+3}.$$

An immediate computation shows that, for fixed $\beta \geq 1$, the function φ takes its maximum on $[2,\infty)$ at $\alpha_0 = \max\{2,\ (16-3\beta)/(2+\frac{1}{s})\}$. Hence,

$$\varphi(\alpha,\beta) \leq \varphi(\alpha_0,\beta) = \begin{cases} \dfrac{(2+\beta s)^2}{s\left(\frac{\beta^2-\beta}{2}s^2+(\beta+2)s+1\right)} & \text{if } \alpha_0 = 2\,, \\ \dfrac{2\beta s-2\beta+16}{(\beta-1)s^2+(7-\beta)s-\frac{1}{4}} & \text{if } \alpha_0 = \dfrac{16-3\beta}{(2+\frac{1}{s})}\,. \end{cases}$$

If $\beta \geq 4$ then $\alpha_0 = 2$. But $\varphi(2,\beta)$ is strictly decreasing for $\beta \geq 4$, $s \geq 1$, and we get

$$\varphi(\alpha_0,\beta) = \varphi(2,\beta) \leq \varphi(2,4) = \frac{4(1+2s)^2}{s\left(6s^2+6s+1\right)}.$$

On the other hand, for $\beta = 1, 2, 3$, we get

$$\varphi(\alpha_0,1) = \frac{s+7}{3s-\frac{1}{8}}, \quad \varphi(\alpha_0,2) = \frac{4(s+3)}{s^2+5s-\frac{1}{4}}, \quad \varphi(\alpha_0,3) = \frac{3s+5}{s^2+2s-\frac{1}{8}}.$$

Since each of these bounds is strictly decreasing for $s \in (1,\infty)$, we can estimate:

	$\varphi(\alpha_0,1)$	$\varphi(\alpha_0,2)$	$\varphi(\alpha_0,3)$	$\varphi(2,\beta)$ for $\beta \geq 4$
if $s \in (1,2)$	$< \frac{64}{23}$	$< \frac{64}{23}$	$< \frac{64}{23}$	$< \frac{36}{13}$
if $s \in [2,4)$	$\leq \frac{72}{47}$	$\leq \frac{16}{11}$	$\leq \frac{88}{63}$	$\leq \frac{50}{37}$
if $s \geq 4$	$\leq \frac{88}{95}$	$\leq \frac{112}{123}$	$\leq \frac{136}{191}$	$\leq \frac{81}{121}$

Case 4. $\alpha \geq 1,\ \beta = 0$. We can estimate (cf. Fig. 1.9, respectively Case 2)

$$\begin{aligned} \deg Z &= \dim_{\mathbb{C}} \mathbb{C}\{x,y\}/\langle I^{es}_{fix}(f)+\langle y^{p-1},x^{\alpha}\rangle\rangle \qquad (1.2.3.6)\\ &\leq \min\left\{\alpha(p-1),\ \alpha p - \sum_{i=1}^{\alpha-1}\left\lfloor\frac{i}{s}\right\rfloor - \lceil s\rceil\right\}. \end{aligned}$$

In any case, the first estimate implies that (1.2.3.5) is bounded by $\alpha p^2 = \frac{\alpha}{s^2}q^2$. Hence, in the following it suffices to consider the following cases: either ($s \in (1,2)$ and $\alpha > 3s^2$), or ($s \in [2,4)$ and $\alpha > 2s^2$), or ($s \geq 4$ and $\alpha > s^2$). The second estimate bounds (1.2.3.5) by $g(\alpha)\cdot q^2$, where

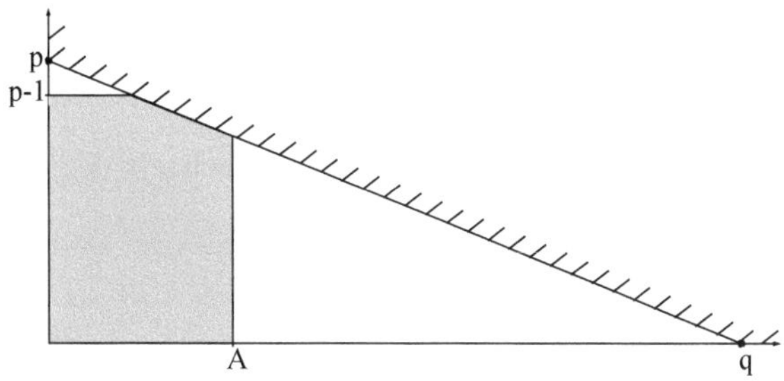

Fig. 1.9 The ideal $I := I^{es}_{fix}(f) + \langle \frac{\partial f}{\partial y} \rangle + \langle x^\alpha \rangle$. The monomials inside the shaded area form a basis for $\mathbb{C}\{x, y\}/I$

$$g(\alpha) := \frac{\alpha^2}{s^2\left(\frac{\alpha(\alpha-1)}{2s} - (\alpha-1) + (s+1)\right)}.$$

Note that the denominator of g exceeds $\frac{1}{B}\alpha^2$ and hence, $g(\alpha) \leq B$, iff

$$\left(\frac{s}{2} - \frac{1}{B}\right) \cdot \alpha^2 - \left(s^2 + \frac{s}{2}\right) \cdot \alpha + (s^3 + 2s^2) \geq 0. \qquad (1.2.3.7)$$

This holds in almost all the cases to be considered:

- $s \geq 4$, $\alpha > s^2$ and $B = 1$;
- $s \in [2, 4)$, $\alpha > 2s^2$ and $B = 2$;
- $s \in (1, 2)$, $\alpha > 3s^2$ and $B = 3$, if we assume that $\alpha \geq 6$.

In all these cases the left-hand side of (1.2.3.7) is strictly increasing in α. In particular, it exceeds the value at $\alpha = s^2, 2s^2, 3s^2$, respectively 6, all of them being positive.

Thus, finally, we are left with the cases $\alpha \in \{4, 5\}$. In both cases the estimate (1.2.3.6) for $\deg Z$ (which yields $4p - 5$, respectively $5p - 8$) in general is not good enough. More precisely, it is sufficiently good iff $s \geq \sqrt{16/15}$, respectively iff $s \geq \sqrt{25/24}$, in particular, for $s \geq 5/4$.

Let us assume $1 < s < 5/4$. Then it is clear that the monomial x^4, respectively x^5, is not only of minimal (p, q)-weighted degree but also of minimal degree among the monomials appearing in g.

Hence, when considering any local degree ordering $<$ instead of the weighted degree ordering $<_{(p,q)}$, we obtain $L(I) \supset I' := \langle x^4, y^{p-1}, y^{p-2}x^3 \rangle$, respectively $L(I) \supset I' := \langle x^5, y^{p-1}, y^{p-2}x^3, y^{p-3}x^4 \rangle$. By Lemma 1.2.30 and property (a) of the slope of the Hilbert–Samuel function (see p. 98), this implies that we have in both cases

$$H^0_{A/I}(d) = H^0_{A/L(I)}(d) \leq H^0_{A/I'}(d) = \begin{cases} \alpha - 2 & \text{if } d = p, \\ 0 & \text{if } d \geq p + 1. \end{cases}$$

Since I is a complete intersection, we can deduce from Lemma 1.2.31 that the inequality for $d = p$ is strict. Thus,

$$\deg(Z) = \sum_{d \geq 0} H^0_{A/I}(d) < \sum_{d \geq 0} H^0_{A/I'}(d) = \dim_{\mathbb{C}}(A/I') = \begin{cases} 4p-5 & \text{if } \alpha = 4\,, \\ 5p-8 & \text{if } \alpha = 5\,. \end{cases}$$

Hence, we get $\deg(Z) \leq 4p-6$, respectively $\deg(Z) \leq 5p-9$, which is sufficient to bound (1.2.3.5) by $3q^2$. □

1.3 Historical Notes and References

We give some historical remarks and references to the different sections of this chapter.

The material of Sect. 1.1.1 is very classical and goes back to the Italian geometers Enriques and Chisini (cf. [EnC]). The notions of cluster, proximity, unloading, etc. originated from their work. We like to mention that in [EnC] the authors introduce another graph, now called *Enriques diagram*, with straight and curved edges, which also keeps record of the proximity relations. For a comprehensive modern treatment clarifying obscure definitions and statements in older sources, which influenced our presentation, see [Cas1, Cas2] and [CGSL] (see also [KPT, Section 2]). In [Pop] the author introduces a "kite" of a constellation of infinitely near points, which embeds both the Enriques diagram and the dual graph of the divisor obtained by blowing up the constellation, allowing to 'see' their relation.

The notion of equisingularity for plane curve singularities used in Sect. 1.1.2 is due to Zariski, who developed the theory in three articles in [Zar2] with different but equivalent definitions. One of these definitions was by successive blowing up the singularities and requiring for two equisingular curve singularities the same multiplicities of the reduced total transform in the resolution process. This definition is used for equisingular families of plane curve singularities in Definition 1.1.27. The same definition was also used by Wahl in [Wah1], who first studied families of algebroid plane curve singularities over algebraically closed fields of characteristic 0 and proved that the equisingularity stratum (μ-constant stratum) in the formal semiuniversal deformation is smooth. Equisingular deformations in the complex analytic setting were studied in [GLS6], where a simpler proof of the smoothness of the μ-constant stratum was given. This smoothness is essential for our approach to equisingular families of global plane curves in Chap. 4. Equisingular deformations of algebroid plane curve singularities in positive characteristic have been studied in [CGL1] and [CGL2]. The investigation of equisingular families of global plane curves in positive characteristic is widely open, see Problem 4.7.7.

The systematic study of equisingular families through zero-dimensional schemes and cohomological vanishing conditions for their ideal sheaves has been developed by the authors in several papers, in particular in [GrK], [GrL1], [GLS3]. In Sect. 1.1.2 we define zero-dimensional schemes, concentrated in a point z in a smooth surface Σ, by their ideal in the analytic local ring $\mathcal{O}_{\Sigma,z}$, related to equisingular deformations of a plane curve singularity $(C, z) \subset (\Sigma, z)$. The different zero-dimensional ideals describe tangent spaces to different classes of deformations of (C, z). The equisingularity ideal I^{es} of Wahl (Definition 1.1.63) is the most basic, as it describes the tangent space to equisingular deformations of (C, z), but it is also the most abstract one. The topological singularity ideal $I^s \subset I^{es}$ (Definition 1.1.31) is important, as it is better suited for numerical estimates guaranteeing H^1-vanishing results as well as for certain theoretical purposes. Further zero-dimensional schemes are related to deformations fixing other invariants (e.g. the conductor ideal is the tangent space to deformations fixing the conductor).

We study in Sect. 1.1.2.4 what we call straight equisingular deformations and determine the base space of their seminuniversal deformation in Proposition 1.1.37 using the ideal I^s, which describes the tangent space of such deformations. The straight es-deformations form a linear subspace (w.r.t. suitable coordinates) of the μ-constant stratum and had been introduced by Wahl [Wah1].

In Sect. 1.1.3 we introduce the Hilbert scheme of a cluster scheme. It is a subscheme of the punctual Hilbert scheme on a smooth surface, which has been intensively studied by Fogarty, Hartshorne, Iarrobino, and Briançon [Fog, HaR1, Iar1, Iar2, Bri]. To be able to use their results, we identify the (algebraically defined) Hilbert scheme with the (analytically defined) Douady space for closed subvarieties in a projective variety (cf. Sect. 2.2.2.1, Proposition 2.2.28).

The punctual Hilbert scheme $\mathcal{H}_0^s(S)$ associated to a topological type S and parameterizing the schemes $Z^s(C, z)$ defined by the ideal I^s, had been introduced by the authors in connection with the irreducibility problem for equisingular families of plane projective curves. We prove in Theorem 1.1.57 that $\mathcal{H}_0^s(S)$ is irreducible, a fact that was already proved by Lossen in his Ph.D. thesis [Los], using ideas from [NoV, Ris1]. $\mathcal{H}_0^s(S)$ is also used to prove in Proposition 1.1.61 that the set $V_d(S)$ of reduced curves $C \subset \mathbb{P}^2$ of degree d having exactly one singular point of topological type S is a quasi-projective variety.

It is interesting to notice that straight es-deformations appear again in connection with $\mathcal{H}_0^s(S)$. Namely, there is a natural morphism $B_{C,z}^{es,fix} \to \mathcal{H}_0^s(S)$ (Remark 1.1.64.2), where $B_{C,z}^{es,fix}$ denotes the base space of the semiuniversal es-deformation of (C, z) along the trivial section, with fibres being the straight equisingular deformations of (C, z) and thus providing a nice formula for the dimension of $\mathcal{H}_0^s(S)$.

We treat analytic types of isolated hypersurface singularities of arbitrary dimension in Sect. 1.1.4 and introduce the ideal I^a, an analog of the ideal I^s for topological types. The punctual Hilbert scheme $\mathcal{H}_0^a(S)$ associated to an analytic type S is used (similar as for a topological type) to prove that the set of reduced curves in P^2 of degree d having exactly one singular point of analytic type S is a quasi-projective variety. The results about quasi-projectivity are used later to prove in general that the set $V_h(S_1, \cdots, S_r)$ of hypersurface in a smooth projective variety Σ with fixed Hilbert polynomial h and r singularities of analytic or topological types $S_1, \cdots, S_r$ is quasi-projective (Theorems 2.2.32 and 2.2.36), a fact that was known before only in special cases (see also Conjecture 4.7.1).

Section 1.2 is devoted to singularity invariants with a rather restricted application range. One of the main goals of our study is to provide sufficient conditions, expressed in terms of singularities and certain discrete characteristics of curves, for the non-emptiness and smoothness of equisingular families of curves and the versality of deformations of curve singularities induced by linear systems of curves on surfaces. The non-classical invariants discussed in Sect. 1.2 were designed just for this purpose.

The finite determinacy is one of the key issues in singularity theory (see e.g. [Arn, AGV, Tou, Mat, Gaf, Wal, Dam, BDW, GLS6]). The explicit bound $\mu + 1$ for the determinacy level of contact and right analytic equivalence has been known for a while, and is attributed to Tougeron [Tou]. In turn, a similar bound for topological equivalence (see Proposition 1.2.8) has been proposed in [Lic], though generally incorrect. The correct form implicitly appeared in [GLS3, Lemma 2.19].

The β-invariant and the isomorphism defect both came from an attempt to deduce the aforementioned sufficient conditions from the H^1-vanishing criteria for ideal sheaves of zero-dimensional singularity schemes which we call "Riemann–Roch" criteria (see Sect. 3.1 below). The β-invariant has been introduced in [Shu4] (originally published in Russian in 1983) for arbitrary topological singularities. An improvement (i.e., smaller invariants) for A_k-singularities has been found in [Shu2]. The isomorphism defect in connection with the contact analytic singularity types has been invented in [GrK] (called there the "local index"). Further development, including an extension to topological singularity types and explicit estimates, has been performed in [GrL1]. The γ-invariant was suggested and estimated in [GLS4] (improved bounds were found in [Los1]). Though the γ-invariant looks a bit artificial, the results of [GLS4] show that this invariant precisely fits the asymptotically proper sufficient conditions for the T-smoothness of equisingular families of plane curves with arbitrary analytic or topological singularities. (see Sects. 4.3.2 and 4.3.3 for details).

References

[GLS6] Greuel, G.-M., Lossen, C., Shustin, E.: Introduction to Singularities and Deformations. Springer, Berlin (2007)

[EnC] Enriques, F., Chisini, O.: Lezioni sulla teoria geometrica delle equazioni e delle funzioni algebriche. Libro IV (1915)

[Wae1] van der Waerden, B.L.: Infinitely near points. Ind. Math. **12**, 401–410 (1950)

[Cas2] Casas-Alvero, E.: Singularities of Plane Curves. LMS, vol. 276. Cambridge University Press, Cambridge (2000)

[BHPV] Barth, W., Hulek, K., Peters, C., van de Ven, A.: Compact Complex Surfaces. Springer, Berlin (1984)

[HaR2] Hartshorne, R.: Algebraic Geometry. Graduate Text in Mathematics, vol. 52. Springer, Berlin (1977)

[Dra] Draper, R.N.: Intersection theory in analytic geometry. Math. Ann. **180**, 175–204 (1969)

[Cas1] Casas-Alvero, E.: Infinitely near imposed singularities and singularities of polar curves. Math. Ann. **287**, 429–454 (1990)

[CGSL] Campillo, A., Gonzalez-Sprinberg, G., Lejeune-Jalabert, M.: Clusters of infinitely near points. Math. Ann. **306**, 169–194 (1996)

[Lip] Lipman, J.: Proximity inequalities for complete ideals in regular two-dimensional local rings. Contemp. Math. **159**, 293–306 (1994)

[NoV] Nobile, A., Villamayor, O.E.: Equisingular stratifications associated to families of planar ideals. J. Algebr. **193**(1), 239–259 (1997)

[Bra] Brauner, K.: Zur Geometrie der Funktionen zweier Veränderlicher. Abh. Math. Sem. Hamburg **6**, 8–54 (1928)

[Zar3] Zariski, O.: Algebraic Surfaces, 2nd edn. Springer, Berlin (1971)

[BrK] Brieskorn, E., Knörrer, H.: Plane Algebraic Curves. Birkhäuser, Basel (1986)

[Zar1] Zariski, O.: Polynomial ideals defined by infinitely near base points. Am. J. Math. **60**, 151–204 (1938)

[GrP] Greuel, G.-M. Pfister, G.: A SINGULAR Introduction to Commutative Algebra. With contributions by Olaf Bachmann, Christoph Lossen and Hans Schnemann. 2nd extended ed. Springer, Berlin (2007)

[DeL] Decker, W., Lossen, C.: Computing in Algebraic Geometry. A Quick Start Using SINGULAR. ACM, vol. 16. Springer, Berlin (2006)

[Wah1] Wahl, J.: Equisingular deformations of plane algebroid curves. Trans. Am. Math. Soc. **193**, 143–170 (1974)

[Mil] Milnor, J.: Singular Points of Complex Hypersurfaces. Annals of Mathematics Studies, vol. 61. Princeton University Press, Princeton (1968)

[HiH] Hironaka, H.: On the arithmetic genera and the effective genera of algebraic curves. Mem. Coll. Sci. Univ. Kyoto **30**, 177–195 (1957)

[HaR1] Hartshorne, R.: Connectedness of the Hilbert scheme. Publ. Math. IHES **29**, 261–304 (1966)

[Fog] Fogarty, J.: Algebraic families on an algebraic surface. Am. J. Math. **90**, 511–521 (1968)

[Iar2] Iarrobino, A.: Hilbert scheme of points: overview of last ten years. In: Proceedings of the Summer Research Institute on Algebraic Geometry, Brunswick/Maine 1985, Part 2, Proceedings of Symposia in Pure Mathematics, vol. 46, pp. 297–320 (1987)

[Dou2] Douady, A.: Le problème des modules locaux pour les espaces $\mathbb{C}$-analytiques compactes. Ann. Sci. ENS **7**, 569–602 (1974)

[Dou1] Douady, A.: Le problème des modules pour les sous-espaces analytiques compacts d'une espace analytique donné. Ann. Inst. Fourier **16**, 1–95 (1966)

[Bin] Bingener, J.: Darstellbarkeitskriterien für analytische Funktoren. Ann. Sci. École Norm. Sup. **13**, 317–347 (1980)

[Mum2] Mumford, D.: Geometric Invariant Theory. Springer, Berlin (1965)

[Bri] Briançon, J.: Description de $\mathrm{Hilb}^n\mathbb{C}\{x, y\}$. Invent. Math. **41**, 45–89 (1977)

[Ris1] Risler, J.-J.: Sur les déformations équisingulières d'idéaux. Bull. Soc. Math. Fr. **101**, 3–16 (1973)

[Ser2] Serre, J.-P.: Algèbre Locale. Multiplicités. SLNM, vol. 11. Springer, Berlin (1965)

[AK] Altman, A., Kleiman, S.: Introduction to Grothendieck Duality Theory. Lecture Notes in Mathematics, vol. 146. Springer, Berlin (1970)

[CGL1] Campillo, A., Greuel, G.-M., Lossen, C.: Equisingular deformations of plane curves in arbitrary characteristic. Compos. Math. **143**, 829–882 (2007)

[BiF] Bingener, J., Flenner, H.: On the fibers of analytic mappings. In: Ancona, V., et al. (eds.) Complex Analysis and Geometry. The University Series in Mathematics, pp. 45–101. Plenum Press, New York (1993)

[MaY] Mather, J.N., Yau, S.-T.: Classification of isolated hypersurface singularities by their moduli algebras. Invent. Math. **69**, 243–251 (1982)

[Lic] Lichtin, B.: Estimates and formulae for the C^0 degree of sufficiency of plane curves. In: Singularities, Part 2 (Arcata 1981). Proceedings of Symposia in Pure Mathematics, vol. 40, pp. 155–160 (1983)

[DiH] Diaz, S., Harris, J.: Ideals associated to deformations of singular plane curves. Trans. Am. Math. Soc. **309**, 433–468 (1988)

[Shu10] Shustin, E.: Smoothness of equisingular families of plane algebraic curves. Int. Math. Res. Not. **2**, 67–82 (1997)

[Sai] Saito, K.: Quasihomogene isolierte Singularitäten von Hyperflächen. Invent. Math. **14**, 123–142 (1971)

[GuS] Gudkov, D.A., Shustin, E.: On the intersection of the close algebraic curves. In: Dold, A., Eckmann, B. (eds.) Topology (Leningrad, 1982). SLN, vol. 1060, pp. 278–289 (1984)

[Iar1] Iarrobino, A.: Punctual Hilbert Schemes. Memoirs of the American Mathematical Society, No. 188 (1977)

[KPT] Kleiman, S., Piene, R., Tyomkin, I.: Enriques diagrams, arbitrarily near points, and Hilbert schemes. Atti Accad. Naz. Lincei, Cl. Sci. Fis. Mat. Nat., IX. Ser., Rend. Lincei, Mat. Appl. **22**(4), 411–451 (2011)

[Pop] Popescu-Pampu, P.: Le cerf-volant dune constellation. Enseign. Math. (2) **57**, 303–347 (2011)

[Zar2] Zariski, O.: Studies in equisingularity I–III. Am. J. Math. **87**, 507–536; 972–1006 (1965), respectively Am. J. Math. **90**, 961–1023 (1968)

[CGL2] Campillo, A., Greuel, G.-M., Lossen, C.: Equisingular calculations for plane curve singularities. J. Symb. Comput. **42**(1–2), 89–114 (2007)

[GrK] Greuel, G.-M., Karras, U.: Families of varieties with prescribed singularities. Compos. Math. **69**, 83–110 (1989)

[GrL1] Greuel, G.-M., Lossen, C.: Equianalytic and equisingular families of curves on surfaces. Manuscr. Math. **91**, 323–342 (1996)

[GLS3] Greuel, G.-M., Lossen, C., Shustin, E.: Plane curves of minimal degree with prescribed singularities. Invent. Math. **133**, 539–580 (1998)

[Los] Lossen, C.: The geometry of equisingular and equianalytic families of curves on a surface. Ph.D. thesis, Universitt Kaiserslautern (1998)

[Arn] Arnold, V.I.: Singularities of smooth mappings. (Russian) Uspehi Mat. Nauk **23**(1), 3–44 (1968)

[AGV] Arnol'd, V.I., Gusein-Zade, S.M., Varchenko, A.N.: Singularities of Differentiable Maps, vol. I. Birkhäuser, Basel (1985)

[Tou] Tougeron, J.C.: Idéaux de fonctions différentiables. Ann. Inst. Fourier (Grenoble) **18**(1), 177–240 (1968)

[Mat] Mather, J.N.: Stability of C^∞ mappings. III. Finitely determined map-germs. Publ. Math. Inst. Hautes Études Sci. **35**, 279–308 (1968)

[Gaf] Gaffney, T.: A note on the order of determination of a finitely determined germ. Invent. Math. **52**(2), 127–130 (1979)

[Wal] Wall, C.T.C.: Finite determinacy of smooth map-germs. Bull. Lond. Math. Soc. **13**(6), 481–539 (1981)

[Dam] Damon, J.: The unfolding and determinacy theorems for subgroups of A and K. Mem. Am. Math. Soc. **50**(306), x+88 pp. (1984)

[BDW] Bruce, J.W., du Plessis, A.A., Wall, C.T.C.: Determinacy and unipotency. Invent. Math. **88**(3), 521–554 (1987)

[Shu4] Shustin, E.: On manifolds of singular algebraic curves. Selecta Math. Sov. **10**, 27–37 (1991)

[Shu2] Shustin, E.: New M-curve of the 8th degree. Math. Notes Acad. Sci. USSR **42**, 606–610 (1987)

[GLS4] Greuel, G.-M., Lossen, C., Shustin, E.: Castelnuovo function, zero-dimensional schemes and singular plane curves. J. Algebr. Geom. **9**(4), 663–710 (2000)

[Los1] Lossen, C.: New asymptotics for the existence of plane curves with prescribed singularities. Commun. Algebr. **27**, 3263–3282 (1999)

Chapter 2
Global Deformation Theory

Global deformation theory serves as a key tool in the study of families of singular algebraic varieties, notably, equisingular families of algebraic curves, the main object of this monograph.

The chapter starts with the introductory section, "Classical global theorems", which, on one hand, makes the exposition self-contained and helps an innocent reader to proceed further and, on the other hand, presents the classical material in a form convenient for our use in the advanced part of the monograph.

In the next section "Families of complex spaces", we introduce various global deformation functors (equisingular, equianalytic, with fixed position of singular points etc.), link them with the Hilbert schemes of curves on surfaces and with the T-smoothness property of families of curves on surfaces (or, more generally, of hypersurfaces in smooth projective varieties). We then compute their tangent spaces and obstructions which finally are expressed via the zero and first cohomology of ideal sheaves of appropriate zero-dimensional schemes on curves and surfaces (Theorems 2.2.38 and 2.2.40). These expressions serve as the key ingredient in our study and are explored further in Chaps. 3 and 4.

In the section "Construction via deformation", we set forth the patchworking construction of hypersurfaces with prescribed isolated singularities. It appears to be the most powerful tool for obtaining examples of curves with a given collection of singularities and it is intensively used in the proof of the existence results in Sect. 4.5. We point out the following important feature of the patchworking construction. In general, to control simultaneous local deformations of all singularities of a curve or hypersurface induced by a given linear system, one demands the joint versatility of the induced global deformation, which is equivalent to the smoothness and expected dimension of the analytic equisingular family of curves (hypersurfaces) in the given linear system. In turn, the patchworking construction provides a large set of controllable simultaneous local deformations (described below via deformation patterns) which require rather less restrictive sufficient conditions, equivalent to the smoothness and expected dimension of the topological equisingular family of curves (hypersurfaces) in the given linear system.

G.-M. Greuel et al., *Singular Algebraic Curves*, Springer Monographs in Mathematics, https://doi.org/10.1007/978-3-030-03350-7_2

Our version of the patchworking grew up from the Viro method invented in 1979–1980 for constructing real nonsingular algebraic varieties with prescribed topology. The main improvement compared to Viro's original method is that the required transversality conditions are purely local and, therefore, we can use the implicit function theorem which gives a lot of additional freedom. A short account from our point of view of the original Viro construction is given in Sect. 2.4, "Appendix" to this chapter. Appendix A at the end of this book presents Viro's original paper "Patchworking real algebraic varieties" which provides all the details of his method.

2.1 Classical Global Theorems

The classical theorems by Bézout, Bertini, Plücker, Noether are often referred to in the further exposition. They refer to algebraic schemes rather than to complex spaces. For the reader's convenience we recall them in the form relevant for our needs and provide proofs or give references, following mainly [FOV, HaR2, Uen, Wae2].

In this section (Sect. 2.1) let Y denote an algebraic scheme.[1] Since Y and its analytification coincide as sets (points are closed points) we may think of Y as a (not necessarily reduced or irreducible) "algebraic variety" with the Zariski topology and the algebraic structure sheaf, which we denote by $\mathcal{O}_Y$ in this section. Y can be covered by finitely many affine open subsets. The *affine coordinate ring* $K[Y]$ of an affine scheme Y is isomorphic to $\mathbb{C}[x_1, \ldots, x_n]/I$ for some n and some ideal I. We have $\Gamma(U, \mathcal{O}_Y) = K[U]$ for U open and affine and the local ring $\mathcal{O}_{Y,y}$ is the localization of $K[U]$ at the maximal ideal $\mathfrak{m}_y$ defining $y \in U$. If $Y \subset \mathbb{P}^n$ is projective, then $K[Y]$ denotes the *homogeneous coordinate ring*, isomorphic to $\mathbb{C}[x_0, \ldots, x_n]/I$ with I a homogeneous ideal. $K[Y]$ is then graded where the homogeneous elements of degree d are represented by homogeneous polynomials of degree d in $\mathbb{C}[x_0, \ldots, x_n]$.

If Y is reduced then, for $U \subset Y$ open, $\Gamma(U, \mathcal{O}_Y)$ can be identified with the set of regular functions on U. If Y is affine (resp. projective), a function $f : U \to \mathbb{C}$ is *regular at* $y \in U$, if there exist $p, q \in K[Y]$ (resp. with p and q homogeneous of the same degree) and a open neighbourhood $V \subset U$ of y, such that $q(y') \neq 0$ and $f(y') = p(y')/q(y')$ for $y' \in V$. f is a *regular function* on Y if it is regular at every point of Y. Then $\mathcal{O}_Y$ is the sheaf of regular functions on Y.

Y is a *curve* resp. *surface* if it is of pure dimension one, resp. two. We say that a property *holds generically* for points $p_1, \ldots, p_r$ on Y, or that the points are *generic*, or *in generic position*, if there exists a (Zariski) open, dense subset U in the Cartesian product Y^r (equivalently in the symmetric product $\mathrm{Sym}^r Y$) such that the property holds for all $(p_1, \ldots, p_r) \in U$. The subset U depends on the property and is in general not further specified.

---

[1]Recall that an algebraic scheme means a separated scheme of finite type over $\mathbb{C}$ and an algebraic variety denotes an algebraic complex space, i.e. the analytification of an algebraic scheme (cf. Notations and Conventions).

2.1.1 Divisors and Linear Systems

Let us recall the notions of divisors and linear systems and prove some of the basic properties. Roughly speaking, a divisor on a smooth algebraic scheme is locally given by the union of the zero set and the pole set of a non-zero rational function, together with multiplicities. A divisor with non-negative multiplicities is called an effective divisor. It is a codimension one subscheme (if non-empty), locally defined by one regular function (also called a hypersurface). A linear system is a family of effective divisors parameterized by a linear space. Linear systems possess some properties not shared by arbitrary algebraic families. It is a classical geometric concept that is nowadays best described by using sheaf theory.

Let Y be an algebraic scheme and $\mathcal{K}_Y$ the sheaf of rational functions on Y. $\mathcal{K}_Y$ is the sheaf associated to the presheaf which associates to an open set $U \subset Y$ the set $\mathcal{K}_Y(U) := \{p/q \mid p, q \in \Gamma(U, \mathcal{O}_Y), q$ a non-zero divisor in $\mathcal{O}_{Y,y}$ for each $y \in U\}$, i.e. the localisation of $\Gamma(U, \mathcal{O}_Y)$ w.r.t the multiplicative set of elements in $\Gamma(U, \mathcal{O}_Y)$ that are locally non-zero divisors on U. If U is affine, then $\mathcal{K}_Y(U) = Q(K[U]) = \{p/q \mid p, q \in K[U], q$ a non-zero divisor of $K[U]\}$, the *total quotient ring* of the affine coordinate ring $K[U]$. (See [Kle], also for the definition of $\mathcal{K}_Y$ on general schemes.) A global section of $\mathcal{K}_Y$ is called a *rational function* on Y and $\Gamma(Y, \mathcal{K}_Y)$ is the *ring of rational functions* on Y.

If Y is reduced, a rational function on Y is given by an equivalence class of pairs (U, f) with $U \subset Y$ open and dense and $f : U \to \mathbb{C}$ a regular function, where two pairs (U, f) and (V, g) are equivalent if $f = g$ on $U \cap V$. The regular function $f : U \to \mathbb{C}$ has a maximal extension $\tilde{f} : \tilde{U} \to \mathbb{C}$ of to an open set $\tilde{U}$ containing U with $\tilde{f}$ regular and $\tilde{f}|U = f$. The set $\tilde{U}$ is called the *definition set* of f and $Y \setminus \tilde{U}$ the *pole set* of f.

If Y is affine then $\Gamma(Y, \mathcal{K}_Y) = Q(K[Y])$ is the total quotient ring of the affine coordinate ring $K[Y]$. If $Y \subset \mathbb{P}^n$ is projective, with homogeneous coordinate ring $K[Y]$, then $\Gamma(Y, \mathcal{K}_Y)$ is the *homogeneous total quotient ring* of $K[Y]$, i.e. $\Gamma(Y, \mathcal{K}_Y) = Q^h(K[Y]) := \{f/g \mid f, g \in K[Y]$ homogeneous of the same degree, g a non-zero divisor of $K[Y]\}$.

If Y is *integral*, i.e. irreducible and reduced, then $\Gamma(Y, \mathcal{K}_Y)$ is a field, denoted by $K(Y)$, the *function field* of Y and $\mathcal{K}_Y$ is the constant sheaf $K(Y)$. If Y is reduced but reducible and $Y = Y_1 \cup \cdots \cup Y_r$ the decomposition of Y into irreducible components, then the ring of rational functions $\Gamma(Y, \mathcal{K}_Y)$ is the direct sum of the fields $K(Y_i), i = 1, \ldots, r$.

2.1.1.1 Cartier and Weil Divisors

Definition 2.1.1 A *Cartier divisor D* on an algebraic scheme Y is a global section of the sheaf $\mathcal{K}_Y^*/\mathcal{O}_Y^*$ where $\mathcal{K}_Y^*$ is the sheaf of invertible elements in the sheaf $\mathcal{K}_Y$ and $\mathcal{O}_Y^*$ the sheaf of invertible elements in $\mathcal{O}_Y$.

In more concrete terms, a Cartier divisor D can be represented by a system $\{(U_i, f_i)\}$ where $\{U_i\}$ is an open covering of Y and $f_i = p_i/q_i \in \mathcal{K}_Y^*(U_i)$, $p_i, q_i \in \Gamma(U_i, \mathcal{O}_Y)$ non-zero divisors in $\mathcal{O}_{Y,y}$ for all $y \in U_i$, such that for each i, j the quotient f_i/f_j is in $\Gamma(U_i \cap U_j, \mathcal{O}_Y^*)$.

Let $U_i' \subset \tilde{U}_i$ the open subset of the definition set $\tilde{U}_i$ of f_i where f_i has no zeros. $\operatorname{supp}(D|U_i) := U_i \setminus U_i'$ is closed in U_i and $\operatorname{supp}(D|U_i) = \operatorname{supp}(D|U_j)$ in $U_i \cap U_j$. Hence the union of the $\operatorname{supp}(D|U_i)$ is well defined and called the *support of the Cartier divisor* D, denoted $\operatorname{supp}(D)$. It is the union of the zeros and poles of D. For a point $y \in \operatorname{supp}(D)$ we also write $y \in D$.

If the f_i and f_j coincide on $U_i \cap U_j$ we can glue them to a global section f of $\Gamma(Y, \mathcal{K}_Y^*)$ and then we say that D is a *principal* divisor, i.e. in the image of the map $\Gamma(Y, \mathcal{K}_Y^*) \to \Gamma(Y, \mathcal{K}_Y^*/\mathcal{O}_Y^*)$. The principal divisor of the invertible rational function $f \in \Gamma(Y, \mathcal{K}_Y^*)$ is denoted by (f).

Two Cartier divisors D and D' given by $\{(U_i, f_i)\}$ resp. $\{(U_i, f_i')\}$ are called *linear equivalent*, written $D \sim D'$, if their difference $D - D'$, given by $\{(U_i, f_i/f_i')\}$, is principal. In other words, f_i/f_i' is independent of i or $f_i/f_j = f_i'/f_j'$ for all i, j.

D is called *effective*, notation $D \geq 0$, if it can be represented by $\{(U_i, f_i)\}$ with $f_i \in \Gamma(U_i, \mathcal{O}_Y)$.[2]

The set of Cartier divisors on Y, $\Gamma(Y, \mathcal{K}_Y^*/\mathcal{O}_Y^*)$, is an abelian group, where $D + D'$ is given by $\{(U_i, f_i f_i')\}$, $-D$ by $\{(U_i, \frac{1}{f_i})\}$, and 0 (the "empty divisor") by any system $\{(U_i, f_i)\}$ with $f_i \in \Gamma(U_i, \mathcal{O}_Y^*)$.

There is another concept of divisors, that of Weil divisors, which we define only for Y integral. Weil divisors are less general but more geometric than Cartier divisors.

Definition 2.1.2 Let Y be an integral algebraic scheme. A *Weil divisor* D on Y is a finite formal sum

$$D = \sum a_j X_j \ , \ a_j \in \mathbb{Z},$$

where $X_j \subset Y$ is a closed, irreducible and reduced subscheme of codimension 1, called *prime divisor*. The *support of the Weil divisor* D is the union of the prime divisors X_j with $a_j \neq 0$, denoted $\operatorname{supp}(D)$.

The set of Weil divisors on Y, written as $\operatorname{Div} Y$, is the free abelian group generated by the prime divisors. D is called *effective*, denoted by $D \geq 0$, if $a_j \geq 0$ for all j. Hence any Weil divisor D can be written as $D' - D''$ with D' and D'' effective.

Definition 2.1.3 Let Y be integral. Then $\mathcal{K}_Y^*$ is the constant sheaf $K(Y)^*$. For $f \in K(Y)^*$ a non-zero rational function on Y and $X \subset Y$ a prime divisor we define the *order* of f at X, denoted by $\operatorname{ord}_X f$, as follows:

Let $R \subset K(Y)$ denote the *local ring of Y along X*, i.e. the localization of the coordinate ring $\Gamma(U, \mathcal{O}_Y)$ at the prime ideal defining X in an affine open set $U \subset Y$ meeting X. Then we can write $f = p/q$ with $p, q \in R$. By setting

---

[2]In the complex analytic situation, when D a complex subspace of the complex space Y, we call D an *(effective) Cartier divisor* if its ideal is generated by a non-zero divisor of $\mathcal{O}_{Y,y}$ for every $y \in D$. If Y is smooth, D is also called a hypersurface.

$$\operatorname{ord}_X \frac{p}{q} = \operatorname{ord}_X p - \operatorname{ord}_X q$$

we reduce to the case $f \in R$. For $f \in R$ we define

$$\operatorname{ord}_X f = \lg(R/f) ,$$

where $\lg(M)$ denotes the length (of a composition series) of the R-module M.

Example 2.1.3.1 For Y an integral curve a prime divisor is just a reduced point $y \in Y$ and the local ring along the prime divisor y is $\mathcal{O}_{Y,y}$. If f is a non-zero divisor of $\mathcal{O}_{Y,y}$, then

$$\operatorname{ord}_y f = \dim_{\mathbb{C}}(\mathcal{O}_{Y,y}/f),$$

since the length coincides with the $\mathbb{C}$-dimension. For non-zero rational function $f = p/q$ on Y, p, q are non-zero divisors in $\mathcal{O}_{Y,y}$ and hence $\operatorname{ord}_y f = \dim_{\mathbb{C}}(\mathcal{O}_{Y,y}/p) - \dim_{\mathbb{C}}(\mathcal{O}_{Y,y}/q)$.

If Y is regular in codimension one,[3] then $X \not\subset \operatorname{Sing}(Y)$ and R is a regular local ring of dimension 1. In this case $f = ut^k$ with $k \in \mathbb{Z}$, u a unit in R and t a generator of the maximal ideal of R. We get $\operatorname{ord}_X f = k$.

We remark that $\operatorname{ord}_X f \neq 0$ for at most finitely many prime divisors $X \subset Y$, which can be seen as follows: The set of points $Z \subset Y$ where f is not a regular function (i.e. where f has poles) is closed of codimension one and contains hence only finitely many prime divisors (since Y is noetherian). For the same reason the set of zeros of $f|_{Y-Z}$ contains only finitely many prime divisors. Since $\operatorname{ord}_X f \neq 0$ only if the prime divisor X is an irreducible component of the set of zeros or of the set of poles of f, the claim follows.

Consider now a Cartier divisor $s \in \Gamma(Y, \mathcal{K}_Y^*/\mathcal{O}_Y^*)$, Y integral, represented by $\{(U_i, f_i)\}$, $f_i \in \Gamma(U_i, \mathcal{K}_Y^*) = K(Y)^*$. If $X \subset Y$ is a prime divisor set $a_X^i := \operatorname{ord}_X f_i$ if $X \cap U_i \neq \emptyset$ and $a_X^i = 0$ otherwise. If $X \cap U_j \neq \emptyset$ for another j, then f_i/f_j is nowhere vanishing on $U_i \cap U_j$, hence $\operatorname{ord}_X(f_i/f_j) = 0$. If follows that $a_X^i = a_X^j =: a_X$ and we define

$$(s) := \sum_{X \subset Y \text{ a prime divisor}} a_X \cdot X,$$

the *Weil divisor associated to the Cartier divisor s*. The support of the Cartier divisor s is the support of the associated Weil divisor.

Definition 2.1.4 Let Y be integral.

(1) Any Weil divisor on Y equal to

$$(f) = \sum_{X \subset Y \text{ a prime divisor}} \operatorname{ord}_X f \cdot X .$$

[3] An algebraic scheme or complex space Y is called *regular* (or *smooth*) *in codimension one* if for all $y \in Y$, $\operatorname{codim}_{(Y,y)}(\operatorname{Sing}(Y), y) \geq 2$. If Y is normal, then it is regular in codimension one.

for some $f \in K(Y)^* = \Gamma(Y, \mathcal{K}_Y^*)$ is called a *principal* (Weil) *divisor*.

(2) Two Weil divisors D and D' are *linear equivalent*, denoted by $D \sim D'$, if $D - D'$ is a principal Weil divisor.

(3) The principal Weil divisors of Y form a subgroup of Div Y and the factor group is called the *divisor class group* of Y.

The following proposition states conditions under which Cartier divisors and Weil divisors coincide.

Proposition 2.1.5 *Let Y be an integral scheme which is* locally factorial *(i.e. each local ring $\mathcal{O}_{Y,y}$ is a unique factorization domain). Then Y is normal and the group* Div Y *of Weil divisors is isomorphic to the group $\Gamma(Y, \mathcal{K}_Y^*/\mathcal{O}_Y^*)$ of Cartier divisors such that principal divisors correspond to each other.*

Proof It is well known that unique factorization domains are normal, hence reduced. Since Y is irreducible, the sheaf of rational functions $\mathcal{K}_Y$ is the constant sheaf $K(Y)$, and we associated already to each Cartier divisor s a Weil divisor (s).

To see the converse, let $D = \sum a_i X_i$ be a Weil divisor and $y \in Y$. Each prime divisor X_i of D passing through y defines a codimension one reduced subscheme and hence defines an ideal $I_i \subset \mathcal{O}_{Y,y}$ which is principal (cf. [HaR2, Proposition 6.2]). Let f_i be a generator of I_i and set $f_y = \prod f_i^{a_i}$, where a_i is the coefficient of X_i in D.

Then there exists a neighbourhood U_y of y such that $f_y \in \Gamma(U_y, \mathcal{K}_Y^*)$ and $(f_y) = D\big|_{U_y}$. Covering Y with such U_y, then the collection $\{(U_y, f_y)\}$ defines a Cartier divisor on Y. If f_i' is a different generator of I_i then f_i and f_i' differ by multiplication with a unit in $\mathcal{O}_{Y,y}$ since normal local rings are integral. Therefore $f_i/f_i' \in \Gamma(U_y, \mathcal{O}_Y^*)$ for sufficiently small U_y and the Cartier divisor obtained from D is well defined. □

Example 2.1.5.1 Weil divisors and Cartier divisors are the same thing on integral, locally factorial schemes, in particular on smooth schemes. We give now an example of an integral singular curve Y, where Weil divisors and Cartier divisors differ. A Weil divisor D on Y is just a finite number of points, each counted with (positive or negative) multiplicity. Every smooth point of Y (with arbitrary multiplicity) is also a Cartier divisor. However, if p is a singular point, then the Weil divisor $D = 1 \cdot p$ need not be a Cartier divisor. Consider the integral curve $Y = V((x^2 + y^2)^3 - 4x^2y^2) \subset \mathbb{C}^2$ (four-leaf clover) at $p = 0$. We show that for $f \in \mathcal{O}_{Y,0} \setminus \{0\}$, $\operatorname{ord}_0 f$ is even. To see this, note that $\operatorname{ord}_0 f = \operatorname{ord}_0 h$, with h the tangent cone (sum of terms of lowest degree) of f. Since h is the product of linear factors l_i, $\operatorname{ord}_0 h = \sum \operatorname{ord}_0 l_i$ and we are reduced to the case that f is linear. For a linear f one computes $\operatorname{ord}_0 f = 6$, if $f = x$ or $f = y$ and $\operatorname{ord}_0 f = 4$ otherwise. It follows that the order at 0 of any Cartier divisor s on Y is even and therefore we get for the associated Weil divisor (s), $(s) \neq D$.

We shall mainly consider Cartier divisors. In the following we describe their relations to invertible sheaves. A sheaf is called *invertible* if it is locally free of rank one.

Definition 2.1.6 If $D \subset Y$ is a Cartier divisor on an algebraic scheme Y given by $\{(U_i, f_i)\}$, $f_i \in \Gamma(U_i, \mathcal{K}_Y^*)$, we define the sheaf $\mathcal{O}_Y(D)$ to be the subsheaf of $\mathcal{K}_Y$ generated as $\mathcal{O}_Y$-module by f_i^{-1} on U_i. It is locally free of rank one.

Since f_i/f_j is in $\Gamma(U_i \cap U_j, \mathcal{O}_Y^*)$, f_i^{-1} and f_j^{-1} generate the same $\mathcal{O}_Y$-module on $U_i \cap U_j$, and hence $\mathcal{O}_Y(D)$ depends only on D. It is called the *invertible sheaf associated with the divisor D*.

Proposition 2.1.7 *Let Y be an algebraic scheme.*

(1) $\mathcal{O}_Y(D)$ *is an invertible sheaf on Y (i.e. locally free of rank* 1*);*
(2) $D \mapsto \mathcal{O}_Y(D)$ *gives a* $1-1$ *correspondence between Cartier divisors on Y and invertible subsheaves of* $\mathcal{K}_Y$*;*
(3) $\mathcal{O}_Y(D-D') \cong \mathcal{O}_Y(D) \otimes_{\mathcal{O}_Y} \mathcal{O}_Y(D')^{-1}$*, where* $\mathcal{L}^{-1}$ *denotes the* $\mathcal{O}_Y$*-dual of an invertible sheaf* $\mathcal{L}$ *on Y;*
(4) $D \sim D'$ *if and only if* $\mathcal{O}_Y(D) \cong \mathcal{O}_Y(D')$ *as abstract sheaves (forgetting the embedding in* $\mathcal{K}_Y$*).*

Proof (1) Multiplication with f_i is an isomorphism $\mathcal{O}_Y(D)|U_i \to \mathcal{O}_Y|U_i$, hence $\mathcal{O}_Y(D)$ is locally free. To get the inverse of the map in (2) let $\mathcal{L} \subset \mathcal{K}_Y$ be an invertible sheaf, trivial on an open covering $\{U_i\}$. Generators f_i of $\mathcal{L}|U_i$ are sections of $\mathcal{K}_Y^*$ over U_i and define a Cartier divisor D. Since two local generators differ by a unit, D is unique and we get $\mathcal{L} = \mathcal{O}_Y(D)$. The rest is straightforward (cf. [HaR2, II Prop. 6.13]). □

We denote by $\mathcal{C}a\,\mathcal{C}\ell\,Y$ the group of Cartier divisors on Y modulo linear equivalence $\sim$, called the *Cartier divisor class group* of Y, and by Pic Y the group of isomorphism classes of invertible sheaves with $\otimes$ as group operation, called the *Picard group* of Y.

Remark 2.1.7.1 By the above proposition we get a monomorphism from the Cartier divisor class group to the Picard group, which is an isomorphism in many cases. E.g. if Y is integral then any invertible sheaf on Y is isomorphic to an invertible subsheaf of $\mathcal{K}_Y$ and we have an isomorphism

$$\mathcal{C}a\,\mathcal{C}\ell\,Y \cong \operatorname{Pic} Y, \quad D \mapsto \mathcal{O}_Y(D). \tag{2.1.1.1}$$

Indeed, if Y is integral, then $\mathcal{K}_Y$ is the constant sheaf $K(Y)$. Moreover, for an invertible sheaf $\mathcal{L}$ on Y, there is an open covering $\{U_i\}$ of Y such that $\mathcal{L}\big|_{U_i} \cong \mathcal{O}_Y\big|_{U_i}$. It follows that $\mathcal{L} \otimes_{\mathcal{O}_Y} \mathcal{K}_Y|U_i \cong \mathcal{K}_{U_i}$ is a constant sheaf on U_i. Hence, $\mathcal{L} \otimes \mathcal{K}_Y$ is in fact a constant sheaf, isomorphic to $\mathcal{K}_Y$. The natural morphism $\mathcal{L} \to \mathcal{L} \otimes \mathcal{K}_Y \cong \mathcal{K}_Y$ shows that $\mathcal{L}$ is isomorphic to a subsheaf of $\mathcal{K}_Y$ and hence (2.1.1.1) is an isomorphism by Proposition 2.1.7(2).

By a result of Nakai, the isomorphism (2.1.1.1) holds for any projective algebraic scheme (cf. [HaR2, II, Rem. 6.14.1]).

Let the Cartier divisor D be effective, represented by $\{(U_i, f_i)\}$ with $f_i \in \Gamma(U_i, \mathcal{O}_Y)$ and f_i a non-zero divisor in $\mathcal{O}_{Y,y}$ for all $y \in U_i$. Then D defines an ideal sheaf $\mathcal{I}_D \subset \mathcal{O}_Y$, generated over U_i by f_i and hence a subscheme of Y, also denoted D, with ideal sheaf $\mathcal{I}_D$. The equations $f_i = 0$ or just f_i are called *local equations* for D. Moreover, by the definition of $\mathcal{O}_Y(D)$ above we get an isomorphism

$$\mathcal{I}_D \cong \mathcal{O}_Y(-D)$$

for any effective Cartier divisor D on Y. In other words, we have an exact sequence of sheaves

$$0 \to \mathcal{O}_Y(-D) \to \mathcal{O}_Y \to \mathcal{O}_D \to 0.$$

Furthermore, if D is effective, then $1 \in \mathcal{O}_Y(D)\big|U_i = \frac{1}{f_i}\mathcal{O}_{U_i}$, since $f_i \in \Gamma(U_i, \mathcal{O}_Y)$, and hence $\mathcal{O}_Y(D)$ contains a *canonical global section*

$$s_D \in \Gamma(Y, \mathcal{O}_Y(D)) \,,$$

corresponding to $1 \in \Gamma(Y, \mathcal{K}_Y)$ if we regard $\mathcal{O}_Y(D)$ as a subsheaf of $\mathcal{K}_Y$. Moreover, $\mathcal{O}_Y = \mathcal{O}_Y \cdot s_D$ is a subsheaf of $\mathcal{O}_Y(D)$, while $\mathcal{O}_Y(-D)$ is a subsheaf of $\mathcal{O}_Y$.

Remark 2.1.7.2 Instead of defining $\mathcal{O}_Y(D)$ as a concrete subsheaf of $\mathcal{K}_Y$ we can define it up to (abstract) isomorphism by gluing the sheaves $\mathcal{O}_{U_i}$ on U_i via transition functions φ_{ij} (cf. [GLS6], A.2). Since the multiplication map $f_i : \frac{1}{f_i}\mathcal{O}_{U_i} \overset{\cong}{\to} \mathcal{O}_{U_i}$ is an isomorphism, the $\mathcal{O}_{U_i}$ are glued by $\varphi_{ij} : \mathcal{O}_{U_j}\big|_{U_i \cap U_j} \overset{\cong}{\to} \mathcal{O}_{U_i}\big|_{U_i \cap U_j}$, with φ_{ij} the multiplication with f_i/f_j (the $\frac{1}{f_i}\mathcal{O}_{U_i}$ are glued inside $\mathcal{K}_Y$ by the identity). Then the local sections $f_i \in \Gamma(U_i, \mathcal{O}_Y)$ glue to a global section of $\mathcal{O}_Y(D)$ since $f_i = \varphi_{ij}(f_i)$ on $U_i \cap U_j$. This is the canonical section s_D.

Note that the ideal sheaf $\mathcal{I}_D = \mathcal{O}_Y(-D)$ can be defined by the transition functions $\varphi_{ij} = f_j/f_i = (f_i/f_j)^{-1}$ which, however, do not give rise to a global section. Indeed, we shall see in Remark 2.1.10.1(1) that $\mathcal{O}_{\mathbb{P}^n}(-dH)$ does not have any global section $\neq 0$ if $d > 0$.

Definition 2.1.8 Let $\mathcal{L}$ be an invertible sheaf on the algebraic scheme Y and $s \in \Gamma(Y, \mathcal{L})$ a global section, which is locally a non-zero divisor, i.e. for any $y \in Y$ and isomorphism $\varphi : \mathcal{L}_y \cong \mathcal{O}_{Y,y}$, $\varphi(s_y)$ is a non-zero divisor. We define an effective Cartier divisor, the *divisor of zeros* of s, as follows:

Y has an open covering $\{U_i\}$ such that there exists $\mathcal{O}_{U_i}$-linear isomorphisms $\varphi_i : \mathcal{L}\big|U_i \overset{\cong}{\to} \mathcal{O}_{U_i}$. Set $f_i = \varphi_i(s|U_i) \in \Gamma(U_i, \mathcal{O}_Y)$, which is locally a non-zero divisor. On $U_i \cap U_j$ we have $f_j = \varphi_j \circ \varphi_i^{-1}(f_i) = \varphi_j \circ \varphi_i^{-1}(1) \cdot f_i$ and therefore f_i and f_j differ by multiplication with an invertible function.

Since φ_i is unique up to multiplication with an element in $\Gamma(U_i, \mathcal{O}^*_{U_i})$, it follows that $\{(U_i, f_i)\}$ defines an effective Cartier divisor on Y, which is independent of the trivialization φ_i. This divisor is denoted by $div_0(s)$,

$$div_0(s) := \text{ divisor represented by } \{(U_i, f_i = \varphi_i(s))\}.$$

and called the *divisor of zeros* of $s \in \Gamma(Y, \mathcal{L})$. For $y \in Y$ let $s(y)$ be the image of s in $\mathcal{L}_Y/\mathfrak{m}_y\mathcal{L}_Y$. Then $y \in div_0(s)$ iff $s(y) = 0$, whence the name divisor of zeros.

If $\mathcal{L} = \mathcal{O}_Y(D)$, D given by $\{U_i, f_i\}$, then φ_i is the multiplication with f_i. If D is effective and s_D the canonical global section, then $D = div_0(s_D)$. Moreover, we may consider $\mathcal{O}_Y(D)$ as a subsheaf of $\mathcal{K}_Y$, and a non-zero section $s \in \Gamma(Y, \mathcal{O}_Y(D))$ is rational function f on Y. Hence s defines the principal divisor (f) on Y, not to be confused with $div_0(f) := div_0(s)$. f has zeros and poles on Y and the set of zeros of f coincides with the support of $div_0(f)$, while the support of $div_0(f^{-1})$ is the set of poles of f.

2.1.1.2 Divisors on Projective Schemes

Consider now divisors on a projective algebraic scheme $Y \subset \mathbb{P}^n$. Then, for any coherent sheaf $\mathcal{F}$ on Y, the cohomology groups $H^i(Y, \mathcal{F})$, in particular $\Gamma(Y, \mathcal{O}_Y(D)) = H^0(Y, \mathcal{O}_Y(D))$, are finite dimensional complex vector spaces.

Let us start with $Y = \mathbb{P}^n$, and let $H \cong \mathbb{P}^{n-1}$ be the hyperplane, given by $x_0 = 0$ for homogeneous coordinates $x_0 : \cdots : x_n$ of $\mathbb{P}^n$. Since $\mathbb{P}^n$ is covered by the affine open sets $U_i = \{x_i \neq 0\}$, the rational functions on $\mathbb{P}^n$ can be identified with

$$K(\mathbb{P}^n) = K(U_i) \cong \mathbb{C}\left(\frac{x_0}{x_i}, \ldots, \frac{x_n}{x_i}\right) \cong \mathbb{C}(y_1, \ldots, y_n),$$

with $\frac{x_i}{x_i}$ deleted. We also have

$$K(\mathbb{P}^n) = \left\{\frac{f}{g} \mid f, g \text{ homogeneous polynomials of the same degree, } g \neq 0\right\},$$

following from the arguments given in Lemma 2.1.9 below.

For $d \in \mathbb{Z}$ the Weil divisor dH is given on U_i, $i \neq 0$, by $\left(\frac{x_0}{x_i}\right)^d \in K(U_i)$. We denote by

$$\mathcal{O}_{\mathbb{P}^n}(d) := \mathcal{O}_{\mathbb{P}^n}(dH), \quad d \in \mathbb{Z},$$

the invertible sheaf defined by the divisor dH on $\mathbb{P}^n$. $\mathcal{O}_{\mathbb{P}^n}(1)$ is called the *twisting sheaf of Serre*.

Any $\mathcal{O}_Y$-module $\mathcal{F}$ on Y is an $\mathcal{O}_{\mathbb{P}^n}$-module and the sheaf

$$\mathcal{F}(d) := \mathcal{F} \otimes_{\mathcal{O}_{\mathbb{P}^n}} \mathcal{O}_{\mathbb{P}^n}(d), \quad d \in \mathbb{Z},$$

is called the dth *twist of* $\mathcal{F}$.

For any divisor $D = \sum_{i=1}^r a_i X_i$ on $\mathbb{P}^n$ (on $\mathbb{P}^n$ Weil divisors and Cartier divisors coincide) define the *degree of D* by

$$\deg D := \sum_{i=1}^{r} a_i \deg(X_i),$$

where $\deg(X_i)$ is the degree of the projective hypersurface $X_i \subset \mathbb{P}^n$.

If D is effective of degree d then it determines a homogeneous polynomial $F \in \mathbb{C}[x_0, \dots, x_n]$ of degree d, unique up to multiplication with a non-zero constant. In fact, the prime divisor X_i is a reduced, irreducible projective hypersurface in $\mathbb{P}^n$, defined by a homogeneous polynomial F_i of positive degree $d_i = \deg X_i$, unique up to multiplication with a non-zero constant. Then $d = \deg D = \sum_{i=1}^{r} a_i \deg X_i$ is the degree of D and $F = F_1^{a_1} \cdot \dots \cdot F_r^{a_r}$ the desired homogeneous polynomial of degree d.

Conversely, any homogeneous polynomial factors as $F = F_1^{a_1} \cdot \dots \cdot F_r^{a_r}$ with F_i irreducible and homogeneous of degree d_i and thus defines the prime divisors $X_i = V(F_i)$ and the effective divisor

$$\operatorname{div}(F) := \sum_{i=1}^{r} a_i X_i,$$

the *divisor of* F. We get the canonical identification

$$\mathbb{C}[x_0, \dots, x_n]_d \setminus \{0\}/\mathbb{C}^* = \{\text{effective divisors of degree } d \text{ on } \mathbb{P}^n\},$$

mapping F to the (divisor of the) projective hypersurface $V(F)$, where $\mathbb{C}[x_0, \dots, x_n]_d$ denotes the homogeneous polynomials of degree d, which is a vector space of dimension $\binom{n+d}{d}$.

Lemma 2.1.9 *The homogeneous polynomials of degree d can be identified with the global sections of* $\mathcal{O}_{\mathbb{P}^n}(d)$, $d \geq 0$,

$$\mathbb{C}[x_0, \dots, x_n]_d \cong H^0(\mathbb{P}^n, \mathcal{O}_{\mathbb{P}^n}(d)),$$

mapping $F \in \mathbb{C}[x_0, \dots, x_n]_d$ *to the rational function* F/x_i^d *on* $U_i = \{x_i \neq 0\}$, *which is a global section of* $H^0(\mathbb{P}^n, \mathcal{O}_{\mathbb{P}^n}(d)) \subset K(\mathbb{P}^n)$. *The inverse maps a global non-zero section of* $\mathcal{O}_{\mathbb{P}^n}(d)$ *to its divisor of zeros. Moreover, for any such F of degree d we have*

$$\operatorname{div}(F) \sim dH.$$

Proof To see this, note first that on U_i a local equation for the divisor dH is x_0^d/x_i^d and that the transition functions $\varphi_{ij} : \mathcal{O}_{U_j}|_{U_i \cap U_j} \to \mathcal{O}_{U_i}|_{U_i \cap U_j}$ of $\mathcal{O}_{\mathbb{P}^n}(dH)$ are multiplications with $\varphi_{ij} = (x_0/x_i)^d/(x_0/x_j)^d = x_j^d/x_i^d$.

Now, if $F \in \mathbb{C}[x_0, \dots, x_n]_d$, then $f_i = F/x_i^d \in \Gamma(U_i, \mathcal{O}_{\mathbb{P}^n}) = \mathbb{C}\left[\frac{x_0}{x_i}, \dots, \frac{x_n}{x_i}\right]$ satisfies $\varphi_{ij}(f_j) = f_i$. Hence $\{(U_i, F/x_i^d)\}$ defines a global section of $\mathcal{O}_{\mathbb{P}^n}(d)$.

Conversely, given a global section s of $\mathcal{O}_{\mathbb{P}^n}(d)$, then its divisor of zeros $div_0(s)$ is a hypersurface in $\mathbb{P}^n$, defined by a homogeneous polynomial of degree d. If fact, s is given on U_i by $f_i \in \mathbb{C}\left[\frac{x_0}{x_i}, \ldots, \frac{x_n}{x_i}\right] \cong \mathbb{C}[y_1, \ldots, y_n]$, $i = 0, \ldots, n$, satisfying $\varphi_{ij}(f_j) = f_i$ i.e. $x_i^d f_i = x_j^d f_j$. Since x_i does not appear as a denominator of f_j for all $j \neq i$, $x_i^d f_i \in \mathbb{C}[x_0, \ldots, x_n]$. Hence f_i has degree $\leq d$ as polynomial in $y_1, \ldots, y_n$ and $x_i^d f_i(x_0/x_i, \ldots, x_n/x_i)$ defines the desired element in $\mathbb{C}[x_0, \ldots, x_n]_d$, independent of i.

It follows that $\operatorname{div}(F) - dH$ is the principal divisor of the rational function F/x_0^d, i.e. $\operatorname{div}(F)$ is linear equivalent to dH if $F \neq 0$.

Proposition 2.1.10

(1) If D is a divisor on $\mathbb{P}^n$ of degree d, then $D \sim dH$, and hence $\mathcal{O}_{\mathbb{P}^n}(D) \cong \mathcal{O}_{\mathbb{P}^n}(d)$.
(2) For a non-zero rational function on $\mathbb{P}^n$, $f \in K(\mathbb{P}^n)^*$, the principal divisor (f) has degree 0.
(3) $\deg : \mathrm{Ca}\,\mathcal{C}\ell\,\mathbb{P}^n \to \mathbb{Z}$ is an isomorphism of abelian groups.
(4) Every invertible sheaf on $\mathbb{P}^n$ is isomorphic to $\mathcal{O}_{\mathbb{P}^n}(d)$ for some $d \in \mathbb{Z}$.

Proof (1) An arbitrary divisor D can be written as $D_1 - D_2$ with D_i effective, hence defined by homogeneous polynomials g_i of degree $d_i = \deg D_i$, and $d = \deg D = d_1 - d_2$. Now g_1 and $x_0^d g_2$ are homogeneous polynomials of the same degree and therefore $f = g_1/x_0^d g_2 = (g_1/g_2)/x_0^d$ defines a rational function on $\mathbb{P}^n$ satisfying

$$D - dH = (f)\,.$$

Hence (1) follows.

(2) If $f = g/h$ is a rational function, g, h homogeneous of the same degree, then (f) has degree $\deg g - \deg h = 0$.

(3) deg is well defined and injective by (1) and (2) and a homomorphism by construction. Surjectivity follows from $\deg(dH) = d$.

(4) The isomorphism (2.1.1.1) and statement (3) imply

$$\operatorname{Pic} \mathbb{P}^n \cong \mathbb{Z},$$

with $d \in \mathbb{Z}$ corresponding to $\mathcal{O}_{\mathbb{P}^n}(d)$. □

Remark 2.1.10.1 The sheaf $\mathcal{O}_{\mathbb{P}^n}(-D)$ has no global sections for a divisor $D \geq 0$ of degree $d > 0$ on $\mathbb{P}^n$: Otherwise D would be defined by local sections $f_i \in \Gamma(U_i, \mathcal{O}_{\mathbb{P}^n}) = \mathbb{C}\left[\frac{x_0}{x_i}, \ldots, \frac{x_n}{x_i}\right]$ satisfying $x_j^d f_i = x_i^d f_j$, which is impossible since f_i involves x_i only with exponents ≤ 0.

Now let $Y \subset \mathbb{P}^n$ be a reduced projective scheme and $F \in \mathbb{C}[x_0, \ldots, x_n]$ a homogeneous polynomial. If F does not vanish identically on any irreducible component of Y, the image of F in the homogeneous coordinate ring $K[Y] = \mathbb{C}[x_0, \ldots, x_n]/I(Y)$

is a non-zero divisor at every point of Y and thus defines an effective Cartier divisor on Y with support $Y \cap V(F)$, the *hypersurface section* of Y, also denoted by $\operatorname{div}(F)\big|_Y$.

Of course, not every effective divisor on Y is a hypersurface section (as follows e.g. from the theorem of Bézout 2.1.16).

Remark 2.1.10.2 A rational function on $\mathbb{P}^n$ may not restrict to a rational function on Y, as its pole set may contain Y. But every rational function on Y is the restriction of a rational function on $\mathbb{P}^n$. This follows, since any rational function on Y is the quotient f/g with f, g homogeneous elements of the same degree, g a non-zero divisor, in the homogeneous coordinate ring $K[Y]$ of Y. Representing f, g by homogeneous polynomials $F, G \in \mathbb{C}[x_0, \ldots, x_n]$ of the same degree, then F/G is a rational function on $\mathbb{P}^n$, whose restriction to Y is f/g.

If Y is integral, then an effective divisor D on Y is the divisor of zeros of a global section of $\mathcal{O}_Y(D)$ by Lemma 2.1.13 below, hence of a rational function $f/g \in K(Y)$ with f, g represented by F, G as above. The hypersurface section $Y \cap V(F)$ contains D and we have

$$\operatorname{div}(F)\big|_Y = D + D'$$

for some effective divisor D', which may be not empty. Take for example $Y = V(x^2 + y^2 - z^2) \subset \mathbb{P}^2$, $F = x, G = y + 1$ and $f/g = F/G\ |Y$. Then $D = div_0$ (f/g) is the point $(0 : 1 : 1)$ while $Y \cap V(F) = \operatorname{div}(F)\big|_Y$ contains in addition the point $(0 : 1 : -1)$.

Definition 2.1.11 An invertible sheaf $\mathcal{L}$ on a projective algebraic scheme Y is called *very ample* if there exists a closed embedding $\varphi : Y \hookrightarrow \mathbb{P}^n$ for some n, such that $\mathcal{L} \cong \varphi^*\mathcal{O}_{\mathbb{P}^n}(1)$. A divisor D on Y is very ample if $\mathcal{O}_Y(D)$ is very ample. An invertible sheaf $\mathcal{L}$ on Y is called *ample* if, for some $m > 0$, the mth tensor power $\mathcal{L}^{\otimes m}$ is very ample, and a divisor D is very ample if mD is very ample.

Every closed embedding φ of Y into some $\mathbb{P}^n$ provides a very ample invertible sheaf on Y. Different embeddings may lead to non-isomorphic very ample sheaves. The following characterization of ampleness is due to Serre.

Theorem 2.1.12 (Serre) *For an invertible sheaf $\mathcal{L}$ on a projective scheme Y the following are equivalent:*

(1) $\mathcal{L}$ is ample.
(2) For every coherent sheaf $\mathcal{F}$ on Y there exists an $n_0 > 0$ such that for each $n \geq n_0$ the sheaf $\mathcal{F} \otimes \mathcal{L}^{\otimes n}$ is generated by its global sections.
(3) For every coherent sheaf $\mathcal{F}$ on Y there exists an $n_0 > 0$ such that

$$H^i(Y, \mathcal{F} \otimes \mathcal{L}^{\otimes n}) = 0 \quad \textit{for all} \quad i > 0,\ n \geq n_0\,.$$

Recall that a sheaf $\mathcal{F}$ on Y is generated by a family of global sections $(s_i \in \Gamma(Y, \mathcal{F}))_{i \in I}$ if the germs $s_{i,y}$ generate the stalk $\mathcal{F}_y$ for all $y \in Y$. For proofs and details we refer to [HaR2, 2.5–7, 3.5]. Since $\mathcal{O}_{\mathbb{P}^n}(1)$ is generated by its global sections, so is every very ample line bundle on Y.

Statement (3) is called the *Serre vanishing theorem*, it is the most basic cohomological vanishing theorem for projective varieties. This and further cohomological vanishing theorems play an important role in this book, for proofs as well as for constructions.

2.1.1.3 Linear Systems

In this section let $Y \subset \mathbb{P}^n$ be a projective scheme, which we assume to be integral. Under this assumption linear equivalence classes of divisors form a projective space and there is a $1-1$ correspondence with isomorphism classes of invertible sheaves, given by $D \mapsto \mathcal{O}_Y(D)$ (Remark 2.1.7.1) By Definition 2.1.8 any non-zero global section $s \in \Gamma(Y, \mathcal{L})$, $\mathcal{L}$ invertible, defines an effective divisor $div_0(s)$, the divisor of zeros of s.

Lemma 2.1.13 *For any integral projective scheme Y, D_0 any divisor on Y and $\mathcal{O}_Y(D_0)$ the associated invertible sheaf the following holds:*

(a) for any $s \in \Gamma(Y, \mathcal{O}_Y(D_0)) \setminus \{0\}$ the effective divisor of zeros $div_0(s)$ is linear equivalent to D_0;
(b) every effective divisor on Y, linear equivalent to D_0, is of the form $div_0(s)$ for some $s \in \Gamma(Y, \mathcal{O}_Y(D_0)) \setminus \{0\}$;
(c) two non-zero sections $s, s' \in \Gamma(Y, \mathcal{O}_Y(D_0))$ satisfy $div_0(s) = div_0(s')$ if and only if there is a $\lambda \in \mathbb{C}^*$ such that $s' = \lambda s$.

Proof We may assume that $\mathcal{O}_Y(D_0) \subset \mathcal{K}_Y$.

(a) Let f be the rational function corresponding to $s \in \Gamma(Y, \mathcal{O}_Y(D_0)) \subset \Gamma(Y, \mathcal{K}_Y) = K(Y)$. If the Cartier divisor D_0 is defined by $\{(U_i, f_i)\}$ then $\mathcal{O}_Y(D_0)$ is generated by $1/f_i$ over U_i and hence multiplication with f_i yields an isomorphism $\varphi_i : \mathcal{O}_Y(D_0)|U_i \overset{\cong}{\to} \mathcal{O}_{U_i}$ with $\varphi_i(s) = f_i f$. Since $div_0(s)$ is given by $\{(U_i, \varphi_i(s))\}$, $div_0(s) = D_0 + (f)$ as required.

(b) If $D \geq 0$ and $D = D_0 + (f)$ with $f \in K(Y)$, then $(f) \geq -D_0$. Hence $s := f \in \Gamma(Y, \mathcal{O}_Y(D_0))$ and $div_0(s) = D$.

(c) Let s resp. s' correspond to f resp. f' in $K(Y)$. By the consideration in (a) we see that $div_0(s) = div_0(s')$ iff $(f/f') = 0 \in \Gamma(Y, \mathcal{K}_Y^*/\mathcal{O}_Y^*)$. Hence $f/f' \in \Gamma(Y, \mathcal{O}_Y^*) = \mathbb{C}^*$ since Y is projective and irreducible and reduced. □

Definition 2.1.14

(1) A *complete linear system* (of divisors) on Y is the set of all effective divisors D on Y with D linear equivalent to some fixed divisor D_0 on Y. It is denoted by $|D_0|$, that is, we have

$$|D_0| = \{D \in \operatorname{Div} Y \mid D \geq 0,\ D \sim D_0\}.$$

By Lemma 2.1.13 we see that

$$\begin{aligned}|D_0| &= \{div_0(s) \mid s \in \Gamma(Y, \mathcal{O}_Y(D_0)) \setminus \{0\}\}\\ &= \Gamma(Y, \mathcal{O}_Y(D_0))\setminus\{0\}/\mathbb{C}^*\\ &= \mathbb{P}(\Gamma(Y, \mathcal{O}_Y(D_0))\end{aligned}$$

is the projective space of the vector space $\Gamma(Y, \mathcal{O}_Y(D_0)$. Note that $|D_0|$ may be empty.

(2) A *linear system* (of divisors) on Y is a projective subspace $\Lambda \subset |D_0|$ of a complete linear system $|D_0|$, i.e. $\Lambda = V \setminus \{0\}/\mathbb{C}^*$ where $V \subset \Gamma(Y, \mathcal{O}_Y(D_0))$ is a vector subspace. The dimension of Λ is its dimension as projective space, i.e. $\dim V - 1$.

(3) Let $\mathcal{L}$ is an invertible sheaf on Y and $V \subset \Gamma(Y, \mathcal{L})$ a vector space. The *linear system belonging to* V is defined as follows. We have $\mathcal{L} \cong \mathcal{O}_Y(D_0)$ for a unique divisor D_0 by Proposition 2.1.7 and Remark 2.1.7.1, and for any global section $s \in \Gamma(Y, \mathcal{L})$ the divisor of zeros $div_0(s)$ is linear equivalent to D_0 by Lemma 2.1.13. Hence $\Gamma(Y, \mathcal{L})$ gives rise to the complete linear system

$$\mathbb{P}(\Gamma(Y, \mathcal{L})) = \{div_0(s) \mid s \in \Gamma(Y, \mathcal{L}) \setminus \{0\}\}$$

coinciding with $|D_0|$, and V to the linear system

$$\Lambda = \mathbb{P}(V) = \{div_0(s) \mid s \in V \setminus \{0\}\}.$$

Remark 2.1.14.1 We have $\mathcal{O}_Y(D_0) \subset \mathcal{K}_Y = K(Y)$ and by Lemma 2.1.13

$$\Gamma(Y, \mathcal{O}_Y(D_0)) = \{f \in K(Y) |\ f = 0 \text{ or } (f) + D_0 \geq 0\}.$$

If Y is smooth and $D_0 = \sum_{i=1}^r a_i X_i$, we have the following geometric interpretation

$$\begin{aligned}\Gamma(Y, \mathcal{O}_Y(D_0)) = \{f \in K(Y) |\ f = 0 \text{ or } \operatorname{ord}_{X_i}(f) \geq -a_i \text{ and } \operatorname{ord}_X(f) \geq 0\\ \text{for all other prime divisors } X \text{ on } Y\}.\end{aligned}$$

That is, $\Gamma(Y, \mathcal{O}_Y(D_0))$ is the vector space of rational functions on Y with no poles except possibly at the X_i and at X_i with poles of order $\leq a_i$ if a_i is positive and zeros of order $\geq -a_i$ if a_i is negative.

Definition 2.1.15 A point $y \in Y$ is called a *base point* of the linear system Λ if y belongs to the support of every divisor $D \in \Lambda$. Λ is *base-point-free* if its *base locus* (i.e. its set of base points) is empty. An effective divisor Z is called a *fixed component* of the linear system Λ if $D - Z$ is effective for any $D \in \Lambda$. If Z is maximal with respect to this property then Z is called the *fixed part* of Λ and $D - Z$ the *variable part* of D.

Hence any $D \in \Lambda$ can be written as

$$D = Z + (D - Z) \text{ with } Z \text{ and } D - Z \text{ effective,}$$

where the fixed part Z is independent of $D \in \Lambda$ and $D - Z$ has no fixed components. The collection of the variable parts of Λ is a linear subsystem of Λ. As the support of Z belongs to the support of every $D \in \Lambda$ the points of Z are base points of Λ. The set of base points of Λ is clearly a closed algebraic subset of Y.

Remark 2.1.15.1 (1) Let $\mathcal{L}$ be an invertible sheaf on Y and Λ a linear system belonging to the subspace $V \subset \Gamma(Y, \mathcal{L})$. Then $y \in Y$ is a base point of Λ if and only if, for every $s \in V \setminus \{0\}$, the germ s_y belongs to $\mathfrak{m}_y\mathcal{L}_y$, where $\mathfrak{m}_y$ denotes the maximal ideal of $\mathcal{O}_{Y,y}$ (c.f. Definition 2.1.8). Since s_y generates $\mathcal{L}_y$ iff it is not contained in $\mathfrak{m}_y\mathcal{L}_y$, we see that Λ is base point free if and only if $\mathcal{L}$ is generated by its global sections belonging to V.

(2) We can describe the base locus of Λ slightly different. If $V \neq \{0\}$, choose a basis $s_0, \ldots, s_r$ of V and consider the rational map $\Phi_\Lambda : \dashrightarrow \mathbb{P}^r$, $z \mapsto (s_0(z) : \cdots : s_r(z))$. The points where Φ_Λ is not defined is exactly the base locus of Λ. In particular, if D is an ample divisor on Y, then mD is very ample for some m and hence basepoint free, so that $\Phi = \Phi_{|mD|}$ is a morphism from Y to some projective space $\mathbb{P}^r$ with $\mathcal{O}_Y(mD) = \Phi^*(\mathcal{O}_{\mathbb{P}^r}(1))$.

(3) A divisor D is called *nef* if the intersection number $D \cdot C$ is non-negative for all effective divisors C on Y. An ample divisor is nef since $mD = \Phi * H, m > 0$, H a hyperplane in $\mathbb{P}^r$, and hence $D \cdot C = \frac{1}{m} H \cdot \Phi_*(C) \geq 0$.

Remark 2.1.15.2 (1) The linear system Λ comes with a "universal family", the *incidence variety*

$$\mathfrak{N}_\Lambda := \{(y, D) \in Y \times \Lambda \mid y \in D\},$$

a subscheme of $Y \times \Lambda$ which is flat over Λ. To see this, let $s_0, \ldots, s_m$ be a basis of V. Then $D = div_0(s)$ for $s \in V \setminus \{0\}$ can be written as

$$D = div_0(\lambda_0 s_0 + \cdots + \lambda_m s_m)$$

with unique $(\lambda_0 : \cdots : \lambda_m) \in \mathbb{P}^m(\mathbb{C})$. We get

$$\mathfrak{N}_\Lambda = \left\{ (y, \lambda_0 : \cdots : \lambda_m) \in Y \times \mathbb{P}^m \mid \sum_{j=0}^{m} \lambda_j \cdot s_j(y) = 0 \right\} .$$

Let $(y, p) \in \mathfrak{N}_\Lambda, y \in U_i$, with $U_i \subset Y$ open such that $\varphi_i \left(s_j\big|_{U_i} \right) = f_{ij} \in \Gamma(U_i, \mathcal{O}_Y)$ as in Definition 2.1.8. For $p \in \mathbb{P}^m$ choose coordinates $\lambda_0 : \cdots : \lambda_m$ such that p is the center of the chart $V_0 = \{\lambda_0 \neq 0\}$. Then, with affine coordinates $\mu_j = \lambda_j / \lambda_0$,

$$\mathfrak{N}_\Lambda \cap U_i \times V_0 = \left\{ (y, \mu_1, \ldots, \mu_m) \in U_i \times V_0 \mid f_{i,0}(y) + \sum_{j=1}^{m} \mu_j \cdot f_{i,j}(y) = 0 \right\} .$$

Since f_0 is a non-zero divisor in $\mathcal{O}_{Y,y}$, it follows that $\mu_1, \ldots, \mu_m, f_0$ is a non-zero divisor sequence in $\mathcal{O}_{\mathfrak{N}_\Lambda,(y,p)}$ and hence $\mathfrak{N}_\Lambda \to \Lambda$ is flat at (y, p).

(2) Let us describe the linear system Λ explicitly if $\mathcal{L} = \mathcal{O}_Y(D_0)$. Let $(U_i)_{i\in I}$ be an open covering of Y and D_0 defined by $\{(U_i, f_{i0})\}$, $f_{i0} \in \Gamma(U_i, \mathcal{K}_Y^*)$. If $D \geq 0$ is given by $f_i \in \Gamma(U_i, \mathcal{O}_Y)$ then, by definition, $D \sim D_0$ iff D transforms in the same way as f_{i0} over U_{ij}, i.e. $\frac{f_i}{f_{i0}} = \frac{f_j}{f_{j0}}$ in $\Gamma(U_{ij}, \mathcal{K}_Y^*)$ for all i, j.

Consider the complete linear system $|D_0| = \mathbb{P}(\Gamma(Y, \mathcal{O}_Y(D_0)))$ and assume that $D_0 \geq 0$. Then $D_0 \in |D_0|$ and let $s_0, \ldots, s_m$ be a basis of $\Gamma(Y, \mathcal{O}_Y(D_0))$ such that $D_0 = div_0(s_0)$. If $div_0(s_l)$ is given by $(f_{il} \in \Gamma(U_i, \mathcal{O}_Y))_{i\in I}$ then any divisor $D = D_\lambda \in |D_0|$ is given by $(\sum_{l=0}^m \lambda_l f_{il})_{i\in I}$ with unique $(\lambda_0 : \cdots : \lambda_m) \in \mathbb{P}^m$, by definition of div_0. Hence we get

$$|D_0| = \left\{ D_\lambda := \left(F_{i,\lambda} := \sum_{l=0}^m \lambda_l f_{il} \right)_{i\in I} \;\middle|\; \begin{array}{c} \lambda = (\lambda_0 : \cdots : \lambda_m) \in \mathbb{P}^m, \\ \frac{f_{il}}{f_{i0}} = \frac{f_{jl}}{f_{j0}} \; \forall i, j \in I, \; l = 1, \ldots, m \end{array} \right\}$$

Note that $F_{i,\lambda}/F_{j,\lambda} = f_{i0}/f_{j0}$ is independent of $\lambda \in \mathbb{P}^m$. If $\Lambda = \mathbb{P}(V) \subset |D_0|$ is a not necessarily complete linear system with $D_0 \in \Lambda$, then we get the same explicit description of Λ if $s_0, \ldots, s_m$ is a basis of V.

It follows from Lemma 2.1.9 and Proposition 2.1.10, that a complete linear system on $\mathbb{P}^n$ can be identified, for some d, with the projective space

$$|dH| = \Gamma(\mathbb{P}^n, \mathcal{O}_{\mathbb{P}^n}(d)) \setminus \{0\}/\mathbb{C}^* = \mathbb{C}[x_0, \ldots, x_n]_d \setminus \{0\}/\mathbb{C}^*,$$

parameterizing all projective hypersurfaces of degree d.

An m-dimensional linear system $\Lambda \subset |dH|$ is hence given by an $(m+1)$-dimensional vector space $V \subset \mathbb{C}[x_0, \ldots, x_n]_d$. Choosing a basis $F_0, \ldots, F_m$ of V, the linear system Λ of divisors on $\mathbb{P}^n$ is given by the family of hypersurfaces

$$F_\lambda = \lambda_0 F_0 + \cdots + \lambda_m F_m, \; \lambda = (\lambda_0 : \cdots : \lambda_m) \in \mathbb{P}^m.$$

2.1.2 *Bézout's Theorem*

The most important theorem relating global and local properties of projective varieties in projective space is the classical theorem of Bézout, which is a generalization of the fundamental theorem of algebra. We start with the case of plane curves. For a version of Bézout's theorem for higher dimensional projective varieties see Theorem 2.1.18.

2.1.2.1 The Planar Bézout's Theorem

Theorem 2.1.16 (Bézout) *Two plane projective curves $C, D \subset \mathbb{P}^2$ of degrees m and n respectively, which have no component in common, intersect at $m \cdot n$ points, counting multiplicities. That is,*

$$m \cdot n = \sum_{w \in C \cap D} i_w(C, D)$$

with $i_w(C, D) := \dim_{\mathbb{C}} \mathcal{O}_{\mathbb{P}^n,w}/\langle f, g\rangle$ the intersection multiplicity of C and D at w if f resp. g are local equations of C resp. D in some affine open neighbourhood of w.

Remark 2.1.16.1 Let, for the moment, $\mathcal{O}^{an}_{\mathbb{P}^n,w}$ denote the analytic local ring of $\mathbb{P}^n$ at w and f', g' be two analytic local equations of C resp. D in some Euclidean open neighbourhood of w. Since C and D have no common component, the complex space germs (C, w) and (D, w) have no common component and hence the ideals $\langle f, g\rangle$ resp. $\langle f', g'\rangle$ contain a power of the maximal ideal of $\mathcal{O}_{\mathbb{P}^n,w}$ resp. $\mathcal{O}^{an}_{\mathbb{P}^n,w}$. This shows that

$$i_w(C, D) = \dim_{\mathbb{C}} \mathcal{O}^{an}_{\mathbb{P}^n,w}/\langle f', g'\rangle.$$

Proof We sketch a proof of Bézout's theorem in the classical manner by using resultants. Let the two plane curves C and D be given by homogeneous polynomials $F, G \in \mathbb{C}[x_0, x_1, x_2]$ of degrees m and n, respectively. By assumption, the curves intersect at finitely many points.

Let $\boldsymbol{p} = (p_0 : p_1 : p_2)$ and $\boldsymbol{q} = (q_0 : q_1 : q_2)$ be generic points of $\mathbb{P}^2$ (such that the line $\overline{\boldsymbol{pq}}$ is neither contained in C nor in D). The point $\boldsymbol{w} = \lambda_0\boldsymbol{p} + \lambda_1\boldsymbol{q}$ runs over $\overline{\boldsymbol{pq}}$ as $(\lambda_0 : \lambda_1)$ runs over $\mathbb{P}^1$. The polynomials $F(\boldsymbol{w})$, $G(\boldsymbol{w})$, considered as elements of $(\mathbb{C}[\boldsymbol{p}, \boldsymbol{q}])[\lambda_0, \lambda_1]$, are homogeneous in λ_0, λ_1, and we consider their resultant $R \in \mathbb{C}[\boldsymbol{p}, \boldsymbol{q}] = \mathbb{C}[p_0, p_1, p_2, q_0, q_1, q_2]$ which has the following properties (cf., e.g., [GKZ, Ch. 12]):

- $R(\boldsymbol{p}, \boldsymbol{q})$ vanishes iff the line $\overline{\boldsymbol{pq}}$ passes through an intersection point $\boldsymbol{w}$ of the curves C and D,
- R does not vanish identically, since there is a line which does not pass through any intersection point of the curves,
- R is a homogeneous polynomial of degree $2mn$.

On the other hand, a point $\boldsymbol{w}$ lies on the line $\overline{\boldsymbol{pq}}$ iff

$$\Delta(\boldsymbol{p}, \boldsymbol{q}, \boldsymbol{w}) := \det \begin{pmatrix} p_0 & p_1 & p_2 \\ q_0 & q_1 & q_2 \\ w_0 & w_1 & w_2 \end{pmatrix} = 0\,.$$

This implies immediately that

$$R(\boldsymbol{p},\boldsymbol{q}) = A \cdot \prod_{\boldsymbol{w}} \Delta(\boldsymbol{p},\boldsymbol{q},\boldsymbol{w})^{\sigma_{\boldsymbol{w}}}, \quad A = \text{const} \neq 0,$$

where $\boldsymbol{w}$ runs over all the intersection points of the given curves. Finally, we notice that $\sigma_{\boldsymbol{w}} = i_{\boldsymbol{w}}(C, D)$ is the intersection number of C and D at w (see, for example, [BrK, Section 6.1] or [GLS6, Proposition I.3.10]).

The latter can be seen as follows: assume that the coordinates are chosen sufficiently general, such that no intersection point lies on the line $\{x = 0\}$, such that the origin $(0\!:\!0\!:\!1)$ neither lies on C nor on D, and such that each line through $(0\!:\!0\!:\!1)$ contains at most one intersection point of C and D. We specialize $\boldsymbol{p} := (1\!:\!u\!:\!0), \boldsymbol{q} = (0\!:\!0\!:\!1)$, and consider $R(\boldsymbol{p},\boldsymbol{q}) = R(u)$ as the resultant of the univariate polynomials $F(1,u,\lambda) = a_0\lambda^m +$ lower terms, $G(1,u,\lambda) = b_0\lambda^n +$ lower terms ($a_0, b_0 \neq 0$ due to our assumptions). Now, suppose that $\boldsymbol{w} = (1\!:\!w_1\!:\!w_2)$ is an intersection point. Then we get over the field of Puiseux series, factorizations into linear factors (cf. [GLS6, Proposition I.3.4]),

$$F(1,u,\lambda) = a_0 \cdot \prod_{i=1}^{m} \big(\lambda - y_i(t)\big), \quad G(1,u,\lambda) = b_0 \cdot \prod_{j=1}^{n} \big(\lambda - y'_j(t)\big),$$

with $t := (u - w_1)^{1/N}$, and $y_1(0), \ldots, y_m(0)$ being the zeros of $F(1, w_1, \lambda)$, resp. $y'_1(0), \ldots, y'_m(0)$ the zeros of $G(1, w_1, \lambda)$. In particular, the well-known formula for the univariate resultant gives

$$R(u) = a_0^n b_0^m \cdot \prod_{i=1}^{m}\prod_{j=1}^{n} \big(y_i(t) - y'_j(t)\big).$$

Since, by assumption, $\boldsymbol{w}$ is the unique intersection point of C and D with coordinates $(1\!:\!w_1\!:\!\lambda)$, Halphen's formula (see [GLS6, Proposition I.3.10]), gives that $R(u)$ vanishes of order $i_{\boldsymbol{w}}(C, D)$ at $u = w_1$. On the other hand, our assumptions imply that $\Delta(\boldsymbol{p},\boldsymbol{q},\boldsymbol{w}) = u - w_1$ and that $u - w_1$ are pairwise different for the intersection point $\boldsymbol{w}$ of C and D. □

2.1.2.2 General Bézout's Theorem

The general (classical) theorem of Bézout is about the intersection of two arbitrary closed subschemes X and Y in $\mathbb{P}^n$. It expresses the product of their degrees as the sum of the degrees of the irreducible components Z_i of their intersection, counted with the intersection multiplicity $i_{Z_i}(X, Y)$, provided the intersection is proper. This generalizes the planar Bézout's Theorem 2.1.16. The main problem is to define the "correct" intersection multiplicity in this generality. We just recall the result and refer to the monograph [FOV], which contains an elegant proof and further generalizations.

Recall that for a local Noetherian ring A with maximal ideal $\mathfrak{m}$, a finitely generated A-module M and an $\mathfrak{m}$-primary ideal $\mathfrak{q} \subset \mathfrak{m}$, the *Hilbert–Samuel function* $HS_{\mathfrak{q}}(M, k) = \lg(M/\mathfrak{q}^{k+1}M)$ (lg denotes length) is a polynomial $P_{\mathfrak{q}}(M, k)$ for large k, the *Hilbert–Samuel polynomial*

$$P_{\mathfrak{q}}(M, t) = a_d t^d + \cdots \text{ lower terms in } d \ \ldots \in \mathbb{Q}[t],$$

a polynomial of degree $d = \dim M$ and leading coefficient a_d. The integer

$$e(M, \mathfrak{q}) := d!\, a_d$$

is called the *(Hilbert–Samuel) multiplicity of M w.r.t.* $\mathfrak{q}$ (cf. e.g. [GrP, Section 5.5, 5.6]).

Remark 2.1.16.2 We mention two important properties of the multiplicity. For proof see [FOV, Proposition 1.2.7 and 1.2.8].

(1) (Additivity) If $0 \to M' \to M \to M'' \to 0$ is an exact sequence of A-modules then

$$e(M, \mathfrak{q}) = e(M', \mathfrak{q}) + e(M'', \mathfrak{q}).$$

(2) (Associativity) $e(M, \mathfrak{q}) = \sum\limits_{\mathfrak{p}} \lg(M_{\mathfrak{p}}) \cdot e(A/\mathfrak{p}, \mathfrak{q})$ where the sum is running through all minimal primes of M with $\dim A/\mathfrak{p} = \dim M$. In particular, the multiplicity does not count embedded or lower dimensional components of M.

Consider now two closed subschemes $X, Y \subset W$ of a pure dimensional smooth algebraic scheme W and let Z be any irreducible component of the intersection $X \cap Y$. To define the intersection multiplicity, we use the "diagonal trick". Denote by Δ the image and by I_Δ the ideal of the diagonal embedding $W \hookrightarrow W \times W$. Note that for $X \cap Y \neq \emptyset$ we have

$$\dim Z \geq \dim_Z X + \dim_Z Y - \dim_Z W \tag{2.1.2.1}$$

for any irreducible component Z of $X \cap Y$, where $\dim_Z$ denotes the dimension at the generic point corresponding to Z. This follows since $X \cap Y = X \times Y \cap \Delta$ and since Δ is locally defined by $\dim W$ equations. The intersection is called *proper at* Z, if equality holds in (2.1.2.1). We consider Z as a subscheme of $X \times Y$ via the diagonal embedding and denote by $\mathcal{O}_{X \times Y, Z}$ the local ring of $X \times Y$ at the generic point corresponding to Z.

Definition 2.1.17 The *intersection multiplicity* of X and Y in W is the integer

$$i_Z(X, Y) := e(\mathcal{O}_{X \times Y}, I_\Delta \cdot \mathcal{O}_{X \times Y}).$$

Recall that the *degree* of a projective subscheme X of $\mathbb{P}^n$ of dimension r is defined as

$$\deg X = r! \cdot \text{ leading coefficient of the Hilbert polynomial of } X.$$

Note that the degree can also be expressed as a multiplicity. Let $R = \mathbb{C}[x_0, \ldots, x_n]$ and and $\mathfrak{m} = \langle x_0, \ldots, x_n \rangle$ the maximal homogeneous ideal of R. For M a finitely generated graded R-module we call $\deg M := e(M_\mathfrak{m}, \mathfrak{m}R_\mathfrak{m})$ the *degree of* M. Taking for M the graded module R/I, with I the homogeneous ideal defining X, we have $\deg X = \deg R/I$, i.e., $\deg X$ is the multiplicity of the affine cone $CX \subset \mathbb{C}^{n+1}$ at 0.

Now we can formulate the general version of Bézout's theorem.

Theorem 2.1.18 (Bézout) *Let $X, Y \subset \mathbb{P}^n$ be closed pure dimensional subschemes with proper intersection, i.e.* $\dim X \cap Y = \dim X + \dim Y - n$. *If $X \cap Y \neq \emptyset$ then*

$$\deg X \cdot \deg Y = \sum_{Z \subset X \cap Y} i_Z(X, Y) \deg Z,$$

where the sum runs through all irreducible components Z of $X \cap Y$.

For the proof we refer to [FOV, Theorem 1.4.4]. A proof of Bézout's theorem in the framework of a general intersection theory is given in [Ful3, Chapter 8.4]. A slightly less general version (X reduced and irreducible, Y a divisor) is proved in [HaR2, Theorem I.7.7].

Remark 2.1.18.1 (1) We like to mention that for X, Y closed subvarieties of $\mathbb{P}^n$ we have

$$\dim X + \dim Y - n \geq 0 \Rightarrow X \cap Y \neq \emptyset.$$

In fact, consider the affine cones $CX, CY \subset \mathbb{C}^{n+1}$ (given by the homogeneous ideals in $\mathbb{C}[x_0, \ldots, x_n]$). Then $\dim CX = \dim X + 1$, $\dim CY = \dim Y + 1$ and by formula (2.1.2.1) we have

$$\dim CX \cap CY \geq \dim X + \dim Y + 2 - (n + 1) \geq 1.$$

Since $C(X \cap Y) = CX \cap CY$ we get $\dim X \cap Y \geq 0$.

(2) Let X and Y meet properly in the smooth algebraic scheme W at an irreducible component Z of $X \cap Y$. It is proved in [FOV, Proposition 1.4.2] that

$$i_Z(X, Y) \leq \lg(\mathcal{O}_{X \cap Y, Z})$$

with equality iff $\mathcal{O}_{X,Z}$ and $\mathcal{O}_{Y,Z}$ are Cohen–Macaulay.

If C, D are two plane curves, then $\mathcal{O}_{C,w}$ and $\mathcal{O}_{D,w}$ are Cohen–Macaulay and the intersection multiplicity of C and D at $w \in C \cap D$ defined here coincides with $i_w(C, D)$ used in Theorem 2.1.16 (see also Example 2.1.3.1). Hence, Theorem 2.1.18 generalizes Theorem 2.1.16.

(3) A useful criterion for the intersection multiplicity being one is proved in [FOV, Proposition 1.4.3]. Let X and Y be closed subschemes of the smooth algebraic

scheme W (not necessarily meeting properly) and let Z be an irreducible component of $X \cap Y$. Assume that $\mathcal{O}_{X,Z}$ and $\mathcal{O}_{Y,Z}$ are pure dimensional and without embedded components. Then

$$i_Z(X, Y) = 1 \Leftrightarrow \text{ the local rings } \mathcal{O}_{X,Z}, \mathcal{O}_{Y,Z}, \mathcal{O}_{X\cap Y,Z} \text{ are regular.}$$

In the case of a proper intersection, $\mathcal{O}_{X\cap Y,Z}$ being regular just means that X and Y meet *transversally* at Z.

(4) Another property of the intersection multiplicity is the following. If $X \subset \mathbb{P}^n$ is not pure dimensional then

$$\deg X = \deg X^*$$

where X^* is the pure dimensional part of X, which follows immediately from Remark 2.1.16.2.

In an affine chart $U \subset \mathbb{P}^n$, the *pure dimensional part* $X^* \cap U$ is the union of the irreducible components of $X \cap U$ defined by the primary ideals belonging to minimal primes of $\langle 0 \rangle \subset \Gamma(U, \mathcal{O}_X)$.

Bézout's theorem has many applications. We just mention the following

Corollary 2.1.19 *Let $X \subset \mathbb{P}^n$ be a closed subscheme and $L \subset \mathbb{P}^n$ a linear subscheme of dimension $n - \dim X$ meeting X in finitely many points. Then*

$$\deg X = \sum_{z \in X^* \cap L} i_z(X^*, L).$$

Hence, if L meets X^* transversally then $\deg X = \#X^* \cap L$.

2.1.3 Bertini's and Noether's Theorems

2.1.3.1 Bertini's Theorems

In general, one uses the notion of *"Bertini theorems"* to refer to statements on the geometry of generic members of linear systems. Here we present a few of them, concerning linear systems on curves and on projective algebraic surfaces. We use the notations of the preceding sections.

Theorem 2.1.20 *Let D_0 be an effective divisor on a reduced, irreducible projective curve C, and let $\Lambda \subset |D_0|$ be a linear subsystem of positive dimension. There exists a Zariski-open dense subset $U \subset \Lambda$ such that $D \in U$ (a generic element) has no multiple points outside* $\operatorname{Sing}(C)$ *and the set of base points of the linear system Λ.*

Proof Let C be embedded in a projective space $\mathbb{P}^n$. We construct a system of hypersurface sections of C containing Λ and prove the desired property for this bigger system.

Choose a basis $s_0, \ldots, s_r \in H^0(C, \mathcal{O}_C(D_0))$, $r = \dim \Lambda \geq 1$, of the linear subspace $L \subset H^0(C, \mathcal{O}_C(D_0))$ such that $(L \setminus \{0\})/\mathbb{C}^* = \Lambda$ and $div_0(s_0) = D_0$. The ratios $s_i/s_0, i = 1, \ldots, r$, are non-trivial rational functions on C. By Remark 2.1.10.2 we can represent them as $s_i/s_0 = (F_i/F_0)|C$, $i = 1, \ldots, r$, with $F_0, F_1, \ldots, F_r \in \mathbb{C}[x_0, \ldots, x_n]$ homogeneous polynomials of the same degree, and we have $\operatorname{div}(F_0)\big|_C = D_0 + D'$ with some effective divisor D'. Setting $F_\lambda := \lambda_0 F_0 + \cdots + \lambda_r F_r$, $\lambda = (\lambda_0 : \cdots : \lambda_r) \in \mathbb{P}^r$, there exists for any $D \in \Lambda$ a unique λ such that $D - D_0 = div_0(s_\lambda/s_0) = div_0((F_\lambda/F_0)|C) = \operatorname{div}(F_\lambda)\big|_C - \operatorname{div}(F_0)\big|_C$. Hence, for any divisor $D \in \Lambda$, there exists a unique element $\lambda = (\lambda_0 : \cdots : \lambda_r) \in \mathbb{P}^r$ such that

$$D + D' = \operatorname{div}(F_\lambda)\big|_C = \operatorname{div}(\lambda_0 F_0 + \cdots + \lambda_r F_r)\big|_C \,.$$

Now let $S = \operatorname{Sing}(C) \cup \{\text{base points of } \Lambda + D'\}$ and consider the projection

$$\pi : \{(z, \lambda) \in (C \setminus S) \times \mathbb{P}^r \,|\, F_\lambda(z) = 0\} \to \mathbb{P}^r,$$

which is flat by Remark 2.1.15.2. Hence (z, λ) is a point of the algebraic scheme $\operatorname{Sing}(\pi)$ iff z is a multiple point of $\operatorname{div}(F_\lambda)\big|_{C\setminus S}$, i.e. $\operatorname{div}(F_\lambda)\big|_{C\setminus S}$ has no multiple points iff $\lambda \in \mathbb{P}^r \setminus \pi(\operatorname{Sing}(\pi))$. If the complement of the Zariski closure of $\pi(\operatorname{Sing}(\pi))$ is non-empty, it is open and dense in $\mathbb{P}^r$ (and corresponds to the generic elements $D \in U$ of the theorem). Therefore it suffices to show the claim for some 1-dimensional subsystem of Λ.

In the following we consider regular functions as complex analytic maps and work with the Euclidean topology. Let $z_0 \in C \setminus S$, and let $t \mapsto z(t)$, $z(0) = z_0$, be a local analytic parametrization of the germ (C, z_0). Without loss of generality we may assume that $F_0(z_0) = 0$ and $F_1(z_0) \neq 0$. The 1-dimensional subsystem $\operatorname{div}(F_0 + \lambda F_1)\big|_C$ of Λ is given as the intersection of the hypersurface $F_0 + \lambda F_1$ with C and satisfies

$$F_0\big(z(t)\big) + \lambda F_1\big(z(t)\big)\big) = 0\,.$$

The analytic function $\lambda(t) := -F_0(z(t))/F_1(z(t))$ vanishes at 0 with some order $p > 0$. Choosing a pth root of unity of λ, we get a continuous inverse $\lambda \mapsto t(\lambda)$, defined in a neighbourhood of 0 and locally analytic outside 0, satisfying $z(t(0)) = z_0$, $F_0\big(z(t(\lambda))\big) + \lambda F_1\big(z(t(\lambda))\big)\big) = 0$.

Differentiating with respect to λ gives

$$F_1\big(z(t)\big) + \sum_{i=0}^{n} \left(\frac{\partial F_0}{\partial x_i}\big(z(t)\big) + \lambda \frac{\partial F_1}{\partial x_i}\big(z(t)\big) \right) \cdot \frac{dx_i}{dt}(t) \cdot \frac{dt}{d\lambda}(\lambda) = 0$$

for λ close to 0. On the other hand, the assumption that $z(t)$ is a multiple point of the divisor $\operatorname{div}(F_0 + \lambda F_1)\big|_C$ for t in an open set (equivalently $z(t(\lambda))$ is a multiple point for λ in an open set) would imply

$$\sum_{i=0}^{n}\left(\frac{\partial F_0}{\partial x_i}\big(z(t)\big)+\lambda\frac{\partial F_1}{\partial x_i}\big(z(t)\big)\right)\cdot\frac{dx_i}{dt}(t) \;=\; 0\,,$$

and the last two equations would give $F_1(z(t)) \equiv 0$.[4] Since $F_1(z_0) \neq 0$, we obtain a contradiction to the assumption that $z(t)$ is a multiple point of $\mathrm{div}(F_0 + \lambda F_1)\big|_C$ for λ in some open set. Hence $\mathrm{div}(F_0 + \lambda F_1)\big|_C$ has multiple points outside S only for finitely many λ. □

Corollary 2.1.21 *Let Λ be a positive dimensional linear system of curves on a smooth projective algebraic surface Σ. If Λ has no fixed part, then a generic element of Λ is reduced.*

Proof This immediately follows from Theorem 2.1.20, if one restricts the linear system to a generic hyperplane section C of Σ. Here C is reduced, irreducible (even smooth), since the space of hyperplanes which are tangent to Σ is a proper closed subset of the space of all hyperplanes. □

Corollary 2.1.22 *Under the hypotheses of Corollary 2.1.21, a generic curve $C \in \Lambda$ is non-singular outside the base points of Λ.*

Proof It is sufficient to consider the case of a one-dimensional system Λ. Then the union of the singular loci $\mathrm{Sing}(C)$, $C \in \Lambda$, is either finite, or a (reduced) algebraic curve $B \subset \Sigma$. In the latter case we would obtain on at least one irreducible component of B a linear system on B, given by the intersection of B with $C \in \Lambda$, with a moving multiple point, contradicting Theorem 2.1.20. □

Corollary 2.1.22 can easily be generalized to higher dimensions:

Proposition 2.1.23 *A generic member of a positive dimensional linear system Λ on a smooth algebraic scheme X is a hypersurface which is non-singular outside the base points of Λ.*

Proof We can assume that $\dim \Lambda = 1$. Let $Y \subset X$ be the locus of all singular points of all the members of Λ, and $B \subset X$ the base locus of Λ. If $Y \setminus B$ is finite, we are done. If $Y \setminus B$ has a positive dimensional component, then there exists a reduced, irreducible curve $C \subset Y \setminus B$, on which Λ cuts off a (one-dimensional) linear system with a moving multiple point along C, in contradiction to Theorem 2.1.20. □

Theorem 2.1.24 *Let Λ be a positive dimensional linear system of curves on a smooth projective algebraic surface Σ. Assume that Λ has no fixed part and all curves $C \in \Lambda$ are either reducible or non-reduced. Then the following holds true:*

(1) Two curves in Λ either have a common component or meet only in the base points of Λ.

[4] Here we use the fact that Λ is a linear system, and not just an algebraic family.

(2) *There exists a Zariski-open dense subset $U \subset \Lambda$ such that the irreducible components of all curves $C \in U$ (generic curves) belong to a one-dimensional algebraic family, which is a linear system if Σ is rational.*

Proof (1) Let $C \in \Lambda$ be a generic element and $C_1, \ldots, C_n$, $n \geq 2$, its irreducible components. Since C is generic, n is the minimal possible number of irreducible components for curves in Λ (because in a given linear system of curves on a surface, the set of curves splitting into at least $k \geq 2$ components is a lower dimensional closed subset). This implies, in particular, that C_i and C_j, $j \neq i$, meet only at base points of Λ, since, by Corollary 2.1.22, C is non-singular outside the base locus.

We prove (1) just for two generic curves. So, let $C' \in \Lambda$ be another generic curve, close to C, splitting into irreducible components $C'_1, \ldots, C'_n$, such that C_i turns into C'_i when deforming C to C' locally in Λ, and we may assume that C and C' have no component in common, i.e. $C_i \neq C'_i$ for all i. Two components C'_i and C'_j intersect at the same (base) points z as C_i and C_j. Moreover, $\mathrm{mt}(C'_i, z) = \mathrm{mt}(C_i, z)$, $\mathrm{mt}(C'_j, z) = \mathrm{mt}(C_j, z)$, in view of the semicontinuity of multiplicity (cf. [GLS6, Theorem I.1.81]). Hence, after blowing up Σ at the intersection points of C_i and C_j, we obtain a linear system Λ^* on the blown-up surface Σ^*, generated by the strict transforms of the curves in Λ close to C. Clearly, $\dim \Lambda^* = \dim \Lambda$, and the system Λ^* has no fixed part. So, we can continue to blow up points in Σ^* infinitely near to points in the base locus until we come to a linear system, still called Λ^*, of curves C^* on the blown up surface Σ^* such that

$$C_i^* \cdot C_j^* = 0\,, \qquad 1 \leq i < j \leq n\,. \tag{2.1.3.1}$$

Since any component C_i^* changes in a non-trivial algebraic family when varying C^* in Λ^*, we have $(C_i^*)^2 \geq 0$. By the Hodge index theorem applied to C_i^* and C_j^*, $i \neq j$,[5] (see [BHPV, Corollary IV.2.16], Theorem 3.2.3)

$$(C_i^*)^2 = 0\,, \qquad i = 1, \ldots, n\,. \tag{2.1.3.2}$$

From (2.1.3.1) and (2.1.3.2) we derive that $D \cdot D' = (C^*)^2 = 0$ for any two curves D, D' in Λ^*. Hence, any two curves in Λ^* do not intersect, which means that the images of these curves in Λ meet only in the base points, implying that (1) holds for generic curves in Λ. But then it holds for any two curves in Λ, since the intersection points of two curves move continuously when varying the curves in Λ.

(2) Let $\Lambda_1 \subset \Lambda$ be a one-dimensional linear subsystem containing C and some other generic curve $C' \in \Lambda$, and assume that, with the above notation, $C_1 \neq C'_1$. The germ of Λ_1 at C generates a germ of a non-trivial one-dimensional algebraic family of irreducible curves on Σ, with fibre over C being C_1. Consider the blown up family Λ_1^*. After trivializing the monodromy by a finite base change, the germ of Λ_1^* at C^* can be extended to all generic curves in Λ_1^*, leading to a one-dimensional family Φ of irreducible curves in Σ, whose base is a finite covering of an open dense subset

[5]Here we need that all curves in Λ are reducible.

of $\Lambda_1 = \mathbb{P}^1$. Since the curves in Λ_1 are moving in Σ, they form a 2-dimensional family which projects onto an open dense subset of Σ. In view of (2.1.3.1), (2.1.3.2), we obtain that *all* irreducible components of generic curves in Λ belong to the same family Φ.

If Σ is rational, then the irreducible components of generic curves in Λ generate a linear system[6] without fixed part, which is one-dimensional in virtue of (2.1.3.1), (2.1.3.2). □

Remark 2.1.24.1 The argument in the proof of Theorem 2.1.24 implies Bertini's theorem in the following traditional formulation:

Let $\Sigma \subset \mathbb{P}^N$ be a smooth algebraic surface, and let Λ be the linear system cut out on Σ by a family of hypersurfaces $\lambda_0 F_0 + \cdots + \lambda_n F_n$, $(\lambda_0 : \ldots : \lambda_n) \in \mathbb{P}^n$, with $F_0, \ldots, F_n \in \mathbb{C}[x_0, \ldots, x_N]$ homogeneous of the same degree.

If all members of Λ are reducible, then the image of the rational map on Σ given by $z \mapsto \big(F_0(z) : \ldots : F_n(z)\big) \in \mathbb{P}^n$ (well-defined on the open dense subset of Σ where not all F_i vanish) is one-dimensional. In other words, if the image has dimension two, then the generic element of Λ is irreducible.

There are many other Bertini type theorems. For a rather exhaustive and general treatment we like to refer to [FOV, Section 3.4].

2.1.3.2 Noether's "$AF + BG$" Theorem

While *Hilbert's Nullstellensatz* answers the question if a polynomial belongs to the radical of a given polynomial ideal, Noether's theorem tells us when a polynomial belongs to the ideal itself. We restrict ourselves to the case of plane curves, although the results easily extend to higher dimensions.

Theorem 2.1.25 (Noether's fundamental theorem)

(1) Let $f, g, h \in \mathbb{C}[x, y]$ be polynomials, and let f and g be coprime. If, for any intersection point $(u, v) \in \mathbb{C}^2$ of the curves $\{f = 0\}$, $\{g = 0\}$, we have

$$h(x, y) = \alpha(x-u, y-v) \cdot f(x, y) + \beta(x-u, y-v) \cdot g(x, y) \qquad (2.1.3.1)$$

with some (formal) power series $\alpha, \beta \in \mathbb{C}[[X, Y]]$, then there exist polynomials $a, b \in \mathbb{C}[x, y]$ such that $h = af + bg$.

In other words, if the germ of h in $\mathbb{C}[[x-u, y-v]]$ belongs to the ideal $\langle f, g\rangle\mathbb{C}[[x-u, y-v]]$ for all intersection points (u, v) of $\{f = 0\}$ and $\{g = 0\}$, then h belongs to the ideal $\langle f, g\rangle \subset \mathbb{C}[x, y]$.

---

[6]It is well-known that homologous curves on a complex rational surface are linearly equivalent, which can easily be derived from the fact that any rational surface is a blow up of the plane or a rational ruled surface (see, for instance, [BHPV, Theorem VI.1.1]).

(2) *Let $F, G, H \in \mathbb{C}[x_0, x_1, x_2]$ be homogeneous polynomials, and let F, G be coprime. If, for any intersection point $z \in \mathbb{P}^2$ of the curves $\{F = 0\}$ and $\{G = 0\}$, there are affine coordinates x, y such that $z = (u, v)$ and the polynomials*

$$f(x, y) = F(x_0(x, y), x_1(x, y), x_2(x, y)),$$
$$g(x, y) = G(x_0(x, y), x_1(x, y), x_2(x, y)),$$
$$h(x, y) = H(x_0(x, y), x_1(x, y), x_2(x, y))$$

satisfy (2.1.3.1), then there are polynomials $A, B \in \mathbb{C}[x_0, x_1, x_2]$ such that

$$H = AF + BG .$$

In other words, H is in the (homogeneous) ideal $\langle F, G\rangle \subset \mathbb{C}[x_0, x_1, x_2]$.

Proof (1) First, we claim that the power series α, β in (2.1.3.1) can be replaced by rational functions, regular at (u, v). Let $\varphi \in \mathbb{C}[x, y]$ vanish at all intersection points of $\{f = 0\}$, $\{g = 0\}$, but not at (u, v). Then, by Hilbert's Nullstellensatz, there is an integer k such that

$$(x-u)^i \cdot (y-v)^j \cdot \varphi^k \in \langle f, g\rangle \subset \mathbb{C}[x, y] \quad \text{for} \quad i + j \geq k . \tag{2.1.3.2}$$

Denote by α_k, β_k the k-jets of α, β, respectively. Since, by assumption, the k-jet of the polynomial

$$h' := h - \alpha_k(x-u, y-v) \cdot f - \beta_k(x - u, y - v) \cdot g \in \mathbb{C}[x-u, y-v]$$

vanishes, (2.1.3.2) implies $h'\varphi^k \in \langle f, g\rangle$, and, hence, $h\varphi^k \in \langle f, g\rangle$. Thus, for any point $z \in M = \{f = 0\} \cap \{g = 0\}$, there are polynomials P_z, Q_z, $R_z \in \mathbb{C}[x, y]$, $R_z(z) \neq 0$, such that $R_z h = P_z f + Q_z g$, proving the claim.

By Hilbert's Nullstellensatz, there exist polynomials ψ_f, ψ_g, $\psi_z \in \mathbb{C}[x, y]$, $z \in M$, such that $1 = \psi_f f + \psi_g g + \sum_{z \in M} \psi_z R_z$. Together, this implies

$$h = \left(\psi_f h + \sum_{z \in M} \psi_z P_z\right) \cdot f + \left(\psi_g h + \sum_{z \in M} \psi_z Q_z\right) \cdot g .$$

(2) Without loss of generality, suppose that all the intersection points lie outside the line $\{x_0 = 0\}$. By (1), there exists polynomials $a, b \in \mathbb{C}[x, y]$ such that $H(1, x, y) = a \cdot F(1, x, y) + b \cdot G(1, x, y)$, which after homogenization turns into

$$x_0^s H = AF + BG , \qquad A, B \in \mathbb{C}[x_0, x_1, x_2] . \tag{2.1.3.3}$$

Assume that $s > 0$. Representing any $P \in \mathbb{C}[x_0, x_1, x_2]$ as $P = x_0 P_0 + P_1$, $P_1 \in \mathbb{C}[x_1, x_2]$, and applying this to (2.1.3.3), we obtain $A_1 F_1 + B_1 G_1 = 0$.

Since the curves $\{F = 0\}$, $\{G = 0\}$ and $\{x_0 = 0\}$ have no intersection, the homogeneous polynomials F_1, G_1 are coprime, and, hence, $A_1 = \varphi \cdot G_1$, $B_1 = -\varphi \cdot F_1$. Plugging the latter formulae into (2.1.3.3), we obtain

$$x_0^s H = x_0 \cdot ((A_0 - G_0)F + (B_0 + F_0)G) ,$$

reducing the exponent of x_0 in the left-hand side of (2.1.3.3), and, by induction, leading to the required statement. □

Remark 2.1.25.1 Let $f, g, h \in \mathbb{C}[x, y]$, and assume that the curves defined by f and g intersect only at non-singular points and all intersections are transversal. Then Noether's fundamental theorem implies that $h \in \langle f, g\rangle \subset \mathbb{C}[x, y]$ iff the curve $\{h = 0\}$ contains $\{f = 0\} \cap \{g = 0\}$. The analogous statement for projective plane curves holds, too.

A useful sufficient condition for (2.1.3.1) is given by the following theorem. Sometimes Theorems 2.1.25 and 2.1.26 together are called Noether's theorem.

Theorem 2.1.26 *Let $f, g, h \in \mathbb{C}[[x, y]]$, and let f and g be coprime. Moreover, let f be reduced, $f = f_1 \cdot \ldots \cdot f_r$, $f_j \in \mathbb{C}[[x, y]]$, its irreducible decomposition, and Π a generic polar of f, that is, $\Pi = a\frac{\partial f}{\partial x} + b\frac{\partial f}{\partial y}$ with $(a, b) \in \mathbb{C}^2 \setminus \{0\}$ such that the line $\{bx - ay = 0\}$ is not tangent to $V(f)$ at the origin. If*

$$i_0(h, f_j) \geq i_0(g, f_j) + i_0(\Pi, f_j) - \mathrm{mt}\ f_j + 1 , \tag{2.1.3.4}$$

for any $j = 1, \ldots, r$, then there exist $\alpha, \beta \in \mathbb{C}[[x, y]]$ such that $h = \alpha f + \beta g$.

Proof We proceed by induction on r, the number of local branches of f.
Step 1. Let $r \geq 2$, and write $f = f_1 \cdot f'$ with $f' = f_2 \cdot \ldots \cdot f_r \in \mathbb{C}[[x, y]]$. Moreover, let Π_1, Π' be generic polars of f_1, respectively f'. By the product rule, we have $\Pi = \Pi_1 f' + \Pi' f_1$ and therefore $i_0(\Pi, f_1) = i_0(\Pi_1 f', f_1) = i_0(\Pi_1, f_1) + i_0(f', f_1)$, and the assumption (2.1.3.4) implies

$$\begin{aligned} i_0(h, f_1) &\geq i_0(g, f_1) + i_0(\Pi_1, f_1) + i_0(f', f_1) - \mathrm{mt}\ f_1 + 1 \\ &= i_0(gf', f_1) + i_0(\Pi_1, f_1) - \mathrm{mt}\ f_1 + 1 . \end{aligned}$$

Hence, by induction hypothesis, there are power series $\alpha_1, \beta_1 \in \mathbb{C}[[x, y]]$ such that $h = \alpha_1 f_1 + \beta_1(gf')$. For $j = 2, \ldots, r$, we obtain

$$\begin{aligned} i_0(\alpha_1, f_j) &= i_0(h, f_j) - i_0(f_1, f_j) \\ &\geq i_0(g, f_j) + i_0(\Pi, f_j) - \mathrm{mt}\ f_j + 1 - i_0(f_1, f_j) \\ &= i_0(g, f_j) + i_0(\Pi', f_j) - \mathrm{mt}\ f_j + 1 , \end{aligned}$$

and, again by induction hypothesis, there are power series $\alpha, \beta_2 \in \mathbb{C}[[x, y]]$ such that

$$h=\alpha_1 f_1+\beta_1(gf')=(\alpha f'+\beta_2 g)f_1+\beta_1(gf')=\alpha f+(\beta_2 f_1+\beta_1 f')g\,.$$

Step 2a. Let $r=1$, and assume that f is non-singular. Then the generic polar $\Pi\in\mathbb{C}[[x,y]]$ is a unit and (2.1.3.4) turns into $i_0(h,f)\geq i_0(g,f)$. After applying a local analytic isomorphism, we can assume that $f=y$ and the latter inequality implies that we can write $g=yg'+g''$, $h=yh'+rg''$ with $g'',r\in\mathbb{C}[[x]]$. Consequently, $h=(h'-rg')y+rg$.
Step 2b. Let $r=1$, $m:=\mathrm{mt}\ f\geq 2$, and assume that the tangent of f is not x. Let $\varphi:\mathbb{C}[[x,y]]\to\mathbb{C}[[x,y]]$ be the covering transformation $(x,y)\mapsto(x^m,y)$. Then

- $\phi:=f\circ\varphi$ decomposes into m non-singular branches, $\phi=\phi_1\cdot\ldots\cdot\phi_m$;
- x is not a tangent of ϕ;
- $i_0(a\circ\varphi,b\circ\varphi)=m\cdot i_0(a,b)$ for any $a,b\in\mathbb{C}[[x,y]]$;
- the power series $h\circ\varphi$, $g\circ\varphi$, $\phi_y=\frac{\partial f}{\partial y}\circ\varphi$ are invariant under transformations of the form $x\mapsto\varepsilon x$, $\varepsilon^m=1$, while the branches ϕ_j are permuted.

In particular, we obtain for $j=1,\ldots,m$

$$\begin{aligned} i_0\big(x^{m-1}(h\circ\varphi),\,\phi_j\big)&=m-1+i_0(h,f)\ \geq\ i_0(g,f)+i_0\big(\tfrac{\partial f}{\partial y},f\big)\\ &=i_0(g\circ\varphi,\,\phi_j)+i_0(\phi_y,\phi_j)\,,\end{aligned}$$

and, by Steps 1 and 2a, we can conclude the existence of $\widetilde{\alpha},\widetilde{\beta}\in\mathbb{C}[[x,y]]$ such that $x^{m-1}(h\circ\varphi)=\widetilde{\alpha}\cdot(f\circ\varphi)+\widetilde{\beta}\cdot(g\circ\varphi)$. Here, we can assume that all monomials in $\widetilde{\alpha},\widetilde{\beta}$ have exponents for x congruent $m-1$ modulo m. Dividing $\widetilde{\alpha}$ and $\widetilde{\beta}$ by x^{m-1} and substituting x^m by x, we come to an equality $h=\alpha f+\beta g$. □

2.1.4 Polar and Dual Curves

2.1.4.1 Hesse Problem

Let $F\in\mathbb{C}[x_0,\ldots,x_n]$, $n\geq 1$, be a homogeneous polynomial. Its *Hessian determinant* is the determinant of the matrix of second derivatives,

$$H_F(x_0,\ldots,x_n):=\det\begin{pmatrix}\frac{\partial^2F}{\partial x_0\partial x_0}&\cdots&\frac{\partial^2F}{\partial x_0\partial x_n}\\ \vdots&&\vdots\\ \frac{\partial^2F}{\partial x_n\partial x_0}&\cdots&\frac{\partial^2F}{\partial x_n\partial x_n}\end{pmatrix}.$$

In his original 1851 paper [Hes], Hesse claims that

> *the equality $H_F\equiv 0$ holds if and only if, after a suitable homogeneous coordinate change, the polynomial F depends on at most n variables.*

In other words, he claims that the Hessian determinant would vanish identically iff the hypersurface $V(F):=\{F=0\}\subset\mathbb{P}^n$ is a *cone*. While the "if" implication is trivial, the “only if” statement is not trivial at all. Actually, it turned out very soon that Hesse’s proof is wrong, and it seems to be Weierstraß who first expressed his doubts on the correctness of Hesse’s claim. Finally, in 1876, Gordan and Noether [GoN] showed that the “only if” statement holds true exactly for $n \leq 3$.

Theorem 2.1.27 (Gordan, Noether) *Let $F \in \mathbb{C}[x_0, \dots, x_n]$, $n \in \{1, 2, 3\}$, be a homogeneous polynomial.*[7] *Then the following are equivalent:*

(a) $H_F \equiv 0$.
(b) The partial derivatives $\frac{\partial F}{\partial x_0}, \dots, \frac{\partial F}{\partial x_n}$ *are linearly dependent (over* $\mathbb{C}$*).*
(c) After a suitable homogeneous coordinate change, F depends on at most n variables.

Note that implications $(b) \Leftrightarrow (c)$ and $(c) \Rightarrow (a)$ hold for any $n \geq 1$. A modern proof of Theorem 2.1.27 was given by Lossen in [Los2].

For higher dimensions, Gordan and Noether gave a concrete construction of families of counterexamples (of degree ≥ 3[8]), see also [Los2]. The most basic counterexample is probably the following

Example 2.1.27.1 Let $F := x_0^2 x_2 + x_0 x_1 x_3 + x_1^2 x_4 + x_0^3 + x_1^3 \in \mathbb{C}[x_0, \dots, x_4]$. Then $H_F \equiv 0$, while the hypersurface $V(F)$ is not a cone.

Let us use SINGULAR to check that condition (a) of Theorem 2.1.27 is fulfilled but not condition (b).

```
ring r = 0,x(0..4),dp;
poly F = x(0)^2*x(2)+x(0)*x(1)*x(3)+x(1)^2*x(4)+x(0)^3+x(1)^3;
ideal j = jacob(F);
det(jacob(j));  //the Hessian determinant
//=> 0
```

Hence the Hessian determinant of F is 0. We show now that the partials of F are linear independent in the space of polynomials of degree 2. We create the *general polar*, i.e. the linear combination of the partials with indeterminates $a(i)$ as coefficients, and show that it vanishes iff all the coefficients vanish.

Create now the ideal C of the coefficients of a general polar of F:

```
ring R = (0,a(0..4)),x(0..4),dp;
matrix A[5][1]=a(0..4);
ideal j = imap(r,j);
poly P = (matrix(j)*A)[1,1];  //the general polar
ideal C;
for (int ii=1; ii<=size(P);ii++)
```

[7] In positive characteristic, the situation is quite different [Per]. There, in general, only the implications $(b) \Rightarrow (a)$ and $(c) \Rightarrow (a)$ are true.

[8] For quadratic forms, Hesse’s statement holds true in any dimension.

```
{
   C[ii] = leadcoef(P[ii]);
}

ring S = 0,a(0..4),dp;
ideal C = imap(R,C);  //coefficient ideal of the general polar
std(C);
//=>_[1]=a(4)
//=>_[2]=a(3)
//=>_[3]=a(2)
//=>_[4]=a(1)
//=>_[5]=a(0)
```

The only zero of C is the origin. Hence the partials of F are linear independent.

In the sequel we present a new proof of Gordan–Noether's theorem in the case $n \leq 2$, making use of Bertini's theorem and the fact that in a non-μ-constant deformation of an ordinary singular point there appear critical points with distinct critical levels by the Lazzeri–Gabrièlov theorem [LaF, Gab] (cf. [GLS6, Theorem I.2.8]). For the proof we shall work in the metric topology in $\mathbb{P}^n = \mathbb{P}^n_{\mathbb{C}}$.

2.1.4.2 Reformulation of the General Hesse Problem

Without any restriction to n, the vanishing of the Hessian determinant translates to the intersecting behaviour of n (generic) polar hypersurfaces $V(F_\alpha)$,

$$F_\alpha = \alpha_0 \frac{\partial F}{\partial x_0} + \cdots + \alpha_n \frac{\partial F}{\partial x_n}, \quad \boldsymbol{\alpha} = (\alpha_0 : \cdots : \alpha_n) \in \mathbb{P}^n .$$

We say that F_α is a *polar hypersurface* of F if F_α is not constant. F_α is *generic* if $\alpha \in \mathbb{P}^n$ is generic.

Proposition 2.1.28 *Let $F \in \mathbb{C}[x_0, \ldots, x_n]$, $n \geq 1$, be a homogeneous polynomial satisfying $H_F \equiv 0$. Then no n distinct polar hypersurfaces of F have an isolated intersection point outside the hypersurface $V(F) \subset \mathbb{P}^n$.*

Remark 2.1.28.1 Let $F \in \mathbb{C}[x_0, \ldots, x_n]$, $n \geq 1$, be any homogeneous polynomial. Then, for each point $z \in \mathbb{P}^n$, such that z is an isolated intersection point of n polar hypersurfaces of F, there are (at least) n linearly independent polar hypersurfaces of F passing through z.

Proof of Proposition 2.1.28. We argue to the contrary. Suppose there exists an isolated intersection point $z \in \mathbb{P}^n \setminus V(F)$ of some n polar hypersurfaces of F. We claim that then there are n polar hypersurfaces of F which intersect at some point $w \in \mathbb{P}^n \setminus V(f)$ with multiplicity 1, that is, they are non-singular at w and intersect transversally.

Indeed, the base point set of the linear system generated by the partials $\frac{\partial F}{\partial x_0}, \dots, \frac{\partial F}{\partial x_n}$ is $\mathrm{Sing}(V(F)) \subset V(F)$. By Remark 2.1.28.1 the linear system has positive dimension. Then, by Bertini's theorem (cf. Proposition 2.1.23), almost all polar hypersurfaces, respectively all close to the given n ones, are non-singular and have isolated intersection points in a neighborhood $U \subset \mathbb{P}^n \setminus V(F)$ of z. In $n = 1$ we are done, otherwise fix one of these non-singular polar hypersurfaces, which we may assume to be $V(\frac{\partial F}{\partial x_n})$. The linear system generated by the remaining partials $\frac{\partial F}{\partial x_0}, \dots, \frac{\partial F}{\partial x_{n-1}}$ has no base points in U; hence, by Proposition 2.1.23 its generic element defines a non-singular hypersurface, crossing $V(\frac{\partial F}{\partial x_n})$ transversally in U. Thus, inductively, we come to the existence of a point $w \in \mathbb{P}^n \setminus V(F)$, where n polar hypersurfaces intersect with multiplicity 1. We may assume these n polar hypersurfaces to be $V(\frac{\partial F}{\partial x_1}), \dots, V(\frac{\partial F}{\partial x_n})$. Observe that $w \notin V(x_0)$, since, otherwise, by Euler's formula, we would have $F(w) = 0$, in contradiction to $w \notin V(F)$.

Setting $x_0 = 1$ in the affine space $\mathbb{A}^n = \mathbb{P}^n \setminus V(x_0)$, we have the well-known formula

$$H_F(1, x_1, \dots, x_n) = \det \begin{pmatrix} m(m-1)F & (m-1)A \\ (m-1)A^t & B \end{pmatrix}\Bigg|_{x_0=1},$$

$$A = \left(\frac{\partial F}{\partial x_1}, \dots, \frac{\partial F}{\partial x_n}\right), \quad B = \left(\frac{\partial^2 F}{\partial x_i \partial x_j}\right)_{1 \le i,j \le n}.$$

However, $w \in \mathbb{A}^n$ is a nondegenerate critical point of $F(1, x_1, \dots, x_n)$, and, hence,

$$\det \begin{pmatrix} m(m-1)F & (m-1)A \\ (m-1)A^t & B \end{pmatrix}\Bigg|_{(1,x_1,\dots,x_n)=q} = m(m-1)F(q) \cdot \det B \neq 0,$$

contradicting $H_F \equiv 0$. □

Let us turn now to the proof of Theorem 2.1.27.

For any n the implications $(c) \Rightarrow (b)$ and $(c) \Rightarrow (a)$ are trivial while $(b) \Rightarrow (c)$ can be seen as follows: If $\partial F/\partial x_0 = \lambda_1 \partial F/\partial x_1 + \dots + \lambda_n \partial F/\partial x_n$, then the linear coordinate change $\Phi(x) = (x_0 - \lambda_1 x_1 - \dots - \lambda_n x_n, x_1, \dots, x_n)$ with matrix A satisfies $D(F \circ \Phi)(x) = (\partial F/\partial x_0(\Phi(x)), \dots, \partial F/\partial x_n(\Phi(x))) \cdot A = (0, \partial F/\partial x_1(\Phi(x)), \dots, \partial F/\partial x_n(\Phi(x)))$. Hence $F \circ \Phi$ is independent of the first variable.

It remains to show that (a) implies (b) resp. (c) if $n \le 2$. We assume that the partial derivatives are linear independent.

The case $\mathbf{n = 1}$. If $F \in \mathbb{C}[x_0, x_1]$ contains at least two distinct linear factors, then the generic polar $V(\lambda_0 \frac{\partial F}{\partial x_0} + \lambda_1 \frac{\partial F}{\partial x_1})$ contains a point $z \in \mathbb{P}^1 \setminus V(F)$, which, of course, is isolated, and hence by Proposition 2.1.28, $H_F \not\equiv 0$.

The case $\mathbf{n = 2}$. Let $F = F_1^{m_1} \cdots F_s^{m_s}$, where $F_1, \dots, F_s \in \mathbb{C}[x_0, x_1, x_2]$ are reduced irreducible homogeneous polynomials. Put

$$F_{red} := F_1 \cdots F_s, \quad \widehat{F} := F_1^{m_1 - 1} \cdots F_s^{m_s - 1},$$

then

$$F_{x_i} = G_i \widehat{F}, \quad G_i := \sum_{j=1}^{s} m_j \frac{\partial F_j}{\partial x_i} \prod_{k \neq j} F_k, \quad i = 0, 1, 2.$$

Lemma 2.1.29 *Let Λ be the linear system generated by G_0, G_1, G_2, and let $C, C' \in \Lambda$ be two generic elements. Then the following holds:*

(a) C and C' have no component in common.
(b) The intersection number $i_z(C, C')$ at a point $z \in \mathrm{Sing}(V(F_{red}))$ does not exceed the Milnor number $\mu(F_{red}, z)$.

Proof First, notice that the linear system Λ has the finite set $\mathrm{Sing}(V(F_{red}))$ as its only base points. Indeed, the base point set of the linear system $\widehat{F} \cdot \Lambda$, generated by the partials $\frac{\partial F}{\partial x_0}, \frac{\partial F}{\partial x_1}, \frac{\partial F}{\partial x_2}$ is $\mathrm{Sing}(V(F_{red})) \cup V(\widehat{F}) \subset V(F)$. The claim follows, since the restriction of Λ to $V(F_j)$ is generated by $\frac{\partial F_j}{\partial x_0}, \frac{\partial F_j}{\partial x_1}, \frac{\partial F_j}{\partial x_2}$, multiplied by $\prod_{k \neq j} F_k$.

Now, by Bertini's theorem, a generic member C of Λ is reduced and has no fixed components. Moreover, a reducible generic member C must consist of distinct irreducible components $C_1, \ldots, C_r$ of the same degree and belonging to a one-dimensional linear system. This immediately implies (a).

To prove statement (b), assume that $z \in \mathbb{A}^2 = \mathbb{P}^2 \setminus V(x_0)$, and introduce coordinates x, y in $\mathbb{A}^2$ such that $z = (0, 0)$ and the line $\{x = 0\}$ is not tangent to $V(F)$ at z. It is sufficient to show that $i_z(G_1, G_2) \leq \mu(F_{red}, z)$. To do so, we use the defining formula $\mu(F_{red}, z) = i_z\big(\frac{\partial F_{red}}{\partial x}, \frac{\partial F_{red}}{\partial y}\big)$.

First, we establish the equalities $i_z(F_j, G_2) = i_z(F_j, \frac{\partial F_{red}}{\partial y})$, $j = 1, \ldots, s$. By definition, $i_z(F_j, G_2) = \sum_\xi \mathrm{ord}_t\big(G_2(x_\xi(t), y_\xi(t))\big)$, where ξ runs over all local branches of $V(F_j)$ at z and $t \mapsto (x_\xi(t), y_\xi(t))$ is a parametrization of ξ, and $i_z(F_{red}, \frac{\partial F_{red}}{\partial y}) = \sum_\xi \mathrm{ord}_t\big(\frac{\partial F_{red}}{\partial y}(x_\xi(t), y_\xi(t))\big)$. Hence, we have to verify only that $\mathrm{ord}_t\big(G_2(x_\xi(t), y_\xi(t))\big) = \mathrm{ord}_t\big(\frac{\partial F_{red}}{\partial y}(x_\xi(t), y_\xi(t))\big)$ for each local branch ξ of $V(F_j)$ at z. But, the identity $F_j(x_\xi(t), y_\xi(t)) \equiv 0$ yields

$$\begin{aligned}\mathrm{ord}_t\big(G_2(x_\xi(t), y_\xi(t))\big) &= \mathrm{ord}_t\left(\frac{\partial F_j}{\partial y}(x_\xi(t), y_\xi(t))\right) + \sum_{k \neq j} \mathrm{ord}_t\big(F_k(x_\xi(t), y_\xi(t))\big) \\ &= \mathrm{ord}_t\left(\frac{\partial F_{red}}{\partial y}(x_\xi(t), y_\xi(t))\right),\end{aligned}$$

and, thus,

$$i_z(F, G_2) = i_z\left(F, \frac{\partial F_{red}}{\partial y}\right), \quad i_z\big(\widehat{F}, G_2\big) = i_z\left(\widehat{F}, \frac{\partial F_{red}}{\partial y}\right). \tag{2.1.4.1}$$

On the other hand, $i_z(F, G_2) = \sum_\eta \mathrm{ord}_t\big(F(x_\eta(t), y_\eta(t))\big)$, where η runs over all local branches of $V(G_2)$ at z. Further on,

$$\begin{aligned}
\sum_{\eta} \mathrm{ord}_t\big(F(x_\eta(t), y_\eta(t))\big) &= \sum_{\eta}\left(\mathrm{ord}_t\left(\frac{d}{dt}F(x_\eta(t), y_\eta(t))\right)+1\right)\\
&= \sum_{\eta}\left(\mathrm{ord}_t\left(\frac{\partial F}{\partial x}(x_\eta(t), y_\eta(t))\cdot\frac{dx_\eta(t)}{dt}+\frac{\partial F}{\partial y}(x_\eta(t), y_\eta(t))\cdot\frac{dy_\eta(t)}{dt}\right)+1\right)\\
&= \sum_{\eta}\Bigg(\mathrm{ord}_t\Bigg(G_1(x_\eta(t), y_\eta(t))\cdot\widehat{F}(x_\eta(t), y_\eta(t))\cdot\frac{dx_\eta(t)}{dt}\\
&\qquad +G_2(x_\eta(t), y_\eta(t))\cdot\widehat{F}(x_\eta(t), y_\eta(t))\cdot\frac{dy_\eta(t)}{dt}\Bigg)+1\Bigg)\\
&= \sum_{\eta}\Big(\mathrm{ord}_t\big(G_1(x_\eta(t), y_\eta(t))\big)+\mathrm{ord}_t\big(\widehat{F}(x_\eta(t), y_\eta(t)\big)+\mathrm{ord}_t\big(x_\eta(t)\big)\Big)\ ,
\end{aligned}$$

and, thus,

$$i_z(F, G_2) = i_z(G_1, G_2) + i_z\big(\widehat{F}, G_2\big) + i_z\big(x, G_2\big)\,. \tag{2.1.4.2}$$

Moreover, by [GLS6, Lemma I.3.37],

$$i_z\left(F_{red}, \frac{\partial F_{red}}{\partial y}\right) = \mu(F_{red}, z) + i_z\left(x, \frac{\partial F_{red}}{\partial y}\right),$$

which together with (2.1.4.1), (2.1.4.2) yields

$$i_z(G_1, G_2) = \mu(F_{red}, z) + i_z\left(x, \frac{\partial F_{red}}{\partial y}\right) - i_z\,(x, G_2) \le \mu(F_{red}, z)\ ,$$

in view of the transversality of $V(x)$ and $V(F_{red})$ at z. □

To complete the proof in the case $n = 2$, assume that F_{red} is not the union of straight lines through one point. Then F_{red} represents a non-equisingular deformation of an ordinary singularity of multiplicity $\deg(F_{red})$ (cf. [Bru]). By [LaF, Gab] (cf. [GLS6, Theorem I.2.8])

$$\sum_{z\in\mathrm{Sing}(F_{red})} \mu(F_{red}, z) < \big(\deg(F_{red})-1\big)^2 .$$

This inequality together with Lemma 2.1.29 (b) and the fact that the base point set of Λ is $\mathrm{Sing}(F_{red})$, imply that two generic members $C, C' \in \Lambda$ intersect at some point $z \in \mathbb{P}^2 \setminus V(F)$, and by Lemma 2.1.29 (a) this intersection point is isolated. Consequently, z is an isolated intersection point of the polar curves $C \cup V\big(\widehat{F}\big)$ and $C' \cup V\big(\widehat{F}\big)$ of F, which in turn contradicts $H_F \equiv 0$ by Proposition 2.1.28.

Remark 2.1.29.1 The above proof generalizes to higher dimensions, provided that the hypersurface $V(F) \subset \mathbb{P}^n$ has *only isolated singularities*. Indeed, pick a hyperplane $L \subset \mathbb{P}^n$, which does not pass through the singular locus of F and crosses $V(F)$ transversally and let $y_1, \ldots, y_n$ be coordinates in $\mathbb{A}^n = \mathbb{P}^n \setminus L$. If $V(F)$ is not a cone, then F induces a non-μ-constant deformation by [LaF, Gab] (cf. [GLS6, Theorem I.2.8]), and, hence, the polynomial $f(y_1, \ldots, y_n)$ induced by F must have an (isolated) critical point on a non-zero critical level. Proposition 2.1.28 shows that this implies $H_F \not\equiv 0$.

2.1.4.3 Application to Polar Curves

Let $C \subset \mathbb{P}^2$ be a reduced plane curve of degree d with (homogeneous) equation $F \in \mathbb{C}[x, y, z]$. Then we define the *polar* of C relative to the point $(\alpha:\beta:\gamma) \in \mathbb{P}^2$ to be the curve $C_{\alpha\beta\gamma}$ (of degree $d-1$) given by $\alpha F_x + \beta F_y + \gamma F_z$.

As in the local case, we use the notion of a *generic polar* C' for $C_{\alpha\beta\gamma}$ with $(\alpha:\beta:\gamma) \in \mathbb{P}^2$ chosen generically. That is, the properties claimed for a generic polar hold for any $C_{\alpha\beta\gamma}$, $(\alpha:\beta:\gamma)$ chosen in some open dense subset $U \subset \mathbb{P}^2$.

Lemma 2.1.30 *Let $C \subset \mathbb{P}^2$ be a reduced plane curve. Then a generic polar C' is irreducible iff C is not the union of $d \geq 3$ lines through the same point.*

Proof Applying Bertini's theorem, we have irreducibility if there is no algebraic relation between the partials $\frac{\partial F}{\partial x}$, $\frac{\partial F}{\partial y}$ and $\frac{\partial F}{\partial z}$ (cf. Remark 2.1.24.1). Considering such a (homogeneous) relation of minimal degree and differentiating, we obtain a system of equations

$$A \cdot \begin{pmatrix} \frac{\partial^2 F}{\partial x \partial x} \\ \frac{\partial^2 F}{\partial x \partial y} \\ \frac{\partial^2 F}{\partial x \partial z} \end{pmatrix} + B \cdot \begin{pmatrix} \frac{\partial^2 F}{\partial y \partial x} \\ \frac{\partial^2 F}{\partial y \partial y} \\ \frac{\partial^2 F}{\partial y \partial z} \end{pmatrix} + \Gamma \cdot \begin{pmatrix} \frac{\partial^2 F}{\partial z \partial x} \\ \frac{\partial^2 F}{\partial z \partial y} \\ \frac{\partial^2 F}{\partial z \partial z} \end{pmatrix} = 0$$

where A, B and Γ generically do not vanish. Now the lemma follows from the Gordan–Noether theorem: the Hessian determinant vanishes identically iff C is a cone, that is, the union of d lines through one point. □

2.1.4.4 Plücker, Riemann–Hurwitz, Genus Formula

Let $C \subset \mathbb{P}^2$ be a reduced, irreducible curve of degree $d > 1$, given by a homogeneous polynomial $F \in \mathbb{C}[x_0, x_1, x_2]$. Denote by $C^* \subset (\mathbb{P}^2)^*$ its *dual curve*, that is,

$$C^* := \left\{ (a_0 : a_1 : a_2) \in (\mathbb{P}^2)^* \;\middle|\; \begin{array}{l} V(a_0x_0 + a_1x_1 + a_2x_2) \text{ is tangent} \\ \text{to } C \text{ at some point } p \in C \end{array} \right\}$$

Here $(\mathbb{P}^2)^*$ is the *(dual) projective 2-space*, whose points $(a_0 : a_1 : a_2)$ are in 1-1 correspondence to lines $V(a_0x_0 + a_1x_1 + a_2x_2) \subset \mathbb{P}^2$.

We have a natural rational duality morphism $\boldsymbol{d} : C \dashrightarrow C^*$. Let $z \in C$ and P an irreducible component of the germ (C, z). In local affine coordinates x, y such that $z = (0, 0)$ and the x-axis is tangent to P, this component admits a parametrization

$$\begin{cases} x = t^p \\ y = \lambda t^q + O(t^{q+1}) \end{cases} \quad 1 \le p < q, \quad \lambda \ne 0, \quad t \in (\mathbb{C}, 0) .$$

and the tangent lines to the points of P are given by equations $y = b(t)x + c(t)$ with

$$b(t) = \frac{\lambda q}{p} t^{q-p} + O(t^{q-p+1}), \quad c(t) = \lambda t^q + O(t^{q+1}), \quad t \in (C, 0) . \quad (2.1.4.3)$$

It follows that generically the duality morphism $\boldsymbol{d}$ is 1-to-1, and hence birational. Furthermore, C^* is an irreducible projective curve of degree $d^* > 1$.[9]

Let $C = V(F)$ with $F(x_0, x_1, x_2) = 0$ a homogeneous polynomial of degree d. At a smooth point $z \in C$, the coefficients of the tangent line are given by

$$a_0 = \frac{\partial F}{\partial x_0}, \quad a_1 = \frac{\partial F}{\partial x_1}, \quad a_2 = \frac{\partial F}{\partial x_2} .$$

Let $t \in (\mathbb{C}, 0)$ parameterize the germ (C, z). Since $F(x_o(t), x_1(t), x_2(t)) \equiv 0$, we have $\frac{\partial F}{\partial x_0} \cdot \frac{dx_0}{dt} + \frac{\partial F}{\partial x_1} \cdot \frac{dx_1}{dt} + \frac{\partial F}{\partial x_2} \cdot \frac{dx_2}{dt} = 0$, that is,

$$a_0(t) \cdot \frac{x_0}{dt} + a_1(t) \cdot \frac{dx_1}{dt} + a_2(t) \cdot \frac{dx_2}{dt} = 0 .$$

Combining this with the Euler formula $a_0(t)x_0(t) + a_1(t)x_1(t) + a_2(t)x_2(t) = 0$, we obtain

$$x_0(t) \cdot \frac{da_0}{dt} + x_1(t) \cdot \frac{da_1}{dt} + x_2(t) \cdot \frac{da_2}{dt} = 0 .$$

Thus, the dual to C^* is the original curve C.

We call a tangent line L to C a *singular tangent*, if

(a) either L is tangent to C at a singular point,
(b) or L is tangent to C at more than one point,
(c) or L intersects C at a non-singular point with multiplicity > 2.

---

[9] An equation F^* for C^* can be obtained as follows: let

$$g(x_1, x_2) := a_0^d \cdot F\left(\frac{-(a_1 x_1 + a_2 x_2)}{a_0}, x_1, x_2, \right) \in K(a_0, a_1, a_2)[x_1, x_2] ,$$

and compute the discriminant $D \in \mathbb{C}[a_0, a_1, a_2] \setminus \{0\}$ of $g(1, x_2)$. D is homogenous of degree $2d^2 - d$ and is the product of F^* with some number of linear factors. Hence, factorizing D and removing all linear factors, we get an equation for C^*.

The set of singular tangents is finite, since the set $\mathrm{Sing}(C)$ is finite, and the conditions (b) and (c) determine L as a singular point of C^* (cf. formula (2.1.4.3)). Hence, there exists a point $\boldsymbol{q} = (q_0 : q_1 : q_2) \in \mathbb{P}^2 \setminus C$ which does not lie on any singular tangent. Denote by $\Lambda_{\boldsymbol{q}}$ the pencil of lines through q. Recall that a line $L \in \Lambda_{\boldsymbol{q}}$ is tangent to C at the (non-singular) point $z \in C$ iff z lies also on the polar curve relative to $\boldsymbol{q}$, that is, iff

$$F(z) = 0 = q_0 \frac{\partial F}{\partial x_0}(z) + q_1 \frac{\partial F}{\partial x_1}(z) + q_2 \frac{\partial F}{\partial x_2}(z) .$$

We observe that d^* is the number of lines $L \in \Lambda_{\boldsymbol{q}}$ tangent to C at non-singular points, and that $\{q_0 \frac{\partial F}{\partial x_0} + q_1 \frac{\partial F}{\partial x_1} + q_2 \frac{\partial F}{\partial x_2} = 0\}$ is a generic polar of C. Applying Bézout's theorem (Theorem 2.1.16) to the non-singular intersection points of $\{F = 0\}$ with a generic polar (which gives d^* as total number) and the singular intersection points (which gives the kappa-invariant κ at each intersection point, cf. [GLS6, Proposition I.3.38]) we obtain the *first Plücker formula*:

$$d^* = d(d-1) - \sum_{z \in \mathrm{Sing}(C)} \kappa(C, z) . \tag{2.1.4.4}$$

In particular, if C has n nodes and k cusps as its only singularities one gets

$$d^* = d(d-1) - 2n - 3k .$$

Now, let $\overline{C} \to C$ be the normalization map. Then the *topological Euler characteristic* of C satisfies (using Mayer–Vietoris)

$$\begin{aligned} \chi_{\mathrm{top}}(C) &= \chi_{\mathrm{top}}\big(\overline{C}\big) - \sum_{z \in \mathrm{Sing}(C)} \big(r(C,z) - 1\big) \\ &= 2 - 2g(C) - \sum_{z \in \mathrm{Sing}(C)} \big(r(C,z) - 1\big) , \end{aligned} \tag{2.1.4.5}$$

where $r(C, z)$ is the number of irreducible branches of the germ (C, z). Besides, considering the projection of C on some straight line $L_0 \not\supset \{\boldsymbol{q}\}$ from the point $\boldsymbol{q}$ leads to the following version of the topological *Riemann–Hurwitz formula*,

$$\chi_{\mathrm{top}}(C) = d \cdot \chi_{\mathrm{top}}(L) - d^* - \sum_{z \in \mathrm{Sing}(C)} \big(\mathrm{mt}(C,z) - 1\big) ,$$

since a line $L \in \Lambda_{\boldsymbol{q}}$, which is tangent to C at a non-singular point, meets C at $d-1$ points, and a line $L \in \Lambda_{\boldsymbol{q}}$ through a point $z \in \mathrm{Sing}(C)$ meets C at $d - \mathrm{mt}(C, z) + 1$ points. Combining the last equation with (2.1.4.4) and (2.1.4.5), we come to the *genus formula*,

$$g(C) = \frac{(d-1)(d-2)}{2} - \sum_{z \in \mathrm{Sing}(C)} \delta(C, z), \tag{2.1.4.6}$$

where $\delta(C, z)$ is the δ-invariant. If C has n nodes and k cusps as its only singularities, (2.1.4.6) turns into

$$g(C) = \frac{(d-1)(d-2)}{2} - n - k\,.$$

Recall that for an arbitrary projective scheme X of dimension n the *arithmetic genus*, is defined as

$$p_a(X) = (-1)^n(\chi(\mathcal{O}_X) - 1).$$

Here, for any coherent sheaf $\mathcal{F}$ on X,

$$\chi(\mathcal{F}) = \sum (-1)^i h^i(X, \mathcal{F}),$$

$h^i(X, \mathcal{F}) = \dim_{\mathbb{C}} H^i(X, \mathcal{F})$, is the *(analytic) Euler characteristic* of $\mathcal{F}$.

For a connected reduced curve C we have for the arithmetic genus

$$p_a(C) = \dim_{\mathbb{C}} H^1(C, \mathcal{O}_C).$$

If C is not connected, use that the Euler characteristic is the sum of the Euler characteristics of the connected components.

The *(geometric) genus* of C is defined as the arithmetic genus of the normalization $\overline{C}$ of C, hence

$$g(C) = p_a(\overline{C}) = 1 - \chi(\mathcal{O}_{\overline{C}}) = p_a(C) - \delta(C),$$

with $\delta(C) = \sum_{z \in \mathrm{Sing}(C)} \delta(C, z)$.

2.2 Equisingular Families of Singular Algebraic Varieties

In this section we introduce definitions and basic properties of the deformation theory of singular complex hypersurfaces such that the *"singularity type"* of specified isolated singularities is not changed during the deformation. Deformations of this kind are called *equisingular*, and families of hypersurfaces inducing equisingular deformations of the given singularities are called *equisingular families* (ESF). In the case of families of projective hypersurfaces we relate them with *equisingular strata* in the corresponding Hilbert scheme of hypersurfaces in projective space.

Strictly speaking we have to consider deformations with sections along which the singularity type is fixed. We are especially interested in deformations with trivial sections, meaning that the positions of the singular points in the ambient space do not change during the deformation.

In most cases we are considering deformations over reduced base spaces. However, we need also non-reduced base spaces, as our most powerful tools for investigating ESF rely on the cohomological interpretation of the tangent and obstruction spaces of such families.

Thus, we determine the tangent space and the obstruction space for equisingular deformations and study the relation between deformations of the global hypersurface and the induced deformation of the singularities. Finally we define *T-smooth equisingular deformations* meaning that the induced equisingular deformations of the singularities are independent in a strong sense and prove necessary and sufficient cohomological conditions for T-smoothness.

In the whole section let Σ denote a smooth complex space, unless we explicitly say otherwise.

2.2.1 Families with Imposed Conditions on Singularities

2.2.1.1 Global Deformation Categories and Functors

A *family of complex spaces* over a complex space T is a flat morphism $\phi : \mathscr{X} \to T$ of complex spaces. A *family of hypersurfaces* in Σ over T is by definition a commutative diagram of complex spaces

$$\begin{array}{ccc} \mathscr{X} & \overset{j}{\hookrightarrow} & \Sigma \times T \\ & {\scriptstyle\phi}\searrow \quad \swarrow{\scriptstyle\pi} & \\ & T & \end{array}$$

such that j is a closed embedding with $j(\mathscr{X}) \subset \Sigma \times T$ a closed subspace whose ideal is locally principal (also called an effective Cartier divisor), π is the projection and ϕ is flat. This implies that the image of the fibres $j(\mathscr{X}_t) \subset \Sigma \times \{t\} = \Sigma$, $\mathscr{X}_t := \phi^{-1}(t)$, are hypersurfaces in Σ for all $t \in T$. A family with sections is a diagram as above, together with sections $\sigma_1, \ldots, \sigma_r : T \to \mathscr{X}$ of ϕ, i.e. $\sigma_i : T \to \mathscr{X}$ are morphisms with $\phi \circ \sigma_i = \mathrm{id}_T$ for $i = 1, \ldots, r$.

If $S \subset \mathscr{X}$ is a closed subspace contained in the singular locus $\mathrm{Sing}(\phi)$ of ϕ such that for each $z \in S$ the induced map of germs $(\mathrm{Sing}(\phi), z) \to (T, \phi(z))$ is finite, we say that $\mathscr{X} \to T$ is a *flat family of hypersurfaces with isolated singularities* along S. For such families the fibres $\mathscr{X}_t$, $t \in T$, have an isolated singularity at each point $z \in \mathscr{X}_t \cap S$ but $\mathscr{X}_t$ may have also other, not necessarily isolated, singularities.

If Σ is algebraic, then a family of hypersurfaces $\phi : \mathscr{X} \to T$ is called a *family of algebraic hypersurfaces* over the complex space T if $\mathscr{X}_t$ is a complex algebraic variety and the restriction $j : \mathscr{X}_t \to \Sigma$ is algebraic for all $t \in T$.

A *pointed complex space* is a complex space T together with a fixed point $t_0 \in T$, denoted by T, t_0. If we fix in the family $\mathscr{X} \to T$ an isomorphism of X with the special fibre $\mathscr{X}_{t_0}$ we obtain the notion of a *deformation* of X over the pointed space T, t_0.

For a general overview of deformation theory we refer to [GLS6, Appendix C] and [HaR3, Se] for deformations of schemes. For local deformation theory (i.e. deformations of singularities) we refer to [GLS6, Sections II.1 and II.2.3]. In this section we consider embedded deformations of (global) varieties and the induced (local) deformations of the singularities.

Definition 2.2.1 Let $X \subset \Sigma$ be a closed complex subspace and T, t_0 a pointed complex space. A *deformation of* X/Σ (or an *embedded deformation* of $X \subset \Sigma$) over T is a triple $(\mathscr{X}, i, j)$ of a complex space $\mathscr{X}$ and closed embeddings i, j, defining a Cartesian diagram

$$\begin{array}{ccc} X & \overset{i}{\hookrightarrow} & \mathscr{X} \\ \cap & & \cap j \\ \Sigma & \hookrightarrow & \Sigma \times T \\ \downarrow & & \downarrow \pi \\ t_0 & \in & T \end{array} \quad \phi \text{ flat} \qquad (2.2.1.1)$$

such that the composition $\phi = \pi \circ j$ is flat ($\Sigma \hookrightarrow \Sigma \times T$ is the canonical embedding with image $\Sigma \times \{t_0\}$ and π the projection). We sometimes simply write $\mathscr{X}$ or $\mathscr{X} \subset \Sigma \times T$ instead of $(\mathscr{X}, i, j)$.

A *deformation of* X/Σ *with sections* is a triple $(\mathscr{X}, i, j)$ as before, together with a set of sections $\sigma = \{\sigma_1, \ldots, \sigma_r\}$ of ϕ. It is called a *deformation with trivial sections* if the compositions $j \circ \sigma_i$ coincide with the trivial sections $t \mapsto (\sigma_i(t_0), t)$ in $\Sigma \times T$.

A *morphism between deformations* $(\mathscr{X}, i, j)$ and $(\mathscr{X}', i', j')$ of X/Σ over T is a morphism $\psi : \mathscr{X} \to \mathscr{X}'$ such that $j = j' \circ \psi$ and $\psi \circ i = i'$. Then $j(\mathscr{X}) = j'(\mathscr{X}')$ and, since j and j' are closed embeddings, ψ is an isomorphism. Note that $\Sigma \times T$ is fixed in this definition. Hence a morphism of deformations of X/Σ is a morphism of diagrams (2.2.1.1) being the identity on $\Sigma \times T$. When considering deformations with sections we require that ψ commutes with the sections.

It follows that deformations of X/Σ over T form a *groupoid*, that is, a category such that every morphism is an isomorphism.

The *category of deformations* of X/Σ over T, resp. of deformations with sections, resp. with trivial sections, is denoted by $\mathcal{D}ef_{X/\Sigma}(T)$, resp. by $\mathcal{D}ef^{sec}_{X/\Sigma}(T)$, resp. by $\mathcal{D}ef^{fix}_{X/\Sigma}(T)$.

If we omit in diagram (2.2.1.1) the middle part $\Sigma \hookrightarrow \Sigma \times T$, we obtain the notion of a (non-embedded) deformation of X over T, resp. of a deformation of X with (trivial) sections. These categories of non-embedded deformations are denoted by $\mathcal{D}ef_X(T)$, resp. $\mathcal{D}ef^{sec}_X(T)$, resp. $\mathcal{D}ef^{fix}_X(T)$.

If $\Phi : \mathscr{X} \to \Sigma \times T \to T$ is a deformation of X/Σ over T, resp. with section $\sigma : T \to \mathscr{X}$, and $\varphi : S \to T$ any morphism, we have the *induced deformation* $\varphi^*\Phi$ of X/Σ over S (with section $\varphi^*\sigma$). It is given by pullback

$$\varphi^*\Phi : \varphi^*\mathscr{X} := \mathscr{X} \times_T S \to (\Sigma \times T) \times_T S = \Sigma \times S \to S$$

resp. with section $\varphi^*\sigma : S \to \mathscr{X} \times_T S, s \mapsto (\sigma(\varphi(s)), s)$. If σ is trivial then $\varphi^*\sigma$ is trivial as well.

Now fix a point $z \in X$ and restrict the diagram (2.2.1.1) to the germs at z resp. t_0. We obtain a diagram of morphisms of germs, called a deformation of $(X, z)/(\Sigma, z)$ over (T, t_0),

$$\begin{array}{ccc} (X,z) & \stackrel{i}{\hookrightarrow} & (\mathscr{X},z) \\ \cap & & \cap j \\ (\Sigma,z) & \hookrightarrow & (\Sigma,z)\times(T,t_0) \\ \downarrow & & \downarrow \pi \\ t_0 & \in & (T,t_0) \end{array} \quad \phi_z \tag{2.2.1.2}$$

where the composition $\pi \circ j$, denoted by ϕ_z or just by ϕ, is flat.

If we forget the middle row $(\Sigma, z) \hookrightarrow (\Sigma, z) \times (T, t_o)$ in (2.2.1.2) we obtain a deformation of the singularity (X, z) over the base space (T, t_0). The diagram (2.2.1.2) and the deformation of (X, z) are both called the *induced local deformation* at z.

The categories of deformations of $(X, z)/(\Sigma, z)$ resp. of (X, z) over (T, t_o) are denoted by $\mathcal{D}ef_{(X,z)/(\Sigma,z)}(T, t_0)$ resp. $\mathcal{D}ef_{X,z}(T, t_0)$. We write a superscript sec resp. fix if we consider deformations with sections resp. trivial sections.

Having defined the deformation categories, we introduce the corresponding deformation functors of isomorphism classes of deformations.

Definition 2.2.2 Let $\underline{\mathcal{D}ef}_{X/\Sigma}$ resp. $\underline{\mathcal{D}ef}^{sec}_{X/\Sigma}$ denote the functor from the category of pointed complex spaces to the category of sets defined as

$$\underline{\mathcal{D}ef}_{X/\Sigma}(T) := \big\{\text{isomorphism classes of deformations of } X/\Sigma \text{ over } T\big\} \text{ resp.}$$

$$\underline{\mathcal{D}ef}^{sec}_{X/\Sigma}(T) := \left\{\begin{array}{c}\text{isomorphism classes of deformations with}\\ \text{sections of } X/\Sigma \text{ over } T\end{array}\right\}.$$

If the sections are all trivial, $\underline{\mathcal{D}ef}^{sec}$ is denoted by $\underline{\mathcal{D}ef}^{fix}$.

There exist a natural (forgetful) transformation $\underline{\mathcal{D}ef}_{X/\Sigma} \to \underline{\mathcal{D}ef}_X$ by forgetting the middle term of (2.2.1.1) and similar for deformations with (trivial) sections.

Furthermore, for each point $z \in X$ we obtain a natural transformation $\underline{\mathcal{D}ef}_X \to \underline{\mathcal{D}ef}_{X,z}$, where $\underline{\mathcal{D}ef}_{X,z}$ denotes the functor of isomorphism classes of deformations of the analytic germ (X, z). In the same way we get natural transformations $\underline{\mathcal{D}ef}^{sec}_X \to \underline{\mathcal{D}ef}^{sec}_{X,z}$ and $\underline{\mathcal{D}ef}^{fix}_X \to \underline{\mathcal{D}ef}^{fix}_{X,z}$.

Remark 2.2.2.1 Recall that each (non-embedded) deformation of an analytic subgerm $(X, z) \subset (\Sigma, z)$ can be embedded, that is, extended to a deformation of $(X, z)/(\Sigma, z)$ (cf. Corollary II.1.6 of [GLS6]). Moreover, if (T, t_0) is the germ of a complex space, we can choose the embedding in such a way that a given section σ maps to the trivial section in $(\Sigma, z) \times (T, t_0)$ by [GLS6, Prop. II.2.2]. In particular, $\underline{\mathcal{D}ef}^{sec}_{X,z} \cong \underline{\mathcal{D}ef}^{fix}_{X,z}$.

Note also that the forgetful morphism of functors $\underline{\mathcal{D}ef}^{sec}_{X,z} \to \underline{\mathcal{D}ef}_{X,z}$ is smooth (cf. [GLS6, Definition C.1.4] for the definition), since any deformation with section is isomorphic to a deformation with trivial section and hence can be extended over extensions of fat points. This implies that the base space of the semiuniversal deformation of $\underline{\mathcal{D}ef}^{sec}_{X,z}$ (if it exists) is isomorphic to the cartesian product of the base space of the semiuniversal deformation of $\underline{\mathcal{D}ef}_{X,z}$ with a smooth factor. If $(X, z) \subset (\Sigma, z)$ is an isolated hypersurface singularity then this factor has dimension $\dim(\Sigma, z)$, by Lemma 2.2.13.

This is, however, in general not true in the global situation. E.g., for $X \subset \Sigma$ a compact subspace, the forgetful morphism $\underline{\mathcal{D}ef}_{X/\Sigma} \to \underline{\mathcal{D}ef}_X$ is in general not surjective.

2.2.1.2 Local Equisingular and VSI Deformations

As in Sect. 2.2.1.1 we consider deformations of the complex hypersurface $X \subset \Sigma$ with isolated singularities $z_1, \ldots, z_r$ (and maybe other singularities, not necessarily isolated).

We are interested in *equisingular deformations* of X/Σ, i.e. deformations such that certain specified invariants of the singularities $z_i \in X$ do not change during the deformation (along some sections). In this subsection we develop the theory of *local equisingular deformations*, i.e. of equisingular deformations of the germs (X, z_i). We introduce in particular the local condition (VSI), which is formalizes the requirement for our approach to work. The attribute "equisingular" is used in this general sense, but when we consider plane curve singularities it means usually that the embedded topological type does not change.

Two important invariants to be fixed during a deformation are the analytic type of the germ $(X, z_i) \subset (\Sigma, z_i)$ or, if Σ is a surface, the embedded topological type of the plane curve singularity (X, z_i). The analytic and the topological type are examples of what we call a *smooth singularity type* (cf. Definition 2.2.42 in Sect. 2.2.3) of (X, z_i).

Fixing the singularity types during the deformation along given sections leads to a subfunctor $\underline{\mathcal{D}ef}'^{sec}_{X/\Sigma}$ of $\underline{\mathcal{D}ef}^{sec}_{X/\Sigma}$ (resp. of $\underline{\mathcal{D}ef}^{fix}_{X/\Sigma}$). The corresponding deformation category $\mathcal{D}ef'^{sec}_{X/\Sigma}(T)$, resp. $\mathcal{D}ef'^{fix}_{X/\Sigma}(T)$, can be described easily if the base space T is reduced. In this case $\mathcal{D}ef'^{sec}_{X/\Sigma}(T)$ is the full subcategory of $\mathcal{D}ef^{sec}_{X/\Sigma}(T)$ with objects $\mathscr{X} \subset \Sigma \times T$ satisfying the following property:

For all $i = 1, \ldots, r$ and all sufficiently small representatives $\phi_i : \mathscr{X}_i \to T$ of the map germ $(\mathscr{X}, z_i) \to (T, t_0)$ and sections $\sigma_i : T \to \mathscr{X}_i$ of ϕ_i, the singularity type $S_i(t)$ of $(\mathscr{X}_i, \sigma_i(t))$ is independent of $t \in T$. For $\mathcal{D}ef'^{fix}_{X/\Sigma}(T)$ we require additionally that the position of the points z_i does not change during the deformation.

We are mainly interested in the full subcategory $\mathcal{D}ef'_{X/\Sigma} \subset \mathcal{D}ef_{X/\Sigma}$ such that there *exist sections* (i.e. the sections are not part of the data) along which the singularity invariants do not change. However, to study this, we have to consider $\mathcal{D}ef'^{sec}_{X/\Sigma}$ and $\mathcal{D}ef'^{fix}_{X/\Sigma}$.

Before we give the formal definitions let us start with some examples, including the two most important examples: deformations which fix the analytic type and deformations which fix the topological type (in the case of plane curve singularities).

Definition 2.2.3 Let $\phi : (\mathscr{X}, z) \to (T, t)$ be a deformation of a complex germ (X, z).

(1) Fixing the analytic type: ϕ is called *equianalytic* if it is isomorphic to the trivial deformation $(X, z) \times (T, t) \to (T, t)$. $\mathcal{D}ef^{ea}_{X,z} \subset \mathcal{D}ef_{X,z}$ denotes the full subcategory of equianalytic deformations and $\underline{\mathcal{D}ef}^{ea}_{X,z}$ its deformation functor.

(2) Fixing the multiplicity: Let $(X, z) \subset (\Sigma, z)$ be a hypersurface singularity given by $f \in \mathcal{O}_{\Sigma,z}$ with multiplicity $m := \mathrm{mt}(f) \geq 2$. A deformation of (X, z) over (T, t) is then given by $F \in \mathcal{O}_{\Sigma\times T,(z,t)}$ with F the ideal of $(\mathscr{X}, (z, t)) \subset (\Sigma \times T, (z, t))$ and $\phi : (\mathscr{X}, (z, t)) \to (T, t)$ the projection. If $\sigma : (T, t) \to (\mathscr{X}, (z, t))$ is a section of ϕ, then ϕ is called *equimultiple* (*along* σ) if $F \in I^m_{\sigma(T)}$, with $I_{\sigma(T)}$ the ideal of $\sigma(T, t)$.
Note that the equimultiplicity of ϕ along σ implies that the multiplicities of the fibres $(\mathscr{X}_{t'}, \sigma(t'))$ of ϕ are equal to m for t' close to t.
If $(X, 0) \subset (\mathbb{C}^n, 0)$ is given by the power series $f = \sum_{|\alpha|\geq m} c_\alpha x^\alpha$ of order m and $(\mathscr{X}, 0) \subset (\mathbb{C}^n \times \mathbb{C}^k, 0)$ by a power series $F = \sum_{\alpha,\beta} c_{\alpha\beta} x^\alpha t^\beta \in \mathbb{C}\{x_1, \ldots, x_n, t_1, \ldots, t_k\}$, then ϕ is equimultiple along the trivial section simply iff $c_{\alpha\beta} = 0$ for all (α, β) such that $|\alpha| < m$.

(3) Fixing the topological type: Let (X, z) be a reduced curve singularity. ϕ is called *equisingular* if the following holds: ϕ admits a section $\sigma : (T, t) \to (\mathscr{X}, z)$ along which it is equimultiple. Moreover, after blowing up σ there exist sections of the blown up family through the non-normal crossing points of the reduced total transform, which are equimultiple for the reduced total transform. These can again been blown up so that equimultiple sections of the blown up family exist. This process can be continued until an embedded normal crossing resolution of the special fibre (X, z) of ϕ is reached (for an explicit formal definition cf. [GLS6, Def. II.2.6]). A system of equimultiple sections of successively blowing ups as above, including an equimultiple section σ of ϕ as initial section, is called a *system of equimultiple sections* of the equisingular deformation ϕ. $\mathcal{D}ef^{es}_{X,z}$ denotes the full subcategory of $\mathcal{D}ef_{X,z}$ of equisingular deformations and $\underline{\mathcal{D}ef}^{es}_{X,z}$ its deformation functor.

(4) Fixing the topological type along a section: Let $\mathcal{D}ef^{es,sec}_{X,z}$ denote the full subcategory of $\mathcal{D}ef^{sec}_{X,z}$ consisting of deformations ϕ of (X, z) with section σ, which are equisingular along σ. This means that ϕ is equisingular with σ as initial equimultiple section. $\mathcal{D}ef^{es,fix}_{X,z}$ denotes the subcategory of $\mathcal{D}ef^{sec}_{X,z}$ of deformations being equisingular along the trivial section.

Equianalytic resp. equisingular is abbreviated by *ea* resp. *es*, and equimultiple by *em*. Note that *em*-deformations can only be defined for deformations with given section.

Remark 2.2.3.1 We shall prove in Proposition 2.2.6 that equianalytic, resp. equisingular deformations have unique equianalytic, resp. equisingular sections. Hence the sections are respected by morphisms and $\mathcal{D}ef^{ea}_{X,x}$ and $\mathcal{D}ef^{es}_{X,x}$ may be considered as subcategories of $\mathcal{D}ef^{sec}_{X,x}$.

For the proof of uniqueness we need the following lemma.

Lemma 2.2.4 *Let $f \in \langle x_1, \dots, x_n\rangle^2 \subset \mathbb{C}\{x_1, \dots, x_n\}$ have an isolated singularity and let*

$$D = \sum g_i \frac{\partial}{\partial x_i}, \quad g_i \in \mathbb{C}\{x_1, \dots, x_n\},$$

be a derivation satisfying $D(f) \in \langle f\rangle$. Then $g_i(0) = 0$ for $i = 1, \dots, n$.

Proof Since the claim is trivial for $n = 1$, we may assume $n \geq 2$.

The assumption and the conclusion are independent of the coordinate system; hence we may choose (generic) coordinates such that $f\big|_{x_i=0}$ has an isolated singularity for $i = 1, \dots, n$. Now assume that there exists an i such that $g_i(0) \neq 0$, and without loss of generality let $i = 1$. Then $\frac{\partial f}{\partial x_1} \in \langle f, \frac{\partial f}{\partial x_2}, \dots, \frac{\partial f}{\partial x_n}\rangle$ and

$$\langle f, \frac{\partial f}{\partial x_1}, \dots, \frac{\partial f}{\partial x_n}\rangle = \langle f, \frac{\partial f}{\partial x_2}, \dots, \frac{\partial f}{\partial x_n}\rangle .$$

It follows that $\dim_{\mathbb{C}} \mathbb{C}\{x_1, \dots, x_n\}/\langle f, \frac{\partial f}{\partial x_2}, \dots, \frac{\partial f}{\partial x_n}\rangle = \tau(f) < \infty$. Since $f \in \langle x_1, \dots, x_n\rangle^2$, $f\big|_{x_1=0} \in \langle x_2, \dots, x_n\rangle^2$ and $\mu(f\big|_{x_1=0}) > 0$. Applying the following lemma, we get

$$\tau(f) = \mu(f) + \mu(f\big|_{x_1=0}) > \mu(f) \geq \tau(f),$$

a contradiction. □

Remark 2.2.4.1 Let J denote the ideal generated by $\frac{\partial f}{\partial x_1}, \dots, \frac{\partial f}{\partial x_n}$ in $R = \mathbb{C}\{x_1, \dots, x_n\}/\langle f\rangle$ and $\mathfrak{m}$ the maximal ideal in R. The proof of Lemma 2.2.4 shows that the minimal number of generators of J equals

$$\dim_{\mathbb{C}} J/\mathfrak{m}J = n,$$

if $f \in \langle x_1, \dots, x_n\rangle^2$ has an isolated singularity.

Lemma 2.2.5 *Let $f \in \mathbb{C}\{x_1, \dots, x_n\}$, $n \geq 2$, and let f and $f\big|_{\{x_1=0\}}$ have an isolated singularity. Then*

$$\dim_{\mathbb{C}} \mathbb{C}\{x_1, \dots, x_n\}/\langle f, \frac{\partial f}{\partial x_2}, \dots, \frac{\partial f}{\partial x_n}\rangle = \mu(f) + \mu(f\big|_{\{x_1=0\}}) .$$

Proof Apply [Gre, Korollar 5.5] to $(f, x_1) : (\mathbb{C}^n, 0) \to (\mathbb{C}^2, 0)$. □

Note that $\mu(f\big|_{\{x_1=0\}}) = \dim_{\mathbb{C}} \mathbb{C}\{x_1, \dots, x_n\}/\langle x_1, \frac{\partial f}{\partial x_2}, \dots, \frac{\partial f}{\partial x_n}\rangle$ is the intersection multiplicity of the curve singularity defined by $\frac{\partial f}{\partial x_2}, \dots, \frac{\partial f}{\partial x_n}$ with the hyperplane $\{x_1 = 0\}$. Hence Lemma 2.2.5 generalizes [GLS6, Lemma I.3.37] from $n = 2$ to arbitrary dimension.

Proposition 2.2.6 *(1) Let $\phi : (\mathscr{X}, z) \to (T, t)$ be an ea-deformation of a non-smooth isolated hypersurface singularity (X, z). Let $(X, z) \times (T, t) \to (T, t)$ be the trivial deformation with trivial section τ. Then there is a unique*[10] *section $\sigma : (T, t) \to (\mathscr{X}, z)$, called the* equianalytic section *of ϕ, satisfying: for each isomorphism $\psi : (X, z) \times (T, t) \xrightarrow{\cong} (\mathscr{X}, z)$ trivializing ϕ, we have $\sigma = \psi \circ \tau$.*

(2) Let (X, z) be a reduced but non-smooth plane curve singularity and $\phi:(\mathscr{X}, z) \to (T, t)$ an es-deformation of (X, z). Then the system of equimultiple sections of ϕ is uniquely determined. In particular, there exists a unique section σ, called the equisingular section of ϕ, along which ϕ is equisingular.

(3) For $'$ being ea or es, the natural maps $\mathcal{D}ef'^{,fix}_{X,z} \to \mathcal{D}ef'^{,sec}_{X,z} \to \mathcal{D}ef'_{X,z}$ induce isomorphisms of functors

$$\underline{\mathcal{D}ef}'^{,fix}_{X,z} \cong \underline{\mathcal{D}ef}'^{,sec}_{X,z} \cong \underline{\mathcal{D}ef}'_{X,z} .$$

Proof (1) Choose any trivialization $\psi : (X, z) \times (T, t) \to (\mathscr{X}, z)$ and set $\sigma := \psi \circ \tau$. The claim is that σ is independent of the trivialization, that is, $\psi \circ \tau = \psi' \circ \tau$ for any other trivialization ψ'. The statement is trivial if (T, t) is reduced. In this case σ is an isomorphism onto the reduced singular locus of Φ, since $\sigma(s)$ is the unique singular point of $\Phi^{-1}(s)$.

If (T, t) is arbitrary we consider morphisms of local rings. We may assume that $(X, z) \subset (\mathbb{C}^n, 0)$ is defined by $f \in \mathbb{C}\{\boldsymbol{x}\}, \boldsymbol{x} = (x_1, \ldots, x_n)$. For simplicity we denote the ring maps by the same letters as the corresponding maps of germs.

It suffices to show that the trivial section τ coincides with $\sigma := \tau \circ \overline{\psi}$, where $\overline{\psi}$ is any automorphism of $\mathcal{O}_{(X,z)\times(T,t)}$ over $\mathcal{O}_{(T,t)}$. If the sections $\sigma, \tau : \mathcal{O}_{(X,z)\times(T,t)} \to \mathcal{O}_{T,t}$ coincide modulo arbitrary powers of $\mathfrak{m}_{T,t}$, they coincide themselves by Krull's intersection theorem. Hence it suffices to consider the case that $\mathcal{O}_{T,t} =: A$ is an Artinian algebra. We argue by induction on $\dim_{\mathbb{C}} A$, the case $\dim_{\mathbb{C}} A = 1$ being trivial.

Let A be a small extension of $\overline{A} = A/\eta, \eta \in \mathfrak{m}_A$ with $\eta \cdot \mathfrak{m}_A = 0$. We have $A = \overline{A} \oplus \eta\mathbb{C}$ and we write $\mathcal{O}_{(X,z)\times(T,t)} \cong \mathbb{C}\{\boldsymbol{x}\}/\langle f\rangle \otimes_{\mathbb{C}} A \cong A\{\boldsymbol{x}\}/\langle f\rangle$ with $A\{\boldsymbol{x}\} = \mathbb{C}\{\boldsymbol{x}\} \otimes_{\mathbb{C}} A \cong \overline{A}\{\boldsymbol{x}\} \oplus \eta\mathbb{C}\{\boldsymbol{x}\}$, where $\oplus$ is the direct sum of vector spaces. The A-isomorphism $\overline{\psi} : A\{\boldsymbol{x}\}/\langle f\rangle \xrightarrow{\cong} A\{\boldsymbol{x}\}/\langle f\rangle$ can be lifted to an isomorphism (cf. [GLS6, Lemma I.1.27])

$$\psi : A\{\boldsymbol{x}\} \xrightarrow{\cong} A\{\boldsymbol{x}\}$$

such that $\psi(f) \in \langle f\rangle$. We can also lift the sections τ and $\sigma := \tau \circ \psi$ to sections of $A \hookrightarrow A\{\boldsymbol{x}\}$,

$$\tau, \sigma : A\{\boldsymbol{x}\} \to A ,$$

[10] Equianalytic sections need not be unique in positive characteristic. For an example consider the trivial family $K[\varepsilon] \to K[\varepsilon][[x, y]]/\langle x^2 + y^3\rangle$, $\varepsilon^2 = 0$, in characteristic 3. The isomorphism $\psi : x \mapsto x, y \mapsto y + \varepsilon, \varepsilon \mapsto \varepsilon$, of this family maps the trivial section τ to the non-trivial section $\tau \circ \psi : x \mapsto 0, y \mapsto \varepsilon, \varepsilon \mapsto \varepsilon$. Note however that the family is not equisingular along the section $\tau \circ \psi$. In fact, *equisingular* sections are always unique, also in positive characteristic by [CGL1, Proposition 2.12].

with $\tau(x_i) = 0$ and $\sigma(x_i) = \tau \circ \psi(x_i) \in A$. We have to show $\sigma(x_i) = 0$.

Denote by $\psi_0 : \mathbb{C}\{\boldsymbol{x}\} \xrightarrow{\cong} \mathbb{C}\{\boldsymbol{x}\}$ the isomorphism induced by ψ modulo $\mathfrak{m}_A$. Then ψ_0 extends to an A-automorphism of $A\{\boldsymbol{x}\}$ with $\psi_0(\langle \boldsymbol{x}\rangle) = \langle \boldsymbol{x}\rangle$. Hence we may replace ψ by $\psi \circ \psi_0^{-1}$ and assume

$$\psi(x_i) = x_i + h_i, \quad h_i = \overline{h}_i + \eta h_i', \quad \overline{h}_i \in \mathfrak{m}_{\overline{A}}\, \overline{A}\{\boldsymbol{x}\},, \quad h_i' \in \mathbb{C}\{\boldsymbol{x}\}\,.$$

Then $\sigma(x_i) = \tau \circ \psi(x_i) = h_i(0) \in A$. By induction we can assume $\sigma = \tau \bmod \langle \eta \rangle$, and hence $\overline{h}_i(0) = 0$. It remains to prove $h_i'(0) = 0$.

Since $\psi(f) \in \langle f \rangle$, there is a unit $u = \overline{u} + \eta u' \in A\{\boldsymbol{x}\}^*$, $\overline{u} \in \overline{A}\{\boldsymbol{x}\}^*$, $u' \in \mathbb{C}\{\boldsymbol{x}\}$ such that

$$\psi(f) = \overline{u} f + \eta u' f\ .$$

Using the Taylor expansion with $\overline{h} = (\overline{h}_1, \ldots, \overline{h}_n)$, $h' = (h_1', \ldots, h_n')$, we get, since $\eta^2 = 0$,

$$\begin{aligned}\psi(f) = f(\boldsymbol{x} + \overline{h} + \eta h) &= f(\boldsymbol{x} + \overline{h}) + \eta \sum_{i=1}^{n} \frac{\partial f}{\partial x_i}(\boldsymbol{x} + \overline{h}) h_i \\ &= f(\boldsymbol{x} + \overline{h}) + \eta \sum_{i=1}^{n} \frac{\partial f}{\partial x_i}(\boldsymbol{x}) h_i\ .\end{aligned}$$

The last equality holds since $f \in \mathfrak{m}^2$ and $h_i \overline{h}_j = 0$.

Since $\psi(f) = f(\boldsymbol{x} + \overline{h}) = \overline{u} f \bmod \langle \eta \rangle$, we get

$$u' f = \sum_{i=1}^{n} \frac{\partial f}{\partial x_i}(\boldsymbol{x}) \cdot h_i'\ .$$

Applying Lemma 2.2.4, we conclude $h_i'(0) = 0$ as required.

(2) This is proved in [GLS6, Prop. II.2.8].

(3) We prove the statement for *es*; for *ea* it is similar but easier. The first isomorphism follows from

$$\underline{\mathcal{D}ef}^{\,fix}_{X,z} \cong \underline{\mathcal{D}ef}^{\,sec}_{X,z}\ ,$$

which holds since every section of a deformation over some germ can be trivialized by [GLS6, Prop. II.2.2].

For the second isomorphism let ϕ and ϕ' be two *es*-deformations of (X, z) with given systems of *em*-sections, which are isomorphic as deformations without sections via some isomorphism ψ. Then ψ maps the initial *em*-section σ of ϕ to some *em*-section $\phi\sigma$ of ϕ'. ψ lifts to an isomorphism of the blowing up of σ and $\phi\sigma$ and we produce by induction a system of *em*-sections of ϕ' being the image of *em*-sections of ϕ under the lifts of ψ. This is another system of *em*-sections of the *es*-deformation ϕ', which must coincide with the given system by the first statement. In particular, ψ maps the initial *em*-section of ϕ to the initial *em*-section of ϕ', showing that ϕ and ϕ' are isomorphic as deformations with sections. This proves the second isomorphism. □

Remark 2.2.6.1 (1) It is easy to see that for equimultiple deformations the sections are in general not unique. It has been pointed by Markwig [Mar] that one has to take special care when applying our method.

(2) Let $\phi : (\mathscr{X}, z) \to (T, t)$ be a deformation of the reduced plane curve singularity $(X, z) \subset (\mathbb{C}^2, 0)$ with (T, t) reduced. Then the following are equivalent (for a sufficiently small representative $\mathscr{X} \to T$):

(a) ϕ is equisingular.
(b) There exists a section $\sigma : (T, t) \to (\mathscr{X}, z)$ of ϕ such that the topological type (cf. beginning of Chap. 1) of $(\mathscr{X}_s, \sigma(s))$ is independent of $s \in T$.
(c) There exists a section σ of ϕ such that the Milnor number $\mu(\mathscr{X}_s, \sigma(s))$ is constant for $s \in T$.
(d) ϕ is (embedded) topologically trivial. That is, there exists a small ball $B \subset \mathbb{C}^2$ around z and a homeomorphism of $B \times T$ over T onto itself, mapping $\mathscr{X} \subset B \times T$ onto $X \times T \subset B \times T$.

For proofs of these statements resp. references see [GLS6, Section II.2.1 and Corollary II.2.68].

For the rest of this section let (Σ, z) be a smooth complex germ, $\mathfrak{m}_z \subset \mathcal{O}_{\Sigma,z}$ the maximal ideal and $f \in \mathfrak{m}_z$ define the hypersurface germ $(X, z) \subset (\Sigma, z)$.

In order to apply infinitesimal methods of deformation theory we have to consider deformations over non reduced base spaces. In particular, we need *first order deformations*, that is, deformations over the *fat point of dimension two*

$$T_\varepsilon := (pt, \mathbb{C}[\varepsilon]) .$$

T_ε is the complex space consisting of one point pt with structure sheaf $\mathbb{C}[\varepsilon] = \mathbb{C}\{t\}/\langle t^2\rangle$. Recall from [GLS6, App. C] that, for any deformation functor $\underline{D}$, the set of isomorphism classes of deformations $\underline{D}(T_\varepsilon)$ over T_ε is called the *tangent space of the functor* $\underline{D}$ and usually denoted by T^1. It carries a natural structure as $\mathbb{C}$-vector space if it satisfies Schlessinger's conditions ([GLS6, Lemma C.1.7]). The following definition contains the well-known description of $T^1_{X,z}$ and $T^{1,sec}_{X,z}$ (for a detailed description of these vector spaces see [GLS6, Section II.1.4 resp. II.2.1]).

If $\underline{D}$ has a semiuniversal deformation, then $T^1 = \underline{D}(T_\varepsilon)$ is the tangent space to the base space of this semiuniversal deformation.

Definition 2.2.7 (1) We define the vector spaces

$$\begin{aligned} T^1_{X,z} &:= \underline{\mathcal{D}\!ef}^{ea}_{X,z}(T_\varepsilon) := \underline{\mathcal{D}\!ef}_{X,z}(T_\varepsilon) = \mathcal{O}_{\Sigma,z}/\langle f, j(f)\rangle, \\ T^{1,sec}_{X,z} &:= \underline{\mathcal{D}\!ef}^{sec}_{X,z}(T_\varepsilon) \cong \underline{\mathcal{D}\!ef}^{fix}_{X,z}(T_\varepsilon), \\ T^{1,fix}_{X,z} &:= \underline{\mathcal{D}\!ef}^{fix}_{X,z}(T_\varepsilon) = \mathfrak{m}_z/\langle f\rangle + \mathfrak{m}_z j(f). \end{aligned}$$

(2) The ideals

$$I^{ea}(X, z) := \langle f, j(f)\rangle, \text{ resp.}$$
$$I^{ea}_{fix}(X, z) := \langle f\rangle + \mathfrak{m}_z j(f)$$

are called the *Tjurina ideal* resp. the *fixed Tjurina ideal* of $(X, z) \subset (\Sigma, z)$. The ideal

$$j(f) := \langle \partial f/\partial x_1, \ldots, \partial f/\partial x_n\rangle,$$

$x_1, \ldots, x_n$ local coordinates of (Σ, z), denotes the *Jacobian ideal* of f.

(3) For (Σ, z) a smooth surface and (X, z) a plane curve singularity we recall (see Definition 1.1.63) the definition of the *equisingularity ideal*,

$$I^{es}(X, z) := \left\{ g \in \mathcal{O}_{\Sigma,z} \;\middle|\; \begin{array}{c} f + \varepsilon g \text{ is an equisingular deformation} \\ \text{of } (X, z) \text{ over } T_\varepsilon \text{ along some section} \end{array} \right\}$$

and of the *fixed equisingularity ideal*

$$I^{es}_{fix}(X, z) := \left\{ g \in \mathcal{O}_{\Sigma,z} \;\middle|\; \begin{array}{c} f + \varepsilon g \text{ is an equisingular deformation} \\ \text{of } (X, z) \text{ over } T_\varepsilon \text{ along the trivial section} \end{array} \right\}.$$

Then we have the isomorphisms

$$\underline{\mathcal{D}ef}^{es}_{X,z}(T_\varepsilon) \cong I^{es}(X, z)/I^{ea}(X, z), \text{ and}$$
$$\underline{\mathcal{D}ef}^{es,fix}_{X,z}(T_\varepsilon) \cong I^{es}_{fix}(X, z)/I^{ea}_{fix}(X, z).$$

For the dimension of these spaces see Definition 2.2.12 and Lemma 2.2.13.

Our approach to deformations of X/Σ with fixed singularity types works not only if the induced local deformations are equianalytic or equisingular, it works whenever the induced deformations of (X, z_i) define a subcategory $\mathcal{D}ef'_{X,z_i}$ of $\mathcal{D}ef_{X,z_i}$ satisfying the following condition (VSIR) for $i = 1, \ldots, r$:

Definition 2.2.8 A subcategory $\mathcal{D}ef'_{X,z}$ of $\mathcal{D}ef_{X,z}$ satisfies *condition (VSI)* if the following holds for the associated deformation functor:

$$\left.\begin{array}{ll} (V) & \underline{\mathcal{D}ef}'_{X,z} \text{ has a versal deformation,} \\ (S) & \underline{\mathcal{D}ef}'_{X,z} \text{ is smooth,} \\ (I) & \underline{\mathcal{D}ef}'_{X,z}(T_\varepsilon) \text{ is an ideal in } \underline{\mathcal{D}ef}_{X,z}(T_\varepsilon). \end{array}\right\} (VSI)$$

The existence of a versal deformation implies the existence of a semiuniversal deformation and smoothness of $\underline{\mathcal{D}ef}'_{X,z}$ means that the base space of the semiuniversal deformation is smooth of dimension $\dim_{\mathbb{C}} \underline{\mathcal{D}ef}'_{X,z}(T_\varepsilon)$. Examples for deformations that satisfy (VSI) are equianalytic and equisingular deformations (with section), cf. Remark 2.2.11.1.

$\mathcal{D}ef'_{X,z}$ satisfies *condition (VSIR)* if moreover the following condition (R) is satisfied: Let $\phi : (\mathscr{X}, z) \to (T, t_0)$ be any deformation of (X, z) and let D'_ϕ be the functor from complex space germs to sets, given by

$$D'_\phi(S, s) = \{\varphi : (S, s) \to (T, t_0) \mid \varphi^*\phi \text{ is a deformation in } \mathcal{D}ef'_{X,z}(S, s)\}.$$

The (representability) condition (R) is

$$(R)\ D'_\phi \text{ is representable by a complex subgerm } (T', t_0) \subset (T, t_0).$$

Note that (R) means the existence of a stratum (T', t_0) such that

(i) the restriction of ϕ to (T', t_0) is in $\mathcal{D}ef'_{X,z}(T', t_0)$,
(ii) for any morphism $\varphi : (S, s) \to (T, t_0)$ such that $\varphi^*\phi$ is in $\mathcal{D}ef'_{X,z}(S, s)$, φ factors through (T', t_0).

It follows that the complex space germ $(T', t_0) \subset (T, t_0)$ is unique, with a unique (not necessarily reduced) analytic structure. It is called the $\mathcal{D}ef'_{X,z}$*-stratum* in (T, t_0).

Definition 2.2.9 We call the category $\mathcal{D}ef'_{X,z}$ *ideal-versal* if it satisfies condition (VSI) and *strictly ideal-versal* if it satisfies (VSIR).

In an analogous way we define conditions (V), (S), (I) and (R) for subcategories $\mathcal{D}ef'^{sec}_{X,z} \subset \mathcal{D}ef^{sec}_{X,z}$ and $\mathcal{D}ef'^{fix}_{X,z} \subset \mathcal{D}ef^{fix}_{X,z}$.

Examples of strictly ideal-versal deformations are equimultiple ($\mathcal{D}ef^{em}_{X,z}$), equisingular ($\mathcal{D}ef^{es}_{X,z}$) and equianalytic ($\mathcal{D}ef^{ea}_{X,z}$) ones as we shall see below (see Example 2.2.10.1, Remark 2.2.11.1 and Example 2.2.20.1).

Remark 2.2.9.1 Let $\mathcal{D}ef'_{X,z}$ be strictly ideal-versal.

(1) If $\Phi : (\mathscr{Y}, z) \to (B, 0)$ is the semiuniversal deformation of (X, z) with $(B', 0) \subset (B, 0)$ the $\mathcal{D}ef'_{X,z}$-stratum, then the restriction of Φ to $(B', 0)$ is the semiuniversal deformation for $\mathcal{D}ef'_{X,z}$. This follows from (R) and the semiuniversality of Φ. Moreover, $(B', 0)$ is smooth by condition (S).

(2) Let $\phi : (\mathscr{X}, z) \to (T, t_0)$ be a deformation in $\mathcal{D}ef'_{X,z}(T, t_0)$. By the semiuniversality of Φ there exists a morphism $\varphi : (T, t_0) \to (B', 0)$ inducing ϕ from Φ via base change. φ is not unique (only its tangent map), but it follows that preimage $\varphi^{-1}(B', 0)$ coincides with the $\mathcal{D}ef'_{X,z}$-stratum (T', t_0) and hence is unique.

(3) For $\mathcal{D}ef^{ea}_{X,z}$ the subspace $(B^{ea}, 0)$ is the *equianalytic stratum* in $(B, 0)$. The restriction of Φ to $(B^{ea}, 0)$ is the trivial deformation of (X, z) and hence can also be induced via the morphism mapping $(B^{ea}, 0)$ to $\{0\}$. It follows from the semiuniversality of Φ that $(B^{ea}, 0)$ is the reduced point $\{0\}$.

(4) For $\mathcal{D}ef^{es}_{X,z}$, (X, z) a plane curve singularity, the subspace $(B^{es}, 0)$ is the *equisingular stratum* in $(B, 0)$. It coincides with the μ-constant stratum and is known to be smooth (cf. [GLS6, Theorem II.2.61, Corollary II.2.67]).

Let $\mathcal{D}ef^{em}_{X,z}$ denote the category of *equimultiple deformations* of (X, z), i.e. deformations with section which are equimultiple along the section.

By [GLS6, Proposition II.2.2] every deformation of (X, z) with section is isomorphic to an embedded deformation with trivial section. Hence we may assume that $(X, 0) \subset (\mathbb{C}^n, 0)$ is given by

$$f(x) = \sum_{|\alpha| \geq m} c_\alpha x^\alpha, \quad m = \mathrm{mt}(f) \geq 2\,,$$

and that the *semiuniversal deformation with section* of (X, z) is given by

$$F(x, t) = f(x) + \sum_{j=1}^{k} t_j g_j(x)\,,$$

with $g_1, \ldots, g_k$ a basis of $\mathfrak{m}/(\langle f \rangle + \mathfrak{m} j(f))$ (cf. Proposition 2.2.46). If we restrict F to the *equimultiple stratum* $\{t_j = 0 \mid \mathrm{mt}(g_j) < m\}$ we get an equimultiple deformation along the trivial section, which is easily seen to be semiuniversal. Hence we get

Proposition 2.2.10 *With the above notation*

$$F^{em}(x, t) := f(x) + \sum_{\mathrm{mt}(g_j) \geq m} t_j g_j(x) \in \mathbb{C}\{x, t'\}$$

is an equimultiple deformation of $(X, 0)$ *along the trivial section. It is a semiuniversal equimultiple deformation of* $(X, 0)$ *over* $(T', 0) = \{t \in \mathbb{C}^k \mid t_j = 0\, for\ \mathrm{mt}(g_j) < m\}$, *i.e. a semiuniversal object of* $\underline{\mathcal{D}ef}^{em}_{(X,0)}(T', 0)$.

Example 2.2.10.1 Equimultiple deformations are strictly ideal-versal: Since $(T', 0) \subset (\mathbb{C}^k, 0)$ is the *em*-stratum of F and smooth, $\mathcal{D}ef^{em}_{X,z}$ satisfies (V), (S), (R). Moreover, $\underline{\mathcal{D}ef}^{em}_{(X,0)}(T_\varepsilon)$ is an ideal in $\underline{\mathcal{D}ef}^{fix}_{X,0} = \mathfrak{m}/\langle f, \mathfrak{m} j(f)\rangle$ generated by g_j such that $\mathrm{mt}(g_j) \geq m$. It follows that equimultiple deformations $\mathcal{D}ef^{em}_{X,z} \subset \mathcal{D}ef^{sec}_{X,z}$ satisfy (VSIR).

More generally, let $(L, z) \subset (\Sigma, z)$ be a smooth germ and consider deformations of (X, z) with section σ such that the intersection multiplicity of $(\mathscr{X}_t, \sigma(t))$ with $(L, \sigma(t))$ is constant. In a similar way as above one shows that such deformations are strictly ideal-versal in $\underline{\mathcal{D}ef}^{sec}_{X,z}$. If (L, z) is transversal to the tangent cone of (X, z), then we get equimultiple deformations of (X, z).

Example 2.2.10.2 Note that our definition for a family being equimultiple requires equimultiplicity along a section, not just along a map. The following example (deformation of a cusp to a node) makes this clear and shows that the semiuniversal equimultiple deformation of $(X, 0)$ is not a stratum in the (usual) semiuniversal deformation of $(X, 0)$ without section.

Let $(X, 0) \subset (\mathbb{C}^2, 0)$ be the ordinary cusp defined by $f = x^2 + y^3$. The semiuniversal deformation of $(X, 0)$ without section is given by

$$\Phi : \mathscr{X} = V(F) = \{(x, y, u, v) \in \mathbb{C}^4 \mid F(x, y, u, v) = 0\} \to T$$

with

$$F(x, y, u, v) = x^2 + y^3 + u + vy$$

and Φ the projection on the (u, v)-plane. The critical locus of Φ is a smooth curve $C \subset \mathscr{X}$ and the discriminant $D = \Phi(C) \subset T$ is the cuspidal cubic $27u^2 + 4v^3 = 0$. Over each point $t \in D$ there is a unique point $p(t) \in C$ and $\mathrm{mt}(\mathscr{X}_t, p(t)) = 2$ for all t. Hence Φ is equimultiple over D along the image of the map $p : D \to C$. However, the map p is not a section of the restriction of Φ over D, since a section would map D isomorphically onto its image which is not the case since C is smooth and D is singular. Here, the only stratum in T over which there exists an equimultiple section is $\{0\}$.

Consider now the semiuniversal deformation with section of the cusp. Since x, y, y^2 is a basis of $\mathfrak{m}/\langle f, \mathfrak{m}j(f)\rangle$ it is given by

$$F(x, y, u, v, w) = x^2 + y^3 + ux + vy + wy^2$$

with $\Phi : \mathscr{X} = V(F) \to T$ the projection to the (u, v, w)-space, and with σ the trivial section. σ is equimultiple along the line $u = v = 0$ and this is the $\mathcal{D}ef^{em}_{X,z}$-*stratum* of Φ introduced in Definition 2.2.8 (R).

Since the canonical morphism $\mathcal{D}ef^{fix}_{X,z} \hookrightarrow \mathcal{D}ef^{sec}_{X,z}$ induces (by Remark 2.2.2.1) an isomorphism $\underline{\mathcal{D}ef}^{sec}_{X,z} \cong \underline{\mathcal{D}ef}^{fix}_{X,z}$ we do sometimes not distinguish between $\underline{\mathcal{D}ef}^{fix}_{X,z}$ and $\underline{\mathcal{D}ef}^{sec}_{X,z}$.

Definition 2.2.11 Let $\mathcal{D}ef'_{X,z}$, resp. $\mathcal{D}ef'^{sec}_{X,z}$, resp. $\mathcal{D}ef'^{fix}_{X,z}$ be a subcategory of $\mathcal{D}ef_{X,z}$, resp. $\mathcal{D}ef^{sec}_{X,z}$, resp. $\mathcal{D}ef^{fix}_{X,z}$.

(1) We set

$$\begin{aligned}
(T^1_z)' \quad &:= \underline{\mathcal{D}ef}'_{X,z}(T_\varepsilon) \subset \underline{\mathcal{D}ef}_{X,z}(T_\varepsilon) = \mathcal{O}_{\Sigma,z}/I^{ea}(X, z)\,,\\
(T^{1,sec}_z)' &:= \underline{\mathcal{D}ef}'^{sec}_{X,z}(T_\varepsilon) \subset \underline{\mathcal{D}ef}^{sec}_{X,z}(T_\varepsilon) \cong \underline{\mathcal{D}ef}^{fix}_{X,z}(T_\varepsilon)\,,\\
(T^{1,fix}_z)' &:= \underline{\mathcal{D}ef}'^{fix}_{X,z}(T_\varepsilon) \subset \underline{\mathcal{D}ef}^{fix}_{X,z}(T_\varepsilon) = \mathfrak{m}_z/I^{ea}_{fix}(X, z).
\end{aligned}$$

(2) Moreover, if condition (I) of Definition 2.2.8 is satisfied, these are ideals and we denote by

$$I'(X, z) = \pi^{-1}((T^1_z)') \text{ resp. } I'_{fix}(X, z) = \pi^{-1}((T^{1,fix}_z)')$$

the ideals in $\mathcal{O}_{\Sigma,z}$, where π is the projection $\mathcal{O}_{\Sigma,z} \to \mathcal{O}_{\Sigma,z}/I^{ea}(X, z)$ resp. $\mathfrak{m}_z \to \mathfrak{m}_z/I^{ea}_{fix}(X, z)$.

Hence

$$(T_z^1)' = I'(X,z)/I^{ea}(X,z),$$

$$(T_z^{1,sec})' \cong (T_z^{1,fix})' = I'_{fix}(X,z)/I^{ea}_{fix}(X,z)$$

with $I^{ea}(X,z) = \langle f, j(f)\rangle \subset I'(X,z)$ and $I^{ea}_{fix}(X,z) = \langle f, \mathfrak{m}_z j(f)\rangle \subset I'_{fix}(X,z)$.

Remark 2.2.11.1 Let $f \in \mathfrak{m}_z \subset \mathcal{O}_{\Sigma,z}$ define $(X,z) \subset (\Sigma,z)$.

(1) The most important examples of strictly ideal-versal deformations are $\mathcal{D}ef^{ea}_{X,z}$, $\mathcal{D}ef^{ea,fix}_{X,z}$, $\mathcal{D}ef^{es}_{X,z}$, and $\mathcal{D}ef^{es,fix}_{X,z}$. This will be proved in Proposition 2.2.21. For the corresponding dimensions of T^1 see Example 2.2.20.1.

(2) It is clear that any deformation of (X,z) over T_ε is given by

$$f + \varepsilon g \in \mathcal{O}_{\Sigma,z} \otimes \mathcal{O}_{T_\varepsilon} = \mathcal{O}_{\Sigma,z} \oplus \mathcal{O}_{\Sigma,z} \cdot \varepsilon$$

for some $g \in \mathcal{O}_{\Sigma,z}$ (resp. $g \in \mathfrak{m}_z$ if we consider deformations with trivial section). Conversely, every such expression defines a deformation of (X,z) over T_ε.

Let $\mathcal{D}ef'_{X,z}$ satisfy condition (I). By definition of $I'(X,z)$ we have for $g \in \mathcal{O}_{\Sigma,z}$

$$f + \varepsilon g \in Def'_{X,z}(T_\varepsilon) \Leftrightarrow g \in I'(X,z)$$

and such a deformation is in $\mathcal{D}ef'^{,fix}_{X,z}(T_\varepsilon)$ iff $g \in I'_{fix}(X,z)$. In particular, $f + \varepsilon g$ is isomorphic to the trivial deformation iff $g \in I^{ea}(X,z)$ (resp. $g \in I^{ea}_{fix}(X,z)$ for deformations with trivial section).

(3) Let $B'_{X,z}$ resp. $B'^{sec}_{X,z}$ denote the base space of the semiuniversal deformation of $\underline{\mathcal{D}ef}'_{X,z}$ resp. $\underline{\mathcal{D}ef}'^{sec}_{X,z}$. These base spaces exist if condition (V) holds and their Zariski tangent spaces can be identified with $(T_z^1)'$ resp. $(T_z^{1,fix})'$ (cf. [GLS6, Lemma II.1.20]). They are smooth if condition (S) holds. If moreover (R) is satisfied, then $B'_{X,z}$ resp. $B'^{sec}_{X,z}$ is representable by a unique stratum in the base space $B_{X,z}$ resp. $B^{sec}_{X,z}$ of the semiuniversal base space of $\underline{\mathcal{D}ef}_{X,z}$ resp. $\underline{\mathcal{D}ef}^{sec}_{X,z}$.

(4) It follows that $\mathcal{D}ef'_{X,z}$ (and similarly $\mathcal{D}ef'^{sec}_{X,z}$) satisfies condition (VSIR) if and only if the following holds:

If $\phi : (\mathscr{X},z) \to B_{X,z}$ denotes the semiuniversal deformation of (X,z), then there exists a unique smooth subgerm $B'_{X,z} \subset B_{X,z}$ such that the restriction of ϕ to $B'_{X,z}$ represents a semiuniversal deformation of $\underline{\mathcal{D}ef}'_{X,z}$. Moreover, the tangent space to $B'_{X,z}$ is an ideal in $\mathcal{O}_{\Sigma,z}/\langle f, j(f)\rangle$.

(5) Compare strictly ideal-versal deformations with families of fixed smooth singularity type S, defined in Sects. 2.3.1 and 2.3.2. There we consider basically the same conditions, but we do not require the existence of an infinitesimal deformation theory (cf. Remark 2.2.50.1). That is, we consider deformations only over reduced base spaces with fixed singularity type S along a given section, which makes the theory a little less powerful, but more easy to apply.

(6) For the definition of (semiuni-)versal deformations see [GLS6, 2.1]; smooth or unobstructed means that the base space of (any) versal deformation is smooth

(see also Remark 2.2.21.1). Recall from [GLS6, Theorem 2.1.13] that the existence of a versal deformation implies the existence of a semiuniversal deformation.

Definition 2.2.12 Let $\langle f\rangle$ be the ideal of (X,z) in (Σ,z) and $(T^1_z)'$ as in Definition 2.2.11. We set

$$\begin{aligned}
\tau(X,z) &:= \dim_{\mathbb{C}} T^1_{X,z} &&= \dim_{\mathbb{C}} \mathcal{O}_{\Sigma,z}/\langle f, j(f)\rangle\,,\\
\tau_{fix}(X,z) &:= \dim_{\mathbb{C}} T^{1,fix}_{X,z}+1 &&= \dim_{\mathbb{C}} \mathcal{O}_{\Sigma,z}/\langle f, \mathfrak{m}_z j(f)\rangle\,,\\
\tau'(X,z) &:= \tau(X,z)-\dim_{\mathbb{C}}(T^1_z)' &&= \dim_{\mathbb{C}} \mathcal{O}_{\Sigma,z}/I'(X,z),\\
\tau'_{fix}(X,z) &:= \tau_{fix}(X,z)-\dim_{\mathbb{C}}(T^{1,fix}_z)' &&= \dim_{\mathbb{C}} \mathcal{O}_{\Sigma,z}/I'_{fix}(X,z).
\end{aligned}$$

In particular, for $'$ being *ea* or *es*, we get

$$\begin{aligned}
\tau^{ea}(X,z) &= \tau(X,z),\\
\tau^{ea}_{fix}(X,z) &= \tau_{fix}(X,z),\\
\tau^{es}(X,z) &= \dim_{\mathbb{C}} \mathcal{O}_{\Sigma,z}/I^{es}(X,z),\\
\tau^{es}_{fix}(X,z) &= \dim_{\mathbb{C}} \mathcal{O}_{\Sigma,z}/I^{es}_{fix}(X,z).
\end{aligned}$$

Since $T^{1,sec}_{X,z}\cong T^{1,fix}_{X,z}$ we have $(T^{1,fix}_z)'\cong(T^{1,sec}_z)'$ and therefore $\tau_{sec}(X,z)=\tau_{fix}(X,z)$ and $\tau'_{sec}(X,z)=\tau'_{fix}(X,z)$. See Example 2.2.20.1 for $'$ being *ea* and *es*.

Let us compare τ and τ_{fix}.

Lemma 2.2.13 *For any non-smooth hypersurface germ* $(X,z)\subset(\mathbb{C}^n,z)$ *with isolated singularity we have*

$$\tau_{fix}(X,z)=\dim_{\mathbb{C}} T^{1,fix}_{X,z}+1=\tau(X,z)+n\,,$$

and for $n=2$ *we have*

$$\tau^{es}_{fix}(X,z)=\tau^{es}(X,z)+2\,.$$

Proof Let $\mathcal{O}_{X,z}=\mathbb{C}\{x_1,\dots,x_n\}/\langle f\rangle$, J resp. $j(f)$ the ideal generated by $\frac{\partial f}{\partial x_1},\dots,\frac{\partial f}{\partial x_n}$ in $\mathcal{O}_{X,z}$ resp. in $\mathbb{C}\{x_1,\dots,x_n\}$ and $\mathfrak{m}=\langle x_1,\dots,x_n\rangle$. Then

$$\tau_{fix}-\tau=\dim_{\mathbb{C}}\left(\langle f,j(f)\rangle/\langle f,\mathfrak{m}j(f)\rangle\right)=\dim_{\mathbb{C}}\left(J/\mathfrak{m}J\right)=n$$

by Lemma 2.2.4. The proof for τ^{es} is given in [GLS6, Proposition II.2.14]. □

Remark 2.2.13.1 The definitions of τ^{ea} and τ^{es} are consistent in the following sense.

$T^1_{X,z}$ is the tangent space of the (smooth) base space $B_{X,z}$ of the semiuniversal deformation of (X,z). $\tau^{ea}(X,z)=\tau(X,z)$ is hence the codimension in $B_{X,z}$ of the base space of the semiuniversal equianalytic deformation of (X,z) (which is the reduced point).

$T^{1,es}_{X,z}=\underline{\mathcal{D}ef}^{es}_{X,z}(T_\varepsilon)$ is the tangent space of the (smooth) base space $B^{es}_{X,z}$ of the semiuniversal equisingular deformation of (X,z). $\tau^{es}(X,z)$ is hence the codimension in $B_{X,z}$ of the base space of the semiuniversal equisingular deformation of (X, z) (which coincides with the μ-constant stratum in $B_{X,z}$, cf. [GLS6, Theorem 2.61, Theorem 2.64 and Proposition 2.63]).

2.2.1.3 Fixing Local Conditions Over Artin Rings

We develop the formal deformation theory of isolated hypersurface singularities over fat points, fixing singularity invariants considered in the previous section. The results, which show that nothing unexpected happens, are essential for the next section.

Let (Σ, z) be again a smooth complex germ, $\mathfrak{m}_z \subset \mathcal{O}_{\Sigma,z}$ the maximal ideal and $f \in \mathfrak{m}_z$ defining a hypersurface germ $(X, z) \subset (\Sigma, z)$. For the whole Sect. 2.2.1.3, let $\mathcal{D}ef'_{X,z}$ be a full subcategory of $\mathcal{D}ef_{X,z}$ and $\underline{\mathcal{D}ef}'_{X,z}$ its functor of isomorphism classes.

By a *fat point* we understand a complex space $T = (\mathrm{pt}, A)$ consisting of one point pt with structure sheaf an analytic Artinian $\mathbb{C}$-algebra A. Morphisms are just morphisms of analytic algebras. An example is the space $T_\varepsilon = (\mathrm{pt}, \mathbb{C}[\varepsilon])$.

An inclusion $\overline{T} \hookrightarrow T$ of fat points corresponds to an extension $A \twoheadrightarrow \overline{A}$ of Artinian algebras and such an *extension* is called *small* if $\overline{A} = A/\eta A$ for some $\eta \in \mathfrak{m}_A$ with $\eta\mathfrak{m}_A = 0$. Instead of $\mathcal{D}ef(T)$ we usually write $\mathcal{D}ef(A)$ for deformations over the fat point $T = (\mathrm{pt}, A)$.

For a small extension $A \to \overline{A} = A/\eta A$ we get, by restriction to $\overline{A}$, a natural map of sets

$$\rho : \underline{\mathcal{D}ef}'_{X,z}(A) \to \underline{\mathcal{D}ef}'_{X,z}(\overline{A}) .$$

In general the fibres of the map ρ may be empty, i.e. for a given deformation $\overline{\xi}$ in $\mathcal{D}ef'_{X,z}(\overline{A})$ there may be no deformation $\xi \in \mathcal{D}ef'_{X,z}(A)$ such that the restriction of ξ to $\overline{A}$ is isomorphic to $\overline{\xi}$ (even if $\underline{\mathcal{D}ef}'_{X,z}$ has a versal deformation, it may be obstructed). However, if versatility holds for $\underline{\mathcal{D}ef}'_{X,z}$ we have the following statement about the non-empty fibres.

Lemma 2.2.14 *If $\underline{\mathcal{D}ef}'_{X,z}$ satisfy (V), then $\underline{\mathcal{D}ef}'_{X,z}(T_\varepsilon)$ acts transitively on the non-empty fibres of ρ.*

Proof If $\underline{\mathcal{D}ef}'_{X,z}$ has a versal deformation, then it has a semiuniversal deformation and satisfies Schlessinger's conditions (H_0)-(H_3) as a functor of Artin rings (cf. [GLS6, Appendix C, Theorem C.1.6]). The claim follows from [GLS6, Remark C.1.7.1]. □

To explain what this means in concrete terms, let us consider embedded deformations. If $(X, z) \subset (\Sigma, z)$ is given by $f \in \mathcal{O}_{\Sigma,z}$ then an *embedded deformation* of (X, z) over (T, t) is defined by an $F \in \mathcal{O}_{(\Sigma,z)\times(T,t)}$ with $f = F \bmod \mathfrak{m}_{T,t}$. That is, the deformation of (X, z) is given by the projection from $V(F) \subset (\Sigma, z) \times (T, t)$ to (T, t). We describe the action for embedded deformations over fat points.

Lemma 2.2.15 *Let $A \to \overline{A} = A/\langle\eta\rangle$ be a small extension with $\eta \cdot \mathfrak{m}_A = 0$. Let $\xi_0 \in \underline{\mathcal{D}ef}'_{X,z}(T_\varepsilon)$, resp. $\xi \in \underline{\mathcal{D}ef}'_{X,z}(A)$ be given by $f + \varepsilon g \in \mathcal{O}_{\Sigma,z} \otimes_{\mathbb{C}} \mathbb{C}[\varepsilon]$, $g \in I'(X,z)$, resp. by $F \in \mathcal{O}_{\Sigma,z} \otimes_{\mathbb{C}} A$.*

(1) $\underline{\mathcal{D}ef}'_{X,z}(T_\varepsilon)$ acts transitively on the fibres of $\rho : \underline{\mathcal{D}ef}'_{X,z}(A) \to \underline{\mathcal{D}ef}'_{X,z}(\overline{A})$ via

$$[\xi_0] \cdot [\xi] = [F + g\eta] \in \underline{\mathcal{D}ef}'_{X,z}(A)\ .$$

(2) If $F, F' \in \mathcal{O}_{\Sigma,z} \otimes A$ define elements in $\underline{\mathcal{D}ef}'_{X,z}(A)$ which have the same image in $\underline{\mathcal{D}ef}'_{X,z}(\overline{A})$, then there exists a $g \in I'(X,z)$ such that

$$[F'] = [F + g\eta]$$

in $\underline{\mathcal{D}ef}'_{X,z}(A)$. This means that there exists an A-automorphism ψ of $\mathcal{O}_{\Sigma,z} \otimes A$ and a unit $G \in (\mathcal{O}_{\Sigma,z} \otimes A)^*$ such that $\psi(F') = GF + h\eta$ for some $h \in I'(X,z)$.

Proof (1) The condition in [GLS6, Remark C.1.7.1] specializes to the given formula in our situation, with $[*]$ denoting the deformation class of $*$. Note that the action does not change the restriction $\overline{F} = F \bmod \langle\eta\rangle$ and hence induces an action on the non-empty fibres of ρ.

(2) It follows from (1) that $\psi(F') = G(F + g\eta) = GF + h\eta$ with $h = G_0 g$, G_0 the image of G in $(\mathcal{O}_{\Sigma,z})^*$. □

Let us consider embedded deformations in more detail, denoting by $\underline{\mathcal{D}ef}_{X/\Sigma,z}$ the functor of embedded deformations of $(X,z) \subset (\Sigma,z)$. Such deformations over (T,t) are given by families $j : (\mathscr{X},(z,t)) \xrightarrow{\cong} V(F) \subset (\Sigma \times T,(z,t))$, $F \in \mathcal{O}_{(\Sigma,z)\times(T,t)}$, $V(F)$ flat over (T,t). Morphisms are the identity on $(\Sigma \times T,(z,t))$, hence fixing the image of j. It is shown in [GLS6, Prop. II.1.5] that the forgetful morphism of functors

$$\underline{\mathcal{D}ef}_{X/\Sigma,z} \to \underline{\mathcal{D}ef}_{X,z}$$

is smooth on the category of complex germs. We need here only the smoothness on the category of fat points.

Remark 2.2.15.1 Recall that a *morphism $F \to G$ of functors* of Artin rings is called *smooth* if for every surjection $A \to \overline{A}$ of Artin rings the following holds: given $\overline{\xi} \in F(\overline{A})$ and $\eta \in G(A)$ such that $\overline{\xi}$ maps to the image of η in $G(\overline{A})$, then there exists a lifting $\xi \in F(A)$ of $\overline{\xi}$ which maps to η in $G(A)$. The *functor F* itself is called *smooth* or *unobstructed* if $F(A) \to F(\overline{A})$ is surjective for every surjection $A \to \overline{A}$ (for details see [GLS6, Appendix C]).

Definition 2.2.16 Let $\underline{\mathcal{D}ef}'_{X,z}$ be a subcategory of $\underline{\mathcal{D}ef}_{X,z}$. Denote by $\underline{\mathcal{D}ef}'_{X/\Sigma,z}$ the full subcategory of $\underline{\mathcal{D}ef}_{X/\Sigma,z}$ whose image in $\underline{\mathcal{D}ef}_{X,z}$ is in $\underline{\mathcal{D}ef}'_{X,z}$ and call it the subcategory of *embedded deformations induced by $\underline{\mathcal{D}ef}'_{X,z}$*.

Since $\underline{\mathcal{D}ef}_{X/\Sigma,z} \to \underline{\mathcal{D}ef}_{X,z}$ is smooth, it follows that

$$\underline{\mathcal{D}ef}'_{X/\Sigma,z} \to \underline{\mathcal{D}ef}'_{X,z}$$

is smooth too. By [GLS6, Prop. II.1.25 and Remark II.1.25.1] we have

$$\underline{\mathcal{D}ef}'_{X/\Sigma,z}(T_\varepsilon) = I'(X,z)/\langle f\rangle \subset \mathcal{O}_{X,z} .$$

Even if $\underline{\mathcal{D}ef}'_{X,z}$ has a semiuniversal deformation, $\underline{\mathcal{D}ef}'_{X/\Sigma,z}$ may not (its tangent space may be infinite dimensional, i.e. Schlessinger's (H_3) may not hold). But we have

Lemma 2.2.17 *If $\underline{\mathcal{D}ef}'_{X,z}$ has a versal deformation, then the functor of embedded deformations $\underline{\mathcal{D}ef}'_{X/\Sigma,z}$ satisfies Schlessinger's conditions (H_0), (H_1), (H_2).*

Proof (H_0) says that $\underline{\mathcal{D}ef}'_{X/\Sigma,z}(\mathbb{C})$ consists of one element only, which is part of the definition of $\underline{\mathcal{D}ef}'$ (the element is given by $(X,z) \hookrightarrow (\Sigma,z) \to \{\text{pt}\}$, where pt is the reduced point).

For (H_1) and (H_2) consider the canonical map

$$\alpha : F(A' \times_A A'') \to F(A') \times_{F(A)} F(A'') ,$$

where we write F for $\underline{\mathcal{D}ef}'_{X/\Sigma,z}$ and where $A' \to A$ and $A'' \to A$ are surjections of Artin rings.

(H_1) says that α is surjective for $A' \to A$ a small extension and (H_2) requires α to be bijective for $A = \mathbb{C}$ and $A' = \mathbb{C}[\varepsilon]$.

We prove first (H_1), writing $\mathcal{O}$ for $\mathcal{O}_{\Sigma,z}$.

Let $\xi' \in F(A')$ resp. $\xi'' \in F(A'')$ be given by $F' \in \mathcal{O} \otimes A'$ resp. by $F'' \in \mathcal{O} \otimes A''$ such that ξ' and ξ'' induce the same element in $F(A)$. If $\overline{F}'$ resp. $\overline{F}''$ denote the images of F' resp. F'' in $\mathcal{O} \otimes A$, then $\overline{F}'' = \overline{G}\overline{F}'$ for some unit $\overline{G} \in (\mathcal{O} \otimes A)^*$. Lift $\overline{G}$ to $G \in (\mathcal{O} \otimes (A' \times_A A''))^*$. Then ξ' is given by GF' and GF' coincides with F'' in $\mathcal{O} \otimes A$. Hence $(GF', F'') \in \mathcal{O} \otimes (A' \times_A A'')$ defines an element in $\underline{\mathcal{D}ef}_{X/\Sigma,z}(A' \times_A A'')$ mapping to (ξ', ξ''). Since $\underline{\mathcal{D}ef}'_{X,z}$ has a semiuniversal deformation by assumption, it satisfies Schlessinger's conditions (H_0)-(H_3) (cf. [GLS6, Theorem C.1.6]). Moreover, since $\underline{\mathcal{D}ef}'_{X/\Sigma,z} \to \underline{\mathcal{D}ef}'_{X,z}$ is smooth, we can choose F' and F'' above in such a way that the embedded deformation given by (GF', F'') maps to $\underline{\mathcal{D}ef}'_{X,z}(A' \times_A A'')$. Hence (GF', F'') defines an element in $\underline{\mathcal{D}ef}'_{X/\Sigma,z}(A' \times_A A'')$ showing that α is surjective.

The fact that α is bijective for $A = \mathbb{C}$ and $A' = \mathbb{C}[\varepsilon]$ follows since $\underline{\mathcal{D}ef}_{X/\Sigma,z}$ satisfies (H_1). This is well known, but let us see how it works.

Given $(f + \varepsilon g, F) \in \mathcal{O} \otimes_{\mathbb{C}} \mathbb{C}[\varepsilon] \times \mathcal{O} \otimes_{\mathbb{C}} A$ defining an element in the target of α, we have to show that the class of $F + \varepsilon g \in \mathcal{O} \otimes_{\mathbb{C}} (\mathbb{C}[\varepsilon] \times_{\mathbb{C}} A) \cong \mathcal{O} \otimes_{\mathbb{C}} A \oplus \varepsilon\mathcal{O}$ (with $\varepsilon^2 = \varepsilon\mathfrak{m}_A = 0$) is the unique preimage of α.

Let $(f' + \varepsilon g', F')$ define the same element in the target of α as $(f + \varepsilon g, F)$. Multiplying F and F' with suitable units we may assume that $f \equiv F \operatorname{mod} \mathfrak{m}_A$ and $f' \equiv F' \operatorname{mod} \mathfrak{m}_A$. There exist units $u \in (\mathcal{O} \otimes_{\mathbb{C}} \mathbb{C}[\varepsilon])^* = \mathcal{O}^* \oplus \varepsilon\mathcal{O}$ and $U \in (\mathcal{O} \otimes A)^* = \mathcal{O}^* + \mathfrak{m}_A\mathcal{O} \otimes A$ such that $u(f + \varepsilon g) = f' + \varepsilon g'$ and $UF = F'$. Writing

$u = u_0 + \varepsilon u_1$ and $U = U_0 + \varepsilon U_1$ with $u_0, U_0 \in \mathcal{O}^*$ and $u_1 \in \mathcal{O}, U_1 \in \mathfrak{m}_A\mathcal{O} \otimes A$, it follows that $u_0 f = f'$ and $\varepsilon UF = \varepsilon U_0 f = \varepsilon f'$. Hence $U_0 f = u_0 f$ and therefore $U_0 = u_0$ since f is a non-zero divisor.

$U + \varepsilon u_1$ is a unit in $\mathcal{O} \otimes A \oplus \varepsilon\mathcal{O}$ and we check that $(U + \varepsilon u_1)(F + \varepsilon g) = F' + \varepsilon g'$. This shows that $F + \varepsilon g$ and $F' + \varepsilon g'$ define isomorphic embedded deformations over $\mathbb{C}[\varepsilon] \times_{\mathbb{C}} A$ and therefore α has a unique preimage. □

Using Lemma 2.2.17 we get the following consequence for induced embedded deformations.

Lemma 2.2.18 *Given a small extension $A \to \overline{A} = A/\langle\eta\rangle$ then $\underline{\mathcal{D}ef}'_{X/\Sigma,z}(T_\varepsilon)$ acts transitively on the non-empty fibres of $\underline{\mathcal{D}ef}'_{X/\Sigma,z}(A) \to \underline{\mathcal{D}ef}'_{X/\Sigma,z}(\overline{A})$ by the same formula as in Lemma 2.2.15.*

In particular, let $F, F' \in \mathcal{O}_{\Sigma,z} \otimes A$ define elements in $\underline{\mathcal{D}ef}'_{X/\Sigma,z}(A)$ with the same image in $\underline{\mathcal{D}ef}'_{X/\Sigma,z}(\overline{A})$. Then there exists a $g \in I'(X, z)$ and a unit $G \in (\mathcal{O}_{\Sigma,z} \otimes A)^*$ such that

$$F' = GF + g\eta\ .$$

2.2.1.4 Global Equisingular and VSI Deformations

Let $X \subset \Sigma$ be a hypersurface in the complex manifold Σ with isolated singularities $z_1, \dots, z_r$. We are now going to define the global deformation category $\mathcal{D}ef'_{X/\Sigma}$, i.e. embedded deformations of X in Σ inducing local deformations in $\mathcal{D}ef'_{X,z_i}$ for $i = 1, \dots, r$, in particular ideal-versal and strictly ideal-versal deformations, i.e. satisfying condition (VSI) and (VSIR).

To any object in $\mathcal{D}ef_{X/\Sigma}$ (resp. in $\mathcal{D}ef^{sec}_{X/\Sigma}$, resp. in $\mathcal{D}ef^{fix}_{X/\Sigma}$), i.e. to any embedded deformation of X, $\mathscr{X} \hookrightarrow \Sigma \times T \to T$ (resp. with sections $\sigma_1, \dots, \sigma_r$, resp. with trivial sections) over the pointed complex space (T, t_0) we associate in a functorial way the deformations

$$(\mathscr{X}, z_i) \to (T, t_0), \quad i = 1, \dots, r,$$

(resp. with sections $\sigma_i : (T, t_0) \to (\mathscr{X}, z_i)$) of the singularity (X, z_i). In particular, we get functors of deformation categories

$$\begin{aligned}
\mathcal{D}ef_{X/\Sigma}(T, t_0) &\to \mathcal{D}ef_{X,z_i}(T, t_0),\\
\mathcal{D}ef^{sec}_{X/\Sigma}(T, t_0) &\to \mathcal{D}ef^{sec}_{X,z_i}(T, t_0),\\
\mathcal{D}ef^{fix}_{X/\Sigma}(T, t_0) &\to \mathcal{D}ef^{fix}_{X,z_i}(T, t_0),
\end{aligned}$$

associating to a deformation of (the global) X, the induced local deformations of the germs (X, z_i), $i = 1, \dots, r$. If the sections σ_i are trivial, then the position of the points $\sigma_i(t)$ in the fibre $\mathscr{X}_t$ is fixed in Σ for each $t \in T$.

Definition 2.2.19 (1) For given local deformation (full) subcategories $\mathcal{Def}'_{X,z_i}$ of $\mathcal{Def}_{X,z_i}$, $i = 1, \ldots, r$, we define $\mathcal{Def}'_{X/\Sigma}$ as the full subcategory of $\mathcal{Def}_{X/\Sigma}$ given by

$$\mathcal{Def}'_{X/\Sigma}(T, t_0) = \left\{ \begin{array}{l} \text{deformations in } \mathcal{Def}_{X/\Sigma}(T, t_0) \text{ which induce} \\ \text{deformations in } \mathcal{Def}'_{X,z_i}(T, t_0) \text{ for } i = 1, \ldots, r \end{array} \right\}.$$

Similarly we define $\mathcal{Def}'^{sec}_{X/\Sigma} \subset \mathcal{Def}^{sec}_{X/\Sigma}$ for given subcategories $\mathcal{Def}'^{sec}_{X,z_i} \subset \mathcal{Def}^{sec}_{X,z_i}$, i.e. for given deformations with sections. If the given sections are trivial we denote $\mathcal{Def}'^{sec}_{X/\Sigma}$ by $\mathcal{Def}'^{fix}_{X/\Sigma}$. The corresponding deformation functors of isomorphism classes are denoted as usual by $\underline{\mathcal{Def}}'_{X/\Sigma}$, $\underline{\mathcal{Def}}'^{sec}_{X/\Sigma}$ and $\underline{\mathcal{Def}}'^{fix}_{X/\Sigma}$.

(2) We say that $\mathcal{Def}'_{X/\Sigma}$ *satisfies (V), (S), (I) or (R)* if this holds for $\mathcal{Def}'_{X,z_i}$ for all $i = 1, \ldots, r$ (cf. Definition 2.2.8). If the conditions (VSI) (resp. (VSIR)) hold, we call $\mathcal{Def}'_{X/\Sigma}$ *ideal-versal* (resp. *strictly ideal-versal*). In the same way we use this notation for $\mathcal{Def}'^{sec}_{X/\Sigma}$ and for $\mathcal{Def}'^{fix}_{X/\Sigma}$.

Of special interest are equianalytic (resp. equisingular) deformations, which fix the analytic (resp. topological) type S_i of the hypersurface (resp. plane curve) singularity (X, z_i), $i = 1, \ldots, r$, in the sense of Definition 2.2.3.

Definition 2.2.20 Let $\mathcal{Def}'_{X,z_i}$ be $\mathcal{Def}^{ea}_{X,z_i}$ (resp. $\mathcal{Def}^{es}_{X,z_i}$) and S_i the analytic (resp. topological) type of (X, z_i). Then we denote $\mathcal{Def}'_{X/\Sigma}$ by

$$\begin{array}{l} \mathcal{Def}^{ea}_{X/\Sigma} = \mathcal{Def}^{ea}_{X/\Sigma}(S_1, \ldots, S_r) \quad \text{resp.} \\ \mathcal{Def}^{es}_{X/\Sigma} = \mathcal{Def}^{es}_{X/\Sigma}(S_1, \ldots, S_r) \end{array}$$

and call it the category of *equianalytic deformations* (resp. *equisingular deformations*) of X/Σ with fixed analytic (resp. topological) type S_i along z_i, $i = 1, \ldots, r$.

Intuitively $\mathcal{Def}^{ea}_{X/\Sigma}$, resp. $\mathcal{Def}^{es}_{X/\Sigma}$, are embedded deformations of X in Σ, such that the analytic (resp. topological) type of the singularities (X, z_i), $i = 1, \ldots, r$, is not changed during the deformation.

Note that in $\mathcal{Def}'_{X/\Sigma}$ we impose only conditions on deformations of the germs (X, z_i), $i = 1, \ldots, r$, while deformations of (X, z), $z \notin \{z_1, \ldots, z_r\}$, can be arbitrary.

Let us consider deformations of $X \subset \Sigma$, which induce locally equianalytic or equisingular deformations. They satisfy (VSIR) by Proposition 2.2.21. We identify the tangent spaces of the local deformation functors $\underline{\mathcal{Def}}'_{X,z}$ and their codimensions $\tau'(X, z)$.

Example 2.2.20.1 (a) Let $\mathcal{Def}'_{X/\Sigma} = \mathcal{Def}^{ea}_{X/\Sigma} \subset \mathcal{Def}_{X/\Sigma}$ be the category of equianalytic deformations. For $z \in \operatorname{Sing} X$, $\mathcal{Def}^{ea}_{X,z}$ has as semiuniversal base space the reduced point (cf. Remark 2.2.9.1) and its tangent space is $\underline{\mathcal{Def}}^{ea}_{X,z}(T_\varepsilon) = 0 \subset \underline{\mathcal{Def}}_{X,z}(T_\varepsilon) = \mathcal{O}_{\Sigma,z}/I^{ea}(X, z)$ with $I^{ea}(X, z) = \langle f, j(f) \rangle$ and $\tau^{ea}(X, z) = \tau(X, z)$ is the Tjurina number.

(b) Let $\underline{\mathcal{Def}}'^{fix}_{X/\Sigma} = \underline{\mathcal{Def}}^{ea,fix}_{X/\Sigma} \subset \underline{\mathcal{Def}}^{fix}_{X/\Sigma}$ denote the category of *equianalytic deformations with fixed position*, that is, with trivial sections. Again, the tangent space is $\underline{\mathcal{D}ef}^{ea,fix}_{X,z}(T_\varepsilon) = 0 \subset \underline{\mathcal{D}ef}^{fix}_{X,z}(T_\varepsilon) = \mathfrak{m}_z/I^{ea}_{fix}(X,z)$, $I^{ea}_{fix}(X,z) = \langle f, \mathfrak{m}j(f)\rangle$, and $\tau^{ea}_{fix}(X,z) = \dim_{\mathbb{C}} \mathcal{O}_{\Sigma,z}/I^{ea}_{fix}(X,z) = \tau_{fix}(X,z)$. By Lemma 2.2.13 we have $\tau_{fix}(X,z) - \tau(X,z) = \dim(\Sigma, z)$.

(c) Let X be a reduced curve in the surface Σ and $\mathcal{D}ef^{es}_{X/\Sigma} \subset \mathcal{D}ef_{X/\Sigma}$ the category of *equisingular deformations.* We have $\underline{\mathcal{D}ef}^{es}_{X,z}(T_\varepsilon) \cong I^{es}(X,z)/I^{ea}(X,z)$ with $I^{es}(X,z)$ the *equisingularity ideal* and $\tau^{es}(X,z) = \dim_{\mathbb{C}} \mathcal{O}_{\Sigma,z}/I^{es}(X,z) = \mu(X,z) - \mathrm{mod}(X,z)$ (cf. Definition 1.1.63).

(d) $\mathcal{D}ef^{es,fix}_{X/\Sigma} \subset \mathcal{D}ef^{fix}_{X/\Sigma}$ denotes the category of *equisingular deformations with fixed position.* $\underline{\mathcal{D}ef}^{es,fix}_{X,z}(T_\varepsilon) \cong I^{es}_{fix}(X,z)/I^{ea}_{fix}(X,z)$ (cf. [GLS6, Proposition II.2.14]) with $I^{es}_{fix}(X,z)$ and $\tau^{es}_{fix}(X,z) = \dim_{\mathbb{C}} \mathcal{O}_{\Sigma,z}/I^{es}_{fix}(X,z)$. By [GLS6, Prop. II.2.14] we have $\tau^{es}_{fix}(X,z) - \tau^{es}(X,z) = 2$ (cf. Definition 1.1.63).

Remark 2.2.20.1 $\tau^{es}(X,z)$ does not admit a simple formula in other basic invariants, it can however be computed. For semiquasi-homogeneous and Newton nondegenerate singularities see [GLS6, Prop. II.2.17], for the general case see [GLS6, Remarks and Exercises to II.2.8, p. 373] and [CGL2, Algorithm 2]. For an explicit formula in terms of the resolution tree of (X,z), see Corollary 1.1.64.

Proposition 2.2.21 *The deformation categories* $\mathcal{D}ef^{ea}_{X,z}$, $\mathcal{D}ef^{ea,fix}_{X,z}$, $\mathcal{D}ef^{es}_{X,z}$, *and* $\mathcal{D}ef^{es,fix}_{X,z}$ *are strictly ideal-versal for each* $z \in X$.

Proof We consider first ideal-versatility and then condition (R). Since the semiuniversal deformations of $\underline{\mathcal{D}ef}^{ea}_{X,z}$ and $\underline{\mathcal{D}ef}^{ea,fix}_{X,z}$ consist of the reduced point, (VSI) follows.

For $\mathcal{D}ef^{es}_{X,z}$ and $\mathcal{D}ef^{es,fix}_{X,z}$ (VSI) follows from [GLS6, Theorems II.2.38 and II.2.64].

The fact that (R) is satisfied follows for $\mathcal{D}ef^{ea}_{X,z}$ from [GrK, Lemma 1.4] (cf. also [FlK, Corollary 0.2]). For $\mathcal{D}ef^{es}_{X,z}$ a proof is given in [GLS6, Corollary II.2.67, Theorem II.2.61]. □

Remark 2.2.21.1 Let $\phi : (\mathscr{X}, z) \to (T, t_0)$ be any deformation of (X,z).

(1) The $\mathcal{D}ef^{ea}_{X,z}$-stratum or *equianalytic stratum* in (T, t_0) exists and is the trivial locus of ϕ, i.e. the unique maximal subgerm $(T^{ea}, t_0) \subset (T, t_0)$ such that the restriction of ϕ over (T^{ea}, t_0) is trivial. If $(B_z, 0)$ denotes the base space of the semiuniversal deformation of (X,z) and $\varphi : (T, t_0) \to (B_z, 0)$ a map inducing ϕ, then $(T^{ea}, t_0) = (\varphi^{-1}(0), t_0)$. It is unique, although φ need not be unique.[11]
(2) The $\mathcal{D}ef^{es}_{X,z}$-stratum or *equisingular stratum* in (T, t_0), the reduction is the *μ-constant stratum* (Δ^μ, t_0). If ϕ is a versal deformation then the equisingular stratum (and hence (Δ^μ, t_0)) is smooth.

Both statements are proved in the references given in the proof of Proposition 2.2.21. The following corollary for embedded deformations of $X \subset \Sigma$ is a consequence of Proposition 2.2.21 for deformations of germs.

[11] In positive characteristic uniqueness may fail and an equianalytic stratum may not exist. The same holds for equisingular deformations (cf. [CGL1]).

Corollary 2.2.22 *The category $\mathcal{D}ef^{ea}_{X/\Sigma}(S_1, \ldots, S_r)$ (resp. $\mathcal{D}ef^{es}_{X/\Sigma}(S_1, \ldots, S_r)$) of equianalytic (resp. equisingular) deformations of $X \subset \Sigma$ with fixed analytic (resp. topological) types $S_1, \ldots, S_r$ along the given points $z_1, \ldots, z_r$, is strictly ideal-versal.*

As before let $X \subset \Sigma$ be a hypersurface and $\{z_1, \ldots, z_r\} \subset \mathrm{Sing}(X)$ a finite set of isolated singularities of X.

Definition 2.2.23 Let $\mathcal{D}ef'_{X/\Sigma}$ resp. $\mathcal{D}ef'^{fix}_{X/\Sigma}$ satisfy (VSI). Using the notations from Definition 2.2.11 we define ideal sheaves of $\mathcal{O}_\Sigma$ as

$$\mathcal{J}_{Z'(X)/\Sigma} := \begin{cases} I'(X,z) & \text{if } z \in \{z_1, \ldots, z_r\} \\ \mathcal{O}_{\Sigma,z} & \text{if } z \notin \{z_1, \ldots, z_r\}, \end{cases}$$

$$\mathcal{J}_{Z'_{fix}(X)/\Sigma} := \begin{cases} I'_{fix}(X,z) & \text{if } z \in \{z_1, \ldots, z_r\} \\ \mathcal{O}_{\Sigma,z} & \text{if } z \notin \{z_1, \ldots, z_r\} \end{cases}.$$

These ideal sheaves define *0-dimensional singularity schemes* $Z'(X)$ and $Z'_{fix}(X)$, which are subschemes of Σ, concentrated on the finitely many points $\{z_1, \ldots, z_r\}$, with

$$\begin{aligned} \mathcal{O}_{Z'(X)} &= \mathcal{O}_\Sigma / \mathcal{J}_{Z'(X)/\Sigma}, \\ \mathcal{O}_{Z'_{fix}(X)} &= \mathcal{O}_\Sigma / \mathcal{J}_{Z'_{fix}(X)/\Sigma}. \end{aligned}$$

Of special interest are the *equianalytic scheme* $Z^{ea}(X)$ and the *equisingular scheme* $Z^{es}(X)$.

The *degree* of $Z'(X)$ at z resp. its *total degree* are given by

$$\begin{aligned} \deg(Z'(X), z) &:= \tau'(X, z) && = \dim_{\mathbb{C}} \mathcal{O}_{\Sigma,z}/I'(X,z) \text{ resp.} \\ \deg Z'(X) &:= \sum_{i=1,\ldots,r} \tau'(X, z_i) && = \dim_{\mathbb{C}} H^0(\Sigma, \mathcal{O}_{Z'(X)}), \end{aligned}$$

and similar for Z'_{fix}.

Note that $Z'(X)$ and $Z'_{fix}(X)$ may be also considered as subschemes of X, since $I'(X,z)$ resp. $I'_{fix}(X,z)$ contains $I^{ea}(X,z)$ resp. $I^{ea}_{fix}(X,z)$ and hence they contain both the generator of the ideal of $(X,z) \subset (\Sigma,z)$. Depending on whether we consider $Z'(X)$ as a subscheme of Σ resp. of X we write

$$Z'(X)/\Sigma \quad \text{resp.} \quad Z'(X)/X$$

for the schemes, and denote by $\mathcal{J}_{Z'(X)/\Sigma}$ resp. $\mathcal{J}_{Z'(X)/X} :=$ image of $\mathcal{J}_{Z'(X)/\Sigma}$ in $\mathcal{O}_X$, the corresponding ideal sheaves in $\mathcal{O}_\Sigma$ resp. $\mathcal{O}_X$.

Let $\mathcal{J}_X = \mathcal{O}_\Sigma(-X)$ denote the ideal sheaf of $X \subset \Sigma$. Then we have an exact sequence

$$0 \to \mathcal{J}_X \to \mathcal{J}_{Z'(X)/\Sigma} \to \mathcal{J}_{Z'(X)/X} \to 0$$

of $\mathcal{O}_\Sigma$-sheaves. Tensoring with $\mathcal{O}_\Sigma(X)$ we get a commutative diagram with exact rows

$$\begin{array}{ccccccccc} 0 \to & \mathcal{J}_X(X) & \to & \mathcal{J}_{Z'(X)/\Sigma}(X) & \to & \mathcal{J}_{Z'(X)/X}(X) & \to 0 \\ & \downarrow\cong & & \downarrow & & \downarrow & \\ 0 \to & \mathcal{O}_\Sigma & \to & \mathcal{O}_\Sigma(X) & \to & \mathcal{O}_X(X) & \to 0 \\ & & & \downarrow & & \downarrow & \\ & & & \mathcal{O}_{Z'(X)/\Sigma} & \xrightarrow{=} & \mathcal{O}_{Z'(X)/X} & \end{array} \qquad (2.2.1.3)$$

which will be used later.

2.2.1.5 Infinitesimal Deformations and Obstructions

Fix isolated singularities $z_1, \ldots, z_r$ of the hypersurface X in the complex manifold Σ and consider deformations $\mathcal{D}ef'_{X/\Sigma}$ as in Definition 2.2.19. Examples are equianalytic or equisingular deformations of X in Σ that induce analytically or topologically trivial deformation of the germs (X, z_i). The relevance of the conditions (V), (S), (I) and (R) from Definition 2.2.8 will become clear in the following results. Let $Z'(X)$ be the 0-dimensional scheme defined in Definition 2.2.23. We determine the tangent space and the obstruction space of the functor $\underline{\mathcal{D}ef}'_{X/\Sigma}$.

Lemma 2.2.24 *Let $\mathcal{D}ef'_{X/\Sigma}$ satisfy condition (I). Then there is a canonical isomorphism of complex vector spaces*

$$\underline{\mathcal{D}ef}'_{X/\Sigma}(T_\varepsilon) \xrightarrow{\cong} H^0(X, \mathcal{J}_{Z'(X)/X}(X)).$$

Hence, if $\mathcal{D}ef'_{X/\Sigma}$ admits a semiuniversal deformation, the Zariski tangent space to its base is isomorphic to $H^0(X, \mathcal{J}_{Z'(X)/X}(X))$.
In particular, if $Z'(X)$ is the empty scheme, we obtain the well known isomorphism

$$\underline{\mathcal{D}ef}_{X/\Sigma}(T_\varepsilon) \cong H^0(X, \mathcal{O}_X(X)) .$$

Proof Since $\mathcal{D}ef'_{X/\Sigma}$ is a subcategory of $\mathcal{D}ef_{X/\Sigma}$, an element of $\underline{\mathcal{D}ef}'_{X/\Sigma}(T_\varepsilon)$ is represented in $\mathcal{D}ef_{X/\Sigma}(T_\varepsilon)$ by local equations $(f_i + \varepsilon g_i = 0)_{i \in I}$, $f_i, g_i \in H^0(U_i, \mathcal{O}_\Sigma)$ for some open covering $(U_i)_{i\in I}$ of Σ, defining deformations in $\mathcal{D}ef'_{X/\Sigma}$, which satisfy:

(1) $(f_i = 0)_{i\in I}$ are local equations of $X \subset \Sigma$,
(2) $(f_i + \varepsilon g_i) = (a_{ij} + \varepsilon b_{ij})(f_j + \varepsilon g_j)$ on $U_{ij} := U_i \cap U_j$ for some $a_{ij}, b_{ij} \in H^0(U_{ij}, \mathcal{O}_\Sigma)$ with a_{ij} invertible,
(3) for each $z \in \{z_1, \ldots, z_r\} \cap U_i$ the germ of g_i at z belongs to $I'(X, z)$.

Properties (1) and (2) follow from the definition of $\mathcal{D}ef_{X/\Sigma}(T_\varepsilon)$. (3) means that $f_i + \varepsilon g_i$ defines an element of $\mathcal{D}ef'_{(X,z)}(T_\varepsilon)$, cf. Remark 2.2.11.1(2).

Property (2) is equivalent to $f_i = a_{ij} f_j$ and $g_i = a_{ij} g_j + b_{ij} f_j$ which shows that the local sections $g_i/f_i \in H^0(U_i, \mathcal{O}_\Sigma(X))$ satisfy

$$\frac{g_i}{f_i} - \frac{g_j}{f_j} = \frac{a_{ij}g_j + b_{ij}f_j}{a_{ij}f_j} - \frac{g_j}{f_j} = f_j\frac{b_{ij}}{f_i} \tag{2.2.1.4}$$

which is zero over U_{ij} as section of $\mathcal{O}_X(X)$. Hence $(g_i/f_i)_{i\in I}$ defines a global section of $\mathcal{O}_X(X)$.

Conversely, every global section of $\mathcal{O}_X(X)$ is given by local sections $g_i/f_i \in \Gamma(U_i, \mathcal{O}_\Sigma(X))$ satisfying $f_i = a_{ij}f_j$ with invertible a_{ij} and (2.2.1.4) for some b_{ij}. This implies condition (2) and we obtain an object $(f_i + \varepsilon g_i)_{i\in I}$ of $\underline{\mathcal{D}ef}_{X/\Sigma}(T_\varepsilon)$. If $f_i + \varepsilon g_i$ is isomorphic to the trivial deformation, we get $f_i + \varepsilon g_i = (a_i + \varepsilon b_i)f_i$ for $a_i, b_i \in H^0(U_i, \mathcal{O}_\Sigma)$ and a_i invertible. Hence $(g_i/f_i = f_i(b_i/f_i))_{i\in I}$ is the zero section of $\mathcal{O}_X(X)$. Since the zero section defines a trivial deformation we get

$$\underline{\mathcal{D}ef}_{X/\Sigma}(T_\varepsilon) \cong H^0(X, \mathcal{O}_X(X)) .$$

This is indeed an isomorphism of vector spaces since the $\mathbb{C}$-linear structure on $\underline{\mathcal{D}ef}_{X/\Sigma}(T_\varepsilon)$ is given by $f_i + \varepsilon(g_i + g_i')$ (addition) and $f_i + \varepsilon\lambda g_i$ (scalar multiplication) for two deformations $f_i + \varepsilon g_i$ and $f_i + \varepsilon g_i'$ and for $\lambda \in \mathbb{C}$.

Now, if $g_i \in I'(X, z)$ for $z \in \{z_1, \dots, z_r\} \cap U_i$ then $g_i/f_i \in \Gamma(U_i, \mathcal{J}_{Z'(X)/X}(X))$ and hence $(g_i/f_i)_{i\in I}$ is a global section of $\mathcal{J}_{Z'(X)/X}(X)$, showing the claimed isomorphism. □

In the following we use the well known regularity criterion for analytic algebras (cf. [Mum1, Lecture 22] and [GLS6, Lemma II.1.30]), which we recall.

Lemma 2.2.25 *An analytic K-algebra R, K a field, is regular iff each surjection $A \to \overline{A}$ of Artinian analytic K-algebras has the property that every morphism $R \to \overline{A}$ lifts through $A \to \overline{A}$.*

For a generalization to morphisms of arbitrary analytic algebras $B \to R$ instead of $K \hookrightarrow R$ and its relation to obstruction theory see [GLS6, Appendix C.2 after Def. C.2.2].

Proposition 2.2.26 *Let $\mathcal{D}ef'_{X/\Sigma}$ be an ideal-versal deformation category. If the natural morphism*

$$H^1\big(\mathcal{J}_{Z'(X)/\Sigma}(X)\big) \to H^1\big(\mathcal{J}_{Z'(X)/X}(X)\big)$$

is the zero morphism then the functor $\underline{\mathcal{D}ef}'_{X/\Sigma}$ is unobstructed on the category of fat points, that is, $\underline{\mathcal{D}ef}'_{X/\Sigma}(A) \to \underline{\mathcal{D}ef}'_{X/\Sigma}(\overline{A})$ is surjective for any surjection $A \to \overline{A}$ of analytic Artinian $\mathbb{C}$-algebras. In particular, if $\mathcal{D}ef'_{X/\Sigma}$ admits a semiuniversal deformation, then its base space is smooth of dimension $h^0(\mathcal{J}_{Z'(X)/X}(X))$.

The special case $Z'(X) = \emptyset$, treated in [Mum1, Lecture 23], shows that the obstructions against lifting a deformation of $X \subset \Sigma$ to higher order lie in the image of $H^1\big(\mathcal{O}_\Sigma(X)\big) \to H^1\big(\mathcal{O}_X(X)\big)$ (cf. diagram (2.2.1.3)). Our result is a generalization of Mumford's.

Proof The last statement is a consequence of the first, using deformation theory (cf. [GLS6, Appendix C2] and Remark 2.2.11.1(3)). Since a smooth germ has the same dimension as its tangent space, the dimension statement follows from Lemma 2.2.24.

Since each surjection of local Artinian $\mathbb{C}$-algebras factors through small extensions, it suffices to consider a small extension $0 \to \mathbb{C}\eta \to A \to \overline{A} \to 0$, $\eta \cdot \mathfrak{m}_A = 0$, to prove the first statement.

Let $\overline{\xi}$ represent an isomorphism class in $\underline{\mathcal{D}ef}'_{X/\Sigma}(\overline{A})$. We have to show that (up to isomorphism) it lifts to an embedded deformation ξ of X/Σ over Spec A representing a class in $\underline{\mathcal{D}ef}'_{X/\Sigma}(A)$.

The deformation $\overline{\xi}$ of X/Σ is given by local equations $\overline{F_i} \in \Gamma(U_i, \mathcal{O}_\Sigma \otimes_{\mathbb{C}} \overline{A})$, where $(U_i)_{i\in I}$ is an open covering of Σ, such that

(1) on $U_{ij} := U_i \cap U_j$, $\overline{F}_i = \overline{G}_{ij} \cdot \overline{F}_j$ with a unit $\overline{G}_{ij} \in \Gamma(U_{ij}, (\mathcal{O}_\Sigma \otimes \overline{A})^*)$,
(2) the image $F_i^{(0)}$ of $\overline{F}_i$ in $\Gamma(U_i, \mathcal{O}_\Sigma \otimes \overline{A}/\mathfrak{m}_{\overline{A}}) = \Gamma(U_i, \mathcal{O}_\Sigma)$ is a generator for the ideal of $U_i \cap X \subset \Sigma$,
(3) the germs of $\overline{F}_i$ at each point $z \in \{z_1, \ldots, z_r\} \cap U_i$ map to classes in $\underline{\mathcal{D}ef}'_{X,z}(\overline{A})$.

We have to show the existence of local equations $F_i \in \Gamma(U_i, \mathcal{O}_\Sigma \otimes A)$ lifting $\overline{F}_i$ and satisfying analogous conditions (1)–(3) for some liftings G_{ij} of $\overline{G}_{ij}$. In the following we assume that each U_i is sufficiently small, in particular, it contains at most one singular point $z \in \{z_1, \ldots, z_r\}$.

Because of the results of section 2.1.3 our argument is now a modification of the argument when $Z'(X) = \emptyset$ (cf. [Mum1, Lecture 23]). Consider the following exact sequence obtained from the cohomology sequence of diagram (2.2.1.3).

$$\begin{aligned} 0 \to H^0(\mathcal{O}_\Sigma) \to H^0(\mathcal{J}_{Z'/\Sigma}(X)) \to H^0(\mathcal{J}_{Z'/X}(X)) \to H^1(\mathcal{O}_\Sigma) \\ \to H^1(\mathcal{J}_{Z'/\Sigma}(X)) \to H^1(\mathcal{J}_{Z'/X}(X)) \to H^2(\mathcal{O}_\Sigma) \to \ldots \end{aligned} \tag{2.2.1.5}$$

Using the smoothness (S) and the versatility (V) of $\mathcal{D}ef'_{X,z}$ we conclude

- there exists a lifting $F_i \in \Gamma(U_i, \mathcal{O}_\Sigma \otimes A)$ of $\overline{F}_i$, $i = 1, \ldots, r$, inducing a deformation in $\mathcal{D}ef'_{X/\Sigma,z}(A)$ for $z \in U_{ij}$,
- for any lifting $G_{ij} \in \Gamma(U_{ij}, (\mathcal{O}_\Sigma \otimes A)^*)$ of $\overline{G}_{ij}$ there exists $h_{ij} \in \Gamma(U_{ij}, \mathcal{J}_{Z'/\Sigma})$ such that
$$F_i = G_{ij} F_j + \eta h_{ij} \,.$$

To see the second statement, note that F_i and F_j define for $z \in U_{ij}$ elements in $\mathcal{D}ef'_{X/\Sigma,z}(A)$ with the same image in $\mathcal{D}ef'_{X/\Sigma,z}(\overline{A})$. Therefore the claim follows from Lemma 2.2.18.

To be able to glue the local liftings, we have to modify the F_i and G_{ij} in a suitable way, such that the h_{ij} become 0. We compute

$$\begin{aligned}\eta \cdot h_{ij} + \eta \cdot G_{ij} \cdot h_{jk} &= F_i - G_{ij} \cdot F_j + G_{ij} \cdot (F_j - G_{jk} \cdot F_k) \\ &= F_i - G_{ij} \cdot G_{jk} \cdot F_k \\ &= \eta \cdot h_{ik} + (G_{ik} - G_{ij} \cdot G_{jk}) \cdot F_k \,.\end{aligned}$$

$\overline{G}_{ik} = \overline{G}_{ij}\overline{G}_{jk}$ implies that $G_{ik} - G_{ij} \cdot G_{jk}$ is a section in $\eta\mathcal{O}_\Sigma$. As $\eta\mathfrak{m}_A = 0$ we obtain in $\mathcal{O}_\Sigma$

$$h_{ij} + G_{ij}^{(0)} \cdot h_{jk} = h_{ik} + h_{ijk} \cdot G_{ik}^{(0)} \cdot F_k^{(0)} \quad \text{with} \quad h_{ijk} = \frac{1 - G_{ij} \cdot G_{jk} \cdot G_{ik}^{-1}}{\eta} \,.$$

Using $G_{ij}^{(0)} = F_i^{(0)}/F_j^{(0)}$, we get a section

$$\frac{h_{ij}}{F_i^{(0)}} + \frac{h_{jk}}{F_j^{(0)}} - \frac{h_{ik}}{F_i^{(0)}} = h_{ijk} \in \Gamma(U_{ijk}, \mathcal{O}_\Sigma) \,.$$

$h_{ij}/F_i^{(0)}$ is a section of $\mathcal{J}_{Z'/\Sigma}(X)$, but the combination $h_{ij}/F_i^{(0)} + h_{jk}/F_j^{(0)} - h_{ik}/F_i^{(0)}$ is a section in $\mathcal{O}_\Sigma$ which maps therefore to 0 in $\mathcal{J}_{Z'/X}(X)$ by (2.2.1.5). Hence $(h_{ij}/F_i^{(0)})_{i,j\in I}$ is 1-Čech cocycle in $\mathcal{J}_{Z'/X}(X)$ and represents a cohomology class $H \in H^1(\mathcal{J}_{Z'/X}(X))$ (cf. [GLS6, Appendix A.9]).

If $H = 0$ then the cocycle is a coboundary and there exist $h_i \in \Gamma(U_i, \mathcal{J}_{Z'/\Sigma})$ such that

$$\frac{h_{ij}}{F_i^{(0)}} = \frac{h_j}{F_j^{(0)}} - \frac{h_i}{F_i^{(0)}} \quad \text{in} \quad \mathcal{J}_{Z'/X}(X) \,.$$

Hence $h_{ij}/F_i^{(0)} + h_i/F_i^{(0)} - h_j/F_j^{(0)}$ is a section in $\mathcal{O}_\Sigma = \mathcal{J}_X(X)$ by sequence (2.2.1.5) and $h_{ij} + h_i - h_j G_{ij}^{(0)} \in \Gamma(U_{ij}, \mathcal{J}_X)$ is divisible by $F_j^{(0)}$. Defining

$$g_{ij} = \frac{h_{ij} + h_i - h_j G_{ij}^{(0)}}{F_j^{(0)}}, \quad \widetilde{F}_i := F_i + \eta h_i, \quad \widetilde{G}_{ij} := G_{ij} + \eta g_{ij} \,,$$

we get $\widetilde{F}_i = \widetilde{G}_{ij}\widetilde{F}_j$ and we can replace F_i by $\widetilde{F}_i$ which glue together to a lifted deformation over A as required.

If $H^1(\mathcal{J}_{Z'/X}(X)) = 0$ we are ready. Note that we did not use characteristic 0 so far.

Otherwise we still have to show that $H = 0$. By assumption $H^1(\mathcal{J}_{Z'/X}(X)) \to H^1(\mathcal{J}_{Z'/\Sigma}(X))$ is zero, and therefore

$$\partial : H^1(\mathcal{J}_{Z'/X}(X)) \to H^2(\mathcal{J}_X(X)) = H^2(\mathcal{O}_\Sigma)$$

is injective. It suffices to show that $\partial H = 0$. But ∂H is represented by the 2-cocycle

$$h_{ijk} = \frac{1 - G_{ij} G_{jk} G_{ik}^{-1}}{\eta} \in \Gamma(U_{ijk}, \mathcal{O}_\Sigma) ,$$

which is zero iff $G_{ik} = G_{ij} G_{jk}$ in $(\mathcal{O}_\Sigma \otimes A)^*$. Note that $(\overline{G}_{ij})_{i,j \in I}$ is a 1-cocycle in the multiplicative group of units $(\mathcal{O}_\Sigma \otimes \overline{A})^*$ and we have to show that it lifts to a 1-cocycle $(G_{ij})_{i,j \in I}$ in $(\mathcal{O}_\Sigma \otimes A)^*$, because then $G_{ik} - G_{ij} G_{jk} = 0$.

That is, we have to show that the map between the groups of 1-cocycles with respect to the covering, $\mathcal{U} = (U_i)_{i \in I}$,

$$Z^1(\mathcal{U}, (\mathcal{O}_\Sigma \otimes A)^*) \to Z^1(\mathcal{U}, (\mathcal{O}_\Sigma \otimes \overline{A})^*) ,$$

is surjective. But this follows from the fact that the exact sequence of abelian groups

$$(\mathcal{O}_\Sigma \otimes A)^* \to (\mathcal{O}_\Sigma \otimes \overline{A})^* \to 1$$

splits in characteristic 0 (cf. [Mum1, Lemma in Lecture 23]). In fact, as $\mathfrak{m}_A \cong \mathfrak{m}_{\overline{A}} \oplus \eta\mathbb{C} \to \mathfrak{m}_{\overline{A}}$ splits as a surjection of vector spaces, $\mathcal{O}_\Sigma \otimes_\mathbb{C} \mathfrak{m}_A \to \mathcal{O}_\Sigma \otimes_\mathbb{C} \mathfrak{m}_{\overline{A}}$ splits as surjection of sheaves of abelian groups. Applying the exponential we get an isomorphism of abelian groups and a commutative diagram

$$\begin{array}{ccc} \mathcal{O}_\Sigma \otimes_\mathbb{C} \mathfrak{m}_A & \longrightarrow & \mathcal{O}_\Sigma \otimes_\mathbb{C} \mathfrak{m}_{\overline{A}} \\ \exp \Big\downarrow \cong & & \cong \Big\downarrow \exp \\ 1 + \mathcal{O}_\Sigma \otimes_\mathbb{C} \mathfrak{m}_A & \longrightarrow & 1 + \mathcal{O}_\Sigma \otimes_\mathbb{C} \mathfrak{m}_{\overline{A}} \end{array}$$

where the horizontal maps split. It follows that

$$(\mathcal{O}_\Sigma \otimes A)^* \cong \mathcal{O}_\Sigma^*(1 + \mathcal{O}_\Sigma \otimes \mathfrak{m}_A) \to \mathcal{O}_\Sigma^*(1 + \mathcal{O}_\Sigma \otimes \mathfrak{m}_{\overline{A}}) \cong (\mathcal{O}_\Sigma \otimes \overline{A})^*$$

splits. □

Remark 2.2.26.1 (1) The result holds also for algebraic schemes with the same proof, but with algebraic cohomology groups H^0 and H^1. If X is projective, the algebraic and analytic cohomology groups coincide by GAGA.

(2) The proof works for any algebraically closed field K of characteristic 0. If $\mathrm{char}(K) > 0$, then the stronger assumption $H^1(\mathcal{J}_{Z'(X)/\Sigma}(X)) = 0$ implies that $\underline{\mathcal{D}ef}'_{X/\Sigma}$ is unobstructed. This follows from the proof, since characteristic 0 was only used for the exponential map in the last step.

From the proof of Proposition 2.2.26 we obtain a useful corollary for which we introduce the following notation.

Let $1 \le p \le r$ and fix local (VSI)-deformation categories $\mathcal{D}ef'_{X,z_i}$ where we assume that

$$\mathcal{D}ef'_{X,z_i} = \mathcal{D}ef^{ea}_{X,z_i} \quad \text{for } i = 1, \ldots, p\,.$$

The 0-dimensional singularity scheme $Z'(X)$ is hence defined by the ideals $I^{ea}(X, z_i)$ for $i = 1, \ldots, p$ and arbitrary ideals $I'(X, z_i)$ belonging to $\mathcal{D}ef'_{X,z_i}$ for $i = p+1, \ldots, r$ (cf. Definition 2.2.23). Let $\mathcal{D}ef'_{X/\Sigma}$ induce $\mathcal{D}ef'_{X,z_i}$ for $i = 1, \ldots, r$. Denote by

$$\mathcal{D}ef''_{X/\Sigma} \subset \mathcal{D}ef_{X/\Sigma}$$

the deformation category inducing $\mathcal{D}ef'_{X,z_i}$ for $i = p+1, \ldots, r$, but no condition for $i = 1, \ldots, p$.

Corollary 2.2.27 *If the image of $H^1(\mathcal{J}_{Z'(X)/\Sigma}(X)) \to H^1(\mathcal{J}_{Z'(X)/X}(X))$ is zero, then $\mathcal{D}ef''_{X/\Sigma}$ is unobstructed and the natural morphism of functors*

$$\underline{\mathcal{D}ef}''_{X/\Sigma} \to \bigoplus_{i=1}^{p} \underline{\mathcal{D}ef}_{X,z_i}$$

is smooth (hence surjective). In particular, if $\underline{\mathcal{D}ef}''_{X/\Sigma}$ admits a semiuniversal deformation, then the map $B_{X/\Sigma} \to \prod_{i=1}^{p} B_{X,z_i}$ of the corresponding semiuniversal base spaces is smooth. Hence, every joint local deformation of the multigerm $(X, z_1) \cup \cdots \cup (X, z_p)$ is induced by a global deformation from $\mathcal{D}ef''_{X/\Sigma}$.

Proof We have to prove the following. Given a small extension $0 \to \mathbb{C}\eta \to A \to \overline{A} \to 0$ of Artin rings and for $i = 1, \ldots, p$ local deformations $\overline{\xi}_i \in \mathcal{D}ef_{X,z_i}(\overline{A})$ together with extensions $\xi_i \in \mathcal{D}ef_{X,z_i}(A)$, then there exists a deformation $\overline{\xi} \in \mathcal{D}ef''_{X/\Sigma}(\overline{A})$ and an extension $\xi \in \mathcal{D}ef''_{X/\Sigma}(A)$ inducing $\overline{\xi}_i$ over $\overline{A}$ resp. ξ_i over A, up to isomorphism.

We prove this by induction on the dimension of $\overline{A}$. If $\overline{A} = \mathbb{C}$ then $\overline{\xi}_i$ is just the germ (X, z_i) and $\overline{\xi}$ being $X \subset \Sigma$ induces $\overline{\xi}_i$. Moreover, we have

$$\xi_i \in \underline{\mathcal{D}ef}_{X,z_i}(T_\varepsilon) = T^1_{X,z_i}\,,$$
$$\xi \in \underline{\mathcal{D}ef}_{X/\Sigma}(T_\varepsilon) = H^0(\mathcal{J}_{Z''(X)/X}(X))$$

by Lemma 2.2.24. Note that $Z'(X) = Z^0(X) \cup Z''(X)$ with $Z^0(X) := \bigcup_{i=1}^{p} Z^{ea}(X, z_i)$. We have a commutative diagram with exact rows

$$\begin{array}{ccccccccc}
0 & \longrightarrow & \mathcal{J}_{Z'(X)/\Sigma}(X) & \longrightarrow & \mathcal{J}_{Z''(X)/\Sigma}(X) & \longrightarrow & \mathcal{O}_{Z^0(X)} & \longrightarrow & 0 \\
 & & \downarrow & & \downarrow & & \| & & \\
0 & \longrightarrow & \mathcal{J}_{Z'(X)/X}(X) & \longrightarrow & \mathcal{J}_{Z''(X)/X}(X) & \longrightarrow & \mathcal{O}_{Z^0(X)} & \longrightarrow & 0 \\
 & & \downarrow & & \downarrow & & & & \\
 & & 0 & & 0 & & & &
\end{array}$$

Note that $H^0(\mathcal{O}_{Z^0(X)}) \cong \bigoplus_{i=1}^{p} T^1_{X,z_i}$ by definition of $I^{ea}(X, z_i)$. The above diagram induces the cohomology diagram with exact rows

$$\begin{array}{ccccccc} H^0(\mathcal{J}_{Z''(X)/\Sigma}(X)) & \to & H^0(\mathcal{O}_{Z^0(X)}) & \to & H^1(\mathcal{J}_{Z'(X)/\Sigma}(X)) & \twoheadrightarrow & H^1(\mathcal{J}_{Z''(X)/\Sigma}(X)) \\ \downarrow & & \| & & \downarrow\alpha & & \downarrow\beta \\ H^0(\mathcal{J}_{Z''(X)/X}(X)) & \to & H^0(\mathcal{O}_{Z^0(X)}) & \to & H^1(\mathcal{J}_{Z'(X)/X}(X)) & \twoheadrightarrow & H^1(\mathcal{J}_{Z''(X)/X}(X)) \end{array}$$

By assumption, α is the zero-map, whence β is the zero-map and $\underline{\mathcal{D}ef}''_{X/\Sigma}$ is unobstructed by Proposition 2.2.26. Moreover, it follows that $H^0(\mathcal{O}_{Z^0(X)}) \to H^1(\mathcal{J}_{Z'(X)/X}(X))$ is also the zero-map, showing that

$$\underline{\mathcal{D}ef}_{X/\Sigma}(T_\varepsilon) \to \bigoplus_{i=1}^{p} \underline{\mathcal{D}ef}_{X,z_i}(T_\varepsilon)$$

is surjective, by the above identification. This proves the claim for $\dim(\overline{A}) = 1$.

For the induction step let $\overline{\xi} \in \underline{\mathcal{D}ef}''_{X/\Sigma}(\overline{A})$ be given, inducing $\overline{\xi}_i \in \underline{\mathcal{D}ef}_{X,z_i}(\overline{A})$, and $\xi_i \in \underline{\mathcal{D}ef}_{X,z_i}(A)$ be any lift of $\overline{\xi}_i$. ξ_i is given by a local equation $F_i \in \Gamma(U_i, \mathcal{O}_\Sigma \otimes A)$ for some neighborhood U_i of z_i.

In the proof of Proposition 2.2.26 we have shown that we can find a lifting $\xi \in \underline{\mathcal{D}ef}''_{X/\Sigma}(A)$ of $\overline{\xi}$ which induces a local deformation $\widetilde{\xi}_i \in \underline{\mathcal{D}ef}_{X,z_i}(A)$ given by $\widetilde{F}_i \in \Gamma(U_i, \mathcal{O}_\Sigma \otimes A)$ satisfying

$$\widetilde{F}_i = F_i + \eta h_i\ ,$$

for some $h \in \Gamma(U_i, \mathcal{J}_{Z'/\Sigma})$. Since $\mathcal{J}_{Z/\Sigma,z_i} = I^{ea}(X, z_i)$ for $i = 1, \ldots, p$, $\widetilde{F}_i$ and F_i define isomorphic deformations in $\underline{\mathcal{D}ef}_{X,z_i}(A)$. This proves the smoothness of

$$\underline{\mathcal{D}ef}''_{X/\Sigma} \to \bigoplus_{i=1}^{p} \underline{\mathcal{D}ef}_{X,z_i}\ .$$

Moreover, since $Z^0(X)$ is 0 -dimensional, the vanishing of $H^1(\mathcal{J}_{Z'(X)/X}(X))$ implies the vanishing of $H^1(\mathcal{J}_{Z''(X),X}(X))$, proving that $\underline{\mathcal{D}ef}''_{X/\Sigma}$ is unobstructed by Proposition 2.2.26. □

Remark 2.2.27.1 A slightly weaker version can be found in [GrK, Proposition 2.3]. There it was proved, using the cotangent complex, for arbitrary complex subspaces $X \subset \Sigma$ with isolated singularities.

2.2.2 Hilbert Schemes of Singular Hypersurfaces

For the whole Sect. 2.2.2 let Σ be a smooth, projective variety, unless we say otherwise. We consider families of projective hypersurfaces $X \subset \Sigma$, where the analytic type resp. the topological type (if X is a curve) of finitely many isolated singularities of X does not change if X varies in Σ. We identify the tangent spaces to such families and prove cohomological conditions under which these families are T-smooth, i.e. smooth of expected dimension (Theorems 2.2.38 and 2.2.40). In sect. 2.2.2.3 we consider more general singularity types which will be needed later.

2.2.2.1 Hilbert Schemes and Embedded Deformations

In this subsection let Σ be an arbitrary projective variety and $X \subset \Sigma$ a fixed closed complex subspace. We compare the concept of embedded deformations with the more classical concept of Hilbert schemes, resp. Douady spaces.

Deformation theory of $X \subset \Sigma$ considers flat families over a pointed complex space T with $t_0 \in T$ the distinguished point. That is, we consider morphisms

$$\phi : \mathscr{X} \overset{j}{\hookrightarrow} \Sigma \times T \overset{\pi}{\longrightarrow} T \,,$$

such that $\phi : \mathscr{X} \to T$ is flat, j is a closed embedding inducing an isomorphism $\mathscr{X}_{t_0} = \phi^{-1}(t_0) \overset{\cong}{\longrightarrow} X$ and π is the projection. Since Σ is projective, ϕ is proper.

When passing to the deformation functor, isomorphic deformations are identified, where "isomorphic" depends on the considered deformation category. In this section we consider the category $\mathcal{D}ef_{X/\Sigma}$, i.e. embedded deformations of $X \subset \Sigma$ with fixed Σ, and then an isomorphism between $\mathscr{X} \hookrightarrow \Sigma \times T$ and $\mathscr{X}' \hookrightarrow \Sigma \times T$ means an isomorphism between $\mathscr{X}$ and $\mathscr{X}'$ being the identity on $\Sigma \times T$. Hence the set of isomorphism classe $\underline{\mathcal{D}ef}_{X/\Sigma}(T)$ is the set of flat families as above where we identify two families if they provide the same subspace in $\Sigma \times T$. In other words, $\underline{\mathcal{D}ef}_{X/\Sigma}(T)$ is the set of closed subspaces $\mathscr{X} \subset \Sigma \times T$ flat over T such that $\mathscr{X}_{t_0} = X$.

On the other hand, the *Douady functor*

$$\mathcal{D}ou_{\Sigma} : (\text{complex spaces}) \to (\text{sets}),$$

associates to a complex space T the set

$$\mathcal{D}ou_{\Sigma}(T) := \{\text{closed subspaces } \mathscr{X} \subset \Sigma \times T, \text{ flat and proper over } T\}.$$

$\mathcal{D}ou_{\Sigma}$ is defined for arbitrary complex spaces Σ, but if Σ is projective then closed subspaces $\mathscr{X} \subset \Sigma \times T$ are proper over T and the functor $\mathcal{D}ou_{\Sigma}$ considers the set of flat families of closed subspaces in $\Sigma \times T$ over T without reference to a special fibre. That is, elements in $\underline{\mathcal{D}ef}_{X/\Sigma}(T)$ are just the elements in $\mathcal{D}ou_{\Sigma}(T)$ together with an isomorphism $\mathscr{X}_{t_0} \cong X$.

The functors $\underline{\mathcal{D}ef}_{X/\Sigma}$ resp. $\mathcal{D}ou_{\Sigma}$ are of particular importance if they admit a semiuniversal deformation resp. a universal family. In our situation, with Σ projective and X compact, this is the case for $\underline{\mathcal{D}ef}_{X/\Sigma}$ by Grauert, [Gra], resp. for $\mathcal{D}ou_{\Sigma}$ by Douady ([Dou2, Dou1], see also [Bin]).

The *universal family* for $\mathcal{D}ou_\Sigma$ is a complex space Dou_Σ and a closed subspace $\mathcal{V} \subset \Sigma \times \text{Dou}_\Sigma$, proper and flat over Dou_Σ, which has the *universal property* that every element of $\mathcal{D}ou_\Sigma(T)$ can be induced from the universal family by a unique morphism $T \to \text{Dou}_\Sigma$ via base change.

The Douady functor is the analytic analog of the Hilbert functor for schemes, $\mathcal{H}ilb_\Sigma$: (schemes) $\to$ (sets), which was introduced and studied by Grothendieck (more generally for Σ a scheme over a locally noetherian scheme S) (cf. [Gro]). It associates to a scheme T the set of all closed subschemes $\mathscr{X} \subset \Sigma \times T$ which are proper and flat over T.

We consider the *Hilbert functor* on the category of algebraic complex spaces

$$\mathcal{H}ilb_\Sigma : (\text{algebraic complex spaces}) \to (\text{sets}),$$

defined as

$$\mathcal{H}ilb_\Sigma(T) := \left\{ \begin{matrix} \text{closed algebraic subspaces } \mathscr{X} \subset \Sigma \times T \\ \text{which are proper and flat over } T \end{matrix} \right\}.$$

In the definition Σ can an arbitrary algebraic complex space. If Σ is projective, say a closed subspace of some $\mathbb{P}^N$, then any closed subspace $\mathscr{X} \subset \Sigma \times T \subset \mathbb{P}^N \times T$ is projective over T. Therefore the fibres $\mathscr{X}_t$ of $\mathscr{X} \to T$ are *projective subvarieties* of $\mathbb{P}^N$ and we can define the *Hilbert polynomial* $h_t = h_{\mathscr{X}_t} \in \mathbb{Q}[x]$ of $\mathscr{X}_t$. It is defined as the analytic Euler characteristic of $\mathcal{O}_{\mathscr{X}_t}(n) = \mathcal{O}_{\mathscr{X}_t} \otimes \mathcal{O}_{\mathbb{P}^N}(n)$ (cf. [HaR2] for alternative definitions),

$$h_t(n) := h_{\mathscr{X}_t}(n) := \chi(\mathcal{O}_{\mathscr{X}_t}(n)) = \sum_i (-1)^i \dim_{\mathbb{C}} H^i(\mathscr{X}_t, \mathcal{O}_{\mathscr{X}_t}(n)), \quad t \in T.$$

We have then the subfunctor $\mathcal{H}ilb^h_\Sigma$ with

$$\mathcal{H}ilb^h_\Sigma(T) = \{\mathscr{X} \in \mathcal{H}ilb_\Sigma(T) \mid \mathscr{X}_t \text{ has Hilbert polynomial } h, \forall\, t \in T\}.$$

If T is connected, h_t is independent of t by the following remark and $\mathcal{H}ilb_\Sigma(T)$ is the disjoint union of the sets $\mathcal{H}ilb^h_\Sigma(T)$, $h \in \mathbb{Q}[x]$.

Remark 2.2.27.2 Let $\mathscr{X} \to T$ be a projective morphism of complex spaces. If X is flat over T, then the function $t \mapsto h_t$ is locally constant on T and the converse is true if T is reduced (cf. [Schn], resp. [HaR3, Theorem 9.9] for morphisms of algebraic schemes).

The fundamental theorem of Grothendieck says that, for Σ a projective scheme, the functor $\mathcal{H}ilb^h_\Sigma$ (on the category of schemes) is representable by a projective scheme ([Gro], see also [Kol2]) and $\mathcal{H}ilb_\Sigma$ is represented by the disjoint union of these schemes.

Thus, the Hilbert functor $\mathcal{H}ilb^h_\Sigma$ on the category of algebraic complex spaces as defined above, is represented by a projective variety Hilb^h_Σ (the analytification of Grothendieck's Hilbert scheme) with a projective subvariety (the universal family) $\mathcal{U}^h \subset \Sigma \times \mathrm{Hilb}^h_\Sigma$, proper and flat over Hilb^h_Σ. We call the (countable) disjoint union of these projective varieties,

$$\mathrm{Hilb}_\Sigma := \coprod_{h\in\mathbb{Q}[x]} \mathrm{Hilb}^h_\Sigma,$$

the *Hilbert scheme* of Σ and

$$\mathcal{U} := \coprod_{h\in\mathbb{Q}[x]} \mathcal{U}^h \subset \Sigma \times \mathrm{Hilb}_\Sigma,$$

its *universal family*. $\mathcal{U} \subset \Sigma \times \mathrm{Hilb}_\Sigma$ represents the functor $\mathcal{H}ilb_\Sigma$.

The same decomposition as for the Hilbert functor holds for the Douady functor $\mathcal{D}ou_\Sigma$ with

$$\mathcal{D}ou^h_\Sigma(T) = \{\mathscr{X} \in \mathcal{D}ou_\Sigma(T) \mid h_{\mathscr{X}_t} = h, \forall\, t \in T\}.$$

The Douady space is then the disjoint union of countably many subspaces,

$$\mathrm{Dou}_\Sigma = \coprod_{h\in\mathbb{Q}[x]} \mathrm{Dou}^h_\Sigma,$$

where Dou^h_Σ represents $\mathcal{D}ou^h_\Sigma$, with $\mathcal{V}^h \subset \Sigma \times \mathrm{Dou}^h_\Sigma$ as universal family and $\mathcal{V} = \coprod_{h\in\mathbb{Q}[x]} \mathcal{V}^h$.

We show now that in our situation the Hilbert scheme and the Douady space coincide.

Proposition 2.2.28 *Let Σ be a projective variety. Then, for each $h \in \mathbb{Q}[x]$, there exist isomorphism of complex spaces $\varphi : \mathrm{Hilb}^h_\Sigma \to \mathrm{Dou}^h_\Sigma$, inducing via base change an isomorphism $\mathcal{U}^h \to \mathcal{V}^h$ over φ.*

Proof $\mathcal{U}^h \subset \Sigma \times \mathrm{Hilb}^h_\Sigma$ is an element of $\mathcal{D}ou^h_\Sigma(\mathrm{Hilb}^h_\Sigma)$ and hence, by the universal property of $\mathcal{D}ou^h_\Sigma$, there is a unique morphism $\varphi : \mathrm{Hilb}^h_\Sigma \to \mathrm{Dou}^h_\Sigma$. We have to show that φ is an isomorphism. The isomorphism $\mathcal{U}^h \to \mathcal{V}^h$ follows, since it is obtained by base change via φ.

Each point of Dou^h_Σ corresponds to a unique closed subvariety of Σ, which is projective, hence a point of Hilb^h_Σ. Thus Hilb^h_Σ and Dou^h_Σ coincide as sets and φ is a set-theoretical bijection.

Let $\mathcal{O}_{D,t}$ resp. $\mathcal{O}_{H,t}$ be the local rings of Dou^h_Σ resp. Hilb^h_Σ at a point t with maximal ideal $\mathfrak{m}$ resp. $\mathfrak{n}$. We show now that φ induces an isomorphism of local rings

$\mathcal{O}_{D,t} \to \mathcal{O}_{H,t}$. If $D_k \subset \mathrm{Dou}^h_\Sigma$ resp. $H_k \subset \mathrm{Hilb}^h_\Sigma$ denote the fat points concentrated in $\{t\}$ defined by $\mathfrak{m}^{k+1}$, resp. $\mathfrak{n}^{k+1}$, $k \geq 0$, then $\varphi_k = \varphi \mid H_k$ maps H_k to D_k. The fat point D_k is algebraic, even projective. The restriction $\mathcal{V}^h_k \to D_k$ of $\mathcal{V}^h \to \mathrm{Dou}^h_\Sigma$ to D_k is an algebraic family for all k, since $\mathcal{V}^h_k$ is a closed subspace of the projective variety $\Sigma \times D_k$, hence algebraic. Therefore it can be induced from $\mathcal{U}^h \to \mathrm{Hilb}^h_\Sigma$ by a unique morphism $\psi_k : D_k \to \mathrm{Hilb}^h_\Sigma$, which factors through H_k. By the universal property of Dou_Σ and Hilb_Σ, $\varphi_k \circ \psi_k$ and $\psi_k \circ \varphi_k$ are the identity. Hence φ_k is an isomorphism for all k, implying that φ induces an isomorphism $\mathcal{O}_{D,t} \to \mathcal{O}_{H,t}$ by the epimorphism theorem [GLS6, Theorem I.1.20] and Krull's intersection theorem. □

Remark 2.2.28.1 It follows that the (algebraic) universal family over Hilb_Σ, Σ projective,

$$\begin{array}{ccc} \mathcal{U} & \hookrightarrow & \Sigma \times \mathrm{Hilb}_\Sigma \\ & {\scriptstyle \Phi}\searrow \quad \swarrow & \\ & \mathrm{Hilb}_\Sigma & \end{array}$$

has the following *universal property:* for any complex space T (e.g. a complex disc) and any complex subspace $\mathscr{X} \subset \Sigma \times T$, flat and proper over T, there is a unique analytic morphism $T \to \mathrm{Hilb}_\Sigma$ inducing $\mathscr{X} \subset \Sigma \times T$ via pullback from $\mathcal{U} \subset \Sigma \times \mathrm{Hilb}_\Sigma$. By this fact we do not need the Douady space in the following and work exclusively with the Hilbert scheme.

The universal property of Hilb_Σ and the definition of $\underline{\mathcal{D}ef}_{X/\Sigma}$, $X \subset \Sigma$ projective, imply that the germ $(\mathrm{Hilb}_\Sigma, X)$ is the base space of the semiuniversal deformation of X/Σ. If Σ is smooth and X a hypersurface, then its tangent space is $H^0(\mathcal{O}_X(X))$ by Lemma 2.2.24. If $H^1(\mathcal{O}_\Sigma(X)) \to H^1(\mathcal{O}_X(X))$ is the zero map $(\mathrm{Hilb}^h_\Sigma, X)$ is smooth by Proposition 2.2.26. Since a smooth space has the same dimension as its tangent space, we have the following classical result.

Lemma 2.2.29 *Let Σ be a smooth projective variety and $X \subset \Sigma$ a projective hypersurface, then $H^0(\mathcal{O}_X(X))$ is the Zariski tangent space to* Hilb_Σ *at X.*

Moreover, if the natural map $H^1(\mathcal{O}_\Sigma(X)) \to H^1(\mathcal{O}_X(X))$ is zero, then Hilb_Σ *is smooth at X of dimension* $h^0(\mathcal{O}_X(X)) := \dim_{\mathbb{C}} H^0(X, \mathcal{O}_X(X))$.

2.2.2.2 Equisingular Families of Hypersurfaces

For the rest of this section let Σ be again a smooth projective variety. We consider families of projective hypersurfaces (effective divisors) X in Σ, parameterized by a complex space, with isolated singularities such that the analytic or topological type of the singularities does not change (along some sections). We call them just *equisingular families (ESF)*. Recall that family means always flat family. When we talk about topological types, we always assume that Σ is a surface. Then a hypersurface on Σ is a curve whose singularities are plane curve singularities.

Such families are strictly ideal-versal (Proposition 2.2.21) and hence we can apply the results of Sect. 2.2.1.5. We prove in Theorem 2.2.32 that for fixed types

$(S_1, \ldots, S_r)$ there exists a universal ESF $V_h(S_1, \ldots, S_r)$ resp. $V_{|X|}(S_1, \ldots, S_r)$, which is a locally closed analytic subspace of the Hilbert scheme Hilb^h_Σ (cf. Sect. 2.2.2.1) resp. of the linear system $|X|$ on Σ. V_h and $V_{|X|}$ carry a unique analytic structure, which is possibly not reduced. We prove, however, that their reduction is quasi-projective, in particular algebraic. Moreover, we give sufficient criteria for the smoothness of these (and related) families.

As introduced in Sect. 2.2.2.1, Hilb^h_Σ is considered as algebraic complex space, the analytification of Grothendieck's Hilbert scheme, which has the universal property for families over arbitrary complex spaces by Proposition 2.2.28 and Remark 2.2.28.1.

Definition 2.2.30 Let $\phi : \mathscr{X} \hookrightarrow \Sigma \times T \to T$ be a proper family of hypersurfaces over the complex space T and assume that for each $t \in T$, r isolated singularities $z_1(t), \ldots, z_r(t) \in \mathscr{X}_t = \phi^{-1}(t)$ are given. We call ϕ an *equianalytic* resp. *equisingular* family along $z_1(t), \ldots, z_r(t)$, if the maps $t \mapsto z_i(t)$, $i = 1, \ldots, r$, are sections of Φ and if the induced local deformations of the germs $(\mathscr{X}_t, z_i(t))$ over (T, t) are equianalytic resp. equisingular for $i = 1, \ldots, r$ and all $t \in T$, with equianalytic resp. equisingular sections $z_i(t)$, $i = 1, \ldots, r$ (cf. Definition 2.2.3 and Remark 2.2.3.1).

By Proposition 2.2.6 the equianalytic resp. equisingular sections are unique. Hence we need to know only that they exist, but we do not need to specify them.

For curves it is allowed that for some singularities the topological type and for others the analytic type is fixed. To simplify notations we call all these families *equisingular along* $z_1(t), \ldots, z_r(t)$, or just *ESF*.

Note that the fibres $\mathscr{X}_t$ of ϕ may have other singularities, isolated or not, which may be deformed in the family ϕ in an arbitrary manner.

Definition 2.2.31 Let $\mathcal{S} = (S_1, \ldots, S_r)$ be a sequence of analytic or topological types. We call an ESF $\phi : \mathscr{X} \hookrightarrow \Sigma \times T \to T$ over the complex space T $\mathcal{S}$-*equisingular* if there exist (disjoint) sections $\sigma_1, \ldots, \sigma_r$ of ϕ such that for $t \in T$ the type of $(\mathscr{X}_t, \sigma_i(t))$ is S_i and ϕ is equisingular along $\sigma_1(t), \ldots, \sigma_r(t)$.

Here and in the following we may mix the types (for curves), that is, some of the S_i are analytic types while the others are topological types. We may also assume that some of the sections are trivial, that is, the position of these points is fixed.

Theorem 2.2.32 *Let Σ be a smooth projective variety and* Hilb^h_Σ *the Hilbert scheme parameterizing proper families of hypersurfaces on Σ with fixed Hilbert polynomial h. Moreover, let $\mathcal{S} = (S_1, \ldots, S_r)$ be a sequence of analytic or topological types.*

Then there exists a unique locally closed analytic subspace $V_h(\mathcal{S}) = V_h(S_1, \ldots, S_r) \subset$ Hilb^h_Σ *representing $\mathcal{S}$-equisingular families. Moreover, the reduction of $V_h(\mathcal{S})$ is a quasi-projective algebraic subvariety of* Hilb^h_Σ.

Note that this means:

(1) The restriction of the universal family $\mathscr{U}_h \hookrightarrow \Sigma \times \mathrm{Hilb}^h_\Sigma \to \mathrm{Hilb}^h_\Sigma$ (cf. Remark 2.2.28.1) to $V_h(\mathcal{S})$ is $\mathcal{S}$-equisingular (along some sections).

(2) If $\phi : \mathscr{X} \to \Sigma \times T \to T$, T a complex space, is any $\mathcal{S}$-equisingular family then the (unique) analytic morphism $\varphi : T \to \mathrm{Hilb}^h_\Sigma$ inducing ϕ via pullback, factors through $V_h(\mathcal{S})$.
(3) For singularities of the fibres of an $\mathcal{S}$-equisingular family different from the given $x_1(t), \dots, x_r(t)$, no conditions are required.

Proof Let $X \subset \Sigma$ be a projective hypersurface with isolated singular points $z_1, \dots, z_r$ of type $S_1, \dots, S_r$. We denote the point in Hilb^h_Σ defined by X with $[X]$ where h is the Hilbert polynomial of X. Then $X \subset \mathscr{U}_h$ is the fibre of the universal family $\mathscr{U}_h \to \mathrm{Hilb}^h_\Sigma$ over $[X]$ and the induced map of germs $(\mathscr{U}_h, z_i) \to (\mathrm{Hilb}^h_\Sigma, [X])$ is a deformation of (X, z_i) for $i = 1, \dots, r$. Since the local *ea*- resp. *es*-deformations are strictly ideal-versal by Proposition 2.2.21, they satisfy (R) and there is a unique closed equianalytic resp. equisingular stratum (a complex germ) in $(\mathrm{Hilb}^h_\Sigma, [X])$ which we denote by $(V_h(S_i), [X]), i = 1, \dots, r$. Define $(V_h(S_1, \dots, S_r), [X])$ as the intersection of the subgerms $(V_h(S_i), [X])$ in $(\mathrm{Hilb}^h_\Sigma, [X])$. Since these subgerms are unique in a neighborhood of each $[X]$ in Hilb^h_Σ, they glue together to a locally closed analytic subspace $V_h(S_1, \dots, S_r)$ of Hilb^h_Σ. Using Proposition 2.2.28 (or Remark 2.2.28.1) it is easy to see that $V_h(S_1, \dots, S_r)$ represents $\mathcal{S}$-equisingular families in Σ.

For the quasi-projectivity note that $V_h(S_i)_{red}$, the reduction of $V_h(S_i)$, is quasi-projective for $i = 1, \cdots, r$. This follows with same proof as in Proposition 1.1.61 resp. Proposition 1.1.72 for hypersurfaces in projective space, replacing $\mathscr{U}_d \to \mathbb{P}\big(H^0(\mathcal{O}_{\mathbb{P}^n}(d))\big)$ by $\mathscr{U}_h \to \mathrm{Hilb}^h_\Sigma$. The fact that we required a unique isolated singularity in the fibres is not relevant, since we consider here deformations along disjoint sections. Hence the reduction of $V_h(S_1, \dots, S_r)$, which is the reduction of the intersection of the $V_h(S_i)_{red}$, is quasi-projective. □

Remark 2.2.32.1 We do not know if $V_h(S_1, \dots, S_r)$ itself with its (possibly) non-reduced structure is quasi-projective, except for the case of plane curves with nodes and cusps (cf. [Wah2, Theorem 3.3.5]). The proof of Wahl does not generalize immediately, since it uses simply a flattening stratification of the space $\mathscr{S}_d$, used in item (2) of the proof of Lemma 1.60, and the characterization of nodes resp. cusps by their Tjurina number (being 1 resp. 2). Nevertheless, we may expect that a more refined analysis of the involved Hilbert schemes would give the quasi-projectivity of $V_h(S_1, \dots, S_r)$.

Example 2.2.32.2 A complex space whose reduction is algebraic is itself not necessarily algebraic. A simple (global) example is the affine line $\mathbb{C}$ with non-reduced structure at the integers.

We like to thank Dorin Popescu for providing the following (local) example of a non-algebraic singularity with algebraic reduction. Let P be a prime ideal in $\mathbb{C}\{x, y\}$, which is not generated by polynomials and consider the ideal $I = \langle z^2, zP\rangle = \langle z\rangle \cap \langle z^2, P\rangle \in C\{x, y, z\}$. Then I is not generated by polynomials but its radical $\langle z\rangle$ is.

Note that isolated singularities are algebraic (by a theorem of M. Artin).

Definition 2.2.33 Let Σ be a smooth projective variety, X a projective hypersurface of Σ with Hilbert polynomial h and $\mathcal{S} = (S_1, \ldots, S_r)$ a sequence of analytic or topological types. We call

$$V_h(\mathcal{S}) = V_h(S_1, \ldots, S_r) \subset \mathrm{Hilb}^h_\Sigma$$

the $\mathcal{S}$-*equisingular Hilbert stratum* and denote by

$$V_h^{irr}(\mathcal{S}) = V_h^{irr}(S_1, \ldots, S_r) \subset V_h(\mathcal{S})$$

the open subspace parameterizing irreducible hypersurfaces.

We often refer to $V_h(S_1, \ldots, S_r)$ (resp. $V_h^{irr}(S_1, \ldots, S_r)$) as the *variety of hypersurfaces* (resp. *irreducible hypersurfaces*) *in Σ with Hilbert polynomial h and with r singularities of type $S_1, \cdots S_r$*.

By Theorem 2.2.32 $V_h(\mathcal{S})$ and $V_h^{irr}(\mathcal{S})$ have a unique (not necessarily reduced) analytic structure. Their reduction is quasi-projective.

Note that $V_h(S_1, \ldots, S_r)$ is the intersection of the (locally closed) analytic subspaces $V_h(S_i)$, $i = 1, \ldots, r$, in Hilb^h_Σ. Intuitively we may think of $V_h(S_1, \ldots, S_r)$ as the set of all hypersurfaces $X \subset \Sigma$ having isolated singularities $z_1, \ldots, z_r$ of prescribed types $S_1, \ldots, S_r$. The position of the singular points $z_1, \ldots, z_r$ is not specified, but they are not randomly distributed. For a hypersurface $X' \subset \Sigma$ in a neighborhood of X in $V_h(S_1, \ldots, S_r)$ with singularities z_i' of type S_i, $i = 1, \ldots, r$, the singularities z_i' and z_i belong to the image of some section σ_i over a neighborhood of X in $V_h(S_1, \ldots, S_r)$.

Classically, instead of the Hilbert scheme Hilb^h_Σ, one considers the complete linear system $|X| = \mathbb{P}\big(H^0(\mathcal{O}_\Sigma(X))\big)$ of the divisor $X \subset \Sigma$, respectively the open subset of $|X|$ corresponding to hypersurfaces with isolated singularities. The two notions are related as follows: The Hilbert polynomial $h = h(X')$ is constant for all divisors X', linear equivalent to X, since the incidence variety

$$\mathfrak{N}_{|X|} = \big\{(x, X') \in \Sigma \times |X| \,\big|\, x \in X'\big\} \subset \Sigma \times |X|$$

is flat over $|X|$ by Remark 2.1.15.2(1).[12]

---

[12] Another way to see this is as follows: having an embedding $\Sigma \subset \mathbb{P}^N$, we write a rational function φ on Σ such that $(\varphi) = X - X'$ as $\varphi = \frac{P}{Q}\big|_\Sigma$ with some coprime homogeneous polynomials $P, Q \in \mathbb{C}[x_0, \ldots, x_N]$ of degree m. P, Q in turn define divisors $(P) = X \cup Y$, $(Q) = X' \cup Y$ on Σ. We then have an exact sequence of ideal sheaves

$$0 \to \mathcal{J}_{\Sigma/\mathbb{P}^N}(n) \to \mathcal{J}_{\Sigma/\mathbb{P}^N}(m+n) \to \mathcal{J}_{X \cup Y/\mathbb{P}^N}(m+n) \to 0$$

and the same for X'. Therefore, $h_{X \cup Y} = h_X + h_Y - h_{X \cap Y} = h_{X' \cup Y} = h_{X'} + h_Y - h_{X' \cap Y}$, and hence $h_X = h_{X'}$, provided that $h_{X \cap Y} = h_{X' \cap Y}$, where $X \cap Y$, $X' \cap Y$ are linear equivalent (with the rational function $\varphi\big|_Y$) divisors on an algebraic variety Y of dimension $\dim \Sigma - 1$. Thus, we complete the proof by induction.

Proposition 2.2.34 *Let Σ be a smooth projective variety and X a projective hypersurface of Σ with Hilbert polynomial h.*

(1) There exists a unique immersion $|X| \hookrightarrow \mathrm{Hilb}^h_\Sigma$ making $|X|$ a subscheme of Hilb^h_Σ. If $H^1(\mathcal{O}_\Sigma) = 0$, (e.g. $\Sigma = \mathbb{P}^N$) then $|X|$ is a connected component of Hilb^h_Σ.

(2) If $\Sigma = \mathbb{P}^N$ then $\mathrm{Hilb}^h_\Sigma = |X|$ and hence $\mathrm{Hilb}^h_{\mathbb{P}^N}$ is a projective space.

Proof (1) By the universal property of Hilb^h_Σ there is a unique morphism $|X| \to \mathrm{Hilb}^h_\Sigma$, which, on the tangent level (cf. Lemma 2.2.29), corresponds to the first inclusion of the exact sequence

$$H^0\big(\mathcal{O}_\Sigma(X)\big)/H^0\big(\mathcal{O}_\Sigma\big) \hookrightarrow H^0\big(\mathcal{O}_X(X)\big) \to H^1(\mathcal{O}_\Sigma)\,,$$

using $|X| = \mathbb{P}(H^0(\mathcal{O}_\Sigma(X)))$ and $H^0\big(\mathcal{O}_\Sigma\big) = \mathbb{C}$. Therefore $|X| \hookrightarrow \mathrm{Hilb}^h_\Sigma$ is an immersion (cf. [GLS6, Theorem I.1.20]) and we may consider it as a subscheme of Hilb^h_Σ. In particular, if $H^1(\mathcal{O}_\Sigma) = 0$, then, since $|X|$ is smooth, $|X|$ has the same dimension as the Zariski tangent space of Hilb^h_Σ at $|X|$ and therefore the above immersion is locally an isomorphism and $|X|$ is a connected component of Hilb^h_Σ (cf. also Remark 2.2.39.1).

(2) By Hartshorne [HaR1] the Hilbert scheme $\mathrm{Hilb}^h_{\mathbb{P}^N}$ is connected for any $h \in \mathbb{Q}[x]$. Since $|X|$ is a connected component of $\mathrm{Hilb}^h_{\mathbb{P}^N}$ by (1), it must be equal.

This can also easily be seen directly. By the following Remark 2.2.34.1 the polynomial h_X is uniquely determined by N and d, i.e. by its degree, which is $N-1$ (the dimension of X) and by its leading coefficient, which is $d/(N-1)!$. Let $Y \subset \mathbb{P}^N$ be any subvariety with Hilbert polynomial h. Then it has dimension $N-1$ and degree d.

Now let I be the intersection of the primary components belonging to the minimal associated primes of the homogeneous ideal J of Y. Then I defines a hypersurface X in $\mathbb{P}^N$ and, if $X \neq Y$, the other components of Y have lower dimension. X is the pure dimensional part of Y of the same dimension and degree as Y (by Remark 2.1.18.1(4)) and the Hilbert polynomials of X and Y coincide. Since the Hilbert polynomial is additive w.r.t. exact sequences (cf. e.g. [GrP, Lemma 5.1.2]) it follows from the exact sequence

$$0 \to J/I \to \mathcal{O}_X \to \mathcal{O}_Y \to 0$$

that the Hilbert polynomial of the graded module J/I is 0, implying $J/I = 0$. That is, $Y = X$ is a hypersurface. (For a slightly different argument see [ACG, Chapter IX, Example 2.3]) □

Remark 2.2.34.1 We have $h_{\mathbb{P}^N}(n) = \binom{N+n}{N}$ and for a hypersurface $X \subset \mathbb{P}^N$ of degree d the *Hilbert polynomial of* X is $h_X(n) = \binom{N+n}{N} - \binom{N-d+n}{N}$. This is well known and follows easily from the cohomology sequence to the exact sequence $0 \to \mathcal{O}_{\mathbb{P}^N}(n-d) \to \mathcal{O}_{\mathbb{P}^N}(n) \to \mathcal{O}_X(n) \to 0$.

Definition 2.2.35 Let Σ be a smooth projective variety, X a projective hypersurface of Σ with Hilbert polynomial h and $\mathcal{S} = (S_1, \ldots, S_r)$ be a sequence of analytic or topological types. We denote by

$$V_{|X|}(\mathcal{S}) = V_{|X|}(S_1, \ldots, S_r) \subset |X|$$

resp.

$$V^{\text{irr}}_{|X|}(\mathcal{S}) = V^{\text{irr}}_{|X|}(S_1, \ldots, S_r) \subset |X|$$

the intersection of $|X|$ with $V_h(S_1, \ldots, S_r)$ resp. with $V_h^{irr}(S_1, \ldots, S_r)$ in Hilb^h_Σ.

We call $V_{|X|}(S_1, \ldots, S_r)$ (resp. $V^{irr}_{|X|}(S_1, \ldots, S_r)$) the *variety of hypersurfaces* (resp. *irreducible hypersurfaces*) *in the linear system* $|X|$ *with* r *singularities of type* $S_1, \cdots S_r$.

If $\Sigma = \mathbb{P}^N$ and $|X| = |dH|$, H a hyperplane in $\mathbb{P}^N$, then $V_{|X|}(S_1, \ldots, S_r)$, resp. $V^{irr}_{|X|}(S_1, \ldots, S_r)$, is denoted by

$$V_d(S_1, \ldots, S_r), \text{ resp. } V_d^{irr}(S_1, \ldots, S_r).$$

$V_d^{(irr)}(S_1, \ldots, S_r)$ is called the *variety of (irreducible) hypersurfaces in* $\mathbb{P}^N$ *of degree* d *with* r *singularities of type* $S_1, \cdots, S_r$.

The hypersurfaces in $V_{|X|}(S_1, \ldots, S_r)$ may have further singularities, outside the considered ones of type $S_1, \ldots, S_r$. In $V_{|X|}(\mathcal{S})$ nothing is required for these. The following theorem is a direct consequence of Theorem 2.2.32.

Theorem 2.2.36 *Let X be a projective hypersurface of the smooth projective variety Σ and $(S_1, \cdots, S_r)$ a sequence of analytic or topological types. $V_{|X|}(S_1, \ldots, S_r)$ resp. $V^{irr}_{|X|}(S_1, \ldots, S_r)$ are (not necessarily reduced) locally closed analytic subspaces of $|X|$, representing $(S_1, \ldots, S_r)$-equianalytic resp. $(S_1, \ldots, S_r)$-equisingular families of (resp. irreducible) hypersurfaces in the linear system $|X|$. Their reduction is a quasi-projective algebraic subvariety of $|X|$.*

In the next section we give a geometric interpretation of the 0th and 1st cohomology group of the ideal sheaves of certain zero-dimensional schemes. To simplify notation we write $X \in V_h(S_1, \ldots, S_r)$ to denote either the point in $V_h(S_1, \ldots, S_r)$ or the hypersurface corresponding to the point, that is, the corresponding fibre of the universal family.

Consider the map

$$\Phi_h : V_h(S_1, \ldots, S_r) \longrightarrow \text{Sym}^r \Sigma\,, \quad X \longmapsto (z_1+\ldots+z_r)\,, \tag{2.2.2.1}$$

where $\text{Sym}^r \Sigma$ is the r-fold symmetric product of Σ and $(z_1+\ldots+z_r)$ the non-ordered tuple of the specified isolated singularities of X. Since any equianalytic resp. equisingular deformation of a germ admits a unique equianalytic resp. equisingular section (cf. Proposition 2.2.6), the universal family

$$\mathscr{U}_h(S_1, \ldots, S_r) \hookrightarrow \Sigma \times V_h(S_1, \ldots, S_r) \to V_h(S_1, \ldots, S_r)$$

admits, locally at X, r unique disjoint analytic sections. Composing these sections with the projections to Σ gives a local description of the map Φ_h and shows in particular that Φ_h is a well-defined analytic morphism, even if $V_h(S_1, \ldots, S_r)$ is not reduced.

Definition 2.2.37 With the assumptions of Definition 2.2.35 and Φ_h as in (2.2.2.1) let $V_{h,fix}(S_1, \ldots, S_r)$ denote the complex subspace of $V_h(S_1, \ldots, S_r)$ consisting of the disjoint union of the fibres of Φ_h. We call $V_{h,fix}(S_1, \ldots, S_r)$ the *fixed* $(S_1, \ldots, S_r)$*-equisingular Hilbert stratum*. We set $V_{|X|,fix}(S_1, \ldots, S_r) := |X| \cap V_{h,fix}(S_1, \ldots, S_r)$.

We also speak about the *variety of hypersurfaces in* Σ resp. *in the linear system* $|X|$ *with* r *singularities of type* $S_1, \cdots S_r$ *with fixed position.*

Remark 2.2.37.1 It follows from the universal property of $V_h(S_1, \ldots, S_r)$ and from the above construction that $V_{h,fix}(S_1, \ldots, S_r)$, resp. $V_{|X|,fix}(S_1, \ldots, S_r)$, together with the induced universal family on each fibre of Φ_h represents the Hilbert functor of hypersurfaces in Σ, resp. in the linear system $|X|$, with given singularity types $S_1, \ldots, S_r$ along trivial sections. The reduction of $V_{h,fix}(S_1, \ldots, S_r)$ resp. of $V_{|X|,fix}(S_1, \ldots, S_r)$ is quasi-projective.

2.2.2.3 T-Smoothness of ESF Through H^1-Vanishing

In this section we derive properties of ESF of projective hypersurfaces and in particular of plane projective curves from the vanishing of H^1 for the ideal sheaves of certain 0-dimensional schemes. This constitutes our basic cohomological approach for the study of ESF. Later we will prove numerical conditions, which imply H^1-vanishing and hence the desired properties.

We use the notations from the previous section. In particular, Σ denotes a smooth projective variety.

Recall from Definition 2.2.23 the 0-dimensional scheme

$$Z'(X) := \bigcup_{i=1}^{r} Z'(X, z_i),$$

concentrated on $\{z_1, \ldots, z_r\}$ with $Z'(X, z_i)$ one of the schemes $Z^{ea}(X, z_i)$, $Z^{ea}_{fix}(X, z_i)$, $Z^{es}(X, z_i)$, $Z^{es}_{fix}(z_i)$, concentrated on z_i. Let $f \in \mathcal{O}_{\Sigma,z_i}$ define (X, z_i), then $Z'(X, z_i)$ is defined by one of the ideals $I^{ea}(X, z_i) = \langle f, j(f)\rangle$, $I^{ea}_{fix}(X, z_i) = \langle f, \mathfrak{m} j(f)\rangle$, $I^{es}(X, z_i)$, $I^{es}_{fix}(X, z_i)$ respectively (see Example 2.2.20.1). Note that for different z_i we allow different kinds of ideals.

Remark 2.2.37.2 We have

$$\deg Z'(X) = \sum \deg Z'(X, z_i)$$

and

$$\begin{aligned}
\deg Z^{ea}(X, z_i) &= \tau(f), \\
\deg Z^{ea}_{fix}(X, z_i) &= \tau_{fix}(f) = \tau(f) + \dim(\Sigma, z_i), \\
\deg Z^{es}(X, z_i) &= \tau^{es}(f) \;\; = \mu(f) - \operatorname{mod}(f), \\
\deg Z^{es}_{fix}(X, z_i) &= \tau^{es}_{fix}(f) = \tau^{es}(f) + 2,
\end{aligned}$$

with $\mu(f)$ the Milnor number, $\tau(f)$ the Tjurina number and $\operatorname{mod}(f)$ the right-modality of f (cf. Definition 1.1.63, Lemma 2.2.13 and Remark 2.2.13.1).

The ideal sheaf of $Z'(X)$ is $\mathcal{J}_{Z'(X)/\Sigma}$ if $Z'(X)$ is considered as a subscheme of Σ resp. $\mathcal{J}_{Z'(X)/X}$ if it is considered as a subscheme of X.

In the following we use the notation $Z'(X)$ if the statement holds for any combination of different kinds of ideals. Otherwise we specify the scheme (e.g. $Z^{ea}(X)$ means $Z'(X, z_i) = Z^{ea}(X, z_i)$ for all $i = 1, \ldots, r$). Moreover, we write V_h to denote $V_h(S_1, \ldots, S_r)$, respectively $V_{h,fix}(S_1, \ldots, S_r)$, and $V_{|X|}$ to denote $V_{|X|}(S_1, \ldots, S_r)$, respectively $V_{|X|,fix}(S_1, \ldots, S_r)$, where some of the S_i can be analytic while the others are topological types.

Theorem 2.2.38 *Let $X \subset \Sigma$ be a projective hypersurface with Hilbert polynomial h and r specified isolated singularities $z_1, \ldots, z_r$ of analytic or topological types $S_1, \ldots, S_r$. Set $V_h = V_h(S_1, \ldots, S_r)$ and $V_{|X|} = V_{|X|}(S_1, \ldots, S_r)$.*

(a) The Zariski tangent space of $V_h \subset \mathrm{Hilb}^h_\Sigma$ at X is $H^0\big(\mathcal{J}_{Z'(X)/X}(X)\big)$, while the Zariski tangent space of $V_{|X|}$ at X is $H^0\big(\mathcal{J}_{Z'(X)/\Sigma}(X)\big)/H^0(\mathcal{O}_\Sigma)$.

(b) $h^0\big(\mathcal{J}_{Z'(X)/X}(X)\big) - h^1\big(\mathcal{J}_{Z'(X)/X}(X)\big) \leq \dim(V_h, X) \leq h^0\big(\mathcal{J}_{Z'(X)/X}(X)\big).$

(c1) If $H^1\big(\mathcal{J}_{Z'(X)/X}(X)\big) = 0$ then V_h is T-smooth at X, that is, smooth of the "expected" dimension $h^0\big(\mathcal{O}_X(X)\big) - \deg Z'(X) = \dim(\mathrm{Hilb}^h_\Sigma, X) - \deg Z'(X)$.

(c2) If $H^1\big(\mathcal{J}_{Z'(X)/\Sigma}(X)\big) = 0$ then $V_{|X|}$ is T-smooth at X of the expected dimension $h^0\big(\mathcal{O}_\Sigma(X)\big) - 1 - \deg Z'(X) = \dim(V_{|X|}, X) - \deg Z'(X)$.

(d) If $H^1\big(\mathcal{J}_{Z^{ea}(X)/X}(X)\big) = 0$ then the morphism of germs

$$\big(\mathrm{Hilb}^h_\Sigma,\, X\big) \longrightarrow \prod_{i=1}^{r} B_{(X,z_i)}$$

is smooth of fibre dimension $h^0\big(\mathcal{J}_{Z^{ea}(X)/X}(X)\big)$. Here, $\prod_{i=1}^r B_{(X,z_i)}$ is the cartesian product of the base spaces of the semiuniversal deformations of the germs (X, z_i) and the morphism exists by versatility of $B_{(X,z_i)}$.

(e) Write $Z_{fix}(X)$ for $Z^{ea}_{fix}(X)$, respectively $Z^{es}_{fix}(X)$. Then $H^0\big(\mathcal{J}_{Z_{fix}(X)/X}(X)\big)$ is the tangent space of $V_{h,fix}(S_1, \ldots, S_r)$ at X and the vanishing of $H^1\big(\mathcal{J}_{Z_{fix}(X)/X}(X)\big)$ implies that the morphism of germs

$$\Phi_h : \big(V_h(S_1, \dots, S_r), X\big) \to \big(\mathrm{Sym}^r \Sigma, (z_1 + \dots + z_r)\big)$$

is smooth (hence surjective) of fibre dimension $h^0\big(\mathcal{J}_{Z_{fix}(X)/X}(X)\big)$.

Recall that a morphism of germs $(X, x) \to (Y, y)$ is smooth if it is locally isomorphic to the projection $(Y, y) \times (Z, z) \to (Y, y)$ with (Z, z) a smooth germ. Note also that for $X \subset \Sigma \subset \mathbb{P}^n$ and for any coherent sheaf $\mathcal{J}$ we have $H^1\big(\mathcal{J}(X)\big) = 0$ if X has sufficiently high degree by Serre-vanishing.

The smoothness of the morphism in (d) says that the germ $\big(\mathrm{Hilb}^h_\Sigma, X\big)$ is a joint versal deformation simultaneously for the germs (X, z_i), $i = 1, \dots, r$. An important consequence of (e) is that $H^1(\mathcal{I}_{Z_{fix}(X)/X}(X)) = 0$ implies that arbitrarily close to X there are hypersurfaces in $V_h(S_1, \dots, S_r)$ having their singularities in generic position in Σ.

In the following supplement we generalize the statement (d) to families where some of the singular points of X are deformed arbitrarily while for others the analytic or topological type is fixed. Moreover, for some points also their position in Σ is fixed. To formulate this, let $1 \le p \le r$ and

$$Z'(X) = \bigcup_{i=1}^{p} Z^{ea}(X, z_i) \cup \bigcup_{i=p+1}^{r} Z'(X, z_i)$$

with $Z'(X, z_i)$ one of the schemes $Z^{ea}(X, z_i)$, $Z^{es}(X, z_i)$, $Z^{ea}_{fix}(X, z_i)$, $Z^{es}_{fix}(X, z_i)$ for $i = p + 1, \dots, r$. Moreover, let S_i, $i = p + 1, \dots, r$, be either an analytic type or a topological type, depending on whether the upper index of $Z'(X, z_i)$ is *ea* or *es*, and set $V'_h(S_i) = V_{h,fix}(S_i)$ if the lower index of $Z'(X, z_i)$ is *fix* and $V'_h(S_i) = V_h(S_i)$ otherwise. Finally define

$$V'_h(S_{p+1}, \dots, S_r) := \bigcap_{i=p+1}^{r} V'_h(S_i) \subset \mathrm{Hilb}^h_\Sigma \ .$$

Supplement 2.2.39 *If* $H^1(\mathcal{J}_{Z'(X)/\Sigma}(X)) = 0$ *then the morphism of germs*

$$\big(V'_h(S_{p+1}, \dots, S_r), X\big) \to \prod_{i=1}^{p} B_{(X, z_i)}$$

is smooth (hence surjective) of fibre dimension $h^0(\mathcal{J}_{Z'(X)/\Sigma}(X))$.

The proof of the supplement is given in the proof of Theorem 2.2.38(d). For the proof of the theorem we use the following commutative diagram with exact rows and columns (see also (2.2.1.3)).

$$\begin{array}{ccccccccc} & & & & 0 & & & & \\ & & & & \downarrow & & & & \\ & & & & \mathcal{O}_\Sigma & & & & \\ & & & & \downarrow & & & & \\ 0 & \to & \mathcal{J}_{Z'/\Sigma}(X) & \to & \mathcal{O}_\Sigma(X) & \to & \mathcal{O}_{Z'/\Sigma} & \to & 0 \\ & & \downarrow & & \downarrow & & \| & & \\ 0 & \to & \mathcal{J}_{Z'/X}(X) & \to & \mathcal{O}_X(X) & \to & \mathcal{O}_{Z'/X} & \to & 0 \\ & & \downarrow & & \downarrow & & & & \\ & & 0 & & 0 & & & & \end{array} \tag{2.2.2.2}$$

Proof of Theorem 2.2.38. By the considerations in the beginning of this section we have that the restriction of the universal family $\mathcal{U}_h \to \mathrm{Hilb}^h_\Sigma$ to the germ (V_h, X) is a semiuniversal object of $\mathcal{D}ef'_{X/\Sigma}$ with $\mathcal{D}ef'_{X/\Sigma}$ being $\mathcal{D}ef^{ea}_{X/\Sigma}$ or $\mathcal{D}ef^{es}_{X/\Sigma}$ (cf. Definition 2.2.20), which is ideal-versal by Proposition 2.2.21. Hence we can use formal deformation theory and apply (the proofs of) Lemma 2.2.24 and Proposition 2.2.26. The difference between V_h and $V_{|X|}$ is that we have to show for $V_{|X|}$ that, in addition, the constructions take place in the linear system $|X| \subset \mathrm{Hilb}^h_\Sigma$.

(a) By the universal property of Hilb^h_Σ and hence of V_h, (V_h, X) is the base space of a semiuniversal deformation of $\mathcal{D}ef'_{X/\Sigma}$. Hence the first part of (a) follows from Lemma 2.2.24.

To see the second part of (a) we analyze the proof of Lemma 2.2.24, using the same notations. In addition we use the fact that the considered deformations belong to the linear system $|X|$. Hence, the deformation given by $(f_i + \varepsilon g_i)_{i\in I}$ is the tangent vector to the pencil given by $(f_i + \lambda g_i)_{i\in I}$, $\lambda \in \mathbb{C} = H^0(\mathcal{O}_\Sigma)$, with $(f_i + \lambda g_i)_{i\in I}$ linear equivalent to X for each $\lambda \in \mathbb{C}$. Since X is given by $(f_i)_{i\in I}$ the divisor given by $(f_i + \lambda g_i)/f_i = 1 + \lambda g_i/f_i$ is principal and therefore $\lambda g_i/f_i - \lambda g_j/f_j$ considered in (2.2.1.4) is zero in $\Gamma(U_i \cap U_j, \mathcal{O}_\Sigma(X))$. This shows that $(\lambda g_i/f_i)_{i\in I}$ is not only a global section of $\mathcal{O}_X(X)$ but of $\mathcal{O}_\Sigma(X)$ which is unique modulo $H^0(\mathcal{O}_\Sigma)$. The rest of the argument is similar as in the proof of Lemma 2.2.24.

(b) The first estimate is due to the fact that (V_h, X) is the germ of the fibre over the origin of a (non-linear) obstruction map $H^0(\mathcal{J}_{Z'(X)/X}(X)) \to H^1(\mathcal{J}_{Z'(X)/X}(X))$ (cf. [Lau, Theorem 4.2.4]). The second estimate follows from (a).

(c1) This is a consequence of Proposition 2.2.26 and the exact sequence (deduced from (2.2.2.2))

$$0 \to H^0(\mathcal{J}_{Z'/X}(X)) \to H^0(\mathcal{O}_X(X)) \to H^0(\mathcal{O}_{Z'}(X)) \to$$
$$\to H^1(\mathcal{J}_{Z'/X}(X)) \to H^1(\mathcal{O}_X(X)) \to 0\,.$$

Note that $H^1(\mathcal{J}_{Z'/X}(X)) = 0$ implies $H^1(\mathcal{O}_X(X)) = 0$ whence Hilb^h_Σ is smooth at X of dimension $h^0(\mathcal{O}_X(X))$. The dimension statement is a consequence of (a).

(c2) The dimension statement follows as in (c1) with the exact cohomology sequence to $\mathcal{J}_{Z'/\Sigma}(X) \hookrightarrow \mathcal{O}_\Sigma(X) \twoheadrightarrow \mathcal{O}_{Z'}(X)$. To show the smoothness of $V_{|X|} = V_h \cap |X|$ at X, we apply Lemma 2.2.25 to the local ring R of $(V_{|X|}, X)$.

Let $A \to \overline{A} = A/\langle \eta \rangle$ be a small extension of Artin rings. We have to show that every morphism $\overline{\varphi} : R \to \overline{A}$ lifts to a morphism $\varphi : R \to A$.

Note first that $H^1(\mathcal{J}_{Z'/\Sigma}(X)) = 0$ implies that V_h is smooth at X by Proposition 2.2.26. Hence, the morphism $\overline{\varphi} : R \to \overline{A}$ extends to a morphism $\varphi_1 : R_1 := \mathcal{O}_{V_h,X} \to A$. This means the following:

Let $\{U_i\}_{i\in I}$ be an open covering of Σ and $\overline{F}_i \in \Gamma(U_i, \mathcal{O}_\Sigma \otimes \overline{A}), i \in I$, the family over $\overline{A}$ induced via $\overline{\varphi}$ from the universal family over V_h. That $\overline{\varphi}$ lifts to φ_1 means the existence of $F_i \in \Gamma(U_i, \mathcal{O}_\Sigma \otimes A)$ lifting $\overline{F}_i$ such that

(1) $F_i = G_{ij} F_j$ on $U_{ij} = U_i \cap U_j$ for some $G_{ij} \in \Gamma(U_{ij}, (\mathcal{O}_\Sigma \otimes A)^*)$,
(2) the image f_i of F_i in $\Gamma(U_i, \mathcal{O}_\Sigma)$ generates the ideal of $X \cap U_i$ in U_i,
(3) the germs of F_i at each $z \in \{z_1, \dots, z_r\}$ define an element in $\underline{\mathcal{D}ef}'_{X,z}(A)$.

However, the family $(F_i)_{i\in I}$ may not be induced from the linear system $|X|$, that is, φ_1 may not define a morphism $\varphi : R := R_1/\underline{I}(|X|, X) \to A$. To do so, we have to change F_i to $\widetilde{F}_i \in \Gamma(U_i, \mathcal{O}_\Sigma \otimes A)$ such that $\widetilde{F}_i$ lifts $\overline{F}_i$, satisfies conditions (1)-(3), and, in addition, $(\widetilde{F}_i)_{i\in I}$ is induced from $|X|$.

Since $A = \overline{A} + \eta\mathbb{C}$, this last condition means the following:

(4) $\widetilde{F}_i = \overline{F}_i + \eta \widetilde{g}_i$ with $\widetilde{g}_i \in \mathcal{O}_\Sigma(U_i)$ such that $\frac{\widetilde{g}_i}{f_i} = \frac{\widetilde{g}_j}{f_j}$ for all i, j, by the explicit description of $|X| = \mathbb{P}^m$ in Remark 2.1.15.2(2)

$$\left\{ \left(F_{i,\lambda} = \sum_{l=0}^{m} \lambda_l f_{il} \in \Gamma(U_i, \mathcal{O}_{\Sigma\times|X|}) \right)_{i\in I} \;\middle|\; \frac{f_{il}}{f_{i0}} = \frac{f_{jl}}{f_{j0}} \right\} .$$

To show (4), let $X \in |X|$ correspond to $(1 : 0 : \dots : 0) \in \mathbb{P}^m$ and $f_{i0} = f_i$. Then $\overline{\varphi}$ is determined by $\overline{\varphi}(\lambda_l) =: \overline{a}_l \in \overline{A}$ and φ by $\varphi(\lambda_l) = \overline{a}_l + \eta\alpha_l$, $\alpha_l \in \mathbb{C}$. Hence $\overline{F}_i = \sum_{l=0}^m \overline{a} f_{il}$ and $\widetilde{F}_i = \overline{F}_i + \eta(\sum_l \alpha_l f_{il}) = \overline{F}_i + \eta\widetilde{g}_i$. We have to choose $\widetilde{g}_i$'s in such a way that (3) is satisfied, i.e. $\widetilde{F}_i$ defines an element in $\underline{\mathcal{D}ef}'_{X,z}(A)$ for each $z \in U_i$. Note that (2) holds by construction and (1) holds then for $\widetilde{F}_i$ with $\widetilde{G}_{ij} = f_i/f_j$ by Remark 2.1.15.2(2).

We start with F_i as above and write $F_i = \overline{F}_i + \eta g_i$, $G_{ij} = \overline{G}_{ij} + \eta g_{ij}$. We have to find $h_i \in \Gamma(U_i, \mathcal{O}_\Sigma)$ such that $\widetilde{g}_i := g_i + h_i$ satisfies $\widetilde{g}_i/f_i = \widetilde{g}_j/f_j$ and $\widetilde{F}_i$ defines an element in $\underline{\mathcal{D}ef}'_{X,z}$, $z \in U_i$, by Lemma 2.2.18.

Since $F_i = G_{ij} F_j$, $\overline{F}_i = \overline{G}_{ij} \overline{F}_j$ and $\eta \cdot (\mathcal{O}_\Sigma \otimes \mathfrak{m}_A) = 0$ implies

$$g_i = \frac{f_i}{f_j} g_j + f_j g_{ij} ,$$

it follows that $g_{ij} = \frac{g_i}{f_i} - \frac{g_j}{f_j} \in \Gamma(U_{ij}, \mathcal{O}_\Sigma)$ is a cocycle in $\mathcal{O}_\Sigma \subset \mathcal{J}_{Z'/\Sigma}(X)$ and defines a cohomology class in $H^1(\Sigma, \mathcal{J}_{Z'/\Sigma}(X))$, which vanishes by assumption. It follows that

$$g_{ij} = \frac{h_j}{f_j} - \frac{h_i}{f_i}$$

for some $h_i \in \Gamma(U_i, \mathcal{J}_{Z'/\Sigma})$. Hence $\widetilde{g}_i := g_i + h_i$ satisfies $\widetilde{g}_i/f_i = \widetilde{g}_j/f_j$ and we are done.

(d) We prove the supplement (statement (d) is the special case $p = r$). The smoothness was proved in Corollary 2.2.27. For the statement about the fibre dimension consider the lower exact sequence in the cohomology diagram of Corollary 2.2.27. We have

$$h^0(\mathcal{O}_{Z^{ea}(X)}) = \sum_{i=1}^{p} \tau(X, z_i) = \dim \prod_{i=1}^{p} B_{(X,z_i)}\,,$$

since $B_{(X,z_i)}$ is smooth. Moreover, for a smooth morphism the fibre dimension equals the dimension of the kernel of its tangent map which is $h^0(\mathcal{J}_{Z'(X)/X}(X))$ in this case.

To see (e), we apply (c1) to $Z_{fix}(X)$ and notice that this implies that Φ_h has a smooth fibre through X of the claimed dimension. Moreover, $\mathcal{J}_{Z_{fix}(X)/X}(X)$ is a subsheaf of $\mathcal{J}_{Z'(X)/X}(X)$ with quotient sheaf concentrated on $\{z_1, \dots, z_r\}$ of finite codimension $2r$ (cf. Example 2.2.20.1(b) and (d)). In particular, the vanishing of $H^1\big(\mathcal{J}_{Z_{fix}(X)/X}(X)\big)$ implies that $H^1\big(\mathcal{J}_{Z'(X)/X}(X)\big) = 0$ and, by (c1), $V_h(S_1, \dots, S_r)$ as well as the fibre $V_{h,fix}(S_1, \dots, S_r)$ of ϕ_h are smooth at X of expected dimension, the fibre having codimension $2r = \dim(\mathrm{Sym}^r\, \Sigma, z_1 + \dots + z_r)$. It follows that Φ_h is flat with smooth fibre, hence smooth. □

Remark 2.2.39.1 Let Σ be a regular surface, that is, $h^1\big(\mathcal{O}_\Sigma\big) = 0$ and $C \subset \Sigma$ a reduced curve. Then the long exact cohomology sequence to

$$0 \longrightarrow \underbrace{\mathcal{J}_{C/\Sigma}(C)}_{\cong\, \mathcal{O}_\Sigma} \longrightarrow \mathcal{J}_{Z'(C)/\Sigma}(C) \longrightarrow \underbrace{\mathcal{J}_{Z'(C)/\Sigma}(C) \otimes \mathcal{O}_C}_{=\, \mathcal{J}_{Z'(C)/C}(C)} \longrightarrow 0\,,$$

implies $H^0\big(\mathcal{J}_{Z'(C)/C}(C)\big) \cong H^0\big(\mathcal{J}_{Z'(C)/\Sigma}(C)\big) / H^0(\mathcal{O}_\Sigma)$, that is, the Zariski tangent spaces to V_h and $V_{|C|}$ are isomorphic, thus, the smoothness and expected dimension of $V_{|C|}$ is equivalent to the smoothness and expected dimension of V_h at C.

If, additionally, $h^2\big(\mathcal{O}_\Sigma\big) = 0$, as, for instance, for a rational surface Σ, then also $H^1\big(\mathcal{J}_{Z'(C)/C}(C)\big)$ and $H^1\big(\mathcal{J}_{Z'(C)/\Sigma}(C)\big)$ are isomorphic.

Let's reformulate and strengthen Theorem 2.2.38 in the *case of plane curves* ($\Sigma = \mathbb{P}^2$), which is of special interest.

Of course, since $h^1\big(\mathcal{O}_{\mathbb{P}^2}\big) = h^2\big(\mathcal{O}_{\mathbb{P}^2}\big) = 0$, there is no difference whether we consider the curves in the linear system $|H^0(\mathcal{O}_{\mathbb{P}^2}(d))| = |dH|$, H a line, or curves with fixed Hilbert polynomial $h(z) = dz - (d^2 - 3d)/2$. We denote the corresponding varieties in $|H^0(\mathcal{O}_{\mathbb{P}^2}(d))|$ by $V_d = V_d(S_1, \dots, S_r)$, respectively $V_{d,\mathrm{fix}}(S_1, \dots, S_r)$, where some of the S_i may be analytic and the others topological types.

We use an analogous notation as in Theorem 2.2.38, with $Z'(C)$ the union of the the zero-dimensional schemes $Z'(C, z_i)$ and $Z'(C, z_i)$ being any of the (fixed) topological or analytic singularity schemes. For $\deg(Z'(C)$ see Remark 2.2.37.2. We obtain:

Theorem 2.2.40 *Let $C \subset \mathbb{P}^2$ be a reduced curve of degree d with specified r singularities $z_1, \ldots, z_r$ of analytic, or topological, types $S_1, \ldots, S_r$ and set $V_d = V_d(S_1, \ldots, S_r)$.*

(a) $H^0\big(\mathcal{J}_{Z'(C)/\mathbb{P}^2}(d)\big)\big/H^0(\mathcal{O}_{\mathbb{P}^2})$ *is isomorphic to the Zariski tangent space of V_d at C.*
(b) $h^0\big(\mathcal{J}_{Z'(C)/\mathbb{P}^2}(d)\big) - h^1\big(\mathcal{J}_{Z'(C)/\mathbb{P}^2}(d)\big) - 1 \le \dim(V_d, C)$ $\le h^0\big(\mathcal{J}_{Z'(C)/\mathbb{P}^2}(d)\big) - 1$.
(c) $H^1\big(\mathcal{J}_{Z'(C)/\mathbb{P}^2}(d)\big) = 0$ *if and only if V_d is T-smooth at C, i.e. smooth and of the expected dimension $d(d+3)/2 - \deg Z'(C)$.*
(d) $H^1\big(\mathcal{J}_{Z^{ea}(C)/\mathbb{P}^2}(d)\big) = 0$ *if and only if the natural morphism of germs* $\big(|H^0(\mathcal{O}_{\mathbb{P}^2}(d))|, C\big) \to \prod_{i=1}^r B_{(C,z_i)}$ *is smooth (hence surjective) of fibre dimension* $h^0\big(\mathcal{J}_{Z^{ea}(C)/\mathbb{P}^2}(d)\big) - 1$.
(e) Write $Z_{fix}(C)$ for $Z^{ea}_{fix}(C)$, respectively $Z^{es}_{fix}(C)$. Then $H^1\big(\mathcal{J}_{Z_{fix}(C)/\mathbb{P}^2}(d)\big) = 0$ implies that the morphism of germs

$$\Phi_d : \big(V_d(S_1, \ldots, S_r), C\big) \to \big(\mathrm{Sym}^r\mathbb{P}^2, (z_1 + \ldots + z_r)\big)$$

is smooth of fibre dimension $h^0\big(\mathcal{J}_{Z_{fix}(C)/\mathbb{P}^2}(d)\big) - 1$. In particular, the vanishing of $H^1\big(\mathcal{J}_{Z_{fix}(C)/\mathbb{P}^2}(d)\big)$ implies that arbitrarily close to C there are curves in $V_d(S_1, \ldots, S_r)$ having their singularities in generic position in $\mathbb{P}^2$.

Proof Most parts of this proposition follow immediately from Theorem 2.2.38 and Remark 2.2.39.1. We need to show only the "if" parts in the statements (c) and (d). Since the expected dimension of V_d is $d(d+3)/2 - \deg Z'(C) = h^0\big(\mathcal{O}_{\mathbb{P}^2}(d)\big) - 1 - h^0\big(\mathcal{O}_{Z'(C)}(d)\big)$, and since the tangent space to V_d has dimension $h^0(\mathcal{J}_{Z'(C)/\mathbb{P}^2}(d)) - 1$ by (a), (c) follows from the exact sequence

$$0 \longrightarrow H^0\big(\mathcal{J}_{Z'(C)/\mathbb{P}^2}(d)\big) \longrightarrow H^0\big(\mathcal{O}_{\mathbb{P}^2}(d)\big) \longrightarrow H^0\big(\mathcal{O}_{Z'(C)}(d)\big)$$
$$\longrightarrow H^1\big(\mathcal{J}_{Z'(C)/\mathbb{P}^2}(d)\big) \longrightarrow 0.$$

(d) follows from the supplement with $p = r$. The last statement of (e) follows since ϕ_d is surjective and since curves in $V_h(S_1, \ldots, S_r)$ with singularities in generic position are an open and dense set. □

Supplement 2.2.41 *Let $1 \le p \le r$ and let $Z'(C) = Z^0(C) \cup Z''(C)$ with $Z^0(C) = \bigcup_{i=1}^p Z^{ea}(C, z_i)$ and $Z''(C) = \bigcup_{i=p+1}^r Z'(C, z_i)$ as well as $V'_d(S_{p+1}, \ldots, S_r) \subset |H^0(\mathcal{O}_{\mathbb{P}^2}(d))|$ be defined as in Supplement 2.2.39.*

Then $H^1(\mathcal{J}_{Z'(C)/\mathbb{P}^2}(d)) = 0$ if and only if $H^1(\mathcal{J}_{Z''(C)/\mathbb{P}^2}(d)) = 0$ and the morphism of germs $(V'_h(S_{p+1}, \ldots, s_r, C) \to \coprod_{i=1}^p B_{(C,z_i)}$ is smooth (and then it is of fibre dimension $h^0(\mathcal{J}_{Z'(C)/\mathbb{P}^2}(d))$).

Proof of the supplement. In view of Supplement 2.2.39 and the proof of Corollary 2.2.27 we have only to show the "if" direction. If $(V'_h(S_{p+1}, \ldots, S_r), C) \to \coprod_{i=1}^p B_{(C,z_i)}$ is smooth, it is surjective on the tangent level. That is, the first morphism in the exact sequence (cf. proof of Corollary 2.2.27)

$$H^0(\mathcal{J}_{Z''(C)/\mathbb{P}^2}(d)) \to H^0(\mathcal{O}_{Z^0(C)} \to H^1(\mathcal{J}_{Z'(C)/\mathbb{P}^2}(d)$$
$$\to H^1(\mathcal{J}_{Z''(C)/\mathbb{P}^2}(d)) \to 0 \qquad (2.2.2.3)$$

is surjective. Hence $H^1(\mathcal{J}_{Z'(C)/\mathbb{P}^2}(d)) = 0$ follows from $H^1(\mathcal{J}_{Z''(C)/\mathbb{P}^2}(d)) = 0$. Note that for $p = r$ we have $Z'' = \emptyset$ and then $H^1(\mathcal{J}_{Z''(C)/\mathbb{P}^2}(d)) = H^1(\mathcal{O}_{\mathbb{P}^2}(d)) = 0$ holds automatically. □

2.2.3 *T-Smooth Families of Isolated Hypersurface Singularities*

The main property of equisingular families of singular hypersurfaces discussed above, the smoothness and expected dimension, is called *T-smoothness* (cf. Definition 2.2.53). In the previous section we proved cohomological conditions for T-smoothness of families with fixed analytic or topological types. We now like to formulate it for more general "smooth singularity types" of isolated hypersurface singularities of any dimension, introduced below.

Moreover, we consider arbitrary families and show that the stratum in the base space over which the singularities of the fibres have fixed smooth singularity types is a locally closed analytic subspace. Moreover, for D a very ample divisor on a smooth projective variety Σ we study the family $V_{|D|/\Sigma}(S_1, \cdots, S_r)$ of projective hypersurfaces in $|D|$ having r isolated singularities of smooth singularity types $S_1, \cdots, S_r$ and prove in Theorem 2.2.55 an analogous H^1-vanishing criterion as in Theorem 2.2.38 for the T-smoothness of these strata.

2.2.3.1 General Smooth Singularity Types

We start with the local situation. In the following $\mathfrak{m}$ denotes the maximal ideal of $\mathbb{C}\{x\} = \mathbb{C}\{x_1, \ldots, x_n\}$ and $\mathfrak{m}_z$ the maximal ideal of $\mathcal{O}_{\Sigma,z}$ where (Σ, z) is a smooth complex germ.

Definition 2.2.42 Let $\mathcal{C} \subset \mathfrak{m}$ be a set of germs of convergent power series, vanishing at 0. An equivalence relation $\sim_s$ on $\mathcal{C}$ is called a *singularity relation*, if any two right equivalent elements of $\mathcal{C}$ are $\sim_s$-equivalent. An equivalence class (of $f \in \mathcal{C}$) is called a *singularity type* (of f) and denoted by S (resp. $S(f)$). It contains the right class (of f). Two elements from S are also called *S-equivalent*. A singularity relation is called *smooth*, and an equivalence class a *smooth singularity type* (*of power series*), if the following holds:

(a) Every element $f \in \mathcal{C}$ is *S-finitely determined*, i.e. there exists a $k \geq 1$ such that for each $g \in \mathfrak{m}$ with $f^{(k)} = g^{(k)}$[13] we have $g \in \mathcal{C}$ and $f \sim_s g$. f is then called *S-k-determined.*
(b) Let $M_S^{(k)} := (M_S^{(k)}, f^{(k)})$ denote the germ at $f^{(k)}$ of the image of the S-equivalence class of $f \in \mathcal{C}$ in $J^{(k)}$. Then, we require that there exists a k_0 such that for all $k \geq k_0$, $M_S^{(k)}$ is a smooth analytic germ in the finite dimensional $\mathbb{C}$-vector space $J^{(k)}$. $M_S^{(k)}$ is called the *S-singularity stratum* of f in $J^{(k)}$. The minimal k_0 such that f is S-k_0-determined and $M_S^{(k)}$ is smooth for $k \geq k_0$ is called the *(smooth) S-determinacy* of f and denoted by $d_S(f)$.

Now let $f \in \mathcal{C}$ and $k \geq d_S(f)$. Then we require:

(c) The tangent space to $M_S^{(k)}$ at $f^{(k)}$ is an ideal

$$I_S^{f,k} \subset \mathbb{C}\{x\}/\mathfrak{m}^{k+1}.$$

(d) The codimension of $M_S^{(k)}$ in $J^{(k)}$ depends only on the type S of f, but not on its representative in S.

If (Σ, z) is a smooth complex space germ of dimension n, then there exists an isomorphism $\psi : \mathcal{O}_{\Sigma,z} \xrightarrow{\cong} \mathbb{C}\{x_1, \ldots, x_n\}$ and we define the *singularity type of* $f \in \mathcal{O}_{\Sigma,z}$ to be that of $\psi(f)$. Since different isomorphisms transform the power series to right-equivalent ones, the singularity type $S(f)$ is well defined. It follows also that the germ $(M_S^{(k)}, f^{(k)}) := \psi^{-1}(M_S^{(k)}, \psi(f)^{(k)})$ in $\mathcal{O}_{\Sigma,z}/\mathfrak{m}_z^{k+1}$ and the tangent space $I_S^{f,k} := \psi^{-1}(I_S^{\psi(f),k}) \subset \mathcal{O}_{\Sigma,z}$ are independent of the choice of ψ.

Instead of functions we may consider their zero sets, i.e. isolated hypersurface singularities (X, z) embedded in (Σ, z). The ideal of (X, z) is a principal ideal in $\mathcal{O}_{\Sigma,z}$ generated by a holomorphic function germ $f \in \mathfrak{m}_z$, and two generators differ by multiplication with a unit. Assume that the singularity relation $\sim_s$ on a set $\mathcal{C} \subset \mathfrak{m}_z$ depends only on the ideal of (X, z) but not on the generator f (also called a *representative*) of the singularity (X, z), i.e. any two contact equivalent elements of $\mathcal{C}$ are $\sim_s$-equivalent. Then $\sim_s$ induces an equivalence relation, also denoted by $\sim_s$, on the set of embedded hypersurface singularities of (Σ, z). It follows easily that the condition of smoothness of $\sim_s$ is independent of the choice of the generator of the ideal of (X, z). In the following we are mainly interested in this case, i.e. the $\sim_s$-equivalence class $S(f)$ contains the contact class of f.

Typical smooth singularity types to be kept in mind are right and contact equivalence, resp. topological equivalence, on the set $\mathcal{C}$ of isolated hypersurface singularities, resp. reduced plane curve singularities (cf. Remark 2.2.44.1).

Remark 2.2.42.1 (i) If $f \in \mathcal{C}$ is S-k-determined then f is also S-ℓ-determined for $\ell > k$ and $f + \mathfrak{m}^\ell \subset \mathcal{C}$ by condition (a). It follows that $I_S^{f,k} = I_S^{f,\ell}/\mathfrak{m}^{k+1}$ for $\ell > k$.

---

[13] $f^{(k)}$ denotes the *k-jet of f*, the image of f in the *jet-space* $J^{(k)} := \mathbb{C}\{x\}/\mathfrak{m}^{k+1}$, i.e. $f^{(k)} = \pi_k(f)$ with $\pi_k : \mathbb{C}\{x\} \to J^{(k)}$ the projection. We usually identify $f^{(k)}$ with the power series expansion of f up to order k.

(ii) Let $\mu < \infty$ be the Milnor number of f. Since $f \overset{r}{\sim} g$ implies $f \sim_s g$, the S-determinacy of f is smaller or equal to the right-determinacy of f, which is itself smaller or equal to $\mu + 1$ (cf. [GLS6, Cor. I.2.34]). We claim that for $k_0 > \mu$, $M_S^{(k_0)}$ is smooth iff $M_S^{(k)}$ is smooth for one (equivalent for all) $k > k_0$. In fact, let $g \in J^{(k)} = J^{(k_0)} \oplus (\mathfrak{m}^{k_0+1}/\mathfrak{m}^{k+1})$ be sufficiently close to f and write $g = g_0 + h$ with $g_0 \in J^{(k_0)}$, $h \in \mathfrak{m}^{k_0+1}/\mathfrak{m}^{k+1}$. Since the Milnor number is semicontinuous, $\mu(g) \leq \mu$ and hence g is right k_0-determined. Then g is right-equivalent to g_0 and hence $g \sim_s g_0$. It follows that $g \sim_s f$ iff $g_0 \sim_s f$. Therefore

$$M_S^{(k)} = M_S^{(k_0)} \times (\mathfrak{m}^{k_0+1}/\mathfrak{m}^{k+1}) ,$$

which implies the claim. In particular, $d_S(f) \leq \mu(f) + 1$.

(iii) If $\mu < \infty$ then the set $M_S^{(k)}$ is analytic for any $k \geq 1$ by Proposition 2.2.48, but not necessarily smooth.

(iv) Let $k \geq k_0 \geq d_S(f)$. The ideal $I^{f,k}$ consists, as a tangent space, of all $g \in \mathfrak{m}J^{(k)}$ such that there exists a holomorphic map germ $\varphi : (\mathbb{C}, 0) \to (M_S^{(k)}, f^{(k)})$, $t \mapsto \varphi(t)$, with $\partial\varphi/\partial t(0) = g$ (equivalently, a morphism $T_\varepsilon \to (M_S^{(k)}, f^{(k)})$ with $\varepsilon \mapsto g$). Since $k_0 \geq d_S(f)$, we have $f \sim_s f + tg$ if $g \in \mathfrak{m}^{k_0+1}$ and therefore

$$I_S^{f,k} = I_S^{f,k_0} \oplus (\mathfrak{m}^{k_0+1}/\mathfrak{m}^{k+1}) .$$

The same results of (i)–(iv) hold for μ replaced by τ if $\sim_s$ is implied by contact equivalence.

We define

$$I_S^f := \pi_k^{-1}(I_S^{f,k}) \subset \mathbb{C}\{x\} ,$$

which is independent of the choice of $k \geq d_S(f)$. I_S^f may be considered as the *tangent space of the smooth singularity type* S at f.

Consider an unfolding $f + \varepsilon g$ over T_ε (cf. Sect. 2.2.3.2). By the proof of Lemma 2.2.45, $g \in \mathfrak{m}j(f)$ iff $f + \varepsilon g$ is trivial. Since unfoldings of hypersurface singularities are unobstructed, $f + \varepsilon g$ is tangent to a "real" trivial unfolding of f over some curve germ $\varphi : (\mathbb{C}, 0) \to (J^{(k)}, f^{(k)})$ for $k \geq d_S(f)$. That is, $\varphi(t) \overset{r}{\sim} f^{(k)}$ for t sufficiently close to 0 and $\partial\varphi/\partial t(0) = g^{(k)}$. Since right equivalence implies S-equivalence, $\varphi(t) \in M_S^{(k)}$ and hence $g^{(k)} \in I_S^{f,k}$ for $k \geq d_S(f)$. We get

$$\mathfrak{m}j(f) \subset I_S^f \subset \mathfrak{m} .$$

Definition 2.2.43 Let $f \in \mathfrak{m}_z$, S a smooth singularity type and $k \geq d_S(f)$.

(i) By assumption

$$\operatorname{codim}(S) := \dim_{\mathbb{C}} \mathbb{C}\{x\}/I_S^f \ = \ \text{codimension of } M_S^{(k)} \text{ in } J^{(k)}$$

is independent of $f \in S$ and we call it the *codimension of the type* S.

(ii) By abuse of notation we denote I_S^f also by I^S or I_z^S and call it the *singularity ideal* of the type S or the *S-singularity ideal.*

Since $\mathfrak{m}^{k+1} \subset I_S^f$ for $k \geq d_S(f)$, I_S^f defines a 0-dimensional scheme in Σ, concentrated on z. This scheme is called the *0-dimensional S-singularity scheme* and denoted by Z_S^f or by Z^S or by Z_z^S. The *degree* of Z^S is by definition the codimension of the type S,

$$\deg Z^S = \dim_{\mathbb{C}} \mathcal{O}_{\Sigma,z}/I^S ,$$

which is independent of $f \in S$ (while the scheme structure of Z^S may depend on $f \in S$).

By Remark 2.2.42.1(iv) and Lemma 2.2.45(i) we have $\deg Z^S \leq \mu(f) + \dim(\Sigma, z)$.

Definition 2.2.44 We say that the relation $\sim_s$ on a set of hypersurface singularities is smooth if it is smooth on $\mathcal{C}$, i.e. on function germs. An equivalence class of a smooth equivalence relation is called a *smooth singularity type* of (embedded) *hypersurface singularities* $(X, z) \subset (\Sigma, z)$.

Remark 2.2.44.1 Let $\mathcal{C} \subset \mathfrak{m} \subset \mathbb{C}\{x\}$ be the set of germs having an isolated singularity. The following equivalence relations are smooth singularity relations on $\mathcal{C}$. Note that in (i) and (ii) the required sections are unique by Proposition 2.2.6.

(i) *Contact equivalence* of function germs having an isolated singularity, which is the same as *analytic equivalence* of the corresponding hypersurface singularities (cf. [GLS6, Def. I.2.9]). Finite determinacy follows from [GLS6, Corollary I.2.24] while smoothness of $M_S^{(k)}$ in Definition 2.2.42(b) holds since $M_S^{(k)}$ is the orbit of $f^{(k)}$ in $J^{(k)}$ under the contact group (cf. [GLS6, section I.2.3]). The corresponding singularity type is called *analytic type* and denoted by *ea*. The tangent space of the contact class in $\mathbb{C}\{x\}$ at f is the fixed Tjurina ideal $\langle f, \mathfrak{m}j(f)\rangle$ and the codimension is the fixed Tjurina number $\tau_{fix}(f)$ (Sect. 1.4.1).
An analogous argument works for $\sim_s$ being *right equivalence.* The tangent space of the right class at f is the fixed Jacobian ideal $\mathfrak{m}j(f)$ and the codimension is the fixed Milnor number $\mu_{fix}(f)$ (Sect. 1.4.1).

(ii) *Embedded topological equivalence* of reduced plane curve singularities. Finite determinacy follows from (i) since analytic equivalence implies topological equivalence. Smoothness of $M_S^{(k)}$ in $J^{(k)}$ can be seen as follows: $M_S^{(k)}$ is known to be the μ-constant stratum in $J^{(k)}$ of $f^{(k)}$ (cf. [Le] and [GLS6, Corollary II.2.68]). If $k \geq \tau_{fix}(f)$, $J^{(k)}$ can be considered as the base space of a versal deformation (with trivial section) of the plane curve singularity $(f^{-1}(0), 0)$ and $M_S^{(k)}$ is then the preimage of the μ-constant stratum Δ^{μ} in the base space $(B, 0)$ of the semiuniversal deformation of $(f^{-1}(0), z)$ under some holomorphic map ψ : $(J^{(k)}, f^{(k)}) \to (B, 0)$. Δ^{μ} is smooth by [Wah1] (see also [GLS6, Thm. II.2.61]), and ψ is a smooth map by [GLS6, Proposition I.1.14], i.e. $M^{(k)}$ is isomorphic to the Cartesian product of Δ^{μ} and a smooth factor, hence smooth. The tangent space of the (embedded) topological type S of f is the *equisingularity ideal*

(Definition 1.1.63 and 2.2.7) $I^{es}(f^{-1}(0), 0)$ and the codimension is denoted by $\tau^{es}(S)$ (Definition 1.1.63) which equals the codimension of Δ^{μ} in $(B, 0)$. The corresponding singularity type is called *topological type* and denoted by *es*.

(iii) For arbitrary isolated singularities, the semiquasi-homogeneous and Newton nondegenerate singularity types, discussed below in Sect. 2.3.4, are smooth in the sense of Definition 2.2.42.

We will be interested in deformations of hypersurfaces $X \subset \Sigma$ that fix smooth singularity types of a finite set of singular points. Such deformations resemble strict ideal versal deformations of X (Definition 2.2.19) considered in Sects. 2.2.1.4 and 2.2.1.5. The difference is that the notion of smooth singularity type refers to an individual singularity but not to a family. Thus we can define families with fixed smooth singularity types of the singularities in general only over a reduced base space. On the other hand, the notion of strict ideal versatility refers to a family over an arbitrary complex germ as base space, e.g. to infinitesimal families, while an infinitesimal deformation theory for a smooth singularity type might not exist. For instance, for semiquasihomogeneous or Newton nondegenerate singularity types we do not consider (and do not try to develop) any infinitesimal version. See Remarks 2.2.50.1 and 2.2.55.1 for a further discussion.

2.2.3.2 General Equisingularity Strata

We consider now arbitrary (flat) families of germs, called *unfoldings* when we consider families of functions and right-equivalence, resp. *deformations with section* when we consider families of hypersurfaces and contact-equivalence. We are interested in the strata in the base space over which the fibres have fixed smooth singularity type (called *S-singularity strata*, cf. Definition 2.2.49 and Proposition 2.2.48). We recall first the basic notions.

Let (Σ, z) be a smooth complex space germ, $f \in \mathfrak{m}_z$ with $\mathfrak{m}_z$ the maximal ideal of $\mathcal{O}_{\Sigma,z}$, and $(T, 0)$ any complex germ. We denote by (X, z) the hypersurface germ $(f^{-1}(0), z) \subset (\Sigma, z)$.

An *unfolding of f over* $(T, 0)$ is given by an $F \in \mathfrak{m}_z\mathcal{O}_{\Sigma\times T,(z,0)}$ such that F mod $\mathfrak{m}_{T,0}\mathcal{O}_{\Sigma\times T,(z,0)}$ coincides with f. A morphism $\varphi : (T', 0) \to (T, 0)$ induces a morphism $\mathcal{O}_{\Sigma\times T,(z,0)} \to \mathcal{O}_{\Sigma\times T',(z,0)}$ and the image of F under this morphism is an element in $\mathfrak{m}_z\mathcal{O}_{\Sigma\times T',(z,0)}$, given by $F(x, \varphi(t))$ and denoted by φ^*F, the pullback of F via φ. Two unfoldings F and F' of f over $(T, 0)$ are isomorphic, if there exists an $\mathcal{O}_{T,0}$-algebra automorphism ψ of $\mathcal{O}_{\Sigma\times T,(z,0)}$, mapping $\mathfrak{m}_z\mathcal{O}_{\Sigma\times T,(z,0)}$ to itself and being the identity modulo $\mathfrak{m}_{T,0}\mathcal{O}_{\Sigma\times T,(z,0)}$ such that $\psi^*F = F'$. Denote by $\underline{\mathcal{U}nf}_f$ the functor from complex germs to sets

$$\underline{\mathcal{U}nf}_f : (T, 0) \to \left\{ \begin{array}{l} \text{isomorphism classes of} \\ \text{unfoldings of } f \text{ over } (T, 0) \end{array} \right\}.$$

Consider now *deformations with sections of* (X, z). Recall from [GLS6, Sect. II.2.1, Prop. II.2.2] that any deformation of (X, z) over $(T, 0)$ with section σ is isomorphic to an embedded *deformation with trivial section*. Hence it is given by a diagram

$$\begin{array}{ccccc} (X,z) & \overset{i}{\hookrightarrow} & (\mathscr{X},z) & \overset{k}{\hookrightarrow} & (\Sigma\times T,(z,0)) \\ \downarrow & & \phi\downarrow\uparrow\sigma & \swarrow \mathrm{pr} & \\ \{0\} & \hookrightarrow & (T,0) & & \end{array}$$

with $k \circ \sigma(t) = (z, t)$ and ϕ flat. Another such deformation is given by the same kind of diagram with different morphisms i', k', ϕ', σ' and an isomorphism between them is given by a morphism of diagrams being the identity on $(\Sigma \times T, (z, 0))$.

$$\underline{\mathcal{D}ef}^{\,\mathrm{sec}} : (T,0) \to \left\{ \begin{array}{c} \text{isomorphism classes of deformations} \\ \text{with section of } (X,z) \text{ over } (T,0) \end{array} \right\}$$

denotes the corresponding deformation functor.

The ideal of $k(\mathscr{X}, z)$ in $\mathcal{O}_{\Sigma\times T,(z,0)}$ is a principal ideal, generated by some $F \in \mathfrak{m}_z\mathcal{O}_{\Sigma\times T,(z,0)}$ such that the residue class of F modulo $\mathfrak{m}_{T,0}\mathcal{O}_{\Sigma\times T,(z,0)}$ is a generator uf, u a unit, of the ideal of (X, z). Hence F is an unfolding of uf. Conversely, any unfolding of f defines, by $(X, z) = (f^{-1}(0), z) \subset (\mathscr{X}, z) = (F^{-1}(0), (z, 0)) \subset (\mathscr{X} \times T, (z, 0))$, a diagram as above and hence an object of $\mathcal{D}ef^{\,\mathrm{sec}}_{X,z}(T, 0)$ with trivial section.

We identify now the tangent spaces of the functors $\underline{\mathcal{U}nf}_f$ and $\underline{\mathcal{D}ef}^{\,\mathrm{sec}}_{X,z}$.

Lemma 2.2.45 *Let $f \in \mathcal{O}_{\Sigma,z}$ have an isolated singularity and $(X, z) = (f^{-1}(0), z)$.*
(i) The tangent space of $\underline{\mathcal{U}nf}_f$ satisfies

$$\underline{\mathcal{U}nf}_f(T_\varepsilon) \cong \mathfrak{m}_z/(\mathfrak{m}_z j(f)) ,$$

which is a vector space of dimension $\mu_{fix}(f) - 1 = \mu(f) + \dim(\Sigma, z) - 1$.
(ii) The tangent space of $\underline{\mathcal{D}ef}^{\,sec}_{X,z}$ satisfies

$$\underline{\mathcal{D}ef}^{\,sec}_{X,z}(T_\varepsilon) \cong T^{1,fix}_{X,z} \cong \mathfrak{m}_z/\langle f, \mathfrak{m}_z j(f)\rangle ,$$

which is a vector space of dimension $\tau_{fix}(f) - 1 = \tau(f) + \dim(\Sigma, z) - 1$.

Proof (i) With local coordinates $x_1, \dots, x_n$ of (Σ, z) we have $j(f) = \langle \partial f/\partial x_1, \dots, \partial f/\partial x_n\rangle$. Since $\mathcal{O}_{T_\varepsilon} = \mathbb{C}[\varepsilon]/\langle\varepsilon^2\rangle$, any unfolding F of f over T_ε is of the form

$$F = f + \varepsilon g, \quad g \in \mathfrak{m} = \langle x_1, \dots, x_n\rangle .$$

An automorphism ψ of unfoldings of f over T_ε is given by

$$\psi(\varepsilon) = \varepsilon, \quad \psi(x_i) = x_i + \varepsilon h_i,\ h_i \in \mathfrak{m},\ i = 1, \dots, n,$$

and F is isomorphic to the trivial unfolding if $\psi(f) = f + \varepsilon g$. By Taylor's formula we get, with $h = (h_1, \dots, h_n)$,

$$\psi(f)(x) = f(x + \varepsilon h) = f(x) + \varepsilon \cdot \sum_{i=1}^{n} \frac{\partial f}{\partial x_i}(x) h_i(x) .$$

Hence $\psi(f) = f + \varepsilon g$ iff $g \in \mathfrak{m} j(f)$, which proves the first part of (i). The dimension statement is easy.

The proof of (ii) is similar, for the dimension statement see Lemma 2.2.13. □

Proposition 2.2.46 *Let $f \in \mathfrak{m} \subset \mathbb{C}\{x_1, \dots, x_n\}$ with $\mu(f) < \infty$ and $(X, 0) = (f^{-1}(0), 0)$.*

(i) If $g_i \in \mathfrak{m}$, $i = 1, \dots, m$, is a basis of $\mathfrak{m}/(\mathfrak{m} j(f))$, then

$$F(x, t) := f(x) + \sum_{i=1}^{m} t_i g_i(x) \in \mathbb{C}\{x, t\}$$

is a semiuniversal unfolding of f over $(\mathbb{C}^m, 0)$. We have $m = \mu(f) + n - 1$.

(ii) If $g_i \in \mathfrak{m}$, $i = 1, \dots, m$, is a basis of $\mathfrak{m}/\langle f, \mathfrak{m} j(f)\rangle$, then

$$F(x, t) := f(x) + \sum_{i=1}^{m} t_i g_i(x) \in \mathbb{C}\{x, t\}$$

defines a semiuniversal deformation with section of $(X, 0)$ over $(\mathbb{C}^m, 0)$. We have $m = \tau(f) + n - 1$ by Lemma 2.2.13.

In case (i) (resp. (ii)) the g_i's may be chosen as monomials and then the degree of g_i is necessarily $\le \mu(f)$ (resp. $\le \tau(f)$).

If the g_i in (i) (resp. (ii)) represent only a generating system, then the corresponding F defines a versal deformation with section of f (resp. $(X, 0)$).

Proof The proof of the semiuniversality is in both cases similar to the proof of [GLS6, Theorem II.1.16 and Corollary II.1.17], where deformations without section were considered.

To show that $\deg g_i \le \mu$ (resp. $\le \tau$) we may assume that $g_i \in \mathfrak{m}^2$, since otherwise the statement is simple. Set, in case (i), $\overline{\mathfrak{m}} = \mathfrak{m}/(\mathfrak{m} j(f))$. Then $\dim_{\mathbb{C}} \overline{\mathfrak{m}}/\overline{\mathfrak{m}}^2 = m$ and $\overline{\mathfrak{m}}^k \neq 0$ implies $0 < \dim_{\mathbb{C}} \overline{\mathfrak{m}}^k \le \dim_{\mathbb{C}} \overline{\mathfrak{m}} - m - (k - 2) = \mu - k + 1$ by Lemma 2.2.45(i). Hence $k \le \mu$ if $\mathfrak{m}^k \not\subset \mathfrak{m} j(f)$, which is fulfilled if $\deg g_i = k$. This proves the claim. The proof of (ii) is similar, using Remark 2.2.4.1 and Lemma 2.2.45(ii). □

An important special unfolding of f is given by a sufficiently high jet space. Let

$$f = \sum_{|\alpha| \ge 0} c_\alpha x^\alpha \in \mathbb{C}\{x\}, \quad c_0 = 0,$$

and $\mu = \mu(f) < \infty$. Let $s = (s_\alpha)_{|\alpha|\le k}$ be coordinates of $J^{(k)}$ such that $f^{(k)}$ corresponds to the point $c = (c_\alpha)_{|\alpha|\le k}$ in $J^{(k)}$. The polynomial

$$G(x,s) := \sum_{1\le|\alpha|\le k} s_\alpha x^\alpha = f^{(k)}(x) + \sum_{1\le|\alpha|\le k} (s_\alpha - c_\alpha)x^\alpha \in \mathbb{C}\{x,s\}$$

is an unfolding of $f^{(k)}$ over the germ $(J^{(k)}, c)$. If $k > \mu$ (resp. $k > \tau = \tau(f)$) then $f^{(k)} \overset{r}{\sim} f$ (resp. $f^{(k)} \overset{c}{\sim} f$) and we may assume in the following that $f = f^{(k)}$. We call G the *k-jet space* or just *jet space unfolding of f*.

Lemma 2.2.47 *For any $k > \mu(f)$ (resp. $k > \tau(f)$) the k-jet space unfolding $G(x,s)$ is a versal unfolding of f (resp. a versal deformation with section of $(X,0) = (f^{-1}(0),0)$).*

Proof Let $F(x,t) = f(x) + \sum t_i x^{\alpha(i)}$, $x^{\alpha(i)}$ a monomial basis of $\mathfrak{m}/(\mathfrak{m}j(f))$ (resp. of $\mathfrak{m}/\langle f, \mathfrak{m}j(f)\rangle$), be the semiuniversal unfolding (resp. deformation with section) of $f = f^{(k)}$. F is obtained from G as pullback via the base change

$$\varphi : (\mathbb{C}^m, 0) \to (J^{(k)}, c), \quad (t_1,\dots,t_m) \mapsto c + \sum_{i=1}^m t_i s_{\alpha(i)},$$

with $s_{\alpha(i)}$ the coordinate of $x^{\alpha(i)}$ in $J^{(k)}$. In fact,

$$G(x,\varphi(t)) = G(x, c + \sum t_i s_{\alpha(i)}) = \sum c_\alpha x^\alpha + \sum t_i x^{\alpha(i)} = F(x,t)\,.$$

This shows that any unfolding (resp. deformation with section) can be induced via base change from $G(x,s)$, the main part of the versatility of G. We omit the proof of extending the inducing map over Artinian base spaces (cf. [GLS6, Definition II.1.8]). □

Remark 2.2.47.1 Let f be an isolated singularity and let $d > \mu(f)$ (resp. $d > \tau(f)$). Proposition 2.2.46 and Lemma 2.2.47 show that any unfolding (resp. deformation with section) of f can be induced from a (semiuni-)versal family given by polynomials of degree $\le d$.

For a given smooth singularity type S and for any unfolding of $f \in \mathcal{O}_{\Sigma,z}$ (resp. a deformation with section of (X,z)) over $(T,0)$ we can now define the S-singularity stratum. We assume f to have an isolated singularity.

Proposition 2.2.48 *Let S be a smooth singularity type.*
(i) Let $F \in \mathfrak{m}_z\mathcal{O}_{\Sigma\times T,(z,0)}$ be an unfolding of f with $\mu(f) < \infty$. For $t \in T$, T a representative of $(T,0)$, set $F_t(x) := F(x,t) \in \mathfrak{m}_z$. Then, for T sufficiently small, the set

$$V_F(S) := \{t \in T \mid S(F_t) = S\}$$

is a closed analytic subset of T. It is smooth if F is versal.

(ii) Let $\Phi : (\mathscr{X}, z) \to (T, 0)$ be a deformation of (X, z) with section $\sigma : (T, 0) \to (\mathscr{X}, z)$. Then $(\mathscr{X}_t, \sigma(t)) = (\Phi^{-1}(t), \sigma(t))$ has an isolated singularity and the set

$$V_{(\mathscr{X},z)/(T,0)}(S) := \{t \in T \mid S(\mathscr{X}_t, \sigma(t)) = S\}$$

is an analytic subset of T, for T sufficiently small. It is smooth if Φ is versal.

Definition 2.2.49 The analytic sets $V_F(S)$ resp. $V_{(\mathscr{X},x)/(T,0)}(S)$ (with their reduced structure) are called the *S-singularity stratum* of F resp. of $(\mathscr{X}, x)/(T, 0)$ in T. If S is an analytic type (resp. a topological type), we call these sets also the *equianalytic strata* (resp. *equisingular strata*).

Proof (of the Proposition) (i) By Lemma 2.2.47, F can be induced from G via some morphism $\varphi : (T, 0) \to (J^{(k)}, c), k > \mu(f)$. Let $M_S^{(k)}$ be the S-singularity stratum of $f^{(k)}$ in $J^{(k)}$ as defined in Definition 2.2.42. φ is in general not unique, but $\varphi^{-1}(M_S^{(k)})$ coincides with $V_F(S)$ and hence is unique as a set. This shows that $V_F(S)$ is analytic in T. Moreover, since G is versal and since two versal unfoldings differ by a smooth factor, $V_F(S)$ is smooth if F is versal.

The proof of (ii) is similar. □

We call F (resp. Φ) an *S-equisingular unfolding* (resp. *S-equisingular deformation*) if $S(F_t) = S$ (resp. $S(\mathscr{X}_t, \sigma(t)) = S$) for all $t \in T$ sufficiently close to 0. In both cases we speak of *S-equisingular families*.

Although we did not make any assumption about the base space T, the condition of being S-equisingular refers only to the points of the set T, that is, to the reduction T_{red}. An unfolding of f over $(T, 0)$ is S-equisingular iff it is S-equisingular over $(T_{red}, 0)$. The same holds for a deformation $\Phi : (\mathscr{X}, z) \to (T, 0)$ with section after base change $(T_{red}, 0) \hookrightarrow (T, 0)$. Therefore, we may always assume that $(T, 0)$ is reduced.

Corollary 2.2.50 *Let F (resp. Φ) as in Proposition 2.2.48 be (semiuni-) versal and denote by F^S the restriction of F to $V_F(S)$ (resp. by Φ^S the restriction of Φ to $V_{(\mathscr{X},z)/(T,0)}(S)$). Then F^S (resp. Φ^S) is a (semiuni-) versal S-equisingular family on the category of reduced complex germs.*

Proof The statement means that any S-equisingular family over a reduced germ (W, w) can be induced from F^S (resp. Φ^S) by some morphism φ from (W, w) to $V_F(S)$ (resp. to $V_{(\mathscr{X},z)/(T,0)}(S)$). The tangent map of φ is unique if the family is semiuniversal. This follows immediately from the properties of (semiuni-) versal unfoldings (resp. deformations). □

Remark 2.2.50.1 (1) We do not require that the notion of smooth singularity type extends to S-equisingular families over non-reduced base spaces, e.g. over fat points, as we did for the local singularity conditions in Sect. 2.2.1.2. Nevertheless we will derive in the next section cohomological smoothness criteria similar to those in Sect. 2.2.2.2.

(2) Of course, there are cases where S-equisingular unfoldings (resp. deformations) can be defined over arbitrary complex germs as base spaces. For example, if the singularity relation is right equivalence or contact equivalence of hypersurface singularities, or embedded topological type of plane curve singularities.

(3) More general, consider any unfolding subcategory $\mathcal{U}nf_f^S$ of $\mathcal{U}nf_f$ (resp. deformations subcategory $\mathcal{D}ef_{X,x}^S$ of $\mathcal{D}ef_{X,x}^{sec}$) satisfying condition (VSIR) from Sect. 2.2.1.2 (cf. Remark 2.2.11.1) and coinciding with S-equisingular unfoldings (resp. deformations) over reduced base spaces. Then the S-singularity strata from Proposition 2.2.48 carry a unique (not necessarily reduced) scheme structure, by property (R) from Definition 2.2.8. A (semiuni-) versal S-equisingular family on the category of arbitrary complex germs is then given as in Corollary 2.2.50.

2.2.3.3 T-Smoothness of General ESF

We consider now a global situation where Σ is a complex manifold and

$$\phi : \mathscr{X} \stackrel{j}{\hookrightarrow} \Sigma \times T \stackrel{\mathrm{pr}}{\longrightarrow} T$$

is a (flat) family of hypersurfaces $\mathscr{X}_t = \phi^{-1}(t) \subset \Sigma$ over a *reduced* complex space T together with finitely many disjoint sections $\sigma_1, \ldots, \sigma_r : T \to \Sigma$. We assume that $\mathscr{X}_t$ has an isolated singularity at $\sigma_i(t)$, $i = 1, \ldots, r$, but may be other singularities elsewhere. This means that we have a closed subspace $\mathrm{Sing} = \bigcup_{i=1}^r \sigma_i(T)$ contained in $\mathrm{Sing}(\phi)$, the singular locus of ϕ, with Sing finite over T and $\mathscr{X}_t \cap \mathrm{Sing} = \{\sigma_1(t), \ldots, \sigma_r(t)\}$.

Definition 2.2.51 We fix a finite sequence of smooth singularity types $\mathcal{S} = (S_1, \ldots, S_r)$ of isolated hypersurface singularities in the sense of Definition 2.2.44 and denote by

$$V_{\mathscr{X}/T}(\mathcal{S}) := V_{\mathscr{X}/T}(S_1, \ldots, S_r) := \{t \in T \mid (\mathscr{X}_t, \sigma_i(t)) \text{ is of type } S_i,\ i = 1, \ldots, r\},$$

the $\mathcal{S}$-*singularity stratum* of $\mathscr{X}/T$ in T. Here, some of the sections may be trivial, that is, the points $\sigma_i(t)$ do not change their position in Σ.

Since the germs of $\sigma_i(T)$ and $\mathrm{Sing}(\phi)$ at $\sigma_i(t)$ coincide with their reduced structure, the sections σ_i are unique. Hence, in the above Definition, we may only assume that there exist sections σ_i (picking the singular points $\sigma_i(t) \in \mathscr{X}_t$) but we need not specify them.

Proposition 2.2.52 *The set $V_{\mathscr{X}/T}(S_1, \ldots, S_r)$ is, with its reduced structure, a locally closed analytic subspace of T.*

Proof By Proposition 2.2.48(i), the germ $V_{(\mathscr{X},\sigma_i(t))/(T,t)}(S_i) \subset (T, t)$ is a unique reduced analytic subgerm of (T, t). Hence, small representatives of this germ glue together to a unique analytic subset $V_{\mathscr{X}/T}(S_i) \subset T$. $V_{\mathscr{X}/T}(S_1, \dots, S_r)$ is the intersection of these analytic sets. □

For the rest of this section let Σ be a smooth *projective variety* of dimension n and D a very ample divisor. We consider the family $V_{|D|/\Sigma}(S_1, \dots, S_r) \subset |D|$ of projective hypersurfaces X in $|D|$ having isolated singularities $z_1, \dots, z_r \in \Sigma$ of smooth singularity types $S_1, \dots, S_r$ (and may be other singularities, not necessarily isolated).

To make this precise consider the incidence variety, which is flat over $|D|$ by Remark 2.1.15.2,

$$\mathfrak{N}_{|D|} = \{(z, [X]) \in \Sigma \times |D| \bigm| z \in X\} \hookrightarrow \Sigma \times |D| \to |D| \,,$$

where $[X] \in |D|$ describes (temporarily) the point belonging to the hypersurface $X \subset \Sigma$, linearly equivalent to D. We assume that there are r disjoint analytic sections $\sigma_1, \dots, \sigma_r$ of $\mathfrak{N}_{|D|} \to |D|$ picking $z_i \in X$, i.e. $\sigma_i([X]) = z_i$. Later, when the context is clear, we shall also write X to describe the point in $|D|$.

If $S_1, \dots, S_r$ are smooth singularity types, $\mathcal{S} = (S_1, \dots, S_r)$, then the $\mathcal{S}$-singularity stratum $V_{\mathfrak{N}_{|D|}/|D|}(\mathcal{S}) \subset |D|$ of $\mathfrak{N}_{|D|} \to |D|$ is by Proposition 2.2.52 a locally closed (in the Euclidean topology) analytic subset of $|D|$. If the S_i are analytic or topological types (or a mixture of them) then $V_{|D|/\Sigma}(S_1, \dots, S_r) \subset |D|$ is quasi-projective by Theorem 2.2.32.

Definition 2.2.53 (i) Denote by

$$V_{|D|}(\mathcal{S}) := V_{|D|/\Sigma}(S_1, \dots, S_r) \subset |D|$$

the $\mathcal{S}$-*singularity stratum*

$$V_{\mathfrak{N}_{|D|}/|D|}(\mathcal{S}) = \{[X] \in |D| \bigm| (X, \sigma_i([X]) \text{ is of type } S_i,\ i = 1, \dots, r\},$$

(also called *equianalytic stratum* resp. *equisingular stratum* if the S_i are analytic resp. topological types). We call it the *equisingular family (ESF) of divisors* $X \in |D|$ with r isolated singularities of types $S_1, \dots, S_r$, respectively.
If $\Sigma = \mathbb{P}^n$, $D = dH$ with H a hyperplane in $\mathbb{P}^n$, then we set

$$V_d^n(S_1, \dots, S_r) := V_{|dH|/\mathbb{P}^n}(S_1, \dots, S_r),$$

and call it the ESF of hypersurfaces X of degree d in $\mathbb{P}^n$ with r isolated singularities $z_i \in X$ of smooth singularity type S_i, $i = 1, \dots, r$.

(ii) We say that $V_{|D|/\Sigma}(S_1, \dots, S_r)$ is T-*smooth* or *smooth of expected dimension* at $[X]$ if the following holds:

(a) choose a positive integer $m \geq d_{S_i}(f)$, $i = 1, \dots, r$, where the representative $f \in \mathcal{O}_{\Sigma,z_i}$ defines the singularity (X, z_i) such that the projection $\pi_i : H^0(\Sigma, \mathcal{O}_\Sigma(D)) \to \mathcal{O}_{\Sigma,z_i}/\mathfrak{m}^{m+1}$ is an embedding; then we require that the smooth germ $M_i := (M_{S_i}^{(m)}, f^{(m)})$ of the S_i-equisingular stratum in $\mathcal{O}_{\Sigma,z_i}/\mathfrak{m}^{m+1}$ intersects the vector space $\pi_i(H^0(\Sigma, \mathcal{O}_\Sigma(D)))$ transversally;

(b) the germ of $V_{|D|/\Sigma}(S_1, \ldots, S_r)$ at $[X]$ is the transverse intersection of the smooth germs $V_{|D|/\Sigma}(S_i)$ at $[X]$, $i = 1, \ldots, r$, in $|D|$.

The *expected dimension* of $V_{|D|/\Sigma}(S_1, \ldots, S_r)$ is

$$\dim V_{|D|/\Sigma} - \sum_{i=1}^{r} \operatorname{codim}(S_i),$$

where $\operatorname{codim}(S_i)$ is the codimension of the type S_i (cf. Definition 2.2.43).

Let us explain the conditions (a) and (b):

(a) The map π_i associates to a section $s \in H^0(\Sigma, \mathcal{O}_\Sigma(D))$ its germ in $\mathcal{O}_\Sigma(D)_{z_i} \cong \mathcal{O}_{\Sigma,z_i}$ and its residue class mod $\mathfrak{m}^{m+1}$. Different choices of π_i differ by an automorphism of $\mathcal{O}_{\Sigma,z_i}$. Such automorphism maps $f_i \in \mathcal{O}_{\Sigma,z_i}$ to another representative f_i' of (X, z_i), it maps $\mathcal{O}_{\Sigma,z_i}/\mathfrak{m}^{m+1}$ to itself and the germ (M_i, f_i) to (M_i, f_i'). Hence, the transversality condition in (a) is independent of the choice of π_i.

To see that π_i may be chosen injective, we may assume that Σ is embedded in some $\mathbb{P}^N$ with $\mathcal{O}_\Sigma(D) = \mathcal{O}_{\mathbb{P}^N}(1) \otimes \mathcal{O}_\Sigma =: \mathcal{O}_\Sigma(1)$, since D is very ample. If $\mathcal{J}_\Sigma$ denotes the ideal sheaf of $\Sigma \subset \mathbb{P}^N$ we have for $n \geq 0$ an exact sequence

$$0 \to \mathcal{J}_\Sigma(n+1) \to \mathcal{O}_{\mathbb{P}^N}(n+1) \to \mathcal{O}_\Sigma(n+1) \to 0\,.$$

By Serre vanishing we have $H^1(\mathbb{P}^N, \mathcal{J}_\Sigma(n+1)) = 0$ for n sufficiently big and hence a surjection

$$\mathbb{C}[x_0, \ldots, x_N]_{n+1} \cong H^0(\mathbb{P}^N, \mathcal{O}_{\mathbb{P}^N}(n+1)) \twoheadrightarrow H^0(\Sigma, \mathcal{O}_\Sigma(n+1))\,.$$

Since $\mathcal{O}_\Sigma \subset \mathcal{O}_\Sigma(n)$ we have $H^0(\Sigma, \mathcal{O}_\Sigma(D)) \subset H^0(\Sigma, \mathcal{O}_\Sigma(n+1))$, and any $s \in H^0(\Sigma, \mathcal{O}(D))$ is represented by a homogeneous polynomial of degree $n+1$. This is mapped by π_i to its germ in $\mathcal{O}_{\Sigma,z_i}/\mathfrak{m}^{m+1}$ showing that π_i is injective if $m > n$.

(b) Due to the smoothness of the types $\mathcal{S} = (S_1, \ldots, S_r)$, the germs M_i in (a) are smooth of expected codimension $= \operatorname{codim}(S_i)$ (cf. Definition 2.2.43). The transversality condition implies that the germs at $[X]$ of $V_{|D|/\Sigma}(S_i) = \pi_i^{-1}\big(M_i \cap \pi_i(H^0(\Sigma, \mathcal{O}_\Sigma(D)))\big)$, $i = 1, \ldots, r$, are smooth of the same codimension $\operatorname{codim}(S_i)$. In this case we call $V_{|D|/\Sigma}(\mathcal{S})$ *T-smooth*.[14] Hence, if $V_{|D|/\Sigma}(S_1, \ldots, S_r)$ is T-smooth, then its codimension in $V_{|D|} = V_{|D|/\Sigma}$ is equal to $\sum_{i=1}^r \operatorname{codim}(S_i)$ which we call the *expected codimension*.

Definition 2.2.54 Let $f_i \in \mathcal{O}_{\Sigma,z_i}$ be a generator of the ideal of $(X, z_i) \subset (\Sigma, z_i)$ and $I_{S_i}^{f_i}$ the S_i-singularity ideal of f_i, defining the 0-dimensional S_i-singularity scheme $Z_{z_i}^{S_i} \subset \Sigma$, concentrated on z_i (cf. Definition 2.2.43). Denote by

[14] The "T" in T-smooth refers to "transversal".

$$Z(X) := Z^{\mathcal{S}}(X) := \bigcup_{i=1}^{r} Z_{z_i}^{S_i} \subset \Sigma$$

the 0-dimensional $\mathcal{S}$-*singularity scheme*, $\mathcal{S} = (S_1, \ldots, S_r)$, and by $\mathcal{J}_{Z(X)/\Sigma} := \mathcal{J}_{Z^{\mathcal{S}}(X)/\Sigma} \subset \mathcal{O}_\Sigma$ the ideal sheaf of $Z^{\mathcal{S}}(X)$.

The cohomological T-smoothness criteria of Theorems 2.2.38 and 2.2.40 and their supplements admit a natural generalization of the following form:

Theorem 2.2.55 *Let Σ be a smooth projective variety of dimension ≥ 2, D a very ample divisor and $X \in |D|$ a projective hypersurface, having isolated singularities $z_1, \ldots, z_r$ of smooth singularity types $S_1, \ldots, S_r$ respectively, and let $Z(X) = Z^{\mathcal{S}}(X)$ be the $\mathcal{S}$-singularity scheme. Then we have:*

(a) the Zariski tangent space to $V_{|D|}(S_1, \ldots, S_r)$ at X can be identified with $H^0(\Sigma, \mathcal{J}_{Z(X)/\Sigma}(D))/H^0(\Sigma, \mathcal{O}_\Sigma)$;
(b) if $H^1(\Sigma, \mathcal{J}_{Z(X)/\Sigma}(D)) = 0$ then $V_{|D|}(S_1, \ldots, S_r)$ is T-smooth at X;
(c) if $H^1(\Sigma, \mathcal{O}_\Sigma(D)) = 0$ (e.g. if $\Sigma = \mathbb{P}^n$), then the T-smoothness of $V_{|D|}(S_1, \ldots, S_r)$ at X is equivalent to $H^1(\Sigma, \mathcal{J}_{Z(X)/\Sigma}(D)) = 0$.
(d) Assume that $S_i = ea$ is an analytic type for $1 \leq i \leq p$ while S_i is arbitrary for $i > p$. Set $Z(X) = Z^0(X) \cup Z^1(X)$ with

$$Z^0(X) = \bigcup_{i=1}^{p} Z_{z_i}^{S_i}, \quad Z^1(X) = \bigcup_{i=p+1}^{r} Z_{z_i}^{S_i}.$$

If $H^1(\Sigma, \mathcal{J}_{Z(X)/\Sigma}(D)) = 0$ then the morphism of germs

$$(V_{|D|}(S_{p+1}, \ldots, S_r), X) \to \prod_{i=1}^{p} B_{(X,z_i)}$$

is smooth of the fibre dimension $h^0(\mathcal{J}_{Z(X)/\Sigma}(D))$, where $B_{(X,z_i)}$ denotes the base space of the semiuniversal deformation of (X, z_i). The morphism exists by the semiuniversality of B_{X,z_i}, its tangent map is unique.
In particular, $V_{|D|}(S_{p+1}, \ldots, S_r)$ is smooth at X and every local deformation of (X, z_i), $i = 1, \ldots, p$, is induced from the family $V_{|D|}(S_{p+1}, \ldots, S_r)$, i.e. from hypersurfaces X' in the linear system $|D|$, close to X, with singularities z_i' close to z_i and singularity type of (X', z_i') equal to S_i for $i = p+1, \ldots, r$. For $p = r$ this means that every local deformation of the germ (X, z_i), $i = 1, \cdots, r$, is induced from the family $V_{|D|}$ of hypersurfaces linear equivalent to D.

Proof The incidence variety

$$\mathfrak{N}_{|D|} = \{(z, X) \in \Sigma \times |D| \mid z \in X\}$$

is flat over $|D|$ by Remark 2.2.27.2 and we consider $|D|$ as a subscheme of Hilb_Σ. By Lemma 2.2.29 the tangent space of $|D|$ at X is $H^0(\Sigma, \mathcal{O}_\Sigma(X))/H^0(\Sigma, \mathcal{O}_\Sigma)$. The condition that a tangent vector of $|D|$ is tangent to the subvariety $V_{|D|}(S_1, \ldots, S_r)$ can be treated in the same way as this was done in the proof of Lemma 2.2.24, using the interpretation of the Zariski tangent space $T_y Y$ as morphisms $T_\varepsilon \to (Y, y)$. This proves (a).

To prove (b) we use the regularity criterion in [GLS6, Lemma II.1.30], in the same way as in the proofs of Lemma 2.2.29 and Proposition 2.2.26. Note that we do not need for this argument deformation theory over fat points, we only need that our family is (semi-)universal, which is a property of the Hilbert scheme. This shows the smoothness of $V_{|D|}(S_1, \ldots, S_r)$.

Denote $Z^{S_i}_{z_i}$ by Z_i and let Z', Z'' be a partition of $\{z_1, \ldots, z_r\}$ with $\mathcal{J}'$ resp. $\mathcal{J}''$ the ideal sheaves of the schemes Z' resp. Z'' and $\mathcal{J} = \mathcal{J}' \cap \mathcal{J}''$ the ideal sheaf of $Z(X)$. Since Z' and Z'' are disjoint we have an exact sequence

$$0 \to \mathcal{J} \to \mathcal{J}' \to \mathcal{O}_{Z''} \to 0\,.$$

Since $H^1(\mathcal{O}_{Z''}(D)) = 0$ we see that $H^1(\mathcal{J}(D)) = 0$ implies $H^1(\mathcal{J}'(D)) = 0$. In particular, the H^1-vanishing assumption implies $H^1(\Sigma, \mathcal{J}_{Z_i}(D)) = 0$ for $i = 1, \ldots, r$. As above we get that $V_{|D|}(S_i)$ is smooth for $i = 1, \ldots, r$.

The fact that $V_{|D|}(S_i)$ and $V_{|D|}(S_1, \ldots, S_r)$ have the expected codimension equal to $\mathrm{codim}(S_i)$ resp. $\sum_{i=1}^r \mathrm{codim}(S_i)$ follows in the same way as in the proof of Theorem 2.2.38. Altogether this implies the transversality conditions required in the definition of T-smoothness.

The proof of (c) follows in the same way as the proof of (c) in Theorem 2.2.40. The proof of the Supplements 2.2.39 and 2.2.41 can be adapted to prove (d). □

Remark 2.2.55.1 Given a hypersurface $X \subset \Sigma$ with isolated singularities $z_1, \ldots, z_r \in X$ of smooth singularity type $\mathcal{S} = (S_1, \ldots, S_r)$ we can talk about $\mathcal{S}$-equisingular deformations of the global hypersurface X over a *reduced* base space T (cf. end of Sect. 2.2.3.2, Remark 2.2.50.1). These are deformations $\phi : \mathscr{X} \to T$ of X such that there exist (local analytic) sections $t \mapsto z_i(t)$ of ϕ s.t. $(\mathscr{X}_t, z_i(t))$ has singularity type S_i for each $t \in T$ and for $i = 1, \ldots, r$. We do not require that we can define $\mathcal{S}$-equisingular deformations over non-reduced bases, e.g. over T_ϵ, and hence we cannot apply infinitesimal methods from deformation theory directly. However, as Theorem 2.2.55 shows, we can nevertheless get results about the tangent space and cohomological conditions for smoothness of such deformations.

On the other hand, not requiring an infinitesimal $\mathcal{S}$-equisingular deformation theory has also advantages, since it is less technical and easier to understand and that an infinitesimal theory might just be unknown for a singularity type S_i.

2.3 Construction via Deformation

2.3.1 General Idea of the Patchworking Construction

In this section we present a method for the *construction* of hypersurfaces belonging to a given linear system on a smooth projective algebraic variety and having prescribed types of isolated singularities. This method, together with another one, based on H^1-vanishing for generic zero-dimensional schemes, will be applied in Sect. 4.5 to solve the existence problem for algebraic curves with prescribed singularities lying in the plane or in another algebraic surface. The main idea of the method is due to Viro, who invented the patchworking construction in the context of real algebraic geometry (see [Vir2, Vir4, Vir5] and the first chapter of Viro's dissertation, reproduced as Appendix A at the end of this book [Vir6]). We also refer to the Sect. 2.4 of this chapter for a very brief introduction to Viro's original method and some further developments.

A commonly used approach to apply deformation theory for the construction of algebraic varieties is to find, first, a (highly) degenerate object and then deform it into an object having the required properties. However, in this way, there is a principal obstacle, a lack of algebraic, geometric, or topological description of the objects appearing in a deformation of complicated singularities. Our idea, developed basically in [Shu5, Shu9, Shu11, Shu12], can be described as follows. We start with a (reduced) *reducible* variety Y_0 and a Cartier divisor $X_0 \subset Y_0$ (the degenerate object) which we find by "guessing". We then construct special deformations in $\mathcal{D}ef_{X_0 \to Y_0}(\mathbb{C}, 0)$, that is, diagrams

$$\begin{array}{ccc} X_0 & \hookrightarrow & X \\ \downarrow & & \downarrow \\ Y_0 & \hookrightarrow & Y \\ \downarrow & & \downarrow \\ \{0\} & \hookrightarrow & (\mathbb{C}, 0) \end{array}$$

where $Y \to (\mathbb{C}, 0)$, resp. $X \to (\mathbb{C}, 0)$, are one-parametric deformations of Y_0, resp. X_0, which are locally topologically trivial over the complement of $0 \in \mathbb{C}$. We construct X and Y such that the fibres Y_t, $t \neq 0$, are irreducible and that $X_t \subset Y_t$ is a hypersurface hypersurface (a Cartier divisor). Usually we fix, in addition to $X_0 \subset Y_0$, also the deformation $Y \to (\mathbb{C}, 0)$ and construct X inside Y.

Depending on the initial data, a variety Y_0 and a hypersurface $X_0 \subset Y_0$, we have the deformation category (cf. [GLS6, Appendix C5])

$$\mathcal{D}ef_{X_0 \to Y_0}(\mathbb{C}, 0) := \left\{ \begin{array}{c} \text{deformations of} \\ X_0 \to Y_0 \text{ over } (\mathbb{C}, 0) \end{array} \right\} .$$

We work with these deformations, and consider the situation when X_0 has isolated singularities outside the singular locus of Y_0. Then any element of $\mathcal{D}ef_{X_0 \to Y_0}(\mathbb{C}, 0)$

naturally induces a deformation of every germ (X_0, z), and we look for elements of $\mathcal{D}ef_{X_0 \to Y_0}(\mathbb{C}, 0)$ which induce equisingular or equianalytic deformations of the germs (X_0, z), $z \in \operatorname{Sing}(X_0) \setminus \operatorname{Sing}(Y_0)$. Hence the collection of singularities of X_t, $t \neq 0$, is topological or analytic equivalent to the collection of singularities of $X_0 \setminus \operatorname{Sing}(Y_0)$.

Geometrically a generic fibre X_t, $t \neq 0$, appears as the result of *"patchworking"* (or *"gluing"*) of the pieces of X_0 lying on separate components of the reducible variety Y_0. This explains the advantage of the method, which allows to construct a complicated singular hypersurface X_t out of relatively simple restrictions of X_0 to components of Y_0. In fact, this method covers a large range of possible values of singularity and hypersurface invariants and appears to be an ultimate tool for constructing controllable deformations. One more important feature of the patchworking construction is that if the initial data are defined over the reals, then so is the resulting singular hypersurface.

Our main result, described below, consists in sufficient conditions for the non-emptiness of the set

$$\left\{ \begin{array}{c} \text{objects in } \ \mathcal{D}ef_{X_0 \to Y_0}(\mathbb{C}, 0) \ \text{ inducing objects in} \\ \bigcup_{z \in \operatorname{Sing}(X_0) \setminus \operatorname{Sing}(Y_0)} \mathcal{D}ef^S_{X_0, z}(\mathbb{C}, 0) \end{array} \right\}$$

where the upper index S stands for a smooth singularity type (cf. Definition 2.2.42). The assumption which guarantees these sufficient conditions is, in fact, the unobstructedness or smoothness (in the sense of Sect. 2.2, cf. Remark 2.2.15.1) or T-smoothness (cf. Definition 2.2.53) of the $\mathcal{S}$-equisingular deformations of the multi-germ $\coprod_{z \in \operatorname{Sing}(X_0) \setminus \operatorname{Sing}(Y_0)} (X_0, z)$ with respect to the adjacency of the components of Y_0 (which later on we call $\mathcal{S}$-transversality).[15]

In Sect. 2.3.3, the main results being presented in Theorems 2.3.4 and 2.3.5, we treat the following form of the general construction. Given a convex piece-wise linear function $\nu : \Delta \to \mathbb{R}$, where $\Delta \subset \mathbb{R}^n$ is a convex lattice polytope and ν is integral valued at integral points, we define $Y \to (\mathbb{C}, 0)$ as a toric variety corresponding to the polytope ([Dan, Ful2, Oda])

$$\widetilde{\Delta} = \left\{ (x, y) \in \mathbb{R}^{n+1} \;\middle|\; x \in \Delta, \quad y \in \mathbb{R}, \quad y \geq \nu(x) \right\},$$

the upper-graph of ν. Then the fibre Y_t is the family of toric hypersurfaces, isomorphic to the toric variety $\operatorname{Tor}(\Delta)$ for $t \neq 0$ and degenerating into the union of divisors in Y corresponding to the faces of the graph of the function ν. This model can be viewed as a modification of the so-called *"Viro method"* invented by Viro in 1979-80 and intended for "patchworking" non-singular real algebraic curves and hypersurfaces (cf. [Ris2, Vir2, Vir4, Vir5, Vir6, ItV]). Further developments, applications,

[15] A more general setting, which does not relate the unobstructedness to the adjacency of the components of Y_0 can be found in [ShT3, ShT4].

and related ideas can be found in [Bir, Ch2, Ch1, ItS1, ItS3, ItS4, Shu2, Shu5, Shu11, Shu12] and [GKZ, Chap. 11]. We provide a brief account of the Viro method in an appendix to this chapter (cf. Sect. 2.4).

The main result of Sect. 2.3.5, formulated in Theorem 2.3.13, is based on another model of the general construction, whose simplest version can be described as follows. Given a hypersurface $X \subset \mathbb{P}^n$ with few ordinary multiple points as singularities, we take Y to be the product $\mathbb{P}^n \times (\mathbb{C}, 0)$ blown up at $\mathrm{Sing}(X) \subset \mathbb{P}^n = \mathbb{P}^n \times \{0\}$, so that Y_0 is the union of $\mathbb{P}^n$ blown up at the given points and the exceptional n-planes. Moreover, X_0 is the union of the strict transform of X and certain projective hypersurfaces of degrees equal to the multiplicities of the singular points of X. The hypersurface $X_t \subset \mathbb{P}^n = \mathbb{P}^n \times \{t\}$, $t \neq 0$, is called a "lower deformation" of X, because with respect to some local coordinates it is determined by monomials lying under the Newton diagrams of the multiple points of X.

In Sect. 2.3.6 we translate the unobstructedness ($\mathcal{S}$-transversality) conditions to cohomology vanishing statements for the ideal sheaves of certain zero-dimensional schemes (Theorem 2.3.13 and Proposition 2.3.14), following the theory presented in Sect. 2.2.2, Theorems 2.2.38 and 2.2.40, and in Sect. 2.2.3, Theorem 2.2.55. Finally, in Sects. 4.4.2.2 and 4.5.1 of Chap. 4, using the results of Chap. 3, we translate the H^1-vanishing conditions into explicit numerical relations. Moreover, we provide there a number of concrete existence results.

2.3.2 Polytopes and $\mathcal{S}$-transversality

All existence statements in this chapter are valid both in the complex and real setting. The reason is that the construction we use commutes with the complex conjugation *conj*. We usually assume that the polynomials and varieties are defined over $\mathbb{C}$, but if the data in any statement is invariant with respect to conj then so are the conclusions.

Convex Polytopes and Convex Subdivisions. Throughout this book the term *polytope* means a convex (not necessarily bounded) polytope

$$\Delta \subset \mathbb{R}^n_+ := \big\{(x_1, \dots, x_n) \in \mathbb{R}^n \,\big|\, x_1, \dots, x_n \geq 0\big\},$$

whose vertices have integral coordinates. A polytope in $\mathbb{R}^n_+$ is called *nondegenerate* if it is n-dimensional. A two-dimensional polytope is called *polygon* and a face of codimension 1 is called *facet*. If $\Delta \subset \mathbb{R}^n_+$ is nondegenerate, put

$$\mathrm{Vol}_n(\Delta) := n! \cdot (\text{Euclidean volume of } \Delta)\,,$$

the *n-dimensional lattice volume* of Δ. If $\Delta \subset \mathbb{R}^n_+$ is k-dimensional, $k < n$, and Δ lies in a k-dimensional affine space $\Lambda \subset \mathbb{R}^n$, put

$$\mathrm{Vol}_k(\Delta) := \mathrm{Vol}_k\big(\pi(\Delta)\big),$$

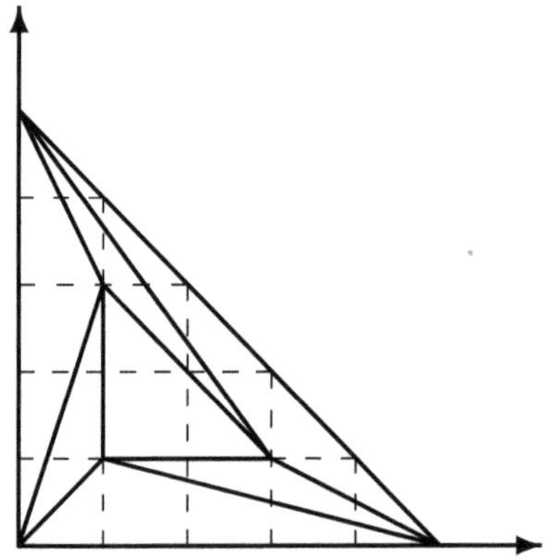

Fig. 2.1 A non-convex subdivision

where $\pi : \Lambda \to \mathbb{R}^k$ is an affine isomorphism which takes $\Lambda \cap \mathbb{Z}^n$ onto $\mathbb{Z}^k \subset \mathbb{R}^k$.

Given a nondegenerate polytope $\Delta \subset \mathbb{R}^n_+$, a set of nondegenerate polytopes $\Delta_1, \ldots, \Delta_N \subset \mathbb{R}^n_+$ is called a *subdivision of* Δ if $\Delta = \Delta_1 \cup \ldots \cup \Delta_N$ and, for any $1 \le i < j \le N$, the intersection $\Delta_i \cap \Delta_j$ is a proper common face of Δ_i and Δ_j, or empty.

A subdivision $\Delta_1, \ldots, \Delta_N$ of Δ is called *convex* if there exists a continuous, convex, piece-wise linear function $\nu : \Delta \to \mathbb{R}$ such that the polytopes $\Delta_1, \ldots, \Delta_N$ are the linearity domains of ν. The geometric meaning of the function ν is that its graph in $\mathbb{R}^{n+1}$ is a convex polyhedral hypersurface, whose facets are projected just onto $\Delta_1, \ldots, \Delta_N$, respectively. We say that the function ν *defines the subdivision* $\Delta = \Delta_1 \cup \ldots \cup \Delta_N$. Note also that the convexity of a subdivision is a non-trivial combinatorial condition.

An example of a non-convex subdivision is shown in Fig. 2.1 (cf. [CoH]). A simple explanation of the lack of convexity is as follows: assuming that the subdivision lifts to a graph of a convex function ν and assuming that the function vanishes on the central triangle, one can see that the function must strictly increase along each edge of the ambient triangle in the counter-clockwise direction contradicting the convexity of ν.

Note that any convex subdivision can be defined by a convex function which is integral-valued at integral points. Indeed, the set of convex piece-wise linear functions, defining the given subdivision, form a convex rational cone which then contains a rational point in its (relative) interior, and the required function is thus obtained by multiplication with an appropriate integer.
Henceforth, we always suppose that the used convex piece-wise linear functions are *integral-valued at integral points*.

Newton Polytopes and PND Polynomials. Given a polynomial $F = \sum_{i \in \mathbb{N}^n} A_i \boldsymbol{x}^{\boldsymbol{i}}$, the convex hull in $\mathbb{R}^n$,

$$\Delta(F) := \operatorname{conv}\{\boldsymbol{i} \in \mathbb{N}^n \mid A_i \neq 0\},$$

is called the *Newton polytope of F* (cf. [GLS6, Definition I.2.14]). Moreover, for a proper face $\sigma \subset \Delta(F)$, we have the *truncation of F at the face σ* which is the polynomial

$$F^\sigma(\boldsymbol{x}) := \sum_{\boldsymbol{i}\in\sigma} A_{\boldsymbol{i}} \boldsymbol{x}^{\boldsymbol{i}} := \sum_{\boldsymbol{i}\in\sigma\cap\mathbb{N}^n} A_{\boldsymbol{i}} \boldsymbol{x}^{\boldsymbol{i}} .$$

Definition 2.3.1 A polynomial F is called *peripherally nondegenerate (PND)* if its Newton polytope is nondegenerate (i.e. not lying in an affine hyperplane) and if all the truncations F^σ at proper faces $\sigma \subset \Delta(F)$ are nondegenerate (here a polynomial is called *nondegenerate* if it defines a non-singular hypersurface in the torus $(\mathbb{C}^*)^n$).

We recall the following result of Kouchnirenko ([Kou, Thm. III'(ii),(iii)]):

Proposition 2.3.2 *If no facet of a nondegenerate polytope $\Delta \subset \mathbb{R}^n_+$ lies on a hyperplane through the origin, then a PND polynomial $F \in \mathbb{C}[x_1, \dots, x_n]$ with Newton polytope Δ has only isolated critical points in $(\mathbb{C}^*)^n$, and their total Milnor number is equal to* $\mathrm{Vol}_n(\Delta(F))$. *Moreover, the set of PND polynomials is Zariski-open and dense in the space of all polynomials with fixed Newton polytope Δ.*

Recall that a *singular point* of F is a point where F and all the partial derivatives of F vanish, that is, a singular point of $\{F = 0\}$, while at a *critical point* all the partial derivatives vanish, but not necessarily F. We denote the sets of singular, respectively critical, points by $\mathrm{Sing}(\{F=0\})$, respectively $\mathrm{Crit}(F)$.

Singular Points and $\mathcal{S}$-transversality. Let $\mathcal{S}$ be a set of smooth singularity types of isolated singular points of hypersurfaces in $(\mathbb{C}^*)^n$ (in the sense of Definition 2.2.42). We say that the singularity type $S \in \mathcal{S}$ is *real* if each singular point of type S has a representative defined over the reals. We assume moreover that the singularity types $S \in \mathcal{S}$ are invariant under the coordinate changes[16] $(x_1, \dots, x_n) \mapsto (\alpha_1 x_1, \dots, \alpha_n x_n)$, $\alpha_1, \dots, \alpha_n \in \mathbb{C}^*$.

The topological or analytic types of singular points, the main examples used in this book, satisfy these requirements. Another example for $\mathcal{S}$ can be a combination of topological or analytic types, or constant intersection number with fixed curves at the singular points (cf. [Shu4, GrL1]).

Given a polynomial $F \in \mathbb{C}[x_1, \dots, x_n]$ such that the hypersurface $\{F=0\}$ has only isolated singular points in $(\mathbb{C}^*)^n$, we denote by $\mathcal{S}_F$ the function

$$\mathcal{S}_F : \mathcal{S} \longrightarrow \mathbb{Z} ,$$

associating to $S \in \mathcal{S}$ the number of singular points of $\{F=0\}$ in $(\mathbb{C}^*)^n$ of type S.

We shall also consider families of hypersurfaces in $\mathbb{P}^n$ with prescribed position of the singular points. This is, however, a non-local condition and can be imposed in addition to any (local) classification of singularities.

[16] If the hypersurface singularities are real in $(\mathbb{R}^n, 0)$ we require coordinate changes with $\alpha_1, \dots, \alpha_n \in \mathbb{R}_{>0}$.

Remark 2.3.2.1 (i) The notations and results from Sect. 2.2.3 are used here for the divisor $D = dH \subset \Sigma = \mathbb{P}^n$, H the hyperplane divisor. Our aim is to prove sufficient conditions for the non-emptiness of $V_{|D|/\Sigma}(\mathcal{S})$ for a given set $\mathcal{S} = (S_1, \dots, S_r)$ of smooth singularity types. We shall construct hypersurfaces in $\mathbb{P}^n$ with isolated singularities of given types in $\mathcal{S}$ which we may assume to lie all in $\mathbb{C}^n$ for some fixed embedding $\mathbb{C}^n \subset \mathbb{P}^n$. In fact, our constructions are all affine, defined by polynomials $F \in \mathbb{C}[x_1, \dots, x_n]$, and the singular points will lie even in $(\mathbb{C}^*)^n$. To prove the existence we use then a toric compactification of $(\mathbb{C}^*)^n$ in a one-parametric family. Nevertheless, if F has degree d, the homogenization $F^h := x_0^d F(x_1/x_0, \dots, x_n/x_0)$ defines a projective hypersurface of degree d in $\mathbb{P}^n$ with the considered singularities in $(\mathbb{C}^*)^n$.

(ii) In some arguments we need that local deformations of a finite set of singularities are induced by polynomials of degree d. This is always possible for d sufficiently large (see Remark 2.2.47.1). We may increase the degree of the projective hypersurface even without change of the affine equation. To achieve this, we use that we have, with respect to the given coordinates, a natural embedding of the space of projective hypersurfaces of degree d into the space of hypersurfaces of degree $d' > d$ by multiplying F^h with $x_0^{d'-d}$. Note that this does not change the affine equation for a given affine hypersurface, and hence not the singularities in $(\mathbb{C}^*)^n$.

At this point, we collect the notations to be used in the remaining part of this section.

Definition 2.3.3 (1) $\mathcal{P}(d) \subset \mathbb{C}[x_1, \dots, x_n]$ denotes the space of polynomials of degree at most d (considered as affine space).

(2) For $\Delta \subset \mathbb{R}^n_+$ a polytope, $\mathcal{P}(\Delta) \subset \mathbb{C}[x_1, \dots, x_n]$ denotes the space of all polynomials with Newton polytope contained in Δ.

(3) Let $F \in \mathbb{C}[x_1, \dots, x_n]$ be a PND polynomial with Newton polytope $\Delta = \Delta(F)$, and let Δ_+ be the union of some facets $\sigma \subset \Delta$ (Δ_+ may be empty). We call (Δ, Δ_+, F) a *Newton triple.*

Then we set, for $n \geq 2$,

$$\mathcal{P}(\Delta, \Delta_+, F) := \left\{ G \in \mathcal{P}(\Delta) \,\middle|\, G^\sigma = F^\sigma \text{ for all } \sigma \subset \Delta_+ \right\},$$

and, for $n = 2$,

$$\widetilde{\mathcal{P}}(\Delta, \Delta_+, F) := \left\{ G \in \mathcal{P}(\Delta) \,\middle|\, \begin{array}{l} \text{for any edge } \sigma \subset \Delta_+, G^\sigma = c \cdot F^\sigma \\ \text{with a non-zero constant } c \end{array} \right\}. \tag{2.3.2.1}$$

The difference between the spaces $\mathcal{P}(\Delta, \Delta_+, F)$ and $\widetilde{\mathcal{P}}(\Delta, \Delta_+, F)$ for $n = 2$ is as follows. If Δ_+ is connected then $\widetilde{\mathcal{P}}(\Delta, \Delta_+, F) \cup \{0\}$ is the linear space spanned by $\mathcal{P}(\Delta, \Delta_+, F)$; if Δ_+ is not connected then $\widetilde{\mathcal{P}}(\Delta, \Delta_+, F) \cup \{0\}$ is larger than the linear span of $\mathcal{P}(\Delta, \Delta_+, F)$. This explains why, below in (5), the weak $\mathcal{S}$-transversality is a less restrictive condition than $\mathcal{S}$-transversality.

(4) If $d \geq \deg F$ we consider F, $\mathcal{P}(\Delta)$ and $\widetilde{\mathcal{P}}(\Delta, \Delta_+, F)$ to be naturally embedded in $\mathcal{P}(d)$ (cf. Remark 2.3.2.1). For a singular point $w \in (\mathbb{C}^*)^n$ of the hypersurface

$\{F = 0\}$ denote by

$$M_d(F, w) := (M_S^{(d)}, (F, w)) \subset \mathcal{P}(d)$$

the germ at F of the set of polynomials $G \in \mathcal{P}(d)$ such that $\{G=0\}$ has a singular point of the same smooth singularity type S as (F, w), in the sense of Definition 2.2.42. $M_d(F, w)$ is an analytic germ by Proposition 2.2.48.
In fact, write $F(\boldsymbol{x}) = \sum_{1\le|\boldsymbol{i}|\le d} A_{\boldsymbol{i}}(\boldsymbol{x} - w)^{\boldsymbol{i}}$ and consider the unfolding $H_t(\boldsymbol{x}) = F(\boldsymbol{x}) + \sum_{1\le|\boldsymbol{i}|\le d} t_{\boldsymbol{i}}(\boldsymbol{x} - w)^{\boldsymbol{i}}$ of F. Then $M_d(F, w)$ is the germ at 0 of the S-analytic stratum $V_H(S) = \{t = (t_{\boldsymbol{i}}) \in \mathcal{P}(d) \mid S(H_t) = S\}$ as defined in Proposition 2.2.48.

(5) Let $\mathcal{S}$ be a set of smooth singularity types associated to the isolated singularities of $\{F = 0\}$ in $(\mathbb{C}^*)^n$. The triple (Δ, Δ_+, F) is called *$\mathcal{S}$-transversal* (respectively *weakly $\mathcal{S}$-transversal*, if $n = 2$) if the following holds:

   (a) the intersection $M_d(F)$ of the germs $M_d(F, w)$ at F, with w running through $\mathrm{Sing}(\{F=0\}) \cap (\mathbb{C}^*)^n$, is a T-smooth germ (cf. Definition 2.2.53, notation $V_{|D|}(\mathcal{S})$) as a complex analytic subset in $\mathcal{P}(d)$,
   (b) $M_d(F)$ intersects $\mathcal{P}(\Delta, \Delta_+, F)$ (respectively $\widetilde{\mathcal{P}}(\Delta, \Delta_+, F)$, if $n = 2$) transversally in $\mathcal{P}(d)$.

In a less formal way, $\mathcal{S}$-transversality (resp. weak $\mathcal{S}$-transversality) means the following. Consider the space $\mathcal{P}(\Delta, \Delta_+, F)$ (resp. $\widetilde{\mathcal{P}}(\Delta, \Delta_+, F)$) of polynomials of degree $\le d$ having fixed Newton polytope Δ with some coefficients fixed and equal to those of F (resp. equal up to multiplication with a constant $\neq 0$), namely, coefficients of those monomials which correspond to points of Δ_+. In this space consider the set of polynomials G having a fixed number of singularities (the same as F) and for each singularity a fixed type (corresponding to an $\mathcal{S}$-type of a singularity of F, for some bijection between the singularities of F and G). This set is $M_d(F) \cap \mathcal{P}(\Delta, \Delta_+, F)$ (resp. $M_d(F) \cap \widetilde{\mathcal{P}}(\Delta, \Delta_+, F)$) which contains F.

$\mathcal{S}$-transversality means: the condition for $G \in \mathcal{P}(\Delta, \Delta_+, F)$ to be in $M_d(F)$ is a locally closed condition on the free coefficients of G and defines a smooth subgerm $(M_d(F) \cap \mathcal{P}(\Delta, \Delta_+, F), F)$ of expected codimension. The smoothness means intuitively that, in $\mathcal{P}(\Delta, \Delta_+, F)$ (and close to F), we have enough freedom so that the conditions imposed by the singularity types are all independent in a rather strong sense. Similarly for $\widetilde{\mathcal{P}}(\Delta, \Delta_+, F)$.

Remark 2.3.3.1 The meaning of $\mathcal{S}$-transversality in the context of the patchworking construction is that the set of parameters describing the whole family of objects considered splits into *independent* groups of parameters, each group being responsible for deformations of certain singularities. This can be compared with the construction suggested by Biran [Bir], which can also be viewed as a kind of "gluing" and which requires some *ampleness* conditions.

2.3.3 Gluing Singular Hypersurfaces

We start with some general statements about the gluing of several hypersurfaces with given Newton polytopes and given singularity types to obtain one hypersurface whose Newton polytope is the union of the given ones and whose singularities are precisely of the given types. The gluing is possible under a certain transversality condition for the given hypersurfaces.

Throughout this section we fix a finite set of smooth singularity types also called a *classification* $\mathcal{S}$ of isolated hypersurface singularities (cf. Sect. 2.3.2).
Assume we are given

- a convex subdivision $\Delta_1, \ldots, \Delta_N$ of a nondegenerate polytope $\Delta \subset \mathbb{R}^n_+$ defined by a convex continuous piece-wise linear function $\nu : \Delta \to \mathbb{R}$,
- a function $\Delta \ni \boldsymbol{i} \mapsto a_{\boldsymbol{i}} \in \mathbb{C}$ such that $a_{\boldsymbol{i}} \neq 0$ if $\boldsymbol{i}$ is a vertex of one of the polytopes $\Delta_1, \ldots, \Delta_N$.

The polynomials in $\mathbb{C}[x_1, \ldots, x_n]$

$$F_k(\boldsymbol{x}) = \sum_{\boldsymbol{i} \in \Delta_k} a_{\boldsymbol{i}} \boldsymbol{x}^{\boldsymbol{i}} , \qquad k = 1, \ldots, N , \tag{2.3.3.1}$$

are called *block polynomials*.

Let G be the *dual graph* of the subdivision $\Delta = \Delta_1 \cup \ldots \cup \Delta_N$, that is, the vertices of G correspond to $\Delta_1, \ldots, \Delta_N$, and the edges correspond to common facets $\sigma = \Delta_i \cap \Delta_j$. Define $\mathcal{G}$ to be the set of oriented graphs $\widetilde{G}$ with support G and without oriented cycles. It is clear[17] that $\mathcal{G} \neq \emptyset$. For any $\widetilde{G} \in \mathcal{G}$ we denote by

$$\Delta_{i,+}(\widetilde{G}) := \left\{ \begin{array}{l} \text{union of facets of } \Delta_i, \text{ which corresponds} \\ \text{to the (oriented) edges of } \widetilde{G} \text{ ending in } \Delta_i \end{array} \right\} \tag{2.3.3.2}$$

and consider the Newton triples $(\Delta_i, \Delta_{i,+}(\widetilde{G}), F_i)$.

The following two theorems show that the transversality conditions of Definition 2.3.3 allow the gluing of the hypersurfaces $\{F_1 = 0\}, \ldots, \{F_N = 0\}$ (in the toric variety defined by the upper-graph of ν, by deformation) to a hypersurface $\{F = 0\}$ while preserving the singularity types.

Of course, it then remains to find effective criteria which guarantee transversality. This problem is reduced, in Sect. 2.3.6 of this chapter, to an H^1-vanishing condition for ideal sheaves of certain 0-dimensional schemes.

With the notations from above and the preceding section we have:

Theorem 2.3.4 *Assume that the block polynomials (2.3.3.1) are PND, and there is a $\widetilde{G} \in \mathcal{G}$ such that, for any $i = 1, \ldots, N$, the triple $(\Delta_i, \Delta_{i,+}(\widetilde{G}), F_i)$ (2.3.3.2) is $\mathcal{S}$-transversal. Then there exists a PND polynomial $F \in \mathcal{P}(\Delta)$ such that*

[17] For example, one can orient the graph G by directing the normals of facets of $\Delta_1, \ldots, \Delta_N$ in such a way that the first non-zero coordinate will be positive.

$$\mathcal{S}_F = \mathcal{S}_{F_1} + \ldots + \mathcal{S}_{F_N} ,$$

and the triple $(\Delta, \emptyset, F)$ *is* $\mathcal{S}$*-transversal.*

In the case $n = 2$ the following claim strengthens Theorem 2.3.4 in many cases.

Theorem 2.3.5 *Let* $n = 2$. *Assume that* $\Delta_{k,+} \neq \partial\Delta_k$, $k = 1, \ldots, N$, *the block polynomials (2.3.3.1) are PND, and there is an oriented graph* $\widetilde{G} \in \mathcal{G}$ *such that, for any* $i = 1, \ldots, N$, *the triple* $(\Delta_i, \Delta_{i,+}(\widetilde{G}), F_i)$ *(2.3.3.2) is weakly* $\mathcal{S}$*-transversal. Then there exists a PND polynomial* $F \in \mathcal{P}(\Delta)$ *such that*

$$\mathcal{S}_F = \mathcal{S}_{F_1} + \ldots + \mathcal{S}_{F_N} ,$$

and the triple $(\Delta, \emptyset, F)$ *is* $\mathcal{S}$*-transversal.*

Remark 2.3.5.1 The shorthand notation $\mathcal{S}_F = \mathcal{S}_{F_1} + \ldots + \mathcal{S}_{F_N}$ means the following. Assume $\{F_i = 0\}$ has precisely r_i isolated singularities in $(\mathbb{C}^*)^n$ of types $S_j^{(i)} \in \mathcal{S}_{F_i}$, $j = 1, \ldots, r_i$, $i = 1, \ldots, N$. Then $\{F = 0\}$ has precisely $r_1 + \ldots + r_N$ singularities in $(\mathbb{C}^*)^n$ and there is a bijection

$$\varphi : \coprod_{i=1}^{N} \mathrm{Sing}(\{F_i = 0\} \cap (\mathbb{C}^*)^n) \to \mathrm{Sing}(\{F = 0\} \cap (\mathbb{C}^*)^n)$$

such that the germs $(\{F_i = 0\}, p)$ and $(\{F = 0\}, \varphi(p))$ have the same singularity type for p in the domain of φ. That is, the given singularities on the different hypersurfaces $\{F_i = 0\}$ are "glued" on one hypersurface $\{F = 0\}$.

Proof of Theorem 2.3.4. Since multiplication of a polynomial by any monomial does not influence its singular points outside the coordinate hyperplanes, we can assume without loss of generality that no facet of the polytopes $\Delta_1, \ldots, \Delta_N$ lies on a hyperplane through the origin. Hence the polynomials $F_1, \ldots, F_N$ satisfy the conditions of Proposition 2.3.2. In particular, the set $\mathrm{Sing}(\{F_i = 0\})$ of singular points of the hypersurface $\{F_i = 0\} \cap (\mathbb{C}^*)^n$ and the set $\mathrm{Crit}(F_i)$ of critical points of F_i in $(\mathbb{C}^*)^n$ are finite for $i = 1, \ldots, N$. By Proposition 2.3.2,

$$\sum_{w \in \mathrm{Crit}(F_i) \cap (\mathbb{C}^*)^n} \mu(F_i, w) = \mathrm{Vol}_n(\Delta_i), \quad i = 1, \ldots, N. \tag{2.3.3.3}$$

We shall look for the desired polynomial F in the following one-parameter family of polynomials:

$$F_t = \sum_{i \in \Delta} A_i(t) t^{\nu(i)} \boldsymbol{x}^{\boldsymbol{i}} , \qquad t > 0, \tag{2.3.3.4}$$

where the A_i, $\boldsymbol{i} \in \Delta$, are analytic functions in t, satisfying $|A_i(t) - a_i| \leq Kt$ with a positive constant K, where the a_i are given by (2.3.3.1) and where ν is the given piecewise linear function defining the convex subdivision. The functions A_i and K will be specified later.

Step 1. Assume that the A_i and K are chosen such that F_t in (2.3.3.4) is defined. Denote by $\ell_k(i_1,\dots,i_n)=\alpha_{0k}+\alpha_{1k}i_1+\ldots+\alpha_{nk}i_n$ the affine function equal to ν on Δ_k, $k=1,\dots,N$. The substitution of $\nu_k=\nu-\ell_k$ for ν in (2.3.3.4) gives the family

$$F_{k,t}(\boldsymbol{x}) \;=\; F_k(\boldsymbol{x})+\sum_{i\notin\Delta_k} A_i(t)\,t^{\nu_k(i)}\boldsymbol{x}^{\boldsymbol{i}}+\sum_{i\in\Delta_k}\big(A_i(t)-a_i\big)\boldsymbol{x}^{\boldsymbol{i}}.$$

This substitution can be expressed as composition of the coordinate change

$$T_k(\boldsymbol{x}) := (x_1t^{\alpha_{1k}},\dots,x_nt^{\alpha_{nk}}),\quad k=1,\dots,N,$$

with the multiplication of the polynomial by a positive real number, since

$$\begin{aligned}F_{k,t}(\boldsymbol{x}) &= \sum_{i\in\Delta} A_i(t)\,t^{\nu(i)-\ell_k(i)}\boldsymbol{x}^{\boldsymbol{i}}\\ &= t^{-\alpha_{0k}}\sum_{i\in\Delta} A_i(t)\,t^{\nu(i)}(x_1t^{-\alpha_{1k}})^{i_1}\cdot\ldots\cdot(x_nt^{-\alpha_{nk}})^{i_n}\\ &= t^{-\alpha_{0k}}F_t\big(T_k^{-1}(\boldsymbol{x})\big).\end{aligned}\tag{2.3.3.5}$$

In particular, this operation does not change the function $\mathcal{S}_F$ but only moves the singular and critical points in $(\mathbb{C}^*)^n$. Recall that our types $S\in\mathcal{S}$ are invariant under coordinate changes T_k, $k=1,\dots,N$.

Step 2. Fix a compact set $Q\subset(\mathbb{C}^*)^n$, whose interior contains the union of the critical sets $\mathrm{Crit}(F_1)\cup\ldots\cup\mathrm{Crit}(F_N)$. Since the difference $F_{k,t}-F_k$ converges to 0 as $t\to 0$, there exists a $t(K)>0$ such that, for any t in the open interval $\big(0,\ t(K)\big)$ and any $k=1,\dots,N$ the following holds:

- $\mathrm{Crit}(F_{k,t})\cap Q$ is finite and located in a neighborhood of $\mathrm{Crit}(F_k)$; moreover, all the critical (respectively singular) points of $F_{k,t}$ in Q appear in the local semiuniversal deformations of the critical (respectively singular) points of F_k,
- no singular, or critical, point of $F_{k,t}$ outside Q comes from a local deformation of singular, or critical, points of F_k in Q,
- no singular point of $F_{k,t}$ in Q comes from a deformation of a critical point of F_k in Q with a non-zero critical value.

In particular, in view of (2.3.3.3), we obtain, for each $k=1,\dots,N$,

$$\sum_{w\in\mathrm{Crit}(F_{k,t})\cap Q}\mu(F_{k,t},w) \;=\; \sum_{w\in\mathrm{Crit}(F_k)\cap(\mathbb{C}^*)^n}\mu(F_k,w) \;=\; \mathrm{Vol}_n(\Delta_k)\,. \tag{2.3.3.6}$$

Due to the convexity of ν, we have $\sum_{j=1}^n|\alpha_{jk}-\alpha_{jl}|>0$ if $k\neq l$. Hence, there exists $t(Q)>0$ such that, for $t\in\big(0,\ t(Q)\big)$ the compact sets $T_1(Q),\dots,T_N(Q)$ are disjoint in $(\mathbb{C}^*)^n$. This follows since Q does not meet any coordinate hyperplane and since, for sufficiently small t, $T_l^{-1}T_k(Q)$ does not meet Q if $l \neq k$. We claim, there are no more singular resp. critical points of F_t in $(\mathbb{C}^*)^n$:

By (2.3.3.5), it follows that for t in the open interval $\big(0, \min\{t(K), t(Q)\}\big)$ the set $\mathrm{Sing}(\{F_t = 0\})$ (respectively $\mathrm{Crit}(F_t)$) contains the disjoint union of the sets $T_k(\mathrm{Sing}(\{F_{k,t} = 0\}) \cap Q)$ (respectively $T_k(\mathrm{Crit}(F_{k,t}) \cap Q)$), $k = 1, \ldots, N$.

We claim that there are no more singular, resp. critical, points of F_t in $(\mathbb{C}^*)^n$:

Lemma 2.3.6 *For* $t \in \big(0, \min\{t(K), t(Q)\}\big)$,

$$\mathrm{Sing}(\{F_t = 0\} \cap (\mathbb{C}^*)^n) = \bigcup_{k=1}^{N} T_k\big(\mathrm{Sing}(\{F_{k,t} = 0\}) \cap Q\big),$$

$$\mathrm{Crit}(F_t) \cap (\mathbb{C}^*)^n = \bigcup_{k=1}^{N} T_k\big(\mathrm{Crit}(F_{k,t}) \cap Q\big).$$

Proof In view of the above assumptions it is sufficient to show that F_t has no critical point in $(\mathbb{C}^*)^n$ outside $T_1(Q) \cup \ldots \cup T_N(Q)$. Note that F_t satisfies the assumptions of Proposition 2.3.2, since the truncation of F_t on Δ_k is, up to a torus action, a small deformation of F_k. Hence the total Milnor number of F_t is equal to $\mathrm{Vol}_n(\Delta(F_t))$. Since $\Delta(F_t) = \Delta$ we get from Proposition 2.3.2 that

$$\sum_{w \in \mathrm{Crit}(F_t) \cap (\mathbb{C}^*)^n} \mu(F_t, w) = \mathrm{Vol}_n(\Delta).$$

The properties of Q and (2.3.3.6) imply

$$\begin{aligned} \sum_{w \in \mathrm{Crit}(F_t) \cap (\bigcup_k T_k(Q))} \mu(F_t, w) &= \sum_{k=1}^{N} \sum_{w \in \mathrm{Crit}(F_{k,t})} \mu(F_{k,t}, w) \\ &= \sum_{k=1}^{N} \mathrm{Vol}_n(\Delta_k) = \mathrm{Vol}_n(\Delta). \end{aligned}$$

Hence there are no further singularities of F_t in $(\mathbb{C}^*)^n$.[18] □

Step 3. To complete the proof of Theorem 2.3.4 we shall define $A_i(t)$ as analytic functions of t such that, $A_i(0) = a_i$, and for any polynomial $F_{k,t}$, its collection of $\mathcal{S}$-types of singular points in Q coincides with the one for F_k, $k = 1, \ldots, N$. Correspondingly, we shall determine the constant K as well.

---

[18] We use here a kind of global argument, showing that the critical points of $F_1, \ldots, F_N$ in the torus provide the maximal possible total Milnor number for critical points in the torus of a polynomial with Newton polytope Δ. However, one may ask for a purely local proof of the fact that the singularities along the boundary divisors are smoothed out in the considered deformation. For $n = 2$ such an argument can be found in [ShT3, ShT4].

For the polynomial F_k, put $M^{(k)} = M_d(F_k) \cap \mathcal{P}(\Delta) \subset \mathcal{P}(d)$, with a sufficiently large d (see Definition 2.3.3). The condition we impose on the functions $A_{\boldsymbol{i}}(t), \boldsymbol{i} \in \Delta$, is that $F_{k,t} \in M^{(k)}$, $k = 1, \ldots, N$.

The $\mathcal{S}$-transversality of the triple $(\Delta_k, \Delta_{k,+}(\widetilde{G}), F_k)$ (2.3.3.2) means that $M^{(k)}$ is the transversal intersection of the linear space $\mathcal{P}(\Delta_k)$ and of $d_k := \operatorname{codim}_{\mathcal{P}(d)} M_d(F_k)$ smooth hypersurface germs at the F_k in the linear space $\mathcal{P}(\Delta)$. Let these d_k hypersurface germs be given by the equations

$$\varphi_r^{(k)}\left(\left\{B_{\boldsymbol{i}} \,\middle|\, \boldsymbol{i} \in \Delta\right\}\right) = 0, \quad r = 1, \ldots, d_k, \tag{2.3.3.7}$$

where $B_{\boldsymbol{i}}$, $\boldsymbol{i} \in \Delta$, denote indeterminates corresponding to coefficients of polynomials in $\mathcal{P}(\Delta)$, considered in a neighborhood of F_k. By transversality of the smooth hypersurfaces there is a subset $\Lambda_k \subset \mathbb{Z}^n \cap \left(\Delta_k \setminus \Delta_{k,+}(\widetilde{G})\right)$ of cardinality d_k such that

$$\det\left(\frac{\partial \varphi_r^{(k)}}{\partial B_{\boldsymbol{i}}}\right)_{\substack{r=1,\ldots,d_k \\ \boldsymbol{i} \in \Lambda_k}} \neq 0 \tag{2.3.3.8}$$

at the point

$$B_{\boldsymbol{i}} = \begin{cases} a_{\boldsymbol{i}} & \text{if } \boldsymbol{i} \in \Delta_k, \\ 0 & \text{if } \boldsymbol{i} \notin \Delta_k. \end{cases} \tag{2.3.3.9}$$

To obtain the functions $A_{\boldsymbol{i}}(t)$, $\boldsymbol{i} \in \Delta$, we substitute

$$B_{\boldsymbol{i}} = A_{\boldsymbol{i}} t^{\nu_k(\boldsymbol{i})} \tag{2.3.3.10}$$

in (2.3.3.7) for $k = 1, \ldots, N$, with indeterminates $A_{\boldsymbol{i}}$.

Lemma 2.3.7

$$\det\left(\left.\frac{\partial \varphi_r^{(k)}}{\partial A_{\boldsymbol{i}}}\right|_{t=0}\right)_{\substack{r=1,\ldots,d_k \\ k=1,\ldots,N \\ \boldsymbol{i} \in \bigcup_k \Lambda_k}} \neq 0. \tag{2.3.3.11}$$

Proof The oriented graph $\widetilde{G}$ turns the set $\{\Delta_1, \ldots, \Delta_N\}$ into a poset: we write $\Delta_i \succ \Delta_j$ if there is an oriented path from Δ_j to Δ_i. Let us renumber $\Delta_1, \ldots, \Delta_N$ such that $\Delta_i \succ \Delta_j$ holds only if $i > j$. We shall prove (2.3.3.11) for $k = 1, \ldots, p \le N$, by induction on p. Let $p = 1$. Then

$$\det\left(\left.\frac{\partial \varphi_r^{(1)}}{\partial A_{\boldsymbol{i}}}\right|_{t=0}\right)_{\substack{r=1,\ldots,d_k \\ \boldsymbol{i} \in \Lambda_1}} \neq 0,$$

since it is equal to the determinant (2.3.3.8) for $k = 1$. Assume $1 < p < N$, and

$$\det\left(\left.\frac{\partial\varphi_r^{(k)}}{\partial A_{\boldsymbol{i}}}\right|_{t=0}\right)_{\substack{r=1,\dots,d_k\\ k=1,\dots,p\\ \boldsymbol{i}\in\Lambda_1\cup\dots\cup\Lambda_p}}\neq 0\,.$$

Consider, in addition, the functions $\varphi_r^{(p+1)}$, $r=1,\dots,d_{p+1}$, and the unknowns $A_{\boldsymbol{i}}$, $\boldsymbol{i}\in\Lambda_{p+1}$. According to the numbering of $\Delta_1,\dots,\Delta_N$, we have $\Lambda_{p+1}\cap(\Delta_1\cup\dots\cup\Delta_p)=\emptyset$. Hence,

$$\left.\frac{\partial\varphi_r^{(k)}}{\partial A_{\boldsymbol{i}}}\right|_{t=0}=0\,,\quad k\le p\,,\quad \boldsymbol{i}\in\Lambda_{p+1}\,,$$

because of (2.3.3.10) and the definition of $\nu_1,\dots,\nu_N$. On the other hand,

$$\det\left(\left.\frac{\partial\varphi_r^{(p+1)}}{\partial A_{\boldsymbol{i}}}\right|_{t=0}\right)_{\substack{r=1,\dots,d_{p+1}\\ \boldsymbol{i}\in\Lambda_{p+1}}}\neq 0$$

by (2.3.3.8), and we are done with the proof of Lemma 2.3.7. □

Now, by means of the implicit function theorem, we derive from Lemma 2.3.7 the existence of the desired functions $A_{\boldsymbol{i}}(t)$, $\boldsymbol{i}\in\Delta$, and the $\mathcal{S}$-transversality of the triple $(\Delta,\emptyset,F_t)$ with F_t defined by (2.3.3.4). □

Proof of Theorem 2.3.5. It suffices to show how to modify Step 3 in the proof of Theorem 2.3.4.

We can pass to the projectivizations of all the spaces of polynomials used. They all are subspaces of the space of polynomials with Newton polygon Δ. Let $A_{\boldsymbol{v}}$, $\boldsymbol{v}\in\Delta\cap\mathbb{Z}^2$, be the coefficients of a polynomial G with Newton polygon Δ, such that $A_{\boldsymbol{v}}\neq 0$ when $\boldsymbol{v}$ is a vertex of one of the $\Delta_1,\dots,\Delta_N$. We shall parameterize the germ of $\mathcal{P}(\Delta)$ at G by the following two groups of parameters:

- In any Δ_k, $1\le k\le N$, we fix a vertex $\boldsymbol{i}_k$ and put

$$C_{\boldsymbol{i}}=\frac{A_{\boldsymbol{i}}}{A_{\boldsymbol{i}_k}}\,,\quad \boldsymbol{i}\in\operatorname{Int}(\Delta_k)\,,\tag{2.3.3.12}$$

  where Int denotes the set of interior integral points.
- Let us orient all edges in $\bigcup_{k=1}^N\partial\Delta_k$. For an edge σ, denote by $\boldsymbol{i}(\sigma)$, $\boldsymbol{j}(\sigma)$ the initial and the terminal point, and put $V(\sigma)=(\sigma\cap\mathbb{Z}^2)\setminus\{\boldsymbol{i}(\sigma)\}$. In addition, for any $k=1,\dots,N$, let us choose an edge $\sigma_k\subset\partial\Delta_k$, $\sigma_k\not\subset\Delta_{k,+}(\widetilde{G})$, which exists by assumption of Theorem 2.3.5, and introduce the parameters (with $A_{\boldsymbol{i}}$ as variables)

$$C_{\boldsymbol{i},\sigma}=\frac{A_{\boldsymbol{i}}}{A_{\boldsymbol{i}(\sigma)}},\quad \boldsymbol{i}\in V(\sigma),\ \sigma\subset\partial\Delta_k,\ \sigma\neq\sigma_k,\tag{2.3.3.13}$$

$$C_{\boldsymbol{i},\sigma_k}=\frac{A_{\boldsymbol{i}}}{A_{\boldsymbol{i}(\sigma_k)}},\quad \boldsymbol{i}\in V(\sigma_k)\setminus\{\boldsymbol{j}(\sigma_k)\}\,.\tag{2.3.3.14}$$

The total number of the parameters (2.3.3.12)–(2.3.3.14) is $\#(\Delta \cap \mathbb{Z}^2) - 1$. To show that these are coordinates in a neighborhood of G in $\mathbb{P}(\mathcal{P}(\Delta))$ it suffices to check that the map, taking a polynomial in a neighborhood of G in $\mathbb{P}(\mathcal{P}(\Delta))$ to the system of parameters (2.3.3.12)–(2.3.3.14), is (locally) surjective. The only possible relation in the image of that neighborhood might be $\prod_\sigma (C_{i(\sigma),\sigma})^{\pm 1} = 1$, where σ runs over certain edges, forming a cycle. Hence it is enough to prove that

$$T = \bigcup_{k=1}^{N} \partial \Delta_k \setminus \bigcup_{k=1}^{N} \mathrm{Int}(\sigma_k)$$

has no cycles. Indeed, suppose that T contains a cycle T_0. It bounds the union of some of $\Delta_1, \dots, \Delta_N$, say, $\Delta_1, \dots, \Delta_m$. The edges $\sigma_1, \dots, \sigma_m$ are embraced by T_0 as well, and they all correspond to (oriented) edges of the graph $\widetilde{G}$. This m-edge subgraph of $\widetilde{G}$ defines a partial order on $\Delta_1, \dots, \Delta_m$, since $\widetilde{G}$ has no oriented cycles. On the other hand, no Δ_k, $1 \le k \le m$, can be a maximal element with respect to such an order, because an edge of $\widetilde{G}$ corresponding to σ_k goes out of Δ_k and ends in Δ_j, $j \ne k$, $1 \le j \le m$. So, one gets a contradiction.

It follows that the formulas (2.3.3.12)–(2.3.3.14) uniquely determine a polynomial $\sum_i A_i \boldsymbol{x}^{\boldsymbol{i}}$ in the (projective) space $\mathbb{P}(\mathcal{P}(\Delta))$ given by the parameters $C_{\boldsymbol{i}}, C_{\boldsymbol{i},\sigma}, C_{\boldsymbol{i},\sigma_k}$. Hence, our problem reduces to finding these parameters as smooth functions of t such that for $t = 0$ one has

$$\begin{aligned}
C_{\boldsymbol{i}}^{(0)} &= \frac{a_{\boldsymbol{i}}}{a_{\boldsymbol{i}_k}}, && \boldsymbol{i} \in \mathrm{Int}(\Delta_k)\,, \\
C_{\boldsymbol{i},\sigma}^{(0)} &= \frac{a_{\boldsymbol{i}}}{a_{\boldsymbol{i}(\sigma)}}, && \boldsymbol{i} \in V(\sigma),\ \sigma \subset \partial\Delta_k,\ \sigma \ne \sigma_k\,, \\
C_{\boldsymbol{i},\sigma_k}^{(0)} &= \frac{a_{\boldsymbol{i}}}{a_{\boldsymbol{i}(\sigma_k)}}, && \boldsymbol{i} \in V(\sigma_k) \setminus \{\boldsymbol{j}(\sigma_k)\}\,,
\end{aligned}$$

$k = 1, \dots, N$. Then we are going to substitute the formulas (2.3.3.10) into the equations (2.3.3.7) and consider $A_{\boldsymbol{i}}$, $\boldsymbol{i} \in \Delta$, as functions of the parameters $C_{\boldsymbol{i}}, C_{\boldsymbol{i},\sigma}, C_{\boldsymbol{i},\sigma_k}$ in a neighbourhood of the point $C_{\boldsymbol{i}}^{(0)}, C_{\boldsymbol{i},\sigma}^{(0)}, C_{\boldsymbol{i},\sigma_k}^{(0)}$. Note that the set of parameters (2.3.3.12)–(2.3.3.14) splits into disjoint subsets

$$\begin{aligned}
&C_{\boldsymbol{i}}, && \boldsymbol{i} \in \mathrm{Int}(\Delta_k), \\
&C_{\boldsymbol{i},\sigma}, && \boldsymbol{i} \in V(\sigma),\ \sigma \subset \partial\Delta_k,\ \sigma \not\subset \Delta_{k,+}(\widetilde{G}),\ \sigma \ne \sigma_k, \\
&C_{\boldsymbol{i},\sigma_k}, && \boldsymbol{i} \in V(\sigma_k) \setminus \{\boldsymbol{j}(\sigma_k)\},
\end{aligned}$$

which, for any $k = 1, \dots, N$, are coordinates in a neighbourhood of F_k in the (projective) space $\mathbb{P}(\widetilde{\mathcal{P}}(\Delta_k, \Delta_{k,+}(\widetilde{G}), F_k))$. Due to the weak $\mathcal{S}$-transversality, for each $k = 1, \dots, N$, there exists a set

$$\Lambda_k \subset \left(\mathrm{Int}(\Delta_k) \cap \mathbb{Z}^2\right) \cup \bigcup_{\substack{\sigma \subset \partial \Delta_k \\ \sigma \not\subset \Delta_{k,+}(\widetilde{G})}} V(\sigma) \cup \left(V(\sigma_k) \setminus \{\boldsymbol{j}(\sigma_k)\}\right),$$

of cardinality d_k such that

$$\det\left(\left(\frac{\partial \varphi_r^{(k)}}{\partial B_j}\right)_{\substack{r=1,\dots,d_k \\ j \in \Lambda_k}} \cdot \left(\frac{\partial B_j}{\partial(C_i, C_{i,\sigma}, C_{i,\sigma_k})}\right)_{\substack{j \in \Lambda_k \\ i,(i,\sigma),(i,\sigma_k) \in \Lambda_k}}\right) \neq 0,$$

at the point where the B_i take the values (2.3.3.9), and the $C_i, C_{i,\sigma}, C_{i,\sigma_k}$ take the values $C_i^{(0)}, C_{i,\sigma}^{(0)}, C_{i,\sigma_k}^{(0)}$, respectively. Finally, one proves that the determinant

$$\det\left(\left(\left.\frac{\partial \varphi_r^{(k)}}{\partial A_j}\right|_{t=0}\right)_{\substack{r=1,\dots,d_k \\ k=1,\dots,N \\ j \in \Delta}} \cdot \left(\left.\frac{\partial A_j}{\partial(C_i, C_{i,\sigma}, C_{i,\sigma_k})}\right|_{t=0}\right)_{\substack{j \in \Delta \\ i,(i,\sigma),(i,\sigma_k) \in \bigcup_k \Lambda_k}}\right)$$

does not vanish, using exactly the same reasoning as in the proof of Lemma 2.3.7. This, together with the implicit function theorem, completes the proof of Theorem 2.3.5. □

2.3.4 SQH and NND Isolated Hypersurface Singularities

In the previous section we considered the gluing of affine hypersurfaces under certain transversality conditions with respect to Newton polygons and singularity types. In this section we develop the local counterpart for germs, that is, for gluing of singularities. We first introduce several notions concerning singularities and their deformations, and then we formulate a transversality condition for deformations of semiquasi-homogeneous and Newton nondegenerate singularities (cf. [GLS6, Definitions I.2.15, I.2.17, respectively I.2.19]).

In this section Σ denotes a smooth complex space of dimension $\dim \Sigma = n \geq 2$ and $X \subset \Sigma$ a hypersurface.

Definition 2.3.8 Let z be an isolated singular point of $X \subset \Sigma$. We say that z is *Newton nondegenerate* (*NND*) if there exists a holomorphic germ $F : (\mathbb{C}^n, 0) \to (\mathbb{C}, 0)$ defining a Newton nondegenerate singularity at the origin such that the analytic local rings $\mathcal{O}_{X,z}$ and $\mathcal{O}_{\mathbb{C}^n,0}/\langle F\rangle$ are isomorphic. Equivalently, there is an isomorphism $\varphi : (\Sigma, z) \to (\mathbb{C}^n, 0)$ such that $F = 0$ is a local equation of the germ $\varphi(X, z)$. We say $F = 0$ is a local equation of (X, z) in the coordinates induced by φ and call F a *representative of the NND singular point* $z \in X$.

Correspondingly, we call a singular point z of a hypersurface $X \subset \Sigma$ *semiquasi-homogeneous* (*SQH*), if it has a semiquasi-homogeneous representative.

Moreover, we call a *monomial* $\boldsymbol{x}^{\boldsymbol{i}} \in \mathbb{C}[x_1, \dots, x_n]$ *upper* (resp., *strictly upper*) if it lies above (resp., above and not on) the Newton diagram $\Gamma = \Gamma(F, 0)$. Similarly, $\boldsymbol{x}^{\boldsymbol{i}}$ is called *lower* (resp. *strictly lower*) if it lies below (resp. below and not on) Γ.

Let us recall some notations for NDD singularities from [GLS6, 1.2]. For $F = \sum_{i \in \mathbb{Z}^n_{\geq 0}} A_i \boldsymbol{x}^{\boldsymbol{i}}$, $A_0 = 0$, let

$$\Delta(F) = \operatorname{conv}\{\boldsymbol{i} \in \mathbb{Z}^n_{\geq} \ : \ A_i \neq 0\}$$

be the *Newton polytope* of F. We set

$$K(F) = \operatorname{conv}\{\Delta(F) \cup \{0\}\}, \quad K_0(F) = \overline{K(F) \backslash \Delta(F)}\,.$$

Then the *Newton diagram* $\Gamma(F, 0)$ of F at 0 is the union of those faces of $K_0(F) \cap \Delta(F)$ through which there are supporting hyperplanes to $\Delta(F)$ with normal vectors having only positive coordinates.

Example 2.3.8.1 (cf. [GLS6, Example II.14.1]) Let $F = x \cdot (y^5 + xy^3 + x^2y^2 - x^2y^4 + x^3y - 10x^4y + x^6)$, then we have

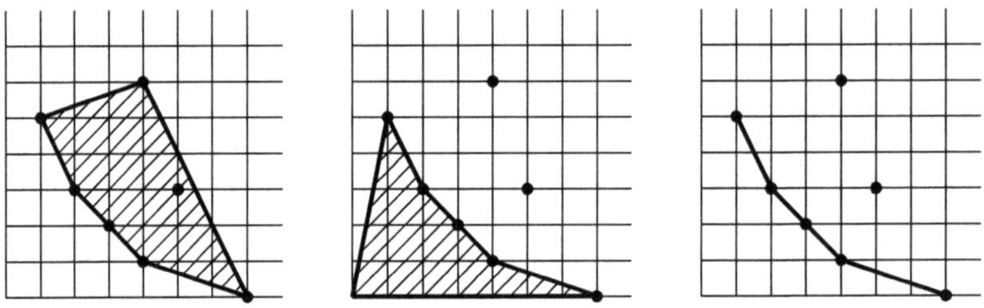

$\Delta(F)$ $\qquad$ $K_0(F)$ $\qquad$ $\Gamma(F,0)$

In particular, the Newton diagram at 0 has three one-dimensional faces, with slopes $-2, -1, -\frac{1}{3}$.

F is called *Newton nondegenerate* (*NND*) if, for all proper faces $\sigma \subset \Gamma(F, 0)$, the hypersurface $\{F^\sigma = 0\}$ has no singular point in $(\mathbb{C}^*)^n$ and if F is *convenient*, i.e. $\Gamma(F, 0)$ meets all coordinate axes.

Note that F in Example 2.3.8.1 is not convenient, hence not NND. But $F + y^N$, $N > \tau(F) + 1$, is NND and is contact equivalent to F at 0, since isolated singularities are contact $(\tau + 1)$-determined ([GLS6, Corollary I.2.24]). Similarly, any isolated singularity F can be made convenient by adding terms of sufficiently high degree N without changing the singularity type: if $N > \tau(F) + 1$ (resp. $N > \mu(F) + 1$) then the contact class (resp. the right class) of F is not changed.

By Kouchnirenko ([Kou, Theorem I]), a NND power series has an isolated singularity at 0 with Milnor number $\mu(F, 0)$ equal to the Newton number of F at 0 and the same holds for its principal part $F^{\Gamma(F,0)} = \sum_{i \in \Gamma(F,0)} A_i \boldsymbol{x}^{\boldsymbol{i}}$. Moreover, NND is a Zariski-open condition in the space of all principal parts with fixed Newton diagram. For F not NND, the Milnor number is bigger or equal to the Newton number.

Zero-Dimensional Schemes Associated with SQH and NND Singular Points. Let $z \in X \subset \Sigma$ be a SQH singular point with a representative $F : (\mathbb{C}^n, 0) \to (\mathbb{C}, 0)$, $F(\boldsymbol{x}) = \sum_i A_i \boldsymbol{x}^i$. Recall from [GLS6] that the *Newton diagram* $\Gamma(F, 0)$ is the union of the compact faces of the *local Newton polytope*

$$N(F,0) = \operatorname{conv}\left(\bigcup_{\boldsymbol{i} \in \operatorname{supp}(F)} \boldsymbol{i} + \mathbb{R}^n_{\geq 0} \right) .$$

F is SQH (at 0) iff there is an $(n-1)$-dimensional (main) face $\sigma \subset \Gamma(F, 0)$ such that the truncation F^σ, the *principal part* of F, has no critical points in $\mathbb{C}^n \backslash \{0\}$. Let $\sigma \subset \Gamma(F, 0)$ be the main face, lying on the hyperplane given by some affine equation $\ell(\boldsymbol{i}) = 0$, $\ell(0) > 0$, i.e.

$$\ell(\boldsymbol{i}) = -\omega_0 + \omega_1 i_1 + \cdots + \omega_n i_n \quad \text{with} \quad \omega_j \in \mathbb{Z}_{<0} \, .$$

Then we define two ideals $I(F)$ and $\widetilde{I}(F)$ in $\mathcal{O}_{\mathbb{C}^n,0}$, generated by the *upper*, respectively, *strictly upper monomials*,

$$I(F) := \left\{ G \in \mathcal{O}_{\mathbb{C}^n,0} \;\middle|\; G = \sum_{\ell(\boldsymbol{i}) \leq 0} \alpha_{\boldsymbol{i}} \boldsymbol{x}^{\boldsymbol{i}} \right\}, \tag{2.3.4.1}$$

$$\widetilde{I}(F) := \left\{ G \in \mathcal{O}_{\mathbb{C}^n,0} \;\middle|\; G = \sum_{\ell(\boldsymbol{i}) < 0} \alpha_{\boldsymbol{i}} \boldsymbol{x}^{\boldsymbol{i}} \right\}.$$

These are vector spaces of finite codimension in $\mathcal{O}_{\mathbb{C}^n,0}$ with quotient spaces having as basis the strictly lower, respectively, lower monomials, such that

$$\dim_{\mathbb{C}} \mathcal{O}_{\mathbb{C}^n,0} / I(F) = \#\{\boldsymbol{i} \mid \ell(\boldsymbol{i}) > 0\} ,$$
$$\dim_{\mathbb{C}} \mathcal{O}_{\mathbb{C}^n,0} / \widetilde{I}(F) = \#\{\boldsymbol{i} \mid \ell(\boldsymbol{i}) \geq 0\} .$$

If $\varphi : (\Sigma, z) \to (\mathbb{C}^n, 0)$ is an isomorphism such that $F = 0$ is an equation of the hypersurface germ $\varphi(X, z) \subset (\mathbb{C}^n, 0)$, then we define the ideals

$$I_z^{sqh} := \varphi^\#\big(I(F)\big) + I_z^{ea}, \text{ resp. } \widetilde{I}_z^{sqh} := \varphi^\#\big(\widetilde{I}(F)\big) + I_z^{ea},$$

in $\mathcal{O}_{\Sigma,z}$, where $\varphi^\# : \mathcal{O}_{\mathbb{C}^n,0} \to \mathcal{O}_{\Sigma,z}$ is the local algebra isomorphism induced by φ and

$$I_z^{ea} = I^{ea}(X, z) = \varphi^\#(\langle F, \frac{\partial F}{\partial x_1}, \ldots, \frac{\partial F}{\partial x_n} \rangle)$$

is the Tjurina ideal of the germ (X, z) (cf. Sect. 1.4.1). The ideals I_z^{sqh}, $\widetilde{I}_z^{sqh}$ define zero-dimensional schemes

$$Z_z^{sqh} = Z^{sqh}(X, z) \subset \Sigma\,, \qquad \widetilde{Z}_z^{sqh} = \widetilde{Z}^{sqh}(X, z) \subset \Sigma\,,$$

respectively, with support $\{z\}$.

Similarly, if $z \in X \subset \Sigma$ is a NND singular point represented by the holomorphic germ $F : (\mathbb{C}^n, 0) \to (\mathbb{C}, 0)$ with Newton diagram $\Gamma(F, 0)$ then we define the ideals $I(F)$, respectively $\widetilde{I}(F)$, in $\mathcal{O}_{\mathbb{C}^n,0}$ as being generated by the monomials $\boldsymbol{x}^{\boldsymbol{i}}$ with $\boldsymbol{i} \in \Delta(F)$, respectively $\boldsymbol{i} \in \Delta(F) \setminus \Gamma(F, 0)$,

$$I(F) := \left\{ G \in \mathcal{O}_{\mathbb{C}^n,0} \;\middle|\; G = \sum_{\boldsymbol{i}\in\Delta(F)} \alpha_{\boldsymbol{i}} \boldsymbol{x}^{\boldsymbol{i}} \right\},$$

$$\widetilde{I}(F) := \left\{ G \in \mathcal{O}_{\mathbb{C}^n,0} \;\middle|\; G = \sum_{\boldsymbol{i}\in\Delta(F)\setminus\Gamma(F,0)} \alpha_{\boldsymbol{i}} \boldsymbol{x}^{\boldsymbol{i}} \right\}.$$

The codimensions of these vector spaces are

$$\dim_{\mathbb{C}} \mathcal{O}_{\mathbb{C}^n,0}\big/ I(F) = \#\big(\mathbb{Z}^n \cap (K_0(F) \setminus \Gamma(F, 0))\big)\,,$$
$$\dim_{\mathbb{C}} \mathcal{O}_{\mathbb{C}^n,0}\big/ \widetilde{I}(F) = \#\big(\mathbb{Z}^n \cap K_0(F)\big)\,.$$

Again, if $\varphi : (\Sigma, z) \to (\mathbb{C}^n, 0)$ is an isomorphism such that $F = 0$ is an equation of the germ $\varphi(X, z)$, we obtain ideals

$$I_z^{nnd} := \varphi^\#\big(I(F)\big) + I_z^{ea}, \text{ resp. } \widetilde{I}_z^{nnd} := \varphi^\#\big(\widetilde{I}(F)\big) + I_z^{ea}$$

in $\mathcal{O}_{\Sigma,z}$. They define zero-dimensional schemes

$$Z_z^{nnd} = Z^{nnd}(X, z) \subset \Sigma\,, \qquad \widetilde{Z}_z^{nnd} = \widetilde{Z}^{nnd}(X, z) \subset \Sigma\,,$$

respectively, with support $\{z\}$.

Deformations of Singularities. In the above notation, let $\varphi : (\Sigma, z) \to (\mathbb{C}^n, 0)$ be an isomorphism defined in an open neighborhood of z. Moreover, let $\overline{B} \subset \Sigma$ be a sufficiently small compact neighborhood of z in Σ (B its interior), such that φ maps $\overline{B}$ diffeomorphically to a closed $2n$-dimensional ball $\overline{B}_{\varepsilon_0} \subset \mathbb{C}^n$ with center 0 and radius $\varepsilon_0 > 0$. Since X has an isolated singularity at z, for every sufficiently small $\varepsilon_0 > 0$, $\overline{X}_0 := X \cap \overline{B}$ is a compact space with boundary $\partial\overline{X}_0 \subset \partial\overline{B}$ a real $(2n-3)$-dimensional C^∞-smooth manifold. We choose $0 < \varepsilon < \varepsilon_0$, denote by B_ε the interior of $\overline{B}_\varepsilon \subset \overline{B}_{\varepsilon_0}$ and redefine $\overline{B} = \varphi^{-1}(\overline{B}_\varepsilon)$ and $\overline{X}_0 = X \cap \overline{B}$, in order that everything is defined in an open neighborhood of $\overline{B}$. Moreover, we set $X_0 := X \cap B$ and denote by ∂B, resp. ∂X_0, the boundary of $\overline{B}$, resp. $\overline{X}_0$. Then X_0 is homeomorphic to the (open) cone over ∂X_0 (cf. [Mil]). It is a complex space with isolated singularity at z, $X_0 \setminus \{z\}$ is smooth and intersects ∂B transversally for all $0 < \varepsilon < \varepsilon_0$.

B is also called a *Milnor ball* and ∂X_0 the *link* of the hypersurface germ $(X, z) \subset (\Sigma, z)$.

We consider now one-parameter (embedded) deformations of the hypersurface germ $(X, z) \subset (\Sigma, z) \cong (\mathbb{C}^n, 0)$. For topological considerations, we need good representatives of such deformations.

Definition 2.3.9 (1) A *good representative* of a *one-parameter deformation* of $(X_0, z) := (X, z) \subset (\Sigma, z)$ (see Fig. 2.2) is given by an analytic hypersurface $\mathscr{X} \subset B \times D_\delta$, defined in an open neighborhood of $\overline{B} \times D_\delta$, B a sufficiently small Milnor ball of (X, z), $D_\delta = \{t \in \mathbb{C} \mid |t| < \delta\}$ an open disc in $\mathbb{C}$, such that (with the above notation):

(a) $X_0 = (\pi|_{\mathscr{X}})^{-1}(0)$, where $\pi : B \times D_\delta \to D_\delta$ is the projection,
(b) $X_t := (\pi|_{\mathscr{X}})^{-1}(t)$ is a complex space with isolated singularities,
(c) The restriction $\pi : \partial B \times D_\delta \to D_\delta$ is submersive, that is, X_t is smooth near ∂B and intersects ∂B transversally in a real $(2n-3)$-dimensional smooth manifold $\partial X_t \subset \partial B$.

If $z, X, \Sigma, \mathscr{X}$ are real, that is, invariant under the complex conjugation conj, then we take only real non-negative values of the parameter t, and denote by D_δ the half open interval $[0, \delta)$.

(2) Let $\mathcal{S}$ be a finite set of smooth singularity types. A good representative $\mathscr{X} = \{X_t,\ t \in D_\delta\}$ of a deformation of (X, z) is called *$\mathcal{S}$-adjacent* if the following holds:

For each $t_0 \neq 0$ in D_δ there exists an open neighborhood $U = U(t_0) \subset D_\delta \setminus \{0\}$ such that the restriction $\pi_U : (\pi|_{\mathscr{X}})^{-1}(U) \to U$ has (disjoint) singular sections $\sigma_1, \dots, \sigma_r : U \to (\pi|_{\mathscr{X}})^{-1}(U)$ and is $\mathcal{S}$-equisingular along these sections. That is, $\{\sigma_1(t), \dots, \sigma_n(t)\} = \mathrm{Sing}(X_t)$ and the $\mathcal{S}$-type of $(X_t, \sigma_i(t))$ is constant for $t \in U$.

Note that for $\mathcal{S}$-adjacent deformations of (X, z) all fibres X_t, $t \neq 0$ have the same number r of singular points, and for $t_1, t_2 \neq 0$ there is a bijection between the singularities of X_{t_1} and X_{t_2} respecting their $\mathcal{S}$-types. We do not require anything about the singularity of X_0, even if $r = 1$.

Given any deformation of the isolated hypersurface singularity (X, z) over the germ $(\mathbb{C}, 0)$, we can always find a good representative (cf. [Mil, Loo]) by choosing $0 < \delta << \varepsilon$ sufficiently small.

Deformation Patterns of SQH Singularities. Let $z \in X \subset \Sigma$ be a semiquasihomogeneous singular point with representative $F : (\mathbb{C}^n, 0) \to (\mathbb{C}, 0)$, $F(\boldsymbol{x}) = \sum_{\boldsymbol{i}} A_{\boldsymbol{i}} \boldsymbol{x}^{\boldsymbol{i}}$. Moreover, let $\sigma \subset \Gamma(F, 0)$ be the main face, lying on the hyperplane given by the affine equation $\ell(\boldsymbol{i}) = 0$, $\ell(0) > 0$, where

$$\ell(\boldsymbol{i}) = -\omega_0 + \omega_1 i_1 + \dots + \omega_n i_n, \quad \omega_j \in \mathbb{Z}_{<0},\ j = 0, \dots, n\ .$$

We call $-\omega_j$ the *weight* of the variable x_j, $j = 1, \dots, n$, and $-\omega_0$ the (weighted) degree of F.

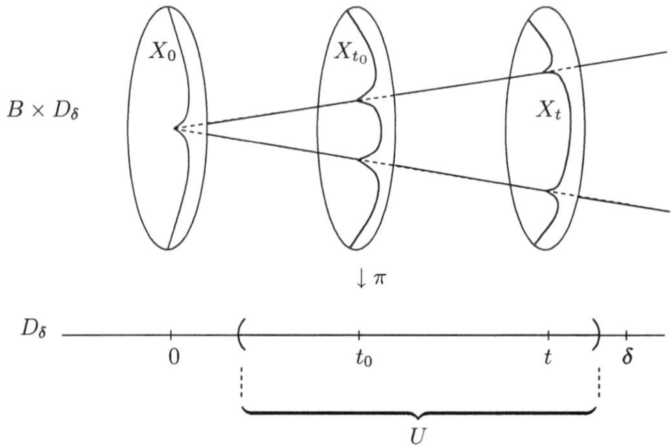

Fig. 2.2 An $\mathcal{S}$-adjacent good representative of a one-parameter deformation

Then we introduce the space of (lower) polynomials

$$\mathcal{P}_+(\sigma) := \left\{ \sum_{\ell(0)\geq\ell(\boldsymbol{i})\geq 0} \alpha_i \boldsymbol{x}^{\boldsymbol{i}} \right\} \subset \mathbb{C}[\boldsymbol{x}] = \mathbb{C}[x_1, \ldots, x_n]$$

with support below or on the affine hyperplane $\{\ell(\boldsymbol{i}) = 0\}$, and the affine subspace

$$\mathcal{P}_+(\sigma, F) = \left\{ P \in \mathcal{P}_+(\sigma) \,\middle|\, P^\sigma = F^\sigma \right\}.$$

By definition, $F^\sigma = \sum_{\ell(\boldsymbol{i})=0} A_i \boldsymbol{x}^{\boldsymbol{i}}$ has 0 as isolated critical point. Using the $\mathbb{C}^*$-action

$$(\boldsymbol{x}, \lambda) \mapsto \lambda\boldsymbol{x} := (\lambda^{-\omega_1} x_1, \ldots, \lambda_n^{-\omega_n} x_n), \quad \lambda \in \mathbb{C}^*,$$

we see that 0 is the only critical point of F^σ in $(\mathbb{C}^*)^n$. We can write $P \in \mathcal{P}_+(\sigma, F)$ as $P = F^\sigma + P'$ with $\ell(\boldsymbol{i}) > 0$ for all $\boldsymbol{i} \in \operatorname{supp}(P')$. Since $P'(\lambda\boldsymbol{x}) = \lambda^{-\omega}\widetilde{P}(\boldsymbol{x}, \lambda)$ for some $\omega < \omega_0$ and some polynomial $\widetilde{P}$, we get

$$P(\lambda\boldsymbol{x}) = \lambda^{-\omega_0}(F^\sigma + \lambda^{\omega_0-\omega}\widetilde{P}(\boldsymbol{x}, \lambda)).$$

Hence, the family of coordinate changes $\boldsymbol{x} \mapsto \lambda\boldsymbol{x}$, $\lambda \neq 0$, induces a family of hypersurfaces $\{P(\lambda\boldsymbol{x}) = 0\} = \{F^\sigma + \lambda^{\omega_0-\omega}P(\boldsymbol{x}, \lambda) = 0\}$ which is a deformation of $\{F^\sigma = 0\}$. It follows that $P(\lambda\boldsymbol{x})$ and therefore $P \in P_+(\sigma, F)$ can have only isolated critical points in $\mathbb{C}^n$.

Definition 2.3.10 Any polynomial $P \in \mathcal{P}_+(\sigma, F)$ is called a *SQH pattern for (a deformation of) the singular point* (X, z) defined by $\{F = 0\}$.

We can assume that the local isomorphism $\varphi : (\Sigma, z) \to (\mathbb{C}^n, 0)$, identifying the germ (X, z) with $(\{F=0\}, 0)$, takes B to a ball $B_\varepsilon \subset \mathbb{C}^n$ centered at the origin with radius $\varepsilon > 0$. By applying the $\mathbb{C}^*$-action, keeping the polynomial P in the space $\mathcal{P}_+(\sigma, F)$, we can move all its critical points to $\mathrm{Int}(B_\varepsilon)$ and obtain that the intersection of $\{P=0\}$ with the boundary of any ball centered at the origin with radius at most ε is transversal. Furthermore, without loss of generality we suppose that $\{F=0\} \cap \partial B_\varepsilon$ and $\{P=0\} \cap \partial B_\varepsilon$ lie in a small tubular neighborhood U of $\{F^\sigma=0\} \cap \partial B_\varepsilon$.

Let P be a SQH pattern of F and $\mathcal{S}$ a collection of smooth singularity types of the singularities of the hypersurface $\{P = 0\}$. We say that an *$\mathcal{S}$-adjacent deformation* $X_t, t \in D_\delta$, of (X, z) *matches the pattern* P, if, for any $t \in D_\delta \setminus \{0\}$ there is a homeomorphism $\psi_t : B_r \to B_r$ such that $\psi_t\big(\varphi(X_t \cap B)\big) = \{P=0\} \cap B_\varepsilon$ and the (via ψ_t) corresponding singular points of $X_t \cap B$ and $\{P=0\} \cap B_\varepsilon$ are $\mathcal{S}$-equivalent.

A pattern P is called *$\mathcal{S}$-transversal*, if the Newton triple $(\Delta(P), \emptyset, P)$ is $\mathcal{S}$-transversal, and P is called *$\mathcal{S}^s$-transversal*, or *strongly $\mathcal{S}$-transversal*, if the triple $(\Delta(P), \sigma, P)$ is $\mathcal{S}$-transversal (cf. Definition 2.3.3).

In the real case, i.e. when the given data are invariant with respect to the complex conjugation, we say that (X, z) matches the pattern P if, in addition to the above requirements, there are real, linearly independent lines $L_1, \ldots, L_n \subset \mathbb{R}^n \subset \mathbb{C}^n$ through the origin, which do not meet U, and the (equivariant) homeomorphisms ψ_t have the fixed point set

$$\mathrm{Fix}(\psi_t) \supset (L_1 \cup \ldots \cup L_n) \cap \partial B_r .$$

The meaning of the fixation of the points $(L_1 \cup \ldots \cup L_n) \cap \partial B_r$ is that ψ_t is, in fact, close to the identity along ∂B_r and, for example, in the case $n = 2$ we allow neither an interchange of components of the link of the given singular point, nor a non-trivial automorphisms of these components.

Deformation Patterns of NND Singularities. Let $z \in X \subset \Sigma$ be a NND singular point, and let $F = \sum_{i \in \Delta(F)} A_i \boldsymbol{x}^{\boldsymbol{i}}$ be a representative of (X, z). Consider the following data

- a convex subdivision of $K(F)$ of the form $K(F) = \Delta(F) \cup \Delta_1 \cup \ldots \cup \Delta_N$ defined by a convex piece-wise linear function ν such that $\nu\big|_{\Delta(F)} = 0$,
- a set of numbers $\big\{A_{\boldsymbol{i}} \in \mathbb{C} \,\big|\, \boldsymbol{i} \in K_0(F)\big\}$ such that $A_{\boldsymbol{i}}$, $\boldsymbol{i} \in \Gamma(F, 0)$, are the (corresponding) coefficients of F, $A_{\boldsymbol{i}} \neq 0$ if $\boldsymbol{i}$ is a vertex of one of the polytopes $\Delta_1, \ldots, \Delta_N$ and, moreover, for any $k = 1, \ldots, N$, the polynomial $F_k := \sum_{\boldsymbol{i} \in \Delta_k} A_{\boldsymbol{i}} \boldsymbol{x}^{\boldsymbol{i}}$ is PND.

It follows that all the polynomials F_k define hypersurfaces in $(\mathbb{C}^*)^n$ with only isolated singularities. This can be seen similarly as in the case of SQH singularities above, since F_k belongs to the space of lower polynomials.

Denote by $\mathrm{Sing}(F_1, \ldots, F_N)$ the disjoint union

$$\big(\mathrm{Sing}(\{F_1=0\}) \cap (\mathbb{C}^*)^n\big) \cup \ldots \cup \big(\mathrm{Sing}(\{F_N=0\}) \cap (\mathbb{C}^*)^n\big) .$$

Definition 2.3.11 The N-tuple of polynomials $(F_1, \ldots, F_N)$ is called a *NND pattern for a deformation of the singular point* (X, z) given by F. Let $\mathcal{S}$ be a collection of smooth singularity types of the singularities of the hypersurfaces $\{F_1 = 0\}$, …, $\{F_N = 0\}$.

We say that an *$\mathcal{S}$-adjacent deformation* X_t, $t \in D_\delta$, of the NND singular point (X, z) *has pattern* $(F_1, \ldots, F_N)$, if there is a continuous family of bijections $\mathrm{Sing}(X_t) \to \mathrm{Sing}(F_1, \ldots, F_N)$, $t \in D_\delta \setminus \{0\}$, which map a singular point of X_t to an $\mathcal{S}$-equivalent singular point of one of the hypersurfaces $\{F_1 = 0\}$, …, $\{F_N = 0\}$.

Remark 2.3.11.1 In fact, similarly as in the case of SQH singular points, one can define here homeomorphisms ψ_t between $X_t \cap \overline{B}$, $t \neq 0$, and some singular hypersurface with boundary, constructed along the procedure described in [ItS4] (see also [Shu12]).

Now, we introduce the $\mathcal{S}$- and $\mathcal{S}^s$-transversality for deformation patterns of NND singularities as defined above.

Definition 2.3.12 Let G be the dual graph of the subdivision $K(F) = \Delta(F) \cup \Delta_1 \cup \cdots \cup \Delta_N$, and $\mathcal{G}$ the set of oriented graphs $\widetilde{G}$ with support G and without oriented cycles. The pattern $(F_1, \ldots, F_N)$ is called *$\mathcal{S}$-transversal*, if there exists a graph $\widetilde{G} \in \mathcal{G}$ such that all triples $(\Delta_i, \Delta_{i,+}(\widetilde{G}), F_i)$, $i = 1, \ldots, N$, (2.3.3.2) are $\mathcal{S}$-transversal. If, in addition, no (oriented) edge of $\widetilde{G}$ has source in $\Delta(F)$ we call the pattern $(F_1, \ldots, F_N)$ *$\mathcal{S}^s$-transversal*. Moreover, if $n = 2$ and $\Delta_{i,+}(G) \neq \partial\Delta_i$, $i = 1, \ldots, N$, we assume in the above definition only (weak) $\mathcal{S}$-transversality for all triples (cf. Definition 2.3.3).

2.3.5 Lower Deformations of Hypersurface Singularities

In Theorems 2.3.4 and 2.3.5 we proved the existence of affine hypersurfaces with the prescribed Newton polygon by gluing and deforming PND block polynomials, provided their Newton polygons satisfied some transversality condition. By prescribing the block polynomials we actually prescribed affine patches with a bunch of singularities.

The transversality condition, defined and discussed in Sect. 2.3.2, allowed to glue the patches to get a single polynomial with Newton polygon the union of the prescribed ones.

In this section we pass from affine patches to analytic local patches and thus we are going to prescribe individual singularities and their types. That is, given a hypersurface X in the complex manifold Σ we wish to prescribe for each $z \in \mathrm{Sing}(X)$ a specific deformation pattern for a one parametric deformation X_t, $t \in D_\delta$, such that for $t \in D_\delta \setminus \{0\}$, X_t has the prescribed singularities (thus proving existence). In our applications we have to consider different kinds of singularity types. Therefore we split $\mathrm{Sing}(X)$ into disjoint subsets S_i, $i = 0, \ldots, 5$, such that

- for the points $z \in S_0$, we prescribe smooth singular types (not further specified),
- for the points $z \in S_1$, we prescribe arbitrary (analytic) singularity types appearing in the semiuniversal deformation of the (X, z),
- for the SQH singular points $z \in S_2 \cup S_3$, we prescribe an SQH deformation pattern,
- for the NND singular points $z \in S_4 \cup S_5$, we prescribe an NND deformation pattern.

Of course, in order to be able to glue the local patches, we have to require again a certain transversality condition. For SQH and NND singularities this can be expressed as an H^1-vanishing condition for the ideal sheaves of the zero-dimensional schemes introduced in Sect. 2.3.4.

Deformations, whose existence is claimed in the following statement are called *lower*, because their local analytic description is a variation of terms lying below the Newton diagrams. For the definition of the zero-dimensional schemes below see Sect. 2.3.4.

Let $\mathcal{S}$ be a set of arbitrary smooth singularity types (Definition 2.2.42) and denote by $H_X^{\mathcal{S}}$, resp. H_X^a, the germs at X of the $\mathcal{S}$-equisingularity stratum, resp. equianalytic stratum, defined in Sect. 2.2.3, Definitions 2.2.49 and 2.2.53 (with the notation $V_{|D|/\Sigma}(\mathcal{S})$).

Theorem 2.3.13 *Let Σ be a smooth projective algebraic variety of dimension $n \geq 2$ and $X \subset \Sigma$ be a hypersurface with a finite singular set* $\mathrm{Sing}(X)$. *Let* $\mathrm{Sing}(X) = \bigcup_{i=0}^{5} S_i$ *be a disjoint splitting such that the singular points in $S_2 \cup S_3$ are SQH, the singular points in $S_4 \cup S_5$ are NND, and the points $z \in S_0$ have singularity types in $\mathcal{S}$.*

Denote by $M_X^{\mathcal{S}}$ the germ at X of the $\mathcal{S}$-equisingular stratum in the linear system $\left|H^0\big(\mathcal{O}_\Sigma(X)\big)\right|$ corresponding to the set S_0, and denote by M_X^a the germ at X of the equianalytic stratum in $\left|H^0\big(\mathcal{O}_\Sigma(X)\big)\right|$ corresponding to the set S_1. Moreover, let $Z = Z' \cup Z''$ be the zero-dimensional scheme such that

$$Z' = \bigcup\nolimits_{z \in S_0} Z_z^{\mathcal{S}} \cup \bigcup\nolimits_{z \in S_1} Z_z^{ea} ,$$

$$Z'' = \bigcup_{z \in S_2} Z_z^{sqh} \cup \bigcup_{z \in S_3} \widetilde{Z}_z^{sqh} \cup \bigcup_{z \in S_4} Z_z^{nnd} \cup \bigcup_{z \in S_5} \widetilde{Z}_z^{nnd}$$

and $\mathcal{J}_{Z/\Sigma}$ its ideal sheaf in $\mathcal{O}_\Sigma$. Assume that

$$H^1\big(\mathcal{J}_{Z/\Sigma}(X)\big) = 0 . \tag{2.3.5.1}$$

Note that $H^1(\mathcal{J}_{Z'/\Sigma}(X)) = 0$ implies that the germs $M_X^{\mathcal{S}}$ and M_X^a are smooth of expected dimension and that $H^1(\mathcal{J}_{Z/\Sigma}(X)) = 0$ implies that the intersection $M_X^{\mathcal{S}} \cap M_X^a \cap \left|H^0\big(\mathcal{J}_{Z''/\Sigma}(X)\big)\right| \subset \left|H^0\big(\mathcal{O}_\Sigma(X)\big)\right|$ is transversal (cf. Theorem 2.2.38). Only these transversality conditions are used in the proof (and not the vanishing of H^1).

Then there exists a family of hypersurfaces $X_t \in |X| = \left|H^0\big(\mathcal{O}_\Sigma(X)\big)\right|$, $t \in D_\delta$, that is, an embedded deformation of X in Σ over D_δ, such that $X_0 = X$, and the following holds. The family X_t induces for $z \in$ $\mathrm{Sing}(X)$ *a deformation of the germ (X, z) of the following types:*

- *for any $z \in S_0$, an $\mathcal{S}$-equisingular family,*
- *for any $z \in S_1$, an arbitrary prescribed deformation of (X, z),*
- *for any $z \in S_2 \cup S_4$ an $\mathcal{S}$-adjacent deformation matching a prescribed $\mathcal{S}^s$-transversal pattern,*
- *for any $z \in S_3 \cup S_5(X)$ an $\mathcal{S}$-adjacent deformation matching a prescribed $\mathcal{S}$-transversal pattern.*

In addition, for any $t \neq 0$, the intersection of the (smooth) germs

$$M_X^{\mathcal{S}}(S_0) \cap M_X^a(S_1) \cap M_{X_t}^{\mathcal{S}}(\mathrm{Sing}'(X_t)) \tag{2.3.5.2}$$

is transversal, where $\mathrm{Sing}'(X_t)$ *is the set of the singular points of X_t in a neighborhood of the set $\bigcup_{i=2}^5 S_i$.*

If all the data are invariant with respect to conj, *then there exists a* conj-*equivariant family X_t, $t \in [0, \delta)$, with the above properties.*

Remark 2.3.13.1 Note that the theorem is a statement about independence and realizability of local deformations of the singularities of X inside the linear system $|X|$. That is, if (2.3.5.1) is satisfied, the following holds:

For $z \in \mathrm{Sing}(X)$ denote by $\phi_z : (Y, z) \to (B, 0)$ the semiuniversal deformation of the germ (X, z), $Y_s = \phi^{-1}(s)$ for $s \in B$, with $\phi : Y \to B$ a good representative of ϕ_z, and for any collection $y = \{y_1, \dots, y_k\}$ of singularities of Y_s denote by (Y, y) the multigerm of Y_s at y.

Then there exists a one-parameter family of hypersurfaces $\mathscr{X} \to D$, $\mathscr{X} = \{X_t \in |X|\}_{t \in D}$, $X_0 = X$, $D = D_\delta$ a small disc in $\mathbb{C}$ centered at 0, and for any $z \in \mathrm{Sing}(X)$ sections $\sigma : D \to \mathscr{X}$ resp. for $z \in S_1(X)$ multisections $\sigma_z = \{\sigma_{1,z}, \dots, \sigma_{p,z}\}$, $\sigma_z(0) = z$, $\sigma_{i,z}(0) = z$, with the following property

- For $z \in S_0$ the germ $(\mathscr{X}, z) \to (D, 0)$ is $\mathcal{S}$-equisingular along σ_z.
- For $z \in S_1$ and any local deformation (Y, y) of (X, z) as above, $(X_t, \sigma_{i,z}(t))$ is analytically isomorphic to (Y, y_i) as $t \neq 0$.
- For $z \in S_2 \cup S_4$ and any $\mathcal{S}^s$-transversal pattern P for (X, z), $(X_t, \sigma_z(t))$ is an $\mathcal{S}$-adjacent deformation, matching P.
- For $z \in S_3(X) \cup S_5(X)$ and any $\mathcal{S}$-transversal pattern P for (X, z), $(X_t, \sigma_z(t))$ is an $\mathcal{S}$-adjacent deformation, matching P.

In particular, for $t \neq 0$ and for each $z \in \mathrm{Sing}(X)$, $X_t \in |X|$ has singularities $z' \in \mathrm{Sing}(X_t)$ close to z, of the types corresponding to z as described above.

Remark 2.3.13.2 (1) When applying Theorem 2.3.13, all simple singularities (of type A, D, or E) of X should be included into S_1, since they are SQH and, by definition, $Z^{sqh} = Z^{ea}$ for a simple singularity.

(2) There is some overlapping between the statements of Theorems 2.3.4, 2.3.5 and 2.3.13. The common situation, when both theorems apply, is when X is a hypersurface in $\mathbb{P}^n$, its SQH and NND singular points are located at the fundamental points of a projective coordinate system, and an equation $F = 0$ of X is a representative for all these singular points.

We should like to point out that in this case Theorem 2.3.13 is *stronger* than Theorems 2.3.4, 2.3.5. Indeed, the $\mathcal{S}$-transversality condition imposed on the triple $(\Delta(F), \Delta_+(\widetilde{G}), F)$ in Theorems 2.3.4, 2.3.5 is equivalent to the H^1-vanishing for the ideal sheaf of a zero-dimensional scheme which is *bigger* than that in Theorem 2.3.13. For instance, for an SQH singular point the ideal sheaf in Theorems 2.3.4, 2.3.5 is given by $I(F)$, resp. by $\widetilde{I}(F)$, whereas in Theorem 2.3.13 we use the ideals $I(F) + I^{ea} \supsetneq I(F)$, $\widetilde{I}(F) + I^{ea} \supsetneq \widetilde{I}(F)$, respectively.

Proof The argument is presented in several steps. First, we treat in detail the case of one SQH or NND singular point with an $\mathcal{S}$-, or $\mathcal{S}^s$-transversal deformation pattern, then we explain how to combine these arguments in the case of several singularities.

Step 1. Assume that $S_2 = \{z\}$, $S_1 = S_3 = S_4 = S_5 = \emptyset$.
1.1. From the exact sequence

$$0 \longrightarrow \mathcal{J}_{Z/\Sigma}(X) \longrightarrow \mathcal{O}_\Sigma(X) \longrightarrow \mathcal{O}_Z \longrightarrow 0 \qquad (2.3.5.3)$$

and, by assumption (a), we obtain

$$\begin{aligned} H^0\big(\mathcal{O}_\Sigma(X)\big) &\cong H^0\big(\mathcal{J}_{Z/\Sigma}(X)\big) \oplus H^0(\mathcal{O}_Z) \\ &= H^0\big(\mathcal{J}_{Z/\Sigma}(X)\big) \oplus \mathcal{O}_{\Sigma,z}\big/I_z^{sqh} . \end{aligned} \qquad (2.3.5.4)$$

For the point z, we have an isomorphism $\varphi = \varphi_z : \mathcal{O}_{\Sigma,z} \to \mathcal{O}_{\mathbb{C}^n,0}$ and a semiquasihomogeneous representative $F : (\mathbb{C}^n, 0) \to (\mathbb{C}, 0)$ such that $F = 0$ is an equation of the germ (X, z) in the coordinates defined in a neighbourhood of z in Σ by φ. We can write

$$F = \sum_{\ell(\boldsymbol{i}) \le 0} A_{\boldsymbol{i}} \boldsymbol{x}^{\boldsymbol{i}} ,$$

where $\ell = 0$ is the equation of a hyperplane containing the main face σ of the Newton diagram of F, $\ell(0) > 0$ and ℓ is integral-valued at integral points. Let

$$P_z = \sum_{\ell(\boldsymbol{i}) \ge 0} A_{\boldsymbol{i}} \boldsymbol{x}^{\boldsymbol{i}} .$$

be a given $\mathcal{S}^s$-transversal deformation pattern for (X, z).
1.2. Let $\Lambda, \Lambda_1, \ldots, \Lambda_n \subset \mathbb{N}^n$ be (finite) sets such that $\{\boldsymbol{x}^{\boldsymbol{i}} \mid \boldsymbol{i} \in \Lambda\}$ is projected to a basis of $\mathcal{O}_{\mathbb{C}^n,0}/I^{sqh}(F)$, where $I^{sqh}(F) := I(F) + \langle \partial F/\partial x_1, \ldots, \partial F/\partial x_n \rangle$, $I(F)$ as in (2.3.4.1), and

$$\big\{\boldsymbol{x}^{\boldsymbol{i}} \,\big|\, \boldsymbol{i} \in \Lambda\big\} \cup \bigcup_{k=1}^{n} \left\{ \boldsymbol{x}^{\boldsymbol{j}} \frac{\partial F}{\partial x_k} \,\middle|\, \boldsymbol{j} \in \Lambda_k \right\}$$

is projected to a basis of $\mathcal{O}_{\mathbb{C}^n,0}/I(F)$. Since, as a vector space, $\mathcal{O}_{\mathbb{C}^n,0}\big/I(F)$ is isomorphic to the direct sum

$$\mathcal{O}_{\mathbb{C}^n,0}\Big/\left(I(F)+\left\langle\frac{\partial F}{\partial x_1},\ldots,\frac{\partial F}{\partial x_n}\right\rangle\right)$$
$$\oplus\bigoplus_{k=1}^{n}\left(\left\langle\frac{\partial F}{\partial x_1},\ldots,\frac{\partial F}{\partial x_k}\right\rangle+I(F)\right)\Big/\left(\left\langle\frac{\partial F}{\partial x_1},\ldots,\frac{\partial F}{\partial x_{k-1}}\right\rangle+I(F)\right),$$

such a basis of $\mathcal{O}_{\mathbb{C}^n,0}/I(F)$ does always exist.

Example 2.3.13.3 Let $F = xy^2 + x^4 \in \mathbb{C}[x, y]$ (a D_5-singularity). Then we obtain $I(F) = \langle x^i y^j \mid 2i + 3j \geq 8\rangle$, while $I^{sqh}(F) = \langle y^2+4x^3,\ xy,\ x^4,\ y^3\rangle$, and $\Lambda \subset \mathbb{Z}^2$ is the set of (multi-)exponents of $\{1, x, y, x^2, x^3\}$ projecting to a basis of $\mathbb{C}\{x, y\}/I^{sqh}(F)$. Moreover, $\{1, x, y, x^2, xy, y^2, x^3, x^2y, xy^2\}$ represents a basis of $\mathbb{C}\{x, y\}/I(F)$. Therefore,

$$\{1,x,y,x^2,x^3\}\cup\left\{\frac{\partial F}{\partial x}=y^2+4x^3,\ x\cdot\frac{\partial F}{\partial x}=xy^2+4x^4\right\}\cup\left\{\frac{\partial F}{\partial y}=2xy,\ x\cdot\frac{\partial F}{\partial y}=2x^2y\right\}$$

projects to a basis of $\mathbb{C}\{x, y\}/I(F)$, whence we can choose $\Lambda_1 = \{(0, 0), (1, 0)\}$, $\Lambda_2 = \{(0, 0), (1, 0)\}$.

Denote by $[Q]$ the sum of those terms of a polynomial Q which do not belong to $I(F)$. By taking germs, global sections of $\mathcal{O}_\Sigma(X)$ are mapped injectively to $\mathcal{O}_{\Sigma,z}$, and then by φ isomorphically to $\mathcal{O}_{\mathbb{C}^n,0}$. Hence, the relation (2.3.5.4) implies that there exists a basis of $H^0(\mathcal{O}_\Sigma(X))$, consisting of elements represented in $\mathcal{O}_{\Sigma,z} = \mathcal{O}_{\mathbb{C}^n,0}$ as

$$F_i = \boldsymbol{x}^i + \sum_{\substack{1\leq k\leq n\\ j\in\Lambda_k}} A_{i,kj}\left[\boldsymbol{x}^j\frac{\partial F}{\partial x_k}\right] + \sum_{\ell(\boldsymbol{m})\leq 0} A_{i,\boldsymbol{m}}\boldsymbol{x}^{\boldsymbol{m}}, \quad \boldsymbol{i}\in\Lambda,$$
$$F_\ell = \sum_{\substack{1\leq k\leq n\\ j\in\Lambda_k}} A_{\ell,kj}\left[\boldsymbol{x}^j\frac{\partial F}{\partial x_k}\right] + \sum_{\ell(\boldsymbol{m})\leq 0} A_{\ell,\boldsymbol{m}}\boldsymbol{x}^{\boldsymbol{m}}, \quad \ell = 0,\ldots,s,$$

where the last sums on the right-hand side represent elements of $I(F)$, and $\{F_0, \ldots, F_s\}$ a basis of $H^0(\mathcal{J}_{Z/\Sigma}(X))$.

Denote by B_i, $\boldsymbol{i} \in \Lambda$, B_ℓ, $\ell = 0, \ldots, s$, the coordinates in $H^0(\mathcal{O}_\Sigma(X))$ with respect to this basis. Since F belongs to $H^0(\mathcal{J}_{Z/\Sigma}(X))$ under this correspondence, we may assume, without loss of generality, that $F = F_0$, that is, F is encoded by $B_i = 0$ for $\boldsymbol{i} \in \Lambda$, and $B_0 = 1$, $B_\ell = 0$ for $\ell = 1, \ldots, s$.
1.3. Consider a deformation of F in $\varphi_z(H^0(\mathcal{O}_\Sigma(X))) \subset \mathcal{O}_{\mathbb{C}^n,0}$ defined by

$$F_t = F + \sum_{\boldsymbol{i}\in\Lambda} a_i F_i + \sum_{\ell=1}^{s} a_\ell F_\ell, \tag{2.3.5.5}$$

and depending a_i, $\boldsymbol{i} \in \Lambda$, a_ℓ, $\ell = 1, \ldots, s$, which will be specified later as smooth functions of a small parameter t vanishing at $t = 0$.

The requirement of Theorem 2.3.13 means that the elements of this deformation realize the given pattern P_z in a neighbourhood of z, and also belong to M, the germ at X of the $\mathcal{S}$-equisingular stratum in $H^0(\mathcal{O}_\Sigma(X))$ corresponding to the points $w \in S_0$ (in the above notation M is the preimage in $H^0(\mathcal{O}_\Sigma(X))$ of the germ $M_X^{\mathcal{S}}(S_0) \subset |X|$).

1.4. As mentioned in the theorem after (2.3.5.1), M is smooth and transversal to $H^0(\mathcal{J}_{Z_z^{sqh}/\Sigma}(X))$. Since the $B_i, \boldsymbol{i} \in \Lambda$, are transversal coordinates to $H^0(\mathcal{J}_{Z_z^{sqh}/\Sigma}(X))$, there is a smooth germ $M' \subset M$ defined by $B_0 = 1$, and $B_\ell = A_\ell + \psi_\ell(\{B_{\boldsymbol{i}} \mid \boldsymbol{i} \in \Lambda\})$, $\ell = 1, \dots, s$, where the ψ_ℓ are analytic functions vanishing at zero. Therefore the elements of (2.3.5.5) belong to M' iff

$$a_\ell = \psi_\ell\big(\{a_{\boldsymbol{i}} \,\big|\, \boldsymbol{i} \in \Lambda\}\big), \quad \ell = 1, \dots, s. \tag{2.3.5.6}$$

1.5. Consider a local coordinate change $T = (T_1, \dots, T_n) : (\mathbb{C}^n, 0) \to (\mathbb{C}^n, 0)$,

$$T_k(\boldsymbol{x}) = x_k + \sum_{\boldsymbol{j} \in \Lambda_k} a_{k,\boldsymbol{j}} \boldsymbol{x}^{\boldsymbol{j}}, \quad k = 1, \dots, n, \tag{2.3.5.7}$$

where $a_{k,\boldsymbol{j}}$, $\boldsymbol{j} \in \Lambda_k$, $k = 1, \dots, n$, are small parameters. One derives

$$\begin{aligned} \widetilde{F}_t := F_t \circ T \;&=\; F_t + \sum_{k=1}^{n} \sum_{\boldsymbol{j} \in \Lambda_k} a_{k,\boldsymbol{j}} \boldsymbol{x}^{\boldsymbol{j}} \frac{\partial F}{\partial x_k} + O(\{a_{\boldsymbol{i}}^2\}, \{a_{k,\boldsymbol{j}}^2\}) \\ &= \sum_{\boldsymbol{k} \in \Lambda} c_{\boldsymbol{k}} \boldsymbol{x}^{\boldsymbol{k}} + \sum_{\ell(\boldsymbol{k}) \le 0} c_{\boldsymbol{k}} \boldsymbol{x}^{\boldsymbol{k}} + \sum_{k=1}^{n} \sum_{\boldsymbol{j} \in \Lambda_k} c_{k,\boldsymbol{j}} \left[\boldsymbol{x}^{\boldsymbol{j}} \frac{\partial F}{\partial x_k} \right], \end{aligned}$$

where the coefficients $c_{\boldsymbol{m}}$, $c_{k,\boldsymbol{j}}$ have the form

$$c_{\boldsymbol{m}} = a_{\boldsymbol{m}} + O(\{a_{\boldsymbol{i}}^2\}, \{a_{k,\boldsymbol{j}}^2\}), \quad \boldsymbol{m} \in \Lambda, \tag{2.3.5.8}$$

$$c_{\boldsymbol{m}} = A_{\boldsymbol{m}} + \psi_{\boldsymbol{m}}^{(1)}(\{a_{\boldsymbol{i}}\}, \{a_{k,\boldsymbol{j}}\}), \quad \ell(\boldsymbol{m}) \le 0, \tag{2.3.5.9}$$

with $\psi_{\boldsymbol{m}}^{(1)}$ being certain smooth functions vanishing at zero, and

$$c_{k,\boldsymbol{j}} = \sum_{\boldsymbol{i} \in \Lambda} a_{\boldsymbol{i}} A_{\boldsymbol{i},k\boldsymbol{j}} + \sum_{\ell=1}^{s} a_\ell A_{\ell,k\boldsymbol{j}} + a_{k,\boldsymbol{j}} + O(\{a_{\boldsymbol{i}}^2\}, \{a_{k,\boldsymbol{j}}^2\}), \tag{2.3.5.10}$$

$\boldsymbol{j} \in \Lambda_k$, $k = 1, \dots, n$.

By the implicit function theorem, the Eqs. (2.3.5.6), (2.3.5.8), (2.3.5.9), and (2.3.5.10) are solvable with respect to $a_{\boldsymbol{i}}$, $\boldsymbol{i} \in \Lambda$, a_ℓ, $\ell = 1, \dots, s$, $a_{k,\boldsymbol{j}}$, $\boldsymbol{j} \in \Lambda_k$, $k = 1, \dots, n$, as functions of

$$\big\{c_{\boldsymbol{m}} \,\big|\, \boldsymbol{m} \in \Lambda\big\} \cup \big\{c_{k,\boldsymbol{j}} \,\big|\, \boldsymbol{j} \in \Lambda_k,\ k = 1, \dots, n\big\}. \tag{2.3.5.11}$$

Hence, the family $\widetilde{F}_t$ can be reparameterized by the independent parameters (2.3.5.11):

$$\widetilde{F}_t = \sum_{m\in\Lambda} c_m x^m + \sum_{k=1}^{n}\sum_{j\in\Lambda_k} c_{k,j}\left[x^j\frac{\partial F}{\partial x_k}\right] + \sum_{\ell(m)\le 0}(A_m + \psi_m^{(2)})x^m$$

with $\psi_m^{(2)}$ being smooth functions in the parameters (2.3.5.11), vanishing at zero. Now we impose the following condition on the parameters (2.3.5.11):

$$\sum_{m\in\Lambda} c_m x^m + \sum_{k=1}^{n}\sum_{j\in\Lambda_k} c_{k,j}\left[x^j\frac{\partial F}{\partial x_k}\right] = \sum_{\ell(m)>0} d_m t^{\ell(m)} x^m,$$

where d_m is close to A_m for any $m \in \mathbb{N}^n$ with $\ell(m) > 0$, and t is a small parameter. The latter relation has a solution with respect to (2.3.5.11) as functions of $\{d_m t^{\ell(m)} \mid \ell(m) > 0\}$. Let $\gamma_1 = -\frac{\partial \ell}{\partial i_1}, \ldots, \gamma_n = -\frac{\partial \ell}{\partial i_n}$. Passing to the parameters $\{d_m\}$ and t and rescaling by the $\mathbb{C}^*$-action induced by the linear function ℓ, we obtain the family

$$\begin{aligned}\hat{F}_t &= \widetilde{F}_t(x_1 t^{\gamma_1}, \ldots, x_n t^{\gamma_n})\, t^{-\ell(0)} \\ &= \sum_{\ell(m)>0} d_m x^m + \sum_{\ell(m)\le 0} t^{-\ell(m)}(A_m + t\psi_m^{(3)})x^m\,,\end{aligned}$$

where the $\psi_m^{(3)}$ are smooth functions of $\{d_m\}$ and t.

1.6. The family $\hat{F}_t$ is contained in the space $\mathcal{F}$ of functions in $x \in \mathbb{C}^n$ analytic in some closed ball B^{2n}, centered at the origin and containing a neighbourhood of $\mathrm{Sing}(\{P_z = 0\})$. The subfamily $\hat{F}_0$ (depending on parameters d_m, $\ell(m) > 0$) is a germ at P_z in the space of polynomials $\mathcal{P}_+(\sigma)$ introduced in Sect. 2.3.4. Due to the strong $\mathcal{S}$-transversality with respect to σ, $\hat{F}_0$ intersects transversally the germ $M_z \subset \mathcal{F}$ at P_z of functions which are $\mathcal{S}$-equisingular to P_z in B^{2n}.

One can choose an index set $\Lambda' \subset \{i \in \mathbb{N}^n \mid \ell(i) > 0\}$ such that the family $\hat{F}'_0$ with fixed parameters $d_m = A_m$, $m \notin \Lambda'$, and free parameters d_m, $m \in \Lambda'$, intersects M_z transversally and only at P_z.

Assume that, for any small fixed t, with fixed parameters $d_m = A_m$, $m \notin \Lambda'$, and free parameters d_m, $m \in \Lambda'$, the family $\hat{F}'_t$ meets M_z at one point as well, and this point corresponds to the parameters $d_m(t)$, $m \in \Lambda'$. Then these values of d_m define the initial parameters a_i, $i \in \Lambda$, as functions of t such that the deformation (2.3.5.5) is $\mathcal{S}$-equisingular outside a neighbourhood of z and realizes the given pattern P_z in a neighbourhood of z.

1.7. It remains to justify the latter argument, since the implicit function theorem in general is not valid in infinite-dimensional spaces. Let e be the maximal finite determinacy index for the singular points $x_1, \ldots, x_q$ of P_z. Hence the condition for an element $\Phi \in \mathcal{F}$, close to P_z, to be $\mathcal{S}$-equisingular to P_z at x_i is expressed by a system of analytic equations

$$\psi^{(4)}_{i,1}\big(\mathrm{jet}_{\boldsymbol{x}_i,e}(\Phi)\big) = \ldots = \psi^{(4)}_{i,s_i}\big(\mathrm{jet}_{\boldsymbol{x}_i,e}(\Phi)\big) = 0\,,$$

whose variables are the coefficients of the e-jet $\mathrm{jet}_{\boldsymbol{x}_i,e}(\Phi)$ of Φ at $\boldsymbol{x}_i$, and whose Jacobian with respect to these variables has rank s_i at $\Phi = P_z$. Introducing qn new variables being the coordinates of points $\overline{\boldsymbol{x}}_1, \ldots, \overline{\boldsymbol{x}}_q \in \mathbb{C}^n$, close to $\boldsymbol{x}_1, \ldots, \boldsymbol{x}_q$, respectively, one can express the relation $\hat{F}'_t \in M_z$ as

$$\psi^{(4)}_{i,j}\big(\mathrm{jet}_{\overline{\boldsymbol{x}}_i,e}(\hat{F}'_t)\big) = 0, \quad j = 1, \ldots, s_i,\ i = 1, \ldots, q. \tag{2.3.5.12}$$

Since the coefficients of $\mathrm{jet}_{\overline{\boldsymbol{x}}_i,e}(\hat{F}'_t)$ are analytic functions in

$$\big\{\overline{\boldsymbol{x}}_1, \ldots, \overline{\boldsymbol{x}}_q\big\} \cup \big\{d_{\boldsymbol{m}} \,\big|\, \boldsymbol{m} \in \Lambda'\big\} \cup \big\{t\big\}, \tag{2.3.5.13}$$

(2.3.5.12) turns into a system of $s_1 + \ldots + s_q$ equations in the $s_1 + \ldots + s_q + 1$ variables (2.3.5.13). The transversality of the intersection of the family $\hat{F}'_0$ with M_z means that, for $t = 0$, the determinant of the Jacobian of functions (2.3.5.12) with respect to variables (2.3.5.13) is not zero. Hence the system (2.3.5.12) has a solution $d_{\boldsymbol{m}} = d_{\boldsymbol{m}}(t)$, $\boldsymbol{m} \in \Lambda'$ (and $\overline{\boldsymbol{x}}_i = \overline{\boldsymbol{x}}_i(t)$, $i = 1, \ldots, q$), which completes the proof of Theorem 2.3.13 in case $S_2 = \{z\}$, $S_3 = \emptyset$.

1.8. Finally, note that the transversality of the intersection of the family $\hat{F}'_t$ with M_z, established in 1.6, 1.7, means that the germ at X_t ($t \neq 0$) of the $\mathcal{S}$-equisingular stratum in $|X|$, corresponding to the singular points of X_t in a neighbourhood of z, is smooth, has the expected dimension and intersects transversally with the image of M' in $|X|$, hence also with $M^{\mathcal{S}}_X(S_0)$.

Step 2. Assume that $S_1 = S_2 = S_4 = S_5 = \emptyset$, $S_3 = \{z\}$.

In this case the proof translates word-by-word as the proof in Step 1 with the following changes:

- substitute $\widetilde{I}^{sqh}_z$ for I^{sqh}_z, $\widetilde{I}(F)$ for $I(F)$,
- substitute in 1.2–1.6 (in summations etc.) the set $\big\{\boldsymbol{i} \in \mathbb{N}^n \,\big|\, \ell(\boldsymbol{i}) > 0\big\}$ by $\big\{\boldsymbol{i} \in \mathbb{N}^n \,\big|\, \ell(\boldsymbol{i}) \geq 0\big\}$, and the set $\big\{\boldsymbol{i} \in \mathbb{N}^n \,\big|\, \ell(\boldsymbol{i}) \leq 0\big\}$ by $\big\{\boldsymbol{i} \in \mathbb{N}^n \,\big|\, \ell(\boldsymbol{i}) < 0\big\}$.

Step 3. Assume that $S_1 = S_2 = S_3 = S_4 = \emptyset$, $S_5 = \{z\}$, and the Newton diagram of a representative of the germ (X, z) has at least two facets (otherwise we obtain the case in Step 2). Here we combine the arguments from Step 1 and from the proof of Theorems 2.3.4, 2.3.5.

3.1. As in Step 1, from the exact sequence (2.3.5.3) we obtain

$$H^0\big(\mathcal{O}_\Sigma(X)\big) \;=\; H^0\big(\mathcal{J}_{Z/\Sigma}(X)\big) \oplus \mathcal{O}_{\Sigma,z}/\widetilde{I}^{nnd}_z\,.$$

For the point z we have an isomorphism $\varphi : \mathcal{O}_{\Sigma,z} \to \mathcal{O}_{\mathbb{C}^n,0}$ and a Newton nondegenerate representative $F : (\mathbb{C}^n, 0) \to (\mathbb{C}, 0)$ such that $F = 0$ is an equation of the germ (X, z) in the coordinates defined in a neighbourhood of z in Σ by φ. We can write

$$F = \sum_{i \in \Delta(F)} A_i \boldsymbol{x}^i .$$

Let $(F_1, \ldots, F_N)$ be an $\mathcal{S}$-transversal deformation pattern for (X, z), given (as described in Sect. 2.3.4) by a convex subdivision $K_0(F) = \Delta_1 \cup \ldots \cup \Delta_N$, a convex, piece-wise linear function $\nu : K(F) \to \mathbb{R}$ with $\nu\big|_{\Delta(F)} = 0$, and a set $\{A_i \in \mathbb{C} \mid \boldsymbol{i} \in K_0(F) \setminus \Gamma(F, 0)\}$.

3.2. We choose (finite) sets $\Lambda, \Lambda_1, \ldots, \Lambda_n \subset \mathbb{N}^n$ as in 1.2 and take a basis of $H^0\big(\mathcal{O}_\Sigma(X)\big)$ consisting of elements represented in $\mathcal{O}_{\Sigma,z} = \mathcal{O}_{\mathbb{C}^n,0}$ by

$$F_i = \boldsymbol{x}^i + \sum_{\substack{1 \le k \le n \\ j \in \Lambda_k}} A_{i,kj} \left[\boldsymbol{x}^j \frac{\partial F}{\partial x_k} \right] + A_{i,0}[F] + \sum_{m \in \Delta(F)} A_{i,m} \boldsymbol{x}^m, \quad \boldsymbol{i} \in \Lambda,$$

$$F_\ell = \sum_{\substack{1 \le k \le n \\ j \in \Lambda_k}} A_{\ell,kj} \left[\boldsymbol{x}^j \frac{\partial F}{\partial x_k} \right] + A_{\ell,0}[F] \sum_{m \in \Delta(F)} A_{\ell,m} \boldsymbol{x}^m, \quad \ell = 0, \ldots, s,$$

(here F is non-zero in the ideal $\widetilde{I}(F) + \langle \partial F/\partial x_1, \ldots, \partial F/\partial x_n \rangle$ if $\Gamma(F, 0)$ contains at least two facets). Denote by B_i, $\boldsymbol{i} \in \Lambda$, B_ℓ, $\ell = 0, \ldots, s$, the coordinates in $H^0\big(\mathcal{O}_\Sigma(X)\big)$ with respect to this basis. Without loss of generality, assume that F is encoded by $B_i = 0$ for $\boldsymbol{i} \in \Lambda$, $B_0 = 1$, $B_\ell = 0$ for $\ell = 1, \ldots, s$.

3.3. Consider a deformation of F in $\varphi_z\big(H^0\big(\mathcal{O}_\Sigma(X)\big)\big) \subset \mathcal{O}_{\mathbb{C}^n,0}$ defined by (2.3.5.5) and depending on small parameters a_i, $\boldsymbol{i} \in \Lambda$, a_ℓ, $\ell = 1, \ldots, s$.

As in 1.4, the elements of (2.3.5.5) belong to a smooth germ $M' \subset M$ if the parameters a_i, $\boldsymbol{i} \in \Lambda$, a_ℓ, $\ell = 1, \ldots, s$, satisfy (2.3.5.6).

3.4. We introduce the local coordinate change (2.3.5.7), but transform the family F_t in a slightly different way:

$$\begin{aligned} \widetilde{F}_t &= (1 + a_0) F_t \circ T \\ &= F_t + a_0 F + \sum_{k=1}^{n} \sum_{j \in \Lambda_k} a_{k,j} \boldsymbol{x}^j \frac{\partial F}{\partial x_k} + O(a_0^2, \{a_i^2\}, \{a_{k,j}^2\}) . \end{aligned} \tag{2.3.5.14}$$

As in 1.5, one can introduce new small parameters (2.3.5.11) and c_0, and reparametrize F_t in the form

$$\widetilde{F}_t = \sum_{m \in \Lambda} c_m \boldsymbol{x}^m + c_0[F] + \sum_{k=1}^{n} \sum_{j \in \Lambda_k} c_{k,j} \left[\boldsymbol{x}^j \frac{\partial F}{\partial x_k} \right] + \sum_{\substack{m \in \Delta(F) \\ m \notin \Gamma(F,0)}} (A_m + \psi_m^{(2)}) \boldsymbol{x}^m$$

with $\psi_m^{(2)}$ being smooth functions of (2.3.5.11) and c_0 vanishing at zero. Then we impose the following condition on the parameters (2.3.5.11) and c_0:

$$\sum_{m\in\Lambda} c_m x^m + c_0[F] + \sum_{k=1}^{n}\sum_{j\in\Lambda_k} c_{k,j}\left[x^j\frac{\partial F}{\partial x_k}\right] = \sum_{m\in K_0(F)} d_m t^{\nu(m)} x^m,$$

where d_m is close to A_m for any $m \in K_0(F)$, and t is a small parameter. The latter relation has a solution with respect to (2.3.5.11) as functions of $\{d_m t^{\nu(m)} \mid m \in K_0(F)\setminus\Gamma(F,0)\}$, and leads us to the family

$$\hat{F}_t = \sum_{m\in K_0(F)} d_m t^{\nu(m)} x^m + \sum_{m\in\Delta(F)\setminus\Gamma(F,0)} (A_m + t\psi_m^{(3)})x^m\ , \tag{2.3.5.15}$$

where the $\phi_m^{(3)}$ are smooth functions of $\{d_m\}$ and t.

To complete the proof in the case considered, we have to show that the parameters $d_m, m \in K_0(F)$, can be defined as functions of t equal to A_m for $t = 0$ and such that the singularities of $\hat{F}_t = 0$ in a neighbourhood of the origin are $\mathcal{S}$-equivalent to $\text{Sing}(F_1, \ldots, F_N)$.

3.5. The proof of the latter claim essentially coincides with the proof of Theorems 2.3.4, 2.3.5: it is indeed the same if F is a polynomial, and it should be completed with the argument from 1.7 if F is a power series.

Remark 2.3.13.3 We should like to comment a little more on the difference between Theorems 2.3.4, 2.3.5 and 2.3.13, pointed out in Remark 2.3.13.2 (2) in the situation considered in Step 3 for $\Sigma = \mathbb{P}^n$.

Taking the Viro-type polynomial in the proof of Theorems 2.3.4, 2.3.5 as a deformation of a polynomial F with a Newton nondegenerate singular point, one would have all the coefficients at $i \in K_0(F)$ as independent free parameters, whereas in our situation, in the beginning, one has the free coefficients a_i, $i \in \Lambda$, only. The remaining necessary free parameters (as in formula (2.3.5.15)) are obtained from the transformation (2.3.5.14).

Step 4. Assume that $S_1 = S_2 = S_3 = S_5 = \emptyset$, $S_4 = \{z\}$. Then the proof goes as in Step 3 with the following changes (cf. Step 2):

- one should get rid of $[F]$, a_0, c_0 in all the formulae,
- the ideals $\widetilde{I}_z^{nnd}$, $\widetilde{I}(F)$ should be replaced by the ideals I_z^{nnd}, $I(F)$, respectively,
- the sets $\Delta(F)\setminus\Gamma(F,0)$, $K_0(F)$ (in summations etc.) should be replaced by the sets $\Delta(F)$, $K_0(F)\setminus\Gamma(F,0)$, respectively.

Step 5. Assume that $S_1 = \emptyset$, but S_2, S_3, S_4, S_5 may be non-empty. Then the exact sequence (2.3.5.3) implies that $H^0(\mathcal{O}_\Sigma(X))$ splits into the direct sum

$$H^0(\mathcal{J}_{Z/\Sigma}(X)) \oplus \bigoplus_{z\in S_2}\frac{\mathcal{O}_{\Sigma,z}}{I_x^{sqh}} \oplus \bigoplus_{z\in S_3}\frac{\mathcal{O}_{\Sigma,z}}{\widetilde{I}_x^{sqh}} \oplus \bigoplus_{z\in S_4}\frac{\mathcal{O}_{\Sigma,z}}{I_x^{nnd}} \oplus \bigoplus_{z\in S_5}\frac{\mathcal{O}_{\Sigma,z}}{\widetilde{I}_x^{nnd}}.$$

This allows us to introduce the following basis in $H^0(\mathcal{O}_\Sigma(X))$:

$$F_i^{(z)} = \boldsymbol{x}^{\boldsymbol{i}} + \sum_{\substack{1 \le k \le n \\ j \in \Lambda_k^{(z)}}} A_{i,kj}^{(z)} \left[\boldsymbol{x}^{j} \frac{\partial F}{\partial x_k} \right]^{(z)} + \sum_{l^{(z)}(\boldsymbol{m}) \le 0} A_{\boldsymbol{im}}^{(z)} \boldsymbol{x}^{\boldsymbol{m}}, \quad \boldsymbol{i} \in \Lambda^{(z)}, \ z \in S_2,$$

$$F_i^{(z)} = \boldsymbol{x}^{\boldsymbol{i}} + \sum_{\substack{1 \le k \le n \\ j \in \Lambda_k^{(z)}}} A_{i,kj}^{(z)} \left[\boldsymbol{x}^{j} \frac{\partial F}{\partial x_k} \right]^{(z)} + \sum_{l^{(z)}(\boldsymbol{m}) < 0} A_{\boldsymbol{im}}^{(z)} \boldsymbol{x}^{\boldsymbol{m}}, \quad \boldsymbol{i} \in \Lambda^{(z)}, \ z \in S_3,$$

$$F_i^{(z)} = \boldsymbol{x}^{\boldsymbol{i}} + \sum_{\substack{1 \le k \le n \\ j \in \Lambda_k}} A_{i,kj}^{(z)} \left[\boldsymbol{x}^{j} \frac{\partial F}{\partial x_k} \right]^{(z)} + \sum_{\boldsymbol{m} \in \Delta^{(z)}(F)} A_{\boldsymbol{im}}^{(z)} \boldsymbol{x}^{\boldsymbol{m}}, \quad \boldsymbol{i} \in \Lambda^{(z)}, \ z \in S_4,$$

$$F_i^{(z)} = \boldsymbol{x}^{\boldsymbol{i}} + \sum_{\substack{1 \le k \le n \\ j \in \Lambda_k}} A_{i,kj}^{(z)} \left[\boldsymbol{x}^{j} \frac{\partial F}{\partial x_k} \right]^{(z)} + A_{i,0}^{(z)}[F]^{(z)} + \sum_{\substack{\boldsymbol{m} \in \Delta^{(z)}(F) \\ \boldsymbol{m} \notin K(\Gamma(F,z))}} A_{\boldsymbol{im}}^{(z)} \boldsymbol{x}^{\boldsymbol{m}},$$
$$\boldsymbol{i} \in \Lambda^{(z)}, \quad z \in S_5,$$

and F_ℓ, $\ell = 0, \ldots, s$, such that $F_0 = F$ and:

(1) each $F_i^{(z)}$ is represented by its image in $\mathcal{O}_{\Sigma,z} = \mathcal{O}_{\mathbb{C}^n,0}$,
(2) the sets $\Lambda^{(z)}, \Lambda_1^{(z)}, \ldots, \Lambda_n^{(z)}$ and the symbols $[*]^{(z)}$ are defined for any point $z \in S_2 \cup \ldots \cup S_5$ as described above,
(3) for any $\ell = 1, \ldots, s$ and any $z \in S_2 \cup \ldots \cup S_5$, $F_\ell \in H^0\big(\mathcal{O}_\Sigma(X)\big)$ has image in $\mathcal{O}_{\Sigma,z} = \mathcal{O}_{\mathbb{C}^n,0}$ of the form

$$\sum_{\substack{1 \le k \le n \\ j \in \Lambda_k}} A_{\ell,kj}^{(z)} \left[\boldsymbol{x}^{j} \frac{\partial F}{\partial x_k} \right]^{(z)} + \sum_{\ell(\boldsymbol{m}) \le 0} A_{\ell \boldsymbol{m}}^{(z)} \boldsymbol{x}^{\boldsymbol{m}}, \quad \text{or}$$

$$\sum_{\substack{1 \le k \le n \\ j \in \Lambda_k}} A_{\ell,kj}^{(z)} \left[\boldsymbol{x}^{j} \frac{\partial F}{\partial x_k} \right]^{(z)} + \sum_{\ell(\boldsymbol{m}) < 0} A_{\ell \boldsymbol{m}}^{(z)} \boldsymbol{x}^{\boldsymbol{m}}, \quad \text{or}$$

$$\sum_{\substack{1 \le k \le n \\ j \in \Lambda_k}} A_{\ell,kj}^{(z)} \left[\boldsymbol{x}^{j} \frac{\partial F}{\partial x_k} \right]^{(z)} + \sum_{\boldsymbol{m} \in \Delta^{(z)}(F)} A_{\ell \boldsymbol{m}}^{(z)} \boldsymbol{x}^{\boldsymbol{m}}, \quad \text{or}$$

$$\sum_{\substack{1 \le k \le n \\ j \in \Lambda_k}} A_{\ell,kj}^{(z)} \left[\boldsymbol{x}^{j} \frac{\partial F}{\partial x_k} \right]^{(z)} + A_{\ell,0}^{(z)}[F]^{(z)} + \sum_{\boldsymbol{m} \in \Delta^{(z)}(F) \setminus \Gamma(F,z)} A_{\ell \boldsymbol{m}}^{(z)} \boldsymbol{x}^{\boldsymbol{m}},$$

according to $z \in S_2$, or S_3, or S_4, or S_5.

The existence of such a basis of $H^0\big(\mathcal{O}_\Sigma(X)\big)$ allows to perform the above construction of the required deformation for all singular points $z \in S_2 \cup \ldots \cup S_5$ independently, and leads to the desired result.

Step 6. Assume that $S_1 \neq \emptyset$.

We construct the required deformation of all the points $z \in S_2 \cup \ldots \cup S_5$, while the points $z \in S_0$ keep their $\mathcal{S}$-type and the points $z \in S_1$ keep their analytic type (may be, slightly moving to another position S_0', resp. S_1'). For an element $X_t, t \neq 0$, of the family constructed, we have the transversal intersection (2.3.5.2). By the infinitesimal versatility criterion (openness of versatility), this means that the intersection

$$M_X^{\mathcal{S}}(S_0) \cap M_{X_t}^{\mathcal{S}}(S') \,, \tag{2.3.5.16}$$

where S' is the set of singular points of X_t in a neighborhood of $S_2 \cup \cdots \cup S_5$, induces a joint versal deformation of the points (X_t, z), $z \in S_1'$. This means that we can deform X_t inside the family (2.3.5.16) realizing any prescribed local deformations of the singular points $z \in S_1'$ simultaneously.

It remains to prove the transversality of intersection (2.3.5.2) for the hypersurfaces in the latter deformation. This follows immediately from (2.3.5.1) and the fact that the deformation patterns for the points (X, z), $z \in S_2 \cup \cdots \cup S_5$ are respectively $\mathcal{S}$- or $\mathcal{S}^s$-transversal (in other words, the latter $\mathcal{S}$- or $\mathcal{S}^s$-transversality of deformation patterns means that the spaces $H^0(\Sigma, \mathcal{J}_{X_w^{\mathcal{S}}/\Sigma}(X))$, $w \in S'$, intersect transversally, and their intersection converges as $t \to 0$ to a subspace of $H^0(\Sigma, \mathcal{O}_\Sigma(X))$ containing the space $H^0(\Sigma, \mathcal{J}_{Z''/\Sigma}(X))$; hence the transversality of intersection (2.3.5.2) follows from the hypothesis (2.3.5.1)). □

2.3.6 Cohomology Vanishing Conditions

We intend to reformulate the hypotheses of Theorems 2.3.4, 2.3.5, 2.3.13 in terms of H^1-vanishing for the ideal sheaves of some zero-dimensional schemes associated to singular points and Newton polytopes. Namely, the geometric property of smoothness and transversality of intersection of various equisingular strata required in Theorems 2.3.4, 2.3.5, 2.3.13 and in the $\mathcal{S}$-transversality condition will be replaced by the H^1-vanishing property.

Let $\mathcal{S}$ be a finite set of smooth singularity types of isolated hypersurface singularities (cf. Definition 2.2.42). Recall that for $S \in \mathcal{S}$ and $f \in \mathcal{O}_{\mathbb{C}^n,z}$ of type S, S is smooth at f (i.e. the analytic germ at f of the image of S in $\mathcal{O}_{\mathbb{C}^n,z}/\mathfrak{m}_z^{k+1}$ is smooth, where f is k-determined). We introduced the S-singularity ideal $I_z^S \subset \mathcal{O}_{\mathbb{C}^n,z}$, the tangent space to the smooth singularity type S, and the 0-dimensional S-singularity scheme Z_z^S defined by I_z^S (cf. Definition 2.2.43). In the following we write also $I_z^{\mathcal{S}}$ resp. $Z_z^{\mathcal{S}}$ where it is clear which type $S \in \mathcal{S}$ at z is considered.

We provide H^1-vanishing criteria, sufficient for the weak, strong and usual $\mathcal{S}$-transversality of Newton triples (Δ, Δ_+, F). Recall that the $\mathcal{S}$- and $\mathcal{S}^s$-transversality of deformation patterns is defined via that of triples (cf. Definitions 2.3.3 and 2.3.12).

Given a triple (Δ, Δ_+, F), assume that F is peripherally nondegenerate (PND), cf. Definition 2.3.1, and that all singular points of the hypersurface $\{F=0\}$ in $(\mathbb{C}^*)^n$ belong to the class $\mathcal{S}$.

Let $X = \mathrm{Tor}(\Delta)$, the toric variety associated with the polytope Δ $X_+ \subset X$ the union of toric divisors corresponding to the facets in Δ_+, and $Y \subset X$ the hypersurface defined by the polynomial F. Introduce the subscheme $Y_+ \subset X$, which is the scheme-theoretic intersection $Y \cap X_+$.

With these objects naturally come the line bundle $\mathcal{L}_\Delta := \mathcal{O}_X(Y)$ (so-called tautological line bundle), and the ideal sheaf $\mathcal{J}_{Y_+/X}$.

Consider also the zero-dimensional subscheme of X:

$$Z(F) := \bigcup_{w \in \mathrm{Sing}(\{F=0\}) \cap (\mathbb{C}^*)^n} Z_w^{\mathcal{S}} ,$$

and its ideal sheaf $\mathcal{J}_{Z(F)/X}$.

Proposition 2.3.14 *In the above notations, assume that*

$$H^1(X, \mathcal{L}_\Delta \otimes \mathcal{J}_{Y_+ \cup Z(F)/X}) = 0 . \tag{2.3.6.1}$$

Then the triple (Δ, Δ_+, F) is weakly $\mathcal{S}$-transversal. Furthermore, if Δ_+ is connected, then the triple (Δ, Δ_+, F) is $\mathcal{S}$-transversal.

Proof Since the weak $\mathcal{S}$-transversality turns to be a usual $\mathcal{S}$-transversality as long as Δ_+ is connected (see descussion in Sect. 2.3.2). By Theorem 2.2.55, the weak $\mathcal{S}$-transversality means (see Definition 2.3.3(5)) that $H^1\big(\mathcal{J}_{Z(F)/\mathbb{P}^n}(d)\big) = 0$ (which is equivalent to the T-smoothness of the intersection of the germ $M_d(F)$ with the linear system $|\mathcal{L}_\Delta|$), and that $H^0\big(X, \mathcal{L}_\Delta \otimes \mathcal{J}_{Z(F)/X}\big)$ (the tangent space to $M_d(F) \cap |\mathcal{L}_\Delta|$) intersects transversally in the linear system $|\mathcal{L}_\Delta|$ with the projective subspace $\widetilde{\mathcal{P}}(\Delta, \Delta_+, F)$ (defined by (2.3.2.1) and coinciding with the linear system $|\mathcal{L}_\Delta \otimes \mathcal{J}_{Y_+/X}|$). The latter transversality of intersection follows from the surjectivity of the morphisms π in the following exact sequences

$$\begin{aligned} 0 &\longrightarrow H^0\big(X, \mathcal{L}_\Delta \otimes \mathcal{J}_{Y_+ \cup Z(F)/X}\big) \longrightarrow H^0\big(X, \mathcal{L}_\Delta \otimes \mathcal{J}_{Y_+/X}\big) \stackrel{\pi}{\longrightarrow} H^0\big(\mathcal{O}_{Z(F)}\big) \\ &\longrightarrow H^1\big(X, \mathcal{L}_\Delta \otimes \mathcal{J}_{Y_+ \cup Z(F)/X}\big) , \end{aligned}$$

where the required surjectivity immediately follows from (2.3.6.1). At last the vanishing of $H^1(X, \mathcal{L}_\Delta \otimes \mathcal{J}_{Z(f)/X})$ follows from the exact sequence

$$H^1\big(X, \mathcal{L}_\Delta \otimes \mathcal{J}_{Y_+ \cup Z(F)/X}\big) \longrightarrow H^1\big(X, \mathcal{L}_\Delta \otimes \mathcal{J}_{Z(F)/X}\big) \longrightarrow H^1\big(\mathcal{O}_{Z(F)}\big) ,$$

in which the extreme terms vanish. □

Example 2.3.14.1 If a deformation pattern for a SQH or NND singular point has no singularities in $(\mathbb{C}^*)^n$ then we call it *a non-singular pattern.* Non-singular deformation patterns are interesting, in particular, when considering real deformations. These applications are discussed in Sect. 4 Appendix.

Lemma 2.3.15 *Any triple* (Δ, Δ_+, F) *with* $\mathrm{Sing}(\{F=0\}) \cap (\mathbb{C}^*)^n = \emptyset$ *is* $\mathcal{S}$*-transversal.*

Proof We have to prove condition (2.3.6.1) in Proposition 2.3.14, which now reads as

$$H^1(X, \mathcal{L}_\Delta \otimes \mathcal{J}_{Y_+/X}) = 0\,. \tag{2.3.6.2}$$

Observe, first, that $H^1(X, \mathcal{L}_\Delta) = 0$. Indeed, this can easily be proved by induction of the dimension of the polytope Δ based on the exact sequence

$$H^1(X, \mathcal{O}_X(Y - \mathrm{Tor}(\sigma))) \to H^1(X, \mathcal{L}_\Delta) \to H^1(\mathrm{Tor}(\sigma), \mathcal{L}_\sigma)\,,$$

where σ is a facet of Δ, and the first term can be regarded as $H^1(\mathrm{Tor}(\Delta^\sigma), \mathcal{L}_{\Delta^\sigma})$ with Δ^σ being the convex hull of $(\Delta \setminus \sigma) \cap \mathbb{Z}^n$. In the same manner we can prove that $H^1(X, \mathcal{O}_X(Y - \mathrm{Tor}(\Delta_+)) = 0$, where $\mathrm{Tor}(\Delta_+)$ denotes the union of the toric divisors in X corresponding to facets $\sigma \subset \Delta_+$.

Finally, we derive (2.3.6.2) by induction on $\dim \Delta$ based on the exact sequence

$$H^1(X_-, \mathcal{L}_{\Delta_-}) \to H^1(X, \mathcal{L}_\Delta \otimes \mathcal{J}_{Y_+/X}) \to H^1(\mathrm{Tor}(\sigma, \mathcal{L}_\sigma)\,,$$

where $\sigma \subset \Delta_+$ is a facet of Δ, Δ_- is the convex hull of $(\Delta \setminus \Delta_+) \cap \mathbb{Z}^n$, X_- is the toric variety associated with Δ_-. □

An important property of non-singular deformation patterns is the following

Proposition 2.3.16 *Any non-singular deformation pattern of a SQH or NND singular point is strongly* $\mathcal{S}$*-transversal.*

Proof In view of Lemma 2.3.15, we have to orient the graph G such that its edges with vertex corresponding to $\Delta(F)$ (the Newton polytope of the representative F of the given singular point) have source in $\Delta(F)$. To do so, assume that the edges of G are orthogonal to the corresponding facets of the Newton polytopes, and orient the edges of G by means of the following order in $\mathbb{R}^n$:

$$\begin{aligned}(x_1, \dots, x_n) \succ (x_1', \dots, x_n') :\Longleftrightarrow\ & x_1 < x_1',\ \text{or}\ (x_1 = x_1',\ x_2 < x_2'),\ \text{or}\ \dots,\\ & \text{or}\ (x_1 = x_1', \dots, x_{n-1} = x_{n-1}',\ x_n < x_n')\,.\end{aligned}$$

Clearly, $\widetilde{G}$ has no cycles and satisfies the requirement. □

2.4 Appendix: The Patchworking Construction

Here we shortly set forth some basics about the original patchworking method as developed by Viro (cf. Appendix A of this book) for the construction of real non-singular hypersurfaces with prescribed topology, and which we generalized for the construction of singular curves and hypersurfaces as in the preceding section. The method is illustrated by examples. We also describe a generalization of Viro's combinatorial patchworking to complete intersections due to Sturmfels as well as some other applications of the patchworking construction.

The Viro construction is a particular case of the following general procedure, described by the commutative diagram

$$\begin{array}{ccc} X & \hookrightarrow & Y \\ \downarrow & & \downarrow \\ (\mathbb{K},0) & = & (\mathbb{K},0) \end{array}$$

where $\mathbb{K}=\mathbb{C}$ or $\mathbb{R}$. That is, we consider a germ of a one-parametric family of algebraic varieties, Y of dimension $n \geq 3$, which is locally trivial over the complement of 0 with an irreducible fibre $Y_t, t \neq 0$, and a reduced reducible fibre Y_0. X is a family of Cartier divisors $X_t \subset Y_t, t \in (\mathbb{K},0)$, locally trivial over the complement of 0. Given a family $Y \to (\mathbb{K},0)$, the procedure consists in constructing a family $X \to (\mathbb{K},0)$ in the following two situations:

- in the *degeneration* procedure, where a Cartier divisor $X_t \subset Y_t, t \neq 0$, is given, and the construction extends it to a family $X \to (\mathbb{K},0)$ with a limit fibre $X_0 \subset Y_0$ splitting into the union of hypersurfaces in the components of Y_0 in a certain way;
- in the *gluing* (or *patchworking*) procedure, a union X_0 of (compatible) hypersurfaces in the components of Y_0 is given, and the construction extends it to a family $X \to (\mathbb{K},0)$ with a generic fibre $X_t \subset Y_t, t \neq 0$, inheriting some properties of the special fibre X_0.

An example of the degeneration procedure can be found in [Ran1, Ran2, Ran3]. An example of the gluing procedure is the Viro construction, where Y is a toric variety built out of a convex polytope and fibred so that the torus $(\mathbb{K}^*)^n \subset X$ splits into a one-dimensional family of isomorphic toric varieties, and the fibre X_0 is the union of some toric divisors.

Here we point out two differences between the construction presented in Sect. 2.3 and the Viro patchworking.

First, practically, for the patchworking construction one chooses an invertible sheaf $\mathcal{L}$ on Y (or, equivalently, a family of invertible sheaves $\mathcal{L}_t$ on $Y_t, t \in (\mathbb{K},0)$) and then constructs a section $\xi \in H^0(Y,\mathcal{L})$ (i.e. a family of sections $\xi_t \in H^0(Y_t,\mathcal{L})$, $t \in (\mathbb{K},0)$) such that $X_t = Z(\xi_t) = Z(\xi) \cap X_t, t \in (\mathbb{K},0)$. In the Viro construction, it is important that $H^0(Y_0,\mathcal{L}_0)$ is isomorphic to $H^0(Y_t,\mathcal{L}_t), t \neq 0$, and $H^1(Y_t,\mathcal{L}_t)=0$ for all $t \in (\mathbb{K},0)$, and both requirements are naturally incorporated into the patchworking data. In turn, in the construction of Sect. 2.3.5 the condition $H^0(Y_0,\mathcal{L}_0) \simeq H^0(Y_t,\mathcal{L}_t), t \neq 0$, is not given in advance, and, moreover, (in the formalism of [ShT3, Section 2.5]) the requirement is that the canonical map $\pi_*\mathcal{L} \otimes K_t \to H^0(Y_t,\mathcal{L}_t)$ is an isomorphism for $t \neq 0$ and an embedding for $t=0$, where $K_t \simeq \mathbb{C}$ is the residue field of the point $t \in (\mathbb{C},0)$.

Second, the properties which we control in the construction considered in Sect. 2.3, i.e. specific singularities and critical points, are not (in general) stable under small deformations, which necessarily leads to appropriate transversality requirements.

On the other hand, the Viro construction fits well for the gluing of the properties of hypersurfaces in toric varieties or polynomials with certain Newton polytopes, which are stable under small deformations and are invariant with respect to the standard real or complex torus action. The embedded topology of the non-singular real point set of projective hypersurfaces, of course, satisfies these demands. A few examples of other stable properties are discussed in Sect. 2.4.5.

2.4.1 Elements of Toric Geometry

Construction of Toric Varieties. A *toric variety* is an irreducible, complex algebraic variety X equipped with an action of the *algebraic torus* $(\mathbb{C}^*)^n$ having an open dense orbit.

In the following we show how such a toric variety can be constructed from a finite set of polyhedral cones. Here, by a *rational, convex, polyhedral cone* in $\mathbb{R}^n$ we mean a set of the form

$$\varrho = \left\{ \sum_{i=1}^{k} r_i \cdot \boldsymbol{v}_i \in \mathbb{R}^n \;\middle|\; r_i \geq 0 \right\},$$

where $\boldsymbol{v}_1, \ldots, \boldsymbol{v}_k$ are vectors in $\mathbb{R}^n$ having integral components. ϱ is called *strongly convex* if ϱ does not contain any linear subspace, that is, if $\{0\}$ is a face of ϱ. The *dual cone* to ϱ is the set

$$\varrho^\vee = \left\{ \varphi \in (\mathbb{R}^n)^* \;\middle|\; \varphi(\boldsymbol{v}) \geq 0 \text{ for all } \boldsymbol{v} \in \varrho \right\}$$

where $(\mathbb{R}^n)^*$ denotes the dual space of linear functionals on $\mathbb{R}^n$.

Lemma 2.4.1 ([Ful2], 1.2). *The dual of a rational, convex, polyhedral cone is again a rational, convex, polyhedral cone.*

Lemma 2.4.2 (Gordan's lemma, [Ful2], 1.2). *Let ϱ be a rational, convex, polyhedral cone. Then $S_\varrho := \varrho^\vee \cap \mathbb{Z}^n$ is a finitely generated semigroup.*

S_ϱ determines a commutative $\mathbb{C}$-algebra $\mathbb{C}[S_\varrho] = \mathbb{C}[\chi^{\boldsymbol{u}} | \boldsymbol{u} \in S_\varrho]$ (with multiplication given by $\chi^{\boldsymbol{u}} \cdot \chi^{\boldsymbol{u}'} = \chi^{\boldsymbol{u}+\boldsymbol{u}'}$), which is finitely generated due to Gordon's lemma and, thus, determines an affine variety U_ϱ which is irreducible and normal ([Oda, 1.2]).

Example 2.4.2.1 Let $\varrho = \{0\}$. Then $S_\varrho = \{0\}^\vee \cap \mathbb{Z}^n = \mathbb{Z}^n$. As a semigroup $\mathbb{Z}^n$ is generated by $e_1, -e_1, \ldots, e_n, -e_n$ where $e_1, \ldots, e_n$ denotes the standard basis of $\mathbb{R}^n$. Thus $\mathbb{C}[S_\varrho] \cong \mathbb{C}[x_1, x_1^{-1}, \ldots, x_n, x_n^{-1}]$ and, hence, $U_\varrho \cong (\mathbb{C}^*)^n$.

For every face $\tau \subseteq \varrho$ the inclusion $\mathbb{C}[S_\varrho] \subseteq \mathbb{C}[S_\tau]$ induces an injective morphism $U_\tau \hookrightarrow U_\varrho$, embedding U_τ as a principal open subset of U_ϱ ([Ful2, 1.3]).

We are always assuming that ϱ is strongly convex (so, $\{0\}$ is a face of ϱ) which implies that $U_{\{0\}}$ is contained in U_ϱ. Thus by the example above U_ϱ contains an algebraic torus as an open dense subset.

In order to glue several affine toric varieties together, the corresponding cones have to be compatible in a straightforward way.

Definition 2.4.3 A *fan* is a collection $\mathcal{F}$ of finitely many rational, strongly convex, polyhedral cones in $\mathbb{R}^n$ satisfying the following two conditions:

1. Each face of a cone in $\mathcal{F}$ is also a cone in $\mathcal{F}$.
2. The intersection of two cones in $\mathcal{F}$ is a face of each.

Given a fan $\mathcal{F}$, the *toric variety of the fan* $\mathcal{F}$, $X(\mathcal{F})$, is constructed out of the disjoint union of all U_ϱ with ϱ in $\mathcal{F}$ and gluing U_ϱ and $U_{\varrho'}$ along $U_{\varrho\cap\varrho'}$, which is a principal open subvariety of each.

Another way to describe the points of an affine toric variety is as follows.

Proposition 2.4.4 *The points of U_ϱ are in one-to-one correspondence with semigroup homomorphisms $S_\varrho \to \mathbb{C} = \mathbb{C}^* \cup \{0\}$, where we consider $\mathbb{C}$ as a semigroup under multiplication. In other words, there are bijections*

$$U_\varrho \longleftrightarrow \mathrm{Hom}_{alg}(\mathbb{C}[S_\varrho], \mathbb{C}), \quad \mathrm{Hom}_{alg}(\mathbb{C}[S_\varrho], \mathbb{C}) \longleftrightarrow \mathrm{Hom}_{sg}(S_\varrho, \mathbb{C}),$$

where $\mathrm{Hom}_{sg}(S_\varrho, \mathbb{C}) = \big\{\varphi\colon S_\varrho \to \mathbb{C} \mid \varphi(0) = 1,\ \varphi(u + u') = \varphi(u)\varphi(u')\big\}$ *is the set of semigroup homomorphisms and* Hom_{alg} *the set of $\mathbb{C}$-algebra homomorphisms.*

In this description, we can define a natural embedding $\varphi\colon (\mathbb{C}^*)^n \hookrightarrow U_\varrho$ by

$$\boldsymbol{x} = (x_1, \ldots, x_n) \mapsto \big(\hat{x} : S_\varrho \to \mathbb{C},\ \ \widehat{\boldsymbol{x}}(u) = x^u := x_1^{u_1} \cdots x_n^{u_n}\big).$$

Furthermore, every $\chi^u \in S_\varrho$ is a regular function $\chi^u : U_\varrho \to \mathbb{C}$ which is given in this new description by $\chi^u(x) = \widehat{x}(u)$. Combined with the embedding $\varphi\colon (\mathbb{C}^*)^n \hookrightarrow U_\varrho$ the map $\chi^u \circ \varphi\colon (\mathbb{C}^*)^n \to \mathbb{C}$ is given by the Laurent monomial $x_1^{u_1} \cdots x_n^{u_n}$.

Proposition 2.4.5 ([Dan, Ful2, Oda]). *The torus $(\mathbb{C}^*)^n$ is acting on $X(\mathcal{F})$ such that it decomposes into a disjoint union of the* orbits *with respect to this action. For every cone $\varrho \in \mathcal{F}$ there is an orbit $O_\varrho \cong (\mathbb{C}^*)^{n-\dim(\varrho)}$ given by* $\mathrm{Hom}_{sg}(\varrho^\perp \cap \mathbb{Z}^n, \mathbb{C}^*)$, *where $\varrho^\perp = \{\varphi \in (\mathbb{R}^n)^* \mid \varphi(\varrho) = 0\}$. The inclusion $O_\varrho \hookrightarrow U_\varrho$ is given by extension by zero outside of $\varrho^\perp$.*

This description gives an intrinsic definition of the *real part* of the toric variety U_ϱ, which is just the subset $U_\varrho(\mathbb{R}) = \mathrm{Hom}_{sg}(S_\varrho, \mathbb{R})$. In the case of a toric variety coming from a fan $\mathcal{F}$ these varieties glue together and give the real part of the toric variety $X(\mathcal{F})$.

Toric Varieties and Convex Polytopes. A toric variety can be constructed out of a convex polytope Δ. One can associate to every such polytope a fan $\mathcal{F}_\Delta$ in the following way. For every face $\Lambda \subseteq \Delta$ consider the cone $C_\Lambda(\Delta) \subseteq (\mathbb{R}^n)^*$ generated by

$$\{\boldsymbol{u} - \boldsymbol{u}' \mid \boldsymbol{u} \text{ is a vertex of } \Delta, \boldsymbol{u}' \text{ is a vertex of } \Lambda\},$$

that is, the cone of vectors pointing from points of Λ to points of Δ.

The fan $\mathcal{F}_\Delta$ consisting of the dual cones to these cones is called *the fan associated to the polytope* Δ. The corresponding toric variety is denoted by X_Δ. Every face $\delta \subset \Delta$ of positive dimension defines a toric subvariety $X_\delta \subset X_\Delta$. The facets (faces of codimension 1) define divisors in X_Δ, which sometimes are called *toric divisors*.

Every $u \in \mathbb{Z}^n$ defines a rational function on X_Δ and so every finite subset $\{u_0, \ldots, u_r\} \subseteq \mathbb{Z}^n$ defines a rational map $\varphi\colon X_\Delta \dashrightarrow \mathbb{C}P^r$. The finite subset $\Delta \cap \mathbb{Z}^n =: \{\boldsymbol{\omega}_0, \ldots, \boldsymbol{\omega}_r\}$ gives a morphism $\varphi\colon X_\Delta \to \mathbb{C}P^r$ ([Ful2, 3.4]), which in general is *not* an embedding (that is, the corresponding divisor is not very ample).

Denote the image by $Y_\Delta = \varphi(X_\Delta)$. It is the closure of the image of the map $(\mathbb{C}^*)^n \to \mathbb{C}P^r$, $\boldsymbol{x} \mapsto (\boldsymbol{x}^{\omega_0} : \ldots : \boldsymbol{x}^{\omega_r})$. Similarly, $Y_\Delta(\mathbb{R}) := \varphi(X_\Delta(\mathbb{R}))$ is the closure (in the metric topology) of the image of $(\mathbb{R}^*)^n \to \mathbb{R}P^r$ defined as above. Both X_Δ and Y_Δ only depend on the affine geometry of Δ.

It should be pointed out that X_Δ is always normal, whereas Y_Δ does not always have this property.

The real torus $(\mathbb{R}^*)^n$ is the union of 2^n orthants

$$\mathbb{R}^n_\varepsilon = \left\{(x_1, \ldots, x_n) \in \mathbb{R}^n \,\middle|\, \varepsilon_i x_i > 0\right\}$$

where $\varepsilon = (\varepsilon_1, \ldots, \varepsilon_n) \in \{+, -\}^n$.

Define $X_\Delta(\mathbb{R}_\varepsilon) \subseteq X_\Delta(\mathbb{R})$ and $Y_\Delta(\mathbb{R}_\varepsilon) \subseteq Y_\Delta$ to be the closure (in the real topology) of the subsets

$$\mathbb{R}^n_\varepsilon \subseteq (\mathbb{R}^*)^n \hookrightarrow X_\Delta(\mathbb{R}), \quad \mathbb{R}^n_\varepsilon \subseteq (\mathbb{R}^*)^n \hookrightarrow Y_\Delta(\mathbb{R})$$

respectively. Thus,.

$$X_\Delta(\mathbb{R}) = \bigcup_{\varepsilon \in \{+,-\}^n} X_\Delta(\mathbb{R}_\varepsilon) \quad \text{and} \quad Y_\Delta(\mathbb{R}) = \bigcup_{\varepsilon \in \{+,-\}^n} Y_\Delta(\mathbb{R}_\varepsilon)$$

Every face $\Lambda \subseteq \Delta$ corresponds to a cone τ in $\mathcal{F}_\Delta$ and so to an orbit of X_Δ which we denote by O_Λ. The intersections $O_\Lambda \cap X_\Delta(\mathbb{R}_\varepsilon)$ and $O_\Lambda \cap Y_\Delta(\mathbb{R}_\varepsilon)$ are denoted by $(O_\Lambda)_\varepsilon$.

Let $\tau^n_d \subset \mathbb{R}^n$, $d, n \geq 1$, be the simplex with vertices $(0, \ldots, 0)$, $(d, 0, \ldots, 0)$, $(0, d, 0, \ldots, 0)$, ..., $(0, \ldots, 0, d)$.

Example 2.4.5.1 Let $\Delta = \tau^n_d$. It is easy to see that $X_\Delta = \mathbb{C}P^n$ and, thus, $X_\Delta(\mathbb{R}) = \mathbb{R}P^n$. The variety $Y_\Delta(\mathbb{R}) \subset \mathbb{R}P^r$ $(r = \#(\Delta \cap \mathbb{Z}^n) - 1)$ is isomorphic to $\mathbb{R}P^n$ in its d-tuple embedding and so in this case the varieties $X_\Delta(\mathbb{R})$ and $Y_\Delta(\mathbb{R})$ are indeed isomorphic.

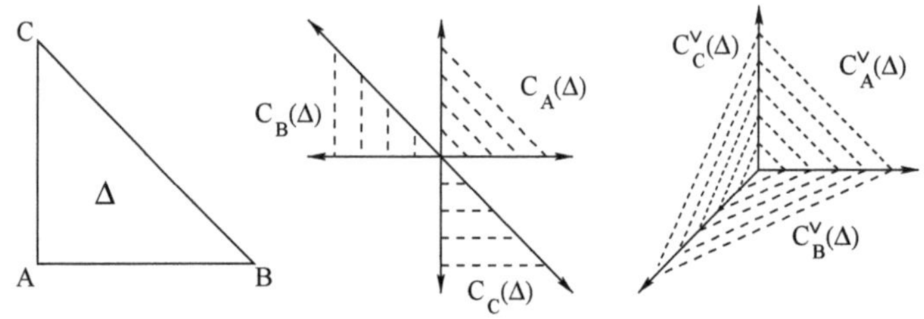

Fig. 2.3 The polytope $\Delta = \tau_d^2$ and its associated cones

Figure 2.3 shows the cones $C_\Lambda(\Delta)$ where Λ is a vertex of Δ and the fan $\mathcal{F}_\Delta$ for $n = 2$.

The Moment Map. Let Δ be an integral, convex, nondegenerate polytope in $\mathbb{R}^n$ and X_Δ the corresponding toric variety. Recall that inside X_Δ we have the real part $X_\Delta(\mathbb{R})$ and the non-negative real part $X_\Delta(\mathbb{R}_{\geq 0})$.

Definition 2.4.6 The *moment map*[19] is the map $\mu := \mu_\Delta : X_\Delta \to \Delta$ with

$$\mu(\boldsymbol{x}) := \frac{1}{\sum\limits_{\boldsymbol{u}\in\Delta\cap\mathbb{Z}^n} |\chi^{\boldsymbol{u}}(\boldsymbol{x})|} \cdot \sum_{\boldsymbol{u}\in\Delta\cap\mathbb{Z}^n} |\chi^{\boldsymbol{u}}(\boldsymbol{x})| \cdot \boldsymbol{u}$$

with the function $\chi^{\boldsymbol{u}} : X_\Delta \to \mathbb{C}$ as introduced above.

The following theorem is the key to a purely combinatorial description of the toric variety $X_\Delta(\mathbb{R})$.

Theorem 2.4.7 ([Ati1, Ati2]) *The moment map μ induces a real analytic homeomorphism from $X_\Delta(\mathbb{R}_{\geq 0})$ onto Δ respecting the stratification of Δ by its faces. More precisely: for any face $\Lambda \subseteq \Delta$, the corresponding orbit $(O_\Lambda)_+$ is mapped by μ diffeomorphically onto the interior of Λ.*

Completely analogous results are true for $X_\Delta(\mathbb{R}_\varepsilon)$ defined by any other orthant $\mathbb{R}^n_\varepsilon$. In particular, Theorem 2.4.7 then implies that from the topological point of view $X_\Delta(\mathbb{R})$ can be thought of as 2^n copies of Δ glued together appropriately.

For every vector $\varepsilon \in \{+,-\}^n$ take a copy Δ_ε of Δ and for every face Λ of Δ denote the corresponding face in Δ_ε by Λ_ε.

Definition 2.4.8 Two vectors $\varepsilon, \boldsymbol{\delta} \in \{+,-\}^n$ are said to *agree on a face* $\Lambda \subseteq \Delta$ if either $\varepsilon^{\boldsymbol{u}} = \boldsymbol{\delta}^{\boldsymbol{u}}$ for all $\boldsymbol{u} \in \Lambda \cap \mathbb{Z}^n$, or $\varepsilon^{\boldsymbol{u}} \neq \boldsymbol{\delta}^{\boldsymbol{u}}$ for all $\boldsymbol{u} \in \Lambda \cap \mathbb{Z}^n$.

[19] Moment maps normally arise when Lie groups are acting on manifolds (more precisely when a Hamiltonian action of a Lie group on a symplectic manifold is given). In our case $(S^1)^r$ is acting on $\mathbb{C}P^r$. For details about relations between general moment maps and the map we define, see [Ful2] or [Ris2].

Proposition 2.4.9 ([Stu]) *The real toric variety $X_\Delta(\mathbb{R})$ is homeomorphic to 2^n copies of Δ glued together by the following recipe: for any $\varepsilon, \boldsymbol{\delta} \in \{+, -\}^n$ and any face $\Lambda \subseteq \Delta$ identify Λ_ε with Λ_δ if ε and δ agree on Λ.*

2.4.2 Viro's Theorem for Hypersurfaces

The Chart of a Real Polynomial. Throughout this section f will denote a real (Laurent) polynomial

$$f = \sum_{\omega \in \mathbb{Z}^n} a_\omega \cdot \boldsymbol{x}^\omega \in \mathbb{R}[x_1, x_1^{-1}, \ldots, x_n, x_n^{-1}].$$

and Δ its Newton polytope, which is always assumed to be nondegenerate, that is, $\dim(\Delta) = n$.

Denote by $Z(f)$ the closure of the image of the zero set $\{f = 0\} \subset (\mathbb{R}^*)^n$ in the toric variety $X_\Delta(\mathbb{R})$ under the embedding $(\mathbb{R}^*)^n \hookrightarrow X_\Delta(\mathbb{R})$. The intersection of this toric hypersurface with each orthant is sent into a separate copy of Δ by means of the moment map, which leads to the definition of the chart of a polynomial.

Definition 2.4.10 Let $f \in \mathbb{R}[x_1, \ldots, x_n]$ and let Δ be its Newton polytope. For every $\varepsilon \in \{+, -\}^n$ let Δ_ε be a copy of Δ in the orthant $\mathbb{R}^n_\varepsilon$ (that is, the image of Δ under the map $\mathbb{R}^n \to \mathbb{R}^n, \boldsymbol{x} \mapsto (\varepsilon_1 x_1, \ldots, \varepsilon_n x_n)$). Define

$$\operatorname{Ch}_\varepsilon(f) = \mu(Z(f) \cap X_\Delta(\mathbb{R}_\varepsilon)) \subseteq \Delta_\varepsilon \,.$$

The *chart of* f is then the union

$$\operatorname{Ch}(f) = \bigcup_{\varepsilon \in \{+,-\}^n} \operatorname{Ch}_\varepsilon(f).$$

Remark 2.4.10.1 $\operatorname{Ch}_\varepsilon(f) \cong \operatorname{Ch}_+(f^\varepsilon)$, $f^\varepsilon(\boldsymbol{x}) := f(\varepsilon_1 x_1, \ldots, \varepsilon_n x_n)$. This implies that it is often enough just to consider the positive orthant.

Example 2.4.10.2 Consider the space of affine linear polynomials, that is, polynomials of the form

$$f = a_0 + \sum_{i=1}^n a_i x_i \,.$$

Assume $a_i \neq 0$ for all i, so that the zero locus of f is a hyperplane in $\mathbb{R}P^n$ transversal to all the coordinate hyperplanes and the Newton polytope $\Delta(f)$ is the simplex

$$\left\{ (t_1, \ldots, t_n) \,\middle|\, t_i \geq 0,\ \sum t_i \leq 1 \right\}.$$

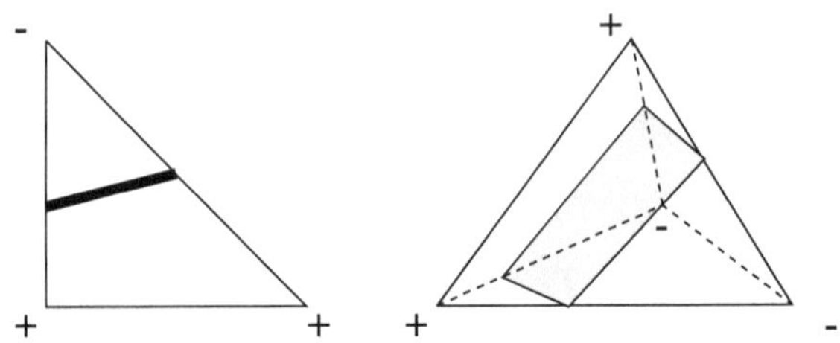

Fig. 2.4 Examples of charts of affine linear polynomials

The corresponding real toric variety is $\mathbb{R}P^n$ and $x_1, \ldots, x_n$ are coordinates in an affine chart. In this chart the moment map is given by

$$\mu(\boldsymbol{x}) = \left(\frac{|x_1|}{1 + \sum |x_i|}, \ldots, \frac{|x_n|}{1 + \sum |x_i|} \right).$$

Consider the intersection of $Z(f)$ with the positive part of $X_\Delta(\mathbb{R}) = \mathbb{R}P^n$, that is, the set of zeros of $f(\boldsymbol{x}) = 0$ with all $x_i > 0$. The image of this set under the moment map is

$$\left\{ (t_1, \ldots, t_n) \in \Delta \,\middle|\, a_0 \left(1 - \sum_{i=1}^{n} t_i \right) + \sum_{i=1}^{n} a_i t_i = 0 \right\},$$

which is a hyperplane section in Δ separating the vertices of Δ with $a_i < 0$ from those with $a_i > 0$.

In particular, the chart of linear polynomials only depends on the signs of the coefficients. Figure 2.4 shows examples of how these charts might look.

Example 2.4.10.2 shows, in fact, how to calculate the charts of a larger class of polynomials because the chart is invariant under certain transformations.

Proposition 2.4.11 [cf. [ItS4], Proposition 1.3 and proof of Theorem 2.1] *Let $A : \mathbb{R}^n \to \mathbb{R}^n$ be a nondegenerate affine-linear map defined over $\mathbb{Z}$, $\lambda = (\lambda_1, \ldots, \lambda_n) \in \mathbb{R}^n_+$, and let*

$$f = \sum_\omega a_\omega \boldsymbol{x} \mapsto f^A = \sum_\omega a_\omega \boldsymbol{x}^{A(\omega)},$$

$$f(x_1, \ldots, x_n)) \mapsto f_\lambda(x_1, \ldots, x_n) = f(\lambda_1 x_1, \ldots, \lambda_n x_n)$$

be the induced actions of A and λ on $\mathbb{R}[x_1, x_1^{-1}, \ldots, x_n, x_n^{-1}]$. Then $\Delta(f_\lambda^A) = A(\Delta)$, and $A(\mathrm{Ch}_+(f))$ is isotopic to $\mathrm{Ch}_+(f_\lambda^A)$ in $A(\Delta(f))$ by an isotopy respecting the faces. Furthermore, if A^{-1} is defined over $\mathbb{Z}$, then $A(\mathrm{Ch}_+(f)) = \mathrm{Ch}_+(f^A)$.

This motivates the definition of the following class of polynomials whose charts are particularly easy.

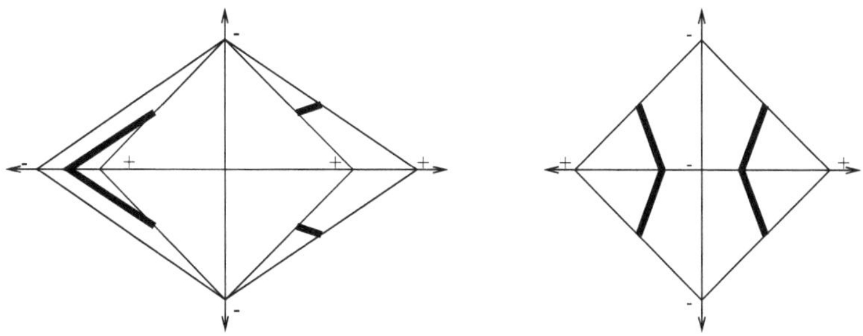

Fig. 2.5 Charts of the polynomials f_1, f_2 from the example

Definition 2.4.12 By an $(n+1)$-*nomial* we mean a polynomial $f \in \mathbb{R}[\boldsymbol{x}]$, $\boldsymbol{x} = (x_1, \dots, x_n)$, whose Newton polytope is an n-dimensional simplex and whose non-zero coefficients correspond to the $n+1$ vertices of $\Delta(f)$ alone.

Since any Newton polytope of an $(n+1)$-nomial can be mapped onto the Newton polytope of an affine linear polynomial by a combination of transformations of the types described in the proposition, the proposition tells us that the chart of an $(n+1)$-nomial is isotopic to the chart of a plane. Note also that by Remark 2.4.10.1, the chart $\mathrm{Ch}_\varepsilon(f)$ is isomorphic to $\mathrm{Ch}_+(f^\varepsilon)$ and, thus, to $\mathrm{Ch}_+\big((f^\varepsilon)^A\big)$. So, the chart $\mathrm{Ch}_\varepsilon(f)$ is obtained by assigning to every vertex $\boldsymbol{\omega}_\varepsilon$ of Δ_ε, corresponding to a vertex $\boldsymbol{\omega} \in \Delta$, the sign $\varepsilon^{\omega} \cdot a_\omega$ and then considering a hyperplane separating the vertices with positive signs from those with negative signs.

Example 2.4.12.1 Figure 2.5 shows the charts of $f_1 = x^3 + x^2 - y^2$ and $f_2 = x^2 - y^2 - 1$.

Viro's Theorem. A polynomial $f \in \mathbb{R}[x_1, \dots, x_n]$ is called *completely nondegenerate* if f and all its truncations f^Λ to faces Λ of $\Delta(f)$ have no singular point in $(\mathbb{R}^*)^n$.

Lemma 2.4.13 *Any $(n+1)$-nomial f is completely nondegenerate.*

The initial data for the Viro construction consist of

- a nondegenerate convex polytope Δ in $\mathbb{R}^n$,
- a regular polyhedral subdivision $\Delta = \Delta_1 \cup \dots \cup \Delta_N$ of Δ given by a convex piece-wise linear function ν,
- real numbers $a_\omega \in \mathbb{R}$ with $\boldsymbol{\omega} \in \Delta \cap \mathbb{Z}^n$, and $a_\omega \neq 0$ whenever $\boldsymbol{\omega}$ is a vertex of some Δ_i.

Define

$$f_k := \sum_{\omega \in \Delta_k} a_\omega \cdot \boldsymbol{x}^\omega , \qquad f^{(t)} = \sum_{\omega \in \Delta} a_\omega \cdot \boldsymbol{x}^\omega \cdot t^{\nu(\omega)} .$$

Theorem 2.4.14 (Viro's theorem) *If all the polynomials f_k are completely non-degenerate then for sufficiently small $t > 0$, the chart $\mathrm{Ch}_\varepsilon(f^{(t)})$ is isotopic to $\bigcup_{k=1}^n \mathrm{Ch}_\varepsilon(f_k)$ by an isotopy respecting the faces of Δ, that is, the chart of $f^{(t)}$ is obtained by gluing the charts of the f_k.*

The proof goes as follows (see [Vir2, Vir6]). Choose the function ν to be integral-valued at integral points and consider the toric variety $X_{\Delta(\nu)}$ determined by the convex polytope

$$\Delta(\nu) = \big\{(\boldsymbol{y}, y_{n+1}) \in \mathbb{R}^{n+1} \,\big|\, \boldsymbol{y} \in \Delta,\ \nu(\boldsymbol{y}) \le y_{n+1} \le M\big\}$$

with an integer $M > \max_{\boldsymbol{y}\in\Delta} \nu(\boldsymbol{y})$.

Then $X_{\Delta(\nu)}(\mathbb{R}) \setminus X_\delta(\mathbb{R})$, $\delta = \Delta(\nu) \cap \{y_{n+1} = M\}$, is fibred over $\mathbb{R}$ by the hypersurfaces $X^{(c)} = Z(t-a)$, $a \in \mathbb{R}$, which, for $a \neq 0$, are isomorphic to $X_\Delta(\mathbb{R})$, and, for $a = 0$, turn into the union of the toric divisors defined by the faces of the graph of ν, that is, $X^{(0)}$ is isomorphic to the union of $X_{\Delta_k}(\mathbb{R})$, $k = 1, \dots, N$, glued up along the common toric divisors. In turn, $Z(f^{(t)})$ is a hypersurface in $X_\Delta(\mathbb{R})$, which, for small positive values of c, intersects transversally with the fibres $X^{(c)}$ so that the intersections $X^{(c)} \cap Z(f^{(t)})$ are isomorphic to the family of isotopic hypersurfaces $Z(f^{(c)}) \subset X_\Delta(\mathbb{R})$, degenerating into the appropriate gluing of $Z(F_k) \subset X_{\Delta_k}$, $k = 1, \dots, N$.

Combinatorial patchworking. An important particular case is that of a triangulation of Δ, that is, the polyhedra Δ_k are simplices. Moreover, assume that $a_\omega \neq 0$ if and only if ω is a vertex of some Δ_k. Thus, all the polynomials f_k are $(n+1)$-nomials.

Denote by $\mathrm{Bar}(\Delta_\nu)$ the first barycentric subdivision of the triangulation Δ_ν. Then every n-simplex ϱ in $\mathrm{Bar}(\Delta_\nu)$ is incident to a unique point ω which is a vertex of some Δ_k. Define the sign of ϱ to be the sign of the real number a_ω. For any lower dimensional cell τ of $\mathrm{Bar}(\Delta_\nu)$ define

$$\mathrm{sign}(\tau) = \begin{cases} + & \text{if } \mathrm{sign}(\varrho) = + \text{ for all facets } \varrho \supseteq \tau \\ - & \text{if } \mathrm{sign}(\varrho) = - \text{ for all facets } \varrho \supseteq \tau \\ 0 & \text{otherwise} \end{cases} \tag{2.4.2.1}$$

Let $Z(\Delta_\nu, f)$ be the subcomplex of $\mathrm{Bar}(\Delta_\nu)$ consisting of all those cells τ with $\mathrm{sign}(\tau) = 0$.

Corollary 2.4.15 *For sufficiently small $t > 0$, the chart $\mathrm{Ch}_+(f^{(t)})$ is homeomorphic to the simplicial complex $Z(\Delta_\nu, f)$.*

For $n = 2$, curves which have been glued by this procedure (that is glued from 3-nomials) are often called *T-curves*.

Example 2.4.15.1 Let $f_1(x, y) = x^3 + x^2 - y^2$ and $f_2(x, y) = x^2 - y^2 - 1$. Define

$$\nu(i, j) = \begin{cases} 0 & \text{if } i + j \ge 2 \\ 2 - i - j & \text{if } i + j \le 2 \end{cases}$$

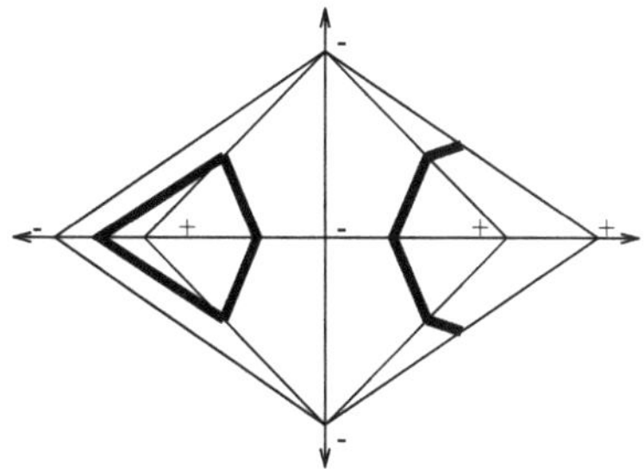

Fig. 2.6 The result of combinatorial patchworking applied to the polynomials f_1, f_2 of Example 2.4.15.1

and so $f^{(t)}(x, y) = x^3 + x^2 - y^2 - t^2$. The chart of $f^{(t)}(x, y)$ for small values of t is homeomorphic to $Z(\Delta_\nu, f)$ (see Fig. 2.6). It is a small deformation of the nodal curve defined by $f^{(0)}(x, y) = f_1(x, y)$.

2.4.3 *Viro's Method in Real Algebraic Geometry*

Isotopy Types of Real Algebraic Curves. The first part of Hilbert's 16th problem [Hil] asks, in particular, for an isotopy classification of real non-singular plane algebraic curves. More precisely: how can the different connected components of such a curve be positioned relative to one another? From the topological viewpoint every connected component has one of the following two types:

1. It divides $\mathbb{R}P^2$ into two parts of which one is orientable and the other one is not. Such a component is called an *oval*.
2. It does not divide $\mathbb{R}P^2$ into two parts. Such a component is called a *one-sided curve*. It is not difficult to see that one-sided curves only occur in odd degree.

At the end of the 19th century, an answer to the problem above was known only up to degree 5. In 1969 Gudkov achieved a complete classification of curves of degree 6 and Viro simplified the constructions of Gudkov and solved the problem for degree 7 by means of his method.

The mutual disposition of the components of a non-singular curve is encoded by its *real scheme* defined by the following procedure:

- A curve consisting of one oval is denoted by the symbol $\langle 1 \rangle$, the empty curve by $\langle 0 \rangle$ and a one-sided curve by $\langle J \rangle$.
- If $\langle A \rangle$, $\langle B \rangle$ are the real schemes of two non-intersecting curves whose ovals do not enclose any oval of the other curve, then the union of these curves is denoted by $\langle A \amalg B \rangle$.
- If $\langle A \rangle$ encodes some set of ovals then the set obtained from it by adjoining one oval enclosing all the ovals from $\langle A \rangle$ is denoted by $\langle 1 \langle A \rangle \rangle$.
- We abbreviate $\langle n \rangle = \langle 1 \amalg \cdots \amalg 1 \rangle$ (n times).

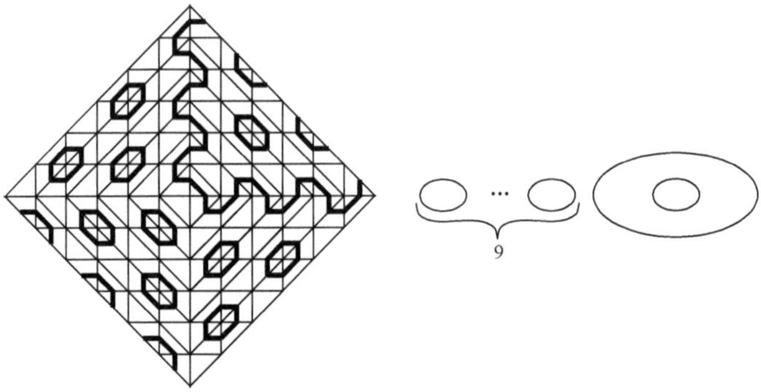

Fig. 2.7 The Harnack curve of degree 6 with real scheme $\langle 9\rangle \amalg 1\langle 1\rangle$

Theorem 2.4.16 ([Gud3]). *Any non-singular curve of degree 6 in $\mathbb{R}P^2$ has one of the following 56 real schemes.*

- $\langle 9 \amalg 1\langle 1\rangle\rangle$, $\langle 5 \amalg 1\langle 5\rangle\rangle$, $\langle 1 \amalg 1\langle 9\rangle\rangle$,
- $\langle 10\rangle$, $\langle 8 \amalg 1\langle 1\rangle\rangle$, $\langle 4 \amalg 1\langle 5\rangle\rangle$, $\langle 1 \amalg 1\langle 8\rangle\rangle$, $\langle 1\langle 9\rangle\rangle$,
- $\langle n \amalg 1\langle m\rangle\rangle$ *with* $n+m \le 8,\ 0 \le n \le 7,\ 1 \le m \le 8$,
- $\langle n\rangle$ *with* $0 \le n \le 9$,
- $\langle 1\langle 1\langle 1\rangle\rangle\rangle$.

Already Hilbert pointed out that the most interesting curves of a given degree are the M-curves, i.e. those having the maximal number of components. The patchworking construction of M-curves of degree 6 is presented below.

There are three different types of M-curves of degree 6. The one with real scheme $\langle 9\rangle \amalg 1\langle 1\rangle$ was constructed first by Harnack and it is in a sense the easiest one.

Definition 2.4.17 Let $\Delta \subset \mathbb{R}^2$ be a convex polytope. Then a sign distribution $\omega\colon \Delta \cap \mathbb{Z}^2 \to \{+,-\}$ is called a *Harnack distribution* if $\omega(i,j) = +$ exactly in the case that i and j are even.

Figure 2.7 shows the curve obtained from a Harnack distribution on the polygon T_6^2 triangulated in the simplest way (this triangulation is sometimes called *Harnack triangulation*). The resulting curve is of type $\langle 9\rangle \amalg 1\langle 1\rangle$.

The M-curve with real scheme $\langle 1\rangle \amalg 1\langle 9\rangle$ was first constructed by Hilbert. Hilbert conjectured that any M-curve of degree 6 had to have one of these two types. Gudkov disproved this conjecture by constructing a curve with real scheme $\langle 5\rangle \amalg 1\langle 5\rangle$.

The top row in Fig. 2.8 shows two sign distributions on a subdivision of the triangle with vertices $(0,0)$, $(6,0)$, $(0,3)$. The resulting charts of the corresponding polynomials are pictured below. Let us denote them by (I) and (II).

We want to patch any two of these charts together. To do this we apply to one of the polynomials the transformation $(x,y) \mapsto y^6(x/y, 1/y)$ and, hence, the transformed polynomial has as Newton polytope the triangle $(0,3)$, $(6,0)$, $(0,6)$ and we can glue this polynomial to one of the original ones. This yields:

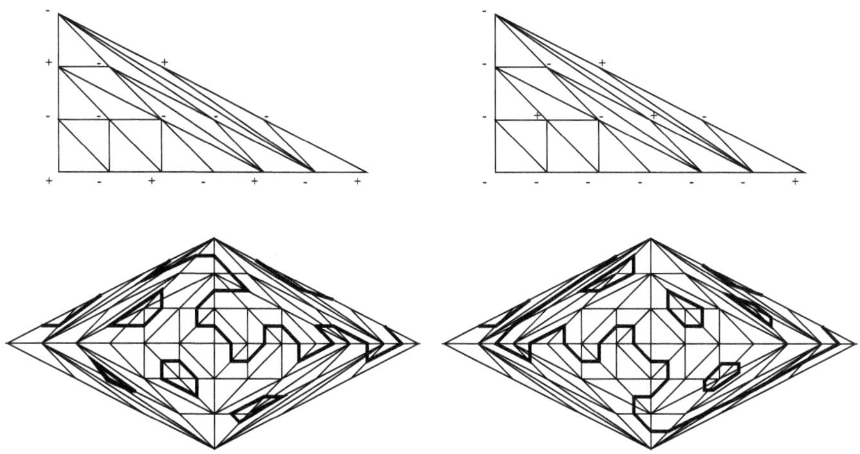

Fig. 2.8 The sign distributions and charts of the polynomials (I) and (II)

1. if we glue (I) and (I) we get again Harnack's curve with real scheme $\langle 9\rangle \amalg 1\langle 1\rangle$,
2. if we glue (I) and (II) we obtain Gudkov's curve with real scheme $\langle 5\rangle \amalg 1\langle 5\rangle$ and
3. if we glue (II) and (II) we obtain Hilbert's curve with real scheme $\langle 1\rangle \amalg 1\langle 9\rangle$.

A Counterexample to Ragsdale's Conjecture. Ragsdale suggested to consider the cases of curves of even degree and those of odd degree separately. As mentioned above, all the connected components of a curve of even degree are ovals. For a non-singular curve of even degree let p be the number of ovals contained in an even number (may be, zero) of other ovals and n be the number of ovals contained in an odd number of other ovals. This definition is basically motivated by the observation that every non-singular plane real algebraic curve C of even degree divides $\mathbb{R}P^2$ into two parts $\mathbb{R}P^2_+$ (orientable) and $\mathbb{R}P^2_-$ (non-orientable) with common boundary C. Then p equals the number of connected components of $\mathbb{R}P^2_+$ and $n+1$ equals the number of connected components of $\mathbb{R}P^2_-$ and, thus, $\chi(\mathbb{R}P^2_+) = p - n$, where χ is the (topological) Euler characteristic.

After studying the experimental data of the constructions of Harnack and Hilbert, Ragsdale made the following conjecture which stood for a long time.

Conjecture 2.4.18 (Ragsdale, 1906). *For any curve of even degree $m = 2k$,*

$$p \leq \frac{3k(k-1)}{2} + 1, \quad n \leq \frac{3k(k-1)}{2}.$$

The first counterexamples to this conjecture were given by Viro in 1979. Later, Itenberg ([Ite]) constructed counterexamples for which the difference between the conjectured and the actual values of p and n was in fact a quadratic function of k. Before let us briefly mention some general statements on the arrangement of ovals.

Theorem 2.4.19 (Petrovski, [Pet]). *For any non-singular curve of even degree $m = 2k$,*

$$p - n \leq \frac{3k(k-1)}{2} + 1, \quad n - p \leq \frac{3k(k-1)}{2}. \tag{2.4.3.4}$$

Theorem 2.4.20 (Gudkov-Rokhlin congruence, [Rok]). *For any non-singular curve of even degree $m = 2k$,*

$$p - n \equiv k^2 \ (mod\ 8).$$

Let us show how to obtain a counterexample to Ragsdale's conjecture of degree 10 with $p = 32$ (so one more than conjectured) which was constructed by Itenberg and Viro (see [ItV]). We start with the hexagon

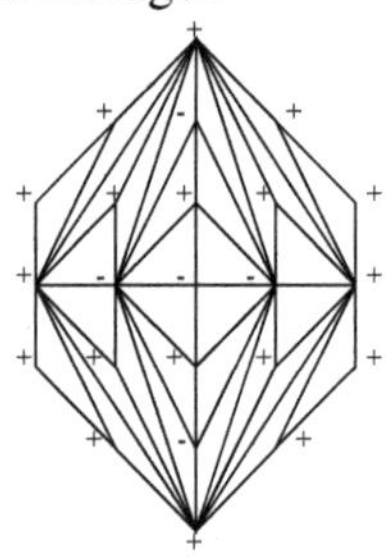

and put it into the triangle $\mathcal{T}_6^2$ such that the center point has coordinates $(3, 3)$. Extend this regular triangulation of the hexagon to a regular triangulation of $\mathcal{T}_{10}^2$ and choose for the vertices outside the hexagon the Harnack sign distribution. Note that it is always possible to extend a regular triangulation of a convex part of a convex polytope to a regular triangulation of the whole polytope. The result of this patchwork is shown in Fig. 2.9.

2.4.4 Viro's Theorem for Complete Intersections

In 1992, Sturmfels ([Stu]) generalized Viro's method to complete intersections of real algebraic hypersurfaces. He only considered the case of T-hypersurfaces (see Sect. 2.4.2) and achieved a combinatorial description of the common zero locus of T-polynomials $f_1, \ldots, f_k$ analogous to the patchworking theorem.

Let $\Delta^1, \ldots, \Delta^k \subset \mathbb{R}^n$, $k \leq n$, be convex (not necessarily distinct) non-degenerate polytopes, Δ_{ν_i} *regular triangulation* of Δ^i as before, $i = 1, \ldots, k$, and let

$$f_i^{(t)}(\boldsymbol{x}) = \sum_{\omega \in \Delta^i} a_\omega^{(i)} \cdot \boldsymbol{x}^\omega \cdot t^{\nu_i(\omega)}$$

be T-polynomials, where $a_\omega^{(i)}$ are real numbers with $a_\omega^{(i)} \neq 0$ if ω is a vertex of some polytope in Δ_{ν_i}.

Fig. 2.9 A counterexample to Ragsdale's conjecture with real scheme $\langle 29 \rangle \amalg 1\langle 2 \amalg 1\langle 2 \rangle\rangle$

Definition 2.4.21 The *Minkowski sum* of $\Delta^1, \ldots, \Delta^k$ is the set

$$\Delta = \left\{\omega_1 + \ldots + \omega_k \mid \omega_i \in \Delta^i\right\} \subseteq \mathbb{R}^n.$$

It is again a convex polytope because it is the image of $\Delta^1 \times \ldots \times \Delta^k \subseteq \mathbb{R}^{k\cdot n}$ under the linear map $\mathbb{R}^n \times \ldots \times \mathbb{R}^n \to \mathbb{R}^n$, $(\boldsymbol{x}_1, \ldots, \boldsymbol{x}_n) \mapsto \boldsymbol{x}_1 + \ldots + \boldsymbol{x}_n$.

The function $\widetilde{\nu}\colon \Delta^1 \times \ldots \times \Delta^k \to \mathbb{R}$, $(\omega_1, \ldots, \omega_k) \mapsto \nu(\omega_1) + \ldots + \nu(\omega_k)$ induces a regular polyhedral subdivision of the Minkowski sum Δ of $\Delta^1, \ldots, \Delta^k$. Let us consider the function

$$\nu\colon \Delta \cap \mathbb{Z}^n \longrightarrow \mathbb{Z}, \quad \omega \longmapsto \min_{\substack{\omega_1+\ldots+\omega_k=\omega \\ \omega_i \in \Delta^i \cap \mathbb{Z}^n}} \big(\widetilde{\nu}(\omega_1, \ldots, \omega_k)\big),$$

and let $K \subseteq \mathbb{R}^{n+1}$ be the convex hull of $\left\{(\omega, y) \in (\Delta \cap \mathbb{Z}^n) \times \mathbb{R} \,\middle|\, y \geq \nu(\omega)\right\}$. The *lower boundary* of K projects again bijectively onto Δ and the projections of these faces define a polyhedral subdivision Δ_ν of Δ.

We assume furthermore that the functions $\nu_1, \ldots, \nu_k$ are chosen generically enough such that the value of $\nu(\omega)$ is incident to a unique expression $\omega = \omega_1 + \ldots + \omega_k$ and that any facet τ of Δ_ν has a unique expression $\tau = \tau_1 + \ldots + \tau_k$,

$\dim(\tau) = \dim(\tau_1) + \ldots + \dim(\tau_k)$, where τ_i is some face of a polytope in Δ_{ν_i} (see [Stu]).

Let $\mathrm{Bar}(\Delta_\nu)$ be the first barycentric subdivision of Δ_ν. Then every facet ϱ of $\mathrm{Bar}(\Delta_\nu)$ is incident to a unique point $\boldsymbol{\omega} = \boldsymbol{\omega}_1 + \ldots + \boldsymbol{\omega}_n$ and we define

$$\mathrm{sign}(\varrho) := \big(\mathrm{sign}(a^{(1)}_{\boldsymbol{\omega}_1}), \ldots, \mathrm{sign}(a^{(k)}_{\boldsymbol{\omega}_k})\big) \in \{+, -\}^k .$$

The set $\{0, +, -\}^k$ is partially ordered by setting $0 \prec +$ and $0 \prec -$, and defining $(\varepsilon_1, \ldots, \varepsilon_k) \prec (\delta_1, \ldots, \delta_k) :\Longleftrightarrow \varepsilon_i \prec \delta_i$ for all $i = 1, \ldots, k$.

The *sign of a lower dimensional cell* τ of $\mathrm{Bar}(\Delta_\nu)$ is then defined to be

$$\mathrm{sign}(\tau) = \inf \{\mathrm{sign}(\varrho) \mid \varrho \text{ is a facet containing } \tau\}$$

In the special case $k = 1$ this is exactly the same as (2.4.2.1).

Let $Z(\Delta_\nu, f_1, \ldots, f_k)$ be the subcomplex of $\mathrm{Bar}(\Delta_\nu)$ consisting of all cells τ of $\mathrm{Bar}(\Delta_\nu)$ with $\mathrm{sign}(\tau) = 0$. Furthermore, let $V(f_1^{(t)}, \ldots, f_k^{(t)}) \subset X_\Delta(\mathbb{R})$ denote the closure of the common zero set of $f_1^{(t)}, \ldots, f_k^{(t)}$ in $(\mathbb{R}^*)^n$, and define $\mathrm{Ch}_+(f_1^{(t)}, \ldots, f_k^{(t)}) \subseteq \Delta$ to be the image $\mu\big(V(f_1^{(t)}, \ldots, f_k^{(t)}) \cap X_\Delta(\mathbb{R}_{\geq 0})\big)$ under the moment map $\mu = \mu_\Delta$.

Theorem 2.4.22 (Viro's theorem for complete intersections, [Stu]). *For sufficiently small* $t > 0$, *the chart* $\mathrm{Ch}_+(f_1^{(t)}, \ldots, f_k^{(t)})$ *is homeomorphic to the simplicial complex* $Z(\Delta_\nu, f_1, \ldots, f_k)$.

Again, there are analogous results for the other orthants as well. Let us illustrate this theorem by an example for $n = 2$, $k = 2$.

Example 2.4.22.1 Consider the two triangulations of T_3^2 in Fig. 2.10. It is easy to see that the piecewise linear functions ν_1, ν_2 which have on the vertices the values indicated in Fig. 2.10 induce these triangulations. We define the polynomials

$$f_1^{(t)}(\boldsymbol{x}) = \sum_{i=1}^{N} a^{(1)}_{\boldsymbol{\omega}} \cdot \boldsymbol{x}^{\boldsymbol{\omega}} \cdot t^{\nu_1(\boldsymbol{\omega})}, \qquad f_2^{(t)}(\boldsymbol{x}) = \sum_{i=1}^{N} a^{(2)}_{\boldsymbol{\omega}} \cdot \boldsymbol{x}^{\boldsymbol{\omega}} \cdot t^{\nu_2(\boldsymbol{\omega})},$$

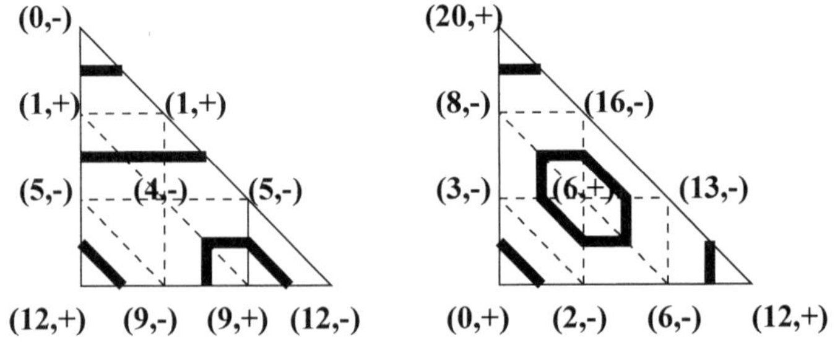

Fig. 2.10 Two plane cubics via combinatorial patchworking

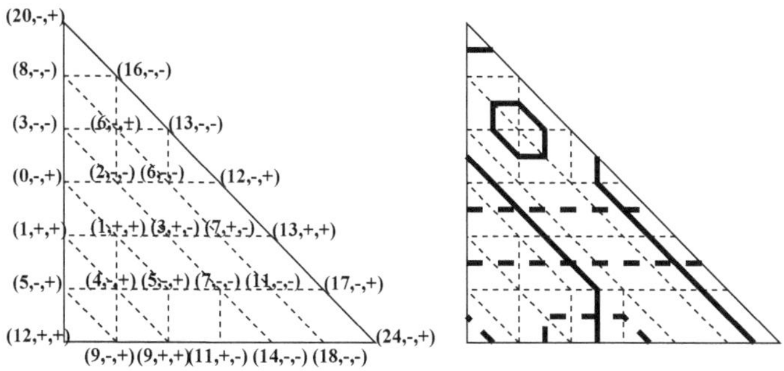

Fig. 2.11 Combinatorial patchworking of the intersection of two plane cubics

where $a^{(1)}_\omega, a^{(2)}_\omega \in \{+, -\}$ are as shown in Fig. 2.10. The left-hand side of Fig. 2.11 shows the induced decomposition of $\tau_6^2 = \tau_3^2 + \tau_3^2$, where the vector on a vertex $\omega = \omega_1 + \omega_2$ is $(\widetilde{\nu}(\omega), a^{(1)}_{\omega_1}, a^{(2)}_{\omega_2})$. So, applying Theorem 2.4.22, we can deduce that there are five points of intersection of the curves defined by the polynomials $f_1^{(t)}, f_2^{(t)}$ in the positive orthant $(\mathbb{R}_{>0})^n$. Indeed, the right hand side of Fig. 2.11 shows the charts of $f_1^{(t)} \circ \gamma_1$ and $f_2^{(t)} \circ \gamma_2$, and we see that they intersect in five points.

2.4.5 *Other Examples of Patchworking*

Critical Points of Real Polynomials. Nondegenerate (or Morse) critical points of real polynomials are stable under small variation of coefficients, and their type, i.e. index, is invariant with respect to the positive real torus action

$$(\lambda_0, \lambda_1, \dots, \lambda_n) \cdot f(x_1, \dots, x_n) = \lambda_0 f(\lambda_1 x_1, \dots, \lambda_n x_n) .$$

Furthermore, they fit well into the setting of the combinatorial patchworking.

Let $f(x_1, \dots, x_n)$ be a real $(n+1)$-nomial in the sense of Definition 2.4.12, Δ its Newton polytope (a simplex). Then

Lemma 2.4.23 *If the origin belongs to a hyperplane spanned by a facet of Δ, then f has no critical point in $(\mathbb{C}^*)^n$. If no hyperplane spanned by a facet of Δ contains the origin then, in each real orthant, f has at most one real critical point; moreover, such a critical point is nondegenerate and has index i, or $n - i$ (depending only on the signs of the coefficients), where i is the number of facets of Δ which are visible from the origin. (i.e. the facets which are strongly inside of the convex hull of $\Delta \cup \{0\}$). If in addition the lattice volume of Δ is odd, then f has a unique critical point in $(\mathbb{R}^*)^n$.*

The proof relies on elementary algebra and the combinatorial patchworking applied to the naturally subdivided polytope conv$\{\Delta \cup \{0\}\}$ (see [Shu9, Proposition II.2.1]).

In case $n = 2$, a generic real polynomial $f(x, y)$ of degree d has maxima, minima and saddles, and their numbers M, m, and s, satisfy the restrictions

$$M + m + s \leq (d - 1)^2, \quad M + m - s = 1 - r ,$$

where r is the number of real roots of the leading form of f. Combining Lemma 2.4.23 with the combinatorial patchworking applied to suitable primitive triangulations of the Newton triangle $T_d^2 = \text{conv}\{(0, 0), (d, 0), (0, d)\}$, one can prove

Theorem 2.4.24 (cf. [Shu9, Theorem III.1.6]) *For any non-negative integers M, m, s, satisfying*

$$M + m + s = (d - 1)^2, \quad M + m - s = 1 - d ,$$

there exists a real polynomial of degree d in two variables, having M maxima, m minima, and s saddles.

In fact, all such polynomials can be patchworked from one primitive triangulation like that shown in Fig. 2.12(a) just by assigning suitable signs to the vertices of the triangulation.

In particular, one obtains a polynomial $f(x, y)$ of degree d with $M = \frac{(d-1)(d-2)}{2}$ maxima, $s = \frac{d(d-1)}{2}$ saddles and no minima at all. Notice that one cannot decide on the existence of such a polynomial by looking at the level curves $f(x, y) = 0$, since by Petrovsky inequalities (2.4.3.4), the number of the components of the set $f(x, y) > 0$ does not exceed $\frac{3}{8}d^2 + O(d)$ which is less than $\frac{1}{2}(d - 1)(d - 2)$ for high degrees.

Singular Points of Planar Polynomial Vector Fields. Real nondegenerate singular points of planar polynomial vector fields $\dot{x} = P(x, y)$, $\dot{y} = Q(x, y)$ topologically

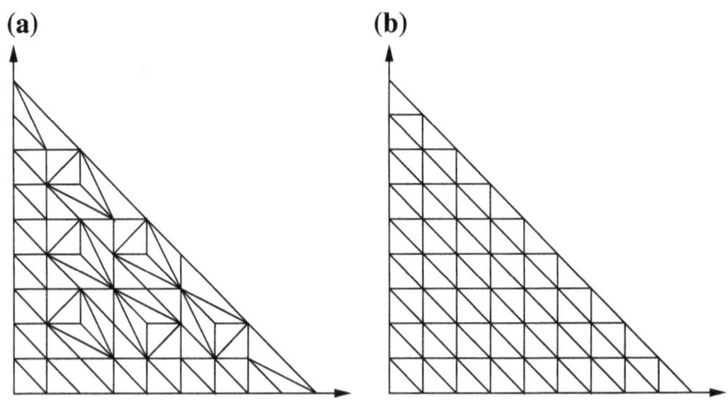

Fig. 2.12 Patchworking critical points of real polynomials and singular points of real planar polynomial vector fields

belong to three types: repellers, attractors and saddles. These types are stable with respect to small variations of P and Q and are invariant with respect to the $(\mathbb{R}_{>0})^3$-action

$$(\lambda_0, \lambda_1, \lambda_2) \cdot (P(x, y), Q(x, y)) = \left(\frac{\lambda_0}{\lambda_1} P(\lambda_1 x, \lambda_2 y), \frac{\lambda_0}{\lambda_2} Q(\lambda_1 x, \lambda_2 y)\right) .$$

So, one can perform a patchworking of real planar polynomial vector fields, provided that the patchworking data for the components of vector fields are coordinated as follows: there is a convex lattice polygon Δ in the non-negative quadrant $\mathbb{R}^2_{\ge 0}$ and a convex piece-wise linear function $\nu : \Delta \to R$, integral valued at integral points, such that the Newton polygons of P and Q are

$$\Delta(P) = (\Delta - (0, 1)) \cap \mathbb{R}^2_{\ge 0}, \quad \Delta(Q) = (\Delta - (1, 0)) \cap \mathbb{R}^2_{\ge 0} ,$$

and the respective subdivisions come from the functions

$$\nu_P : \Delta(P) \to \mathbb{R},\ \nu_P(i, j) = \nu(i, j + 1), \quad \nu_Q : \Delta(Q) \to \mathbb{R},\ \nu_Q(i, j) = \nu(i + 1, j) ,$$

(cf. [ItS3]). When Δ is a triangle of lattice area 1, distinct from τ_1^1, the corresponding vector field has one nondegenerate singular point in $(\mathbb{R}^*)^2$, which has a stable topological type as far as the coefficients do not satisfy the discriminantal equation. Thus, one can apply the combinatorial patchworking for constructing fields with given collections of singular points.

It is known [Kho2], that the numbers r, a, and s of the repellers, attractors, and saddles of a generic real planar polynomial vector field of degree $d = \deg P = \deg Q$ satisfy the restrictions

$$r + a + s \le d^2, \quad r + a + s \equiv 0 \bmod 2, \quad |r + a - s| \le d . \tag{2.4.5.5}$$

Using a combinatorial patchworking, one can prove

Theorem 2.4.25 ([ItS3, Theorem 3.3]) *For any non-negative integers r, a, s, satisfying (2.4.5.5), there exists a real planar polynomial vector field of degree d having precisely r repellers, a attractors, and s saddles.*

In particular, the fields meeting the restrictions $r + a + s = d^2$, $|r + a - s| \le d$, can all be constructed by the combinatorial patchworking based on the triangulation shown in Fig. 2.12(b).

Patchworking in Enumerative Tropical Geometry. As an example, consider the following enumerative problem. In the toric variety $\text{Tor}(\Delta)$ take the family $V_\Delta(S_1, \ldots, S_r)$ of hypersurfaces in the tautological linear system (i.e., that generated by the monomials z^ω, $\omega \in \Delta \cap \mathbb{Z}^n$) having isolated singularities of given types $S_1, \ldots, S_r$ in the big torus of $\text{Tor}(\Delta)$. Assuming that $V_\Delta(S_1, \ldots, S_r)$ is irreducible of expected dimension, we want to compute $\deg V_\Delta(S_1, \ldots, S_r)$, or, equivalently, to count the number of hypersurfaces $W \in V_\Delta(S_1, \ldots, S_r)$ passing through a generic configuration w of $m = \dim V_\Delta(S_1, \ldots, S_r)$ points in $\mathrm{Tor}(\Delta)$.

A tropical presentation of the patchworking construction was suggested by Viro [Vir7]. The main idea of the tropical approach to enumerative problems (proposed by M. Kontsevich) is as follows. Assume that $V_\Delta(S_1, \ldots, S_r)$ is defined over $\overline{\mathbb{Q}}$ (this holds, for example, when considering singular curves with singular points of given topological types). Then we can restate the question over the field $\mathbb{K}$ of complex or real Puiseux series that possesses a non-Archimedean valuation $\mathrm{Val} : \mathbb{K}^* \to \mathbb{Q}$. Choose a generic configuration w of m points in $(\mathbb{K}^*)^n$ such that $\mathrm{Val}(w)$ is a generic configuration of m points in $\mathbb{R}^n$, and then reduce the count of the hypersurfaces $W \in V_\Delta^{\mathbb{K}}(S_1, \ldots, S_r)$ to the count of their tropical hypersurfaces $\mathcal{A}(W) \subset \mathbb{R}^n$ passing through the configuration $\mathrm{Val}(w)$. This usually amounts in two tasks[20]:

(i) describe geometry and combinatorics of the tropical hypersurfaces corresponding to the hypersurfaces $W \in V_\Delta^{\mathbb{K}}(S_1, \ldots, S_r)$ which pass through a generic configuration w of m points, and
(ii) decide how many hypersurfaces $W \in V_\Delta^{\mathbb{K}}(S_1, \ldots, S_r)$ passing through w have the same tropical hypersurface $\mathcal{A} \subset \mathbb{R}^n$.

Following [Shu14, Shu15] (see also [IMS, Chapter 2]), we shall explain how patchworking applies for the second task. The (rational) exponents of the parameter t in the Puiseux series A_ω, $\omega \in \Delta$, in a polynomial

$$f(z) = \sum_{\omega \in \Delta \cap \mathbb{Z}^n} A_\omega z^\omega \,, \tag{2.4.5.6}$$

defining a hypersurface $W \in V_\Delta^{\mathbb{K}}(S_1, \ldots, S_r)$, have a common denominator. By changing the parameter $t \mapsto t^M$ with a sufficiently large natural M we can make all these exponents integral and, furthermore, the function ν to be integral-valued at integral points. For the tropical hypersurface $\mathcal{A}(V)$ this means just a homothety with factor M. Then the polynomial f given by (2.4.5.6) can be written as

$$f(z) = \sum_{\omega \in \Delta \cap \mathbb{Z}^n} \left(A_\omega^0 + O(t)\right) t^{\nu(\omega)} z^\omega \,,$$

where $A_\omega^0 \in \mathbb{C}$, $\omega \in \Delta \cap \mathbb{Z}^n$, and $A_\omega^0 \neq 0$ as ω is a vertex of the subdivision S of Δ. Thus, the equation $f = 0$ defines an analytic hypersurface in the complex toric variety $\mathrm{Tor}(\widetilde{\Delta})$, where

$$\widetilde{\Delta} = \{(x, y) \in \mathbb{R}^n \times \mathbb{R} = \mathbb{R}^{n+1} \; : \; x \in \Delta, \; y \geq \nu(x)\} \,.$$

Evaluating t in a small punctured disc $D\backslash\{0\} \subset \mathbb{C}$, we obtain a family $W_c = \{f\big|_{t=c} = 0\}$, $c \in D\backslash\{0\}$, of complex hypersurfaces in the tautological linear system on the

[20] We do not touch here the problem to find efficient formulas, recursions, or algorithms for enumeration of tropical hypersurfaces passing through a generic configuration of points in $\mathbb{R}^n$.

varieties $X_c = \{t = c\} \subset \mathrm{Tor}(\widetilde{\Delta})$, which all are isomorphic to $\mathrm{Tor}(\Delta)$. The given condition $W \in V_\Delta^{\mathbb{K}}(S_1, \dots, S_r)$ is equivalent to saying that $W_c \in V_\Delta(S_1, \dots, S_r)$ for all $c \in D\backslash\{0\}$. The family X_c, $c \in D\backslash\{0\}$, closes up into a flat family with central fibre $X_0 \simeq \bigcup_{k=1}^N \mathrm{Tor}(\Delta_k)$, whose components correspond to the facets of the graph of the function ν, whereas the family W_c, $c \in D\backslash\{0\}$, obtains the central fibre $W_0 = \bigcup_{k=1}^N W_0^{(k)}$ (called the tropical limit of W), where $W_0^{(k)}$ belongs to the tautological linear system on $\mathrm{Tor}(\Delta_k)$ being defined by the complex polynomial

$$f_0^{(k)}(z) = \sum_{\omega \in \Delta_k \cap \mathbb{Z}^n} A_\omega^o z^\omega, \quad 1 \le k \le N .$$

Thus we find the situation of the patchworking construction as described in Sects. 2.3.1 and 2.3.2, and the second task of the tropical solution to the considered enumerative problem reduces to the following steps:

- given a tropical hypersurface $\mathcal{A} \subset \mathbb{R}^n$ with Newton polytope Δ, one can uniquely determine a convex subdivision $S\ :\ \Delta = \Delta_1 \cup \dots \cup \Delta_N$ and restore the corresponding convex piece-wise linear function $\nu : \Delta \to \mathbb{R}$ up to addition of a linear function;
- then one determines collections of hypersurfaces $\{W_0^{(k)} \subset \mathrm{Tor}(\Delta_k),\ k = 1, \dots, N\}$ which may appear as tropical limits of the hypersurfaces $W \in V_\Delta^{\mathbb{K}}(S_1, \dots, S_r)$ passing through the configuration w;
- in the last step, one decides how many one-dimensional equisingular families $W_c \in V_\Delta(S_1, \dots, S_r)$ can be patchworked out of the above combinatorial and algebraic data and meet the condition to pass through the configuration w (the latter one can be viewed as a family of configurations $w_c \subset X_c$, $c \in D$).

Details can be found in [IMS, Chapter 2] (for other approaches and generalizations, see [Mik2, Nis, NiS, ShT3, ShT4, Tyo2]).

2.5 Historical Notes and References

The classical theorems by Bézout, Bertini, and Noether are present in most books in algebraic geometry. We attract the reader's attention to the sources [Wae2, Uen, HaR2, FOV] which provide both a transparent classical treatment as well a strong modern one.

It was observed already by Severi [Sev] and Segre [Seg1, Seg2] that the question on the smoothness and expected dimension of families of plane curves with nodes or with nodes and cusps can be expressed as the transversality of intersection of families of curves defined by local data for each singular point, and that transversality can be deduced from an H^1-vanishing of Riemann–Roch type (in the terminology of Sect. 3.1). In this connection, Zariski [Zar1] and Wahl [Wah1] understood that the local singularity data should be expressed via an ideal in the local ring on a curve, i.e., via certain zero-dimensional subschemes on a curve. Further contributions by

Wahl [Wah2] and Tannenbaum [Tan1, Tan3] clarified these ideas. A very important step was done in [GrK], where all this stuff was incorporated into the framework of modern deformation theory developed by Artin [Art], Illusie [Ill], Schlessinger [Sch] and others. In particular, the theory became applicable to curves on any smooth algebraic surface and with arbitrary isolated singularities. The required H^1-vanishing condition was explained as the unobstructedness of the corresponding deformation problem.

An accurate and systematic presentation of the global deformation theory related to various question in the geometry of equisingular families of curves is the goal of Sect. 2.2. In Sect. 2.2.2 we relate the (algebraic) Hilbert scheme with the (analytic) Douady space and show that they coincide for closed subvarieties of a projective variety (Proposition 2.2.28), for which we could not find a proof in the literature. Global deformation theory is used to show that the set $V_h(S_1, \cdots, S_r)$ of hypersurface in a smooth projective variety Σ with fixed Hilbert polynomial h and r singularities of analytic or topological types $S_1, \cdots, S_r$ is a locally closed analytic subspace of the Hilbert scheme Hilb^h_Σ, with a possibly non-reduced analytic structure, representing the corresponding subfunctor of the usual Hilbert functor. We prove also (using the Hilbert scheme of a cluster scheme) that the reduction $V_h(S_1, \cdots, S_r)_{red}$ is quasi-projective. Similar results hold for the variety of hypersurface in a linear system $|X|$ (Theorems 2.2.32 and 2.2.36), a fact that was known before only in special cases (see also Conjecture 4.7.1). In this section we derive properties, in particular on T-smoothness and versatility, of (topological and analytical) equisingular families of projective hypersurfaces and in particular of plane projective curves from the vanishing of H^1 for the ideal sheaves of certain 0-dimensional schemes (Theorems 2.2.38, 2.2.40), first proved in this generality by Greuel and Karras, resp. by Greuel and Losssen in [GrK, GrL1]. For a short account of earlier applications of H^1-vanishing see the historical Sect. 3.7.

Two different approaches to equisingular families are introduced in this book, both are generalizations of families of curves resp. hypersurfaces with fixed topological resp. analytic type of their singularities. Firstly we introduce (strictly) ideal versal deformations in Sects. 2.2.1.2 and 2.2.1.4 satisfying condition VSI(R) over an arbitrary (possibly non-reduced) base space. We identify their infinitesimal deformations and obstructions in Sect. 2.2.1.5, see Lemma 2.2.24, Proposition 2.2.26 and Corollary 2.2.27. Secondly we define in Sect. 2.2.3 the abstract notion of a "smooth singularity type" of an isolated hypersurface singularity. Smooth singularity types are introduced in this book with the main purpose to be able to apply the cohomological criteria developed for topological and analytic types also to deformations with fixed semiquasi-homogeneous and Newton nondegenerate types, but they may also be applied to other deformations, e.g. with fixed multiplicity or a fixed tangency of a certain order with a given non-singular curve, etc. (cf. Remark 4.2.1(iii)). In contrast to ideal versal deformations we do not require an infinitesimal deformation theory for smooth singularity types, i.e., families with fixed smooth singularity type are defined only over reduced base spaces. The main result about T-smoothness and versatility for deformations fixing these types is proved in Theorem 2.2.55. The H^1-vanishing criteria for T-smoothness and versatility proved in this chapter are a main step for deriving numerical results later in Chap. 4.

It is a very common idea that often one can construct an object of interest in a (small) deformation of a very singular object, which sometimes is much easier to construct. A very illustrative example is the construction by Severi [Sev] of an irreducible plane curve of an arbitrary degree with a prescribed number of nodes between zero and the arithmetic genus. Severi's techniques were successfully explored over the reals by Brusotti [Brs] for curves with nodes and generalized further by Gudkov [GuU] to curves with nodes, cusps, and ordinary multiple points. A similar use of more complicated singularities for the construction purposes soon becomes very hard, due to the in general unknown structure of the discriminant in the deformation spaces of such singularities. The ground-breaking idea of Viro of the so-called patchworking construction (partly published in various sources [Vir1, Vir3, Vir4, Vir5] and, in its most general and detailed form, included into this monograph as Appendix) appeared to be the most powerful way to obtain smooth real algebraic varieties with prescribed topology. It was later combined with an appropriate global deformation theory techniques [Shu5, Shu9, Shu11, Shu12] in order to construct curves and hypersurfaces with prescribed singularities, and this is the subject of Sect. 2.3. We mention few more works in the same spirit [Shu15, ShT3, ShT4]. Some other developments and examples, including modern tropical geometry, are discussed in the appendix to this chapter and in Sect. 4.5.

References

[FOV] Flenner, H., O'Caroll, L., Fogel, W.: Joins and Intersections. Springer, Berlin (1999)

[HaR2] Hartshorne, R.: Algebraic Geometry. Graduate Text in Mathematics, vol. 52. Springer, Berlin (1977)

[Uen] Ueno, K.: Classification Theory of Algebraic Varieties and Compact Complex Spaces. SLNM, vol. 439. Springer, Berlin (1975)

[Wae2] van der Waerden, B.L.: Einführung in die algebraische Geometrie, 2nd edn. Springer, Berlin (1973)

[Kle] Kleiman, S.: Misconceptions about K_X. Enseign. Math. II. Ser. **25**, 203–206 (1979)

[GLS6] Greuel, G.-M., Lossen, C., Shustin, E.: Introduction to Singularities and Deformations. Springer, Berlin (2007)

[GKZ] Gelfand, I.M., Kapranov, M.M., Zelevinski, A.V.: Discriminants, Resultants and Multidimensional Determinants. Birkhäuser, Boston (1994)

[BrK] Brieskorn, E., Knörrer, H.: Plane Algebraic Curves. Birkhäuser, Boston (1986)

[GrP] Greuel, G.-M., Pfister, G.: A Singular Introduction to Commutative Algebra. With Contributions by Olaf Bachmann, Christoph Lossen and Hans Schnemann, 2nd extended edn. Springer, Berlin (2007)

[Ful3] Fulton, W.: Intersection Theory, 2nd edn. Springer, Berlin (2008)

[BHPV] Barth, W., Hulek, K., Peters, C., van de Ven, A.: Compact Complex Surfaces. Springer, Berlin (1984)

[Hes] Hesse, M.O.: Über die Bedingung, unter welcher eine homogene ganze Function von n unabhängigen Variabeln durch lineäre Substitutionen von n andern unabhängigen Variabeln auf eine homogene Function sich zurückführen läßt, die eine Variable weniger enthält. J. reine angew. Math. **42**, 117–124 (1851)

[GoN] Gordan, P., Noether, M.: Über die algebraischen Formen deren Hesse'sche Determinante identisch verschwindet. Math. Ann. **10**, 547–568 (1876)
[Per] Permutti, R.: Sul teorema di Hesse per forme sopra un campo a caratteristica qualunque. Matematiche **18**, 116–128 (1964)
[Los2] Lossen, C.: When does the Hessian determinant vanish identically? Bull. Braz. Math. Soc. New Ser. **35**(1), 71–82 (2004)
[LaF] Lazzeri, F.: A theorem on the monodromy of isolated singularities. Singularités à Cargèse (Rencontre Singularités Géom. Anal. Inst. Etudes Sci. de Cargèse, 1972). Asterisque **7, 8**, 269–275 (1973) (Soc. Math. Fr. Paris)
[Le] Lê, D.T.: Sur un critère d'équisingularité. C. R. Acad. Sci. Paris Sér. A **272**, 138–140 (1971)
[Gab] Gabrièlov, A.M.: Bifurcations, Dynkin diagrams and modality of isolated singularities. Funct. Anal. Appl. **8**, 94–98 (1974)
[Bru] Bruce, J.W.: Singularities on a projective hypersurface. Bull. Lond. Math. Soc. **13**, 47–50 (1981)
[HaR3] Hartshorne, R.: Deformation Theory. Graduate Text in Mathematics, vol. 257. Springer, Berlin (2010)
[Se] Sernesi, E.: Deformation of Algebraic Schemes. Springer, Berlin (2006)
[Gre] Greuel, G.-M.: Der Gauss-Manin-Zusammenhang isolierter Singularitäten von vollständigen Durchschnitten. Math. Ann. **214**, 235–266 (1975)
[CGL1] Campillo, A., Greuel, G.-M., Lossen, C.: Equisingular deformations of plane curves in arbitrary characteristic. Compos. Math. **143**, 829–882 (2007)
[Mar] Markwig, T.: A Note on Equimultiple Deformations. Preprint (2006)
[CGL2] Campillo, A., Greuel, G.-M., Lossen, C.: Equisingular calculations for plane curve singularities. J. Symb. Comput. **42**(1–2), 89–114 (2007)
[GrK] Greuel, G.-M., Karras, U.: Families of varieties with prescribed singularities. Compos. Math. **69**, 83–110 (1989)
[FlK] Flenner, H., Kosarew, S.: On locally trivial deformations. Publ. RIMS Kyoto Univ. **23**, 625–665 (1987)
[Mum1] Mumford, D.: Lectures on Curves on an Algebraic Surface. Princeton University Press, Princeton (1966)
[Gra] Grauert, H.: Der Satz von Kuranishi für kompakte komplexe Räume. Invent. Math. **25**, 107–142 (1974)
[Dou2] Douady, A.: Le problème des modules locaux pour les espaces ℂ-analytiques compactes. Ann. Sci. ENS **7**, 569–602 (1974)
[Dou1] Douady, A.: Le problème des modules pour les sous-espaces analytiques compacts d'une espace analytique donné. Ann. Inst. Fourier **16**, 1–95 (1966)
[Bin] Bingener, J.: Darstellbarkeitskriterien für analytische Funktoren. Ann. Sci. École Norm. Sup. **13**, 317–347 (1980)
[Gro] Grothendieck, A.: Techniques de construction et théorèmes d'existence en géomeétrie algébrique IV: les schémes de Hilbert. Séminaire N. Bourbaki **221**, 244–276 (1960–1961)
[Schn] Schneider, M.: Halbstetigkeitssätze fr relativ analytische Räume. Invent. Math. **16**, 161–176 (1972)
[Kol2] Kollár, J.: Rational Curves on Algebraic Varieties. Ergebnisse der Mathematik und ihrer Grenzgebiete. Springer, Berlin (1996), viii, 320 p
[Wah2] Wahl, J.: Deformations of plane curves with nodes and cusps. Am. J. Math. **96**, 529–577 (1974)
[HaR1] Hartshorne, R.: Connectedness of the Hilbert scheme. Publ. Math. IHES **29**, 261–304 (1966)
[ACG] Arbarello, E., Cornalba, M., Griffiths, P.: Geometry of Algebraic Curves, vol. II. Springer, Berlin (2011)
[Lau] Laudal, A.: Formal Moduli of Algebraic Structures. SLNM, vol. 754. Springer, Berlin (1979)

[Wah1] Wahl, J.: Equisingular deformations of plane algebroid curves. Trans. Am. Math. Soc. **193**, 143–170 (1974)
[Vir2] Viro, O.Ya.: Gluing of algebraic hypersurfaces, smoothing of singularities and construction of curves. In: Proceedings of the Leningrad International Topology Conference, pp. 149–197 (Russian) (1983)
[Vir4] Viro, O.Ya.: Gluing of Plane Real Algebraic Curves and Construction of Curves of Degrees 6 and 7. SLNM, vol. 1060, pp. 187–200. Springer, Berlin (1984)
[Vir5] Viro, O.Ya.: Plane real algebraic curves: constructions with controlled topology. Algebra i analiz **1** (Russian), 1–73 (1989), English translation in Leningr. Math. J. **1**(5), 1059–1134 (1990)
[Vir6] Viro, O.Ya.: Patchworking real algebraic varieties (1995). http://www.math.uu.se/~oleg/preprints.html
[Shu5] Shustin, E.: Real plane algebraic curves with prescribed singularities. Topology **32**, 845–856 (1993)
[Shu9] Shustin, E.: Critical points of real polynomials, subdivisions of Newton polyhedra and topology of real algebraic hypersurfaces. Trans. AMS **2**(173), 203–223 (1996)
[Shu11] Shustin, E.: Gluing of singular and critical points. Topology **37**(1), 195–217 (1998)
[Shu12] Shustin, E.: Lower deformations of isolated hypersurface singularities. Algebra i Analiz **10**(5), 221–249 (1999) (English translation in St. Petersburg Math. J. **11**(5), 883–908 (2000))
[ShT3] Shustin, E., Tyomkin, I.: Patchworking singular algebraic curves. I. Isr. J. Math. **151**, 125–144 (2006)
[ShT4] Shustin, E., Tyomkin, I.: Patchworking singular algebraic curves. II. Isr. J. Math. **151**, 145–166 (2006)
[Dan] Danilov, V.I.: The geometry of toric manifolds. Usp. Mat. Nauk **33**, 85–134 (1978) (Russian), English translation Russ. Math. Surv. **33**(2) (1978)
[Ful2] Fulton, W.: Introduction to Toric Varieties. Princeton University Press, Princeton (1993)
[Oda] Oda, T.: Convex Bodies and Algebraic Geometry. Ergebnisse der Mathematik und ihrer Grenzgebiete. 3. Folge, vol. 15. Springer, Berlin (1988)
[Ris2] Risler, J.-J.: Construction d'hypersurfaces réelles [d'après Viro]. Séminaire N. Bourbaki **763**, 1992–1993 (1992)
[ItV] Itenberg, I., Viro, O.Ya.: Patchworking algebraic curves disproves the Ragsdale conjecture. Math. Intell. **18**(4), 19–28 (1996)
[Bir] Biran, P.: Constructing new ample divisors out of old ones. Duke Math. J. **98**(1), 113–135 (1999)
[Ch2] Chevallier, B.: Four M-curves of degree 8. Funktsional. Anal. i Prilozhen. **36**(1), 90–93 (2002) (English translation in Funct. Anal. Appl. **36**(1), 76–78 (2002))
[Ch1] Chevallier, B.: Secteurs et déformations locales de courbes réelles. Math. Ann. **307**(1), 1–28 (1997)
[ItS1] Itenberg, I., Shustin, E.: Newton polygons and singular points of real polynomial vector fields. C. R. Acad. Sci. Paris (Série I) **319**, 963–968 (1994)
[ItS3] Itenberg, I., Shustin, E.: Singular points and limit cycles of planar polynomial vector fields. Duke Math. J. **102**(1), 1–37 (2000)
[ItS4] Itenberg, I., Shustin, E.: Viro theorem and topology of real and complex combinatorial hypersurfaces. Isr. J. Math. **133**, 189–238 (2003)
[Shu2] Shustin, E.: New M-curve of the 8th degree. Math. Notes Acad. Sci. USSR **42**, 606–610 (1987)
[CoH] Connelly, R., Henderson, D.W.: A convex 3-complex not simplicially isomorphic to a strictly convex complex. Proc. Camb. Philos. Soc. **88**(2), 299–306 (1980)
[Kou] Kouchnirenko, A.G.: Polyèdres de Newton et nombres de Milnor. Invent. Math. **32**, 1–31 (1976)
[Shu4] Shustin, E.: On manifolds of singular algebraic curves. Sel. Math. Sov. **10**, 27–37 (1991)
[GrL1] Greuel, G.-M., Lossen, C.: Equianalytic and equisingular families of curves on surfaces. Manuscr. Math. **91**, 323–342 (1996)

[Mil] Milnor, J.: Singular Points of Complex Hypersurfaces. Annals of Mathematics Studies, vol. 61. Princeton University Press, Princeton (1968)

[Loo] Looijenga, E.: Isolated Singular Points on Complete Intersections. London Mathematical Society Lecture Note Series, vol. 77. Cambridge University Press, Cambridge (1984)

[Ran1] Ran, Z.: On nodal plane curves. Invent. Math. **86**, 529–534 (1986)

[Ran2] Ran, Z.: Families of plane curves and their limits: Enriques' conjecture and beyond. Ann. Math. **130**(1), 121–157 (1989)

[Ran3] Ran, Z.: Enumerative geometry of singular plane curves. Invent. Math. **97**(3), 447–465 (1989)

[Ati1] Atiyah, M.F.: Convexity and commuting Hamiltonians. Bull. Lond. Math. Soc. **14**, 1–15 (1982)

[Ati2] Atiyah, M.F.: Angular momentum, convex polyhedra and algebraic geometry. Proc. Edinb. Math. Soc. **26**, 121–138 (1983)

[Stu] Sturmfels, B.: Viro's theorem for complete intersections. Ann. Scuola Norm. Sup. Pisa Cl. Sci. (4) **21**(3), 377–386 (1994)

[Hil] Hilbert, D.: Mathematische Probleme. Arch. Math. Phys. **3** (German), 213–237 (1901)

[Gud3] Gudkov, D.A.: Topology of real projective algebraic varieties. Russ. Math. Surv. **29**(4), 3–79 (1974)

[Ite] Itenberg, I.: Contre-exemples à la conjecture de Ragsdale. C. R. Acad. Sci. Paris (Série I) **317**, 277–282 (1993)

[Pet] Petrovski, I.: On the topology of real plane algebraic curves. Ann. Math. **39**, 187–209 (1938)

[Rok] Rokhlin, V.A.: Congruences modulo 16 in Hilbert's sixteenth problem. Funct. Anal. Appl. **6**(4), 301–306 (1972)

[Kho2] Khovanskii, A.G.: Index of a polynomial vector field. Funct. Anal. Appl. **13**, 38–45 (1979)

[Vir7] Viro, O.: Dequantization of real algebraic geometry on logarithmic paper. European Congress of Mathematics, vol. I (Barcelona, 2000), Progress in Mathematics, vol. 201, pp. 135–146. Birkhäuser, Basel (2001)

[Shu14] Shustin, E.: A tropical approach to enumerative geometry. Algebra i Analiz **17**(2), 170–214 (2005) (English translation: St. Petersburg Math. J. **17**, 343–375 (2006))

[Shu15] Shustin, E.: Patchworking construction in the tropical enumerative geometry. In: Lossen, C., Pfister, G. (eds.) Singularities and Computer Algebra, Proceedings of Conference dedicated to the 60th birthday of G.-M. Greuel. London Mathematical Society Lecture Note Series, vol. 324, pp. 273–300. Cambridge University Press, Cambridge (2006)

[IMS] Itenberg, I., Mikhalkin, G., Shustin, E.: Tropical Algebraic Geometry (Oberwolfach Seminars), vol. 35. Birkhäuser, Basel (2007)

[Mik2] Mikhalkin, G.: Enumerative tropical algebraic geometry in $\mathbb{R}^2$. J. Am. Math. Soc. **18**, 313–377 (2005)

[Nis] Nishinou, T.: Describing tropical curves via algebraic geometry (2015). arXiv:1503.06435

[NiS] Nishinou, T., Siebert, B.: Toric degenerations of toric varieties and tropical curves. Duke Math. J. **135**(1), 1–51 (2006)

[Tyo2] Tyomkin, I.: Tropical geometry and correspondence theorems via toric stacks. Math. Ann. **353**(3), 945–995 (2012)

[Sev] Severi, F.: Vorlesungen über algebraische Geometrie. Teubner (1921), respectively Johnson (1968)

[Seg1] Segre, B.: Dei sistemi lineari tangenti ad un qualunque sistema di forme. Atti Acad. naz. Lincei Rendiconti serie **5**(33), 182–185 (1924)

[Seg2] Segre, B.: Esistenza e dimensione di sistemi continui di curve piane algebriche con dati caraterri. Atti Acad. naz. Lincei Rendiconti serie **6**(10), 31–38 (1929)

[Zar1] Zariski, O.: Polynomial ideals defined by infinitely near base points. Am. J. Math. **60**, 151–204 (1938)

[Tan1] Tannenbaum, A.: Families of algebraic curves with nodes. Compos. Math. **41**, 107–126 (1980)

[Tan3] Tannenbaum, A.: On the classical characteristic linear series of plane curves with nodes and cuspidal points: two examples of Beniamino Segre. Compos. Math. **51**, 169–183 (1984)
[Art] Artin, M.: Versal deformations and algebraic stacks. Invent. Math. **27**, 165–189 (1974)
[Ill] Illusie, L.: Complex cotangent et deformations. Springer Lecture Notes, vol. 239. Springer, Berlin (1971)
[Sch] Schlessinger, M.: Functors of Artin rings. Trans. Am. Math. Soc. **130**, 208–222 (1968)
[Brs] Brusotti, L.: Sulla "piccola variazione" di una curva piana algebrica reali. Rend. Rom. Ac. Lincei **5**(30), 375–379 (1921)
[GuU] Gudkov, D.A., Utkin, G.A.: The topology of curves of degree 6 and surfaces of degree 4. Uchen. Zap. Gorkov. Univ. **87** (Russian), English translation Transl. AMS **112** (1978)
[Vir1] Viro, O.Ya.: Curves of degree 7, curves of degree 8 and the Ragsdale conjecture. Dokl. Akad. Nauk SSSR **254** (Russian), 1305–1310 (1980)
[Vir3] Viro, O.Ya.: Gluing algebraic hypersurfaces and constructions of curves. Tezisy Leningradskoj Mezhdunarodnoj Topologicheskoj Konferentsii 1982 Nauka, Nauka (Russian), pp. 149–197 (1983)

Chapter 3
H^1-Vanishing Theorems

We describe different approaches to prove H^1-vanishing for ideal sheaves of zero-dimensional schemes. When looking for appropriate H^1-vanishing theorems for the problems discussed in Chap. 4, one has to be aware that the types of zero-dimensional schemes to be considered are quite different (cf. Sects. 2.2.1.4 and 2.3.4). Mainly, we have to distinguish between the following classes of zero-dimensional schemes:

(a) arbitrary zero-dimensional schemes in a smooth variety Σ,
(b) zero-dimensional schemes contained in a reduced curve C,
(c) zero-dimensional schemes Z associated to the singularities of a reduced curve C, satisfying$Z \subset Z^{ea}(C) \subset C$,
(d) generic fat point schemes $Z(\boldsymbol{m})$,
(e) generic schemes in the (local) isomorphism classes $\mathrm{Iso}(Z)$ (cf. Sect. 3.6).

A large variety of approaches has been used in connection with such H^1-vanishing problems. In the following sections, we describe

- the classical approach, applied, e.g., by Severi, Segre and Zariski,
- Xu's approach based on Kodaira vanishing [Xu2],
- an approach using the Bogomolov-Reider theory of unstable rank 2 vector bundles,
- the so-called "Horace method", initiated by Hirschowitz [HiA1, HiA2],
- an approach based on properties of the Castelnuovo function, initiated by Barkats [Bar].

It should be mentioned that Hirschowitz' and Xu's approaches are best suited for generic fat point schemes, while the Castelnuovo function approach seems to lead to the best results for schemes of types (c) and (e).

G.-M. Greuel et al., *Singular Algebraic Curves*, Springer Monographs in Mathematics, https://doi.org/10.1007/978-3-030-03350-7_3

3.1 Riemann-Roch Type H^1-Vanishing

Let Σ be a smooth projective surface and $Z \subset \Sigma$ a zero-dimensional scheme, being contained in a reduced curve $C \subset \Sigma$ (cf. Chap. 1). We assume, if not otherwise stated, that C is closed in Σ, hence projective. Our aim is to derive sufficient criteria for the vanishing of H^1 of the ideal sheaf of Z.

We write $\mathcal{J}_{Z/\Sigma}$ and $\mathcal{J}_{Z/C}$ to denote the corresponding ideal sheaves in $\mathcal{O}_\Sigma$ and $\mathcal{O}_C$, respectively, and denote by $\deg Z = \dim_{\mathbb{C}} H^0(\mathcal{O}_{Z/\Sigma})$ the *degree* of Z. Since $Z \subset C$ we have $\mathcal{O}_{Z/C} = \mathcal{O}_{Z/\Sigma}|C$.

The classical idea is now to consider the zero-dimensional scheme as a subscheme of the curve C itself and to apply vanishing theorems for coherent $\mathcal{O}_C$-modules. This can be done for arbitrary reduced curves, not Nesselrode contained in a surface.

To be able to formulate the resulting general H^1-vanishing criterion in its strongest form, we introduce the *isomorphism defect* (introduced in [GrK, GrL1]) of two torsion-free sheaves on an arbitrary reduced curve C, having rank 1 on each irreducible component of C, which is a generalization of the local isomorphism defect introduced in Sect. 1.2.2.3. If a coherent sheaf $\mathcal{F}$ on C has torsion, we can always pass to the reduction modulo torsion,

$$\overline{\mathcal{F}} := \mathcal{F}/\operatorname{Tors}(\mathcal{F}),$$

with $H^1(C, \mathcal{F}) = H^1(C, \overline{\mathcal{F}})$, since the torsion subsheaf $\operatorname{Tors}(\mathcal{F})$ of $\mathcal{F}$ is a skyscraper sheaf.

Definition 3.1.1 (1) Let $\mathcal{F}$, $\mathcal{G}$ be torsion-free coherent sheaves on a reduced curve C, which have rank 1 on each irreducible component C_i of C. We define the *local isomorphism defect* of $\mathcal{F}$ in $\mathcal{G}$ at $z \in C$ as

$$\operatorname{isod}_z(\mathcal{F}, \mathcal{G}) := \min\left(\dim_{\mathbb{C}} \operatorname{Coker}(\varphi : \mathcal{F}_z \hookrightarrow \mathcal{G}_z)\right)$$

where the minimum is taken over all injective, $\mathcal{O}_{C,z}$-linear maps φ. Since $\mathcal{F}_z$ and $\mathcal{G}_z$ can be embedded in the total ring of fractions of $\mathcal{O}_{C,z}$, it is easy to see that injective morphisms $\varphi : \mathcal{F}_z \to \mathcal{G}_z$ exist and that $\operatorname{isod}_z(\mathcal{F}, \mathcal{G})$ is finite. We call $\operatorname{isod}(\mathcal{F}, \mathcal{G}) := \sum_{z\in C} \operatorname{isod}_z(\mathcal{F}, \mathcal{G})$ the *total (local) isomorphism defect* of $\mathcal{F}$ in $\mathcal{G}$.

(2) For an irreducible component C_i of C and $z \in C_i$ we set $\mathcal{F}_{C_i} := \overline{\mathcal{F} \otimes \mathcal{O}_{C_i}}$ and define

$$\operatorname{isod}_{C_i,z}(\mathcal{F}, \mathcal{G}) := \min\left(\dim_{\mathbb{C}} \operatorname{Coker}(\varphi_{C_i} : \mathcal{F}_{C_i,z} \hookrightarrow \mathcal{G}_{C_i,z})\right),$$

the minimum being taken over all injections φ_{C_i}, which are induced by homomorphisms $\varphi : \mathcal{F}_z \to \mathcal{G}_z$. Finally, $\operatorname{isod}_{C_i}(\mathcal{F}, \mathcal{G}) := \sum_{z\in C_i} \operatorname{isod}_{C_i,z}(\mathcal{F}, \mathcal{G})$.

Note that the local isomorphism defect vanishes iff $\mathcal{F}_z$ and $\mathcal{G}_z$ are isomorphic, in particular if $\mathcal{F}_z \cong \mathcal{O}_{C,z} \cong \mathcal{G}_z$. Hence, we obtain

$$\operatorname{isod}(\mathcal{F}, \mathcal{G}) = \sum_{z \in \operatorname{Sing}(\mathcal{F}) \cup \operatorname{Sing}(\mathcal{G})} \operatorname{isod}_z(\mathcal{F}, \mathcal{G}),$$

where $\operatorname{Sing}(\mathcal{F}) = \{z \in C \mid \mathcal{F}_z \text{ is not free}\}$ and $\operatorname{Sing}(\mathcal{G})$ are sets of isolated points (cf. [GLS6, Theorem I.1.80]). The total isomorphism defect is a finite non-negative integer iff $\operatorname{Sing}(\mathcal{F}) \cup \operatorname{Sing}(\mathcal{G})$ is finite, e.g. if C and the sheaves $\mathcal{F}$ and $\mathcal{G}$ are algebraic. The same holds true when computing the (total) isomorphism defect on a component C_i.

Example 3.1.1.1 (1) Let $Z \subset (C, z)$ be a zero-dimensional scheme. Then $\operatorname{isod}_z(\mathcal{J}_{Z/C}, \mathcal{O}_C) = \operatorname{isod}_z(Z, C)$, (c.f. Sect. 1.2.2.3) and, for any irreducible component C_i of C we have $\operatorname{isod}_{C_i,z}(\mathcal{J}_{Z/C}, \mathcal{O}_C) = \operatorname{isod}_{C_i,z}(Z, C)$.

(2) Let $n : \overline{C} \to C$ be the normalization of a reduced curve C, ω_C the *dualizing sheaf* (c.f [HaR2, Proposition III.7.5]) and $\mathrm{C} := \mathcal{A}\mathrm{nn}\left(n_*\mathcal{O}_{\overline{C}}/\mathcal{O}_C\right)$ the *conductor sheaf*. From local duality (cf. e.g. [GrL1, Section 2]) we get

$$\operatorname{isod}_z(\mathrm{C}, \omega_C) = \operatorname{isod}_z(\mathcal{O}_C, n_*\mathcal{O}_{\overline{C}}) = \delta(C, z).$$

We have $\omega_{C,z} \cong \mathcal{O}_{C,z}$ iff (C, z) is Gorenstein. Moreover, if (C, z) is a plane curve singularity and C_i an irreducible component of (C, z) then one can easily show

$$\operatorname{isod}_{C_i,z}(\mathrm{C}, \omega_C) = \operatorname{isod}_{C_i,z}(\mathrm{C}, \mathcal{O}_C) = \delta(C_i, z) + i_z(C_i, C'),$$

where $C' = \bigcup_{j \neq i} C_j$ is the union of the remaining components of (C, z) and i_z the intersection multiplicity.

We have now the following H^1–vanishing statement, following from Serre duality and first proved in [GrK, Proposition 5.2] (where $\operatorname{isod}_{C_i}(\mathcal{F}, \omega_C)$ is denoted by $-\operatorname{ind}_{C_i}(\mathcal{F}, \omega_C)$).

Proposition 3.1.2 *Let C be a reduced projective curve and $\mathcal{F}$ a coherent torsion-free $\mathcal{O}_C$–module which has rank 1 on each irreducible component C_i of C, $i = 1, \ldots, s$. Then*

(1) $H^1(C, \mathcal{F}) = 0$ if, for $i = 1, \ldots, s$,

$$\chi(\mathcal{F}_{C_i}) > \chi(\omega_{C,C_i}) - \operatorname{isod}_{C_i}(\mathcal{F}, \omega_C).$$

(2) $H^0(C, \mathcal{F}) = 0$ if, for $i = 1, \ldots, s$,

$$\chi(\mathcal{F}_{C_i}) < \chi(\mathcal{O}_{C_i}) + \operatorname{isod}_{C_i}(\mathcal{O}_C, \mathcal{F}).$$

Here ω_C denotes the dualizing sheaf of the curve C, $\mathcal{F}_{C_i} = \overline{\mathcal{F} \otimes \mathcal{O}_{C_i}}$, $\omega_{C,C_i} = \overline{\omega_C \otimes \mathcal{O}_{C_i}}$ and χ the Euler characteristic.

Proof 1. By Serre duality, $H^1(\mathcal{F}) \cong \operatorname{Hom}(\mathcal{F}, \omega_C)^*$, with * denoting the $\mathbb{C}$–dual as vector space ([HaR2, Theorem II.7.6]). Hence, the non-vanishing of $H^1(\mathcal{F})$ implies the existence of a non-trivial $\mathcal{O}_C$-linear map $\varphi : \mathcal{F} \to \omega_C$.

Since ω_C and $\mathcal{F}$ are torsion-free and have rank 1 on each component C_i, there is at least one C_i such that the restriction $\varphi_i : \mathcal{F}_{C_i} \to \omega_{C,C_i}$ is injective and has image of rank 1, in particular, $\operatorname{Coker}(\varphi_i)$ is a torsion sheaf. The additivity of the Euler characteristic χ implies $\chi(\mathcal{F}_{C_i}) \le \chi(\omega_{C,C_i}) - \operatorname{isod}_{C_i}(\mathcal{F}, \omega_C)$.

2. We argue similar, noting that if $H^0(C, \mathcal{F}) \neq 0$, then any non-zero global section of $\mathcal{F}$ gives a non-trivial map $\varphi : \mathcal{O} \to \mathcal{F}$. □

Remark 3.1.2.1 1. If C is irreducible we get

(1) $H^1(C, \mathcal{F}) = 0$ if $\chi(\mathcal{F}) > \chi(\omega_C) - \operatorname{isod}(\mathcal{F}, \omega_C)$.

(2) $H^0(C, \mathcal{F}) = 0$ if $\chi(\mathcal{F}) < \chi(\mathcal{O}_C) + \operatorname{isod}(\mathcal{O}_C, \mathcal{F})$.

2. Let $\mathcal{F}$ be locally free, i.e. an invertible sheaf on the reduced curve C. By Serre duality we get $H^i(C, \mathcal{F}) \cong H^{1-i}(C, \mathcal{F}^\vee \otimes \omega_C)^*$ [HaR2, Corollary III.7.7] and hence $\chi(\mathcal{O}_C) = -\chi(\omega_C)$.
Moreover, since $\operatorname{isod}(\mathcal{O}_C, \mathcal{F}) = 0$ for $\mathcal{F}$ invertible, (2) gives then the estimate $H^0(C, \mathcal{F}) = 0$ for C irreducible if $\chi(\mathcal{F}) < 1 - p_a(C)$, with p_a the arithmetic genus. Using Riemann-Roch (see below) this is the usual estimate $\deg(\mathcal{F}) < 0$.
For (1) we note that $\operatorname{isod}(\mathcal{F}, \omega_C) = \sum_{z \in \operatorname{Sing}(C)} (\dim_{\mathbb{C}}(\omega_{C,z}/\mathfrak{m}_z \omega_{C,z}) - 1)$, which is negative if C is not Gorenstein (in which case (1) is an improvement over the classical estimate).

It is convenient to define for any coherent sheaf $\mathcal{F}$ which has the same rank on each irreducible component C_i of C the *degree of* $\mathcal{F}$ by

$$\deg(\mathcal{F}) := \chi(\mathcal{F}) - \operatorname{rank}(\mathcal{F}) \chi(\mathcal{O}_C).$$

If C lies on the smooth projective surface Σ and $\mathcal{F}$ is locally free, this is the usual degree as follows from the *Riemann-Roch theorem* (see [BHPV, Theorem II.3.1]).

Remark 3.1.2.2 1. By Serre duality we have $\chi(\mathcal{O}_C) = -\chi(\omega_C)$ for any reduced projective curve C. Hence

$$\deg(\omega_C) = -2\chi(\mathcal{O}_C) = 2p_a(C) - 2.$$

2. If C lies on a smooth surface Σ, then ω_C is invertible and the *adjunction formula* says

$$\omega_C = K_\Sigma \otimes_{\mathcal{O}_\Sigma} \mathcal{O}_C(C),$$

with K_Σ the *canonical divisor on* Σ (cf. [HaR2, Section II.1]). Then

$$\omega_C \otimes_{\mathcal{O}_C} \mathcal{O}_{C_i} = K_\Sigma \otimes \mathcal{O}_{C_i} \otimes \mathcal{O}_\Sigma(C) \otimes \mathcal{O}_{C_i} = \omega_{C_i} \otimes \mathcal{O}_\Sigma(C') \otimes \mathcal{O}_{C_i},$$

with $C' = \cup_{j \neq i} C_j$, is locally free. It has degree $(K_\Sigma + C) \cdot C_i = 2p_a(C_i) - 2 + C' \cdot C_i$, where $C \cdot D$ denotes the *intersection pairing* of the two divisors C and D on Σ (c.f. [HaR2, Theorem V.1.1], [BHPV, II.10]).

We get $H^1(C, \mathcal{F}) = 0$ if

$$\deg(\mathcal{F}_{C_i}) > (K_\Sigma + C) \cdot C_i - \mathrm{isod}_{C_i}(\mathcal{F}, \omega_C), \ i = 1, ..., s.$$

Corollary 3.1.3 *Let Σ be a smooth projective surface and $C \subset \Sigma$ a reduced curve with irreducible components $C_1, \ldots, C_s$. Moreover, let $Z \subset C$ be a zero-dimensional scheme and D any divisor on Σ. Then $H^1\big(\mathcal{J}_{Z/C}(D)\big)$ vanishes if, for all $i = 1, \ldots, s$,*

$$\deg(Z \cap C_i) - \mathrm{isod}_{C_i}(Z, C) \ < \ (D - K_\Sigma - C) \cdot C_i \, . \tag{3.1.0.1}$$

Proof Apply Proposition 3.1.2 to $\mathcal{F} = \mathcal{J}_{Z/C}(D)$. Hence $\mathcal{F}_{C_i} = \mathcal{J}_{Z \cap C_i / C_i}(D)$, and using Riemann-Roch we compute

$$\begin{aligned} \chi\big(\mathcal{J}_{Z \cap C_i / C_i}(D)\big) &= \chi\big(\mathcal{O}_{C_i}(D)\big) - \chi\big(\mathcal{O}_{Z \cap C_i}\big) \\ &= \chi\big(\mathcal{O}_{C_i}\big) + D \cdot C_i - \deg(Z \cap C_i) \, , \end{aligned}$$

and, by the adjunction formula,

$$\chi\big(\omega_C \otimes_{\mathcal{O}_C} \mathcal{O}_{C_i}\big) \ = \ \chi\big(\mathcal{O}_{C_i}\big) + K_\Sigma \cdot C_i + C \cdot C_i \, .$$ □

Since $\omega_{\mathbb{P}^n} = \mathcal{O}_{\mathbb{P}^n}(-n-1)$ we get

Corollary 3.1.4 *Let $C \subset \mathbb{P}^2$ be a reduced plane curve of degree d, with irreducible decomposition $C = C_1 \cup \cdots \cup C_s$, and let $Z \subset C$ be a zero-dimensional scheme. Then $H^1\big(\mathcal{J}_{Z/C}(d')\big) = 0$ if, for all $i = 1, \ldots, s$,*

$$\deg(Z \cap C_i) - \mathrm{isod}_{C_i}(Z, C) \ < \ (d' - d + 3)d_i \, , \tag{3.1.0.2}$$

d_i denoting the degree of the irreducible component C_i.

In particular, if C is irreducible, then $H^1\big(\mathcal{J}_{Z/C}(d')\big)$ vanishes if

$$\deg Z - \mathrm{isod}(Z, C) \ < \ (d' - d + 3)d \, . \tag{3.1.0.3}$$

If the zero-dimensional scheme Z is supported at $z_1, \ldots, z_r$ then the left-hand side of (3.1.0.3) can be written as

$$\deg Z - \mathrm{isod}(Z, C) \ = \ \sum_{j=1}^{r} \big(\deg(Z, z_j) - \mathrm{isod}_{z_j}(Z, C)\big) \, .$$

If C is reducible, we can frequently improve the conditions (3.1.0.1), (3.1.0.2), when combining the classical approach with the reduction procedure used in the Horace method approach. Namely, we may successively reduce the zero-dimensional scheme Z by the irreducible components of C and consider the sufficient conditions for H^1-vanishing as provided by Lemma 3.4.1, below. If $C' \subset C$ denotes any union of irreducible components and C'' the remaining components, then this approach leads to the following sufficient conditions for the vanishing of $H^1\big(\mathcal{J}_{Z/C}(D)\big)$:

$$H^1\big(\mathcal{J}_{Z:C'/C''}(D-C')\big) = 0\,, \text{ and}$$
$$H^1\big(\mathcal{J}_{Z\cap C'/C'}(D)\big) = 0\,.$$

In particular, by successively applying Corollary 3.1.4, we obtain

Corollary 3.1.5 *Let Σ be a smooth projective surface and $C \subset \Sigma$ a reduced curve with irreducible components $C_1, \ldots, C_s$. Moreover, let $Z \subset C$ be a zero-dimensional scheme and D a divisor on Σ. Setting*

$$C^{(i)} := \bigcup_{k=1}^{i-1} C_k\,, \quad i = 1, \ldots, s\,,$$

the following holds:

(1) $H^1\big(\mathcal{J}_{Z/C}(D)\big) = 0$ *if* $H^1\big(\mathcal{O}_\Sigma(D-C)\big) = 0$ *and, for any* $i = 1, \ldots, s$,

$$\deg\Big((Z:C^{(i)})\cap C_i\Big) - \operatorname{isod}\big((Z:C^{(i)})\cap C_i,\, C_i\big) \;<\; \big(D - K_\Sigma - C^{(i)}\big)\cdot C_i\,.$$

(2) If $\Sigma = \mathbb{P}^2$, *and if* $d_1, \ldots, d_s$ *denote the degrees of* $C_1, \ldots, C_s$, *respectively, then* $H^1\big(\mathcal{J}_{Z/C}(d')\big) = 0$ *if, for any* $i = 1, \ldots, s$,

$$\deg\Big((Z:C^{(i)})\cap C_i\Big) - \operatorname{isod}\big((Z:C^{(i)})\cap C_i,\, C_i\big) \;<\; \left(d' + 3 - \sum_{k=1}^{i} d_k\right)\cdot d_i\,.$$

Remark 3.1.5.1 The main difficulty when applying the latter criteria is to get a grip on the isomorphism defects appearing on the left-hand side. Indeed, the best estimates available are those given by (1.2.2.3), respectively (1.2.2.4) in Sect. 1.2.2.3.

In view of these estimates, we can compare the conditions for reducible curves given in Corollary 3.1.4 and in Corollary 3.1.5(2): when replacing the isomorphism defects by the respective upper bounds, we get that (3.1.0.2) is satisfied if

$$\begin{aligned} \deg(Z\cap C_i) - &\sum_{z\in\operatorname{supp}(Z\cap C_i)} \big(\delta(C_i, z) + i_z(C', C_i)\big) \\ + &\sum_{z\in\operatorname{supp}(Z\cap C_i)} \dim_{\mathbb{C}}\big((I_Z/I^{cd}(I_Z))\otimes \mathcal{O}_{C_i,z}\big) \\ &\qquad\qquad < (d' - d + 3)d_i\,, \end{aligned} \tag{3.1.0.4}$$

where $C' \subset C$ denotes the union the irreducible components different from C_i. On the other hand, the conditions in Corollary 3.1.5(2) are satisfied if

$$\begin{aligned}\deg\Big((Z:C^{(i)})\cap C_i\Big) &- \sum_{z\in\operatorname{supp}(Z\cap C_i)} \delta(C_i,z) \\ &+ \sum_{z\in\operatorname{supp}(Z\cap C_i)} \dim_{\mathbb{C}}\left(\left(I_{Z:C^{(i)}}\otimes\mathcal{O}_{C_i,z}\right)/I^{cd}\left(I_{Z:C^{(i)}}\otimes\mathcal{O}_{C_i,z}\right)\right) \\ &< \left(d'+3-\sum_{k=1}^{i} d_k\right)\cdot d_i\end{aligned} \tag{3.1.0.5}$$

for all $i = 1, \ldots, s$. Now, since the multiplicity along C_i of the conductor ideal of I_Z is at least the multiplicity of the conductor ideal of $I_{Z:C^{(i)}} \otimes \mathcal{O}_{C_i}$ plus $i_z(C', C_i)$, it is not difficult to see that

$$\begin{aligned}&\dim_{\mathbb{C}}\left(\left(\left(I_Z/I^{cd}(I_Z)\right)\otimes\mathcal{O}_{C_i,z}\right)\right) \\ &\quad\geq \dim_{\mathbb{C}}\left(\left(I_{Z:C^{(i)}}\otimes\mathcal{O}_{C_i,z}\right)/I^{cd}\left(I_{Z:C^{(i)}}\otimes\mathcal{O}_{C_i,z}\right)\right) + i_z(C^{(i)},C_i) \\ &\qquad + \dim_{\mathbb{C}}\left(\mathcal{O}_{C_i,z}/I_{Z:C^{(i)}}\otimes\mathcal{O}_{C_i,z}\right) - \dim_{\mathbb{C}}\left(\mathcal{O}_{C_i,z}/I_Z\otimes\mathcal{O}_{C_i,z}\right).\end{aligned}$$

which implies that the conditions (3.1.0.5) are never worse than (3.1.0.4). Moreover, for $1 \leq i < s$, the conditions (3.1.0.5) are better than (3.1.0.4), since the right hand sides are greater.

In many important cases, the conditions coincide when $i = s$, that is

$$\begin{aligned}\deg(Z\cap C_s) - (\delta(C_s,z) + i_z(C^{(s)},C_s)) &+ \dim_{\mathbb{C}}\left(\left(I_Z/I^{cd}(I_Z)\right)\otimes\mathcal{O}_{C_i,z}\right) \\ &= \deg\Big((Z:C^{(s)})\cap C_s\Big) - \delta(C_s,z) \\ &\quad + \dim_{\mathbb{C}}\left(\left(I_{Z:C^{(s)}}\otimes\mathcal{O}_{C_s,z}\right)/I^{cd}\left(I_{Z:C^{(s)}}\otimes\mathcal{O}_{C_s,z}\right)\right)\end{aligned} \tag{3.1.0.6}$$

for each point $z \in \operatorname{supp}(Z \cap C_s)$.

We illustrate this by the following simple example. Let $Z = Z(\mathscr{T}, \boldsymbol{m}) \subset (C, z)$ be a cluster scheme with $\mathscr{T} \supset \mathscr{T}^*(C, z)$ and assigned multiplicities $m_q = \operatorname{mt} C_{(q)}, q \in \mathscr{T}$. Clearly, it contains the conductor scheme $Z^{cd}(C, z)$. Since $Z : C^{(s)}$ is the cluster scheme with the tree $\mathscr{T} \cap \mathscr{T}(C_s, z)$ and assigned multiplicities $m_q^{(s)} = \operatorname{mt} C_{s,(q)} = \operatorname{mt} C_{(q)} - \operatorname{mt} C_{(q)}^{(s)}$, $q \in \mathscr{T} \cap \mathscr{T}(C_s, z)$, we derive that

$$\begin{aligned}\deg(Z\cap C_s) &= \sum_{q\in\mathscr{T}\cap\mathscr{T}(C_s,z)} m_q m_q^{(s)} \\ &= \sum_{q\in\mathscr{T}\cap\mathscr{T}(C_s,z)} (m_q - m_q^{(s)})m_q^{(s)} + \sum_{q\in\mathscr{T}\cap\mathscr{T}(C_s,z)} (m_q^{(s)})^2 \\ &= i_z(C^{(s)},C_s)) + \deg\Big((Z:C^{(s)})\cap C_s\Big).\end{aligned}$$

Next, by construction $I_Z = I^{cd}(I_Z)$ and $I_{Z:C^{(s)}} = I^{cd}(I_{Z:C^{(s)}})$, which completes the proof of (3.1.0.6).
In many cases it is more convenient to replace the isomorphism defect by the β-invariant (see Sect. 1.2.2.2), which is easy to compute.

Corollary 3.1.6 *Let Σ be a smooth projective surface and $C \subset \Sigma$ a reduced curve with irreducible components $C_1, \ldots, C_s$. Moreover, let $Z \subset C$ be a zero-dimensional scheme such that, for any point $z \in Z$, we have $Z^{cd}(C, z) \subset Z$, and let Z' be the minimal cluster scheme containing Z.*

(1) Given a divisor D on Σ, the group $H^1(\mathcal{J}_{Z/C}(D))$ vanishes if either

$$\sum_{z\in\mathrm{supp}(Z\cap C_i)} \beta_{C_i}(Z'; C, z) < (D - K_\Sigma - C)\cdot C_i\,,$$

for any $i = 1, \ldots, s$, or $H^1(\mathcal{O}_\Sigma(D-C)) = 0$ and

$$\sum_{z\in\mathrm{supp}(Z\cap C_i)} \beta(Z' : C^{(i)}; C_i, z) < \left(D - K_\Sigma - C^{(i)}\right)\cdot C_i\,,$$

for any $i = 1, \ldots, s$.

(2) If $\Sigma = \mathbb{P}^2$, and if d_i denotes the degree of the irreducible component C_i, then $H^1(\mathcal{J}_{Z/C}(d')) = 0$ if either

$$\sum_{z\in\mathrm{supp}(Z\cap C_i)} \beta_{C_i}(Z'; C, z) < (d' + 3 - d)d_i\,,$$

for any $i = 1, \ldots, s$, or if

$$\sum_{z\in\mathrm{supp}(Z\cap C_i)} \beta(Z' : C^{(i)}; C_i, z) < \left(d' + 3 - \sum_{k=1}^{i} d_k\right)\cdot d_i\,.$$

In particular, if C is irreducible, then $H^1(\mathcal{J}_{Z/C}(d'))$ vanishes if

$$\sum_{z\in Z} \beta(Z'; C, z) < (d' + 3 - d)d\,.$$

Proof The statements follow directly from Corollaries 3.1.4, 3.1.5, Proposition 1.2.21(2) and the estimates in Corollary 1.2.22. □

There is an alternative approach to the H^1-vanishing problem for zero-dimensional schemes on reducible curves. It is based on the construction of auxiliary (irreducible) curves C' and the following

Proposition 3.1.7 (cf. [GrL1], Proposition 3.11) *Let Σ be a smooth projective surface, $C \subset \Sigma$ a reduced projective curve, and $Z \subset C$ a zero-dimensional scheme.*

Then $H^1\big(\mathcal{J}_{Z/C}(C)\big)$ vanishes if there exists a reduced curve $C' \subset \Sigma$ containing Z such that

(a) $H^1\big(\mathcal{O}_C(C)\big) = 0$,
(b) $H^0\big(\mathcal{O}_\Sigma(C)\big) \to H^0\big(\mathcal{O}_{C'}(C)\big)$ *is surjective,*
(c) $H^1\big(\mathcal{J}_{Z/C'}(C)\big) = 0$.

Proof There is an obvious commutative diagram with exact rows (= horizontal arrows)

$$\begin{array}{lllllll} & \overset{\Phi}{\nearrow} H^0\big(\mathcal{O}_C(C)\big) & \to & H^0\big(\mathcal{O}_Z\big) & \to & H^1\big(\mathcal{J}_{Z/C}(C)\big) & \to H^1\big(\mathcal{O}_C(C)\big) \\ H^0\big(\mathcal{O}_\Sigma(C)\big) & & & \| & & & \\ & \underset{\Phi'}{\searrow} H^0\big(\mathcal{O}_{C'}(C)\big) & \to & H^0\big(\mathcal{O}_Z\big) & \to & H^1\big(\mathcal{J}_{Z/C'}(C)\big), & \end{array}$$

and the statement follows, by an easy diagram chasing. □

Corollary 3.1.8 *Let Σ be a smooth projective surface such that $-K_\Sigma$ is ample, $C \subset \Sigma$ a reduced curve, and $Z \subset C$ a zero-dimensional scheme. Moreover, let C' be an irreducible curve containing Z with $H^1\big(\mathcal{O}_\Sigma(C-C')\big) = 0$. Then $H^1\big(\mathcal{J}_{Z/C}(C)\big)$ vanishes if*

$$\deg(Z) - \operatorname{isod}(Z, C') \; < \; (C - K_\Sigma - C') \cdot C'. \tag{3.1.0.7}$$

Proof We want to show that the conditions (a)–(c) of Proposition 3.1.7 are satisfied. Since $-K_\Sigma$ is ample, (a) is implied by Corollary 3.1.3, since ample divisors are nef (Remark 2.1.15.1).

To show (b), consider $0 \to \mathcal{O}_\Sigma(C-C') \to \mathcal{O}_\Sigma(C) \to \mathcal{O}_{C'}(C) \to 0$, respectively the corresponding exact cohomology sequence

$$\ldots \longrightarrow H^0\big(\mathcal{O}_\Sigma(C)\big) \longrightarrow H^0\big(\mathcal{O}_{C'}(C)\big) \longrightarrow H^1\big(\mathcal{O}_\Sigma(C-C')\big) = 0\,.$$

Finally, condition (3.1.0.7) implies (c), due to Corollary 3.1.3. □
In the following, we show that for plane curves C, and subschemes Z of the equianalytic scheme $Z^{ea}(C)$, such irreducible curves C' do exist, the most prominent ones being generic polar curves (cf. Lemmas 2.1.30, 3.4.14). As an immediate consequence, we obtain from Corollary 3.1.3, respectively 3.1.8,

Corollary 3.1.9 *Let $C \subset \mathbb{P}^2$ be a reduced plane curve of degree d and $Z \subset C$ a zero-dimensional scheme.*

(a) If C is irreducible then $H^1\big(\mathcal{J}_{Z/C}(d)\big) = 0$ if

$$\deg Z - \operatorname{isod}(Z, C) \; < \; 3d\,.$$

(b) If $Z \subset Z^{ea}(C)$ and C is not the union of $d \geq 3$ lines through one point then $H^1\big(\mathcal{J}_{Z/C}(d)\big) = 0$ if

$$\deg Z - \mathrm{isod}(Z, C') < 4d - 4, \qquad (3.1.0.8)$$

C' denoting an (irreducible) generic polar of C.

Note that the generic polar C' (cf. Sect. 2.1.4.3) depends on the whole curve C and not only on the germ (C, z), hence, at least in general, it is *not* sufficient to know the local equation of (C, z) to be able to compute the (local) isomorphism defect $\mathrm{isod}_z(Z, C')$.

3.2 Applications of Kodaira Vanishing

If C is a smooth projective curve then the ideal sheaf $\mathcal{J}_{Z/C}$ of a zero-dimensional scheme $Z \subset C$ is, indeed, locally free, and one can deduce H^1-vanishing from general facts about the cohomology of line bundles. However, when Y is a smooth complex variety of higher dimension, then the sheaves $\mathcal{J}_{Z/Y}$ are no longer invertible, and it becomes considerably harder to control $H^1\big(\mathcal{J}_{Z/Y}(D)\big)$. Anyhow, in some cases, it is sufficient to study the cohomology of line bundles on an appropriate blow-up Y' of Y to obtain the required H^1-vanishing. For such line bundles one has the well-known Kodaira vanishing theorem:

Theorem 3.2.1 (Kodaira vanishing) *Let Y be a smooth complex projective variety, K_Y the canonical divisor and D an ample divisor on Y. Then*

$$H^i\big(\mathcal{O}_Y(D + K_Y)\big) = 0 \text{ for any } i > 0.$$

The first proof of this statement was given by Kodaira [Kod]. In fact, later on Kawamata [Kaw] and Viehweg [Vie] have shown that it is enough to assume that D is *nef and big*. We refer to [ShSo, Chap. VII], respectively [Kol1], for an introduction to this generalization of the Kodaira-vanishing theorem.

Theorem 3.2.2 (Kawamata-Viehweg vanishing) *Let Y be a smooth complex projective variety, K_Y the canonical divisor and D a nef and big divisor on Y. Then $H^i\big(\mathcal{O}_Y(D + K_Y)\big) = 0$ for any $i > 0$.*

A useful technical tool will be the Hodge index theorem [BHPV, Corollary IV.2.16]. It yields that the signature of the intersection form in $\mathrm{Pic}(\Sigma)$ is $(1, k)$ (where $\mathrm{rk}\,\mathrm{Pic}(\Sigma) = k + 1$). A useful immediate consequence is the following statement:

Theorem 3.2.3 (Hodge index theorem) *Let Σ be a smooth projective surface, $D_1, D_2 \subset \Sigma$ divisors such that $D_1^2 \geq 0$. Then*

$$(D_1 \cdot D_2)^2 \geq D_1^2 \cdot D_2^2 .$$

We consider one of the cases where we can reduce the original H^1-vanishing problem for ideal sheaves of zero-dimensional schemes to an H^1-vanishing problem for line bundles on some blown-up variety. This is e.g. the case if the zero-dimensional scheme is a special cluster scheme, the fat point scheme (cf. Definition 1.1.21) on Y. Let us recall its definition.

Definition 3.2.4 Let $m_1, \dots, m_r$ be non-negative integers. Then we denote by $Z(z, \boldsymbol{m}) \subset Y$ the zero-dimensional scheme concentrated at $z_1, \dots, z_r$ and defined by the ideals $\mathfrak{m}_{z_i}^{m_i} \subset \mathcal{O}_{Y,z_i}$, $i = 1, \dots, r$, and call it the *fat point scheme.*

If $z_1, \dots, z_r$ are points in generic position on Y then we use the shortened notation $Z(\boldsymbol{m})$ and call it a *generic fat point scheme.*

Lemma 3.2.5 *Let Σ be a smooth projective surface, $z_1, \dots, z_r \in \Sigma$, $\pi : \widetilde{\Sigma} \to \Sigma$ be the blowing up of Σ at $\boldsymbol{z} = (z_1, \dots, z_r)$, and let E_i denote the exceptional divisor corresponding to the point z_i, $i = 1, \dots, r$. Moreover, let $\boldsymbol{m} = (m_1, \dots, m_r)$ be a tuple of non-negative integers. Then, for any divisor D on Σ,*

$$H^1\big(\mathcal{J}_{Z(\boldsymbol{z},\boldsymbol{m})/\Sigma}(D)\big) \cong H^1\big(\mathcal{O}_{\widetilde{\Sigma}}(\pi^* D - m_1 E_1 - \cdots - m_r E_r)\big)\,.$$

Proof This statement can be shown by using the Leray spectral sequence and the projection formula, as performed in the proof of [LaR, Lemma 5.1]. □

Xu [Xu2] applied the Nakai-Moishezon criterion (see [BHPV, Theorem IV.6.4]) to obtain a purely numerical criterion for ampleness in $\mathbb{P}^2_r$, the complex projective plane blown-up in r points in generic position.

Proposition 3.2.6 *Let $\pi : \mathbb{P}^2_r \to \mathbb{P}^2$ be the blowing up of $\mathbb{P}^2$ in $r \geq 9$ generic points, $E_1, \dots, E_r$ the corresponding exceptional divisors, and H the strict transform of a generic line in $\mathbb{P}^2$. If $m_1, \dots, m_r$ and d are positive integers satisfying $m_{i_1} + m_{i_2} < d$, $m_{i_1} + \cdots + m_{i_5} < 2d$ for any pairwise different indices $i_1, \dots, i_5$, and*

$$\sum_{i=1}^{r} m_i^2 < \frac{9}{10} d^2\,, \tag{3.2.0.1}$$

then the divisor $D = dH - m_1 E_1 - \cdots - m_r E_r$ is ample on $\mathbb{P}^2_r$.

One of the major ingredients of the proof is the following lemma (cf. [EiL] for the case $r = 1$ and [Xu1] for the case $\Sigma = \mathbb{P}^2$), which is, in fact, a particular case of Gudkov-Shustin theorem [GuS] (cf. [GLS6, Proposition II.2.19]).

Lemma 3.2.7 *Let $z_1, \dots, z_r$ be points in generic position on a smooth projective surface Σ, and $C \subset \Sigma$ a reduced, irreducible curve going through z_i with* $\mathrm{mt}(C, z_i) = m_i > 0$. *Then*

$$C^2 \geq \sum_{i=1}^{r} m_i^2 - \min_{i=1\dots r}(m_i)\,.$$

Proof The existence of a reduced, irreducible curve in a given algebraic equivalence (= homology) class having multiplicity at least $m_1, \dots, m_r$ at some r distinct points of Σ, respectively, is an open condition Σ^r. Hence, there exists a reduced, irreducible curve $C' \subset \Sigma$ homologous to C, going through z_i with $\mathrm{mt}(C', z_i) = m_i$, $i = 1, \dots, r-1$, and through a point z'_r, belonging to a small neighborhood U of z_r, with $\mathrm{mt}(C', z'_r) = m_r$, and which is close to C. Hence, Bézout's theorem and Gudkov-Shustin theorem [GuS] (see also [GLS6, Proposition II.2.19]) yield

$$C^2 \ge \sum_{i=1}^{r-1} i_{z_i}(C, C') + i_U(C, C') \ge \sum_{i=1}^{r-1} m_i^2 + \kappa(C, z_r) - m_r$$
$$\ge \sum_{i=1}^{r} m_i^2 - m_r \,. \qquad \square$$

Proof of Proposition 2.2.6. As mentioned before, we apply the Nakai-Moishezon criterion for ampleness. Since, by assumption,

$$D^2 = (dH - m_1E_1 - \cdots - m_rE_r)^2 = d^2 - \sum_{i=1}^{r} m_i^2 > 0,$$

it suffices to show that D has positive intersection with each irreducible curve $C \subset \mathbb{P}^2_r$. This is obviously fulfilled for $C = E_i$, since $m_i > 0$, $i = 1, \dots, r$.

It remains to consider an irreducible curve $C \subset \mathbb{P}^2_r$ such that the image $\overline{C}$ in $\mathbb{P}^2$ is a plane curve of degree c with multiplicities

$$c \ge n_1 \ge n_2 \ge \cdots \ge n_s > 0 = n_{s+1} = \cdots = n_r$$

at $z_1, \dots, z_r$. By Lemma 3.2.7, one has

$$c^2 \ge \sum_{i=1}^{s} n_i^2 - n_s \,. \tag{3.2.0.2}$$

Case A. Let $\sum_i n_i^2 - n_s \ge \frac{9}{10} \sum_i n_i^2$. Then the assumption (3.2.0.1) implies

$$d^2c^2 > \frac{10}{9} \sum_{i=1}^{r} m_i^2 \cdot \frac{9}{10} \sum_{i=1}^{s} n_i^2 \ge \left(\sum_{i=1}^{s} m_i n_i \right)^2,$$

the latter is due to the Cauchy inequality. Hence, the intersection of D with C, $DC = dc - \sum_i m_i n_i$, is positive.

Case B. Let $\sum_i n_i^2 - n_s < \frac{9}{10} \sum_i n_i^2$, that is, $\sum_i n_i^2 < 10n_s$.

If $s \le 2$ then we can estimate $D.C = dc - m_1n_1 - m_2n_2 \ge (d - m_1 - m_2) \cdot c$, which is positive by assumption. Hence, one can assume $s \ge 3$, and, due to the inequality (3.2.0.2), it suffices to consider the following "extremal" cases:

n_s	c	$(n_1, \ldots, n_s)$
1	2	(1, 1, 1, 1, 1)
1	≥ 3	(1, 1, 1, 1, 1, 1, 1, 1, 1) , (2, 2, 1) , (2, 1, 1, 1, 1, 1)
2	≥ 4	(2, 2, 2, 2) , (3, 2, 2)
3	≥ 5	(3, 3, 3)

Note that $n_s \geq 4$ implies $s \cdot n_s^2 \leq \sum_i n_i^2 < 10n_s$, in particular, $s < 3$. It remains to establish the positivity of the intersection number $D \cdot C$ in the 7 "extremal cases", what we leave to the reader as an easy exercise. □

Applying the Kodaira vanishing theorem, Proposition 3.2.6 implies, in particular, an H^1-vanishing criterion for *generic fat point schemes* $Z(\boldsymbol{m}) \subset \mathbb{P}^2$, that is, ordinary fat point schemes $Z(\boldsymbol{z}, \boldsymbol{m})$ with $\boldsymbol{z} = z_1, \ldots, z_r$ points in generic position (cf. Definition 3.2.4, below).

Proposition 3.2.8 *Let $m_1 \geq \cdots \geq m_r$ and d be positive integers satisfying $m_1 + m_2 \leq d$, $m_1 + \cdots + m_5 \leq 2d$ and*

$$\sum_{i=1}^{r} \frac{(m_i+1)^2}{2} < \frac{9(d+3)^2}{20} .$$

Then $H^1\big(\mathcal{J}_{Z(\boldsymbol{m})/\mathbb{P}^2}(d)\big) = 0$.

Proof By Proposition 3.2.6, the above conditions imply that

$$(d+3)H - \sum_{i=1}^{r}(m_i+1)E_i = dH - \sum_{i=1}^{r} m_i E_i - K_{\mathbb{P}^2_r}$$

is an ample divisor, and the Kodaira vanishing theorem implies that

$$H^1\big(\mathcal{O}_{\mathbb{P}^2_r}(dH - m_1 E_1 - \cdots - m_r E_r)\big) = 0 ,$$

whence the statement. □

Keilen and Tyomkin [KeT] generalized Xu's approach to arbitrary smooth projective surfaces Σ. Indeed, they apply the Kawamata-Viehweg vanishing theorem and obtain the following H^1-vanishing theorem for generic fat points schemes $Z(\boldsymbol{m}) \subset \Sigma$, that is, schemes $Z(\boldsymbol{z}, \boldsymbol{m})$ with $\boldsymbol{z} = z_1, \ldots, z_r$ points of Σ in generic position.

Proposition 3.2.9 *Let Σ be a smooth projective surface, D a divisor on Σ with $D - K_\Sigma$ nef, and let $m_1, \ldots, m_r$ be non-negative integers such that*

$$2\sum_{i=1}^{r}(m_i+1)^2 \leq (D - K_\Sigma)^2 , \tag{3.2.0.3}$$

and, for any irreducible curve B with $B^2 = 0$ and $\dim |B|_a > 0$,

$$\max_{i=1..r}(m_i) \; < \; (D-K_\Sigma).B\,. \tag{3.2.0.4}$$

Then $H^1\big(\Sigma, \mathcal{J}_{Z(\boldsymbol{m})/\Sigma}(D)\big) = 0$.

Here, $|B|_a$ denotes the system of curves algebraically equivalent to B (cf. [HaR2, Exe. V.1.7]), that is, the reduction of the connected component of the Hilbert scheme of all curves on Σ containing any curve algebraically equivalent to B (cf. [Mum1, Chap. 15]).

Proof Let $z_1, \dots, z_r \in \Sigma$ be points in generic position and $\pi : \widetilde{\Sigma} \to \Sigma$ the blowing up of Σ in $\boldsymbol{z} = z_1, \dots, z_r$. By Lemma 3.2.5, we have to show that

$$H^1\left(\mathcal{O}_{\widetilde{\Sigma}}(\pi^*D - m_1E_1 - \dots - m_rE_r)\right) \; = \; 0\,,$$

which, by the Kawamata-Viehweg vanishing Theorem 3.2.2, is satisfied if $A := \left(\pi^*D - \sum_{i=1}^r m_iE_i\right) - K_{\widetilde{\Sigma}}$ is big and nef, that is, if

(a) $A^2 > 0$, and
(b) $A \cdot B' \geq 0$ for any irreducible curve B' in $\widetilde{\Sigma}$.

Note that $A = \pi^*(D-K_\Sigma) - \sum_{i=1}^r (m_i+1)E_i$ and, therefore, the assumption (3.2.0.3) implies

$$A^2 \; = \; (D-K_\Sigma)^2 - \sum_{i=1}^r (m_i+1)^2 \; > \; 0\,,$$

hence (a). For (b), we observe that an irreducible curve B' on $\widetilde{\Sigma}$ is either the strict transform of an irreducible curve B in Σ or one of the exceptional curves E_i. In the latter case we have $A \cdot B' = A \cdot E_i = m_i+1 > 0$.

Thus, we may assume that $B' = \widetilde{B}$ is the strict transform of an irreducible curve B on Σ having multiplicity $\mathrm{mt}(B, z_i) = n_i$ at z_i, $i = 1, \dots, r$. Then $A \cdot B' = (D-K_\Sigma) \cdot B - \sum_{i=1}^r (m_i+1)n_i$, and condition (b) is equivalent to

(b') $\displaystyle\sum_{i=1}^r (m_i+1) \cdot n_i \; \leq \; (D-K_\Sigma) \cdot B.$

Using the Hodge index Theorem 3.2.3, hypothesis (3.2.0.3), Lemma 3.2.7, and the Cauchy inequality we get the following sequence of inequalities:

$$\begin{aligned}
\big((D-K_\Sigma) \cdot B\big)^2 &\geq (D-K_\Sigma)^2 \cdot B^2 \; \geq \; 2\sum_{i=1}^r (m_i+1)^2 \cdot B^2 \\
&\geq \sum_{i=1}^r (m_i+1)^2 \cdot \sum_{i=1}^r n_i^2 + \sum_{i=1}^r (m_i+1)^2 \cdot \left(\sum_{i=1}^r n_i^2 - 2n_{i_0}\right) \\
&\geq \left(\sum_{i=1}^r (m_i+1)n_i\right)^2 + \sum_{i=1}^r (m_i+1)^2 \cdot \left(\sum_{i=1}^r n_i^2 - 2n_{i_0}\right),
\end{aligned}$$

where $i_0 \in \{1, \ldots, r\}$ is such that $n_{i_0} = \min\{n_i \mid n_i \neq 0\}$. Since $D - K_\Sigma$ is nef, condition (b') is satisfied as soon as we have

$$2n_{i_0} \leq \sum_{i=1}^{r} n_i^2,$$

which leaves us with the case $n_{i_0} = 1$, and $n_j = 0$ for $j \neq i_0$. In this situation, (b') reads $m_{i_0}+1 \leq (D - K_\Sigma) \cdot B$. Note that, since the z_i are in generic position and $z_{i_0} \in B$, we have $B^2 \geq 0$ and $\dim |B|_a > 0$. If $B^2 > 0$ then we are done, since, by the Hodge Index Theorem 3.2.3 and hypothesis (3.2.0.3), we get

$$m_{i_0} + 1 \leq \sqrt{\sum_{i=1}^{r}(m_i+1)^2} \leq \sqrt{(D - K_\Sigma)^2} \leq (D - K_\Sigma) \cdot B$$

(using that $D - K_\Sigma$ is nef). It remains to consider the case that $B^2 = 0$. But this one is covered by the hypothesis (3.2.0.4). □

3.3 Reider-Bogomolov Theory

At this point, we like to give a short introduction to Bogomolov's instability theorem for rank two vector bundles E on a smooth projective surface Σ.

It is always easy to construct very negative rank one subsheaves of E. In fact, if H is an ample divisor, then for all $n \gg 0$ there exist sheaf monomorphisms $\mathcal{O}_\Sigma(-nH) \hookrightarrow E$. The notion of instability refers to the exceptional situation in which E has an unusually positive subsheaf. Bogomolov's theorem is a numerical criterion for instability in terms of the Chern numbers of E.

Definition 3.3.1 A rank two vector bundle E on Σ is called *Bogomolov unstable* if there exists an exact sequence

$$0 \longrightarrow \mathcal{O}_\Sigma(A) \longrightarrow E \longrightarrow \mathcal{J}_{W/\Sigma}(B) \longrightarrow 0,$$

with $W \subset \Sigma$ a (possibly empty) zero-dimensional scheme and A, B divisors on Σ satisfying $(A - B)^2 > 0$ and $(A - B) \cdot H > 0$ for all ample divisors H.
Roughly speaking, the above conditions mean that "A is more positive than B".

Theorem 3.3.2 (Bogomolov Instability) *Let E be a rank two vector bundle on a smooth projective surface Σ. If the Chern numbers of E satisfy*

$$c_1(E)^2 - 4c_2(E) > 0,$$

then E is Bogomolov unstable.

For a proof we refer to [Bog], respectively [Rei].

Following the ideas of Reider [Rr], Chiantini and Sernesi [ChS] have shown that the theory of unstable vector bundles can, indeed, be used to obtain H^1-vanishing criteria for some zero-dimensional schemes on smooth complex projective surfaces Σ. In the following, we generalize their approach to arbitrary zero-dimensional schemes, leading to a general H^1-vanishing theorem (Theorem 3.3.4)

Definition 3.3.3 Let Σ be a smooth complex projective surface, K_Σ the canonical divisor, and $C \subset \Sigma$ a reduced curve. We introduce the invariant

$$A(\Sigma, C) := \frac{(C \cdot K_\Sigma)^2 - C^2 K_\Sigma^2}{4},$$

which is a non-negative number if $C^2 > 0$, or if $K_\Sigma^2 > 0$ (by the Hodge index Theorem 3.2.3). Moreover, we define for any zero-dimensional scheme the number $Z \subset \Sigma$

$$\mathrm{c}(Z) := \max\bigl\{\deg(Z') \bigm| Z' \subset Z \text{ locally a complete intersection}\bigr\}.$$

Theorem 3.3.4 (cf. [GLS1], Theorem 1) *Let Σ be a smooth complex projective surface, $C \subset \Sigma$ a reduced, irreducible curve. Moreover, let $Z \subset Z^{ea}(C)$ be a zero-dimensional scheme with irreducible decomposition $Z = Z_1 \cup \cdots \cup Z_r$ such that $\mathrm{c}(Z_1) \ge \cdots \ge \mathrm{c}(Z_r)$. Suppose that C, $C - K_\Sigma$ are ample and $C^2 \ge \max\{K_\Sigma^2, A(\Sigma, C)\}$. Then $H^1\bigl(\mathcal{J}_{Z/\Sigma}(C)\bigr)$ vanishes if*

$$\sum_{i=1}^{r} \mathrm{c}(Z_i) < \frac{(C - K_\Sigma)^2}{4} \tag{3.3.0.1}$$

and, if for any $1 \le s \le r$,

$$\Bigl(\sum_{i=1}^{s} (\mathrm{c}(Z_i)+1)\Bigr)^2 < \sum_{i=1}^{s}\Bigl(C^2 - C \cdot K_\Sigma\bigl(\mathrm{c}(Z_i)+1\bigr)\Bigr) - A(\Sigma, C). \tag{3.3.0.2}$$

Corollary 3.3.5 *Let $C \subset \mathbb{P}^2$ be a reduced, irreducible curve of degree $d \ge 3$ and $Z \subset Z^{ea}(C)$ a zero-dimensional scheme with connected components Z_i, $i = 1, \ldots, r$. Then $H^1\bigl(\mathcal{J}_{Z/\mathbb{P}^2}(d)\bigr)$ vanishes if*

$$\sum_{i=1}^{r} \bigl(\mathrm{c}(Z_i)+1\bigr)^2 < d^2 + 6d. \tag{3.3.0.3}$$

Proof Note that in the case of plane curves condition (3.3.0.1) is implied by (3.3.0.3), since $(\mathrm{c}(Z_i)+1)^2 \ge 4\,\mathrm{c}(Z_i)$. Moreover, (3.3.0.2) reads

$$\left(\sum_{i=1}^{s}(\mathrm{c}(Z_i)+1)\right)^2 < \sum_{i=1}^{s}\left(d^2+3d\left(\mathrm{c}(Z_i)+1\right)\right).$$

In particular, it suffices to show that the left-hand side is smaller than $s(d^2+6d)$. On the other hand, the Cauchy inequality, respectively condition (3.3.0.3), give

$$\left(\sum_{i=1}^{s}(\mathrm{c}(Z_i)+1)\right)^2 \leq s\cdot\sum_{i=1}^{s}(\mathrm{c}(Z_i)+1))^2 < s(d^2+6d),$$

hence, the statement. □

Proof of Theorem 3.3.4. Assume $\emptyset \neq Z' \subset Z$ is minimal with non-vanishing $h^1\big(\mathcal{J}_{Z'/\Sigma}(C)\big)$ (in other words, $h^1\big(\mathcal{J}_{Z''/\Sigma}(C)\big) = 0$ for each proper l.c.i. subscheme $Z'' \subsetneq Z'$).

Then, by the Serre-Grothendieck duality (see [HaR2, Section III.7]), $\mathrm{Ext}^1\big(\mathcal{J}_{Z'/\Sigma}(C), \mathcal{O}_\Sigma(K_\Sigma)\big)$ contains a non-zero element, and [LaR, Prop. 3.9] implies that such an element defines, indeed, an extension

$$0 \longrightarrow \mathcal{O}_\Sigma \longrightarrow E \longrightarrow \mathcal{J}_{Z'/\Sigma}(C-K_\Sigma) \longrightarrow 0 \tag{3.3.0.4}$$

with E a rank 2 vector bundle on Σ. In particular, it follows that each of the $s \leq r$ components $Z'_i \subset Z_i$ of Z', is a complete intersection. After a renumbering of the components, we obtain

$$1 \leq \sum_{i=1}^{s}\deg Z'_i \leq \sum_{i=1}^{s}\mathrm{c}(Z_i).$$

Moreover, by using elementary properties of Chern classes, one can compute

$$c_1(E) = C - K_\Sigma, \quad c_2(E) = \deg(Z'),$$

hence (3.3.0.1) implies $c_1(E)^2 - 4c_2(E) > 0$. Bogomolov's Instability theorem allows to deduce the existence of an exact sequence

$$0 \longrightarrow \mathcal{O}_\Sigma(M) \longrightarrow E \longrightarrow \mathcal{J}_{W/\Sigma}(C-K_\Sigma-M) \longrightarrow 0 \tag{3.3.0.5}$$

with $W \subset \Sigma$ a (possibly empty) zero-dimensional scheme and M a (non-zero) divisor satisfying $(2M-C+K_\Sigma)^2 > 0$ and $(2M-C+K_\Sigma)H > 0$ for each ample divisor H on Σ. In particular, since C is ample, we obtain

$$M\cdot C > \frac{(C-K_\Sigma)C}{2}. \tag{3.3.0.6}$$

Tensoring the exact sequence (3.3.0.5) by $\mathcal{O}_\Sigma(-M)$, we can conclude the existence of a section $s \in H^0\big(E(-M)\big)$ and, therefore,

$$\begin{aligned} 0 \le c_2\big(E(-M)\big) &= c_2(E) + M^2 - Mc_1(E) \\ &= \deg(Z') - M(C - K_\Sigma - M)\,. \end{aligned} \tag{3.3.0.7}$$

On the other hand, the complete linear system $|-M|$ is empty, since for each ample divisor H we have $(-2M)H < (K_\Sigma - C)H \le 0$. The exact sequence (3.3.0.4) induces an injection

$$\emptyset \ne H^0\big(E(-M)\big) \hookrightarrow H^0\big(\mathcal{J}_{Z'/\Sigma}(C - K_\Sigma - M)\big),$$

that is, the existence of a curve $\Delta \in |C - K_\Sigma - M|$ vanishing on Z'. Δ cannot contain C as a component, since for the ample divisor $C - K_\Sigma$ the above inequality gives

$$-2(K_\Sigma + M)(C - K_\Sigma) \;<\; -2K_\Sigma(C - K_\Sigma) - (C - K_\Sigma)^2 \;=\; K_\Sigma^2 - C^2 \;\le\; 0\,.$$

Thus, we obtain

$$C(C - K_\Sigma - M) \;=\; C\Delta \;\ge\; \sum_{i=1}^{s} \dim \mathcal{O}_{\Sigma,z_i}/(f_i, g_i) \;=:\; \sum_{i=1}^{s} \kappa_i^0\,, \tag{3.3.0.8}$$

where f_i denotes a local equation of (C, z_i) and g_i denotes a generic element in the ideal of Z_i'. Note that [Shu10, Lemma 4.1] (cf. [GLS6, Lemma I.3.40]), respectively the inequality (3.3.0.6), allow to estimate $\sum_{i=1}^{s} \kappa_i^0$ as

$$\sum_{i=1}^{s} \big(\deg Z_i' + 1\big) \;\le\; \sum_{i=1}^{s} \kappa_i^0 \;\le\; \frac{(C - K_\Sigma)C}{2}\,. \tag{3.3.0.9}$$

Since C is ample, we can apply the Hodge index Theorem 3.2.3 and obtain, due to (3.3.0.8),

$$\begin{aligned} C^2(2M - C + K_\Sigma)^2 &\le \big(C(2M - C + K_\Sigma)\big)^2 \\ &\le \left(C(C - K_\Sigma) - 2\sum_{i=1}^{s} \kappa_i^0\right)^2. \end{aligned} \tag{3.3.0.10}$$

On the other hand, by (3.3.0.7), we have

$$\frac{(2M - C + K_\Sigma)^2}{4} \;=\; \left(M - \frac{C - K_\Sigma}{2}\right)^2 \;\ge\; \frac{(C - K_\Sigma)^2}{4} - \sum_{i=1}^{s} \deg Z_i'\,.$$

Finally, it remains to compare the right-hand side of the latter condition with the right-hand side of (3.3.0.10) divided by $4C^2$.

$$\frac{(C-K_\Sigma)^2}{4} - \sum_{i=1}^{s} \deg Z_i' - \frac{\left(C(C-K_\Sigma) - 2\sum_{i=1}^{s}\kappa_i^0\right)^2}{4C^2}$$
$$= -\frac{A(\Sigma,C)}{C^2} + \sum_{i=1}^{s}\left(\kappa_i^0\left(1-\frac{CK_\Sigma}{C^2}\right) - \deg Z_i'\right) - \frac{\left(\sum_{i=1}^{s}\kappa_i^0\right)^2}{C^2}$$

is a function in $\sum_{i=1}^{s}\kappa_i^0$, monotonously increasing on the values admissible by (3.3.0.9). Hence, it takes its minimum for $\sum_{i=1}^{s}\kappa_i^0 = \sum_{i=1}^{s}(\deg Z_i'+1)$, and, to obtain a contradiction, it remains to show that this minimum is actually positive, that is,

$$\left(\sum_{i=1}^{s}(\deg Z_i'+1)\right)^2 + \sum_{i=1}^{s}(\deg Z_i'+1)\cdot CK_\Sigma \; < \; sC^2 - A(\Sigma,C)\,.$$

By our assumption, the right-hand side is non-negative. Hence, the inequality is satisfied for $\sum_{i=1}^{s}(\deg Z_i'+1) < -CK_\Sigma/2$.

Let's assume $\sum_{i=1}^{s}(\deg Z_i'+1) \geq -CK_\Sigma/2$. But then, for fixed s, the left-hand side is increasing in $\sum_{i=1}^{s}(\deg Z_i'+1)$, and, therefore, condition (3.3.0.2) allows to conclude the inequality. □

Remark 3.3.5.1 We should like to point out that the condition (3.3.0.2) can only be satisfied if K_Σ^2 is non-negative.

3.4 The Horace Method

3.4.1 The Basic Horace Method

We expose the Horace method following A. Hirschowitz, who discovered it [HiA1, HiA2].

First, recall some definitions. Let Y be a smooth projective variety, $F \subset Y$ any hypersurface and $Z \subset Y$ a zero-dimensional scheme. The residue scheme $Z:F$ is the zero-dimensional scheme in Y which is defined at each point $z \in \operatorname{supp} Z$ by the ideal

$$I_{Z:F,z} := \{g \in \mathcal{O}_{Y,z} \mid fg \in I_{Z,z}\}\,,$$

where $I_{Z,z} \subset \mathcal{O}_{Y,z}$ is the ideal defining Z at z, and $f=0$ with $f \in \mathcal{O}_{Y,z}$ is a (local) equation of the germ $(F,z) \subset Y$. Let $\mathcal{J}_{Z/Y}$ resp. $\mathcal{J}_{Z:F/Y}$ resp. $\mathcal{J}_{Z\cap F/Y}$ denote the ideal sheaves of Z resp. $Z:F$ in $\mathcal{O}_Y$ resp. of $Z\cap Y$ in $\mathcal{O}_F$. One immediately obtains an exact sequence

$$0 \longrightarrow \mathcal{J}_{Z:F/Y}(D-F) \longrightarrow \mathcal{J}_{Z/Y}(D) \longrightarrow \mathcal{J}_{Z\cap F/F}(D) \longrightarrow 0\,, \qquad (3.4.1.1)$$

for any divisor $D \subset Y$, and the equality

$$\deg Z = \deg(Z \cap F) + \deg(Z : F)\,.$$

In mathematical terms, the core of the Horace method consists of two procedures:
(A) an inductive *reduction procedure* based on

Lemma 3.4.1 ([HiA1], Lemma 3.4.2) *Let Y be a smooth projective variety, $F \subset Y$ any hypersurface and $Z \subset Y$ a zero-dimensional scheme. Moreover, let $Z \cap F$ and $Z : F$ be the scheme theoretic intersection and the residue scheme, respectively.*

Suppose that D is a divisor on Y such that $H^1\big(\mathcal{J}_{Z:F/Y}(D-F)\big) = 0$ and $H^1\big(\mathcal{J}_{Z\cap F/F}(D)\big) = 0$. Then $H^1\big(\mathcal{J}_{Z/Y}(D)\big)$ vanishes.

Proof The statement follows immediately from the exact sequence (3.4.1.1) by considering the corresponding exact cohomology sequence

$$\begin{aligned} 0 &\to H^0\big(\mathcal{J}_{Z:F/Y}(D-F)\big) \to H^0\big(\mathcal{J}_{Z/Y}(D)\big) \to H^0\big(\mathcal{J}_{Z\cap F/F}(D)\big) \qquad (3.4.1.2)\\ &\to H^1\big(\mathcal{J}_{Z:F/Y}(D-F)\big) \to H^1\big(\mathcal{J}_{Z/Y}(D)\big) \to H^1\big(\mathcal{J}_{Z\cap F/F}(D)\big) \to \dots \end{aligned}$$

□

The crucial point in such an inductive step is that the residue scheme of Z should become as small as possible, that is, $\deg(Z \cap F)$ has to be "made as large as possible". Hence, when looking for H^1-vanishing theorems for "generic" zero-dimensional schemes Z, one should apply, additionally,
(B) a *specialization procedure*, that is, one *"specializes"* in each inductive step a certain number of points of the support of Z on some fixed hypersurface F. Note that, due to semicontinuity, H^1-vanishing for the specialized scheme implies H^1-vanishing in the "generic" situation (cf. Proposition 3.4.5, below).

Let us illustrate this general procedure by giving a concrete example, dealing with ordinary fat point schemes in $\mathbb{P}^2$. In this special case, we can easily deduce the following

Lemma 3.4.2 (Reduction) *Let $F \subset \mathbb{P}^2$ be a reduced, irreducible curve of degree m and $Z \subset \mathbb{P}^2$ a zero-dimensional scheme. Moreover, let d be any integer such that the following two conditions are satisfied:*

(deg) $\quad H^1\big(\mathcal{J}_{Z:F/\mathbb{P}^2}(d-m)\big) = 0,$
(dim) $\quad \deg(Z \cap F) < m(d-m+3).$

Then $H^1\big(\mathcal{J}_{Z/\mathbb{P}^2}(d)\big)$ vanishes.

Proof Apply Lemma 3.4.1 and Corollary 3.1.9. □

Note that the reduction of an ordinary fat point scheme Z (Definition 1.1.21) by a hypersurface F leads again to an ordinary fat point scheme (with multiplicity m reduced by $\min\{m, \mathrm{mt}(F, z)\}$ at each point z in the support of $Z \cap F$).

Example 3.4.2.1 Let $m_1 = m_2 = m_3 = 3$ and $m_4 = m_5 = 2$. Moreover, let $z_1, \dots, z_5$ be (distinct) points in $\mathbb{P}^2$. We denote by $Z = Z(\boldsymbol{z}, \boldsymbol{m}) \subset \mathbb{P}^2$ the ordinary fat point scheme concentrated at $z_1, \dots, z_5$ and defined by the ideals $\mathfrak{m}_{z_i}^{m_i} \subset \mathcal{O}_{\mathbb{P}^2, z_i}$, $i = 1, \dots, 5$. We claim

If $z_1, \dots, z_5$ are in generic position then $H^1\big(\mathcal{J}_{Z/\mathbb{P}^2}(6)\big)$ vanishes.

Actually, we could weaken the condition "$z_1, \dots, z_5$ in generic position": let L_0 be a fixed straight line, then $H^1\big(\mathcal{J}_{Z/\mathbb{P}^2}(6)\big)$ vanishes whenever a certain number r' of the points are on L_0, the remaining points are in generic position on $\mathbb{P}^2$ (in particular, not on L_0), and

$$\sum_{z_i \in L_0} m_i \le 6 + 1 = 7 .$$

Now, we start with the inductive procedure. For instance, we can specialize the points z_1, z_2 on L_0. Then $\deg(Z \cap L_0) = 3 + 3 = 6$,

$$Z : L_0 = Z(\boldsymbol{z}, (2, 2, 3, 2, 2)) =: Z(\boldsymbol{z}, \boldsymbol{m}') ,$$

and it suffices to show that $H^1\big(\mathcal{J}_{Z(\boldsymbol{z},\boldsymbol{m}')/\mathbb{P}^2}(5)\big)$ vanishes by (3.4.1.1).

We see that $\deg(Z : L_0) \cap L_0 = 2 + 2 = 4$, whence we can consider the more special situation of points $z_1', \dots, z_5'$, all, except z_3', z_5', lying on the line L_0. Let $Z' := Z(\boldsymbol{z}', \boldsymbol{m}')$. Then $\deg(Z' \cap L_0) = 2 + 2 + 2 = 6$,

$$Z' : L_0 = Z(\boldsymbol{z}', (1, 1, 3, 1, 2)) =: Z'' ,$$

and the problem is reduced to the vanishing of $H^1\big(\mathcal{J}_{Z''/\mathbb{P}^2}(4)\big)$.

In the next step we reduce by the line L_1 through z_3', z_5'. Then we have $\deg(Z'' \cap L_1) = 3 + 2 = 5$,

$$Z'' : L_1 = Z(\boldsymbol{z}', (1, 1, 2, 1, 1)) =: Z(\boldsymbol{z}', \boldsymbol{m}'') ,$$

and it suffices to show the vanishing of $H^1\big(\mathcal{J}_{Z(\boldsymbol{z}',\boldsymbol{m}'')/\mathbb{P}^2}(3)\big)$.

We see that $\deg(Z'' : L_1) \cap L_0 = 1 + 1 + 1 = 3$, whence we can consider the more special situation of points $z_1'', \dots, z_5''$, all, except z_3'', lying on the line L_0. Let $Z''' := Z(\boldsymbol{z}'', \boldsymbol{m}'')$. Then $\deg(Z''' \cap L_0) = 1 + 1 + 1 + 1 = 4$ and

$$Z''' : L_0 = Z(z_3'', (2)) .$$

Now, the claim follows, since $H^1\big(\mathcal{J}_{Z(z_3'',2)/\mathbb{P}^2}(2)\big)$ vanishes (which can be seen by reducing $Z(z_3'', (2))$ by a double line through z_3'' and using $H^1(\mathcal{O}_{\mathbb{P}^2}(d)) = 0$ for all $d \in \mathbb{Z}$ by [HaR2, Theorem III.5.1]).

At this point we should like to show how the general semicontinuity theorem (see [HaR2, Section III.12] for the algebraic and [Schn] for the analytic case) applies to the cohomology of ideal sheaves of (ordinary) fat point schemes.

Definition 3.4.3 Let X be a smooth complex space. A flat family of zero-dimensional schemes on X over T,

$$\begin{array}{ccc} \mathcal{Z} & \stackrel{\iota}{\hookrightarrow} & X \times T\,, \\ & \varphi \searrow \quad \swarrow \mathrm{pr} & \\ & T & \end{array}$$

is called a *family of (ordinary) fat point schemes of multiplicity* $\boldsymbol{m}$ if the fibres of φ are ordinary fat point schemes of multiplicity $\boldsymbol{m} = (m_1, \ldots, m_r)$ (that is, supported at r points and locally given by the m_i-th power of the maximal ideal, $i = 1, \ldots, r$).

We define the *(generalized) diagonal* in X^r to be the union

$$\Delta ::= \bigcup_{i<j} \Delta_{i,j} \subset X^r$$

where $\Delta_{i,j}$ denotes the image of the "diagonal" morphism

$$\begin{aligned} X \times X^{r-2} &\longrightarrow X^r \\ (z; w_1, \ldots, w_{r-2}) &\longmapsto (w_1, \ldots, w_{i-1}, z, w_i, \ldots, w_{j-2}, z, w_{j-1}, \ldots, w_{r-2}). \end{aligned}$$

Let $Y := X^r \setminus \Delta$. The projection $\mathrm{pr} : X \times Y \to Y$ has r natural sections

$$\gamma_i : Y \longrightarrow X \times Y\,, \quad z \longmapsto (z_i, z)\,, \qquad i = 1, \ldots, r.$$

Let $\Gamma_i \subset X \times Y$ denote the image of γ_i and $\mathcal{J}_{\Gamma_i} \subset \mathcal{O}_X \otimes_{\mathbb{C}} \mathcal{O}_Y$ its defining ideal sheaf. Then the intersection of the sheaves $\mathcal{J}_{\Gamma_i}^{m_i}$, $i = 1, \ldots, r$, defines a family of (ordinary) fat point schemes of multiplicity $\boldsymbol{m}$ on X,

$$\begin{array}{ccc} \mathcal{Z} = \mathcal{Z}(\boldsymbol{m}) & \hookrightarrow & X \times Y =: \mathscr{X} \\ & \varphi \searrow \quad \swarrow \mathrm{pr} & \\ & Y\,. & \end{array}$$

Lemma 3.4.4 *With the notion above, the ideal sheaf $\mathcal{J}_{\mathcal{Z}}$ of $\mathcal{Z} = \mathcal{Z}(\boldsymbol{m})$ is a coherent sheaf on the product $\mathscr{X} = X \times Y$, being flat over Y.*

Proof The coherence of $\mathcal{J}_{\mathcal{Z}}$ follows from [GLS6, Corollary I.1.64]. Moreover, consider the exact sequence of sheaves $0 \to \mathcal{J}_{\mathcal{Z}} \to \mathcal{O}_{\mathscr{X}} \stackrel{\mathrm{pr}}{\to} \mathcal{O}_{\mathcal{Z}} \to 0$. We have

- $\mathcal{O}_{\mathscr{X}} = \mathcal{O}_X \otimes_{\mathbb{C}} \mathcal{O}_Y$ is flat over Y,
- $\mathcal{O}_{\mathcal{Z}}$ is flat over Y, by [HaR1, Thm. III.9.9] and [Schn], since the fibre $\mathcal{Z}_{\boldsymbol{w}}$ of $\mathcal{Z}$ over any point $\boldsymbol{w} = (w_1, \ldots, w_r) \in Y$ satisfies

$$\dim_{\mathbb{C}} H^0(\mathcal{Z}_{\boldsymbol{w}}, \mathcal{O}_{\mathcal{Z}_{\boldsymbol{w}}}) = \sum_{i=1}^{r} \binom{n + m_i - 1}{n}.$$

Finally, [HaR2, III.9.1A(e) and 9.2(d)] imply that $\mathcal{J}_{\mathcal{Z}}$ is flat over Y. □

Proposition 3.4.5 *Let X be a smooth complex space and $\mathcal{L}$ an invertible sheaf on X. Then the function*

$$X^r \setminus \Delta \longrightarrow \mathbb{N}, \quad z \longmapsto \dim_{\mathbb{C}} H^i\big(X, \mathcal{J}_{Z(z,\boldsymbol{m})} \otimes_{\mathcal{O}_X} \mathcal{L}\big)$$

is upper semi-continuous.

Proof This follows immediately from Lemma 3.4.4 and the semicontinuity theorem (see [HaR2, Section III.12] for X projective and [Schn] in general). □

3.4.2 Applications of the Basic Horace Method

In the following, we apply the Horace method to obtain several H^1-vanishing criteria for ideal sheaves of zero-dimensional schemes in $\mathbb{P}^n$, $n \geq 1$. Actually, we consider different classes of zero-dimensional schemes:

- zero-dimensional schemes of arbitrary type,
- generic fat point schemes,
- zero-dimensional schemes associated to singularities.

H^1-Vanishing for Arbitrary Zero-Dimensional Schemes. The following statement is classically known (see, e.g., [DaG]).

Proposition 3.4.6 *Let $Z \subset \mathbb{P}^n$, $n \geq 1$, be a zero-dimensional scheme of degree* $\deg Z \leq d+1$. *Then* $H^1\big(\mathcal{J}_{Z/\mathbb{P}^n}(d)\big) = 0$.

Proof We use induction on $n+d$. If $n=1$ then the statement follows from Corollary 3.1.9. If $d=-1$ then $Z=\emptyset$, whence $\mathcal{J}_{Z/\mathbb{P}^n}(-1) = \mathcal{O}_{\mathbb{P}^n}(-1)$ and the statement follows, since $H^1(\mathcal{O}_{\mathbb{P}^n}(-1)) = 0$ (see [HaR2, III.5.1]).

If $d \geq 0$, $n \geq 2$, take a hyperplane $\mathbb{P}^{n-1} \subset \mathbb{P}^n$ with a non-empty intersection $Z \cap \mathbb{P}^{n-1}$. Then $\deg(Z : \mathbb{P}^{n-1}) = \deg Z - \deg(Z \cap \mathbb{P}^{n-1}) < \deg Z$, and one derives the vanishing of $H^1\big(\mathcal{J}_{Z/\mathbb{P}^n}(d)\big)$ from the exact sequence

$$H^1\big(\mathcal{J}_{Z:\mathbb{P}^{n-1}/\mathbb{P}^n}(d-1)\big) \to H^1\big(\mathcal{J}_{Z/\mathbb{P}^n}(d)\big) \to H^1\big(\mathcal{J}_{Z\cap\mathbb{P}^{n-1}/\mathbb{P}^{n-1}}(d)\big),$$

the first and last term being 0 by the induction hypothesis. □

Combining successive reduction by the components and Corollary 3.1.9, (respectively Corollary 3.1.8 in the case $\Sigma = Q$, a quadric surface in $\mathbb{P}^3$) we obtain the following H^1-vanishing criterion for zero-dimensional schemes on reduced curves.

Proposition 3.4.7 ([ShT2], Lemma 2.2.4)

(i) *Let $d \geq 2$ and $Z \subset \mathbb{P}^2$ a zero-dimensional scheme. Suppose Z is contained in a reduced curve $C \subset \mathbb{P}^2$ of degree $d-1$ with irreducible decomposition $C =$*

$C_1 \cup \cdots \cup C_s$, all components C_i having the same degree k ($ks = d-1$) and satisfying

$$\deg(Z \cap C_i) \leq \deg C_i \cdot (d-1) = k \cdot (d-1), \quad i = 1 \ldots s. \tag{3.4.2.1}$$

Then we obtain the following implications

(a) $\deg Z < 4d - 4 \implies H^1\big(\mathcal{J}_{Z/\mathbb{P}^2}(d)\big) = 0,$
(b) $\deg Z < 3d - 3 \implies H^1\big(\mathcal{J}_{Z/\mathbb{P}^2}(d-1)\big) = 0,$
(c) $\deg Z < 2d - 2 \implies H^1\big(\mathcal{J}_{Z/\mathbb{P}^2}(d-2)\big) = 0.$

(ii) Let $Q \subset \mathbb{P}^3$ be a smooth quadric surface with hyperplane section H, $Z \subset Q$ a zero-dimensional scheme and $d \geq 2$. Suppose Z is contained in a reduced curve $C \in \big|H^0\big(\mathcal{O}_Q((d-1)H)\big)\big|$ with irreducible components $C_1, \ldots, C_s \in \big|H^0\big(\mathcal{O}_Q(kH)\big)\big|$ satisfying

$$\deg(Z \cap C_i) \leq (d-1) \cdot \deg C_i = 2k(d-1), \quad i = 1 \ldots s.$$

Then $H^1\big(\mathcal{J}_{Z/Q}(dH)\big)$ vanishes if $\deg Z < 6d - 6$.

Proof We introduce zero-dimensional schemes Z_i, Z'_i, setting $Z'_0 := Z$,

$$Z_i := Z'_{i-1} \cap C_i, \quad Z'_i = Z'_{i-1} : C_i, \quad i = 1, \ldots s-1,$$

and $Z_s := Z'_{s-1}$ which, by construction, is contained in C_s. After a suitable renumbering of the components $C_1, \ldots, C_s$ we can assume

$$\deg Z_1 \geq \deg Z_2 \geq \cdots \geq \deg Z_s, \qquad \sum_{i=1}^{s} \deg Z_i = \deg Z. \tag{3.4.2.2}$$

(i) To be able to prove the implications (a)–(c) simultaneously, we introduce $\varepsilon \in \{0, 1, 2\}$. We proceed inductively, showing

$$H^1\big(\mathcal{J}_{Z'_i/\mathbb{P}^2}(d - \varepsilon - ik)\big) = 0, \qquad i = s-1, \ldots, 0.$$

Since $Z'_{s-1} : C_s = \emptyset$ and $H^1\big(\mathcal{O}_{\mathbb{P}^2}(1-\varepsilon)\big) = 0$, it suffices to show

$$\deg Z_i < k \cdot (d - \varepsilon - ik + 3) = k \cdot \big(k(s-i) + 4 - \varepsilon\big), \tag{3.4.2.3}$$

$i = 1, \ldots, s$ (cf. Lemma 3.4.2). We know, by construction and the assumptions (3.4.2.1), respectively (3.4.2.2),

$$\deg Z_i \leq \deg Z \cap C_i \leq k(d-1) = sk^2,$$
$$\deg Z_i \leq \frac{\deg Z}{i} < \frac{(4-\varepsilon)(d-1)}{i} = \frac{(4-\varepsilon)sk}{i}.$$

Note that the second estimate implies (3.4.2.3) whenever $i = s$ or $4-\varepsilon \leq ik$. In the remaining cases, (3.4.2.3) is obviously implied by the first estimate.
(ii) As before, we proceed inductively, showing

$$H^1\big(\mathcal{J}_{Z_i'/Q}((d-ik)H)\big) = 0\,, \qquad i = s-1, \dots, 0\,.$$

Note that $K_Q \in \big|H^0\big(\mathcal{O}_Q(-2H)\big)\big|$ and $h^1\big(\mathcal{O}_Q(H)\big) = h^0\big(\mathcal{O}_Q(-3H)\big) = 0$. Hence, in view of Lemma 3.4.1 and Corollary 3.1.8, it suffices to show

$$\deg Z_i \;<\; kH \cdot \big((d-ik)H \,-\, K_Q\big) \;=\; 2k \cdot \big(k(s-i)+3\big)\,,$$

$i = 1, \dots, s$. Now, the above reasoning applies. Actually, all appearing inequalities coincide with those in the case $\varepsilon = 1$ up to a factor 2. □

H^1**-Vanishing for Generic Fat Point Schemes**. The Horace method turns out to be particularly effective when dealing with generic fat point schemes on a projective variety Y (cf. Definition 3.2.4 and Example 3.4.2.1, above).

Recall that the class of fat point schemes in Y is closed under reduction. More precisely, reducing $Z(\boldsymbol{z}, \boldsymbol{m})$ by a hypersurface F leads to a scheme $Z(\boldsymbol{z}, \boldsymbol{m}')$ with $m_i' = m_i - \mathrm{mt}(F, z_i)$.

In the following, we present a generalization of Hirschowitz' vanishing theorem (Corollary 3.4.11, respectively [HiA2]) to generic fat point schemes in $\mathbb{P}^n$, $n \geq 2$.

Definition 3.4.8 Let $n \geq 1$ and $d \geq 1$. We define

$$M(n,d) \;=\; 2 \cdot \binom{k+n}{n} \quad \text{for } d = 2k \text{ and}$$
$$M(n,d) \;=\; \binom{k+n}{n} + \binom{k+n-1}{n} \quad \text{for } d = 2k-1.$$

In particular, $M(1,d) = d+2$. We start by proving a simple inequality which will be useful later on.

Lemma 3.4.9 *Let $n \geq 2$, $d \geq 2$. Then*

$$\frac{d-1}{d+n-1} \cdot M(n,d) \;\leq\; M(n,d-1).$$

Proof If $d = 2k$, $k \geq 1$ then

$$\begin{aligned}
\frac{d-1}{d+n-1} \cdot M(n,d) &= \frac{4k-2}{2k+n-1} \cdot \binom{k+n}{n} \;=\; \frac{(4k-2)(k+n)}{(2k+n-1)(k+n)} \cdot \binom{k+n}{n} \\
&= \frac{(2k+n-1)(2k+n) - n^2 - n}{(2k+n-1)(k+n)} \cdot \binom{k+n}{n} \\
&\leq \frac{2k+n}{k+n} \cdot \binom{k+n}{n} \;=\; \binom{k+n}{n} + \binom{k+n-1}{n} \\
&= M(n,d-1)\,.
\end{aligned}$$

Analogously, if $d = 2k+1$, $k \geq 1$ then

$$\begin{aligned} \frac{d-1}{d+n-1} \cdot M(n,d) &= \frac{2k}{2k+n}\left(\binom{k+1+n}{n} + \binom{k+n}{n}\right) \\ &= \frac{2k}{2k+n} \cdot \frac{2k+n+2}{k+1} \cdot \binom{k+n}{n} \\ &\leq 2 \cdot \binom{k+n}{n} = M(n,d-1)\,. \end{aligned}$$

□

Theorem 3.4.10 ([Shu12], Proposition 10) *For a generic fat point scheme on $\mathbb{P}^n$ the following H^1-vanishing holds. Let $n \geq 1$, $m_1, \ldots, m_r$ positive integers such that*

$$\sum_{i=1}^{r} \binom{m_i+n-1}{n} < M(n,d)\,. \tag{3.4.2.4}$$

Then $H^1\big(\mathcal{J}_{Z(\boldsymbol{m})/\mathbb{P}^n}(d)\big) = 0$.
More precisely, let $\mathbb{P}^{n-1} \cong H \subset \mathbb{P}^n$ be a hyperplane and $0 \leq r' \leq r$. If

$$\sum_{i=1}^{r'} \binom{m_i+n-2}{n-1} < M(n-1,d)\,. \tag{3.4.2.5}$$

and condition (3.4.2.4) is satisfied, then there exists an open dense subset $U \subset H^{r'} \times (\mathbb{P}^n)^{r-r'}$ such that $H^1\big(\mathcal{J}_{Z(\boldsymbol{z},\boldsymbol{m})/\mathbb{P}^n}(d)\big) = 0$ for $(z_1, \ldots, z_r) \in U$.

Proof Let $\mathbb{P}^{n-1} \cong H_n \subset \mathbb{P}^n$ be fixed hyperplanes. We show by double induction on the dimension n and on the degree d that, for any $0 \leq r' \leq r$ and $\boldsymbol{m}$ satisfying (3.4.2.5), (3.4.2.4), there exists a Zariski open dense subset

$$U_n = U_n(r', \boldsymbol{m}) \subset (H_n)^{r'} \times (\mathbb{P}^n)^{r-r'}$$

such that $H^1\big(\mathcal{J}_{Z(\boldsymbol{z},\boldsymbol{m})/\mathbb{P}^n}(d)\big) = 0$ for $\boldsymbol{z} \in U_n$.

Throughout the following, we assume that the points z_i, $i = 1, \ldots, r$, are pairwise distinct, that is, $\boldsymbol{z} \in U := (\mathbb{P}^n)^r \setminus \big\{\boldsymbol{z} \;\big|\; \exists\, i \neq j : z_i = z_j\big\}$, and write $Z = Z(\boldsymbol{z}, \boldsymbol{m})$.
Step 1a. If $n = 1$, $d \geq 1$, then condition (3.4.2.4) reads

$$\sum_{i=1}^{r} m_i < d+2\,,$$

which is sufficient for the vanishing of $H^1\big(\mathcal{J}_{Z/\mathbb{P}^1}(d)\big)$ (Corollary 3.1.9).
Step 1b. For $d = 1$, $n \geq 1$, the condition (3.4.2.4) becomes

$$\sum_{i=1}^{r} \binom{m_i+n-1}{n} < n+2\,,$$

which is only possible for ($r = 1, \boldsymbol{m} = 2$), or for ($r \le n+1, \boldsymbol{m} = 1, \ldots, 1$).

In the first case, $h^0\big(\mathcal{J}_{Z/\mathbb{P}^n}(1)\big) = 0$ and $h^0\big(\mathcal{O}_Z\big) = n+1 = h^0\big(\mathcal{O}_{\mathbb{P}^n}(1)\big)$, and the obvious exact cohomology sequence

$$0 \to H^0\big(\mathcal{J}_{Z/\mathbb{P}^n}(1)\big) \to H^0\big(\mathcal{O}_{\mathbb{P}^n}(1)\big) \to H^0\big(\mathcal{O}_Z\big) \to H^1\big(\mathcal{J}_{Z/\mathbb{P}^n}(1)\big) \to 0$$

allows to deduce H^1-vanishing for $\mathcal{J}_{Z/\mathbb{P}^n}(1)$.

In the second case, we can conclude in the same manner, since then $h^0\big(\mathcal{J}_{Z/\mathbb{P}^n}(1)\big) = n+1-k$ and $h^0\big(\mathcal{O}_Z\big) = k$.

Step 2. Before proceeding with the general case, let us consider the special case $r = 1$, that is, $Z = Z(z_1; m_1) \subset \mathbb{P}^n$. In particular, $H^1\big(\mathcal{J}_{Z/\mathbb{P}^n}(d)\big)$ vanishes iff $m_1 < d+2$.

Step 3. Let $H = H_n$. After shrinking U, we can assume that $z_1, \ldots, z_{r'}$ are on the hypersurface H, while $z_{r+1}, \ldots, z_r \in \mathbb{P}^n \setminus H$. When reducing by H we have to consider the trace $Z \cap H$, that is, the ordinary fat point scheme supported at $\{z_1, \ldots, z_{r'}\} \subset H$ with multiplicities $m_1, \ldots, m_{r'}$, and the residual scheme $Z : H$, that is, the ordinary fat point scheme supported at a subset of $\{z_1, \ldots, z_r\} \subset \mathbb{P}^n$ with multiplicities $m_1 - 1, \ldots, m_{r'} - 1, m_{r'+1}, \ldots, m_r$.

Note that, whenever we apply an induction hypothesis in the following, we are actually shrinking the open dense subset U by intersecting it with another open dense subset U'.

Case 1. $r' = r$.
Then condition (3.4.2.5) reads

$$\deg(Z \cap H) = \sum_{i=1}^{r} \binom{m_i + n - 2}{n-1} < M(n-1, d)\,.$$

Hence, by the induction hypothesis in dimension $n-1$ we can assume that $H^1\big(\mathcal{J}_{Z\cap H/H}(d)\big)$ vanishes. It remains to show $H^1\big(\mathcal{J}_{Z:H/\mathbb{P}^n}(d-1)\big) = 0$. This follows from the induction hypothesis in degree $d-1$, since

$$\begin{aligned}\deg(Z:H) &= \sum_{i=1}^{r} \binom{m_i + n - 2}{n} \le \frac{d-1}{d+n-1} \cdot \sum_{i=1}^{r} \binom{m_i + n - 1}{n} \\ &\overset{(3.2.4)}{<} \frac{d-1}{d+n-1} \cdot M(n,d) \le M(n, d-1)\,,\end{aligned}$$

the latter inequality being due to Lemma 3.4.9, and since

$$\begin{aligned}\deg\big((Z:H) \cap H\big) &= \sum_{i=1}^{r} \binom{m_i + n - 3}{n-1} \le \frac{d-1}{d+n-2} \cdot \sum_{i=1}^{r} \binom{m_i + n - 2}{n-1} \\ &< \frac{d-1}{d+n-2} \cdot M(n-1, d) \le M(n-1, d-1)\,,\end{aligned}$$

again, by Lemma 3.4.9.

Case 2. $r' < r$ and $\deg(Z \cap H) \ge M(n,d) - M(n,d-1)$.
Again, the induction hypothesis and (3.4.2.5) imply $H^1\big(\mathcal{J}_{Z\cap H/H}(d)\big) = 0$. Moreover, by (3.4.2.4), $\deg(Z:H) = \deg Z - \deg(Z \cap H) < M(n,d-1)$ and, as in the first case, $\deg\big((Z:H)\cap H\big) < M(n-1,d-1)$. Therefore, the induction hypothesis in degree $d-1$ applies and $H^1\big(\mathcal{J}_{Z:H/\mathbb{P}^n}(d-1)\big) = 0$.
Case 3. $r' < r$ and $\deg(Z \cap H) < M(n,d) - M(n,d-1)$, that is,

$$\deg(Z \cap H) < \binom{k+n-1}{n-1} \text{ if } d = 2k, \text{ respectively} \tag{3.4.2.6}$$

$$\deg(Z \cap H) < \binom{k+n}{n-1} \text{ if } d = 2k+1. \tag{3.4.2.7}$$

This is the situation where we cannot simply reduce by H. Either, we have to use another reduction hypersurface, or we have to apply a *specialization* argument.

Let's make precise what we mean by a "specialization argument": we apply descending induction on r'. Then "specializing a point on H" means that we apply the induction hypothesis in $r'+1$ which gives H^1-vanishing for $z \in U' = U_n(r'+1, \boldsymbol{m}) \subset (H_n)^{r'+1} \times (\mathbb{P}^n)^{r-r'-1}$ and, by semicontinuity (Proposition 3.4.5), for z in a Zariski neighbourhood U of U' in $(H_n)^{r'} \times (\mathbb{P}^n)^{r-r'}$.
Case 3a. Assume that $d = 2k$, $k \ge 1$, and there exists $j \in \{r'+1, \dots, r\}$ with $m_j \le k+1$. Without loss of generality assume $j = r'+1$. Then we can specialize the point $z_{r'+1}$ onto H without violating (3.4.2.5), since

$$\deg(Z \cap H) + \binom{m_{r'+1}+n-2}{n-1} < 2 \cdot \binom{k+n-1}{n-1} = M(n-1,d)\,.$$

Case 3b. $d = 2k$ and $m_j > k+1$ for all $j > r'$, in particular, $m_r > k+1$.
In this case, we can reduce by a generic hyperplane H' through z_r instead of H. The support of $Z \cap H'$ being just the point z_r having multiplicity $m_r \le d$, we know $H^1\big(\mathcal{J}_{Z\cap H'/H'}(d)\big) = 0$. Moreover, by (3.4.2.4),

$$\begin{aligned}\deg(Z:H') = \deg Z - \deg(Z \cap H') &= \sum_{i=1}^{r} \binom{m_i+n-1}{n} - \binom{m_r+n-1}{n-1} \\ &< M(n,d) - \big(M(n,d) - M(n,d-1)\big) = M(n,d-1)\,,\end{aligned}$$

and $\deg\big((Z:H')\cap H\big) = \deg(Z \cap H) < \binom{k+n-1}{n-1} \le M(n-1,d-1)$. Thus, by induction, $H^1\big(\mathcal{J}_{Z:H'/\mathbb{P}^n}(d-1)\big) = 0$ and, hence, $H^1\big(\mathcal{J}_{Z/\mathbb{P}^n}(d)\big) = 0$.
Case 3c. $d = 2k+1$ and there exists $j \in \{r'+1, \dots, r\}$ with

$$\sum_{i=1}^{r'} \binom{m_i+n-2}{n-1} + \binom{m_j+n-2}{n-1} < M(n-1,d).$$

Then, we can proceed as in Case 3a, specializing z_j on H.
Case 3d. $d = 2k+1$ and

$$\sum_{i=1}^{r'} \binom{m_i+n-2}{n-1} + \binom{m_r+n-2}{n-1} \geq M(n-1,d). \tag{3.4.2.8}$$

Let, again, H' be a generic hyperplane through z_r. As in Case 3b, we obtain $H^1\big(\mathcal{J}_{Z\cap H'/H'}(d)\big) = 0$, and it remains to show $H^1\big(\mathcal{J}_{Z:H'/\mathbb{P}^n}(d-1)\big) = 0$. By (3.4.2.7) and (3.4.2.8), we know

$$\binom{m_r+n-2}{n-1} > M(n-1,d) - \binom{n+k}{n-1} = \binom{k+n-1}{n-1}$$

and, thus, $m_r \geq k+2$. Together with (3.4.2.4), this implies that

$$\binom{k+n+1}{n} \leq \binom{m_r+n-1}{n} < M(n,d) = \binom{k+n+1}{n} + \binom{k+n}{n}.$$

Hence,

$$\begin{aligned}\deg(Z\cap H) &= \sum_{i=1}^{r'} \binom{m_i+n-2}{n-1} \leq \sum_{i=1}^{r'} \binom{m_i+n-1}{n} \\ &< M(n,d) - \binom{m_r+n-1}{n} \leq \binom{k+n}{n}.\end{aligned}$$

We conclude that

$$\begin{aligned}\deg(Z:H') &= \sum_{i=1}^{r} \binom{m_i+n-1}{n} - \binom{m_r+n-2}{n-1} \\ &< M(n,d) - \binom{k+n}{n-1} = M(n,d-1),\end{aligned}$$

and, by (3.4.2.7), respectively (3.4.2.9),

$$\begin{aligned}\deg\big((Z:H')\cap H\big) = \deg(Z\cap H) &< \min\left\{\binom{k+n}{n-1}, \binom{k+n}{n}\right\} \\ &\leq 2\cdot\binom{k+n-1}{n-1} = M(n-1,d-1).\end{aligned}$$

Thus, by the induction hypothesis, $H^1\big(\mathcal{J}_{Z:H'/\mathbb{P}^n}(d-1)\big) = 0$, which completes the proof. $\square$

Note that the inequality (3.4.2.4) is sharp, as the following example shows:

Example 3.4.10.1 Let $r = 2$ and $m_1 = \lfloor \frac{d}{2} \rfloor + 1$, $m_2 = d + 1 - \lfloor \frac{d}{2} \rfloor$, in particular

$$\binom{m_1+n-1}{n} + \binom{m_2+n-1}{n} = M(n,d)\,.$$

On the other hand, $H^1\big(\mathcal{J}_{Z(\boldsymbol{z},\boldsymbol{m})/\mathbb{P}^n}(d)\big) \neq 0$ for all $z_1 \neq z_2 \in \mathbb{P}^n$. This can be seen as follows: let $Z = Z(\boldsymbol{z}, \boldsymbol{m})$, and let L be the line through z_1, z_2. Then $\deg(Z \cap L) = d+2$, whence $h^1\big(\mathcal{J}_{Z\cap L/L}(d)\big) = h^1\big(\mathcal{O}_{\mathbb{P}^1}(d - \deg(Z\cap L))\big) > 0$. Consider $L = \mathbb{P}^1 \subset \mathbb{P}^2 \subset \ldots \subset \mathbb{P}^n$. Then the exact sequences

$$0 \to \mathcal{O}_{\mathbb{P}^m}(d-1) \to \mathcal{J}_{Z\cap\mathbb{P}^{m-1}/\mathbb{P}^m}(d) \to \mathcal{J}_{Z\cap\mathbb{P}^{m-1}/\mathbb{P}^{m-1}}(d) \to 0\,,$$

$m = 2, \ldots, n$, respectively the corresponding exact cohomology sequence, imply inductively the non-vanishing of $H^1\big(\mathcal{J}_{Z\cap\mathbb{P}^{n-1}/\mathbb{P}^n}(d)\big)$. This allows to conclude the above non-vanishing statement.

Corollary 3.4.11 ([HiA2]) *Let $m_1, \ldots, m_r$ and d be positive integers such that*

$$\sum_{i=1}^{r} \frac{m_i(m_i+1)}{2} < \left\lfloor \frac{(d+3)^2}{4} \right\rfloor . \tag{3.4.2.9}$$

Then $H^1\big(\mathcal{J}_{Z(\boldsymbol{m})/\mathbb{P}^2}(d)\big)$ vanishes.
More precisely, let $L \subset \mathbb{P}^2$ be a line and $0 \leq r' \leq r$. If

$$\sum_{i=1}^{r'} m_i \leq d+1\,,$$

and condition (3.4.2.9) is satisfied, then there exists an open dense subset $U \subset L^{r'} \times (\mathbb{P}^2)^{r-r'}$ such that $H^1\big(\mathcal{J}_{Z(\boldsymbol{z},\boldsymbol{m})/\mathbb{P}^2}(d)\big) = 0$ for $(z_1, \ldots, z_r) \in U$.

Proof Apply Theorem 3.4.10, and compute $M(2,d) = \left\lfloor \frac{(d+3)^2}{4} \right\rfloor$. □

The following H^1-vanishing theorem is due to Ballico [Bal].

Proposition 3.4.12 *Let $\boldsymbol{m} = (m_1, \ldots, m_r)$ and $d > 0$ be such that*

$$\sum_{i=1}^{r} \frac{m_i(m_i+1)}{2} \leq \frac{(d+2)(d+1)}{2} - d \cdot \max_{i=1..r}(m_i - 1)\,, \tag{3.4.2.10}$$

then $H^1\big(\mathcal{J}_{Z(\boldsymbol{m})/\mathbb{P}^2}(d)\big)$ vanishes.

Proof Let $L \subset \mathbb{P}^2$ be a fixed line. We proceed by induction on d, proving that for any $0 \leq r' \leq r$ and $\boldsymbol{m}$ satisfying (3.4.2.10) and

$$\sum_{i=1}^{r'} m_i \;<\; d+2\,, \tag{3.4.2.11}$$

there is an open dense subset $U \subset L^{r'} \times (\mathbb{P}^2)^{r-r'}$ such that $H^1\big(\mathcal{J}_{Z(z,\boldsymbol{m})/\mathbb{P}^2}(d)\big)$ vanishes for $(z_1,\dots,z_r) \in U$ (cf. the proof of Theorem 3.4.10).

We write $Z = Z(z,\boldsymbol{m})$ and consider the residue scheme $Z:L = Z(z,\boldsymbol{m}')$, where $m'_i = m_i - 1$ if $z_i \in L$, and $m'_i = m_i$ otherwise. Note that (3.4.2.11) implies that $H^1\big(\mathcal{J}_{Z\cap L/L}(d)\big)$ and $H^1\big(\mathcal{J}_{(Z:L)\cap L/L}(d-1)\big)$ vanish.

Case 1. $r = r'$. Then we obtain, by the assumption (3.4.2.10)

$$\begin{aligned}\deg(Z:L) \le \deg Z - 1 \;&\le\; \frac{(d+2)(d+1)}{2} - d\cdot \max_{i=1..r}\,(m_i - 1) - 1\\ &= \frac{(d+1)d}{2} - d\cdot\max_{i=1...r}\,(m'_i - 1)\,,\end{aligned}$$

hence, by the induction hypothesis, $H^1\big(\mathcal{J}_{Z:L/\mathbb{P}^2}(d-1)\big) = 0$.

Case 2. $r < r'$ and $\deg(Z\cap L) > d+1-\max_i m_i$. Then, by (3.4.2.10),

$$\deg(Z:L) \;=\; \deg Z - \deg(Z\cap L) \;\le\; \frac{(d+1)d}{2} - (d-1)\cdot \max_{i=1..r}\,(m_i - 1)\,,$$

and, since $m'_i \le m_i$, the induction hypothesis applies again, and we obtain $H^1\big(\mathcal{J}_{Z:L/\mathbb{P}^2}(d-1)\big) = 0$.

Case 3. $r < r'$ and $\deg(Z\cap L) \le d+1-\max_i m_i$. In this case, we specialize some point z_j on L (cf. Case 3 in the proof of Theorem 3.4.10). □

Remark 3.4.12.1 If Z is a generic reduced zero-dimensional scheme, that is, a generic fat point scheme $Z = Z(\boldsymbol{m})$ with all multiplicities $m_i = 1, i = 1,\dots r$, then $H^1\big(\mathcal{J}_{Z/\mathbb{P}^2}(d)\big)$ vanishes exactly if the expected dimension $(d+2)(d+1)/2 - \deg Z$ is non-negative.

H^1-Vanishing for Zero-Dimensional Schemes Associated to Singularities. The following result is due to du Plessis and Wall [DPW2]. The first relation has been obtained independently in [ShT2]. The proof presented here follows [ShT2] and is based on the Horace method.

Theorem 3.4.13 *Let $W \subset \mathbb{P}^n$, $n \ge 3$, be a hypersurface of degree $d \ge 2$ with only isolated singularities, and let $Z \subset Z^{ea}(W)$ a zero-dimensional scheme. Then the following implications hold true*

(a) $\deg Z < 4d - 4 \implies H^1\big(\mathcal{J}_{Z/\mathbb{P}^n}(d)\big) = 0,$
(b) $\deg Z < 3d - 3 \implies H^1\big(\mathcal{J}_{Z/\mathbb{P}^n}(d-1)\big) = 0,$
(c) $\deg Z < 2d - 2 \implies H^1\big(\mathcal{J}_{Z/\mathbb{P}^n}(d-2)\big) = 0.$

Here $Z^{ea}(W)$ denotes the equianalytic scheme (cf. Definition 2.2.23).

Remark 3.4.13.1 For small degrees the conditions in (a)–(c) can be weakened (see [DPW2]): $H^1(\mathcal{J}_{Z/\mathbb{P}^n}(d))$ vanishes if

$$(d = 4,\ \deg Z < 18)\,,\ \text{or}\ (d = 3,\ \deg Z < 16)\,,\ \text{or}\ (d = 2,\ \deg Z < 8)\,.$$

The sharpness of (a)–(c) is discussed in Sect. 6.4.3.4.
The proof of Theorem 3.4.13. Is organized in three parts: preparatory lemmas, the case $n = 3$, and, at last, we treat the case $n > 3$.

Part 1: Preparatory Lemmas

First, we construct curves as required in Proposition 3.4.7. The statement interferes with Lemma 2.1.30, which is proven in a different way.

Lemma 3.4.14 *Let $F \in \mathbb{C}[x_0, \dots, x_n]$, $n \geq 3$, be a homogeneous polynomial of degree $d \geq 2$, defining a hypersurface $W \subset \mathbb{P}^n$ having only isolated singular points. Then*

(i) for any 2-plane $\Pi \subset \mathbb{P}^n$, a generic member of the linear system of curves $C \subset \Pi$ with equations

$$(\lambda_0 F_{x_0} + \dots + \lambda_n F_{x_n})\big|_{\Pi} = 0\,, \qquad (\lambda_0 : \dots : \lambda_n) \in \mathbb{P}^n \tag{3.4.2.12}$$

is a curve of degree $d-1$, either reduced and irreducible, or consisting of distinct reduced, irreducible components $C_1, \dots, C_s$ of the same degree such that

$$\deg\big(Z^{ea}(W) \cap C_i\big) \leq (d-1) \cdot \deg C_i\,, \qquad i = 1, \dots, s\,; \tag{3.4.2.13}$$

(ii) for any non-singular two-dimensional quadric $Q \subset \mathbb{P}^n$, a generic member $C \subset Q$ of the linear system of curves $C \subset Q$ with equations

$$(\lambda_0 F_{x_0} + \dots + \lambda_n F_{x_n})\big|_{Q} = 0\,, \qquad (\lambda_0 : \dots : \lambda_n) \in \mathbb{P}^n,$$

is a curve of bidegree $(d-1, d-1)$, either reduced and irreducible, or consisting of distinct reduced irreducible components $C_1, \dots, C_s$ of the same bidegree and satisfying (3.4.2.13).

Proof First, note that the linear system of polar hypersurfaces in $\mathbb{P}^n$ (of degree $d-1$), with equations $\lambda_0 F_{x_0} + \dots + \lambda_n F_{x_n} = 0$, $(\lambda_0 : \dots : \lambda_n) \in \mathbb{P}^n$, has the singular points of W as its only base points. Hence, by Corollary 2.1.21, a generic C in the linear system (3.4.2.12) is reduced and has no fixed components. Moreover, by Theorem 2.1.24, a reducible C must consist of distinct irreducible components $C_1, \dots, C_s$, $s \geq 2$, of the same degree. At last, (3.4.2.13) follows from Bézout's theorem, since $Z^{ea}(W)$ is contained in any polar hypersurface.

The proof of the second statement is a word-by-word copy of the previous reasoning with the only difference being that $\mathrm{Pic}(Q) = \mathbb{Z}^2$. □

Now we estimate intersections of hyperplanes and quadrics with zero-dimensional schemes from below.

Lemma 3.4.15 *For any zero-dimensional scheme $Z \subset \mathbb{P}^n$, $n \geq 3$, there exists a hyperplane $P \subset \mathbb{P}^n$ such that*

$$\deg(Z \cap P) \geq \min\{n,\ \deg Z\}.$$

Proof This follows from a general elementary statement that in an m-dimensional linear family of hypersurfaces W_λ with equations

$$\lambda_0 M_0 + \cdots + \lambda_m M_m = 0\,, \qquad (\lambda_0 : \cdots : \lambda_m) \in \mathbb{P}^m,$$

there exists a hypersurface W_* such that $\deg(Z \cap W_*) \geq \min\{m,\ \deg Z\}$.

Clearly, it suffices to consider the case of Z concentrated at one point z. Let $I \subset \mathcal{O}_{\mathbb{P}^n,z}$ define the germ (Z, z), in particular, $p := \dim \mathcal{O}_{\mathbb{P}^n,z}/I = \deg Z$. Take a filtration of I by ideals $I = I_p \subsetneq I_{p-1} \subsetneq \cdots \subsetneq I_1 \subsetneq \mathcal{O}_{\mathbb{P}^n,z}$.

If $q = \min\{m, p\}$, then $M_0, \ldots, M_m$ define linearly dependent elements in $\mathcal{O}_{\mathbb{P}^n,z}/I_q$. Hence there exists a linear combination M_* defining a hypersurface W_* which contains the scheme $Z_q \subset \mathbb{P}^n$ of degree q, defined by I_q. □

Lemma 3.4.16 *If a zero-dimensional scheme $Z \subset \mathbb{P}^3$ of degree* $\deg Z \geq 7$ *satisfies* $\deg(Z \cap \Pi) \leq 3$ *for each plane $\Pi \subset \mathbb{P}^3$, then there exists a non-singular quadric $Q \subset \mathbb{P}^3$ such that* $\deg(Z \cap Q) \geq \min\{8,\ \deg Z\}$.

Proof Since $\dim \left|H^0\big(\mathcal{O}_{\mathbb{P}^3}(2)\big)\right| = 9$ and by the preceding argument, there exists a one-dimensional linear family $\lambda Q_1 + \mu Q_2$, $(\lambda : \mu) \in \mathbb{P}^1$, of quadrics intersecting Z with multiplicity at least $\min\{8,\ \deg Z\}$. They all are irreducible, since $\deg Z \geq 7 > \deg\big(Z \cap (\Pi_1 \cup \Pi_2)\big)$ for any two planes Π_1, Π_2. Suppose that all the quadrics $\lambda Q_1 + \mu Q_2$ are singular. Then they have a common singular point, but such a linear family of quadrics contains a reducible quadric, which is impossible. □

Lemma 3.4.17 *Let $Z \subset \mathbb{P}^3$ be a zero-dimensional scheme such that*

(a) $\deg Z \leq 3k - 1$, *and*
(b) $\deg(Z \cap \Pi) \leq 3$, *for each plane $\Pi \subset \mathbb{P}^3$.*

Then $H^1\big(\mathcal{J}_{Z/\mathbb{P}^3}(k)\big) = 0$.

This can be proven in the same manner as Proposition 3.4.6 with the use of Lemma 3.4.15, which claims that, for $\deg Z \geq 3$ there exists a plane $\Pi \subset \mathbb{P}^3$ such that $\deg(Z \cap \Pi) = 3$.

Part 2: Proof of Theorem 3.4.13 in the Case n = 3

The following statement gives the required statement for $n = 3$ and provides an induction step for higher dimensions.

Lemma 3.4.18 *Let $P \subset \mathbb{P}^n$, $n \geq 3$ be a 3-plane. Moreover, let $W \subset \mathbb{P}^n$ be a hypersurface of degree $d \geq 2$ with only isolated singularities and $Z \subset Z^{ea}(W)$. Then*

(a) $\deg(Z \cap P) < 4d - 4 \implies H^1\big(\mathcal{J}_{(Z\cap P)/P}(d)\big) = 0,$
(b) $\deg(Z \cap P) < 3d - 3 \implies H^1\big(\mathcal{J}_{(Z\cap P)/P}(d-1)\big) = 0,$
(c) $\deg(Z \cap P) < 2d - 2 \implies H^1\big(\mathcal{J}_{(Z\cap P)/P}(d-2)\big) = 0.$

Proof Without restriction, we can assume that the zero-dimensional scheme Z is already contained in the 3-plane P, that is, $Z = Z \cap P$. We divide our reasoning into several steps.

Step 1. If $d = 2$ then the implications (a)–(c) are special cases of Proposition 3.4.6. Let $d = 3$.

If $\deg Z \leq 2d-3 = 3$ then, due to Lemma 3.4.15, Z is contained in a 2-plane Π. Hence, the generic element C of the linear system of curves on Π cut out by the polar hypersurfaces $W_\lambda \subset \mathbb{P}^n$ contains Z. On the other hand, by Lemma 3.4.14, C satisfies the assumptions of Proposition 3.4.7(i), whence (c).

If $\deg Z \leq 3d-4 = 5$, then by Lemma 3.4.15 there is a 2-plane Π_1 such that $\deg(Z \cap \Pi_1) \geq 3$. On the other hand, $Z \cap \Pi_1$ is contained in the intersection of Π_1 with any polar hypersurface $W_\lambda \subset \mathbb{P}^n$. Hence, by Bézout's theorem, $\deg(Z \cap \Pi_1) \leq (d-1)^2 = 4$, and applying Lemma 3.4.14 and Proposition 3.4.7(i) gives the vanishing of $H^1\big(\mathcal{J}_{Z\cap\Pi_1/\Pi_1}(2)\big)$. The residue scheme satisfies $\deg(Z : \Pi_1) \leq 2$, hence it is contained in a 2-plane Π_2 and $H^1\big(\mathcal{J}_{Z:\Pi_1/\Pi_2}(1)\big) = 0$. Successive reduction by Π_1, respectively Π_2, gives (b).

If $\deg Z \leq 8$ then, in the same manner, we obtain planes Π_1, Π_2, Π_3 satisfying $\deg(Z \cap \Pi_1) \in \{3, 4\}$, $\deg\big((Z : \Pi_1) \cap \Pi_2\big) \in \{3, 4\}$, and the residue scheme $(Z : \Pi_1) : \Pi_2$ is contained in Π_3 (and of degree at most 2). Successively reducing by Π_i and applying Propositions 3.4.6, respectively 3.4.7(i), gives (a).

So, further on we suppose $d \geq 4$.

Step 2. Let us prove (b). We construct inductively a sequence $\Pi_1, \ldots, \Pi_m$ of 2-planes in P as follows. Denote $Z_1 := Z$, and let $k \geq 1$. If $Z_k = \emptyset$ then the sequence is complete, i.e., $m = k-1$. If $Z_k \neq \emptyset$ then we define Π_k to be any 2-plane realizing the maximal possible intersection with Z_k, that is,

$$\deg(Z_k \cap \Pi_k) = \max_{\Pi\in|H^0(\mathcal{O}_P(1))|} \deg(Z_k \cap \Pi)\,,$$

and set $Z_{k+1} = Z_k : \Pi_k$. By Lemma 3.4.15, $m \leq d-1$. Hence, successively reducing by Π_i, (b) will follow from

$$H^1\big(\mathcal{J}_{Z_k\cap\Pi_k/\Pi_k}(d-k)\big) = 0\,, \qquad k = 1, \ldots, m. \tag{3.4.2.14}$$

Observe that, as before, the vanishing of $H^1\big(\mathcal{J}_{Z_1\cap\Pi_1/\Pi_1}(d-1)\big)$ follows from Lemma 3.4.14 and Proposition 3.4.7(i). In the same manner, we obtain the vanishing of $H^1\big(\mathcal{J}_{Z_2\cap\Pi_2/\Pi_2}(d-2)\big)$, since $\deg(Z_2 \cap \Pi_2) \leq (\deg Z)/2 < 2d-2$. The remaining equalities in (3.4.2.14) follow from Proposition 3.4.6, since for $k \geq 3$, we can easily estimate $\deg(Z_k \cap \Pi_k) \le d - k + 1$. Indeed, by construction,

$$\deg(Z_1 \cap \Pi_1) \ge \cdots \ge \deg(Z_m \cap \Pi_m), \tag{3.4.2.15}$$

and $\deg(Z_k \cap \Pi_k) \ge d-k+2$ for some $k \ge 3$ would imply that $\deg Z$ is at least $k(d-k+2) \ge 3d-3$, contradicting the assumption.
Step 3. Let us prove (c). By Lemma 3.4.15, in the above sequence $\Pi_1, \ldots, \Pi_m$, one has $m < (2d-2)/3 < d-1$. Hence, successively reducing by the Π_i, (c) will follow from

$$H^1\big(\mathcal{J}_{Z_k \cap \Pi_k/\Pi_k}(d-k-1)\big) = 0\,, \qquad k = 1, \ldots, m. \tag{3.4.2.16}$$

Again, Lemma 3.4.14 and Proposition 3.4.7(i) imply $H^1\big(\mathcal{J}_{Z_1 \cap \Pi_1/\Pi_1}(d-2)\big) = 0$, while the remaining equalities in (3.4.2.16) follow from Proposition 3.4.6. Indeed, (3.4.2.15) and $\deg(Z_k \cap \Pi_k) \ge d-k+1$ for some $k \ge 2$ would lead to the contradiction: $\deg Z \ge k(d-k+1) \ge 2d-2$.
Step 4. Suppose that $\deg Z < 4d-4$, and the 2-plane Π_1 from the above construction satisfies $\deg(Z \cap \Pi_1) \ge d-1$.

Then $\deg(Z : \Pi_1) < 3d-3$ and (b) implies $H^1\big(\mathcal{J}_{Z:\Pi_1/P}(d-1)\big) = 0$. Moreover, by Lemma 3.4.14 and Proposition 3.4.7(i), $H^1\big(\mathcal{J}_{Z\cap\Pi_1/\Pi_1}(d)\big) = 0$, and, as usually, we can conclude $H^1\big(\mathcal{J}_{Z/P}(d)\big) = 0$.
Step 5. Suppose that $\deg Z < 4d-4, d \ge 6$, and the 2-planes Π_1, Π_2, Π_3 constructed in Step 2 satisfy

$$\deg(Z_1 \cap \Pi_1) \;=\; \deg(Z_2 \cap \Pi_2) \;=\; \deg(Z_3 \cap \Pi_3) \;=\; d-2\,.$$

In particular, $\deg Z_4 \le d+1$, whence Lemma 3.4.15 implies that $m \le d$ (as $d \ge 6$). Hence, (a) follows if

$$H^1\big(\mathcal{J}_{Z_k \cap \Pi_k/\Pi_k}(d-k+1)\big) = 0\,, \qquad k = 1, \ldots m\,. \tag{3.4.2.17}$$

But this is satisfied by Proposition 3.4.6, since $\deg(Z_k \cap \Pi_k) \le d-2$ if $k \le 4$, and

$$\deg(Z_k \cap \Pi_k) \le \frac{d+1}{k-3} < d-k+3\,, \text{ if } 5 \le k \le 3 + \frac{d+1}{3}\,, \quad (d \ge 6).$$

Step 6. Suppose that $3d-3 \le \deg Z < 4d-4$, $\deg(Z \cap \Pi_1) \le d-2$, and each nonsingular quadric $Q \subset P$ satisfies $\deg(Z \cap Q) \le 7$.
We construct a sequence of planes $\Pi_1, \ldots, \Pi_m$ as in Step 2 but stop as soon as $\deg Z_k \le 7$. Then, by Lemma 3.4.16 and our assumption, $\deg(Z_k \cap \Pi_k) \ge 4$ for all $k = 1, \ldots, m$.

In particular, $\deg Z_{m+1} \le 7$ and $4m \le 4d-5-\deg Z_{m+1}$, and (a) reduces to (3.4.2.17) and $H^1\big(\mathcal{J}_{Z_{m+1}/P}(d-m)\big) = 0$. The latter follows from Lemma 3.4.17, since we can easily deduce that $d-m \ge 3$ if $4 \le \deg Z_{m+1} \le 7$, and $d-m \ge 2$ if $\deg Z_{m+1} \le 3$.

The first 3 equalities of (3.4.2.17) are obtained from Lemma 3.4.14 and Proposition 3.4.7(i), since $\deg(Z_1 \cap \Pi_1) \le d-2 < 2d-2$. For $k \ge 4$, one can apply Proposition 3.4.6, since

$$\deg(Z_k \cap \Pi_k) < \frac{4d-4}{k} \le d-k+3\,, \text{ if } 4 \le k \le d-1.$$

Step 7. Suppose that $3d-3 \le \deg Z < 4d-4$, $\deg(Z \cap \Pi_1) \le d-2$, and there is a non-singular quadric $Q \subset P$ satisfying $\deg(Z \cap Q) \ge 8$. Choose Q_0 such that it realizes the maximal possible intersection with Z.

We construct successively sequences of 2-planes $\Pi_1, \ldots, \Pi_s$ and quadrics $Q_1, \ldots, Q_q$, by the following procedure. Set $Z_1 := Z : Q_0$, and let $k \ge 1$. As long as there is a 2-plane Π with $\deg(Z_k \cap \Pi) \ge 4$, choose Π_k such that

$$\deg(Z_k \cap \Pi_k) = \max_{\Pi \in |H^0(\mathcal{O}_P(1))|} \deg(Z_k \cap \Pi)\,,$$

and set $Z_{k+1} = Z_k : \Pi_k$. Now, let $Z'_1 := Z_{s+1}$ and $j \ge 1$. As long as there is a non-singular quadric $Q \subset P$ with $\deg(Z'_j \cap \Pi) \ge 8$, choose Q_j such that

$$\deg(Z'_j \cap Q_j) = \max_{\Pi \in |H^0(\mathcal{O}_P(2))|} \deg(Z'_j \cap Q)\,,$$

and set $Z'_{j+1} = Z'_j : Q_j$. By Lemma 3.4.16, it follows that $p := \deg Z'_{q+1} \le 7$. Moreover, note that

$$2+s+2q \le 2+\frac{4d-5-\deg(Z \cap Q_0)-p}{4} \le d-\frac{p+5}{4}\,. \qquad (3.4.2.18)$$

Successively reducing by Q_0, $\Pi_k, k = 1, \ldots, s$, and $Q_j, j = 1, \ldots, q$, it suffices to show

(7.a) $H^1\big(\mathcal{J}_{Z\cap Q_0/Q_0}(d)\big) = 0$
(7.b) $H^1\big(\mathcal{J}_{Z_k\cap \Pi_k/\Pi_k}(d-1-k)\big) = 0\,, \quad k = 1, \ldots, s\,,$
(7.c) $H^1\big(\mathcal{J}_{Z'_j\cap Q_j/Q_j}(d-s-2j)\big) = 0\,, \quad j = 1, \ldots, q\,,$
(7.d) $H^1\big(\mathcal{J}_{Z'_{q+1}/P}(d-s-2q-2)\big) = 0\,.$

Since $\deg Z \cap Q_0 \le 4d-4$, (7.a) is implied by Lemma 3.4.14 and Proposition 3.4.7(ii). The equalities (7.b) follow from Proposition 3.4.6, since we can assume

$$\deg(Z_2 \cap \Pi_2) \le \deg(Z_1 \cap \Pi_1) \le d-2, \quad \deg(Z_3 \cap \Pi_3) \le d-3$$

(note that $\deg(Z_3 \cap \Pi_3) = d-2$ would mean that we have been in the situation already considered in Step 5), and, for $k = 4, \ldots, s$,

$$\deg(Z_k \cap \Pi_k) \le \frac{4d-5-\deg(Z \cap Q_0)-p}{k} < d+1-k\,.$$

To see the latter inequalities, note that it suffices to consider the two extremal cases, $k = 4$, respectively $k = s \leq d - \frac{p+13}{4}$. But these are obvious, applying that $\deg(Z \cap Q_0) \geq 8$.

The equalities (7.c) are implied by Lemma 3.4.17, since, by construction,

$$\deg(Z'_j \cap Q_j) < \frac{4d-4-4s}{j+1} \leq 3(d-s-2j), \quad 1 \leq j \leq q,$$

which has to be verified only for $j = 1$ and some $j \geq q$. For $j = 1$, it is satisfied as $d-s \geq 4$, by the above. On the other hand, the inequality is obviously satisfied for $j = \frac{d-1-s}{2} - 1$, which exceeds q, by (3.4.2.18).

Finally, since $p \leq 7$, (3.4.2.18) implies that $p < 3(d-s-2q-2)$, and (7.d) follows from Lemma 3.4.17. □

Part 3: Proof of Theorem 3.4.13 **in the Case n** > 3

Assume that $n \geq 4$ and show that the implications (a)–(c) in Lemma 3.4.18 hold for any k-plane $P \subset \mathbb{P}^n$, $k \geq 3$.

We use induction on k. The case $k = 3$ was treated in Lemma 3.4.18. Assume that $3 \leq k < n$, and let $P \subset \mathbb{P}^n$ be a $(k+1)$-plane.

The case $d \leq 3$ and implications (b), (c) can be treated just as those in Steps 1, 2, 3 of the proof of Lemma 3.4.18.

In the remaining case $d \geq 4$, $3d - 3 \leq \deg Z < 4d - 4$, we construct the following sequence $\Pi_1, \ldots, \Pi_m$ of k-planes. Denote $Z_1 = Z$ and let $j \geq 1$. If $Z_j = \emptyset$ then the sequence is complete, i.e., $m = j-1$. If $Z_j \neq \emptyset$ then we define Π_j to be any k-plane realizing the maximal possible intersection with Z_j, that is,

$$\deg(Z_j \cap \Pi_j) = \max_{\Pi \in |H^0(\mathcal{O}_P(1))|} \deg(Z_j \cap \Pi),$$

and set $Z_{j+1} = Z_j : \Pi_j$. The case $\deg(Z \cap \Pi_1) \geq d-1$ can be treated as in Step 4 of the proof of Lemma 3.4.18. In the case $\deg(Z \cap \Pi_1) \leq d-2$ we note that, by Lemma 3.4.15, $\deg(Z_j \cap \Pi_j) \geq \min\{4,\ \deg Z_j\}$. Hence, we complete the proof as in Step 6 of the proof of Lemma 3.4.18. □

3.4.3 Variations of the Horace Method

Differential Horace Method. In the following, we introduce a further developed, generalized Horace method, the so-called *"differential" Horace method* due to Alexander and Hirschowitz [AH1, AH2].

The main difference to the basic Horace method as described in Sect. 3.4.1 is that we consider a new, infinitesimal type of conditions imposed on the elements of the linear systems, which raises the flexibility when trying to keep the necessary balance between the degree of the trace and residual scheme.

Throughout the following let X denote a projective variety of dimension n and $H \subset X$ a hypersurface.

Definition 3.4.19 Let z be a smooth point of both X and H. Then we call a zero-dimensional scheme $Z \subset X$ *vertically graded with support* z *and basis* H iff there exist local coordinates $x_1, \dots, x_{n-1}, y$ for X at z such that $H = \{y = 0\}$ and the defining ideal $I \subset \mathbb{C}\{\boldsymbol{x}, y\} = \mathbb{C}\{x_1, \dots, x_{n-1}, y\}$ of Z is of the form

$$I = I_0 \oplus yI_1 \oplus \cdots \oplus y^{m-1} I_{m-1} \oplus y^m \mathbb{C}\{\boldsymbol{x}, y\}, \quad I_j \subset \mathbb{C}\{\boldsymbol{x}\} \text{ ideals.}$$

Note that necessarily $I_0 \subset I_1 \subset \cdots \subset I_{m-1}$.

Let $Z \subset X$ be a vertically graded scheme with support z and basis H. Let $\mathcal{J}_{Z/X}$ denote the ideal sheaf of Z in $\mathcal{O}_X$ and $\mathcal{J}^k_{H/X}$ the k-th power of the ideal sheaf of H. Then we introduce for any non-negative integer k

(a) the *k-trace $T^k_H(Z)$ of Z along H*, given by the ideal sheaf

$$\mathcal{J}_{T^k_H(Z)/X} := \bigl(\mathcal{J}_{Z/X} : \mathcal{J}^k_{H/X}\bigr) \big/ \bigl(\mathcal{J}_{Z/X} : \mathcal{J}^{k+1}_{H/X}\bigr) \cdot \mathcal{J}_{H/X}\,,$$

(b) the *k-th residual $D^k_H(Z)$ of Z with respect to H*, given by the ideal sheaf

$$\mathcal{J}_{D^k_H(Z)/X} := \mathcal{J}_{Z/X} + \bigl(\mathcal{J}_{Z/X} : \mathcal{J}^{k+1}_{H/X}\bigr) \cdot \mathcal{J}^k_{H/X}\,.$$

More generally, we study zero-dimensional schemes $Z \subset X$ with support $z = \{z_1, \dots, z_r\}$, obtained as the disjoint union of vertically graded zero-dimensional schemes $Z_i \subset X$ with support z_i and basis H. For any r-tuple $\boldsymbol{k} = (k_1, \dots, k_r)$ of non-negative integers we introduce

$$T^{\boldsymbol{k}}_H(Z) := \bigcup_{i=1}^r T^{k_i}_H(Z_i)\,, \qquad D^{\boldsymbol{k}}_H(Z) := \bigcup_{i=1}^r D^{k_i}_H(Z_i)\,.$$

The following proposition is the core of the differential Horace method approach, generalizing Lemma 3.4.1 (the fundamental lemma of the basic Horace method).

Proposition 3.4.20 *Let X be a smooth projective variety of dimension n, $H \subset X$ a reduced, irreducible hypersurface and $\mathcal{L}$ an invertible sheaf on X. Let $Z_0 \subset X$ be any zero-dimensional scheme and $Z = Z_1 \cup \cdots \cup Z_r$ the disjoint union of vertically graded subschemes of X with basis H and support disjoint from the support of Z_0. If there exists an r-tuple $\boldsymbol{k} = (k_1, \dots, k_r)$ of non-negative integers such that the two conditions*

$$\begin{aligned} H^0\bigl(\mathcal{J}_{(Z_0 \cap H) \cup T^{\boldsymbol{k}}_H(Z)/X} \otimes \mathcal{L}|_H\bigr) &= 0\,, \\ H^0\bigl(\mathcal{J}_{(Z_0 : H) \cup D^{\boldsymbol{k}}_H(Z)/X} \otimes \mathcal{L}(-H)\bigr) &= 0\,, \end{aligned}$$

are satisfied, then $H^0\bigl(\mathcal{J}_{Z_0 \cup Z_{gen}/X} \otimes \mathcal{L}\bigr) = 0$, where Z_{gen} denotes the generic translation of Z.

Proof Cf. [AH2, Lemma 2.3]. □

An important application of the differential Horace method is to obtain asymptotic results confirming the *Harbourne-Hirschowitz conjecture* for ideal sheaves of generic fat point schemes in $\mathbb{P}^2$ (see [HaB1, HaB2, HiA2]):

Conjecture 3.4.21 (Harbourne/Hirschowitz) *Let $m_1 \geq m_2 \geq \cdots \geq m_r$, $r \geq 3$, d be non-negative integers and $\boldsymbol{m} := (m_1, \ldots, m_r)$. Suppose that*

$$m_1 + m_2 + m_3 \leq d .$$

Denote by $Z(\boldsymbol{m})$ the zero-dimensional subscheme of $\mathbb{P}^2$ of r fat points in general position of multiplicities $m_1, ..., m_r$, respectively. Then $H^1\big(\mathcal{J}_{Z(\boldsymbol{m})/\mathbb{P}^2}(d)\big)$ vanishes whenever the expected dimension is non-negative, that is, whenever

$$\sum_{i=1}^{r} \frac{m_i(m_i+1)}{2} \leq \frac{(d+1)(d+2)}{2} . \tag{3.4.3.1}$$

The following result is due to Alexander and Hirschowitz [AH2].

Theorem 3.4.22 *Let X be a smooth projective variety, and let $\mathcal{M}, \mathcal{L}$ be invertible sheaves on X with $\mathcal{L}$ ample. Moreover, let $m > 0$. Then there exists a positive integer $d_0 = d_0(m)$ (depending also on $X, \mathcal{M}, \mathcal{L}$) such that for any $d \geq d_0$ and any r-tuple $\boldsymbol{m} = (m_1, \ldots, m_r)$ of multiplicities $m_i \leq m$ the canonical map $H^0\big(\mathcal{M} \otimes \mathcal{L}^d\big) \to H^0\big(\mathcal{O}_{Z(\boldsymbol{m})} \otimes \mathcal{M} \otimes \mathcal{L}^d\big)$ has maximal rank. In particular, $H^1\big(\mathcal{J}_{Z(\boldsymbol{m})/X} \otimes \mathcal{M} \otimes \mathcal{L}^d\big)$ vanishes, whenever the expected dimension $H^0\big(\mathcal{M} \otimes \mathcal{L}^d\big) - \deg Z(\boldsymbol{m})$ is non-negative.*

Geometric Horace Method. Next step in the generalization of the Horace method was done by Évain [Eva1, Eva2].

Let X be a quasi-projective variety, H a hypersurface in X, $\mathcal{L}$ a locally free sheaf of rank one on X, $L \subset H^0(X, \mathcal{L})$ a linear subspace. For a closed subscheme Y of X, $L(-Y)$ denotes $L \cap \varphi(H^0(X, \mathcal{L} \otimes \mathcal{J}_{Y/X}))$, where $\varphi : H^0(X, \mathcal{L} \otimes \mathcal{J}_{Y/X}) \to H^0(X, \mathcal{L})$ is the natural inclusion.

Theorem 3.4.23 *With the above notations and assumptions, let $Z = Z_1 \cup ... \cup Z_r$ be the disjoint union of vertically graded zero-dimensional subschemes of X with base H and support at an r-tuple $z = (z_1, ..., z_r)$ of smooth points of H and X. Let $\boldsymbol{v} = (v_1, ..., v_s) \in \mathbb{N}^s$, and let the zero-dimensional scheme $Z_{t,\boldsymbol{v}}$ be obtained from Z by the translation $x_1 \mapsto x_1 - t^{v_i}$ of each component Z_i in the corresponding local coordinates in a neighborhood of z_i, $i = 1, ..., r$, $t \in (\mathbb{C}, 0)$. Suppose that there are integers $n_1 > ... > n_s > n_{s+1} = 0$ such that*

- *$n_j - n_{j+1} \geq \max_i v_i$ for all $j = 1, ..., s$,*
- *$L(-(j-1)H - T_H^{\boldsymbol{k}_j}(Z)) = L(-jH)$ for all $j = 1, ..., s$, where*

$$\boldsymbol{k}_j = (k_{1j}, ..., k_{rj}), \quad k_{ij} = \left[\frac{n_j}{v_i}\right], \ i = 1, ..., r .$$

Then

$$\lim_{t\to 0} L(-Z_{t,v}) \subset L(-sH - \widetilde{Z}),$$

where $\widetilde{Z} = (D_H^{k_s} \circ \ldots \circ D_H^{k_1})(Z)$.

This generalizes Proposition 3.4.20, which essentially corresponds to the case $s = 1$. Furthermore, let Z' be a generic translate of Z (along X), then the following bound holds true.

Corollary 3.4.24 *Under the hypotheses of Theorem 3.4.23, assume, in addition, that X is irreducible. Then*

$$\dim L(-Z') \leq \dim L(-sH - \widetilde{Z}).$$

Using Theorem 3.4.23, Évain [Eva2] established the Harbourne-Hirschowitz Conjecture 3.4.21 for the number of points $r = k^2$ and equal multiplicities $m_1 = \ldots = m_r$. Other proofs of this result were found by Ciliberto and Miranda [CiM] and Roé [Roe1]. For an arbitrary r and equal multiplicities, the H^1-vanishing condition in Conjecture 3.4.21 was established by Roé [Roe] under condition slightly stronger than (3.4.3.1). For a priori bounded multiplicities the required H^1-vanishing holds when degree is sufficiently high (see [Mig1, Mig2]).

An interesting extension of Theorem 3.4.23 to cluster schemes was found by J. Roé [Roe1], who proved H^1-vanishing results for A-D-E singularity schemes (see also [Roe2] for H^1-vanishing results for A_k singularity schemes).

3.5 The Castelnuovo Function

Let $Z \subset \mathbb{P}^2$ be a zero-dimensional scheme and $\mathcal{J}_{Z/\mathbb{P}^2} \subset \mathcal{O}_{\mathbb{P}^2}$ the corresponding ideal sheaf.

Definition 3.5.1 The *Castelnuovo function* of Z is defined as

$$\mathcal{C}_Z : \mathbb{Z}_{\geq 0} \to \mathbb{Z}_{\geq 0}, \quad d \mapsto h^1\big(\mathcal{J}_{Z/\mathbb{P}^2}(d-1)\big) - h^1\big(\mathcal{J}_{Z/\mathbb{P}^2}(d)\big).$$

Remark 3.5.1.1 Let $Z \subset \mathbb{P}^2$ be a zero-dimensional scheme and $L \subset \mathbb{P}^2$ be a line not passing through the support of Z. Then we have an exact reduction sequence (3.4.1.1), respectively the corresponding exact cohomology sequence

$$H^0\big(\mathcal{J}_{Z/\mathbb{P}^2}(d)\big) \xrightarrow{\pi_L} H^0\big(\mathcal{O}_L(d)\big) \to H^1\big(\mathcal{J}_{Z/\mathbb{P}^2}(d-1)\big) \to H^1\big(\mathcal{J}_{Z/\mathbb{P}^2}(d)\big) \to 0.$$

In particular, $\mathcal{C}_Z(d) = h^0\big(\mathcal{O}_L(d)\big) - \dim_{\mathbb{C}} \pi_L\big(H^0\big(\mathcal{J}_{Z/\mathbb{P}^2}(d)\big)\big)$.

In the following, we recall some basic properties of the Castelnuovo function, basically following [Dav]. For an extension of this theory to zero-simensional schemes in the projective spaces of arbitrary dimension, see [BGM].

We introduce the *order* $\mathrm{ord}(Z)$, respectively the *order of regularity* $\mathrm{reg}(Z)$, of Z,

$$\begin{aligned}\mathrm{ord}(Z) &:= \min\left\{d \in \mathbb{Z} \,\middle|\, h^0\big(\mathcal{J}_{Z/\mathbb{P}^2}(d)\big) > 0\right\}\\ \mathrm{reg}(Z) &:= \min\left\{d \in \mathbb{Z} \,\middle|\, H^1\big(\mathcal{J}_{Z/\mathbb{P}^2}(d)\big) = 0\right\}.\end{aligned}$$

Moreover, we introduce

$$b(Z) := \min\left\{d \in \mathbb{Z} \,\middle|\, \big|\big(H^0(\mathcal{J}_{Z/\mathbb{P}^2}(d)\big)\big| \text{ has no fixed component}\right\}.$$

Obviously, $\mathrm{ord}(Z) \le b(Z)$. On the other hand, $b(Z) \le \mathrm{reg}(Z)+1$, since for $d \ge \mathrm{reg}(Z)+1$ and any line $L \subset \mathbb{P}^2$ not meeting the support of Z the morphism

$$\pi_L : H^0\big(\mathcal{J}_{Z/\mathbb{P}^2}(d)\big) \longrightarrow H^0\big(\mathcal{O}_L(d)\big) \cong H^0\big(\mathcal{O}_{\mathbb{P}^1}(d)\big)$$

induced by the restriction to L is surjective (in particular $\big|\big(H^0(\mathcal{J}_{Z/\mathbb{P}^2}(d)\big)\big|$ cannot have a fixed component).

Lemma 3.5.2 (cf. [Dav], Claim 2.1) *Let $Z' \subset Z$ be zero-dimensional schemes in $\mathbb{P}^2$ and let d be a non-negative integer then (see Fig. 3.1)*

(a) $\mathcal{C}_{Z'}(d) \le \mathcal{C}_Z(d)$,
(b) $h^0\big(\mathcal{J}_{Z/\mathbb{P}^2}(d)\big) = h^0\big(\mathcal{J}_{Z'/\mathbb{P}^2}(d)\big)$ *iff* $\mathcal{C}_{Z'}(i) = \mathcal{C}_Z(i)$ *for each* $0 \le i \le d$,
(c) $\mathcal{C}_Z(0) + \cdots + \mathcal{C}_Z(d) = \deg Z - h^1\big(\mathcal{J}_{Z/\mathbb{P}^2}(d)\big)$,
(d) $\mathcal{C}_Z(d) = 0$ *iff* $d \ge \mathrm{reg}(Z)+1$,
(e) $\mathcal{C}_Z(d) \le d+1$ *with equality iff* $h^0\big(\mathcal{J}_{Z/\mathbb{P}^2}(d)\big) = 0$, *i.e., if* $d \le \mathrm{ord}(Z)-1$,
(f) if $d \ge \mathrm{ord}(Z)$ *then* $\mathcal{C}_Z(d) \le \mathcal{C}_Z(d-1)$.

Proof Since $\mathcal{J}_{Z/\mathbb{P}^2}(d) \subset \mathcal{J}_{Z'/\mathbb{P}^2}(d)$, (a) follows immediately from Remark 3.5.1.1.
(b) The above exact cohomology sequence implies that

$$\begin{aligned}&h^0\big(\mathcal{J}_{Z'/\mathbb{P}^2}(d)\big) - h^0\big(\mathcal{J}_{Z/\mathbb{P}^2}(d)\big)\\ &\qquad = h^0\big(\mathcal{J}_{Z'/\mathbb{P}^2}(d-1)\big) - h^0\big(\mathcal{J}_{Z/\mathbb{P}^2}(d-1)\big) + \mathcal{C}_Z(d) - \mathcal{C}_{Z'}(d),\end{aligned}$$

whence $h^0\big(\mathcal{J}_{Z'/\mathbb{P}^2}(d)\big) - h^0\big(\mathcal{J}_{Z/\mathbb{P}^2}(d)\big) = \sum_{i=0}^{d}\big(\mathcal{C}_Z(i) - \mathcal{C}_{Z'}(i)\big)$, by induction, and the statement follows from (a).
(c) $\mathcal{C}_Z(0) + \cdots + \mathcal{C}_Z(d) = h^1\big(\mathcal{J}_{Z/\mathbb{P}^2}(-1)\big) - h^1\big(\mathcal{J}_{Z/\mathbb{P}^2}(d)\big)$, and the statement follows from the exact sequence

$$0 \to H^0\big(\mathcal{O}_Z(-1)\big) \to H^1\big(\mathcal{J}_{Z/\mathbb{P}^2}(-1)\big) \to H^1\big(\mathcal{O}_{\mathbb{P}^2}(-1)\big) = 0.$$

(d) Assume that $\mathcal{C}_Z(d) = 0$, that is, for any generic line $L \subset \mathbb{P}^2$ we have an exact sequence $H^0\big(\mathcal{J}_{Z/\mathbb{P}^2}(d)\big) \xrightarrow{\pi_L} H^0\big(\mathcal{O}_L(d)\big) \to 0$. It follows that, for a generic line $L' \neq L$, we have

$$\pi_L\left(H^0\big(\mathcal{J}_{Z/\mathbb{P}^2}(d+1)\big)\right) \supset \pi_L\left(L' \cdot H^0\big(\mathcal{J}_{Z/\mathbb{P}^2}(d)\big)\right) = \pi_L(L') \cdot H^0\big(\mathcal{O}_L(d)\big),$$

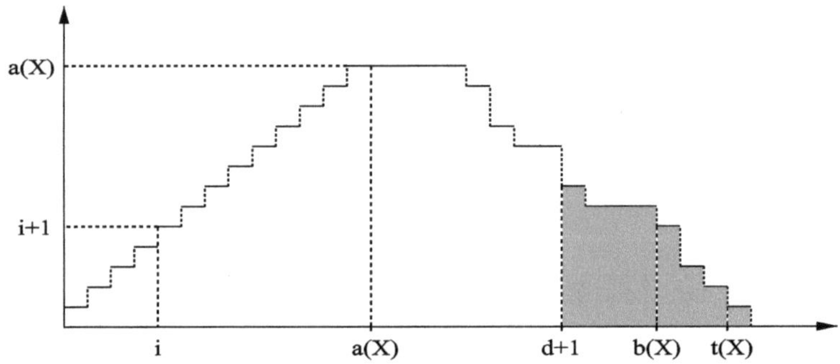

Fig. 3.1 The graph of a Castelnuovo function (considered as a function on $\mathbb{R}_{\geq 0}$ given by $\mathcal{C}_Z(t) = \mathcal{C}_Z(\lfloor t \rfloor)$). The area of the shaded region is $h^1(\mathcal{J}_{Z/\mathbb{P}^2}(d))$

which is of codimension one in $H^0(\mathcal{O}_L(d+1))$. On the other hand, let $L'' \subset \mathbb{P}^2$ be another generic line, then

$$\pi_L(L') \cdot H^0(\mathcal{O}_L(d)) \neq \pi_L(L'') \cdot H^0(\mathcal{O}_L(d)) \subset \pi_L\left(H^0(\mathcal{J}_{Z/\mathbb{P}^2}(d+1))\right),$$

whence the surjectivity of $H^0(\mathcal{J}_{Z/\mathbb{P}^2}(d+1)) \xrightarrow{\pi_L} H^0(\mathcal{O}_L(d+1))$. In particular, $\mathcal{C}_Z(d+1)$ vanishes, and the statement follows by induction.
(e), (f) By Remark 3.5.1.1, we have $\mathcal{C}_Z(d) \leq h^0(\mathcal{O}_L(d)) = d+1$, with equality iff $h^0(\mathcal{J}_{Z/\mathbb{P}^2}(d-1)) = h^0(\mathcal{J}_{Z/\mathbb{P}^2}(d)) = 0$. On the other hand, suppose that $h^0(\mathcal{J}_{Z/\mathbb{P}^2}(d-1)) > 0$ and choose generic lines L, L', L''. Then

$$\pi_L(L') \cdot \pi_L(H^0(\mathcal{J}_{Z/\mathbb{P}^2}(d-1))) \neq \pi_L(L'') \cdot \pi_L(H^0(\mathcal{J}_{Z/\mathbb{P}^2}(d-1))),$$

both being contained in $\pi_L\left(H^0(\mathcal{J}_{Z/\mathbb{P}^2}(d))\right)$. It follows that

$$\begin{aligned}\mathcal{C}_Z(d) - \mathcal{C}_Z(d-1) = h^0(\mathcal{O}_L(d)) - h^0(\mathcal{O}_L(d-1)) - \dim_{\mathbb{C}} \pi_L(H^0(\mathcal{J}_{Z/\mathbb{P}^2}(d))) \\ + \dim_{\mathbb{C}} \pi_L(H^0(\mathcal{J}_{Z/\mathbb{P}^2}(d-1))),\end{aligned}$$

is a non-positive number. □

Example 3.5.2.1 The graphs of the Castelnuovo function of the zero-dimensional schemes $Z \subset \mathbb{P}^2$ supported at $z \in \mathbb{P}^2$, respectively at $z_1, \ldots, z_9 \in \mathbb{P}^2$, and given by the ideals

(a) $\mathcal{J}_{Z,z} = \mathfrak{m}_z^m$,

respectively given by $\mathcal{J}_{Z,z_i} = \mathfrak{m}_{z_i}$, $i = 1, \ldots, 9$, and

(b) $z_1, \ldots, z_9$ in generic position,
(c) $z_1, \ldots, z_6$ on a line, z_7, z_8, z_9 in generic position,
(d) $z_1, \ldots, z_7$ on a line, z_8, z_9 in generic position,

are the following:

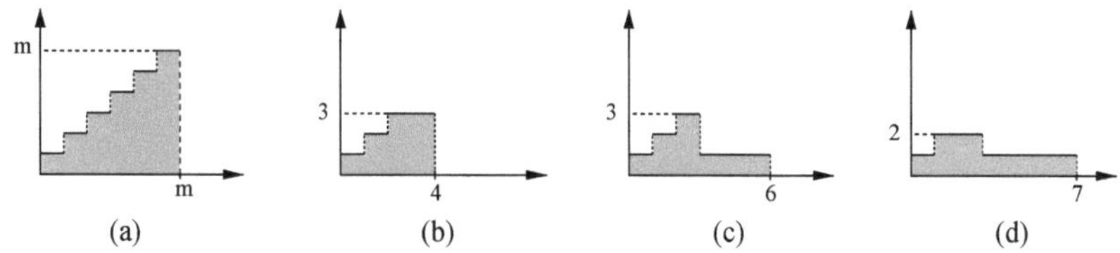

Lemma 3.5.3 ([Dav], Claims 2.3 and 2.4) *Let $Z \subset \mathbb{P}^2$ be a zero-dimensional scheme, and $d_0 \geq \operatorname{ord}(Z)$ such that $\mathcal{C}_Z(d_0) = \mathcal{C}_Z(d_0+1)$. Then there exists a fixed curve D of degree $\mathcal{C}_Z(d_0)$ in the complete linear system $\left|H^0\big(\mathcal{J}_{Z/\mathbb{P}^2}(d_0)\big)\right|$ with the additional property that for each $d \geq 0$*

$$\mathcal{C}_{Z\cap D}(d) = \min\left\{\mathcal{C}_Z(d), \mathcal{C}_Z(d_0)\right\}. \tag{3.5.0.2}$$

In particular, $\mathcal{C}_Z(d) < \mathcal{C}_Z(d-1)$ if $b(Z) \leq d \leq \operatorname{reg}(Z)+1$.

Proof Let $L \subset \mathbb{P}^2$ be a generic line, and $\pi_L\big(H^0\big(\mathcal{J}_{Z/\mathbb{P}^2}(d_0)\big)\big) \subset H^0\big(\mathcal{O}_L(d_0)\big)$ (minimally) generated by $f_1, \ldots, f_m$. After choosing an affine coordinate on L we can identify $f_1, \ldots, f_m$ with polynomials $f_1(x), \ldots, f_m(x) \in \mathbb{C}[x]$. Without restriction, we can suppose that $\deg(f_1) < \deg(f_2) < \cdots < \deg(f_m) = d_0$. Now, $f_1(x), \ldots, f_m(x);\ xf_1(x), \ldots, xf_m(x)$ can be identified with elements in $\pi_L\big(H^0\big(\mathcal{J}_{Z/\mathbb{P}^2}(d_0+1)\big)\big)$.

Since, by our assumption, $\dim_{\mathbb{C}} \pi_L\big(H^0\big(\mathcal{J}_{Z/\mathbb{P}^2}(d_0+1)\big)\big) = m+1$, and since $f_1, \ldots, f_m$ are linearly independent, we obtain relations

$$x \cdot f_j = \alpha_{j,1} f_1 + \alpha_{j,2} f_2 + \cdots + \alpha_{j,j+1} f_{j+1}, \quad \alpha_{j,i} \in \mathbb{C},\ \alpha_{j,j+1} \neq 0,$$

$j = 1, \ldots, m-1$. Hence, (after base change) we can suppose $f_j = x^{j-1} \cdot f_1$, $j = 2, \ldots, m$. In particular, the roots of f_1 are base points of the linear system $\left|\big(\pi_L\big(H^0\big(\mathcal{J}_{Z/\mathbb{P}^2}(d_0)\big)\big)\right|$. Varying L, we obtain a fixed component D in $\left|H^0\big(\mathcal{J}_{Z/\mathbb{P}^2}(d_0)\big)\right|$ of degree

$$\begin{aligned} \deg D = \deg(f_1) &= d_0 - m + 1 \\ &= h^0\big(\mathcal{O}_L(d_0)\big) - \dim_{\mathbb{C}} \pi_L\big(H^0\big(\mathcal{J}_{Z/\mathbb{P}^2}(d_0)\big)\big) = \mathcal{C}_Z(d_0). \end{aligned}$$

Now, we prove that D satisfies (3.5.0.2). Since D is a fixed component of degree $\mathcal{C}_Z(d_0)$ in $\left|H^0\big(\mathcal{J}_{Z/\mathbb{P}^2}(d_0)\big)\right|$, we have

$$\operatorname{ord}(Z \cap D) = \min\left\{d \in \mathbb{Z} \;\middle|\; h^0\big(\mathcal{J}_{Z\cap D/\mathbb{P}^2}(d)\big) > 0\right\} = \mathcal{C}_Z(d_0).$$

By Lemma 3.5.2(e),(f), it follows that $\mathcal{C}_{Z\cap D}(d) \leq \mathcal{C}_Z(d_0)$ for any $d \geq 0$ and to obtain (3.5.0.2) it suffices to show that $\mathcal{C}_{Z\cap D}(d) = \mathcal{C}_Z(d)$ for any $d \geq d_0$ (Fig. 3.2). We know that $H^0\big(\mathcal{J}_{Z/\mathbb{P}^2}(d_0)\big) = D \cdot H^0\big(\mathcal{J}_{Z:D/\mathbb{P}^2}(d_0 - \mathcal{C}_Z(d_0))\big)$, whence

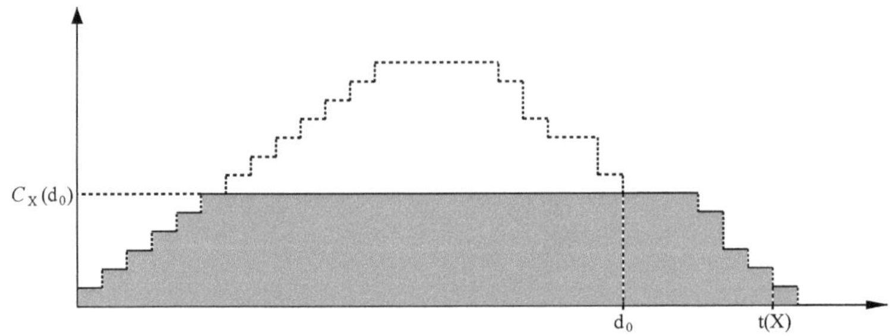

Fig. 3.2 The graph of the Castelnuovo function $\mathcal{C}_{Z\cap D}$, where D is the fixed curve in $|H^0(\mathcal{J}_{Z/\mathbb{P}^2}(d_0))|$ given by Lemma 3.5.3. The area of the shaded region is $\deg(Z \cap D)$

$$\begin{aligned}\dim_{\mathbb{C}} \pi_L\bigl(H^0\bigl(\mathcal{J}_{Z:D/\mathbb{P}^2}(d_0-\mathcal{C}_Z(d_0))\bigr)\bigr) &= \dim_{\mathbb{C}} \pi_L\bigl(H^0\bigl(\mathcal{J}_{Z/\mathbb{P}^2}(d_0)\bigr)\bigr) \;=\; m \\ &= d_0+1-\mathcal{C}_Z(d_0)\,,\end{aligned}$$

that is, $\mathcal{C}_{Z:D}(d_0-\mathcal{C}_Z(d_0)) = 0$. By Lemma 3.5.2(d), this implies the vanishing of $H^1\bigl(\mathcal{J}_{Z:D/\mathbb{P}^2}(d-\mathcal{C}_Z(d_0))\bigr)$ for $d \geq d_0-1$, and we deduce from the exact sequence

$$0 \longrightarrow \mathcal{J}_{Z:D/\mathbb{P}^2}(d-\mathcal{C}_Z(d_0)) \xrightarrow{\cdot D} \mathcal{J}_{Z/\mathbb{P}^2}(d) \longrightarrow \mathcal{J}_{Z\cap D/D}(d) \longrightarrow 0\,,$$

respectively the corresponding cohomology sequence, that

$$h^1\bigl(\mathcal{J}_{Z/\mathbb{P}^2}(d)\bigr) - h^1\bigl(\mathcal{J}_{Z\cap D/D}(d)\bigr) \;\leq\; 0\,. \tag{3.5.0.3}$$

On the other hand, the exact cohomology sequences corresponding to

$$\begin{aligned}&0 \longrightarrow \mathcal{O}_{\mathbb{P}^2}(d-\mathcal{C}_Z(d_0)) \longrightarrow \mathcal{J}_{Z\cap D/\mathbb{P}^2}(d) \longrightarrow \mathcal{J}_{Z\cap D/D}(d) \longrightarrow 0\,,\\ &0 \longrightarrow \mathcal{J}_{Z/\mathbb{P}^2}(d) \longrightarrow \mathcal{J}_{Z\cap D/\mathbb{P}^2}(d) \longrightarrow \bigl(\mathcal{J}_{Z\cap D/\mathbb{P}^2}\big/\mathcal{J}_{Z/\mathbb{P}^2}\bigr)(d) \longrightarrow 0\,,\end{aligned}$$

show that $h^1\bigl(\mathcal{J}_{Z\cap D/D}(d)\bigr) = h^1\bigl(\mathcal{J}_{Z\cap D/\mathbb{P}^2}(d)\bigr) \leq h^1\bigl(\mathcal{J}_{Z/\mathbb{P}^2}(d)\bigr)$. This, together with (3.5.0.3) implies that $\mathcal{C}_{Z\cap D}(d) = \mathcal{C}_Z(d)$ for any $d \geq d_0$. □

Applying Bézout's Theorem, we can compute the Castelnuovo function of a (global) complete intersection $Z = C_d \cap C_k$, C_d, C_k plane curves of degree d, respectively k, without common components.

Lemma 3.5.4 *Let $Z = C_d \cap C_k$ be the intersection of two plane curves C_d, C_k of degree d, respectively k, without common components. Suppose $k \leq d$. Then $\mathcal{C}_Z(i) \leq k$ for $i \geq 0$, and $\mathcal{C}_Z(d+k-j) = j-1$ for $j = 1, \ldots, k+1$ (see Fig. 3.3).*

Proof Bézout's Theorem implies that $\deg Z = kd$, $\operatorname{ord}(Z) = \min\{k,d\} = k$, and $b(Z) = \max\{k,d\} = d$. In particular, by Lemma 3.5.2, we obtain that $\mathcal{C}_Z(i) \leq k$ for all $i \geq 0$ and

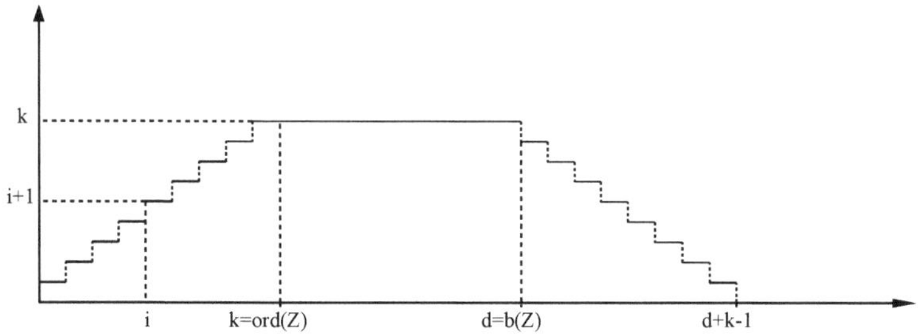

Fig. 3.3 The graph of the Castelnuovo function for the (global) complete intersection scheme $Z = C_d \cap C_k$

$$
\begin{aligned}
kd = \deg Z &= \sum_{i=0}^{\operatorname{ord}(Z)-1} c_Z(i) + \sum_{i=\operatorname{ord}(Z)}^{b(Z)-1} c_Z(i) + \sum_{i=b(Z)}^{\operatorname{reg}(Z)} c_Z(i) \\
&\le \frac{k(k+1)}{2} + \big(d-k\big)\cdot k + \sum_{i=b(Z)}^{\operatorname{reg}(Z)} c_Z(i) = kd - \frac{k^2-k}{2} + \sum_{i=b(Z)}^{\operatorname{reg}(Z)} c_Z(i)\,.
\end{aligned}
$$

Note that, by Lemma 3.5.3, we have, for any $i \ge 0$,

$$
\begin{aligned}
c_Z(b(Z)+i) &\le \max\big\{0,\, c_Z(b(Z)) - i\big\}\,, \\
c_Z(b(Z)) &\le c_Z(b(Z)-1) - 1 \le k-1\,.
\end{aligned}
$$

If any of these inequalities would be strict, from the above we would obtain

$$
\frac{k^2-k}{2} \le \sum_{i=b(Z)}^{\operatorname{reg}(Z)} c_Z(i) < \sum_{i=1}^{k-1} i = \frac{k^2-k}{2}\,,
$$

whence a contradiction. □

Definition 3.5.5 We call a zero-dimensional scheme $Z \subset \mathbb{P}^2$ *decomposable* if there exists a positive integer d_0 such that

$$
c_Z(d_0-1) > c_Z(d_0) = c_Z(d_0+1) > 0\,.
$$

Considering the above properties of the Castelnuovo function, it is not difficult to prove the following lemma, which is basically due to Barkats [Bar].

Lemma 3.5.6 *Let $C_d \subset \mathbb{P}^2$ be an irreducible curve of degree $d > 0$, and $Z \subset C_d$ be a zero-dimensional scheme such that $h^1\big(\mathcal{J}_{Z/\mathbb{P}^2}(d)\big) > 0$. Suppose, moreover, $d >$ $\operatorname{ord}(Z)$. Then there exists a curve C_k of degree $k \ge 3$ such that $Z \cap C_k \subset C_d \cap C_k$ is non-decomposable and satisfies*

(a) $h^1\big(\mathcal{J}_{Z\cap C_k/\mathbb{P}^2}(d)\big) = h^1\big(\mathcal{J}_{Z/\mathbb{P}^2}(d)\big)$,

(b) $\deg(Z \cap C_k) \geq k \cdot (d+3-k)$ *if* $k \leq \frac{d+3}{2}$, *and* $\deg(Z \cap C_k) > \frac{(d+3)^2}{4}$ *if* $k > \frac{d+3}{2}$.

Proof Case 1. Suppose Z to be decomposable and let $d_0 > 0$ be maximal with the property $\mathcal{C}_Z(d_0) = \mathcal{C}_Z(d_0+1) > 0$.

By Lemma 3.5.3, there exists a curve C_k of degree $k = \mathcal{C}_Z(d_0) < \mathrm{ord}(Z)$ such that $\mathcal{C}_{Z\cap C_k}(i) = \min\{\mathcal{C}_Z(i), \mathcal{C}_Z(d_0)\}$. In particular, $Z \cap C_k$ is non-decomposable. Note that C_d is irreducible, and $k < \mathrm{ord}(Z) < d$ by assumption. Hence, $C_d \cap C_k$ is a (global) complete intersection, and it follows (by applying Lemma 3.5.4)

$$1 \leq \mathcal{C}_{Z\cap C_k}(d+1) \leq \mathcal{C}_{C_d \cap C_k}(d+1) = k-2 < \mathcal{C}_Z(d_0). \tag{3.5.0.4}$$

In particular, $k \geq 3$, $d_0 \leq d$, and

$$h^1\big(\mathcal{J}_{Z\cap C_k/\mathbb{P}^2}(d)\big) = \sum_{i=d+1}^{\infty} \mathcal{C}_{Z\cap C_k}(i) = \sum_{i=d+1}^{\infty} \mathcal{C}_Z(i) = h^1\big(\mathcal{J}_{Z/\mathbb{P}^2}(d)\big) > 0.$$

Since $\mathcal{C}_{Z\cap C_k}(i) \leq k$ for $i \geq 0$, $\mathcal{C}_{Z\cap C_k}(k-1) = k$, and since $Z \cap C_k$ is non-decomposable, we have

$$\deg(Z \cap C_k) \geq \sum_{i=0}^{d+1} \mathcal{C}_{Z\cap C_k}(i) \geq \sum_{j=1}^{k} \max\{1,\, d+2-2(j-1)\}$$
$$\begin{cases} = k(d+2-k+1) & \text{if } 2k \leq d+3\,, \\ > \dfrac{(d+3)^2}{4} & \text{if } 2k > d+3\,, \end{cases}$$

whence the statement of the lemma.

Case 2. If Z is a non-decomposable scheme, we can choose $k := \mathrm{ord}(Z) < d$ and $C_k \in H^0\big(\mathcal{J}_{Z/\mathbb{P}^2}(k)\big)$. Again, the above reasoning shows that $k \geq 3$ and the inequality in (b) is satisfied. □

Remark 3.5.6.1 The zero-dimensional scheme $Y := Z \cap C_k$ in Lemma 3.5.6 satisfies

$$h^1\big(\mathcal{J}_{Y/\mathbb{P}^2}(d)\big) = \sum_{i=d+1}^{\infty} \mathcal{C}_Y(i) \leq \frac{r_0(r_0+1)}{2}\,,$$

where $r_0 := \mathcal{C}_Y(d+1) \leq k-2$ (cf. (3.5.0.4)), and (cf. Fig. 3.4)

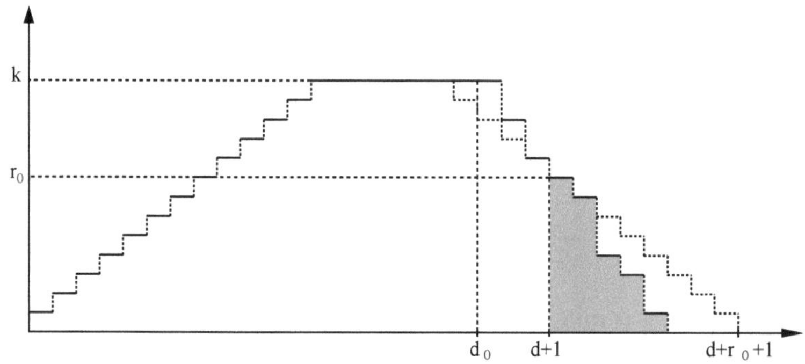

Fig. 3.4 The graph of a Castelnuovo function $c_{Z\cap C_k}$. The area of the shaded region is $h^1\big(\mathcal{J}_{Z\cap C_k/\mathbb{P}^2}(d)\big)$

$$\begin{aligned}\deg Y &= h^1\big(\mathcal{J}_{Y/\mathbb{P}^2}(d)\big) + \sum_{i=0}^{d} c_Y(i) \\ &\geq h^1\big(\mathcal{J}_{Y/\mathbb{P}^2}(d)\big) - \frac{r_0(r_0+1)}{2} + \sum_{j=1}^{k} \max\big\{1,\ d+r_0+1-2(j-1)\big\} \qquad (3.5.0.5) \\ &\geq h^1\big(\mathcal{J}_{Y/\mathbb{P}^2}(d)\big) - \frac{r_0(r_0+1)}{2} + \begin{cases} (d+2-k+r_0)\cdot k & \text{if } 2k \leq d+2+r_0\,, \\ \dfrac{(d+2+r_0)^2}{4} & \text{if } 2k > d+2+r_0\,. \end{cases}\end{aligned}$$

H^1-Vanishing for Zero-Dimensional Subschemes of $Z^{ea}(C)$

The above preparatory considerations can be used to derive an H^1-vanishing theorem for zero-dimensional subschemes of the equianalytic scheme $Z^{ea}(C)$ corresponding to a reduced plane curve C. Here, the γ-invariant appears, which has been introduced and studied in Sect. 1.2.3.

Theorem 3.5.7 ([GLS4], Proposition 4.1) *Let $C \subset \mathbb{P}^2$ be an irreducible curve of degree $d \geq 6$ with r singular points $z_1, \ldots, z_r$, and let $Z_i \subset Z^{ea}(C, z_i)$, $i = 1, \ldots, r$, be any zero-dimensional schemes. Moreover, let Z be the (disjoint) union of the schemes $Z_1, \ldots, Z_r$. If*

$$\sum_{i=1}^{r} \gamma(C; Z_i) \leq (d+3)^2\,, \qquad (3.5.0.6)$$

then $H^1\big(\mathcal{J}_{Z/\mathbb{P}^2}(d)\big) = 0$.

Proof Assume $h^1\big(\mathcal{J}_{Z/\mathbb{P}^2}(d)\big) > 0$, and, as in the proof of Theorem 3.3.4, let Z' be a minimal subscheme of Z with non-vanishing $h^1\big(\mathcal{J}_{Z'/\Sigma}(d)\big)$.

By the Serre-Grothendieck duality (see [HaR2, Section III.7]), $\mathrm{Ext}^1\big(\mathcal{J}_{Z'}(d), \mathcal{O}_{\mathbb{P}^2}(-3)\big) \neq 0$, and a generic element defines an extension $0 \to \mathcal{O}_{\mathbb{P}^2} \to E \to$

$\mathcal{J}_{Z'}(d+3) \to 0$, where E is a rank 2 vector bundle on $\mathbb{P}^2$. In particular, it follows that Z' is a local complete intersection, that is, the local ring of each of the $r' \le r$ components $Z'_i \subset Z_i$ of Z' is a complete intersection ring.

By condition (3.5.0.6) and since $d \ge 6$, we can estimate

$$\deg Z' \;\le\; \sum_{i=1}^{r} \frac{\gamma_{ci}(C; Z_i)}{4} \;\le\; \frac{(d+3)^2}{4} \;<\; \frac{d(d+1)}{2}\,.$$

Hence, $d > \mathrm{ord}(Z') = \min\left\{j \mid h^0\big(\mathcal{J}_{Z'/\mathbb{P}^2}(j)\big) > 0\right\}$, and Lemma 3.5.6 gives the existence of a curve D of degree $k \ge 3$ such that the trace $Z' \cap D$ is non-decomposable and satisfies $h^1\big(\mathcal{J}_{Z'\cap D/\mathbb{P}^2}(d)\big) = h^1\big(\mathcal{J}_{Z'/\mathbb{P}^2}(d)\big) > 0$. Hence, by the minimality assumption for Z', $Z' \cap D = Z'$. Moreover, by Lemma 3.5.6,

$$\deg Z' \;\ge\; k \cdot (d+3-k)\,. \tag{3.5.0.7}$$

Consequently, we can estimate k as

$$k \;\le\; \frac{2 \cdot \deg Z'}{d+3+\sqrt{(d+3)^2 - 4\deg Z'}}\,. \tag{3.5.0.8}$$

On the other hand, let $Z' \cap D = Z''_1 \cup \cdots \cup Z''_{r''}$, $r'' := \#(\mathrm{supp}(Z'))$, be the decomposition of the zero-dimensional scheme Z' into its irreducible components (without loss of generality, we may assume that Z''_i is supported at z_i for $i = 1, \ldots, r'' \le r'$). Note that, due to [Shu10, Lemma 4.1] (cf. [GLS6, Lemma I.3.40]), we have

$$\Delta_i \;:=\; \min\left\{i_{z_i}(C, D) - \deg Z''_i\,,\ \deg Z''_i\right\} \;\ge\; 1\,, \quad i = 1, \ldots, r''\,,$$

and Bézout's Theorem, respectively (3.5.0.7), implies that

$$k \cdot d \;\ge\; \sum_{i=1}^{r''} i_{z_i}(C, D) \;\ge\; \deg Z' + \sum_{i=1}^{r''} \Delta_i \;\ge\; k \cdot (d+3-k) + \sum_{i=1}^{r''} \Delta_i\,.$$

Thus, by (3.5.0.8), we can estimate

$$\sum_{i=1}^{r''} \Delta_i \;\le\; k(k-3) \;<\; k^2 \;\le\; \left(\frac{2 \cdot \deg Z'}{d+3+\sqrt{(d+3)^2 - 4\deg Z'}}\right)^2.$$

In particular, applying the Cauchy inequality, we obtain

$$\sum_{i=1}^{r''} \frac{(\deg Z''_i)^2}{\Delta_i} \;\ge\; \frac{(\deg Z')^2}{\sum_{i=1}^{r''} \Delta_i} \;>\; \left(1 + \sqrt{1 - \frac{4\deg Z'}{(d+3)^2}}\right)^2 \cdot \frac{(d+3)^2}{4}\,.$$

Setting

$$\alpha := \frac{\sum_{i=1}^{r''} \frac{(\deg Z_i'')^2}{\Delta_i}}{(d+3)^2}, \quad \beta := \frac{\sum_{i=1}^{r''} \frac{(\deg Z_i'')^2}{\Delta_i}}{\deg Z'},$$

this implies $\alpha > \frac{1}{4} \cdot \left(1 + \sqrt{1 - \frac{4\alpha}{\beta}}\right)^2$, that is, $\alpha > \left(\frac{\beta}{\beta+1}\right)^2$. Finally, we have

$$\begin{aligned}(d+3)^2 = \frac{\beta}{\alpha} \cdot \deg Z' &< \left(\beta + 2 + \frac{1}{\beta}\right) \cdot \deg Z' \\ &\leq \sum_{i=1}^{r''} \left(\frac{(\deg Z_i'')^2}{\Delta_i} + 2 \cdot \deg Z_i'' + \Delta_i \right), \qquad (3.5.0.9)\end{aligned}$$

the latter again due to the Cauchy inequality. Recall that Z_i'' is contained in the (local) complete intersection Z_i'. Hence (3.5.0.9) contradicts the assumption (3.5.0.6). □

3.6 H^1-Vanishing for Generic Zero-Dimensional Schemes

The content of this section follows [Shu13] with certain improvements.

Let $Z \subset \mathbb{P}^2$ be a zero-dimensional scheme of degree n. We introduce in the following the subvariety $\mathrm{Iso}(Z) \subset \mathrm{Hilb}^n_{\mathbb{P}^2}$ of zero-dimensional schemes which are locally isomorphic to Z (as subschemes of $\mathbb{P}^2$). To be more precise, we consider the germ of $\mathrm{Iso}(Z)$ at the point corresponding to Z (which is usually denoted by $\mathrm{Iso}(Z)$, too).

First, assume that Z is irreducible, that is, supported at a unique point $z \in \mathbb{P}^2$. Let $\mathrm{Aut}(\mathbb{C}^2)$ be the group of algebraic automorphisms of some fixed affine plane $\mathbb{C}^2 \subset \mathbb{P}^2$ containing $z = (0, 0)$. The action of $\mathrm{Aut}(\mathbb{C}^2)$ on $\mathbb{C}^2$ induces an (algebraic) action of $\mathrm{Aut}(\mathbb{C}^2)$ on (the germ at Z of) $\mathrm{Hilb}^n_{\mathbb{P}^2}$, and we define $\mathrm{Iso}(Z) \subset \mathrm{Hilb}^n_{\mathbb{P}^2}$ to be the orbit of Z under this action.

Now, let Z decompose into irreducible components $Z_1, \ldots, Z_r$ of degree $n_1, \ldots, n_r$, respectively. Then, locally at Z, there is an isomorphism

$$\zeta : \mathrm{Hilb}^n_{\mathbb{P}^2} \xrightarrow{\cong} \prod_{i=1}^{r} \mathrm{Hilb}^{n_i}_{\mathbb{P}^2}$$

(cf., e.g., [Iar1]), and we define $\mathrm{Iso}(Z)$ to be the preimage

$$\mathrm{Iso}(Z) := \zeta^{-1}\left(\prod_{i=1}^{r} \mathrm{Iso}(Z_i)\right).$$

Moreover, we introduce the following invariants:

- the *analytic order* of Z,

$$\text{ord}^{\text{an}}(Z) := \min\bigl\{d \in \mathbb{Z} \bigm| H^0\bigl(\mathcal{J}_{Y/\mathbb{P}^2}(d)\bigr) \neq 0 \text{ for all } Y \in \text{Iso}(Z)\bigr\},$$

- the *analytic order of regularity* of Z,

$$\text{reg}^{\text{an}}(Z) := \min\bigl\{d \in \mathbb{Z} \bigm| H^1\bigl(\mathcal{J}_{Y/\mathbb{P}^2}(d)\bigr) = 0 \text{ for some } Y \in \text{Iso}(Z)\bigr\}.$$

Note that $\text{ord}^{\text{an}}(Z) = \text{ord}(Y)$, $\text{reg}^{\text{an}}(Z) = \text{reg}(Y)$ for a generic $Y \in \text{Iso}(Z)$.

Example 3.6.0.1 If Z is a reduced zero-dimensional scheme, or, more generally, if the generic element of I_Z is non-singular at each point of the support of Z, then $\text{reg}^{\text{an}}(Z)$ equals the smallest integer d satisfying

$$\frac{(d+1)(d+2)}{2} \geq \deg(Z)$$

(cf. Remark 3.4.12.1 and Fig. 3.5), that is,

$$\text{reg}^{\text{an}}(Z) = \left\lceil \frac{\sqrt{1+8\deg(Z)}-3}{2} \right\rceil.$$

In the following, we shall estimate the analytic order, respectively order of regularity, of Z, using the above properties of the Castelnuovo function of Z. Combining [Shu13, Propositions 8 and 10], we obtain the following statement (it can be also deduced from the results of [BGM]).

Proposition 3.6.1 *Let $Z \subset \mathbb{P}^2$ be any zero-dimensional scheme and Z_{cl} the maximal cluster scheme in Z. Moreover, let*

$$M_2(Z) := \sum_{q \in \mathcal{T}(Z)} \text{mt}(Z_{\text{cl}}, q)^2,$$

where $\mathcal{T}(Z) = \mathcal{T}(Z_{\text{cl}})$ denotes the constellation of the cluster defined by Z (Definition 1.1.25). Then

(a) $\dfrac{\deg(Z)}{\sqrt{2M_2(Z)}} \leq \text{ord}^{\text{an}}(Z) \leq \sqrt{2\deg(Z)}$,

(b) $\text{reg}^{\text{an}}(Z) \leq \dfrac{\deg(Z)}{\text{ord}^{\text{an}}(Z)} + \text{ord}^{\text{an}}(Z) - 2$.

Proof Throughout this proof, let $Y \in \text{Iso}(Z)$ be generic, in particular, Y has the same degree as Z, $M_2(Y) = M_2(Z)$, and we have $\text{ord}(Y) = \text{ord}^{\text{an}}(Z)$, $\text{reg}(Y) = \text{reg}^{\text{an}}(Z)$.

(a) The upper bound for $\text{ord}(Y)$ can be easily read off from the graph of the Castelnuovo function $\mathcal{C}_Y$. Hence, it is only left to prove the lower bound for $d := \text{ord}(Y)$.

Assume for the moment that $\alpha \geq 1$ satisfies the following property: if $C_1, \ldots, C_s$ denote the reduced, irreducible components of a generic curve $C \in \bigl|H^0\bigl(\mathcal{J}_{Y/\mathbb{P}^2}(d)\bigr)\bigr|$

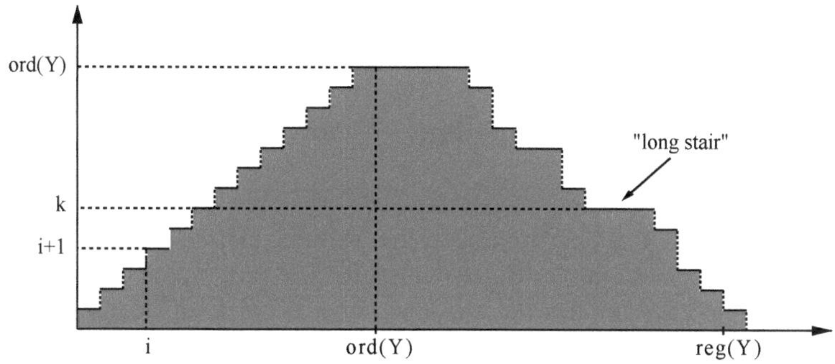

Fig. 3.5 The graph of the Castelnuovo function $\mathcal{C}_Y$. The area of the shaded region is $\deg(Y)$

(which appear with multiplicities $\ell_1, \ldots, \ell_s$, respectively, in the irreducible decomposition of C) then

$$\deg(C_i)^2 \geq \frac{1}{\alpha} \cdot \sum_{q \in \mathcal{T}(Y)} \mathrm{mt}(C_i, q)^2, \quad i = 1, \ldots, s. \tag{3.6.0.10}$$

Let's estimate the degree of the intersections scheme $Y \cap C$ under these assumptions. Certainly, it is at most the (total) intersection number of C with a generic element $D \in \left|H^0\big(\mathcal{J}_{Y/\mathbb{P}^2}(d')\big)\right|$, $d' \gg d$, that is,

$$\begin{aligned}\deg(Y \cap C) &\leq \sum_{z \in \mathrm{supp}(Y)} i_z(D, C) = \sum_{q \in \mathcal{T}(Y)} \mathrm{mt}(Y, q) \cdot \mathrm{mt}(C, q) \\ &\leq \sqrt{M_2(Y)} \cdot \sqrt{\sum_{q \in \mathcal{T}(Y)} \Big(\sum_{i=1}^{s} \ell_i \cdot \mathrm{mt}(C_i, q)\Big)^2},\end{aligned}$$

the latter due to the Cauchy inequality. Let $\boldsymbol{m}_i := \big(\mathrm{mt}(C_i, q)\big)_{q \in \mathcal{T}(Y)}$, then the triangular inequality implies $\| \sum_i \ell_i \boldsymbol{m}_i \| \leq \sum_i \ell_i \|\boldsymbol{m}_i\|$, that is,

$$\begin{aligned}\sqrt{\sum_{q \in \mathcal{T}(Y)} \Big(\sum_{i=1}^{s} \ell_i \,\mathrm{mt}(C_i, q)\Big)^2} &\leq \sum_{i=1}^{s} \ell_i \cdot \sqrt{\sum_{q \in \mathcal{T}(Y)} \mathrm{mt}(C_i, q)^2} \\ &\leq \sum_{i=1}^{s} \ell_i \cdot \sqrt{\alpha} \cdot \deg(C_i) = \sqrt{\alpha} \cdot \deg(C),\end{aligned}$$

the latter inequality due to the definition of α. Hence,

$$\deg(Y \cap C) \leq \sqrt{\alpha M_2(Y)} \cdot \mathrm{ord}(Y),$$

and the statement follows from Proposition 3.6.5, below, which shows that $\alpha = 2$ satisfies the above property (3.6.0.10).

(b) If Y is non-decomposable, that is, the graph of $\mathcal{C}_Y$ has no "long stairs", then the basic properties of the Castelnuovo function (cf. Lemma 3.5.2) imply immediately

$$\operatorname{ord}(Y) \cdot \big(\operatorname{reg}(Y)+1\big) \leq \deg(Y) + \operatorname{ord}(Y) \cdot \big(\operatorname{ord}(Y)-1\big)\,,$$

hence the statement.

Now, assume that Y is decomposable, that is, there exists some positive integer d_0 such that $\operatorname{ord}(Y) > \mathcal{C}_Y(d_0) = \mathcal{C}_Y(d_0+1) > 0$.

Then Lemma 3.5.3 implies the existence of a curve C of degree $k = \mathcal{C}_Y(d_0)$ satisfying $\mathcal{C}_{Y\cap C}(d) = \min\big\{k,\, \mathcal{C}_Y(d)\big\}$. In particular, if d_0 was chosen to be maximal with the above property then the graph of $Y \cap C$ is non-decomposable. Since $\operatorname{ord}(Y \cap C) = k$ and $\operatorname{reg}(Y \cap C) = \operatorname{reg}(Y)$, we obtain, as before,

$$k \cdot \big(\operatorname{reg}(Y)+1\big) \leq \deg(Y \cap C) + k \cdot (k-1)\,. \tag{3.6.0.11}$$

If $k \leq \sqrt{\deg(Y \cap C)}$ then we can easily deduce that

$$\operatorname{reg}(Y) \;\leq\; \frac{\deg(Y\cap C)}{k} + k - 2 \;\leq\; \frac{\deg(Y\cap C)}{\operatorname{ord}(Y)} + \operatorname{ord}(Y) - 2\,,$$

implying the statement as $\deg(Y \cap C) \leq \deg(Y)$.

Finally, let us assume $k > \sqrt{\deg(Y \cap C)}$. Then the inequality (3.6.0.11) implies $\operatorname{reg}(Y) < 2k-2$, that is, $k > (\operatorname{reg}(Y)/2)+1$. In particular,

$$\begin{aligned}\deg(Y\cap C) &\geq \frac{k(k+1)}{2} + \frac{(\operatorname{reg}(Y)-k+1)(\operatorname{reg}(Y)-k+2)}{2}\\ &> \frac{(\operatorname{reg}(Y)+2)(\operatorname{reg}(Y)+4)}{8} + \frac{\operatorname{reg}(Y)(\operatorname{reg}(Y)+2)}{8}\\ &= \frac{(\operatorname{reg}(Y)+2)^2}{4}\,.\end{aligned} \tag{3.6.0.12}$$

Hence, $\operatorname{reg}(Y) < 2\sqrt{\deg(Y\cap C)} - 2 \leq 2\sqrt{\deg(Y)} - 2$, which is the minimal possible value for the right-hand side of (b). □

Corollary 3.6.2 *With the notations of Proposition 3.6.1,*

$$\operatorname{reg}^{\mathrm{an}}(Z) \leq \frac{\deg(Z)}{\sqrt{2M_2(Z)}} + \sqrt{2M_2(Z)} - 2\,.$$

Proof If $\operatorname{ord}^{\mathrm{an}}(Z) \leq \sqrt{\deg(Z)}$ the statement is an immediate consequence of Proposition 3.6.1(a) and (b). Hence, assume the opposite inequality to hold.

Let $Y \in \operatorname{Iso}(Z)$ be generic. As shown in the proof of Proposition 3.6.1(b), in any case there is a curve C such that $Y \cap C$ is non-decomposable and satisfies $\operatorname{reg}(Y \cap C) = \operatorname{reg}(Y)$. Moreover, (3.6.0.12) applies, and we obtain

$$\begin{aligned}\mathrm{reg}(Y) &< 2\sqrt{\deg(Y)} - 2 = 2 \cdot \sqrt{\frac{\deg(Y)}{\sqrt{2M_2(Z)}} \cdot \sqrt{2M_2(Z)}} - 2 \\ &\leq \frac{\deg(Y)}{\sqrt{2M_2(Z)}} + \sqrt{2M_2(Z)} - 2\,.\end{aligned}$$

□

Using a more thorough reasoning we can actually improve this bound.

It turns out that some easy reduction steps (cf. Lemma 3.4.2) enable us to weaken the definition of α in the proof of Proposition 3.6.1(a), by demanding the following property to hold: let $Y \in \mathrm{Iso}(Z)$ be a generic element, and let $C_i, i = 1, \ldots, s$, denote the reduced, irreducible components of a generic curve $C \in \left|H^0\big(\mathcal{J}_{Y/\mathbb{P}^2}(d)\big)\right|$ (with $d \geq \mathrm{ord}(Y)$ any integer). Then *either C_i is a straight line containing exactly two (infinitely near) points of $\mathscr{T}(Y_{\mathrm{cl}})$, or*

$$\deg(C_i)^2 \geq \frac{1}{\alpha} \cdot \sum_{q \in \mathscr{T}(Y)} \mathrm{mt}(C_i, q)^2\,, \quad i = 1, \ldots, s\,. \tag{3.6.0.13}$$

Proposition 3.6.3 *Let $Z \subset \mathbb{P}^2$ be any zero-dimensional scheme, $M_2(Z)$ as in Proposition 3.6.1, and $\alpha \geq 1$ satisfying property (3.6.0.13). Then either*

$$\mathrm{reg}^{\mathrm{an}}(Z) \leq \frac{\deg(Z)}{\sqrt{\alpha M_2(Z)}} + \sqrt{\alpha M_2(Z)} - 2\,,$$

or Z is a (generic) fat point scheme supported at exactly two points with multiplicities $\boldsymbol{m} = (1, 1)$, respectively $\boldsymbol{m} = (2, 2)$, or Z is the singularity scheme of an A_3-singularity. In the latter cases we have

$$\mathrm{reg}^{\mathrm{an}}\big(Z(1,1)\big) = 1\,, \quad \mathrm{reg}^{\mathrm{an}}\big(Z(2,2)\big) = \mathrm{reg}^{\mathrm{an}}\big(Z^s(A_3)\big) = 3\,.$$

Corollary 3.6.4 (cf. [Shu13], Proposition 10) *With the notations of Proposition 3.6.3,*

$$\mathrm{reg}^{\mathrm{an}}(Z) \leq \frac{\deg(Z)}{\sqrt{\frac{4}{3}M_2(Z)}} + \sqrt{\frac{4}{3}M_2(Z)} - 2\,,$$

unless $Z = Z(1,1)$.

Proof This follows immediately from Proposition 3.6.3 in view of the possible choice $\alpha = 4/3$ (Proposition 3.6.5, below). □

Proof of Proposition 3.6.3. Let $Y \in \mathrm{Iso}(Z)$ be generic. The proof of Corollary 3.6.2 shows that it would be sufficient to have

$$\mathrm{ord}(Y) \geq \frac{\deg(Y)}{\sqrt{\alpha M_2(Y)}}\,.$$

Applying the method used in the proof of Proposition 3.6.1(a), we obtain this inequality, unless there exists a line L which contains exactly two points q_1, q_2 of $\mathcal{T}(Y)$ and satisfies

$$\deg(Y \cap L) = \mathrm{mt}(Y, q_1) + \mathrm{mt}(Y, q_2) > \sqrt{\alpha M_2(Y)}\,. \qquad (3.6.0.14)$$

In particular this allows us to proceed by induction on $\deg(Y)$ (the base being the case that no such line exists). Let

$$d := \left\lfloor \sqrt{\alpha M_2(Y)} + \frac{\deg Y}{\sqrt{\alpha M_2(Y)}} - 2 \right\rfloor,$$

and consider the exact (reduction) sequence

$$H^1\big(\mathcal{J}_{Y:L/\mathbb{P}^2}(d-1)\big) \longrightarrow H^1\big(\mathcal{J}_{Y/\mathbb{P}^2}(d)\big) \longrightarrow H^1\big(\mathcal{J}_{Y\cap L/L}(d)\big)\,.$$

Setting $m_1 := \mathrm{mt}(Y, q_1)$ and $m_2 := \mathrm{mt}(Y, q_2)$, it suffices to show (cf. Lemma 3.4.2)

(a) $m_1 + m_2 \leq d + 1$,
(b) $H^1\big(\mathcal{J}_{Y:L/\mathbb{P}^2}(d-1)\big) = 0$, that is, $\mathrm{reg}(Y : L) \leq d - 1$.

Indeed, (a) holds true, since

$$\begin{aligned}\sqrt{\alpha M_2(Y)} + \frac{\deg Y}{\sqrt{\alpha M_2(Y)}} - 2 &\geq \sqrt{\alpha M_2(Y)} + \frac{m_1^2 + m_1 + m_2^2 + m_2}{2\sqrt{\alpha M_2(Y)}} - 2 \\ &\geq \sqrt{\alpha(m_1^2 + m_2^2)} + \frac{m_1^2 + m_1 + m_2^2 + m_2}{2\sqrt{\alpha(m_1^2 + m_2^2)}} - 2 \geq m_1 + m_2 - 1\,,\end{aligned}$$

where the latter inequality holds, because fixing $m := m_1 + m_2$ the left-hand side takes its minimum for $m_1 = m_2 = m/2$, and we obtain an inequality

$$\frac{2\alpha + 1}{2\sqrt{2\alpha}} \cdot m + \frac{1}{\sqrt{2\alpha}} - 2 \geq m - 1\,,$$

which is satisfied as $m \geq 5$. Moreover, if $m \leq 4$, the above inequality (3.6.0.14) restricts ourselves to finitely many cases:

(m_1, m_2)	$M_2 = M_2(Y)$	$(\mathrm{mt}(Y, q))_{q\in\mathcal{T}(Y)}$	$d+1$
(2, 2)	8	(2, 2)	$\to$ exceptional case
(2, 2)	$\in \{9, \dots, 15\}$	$(2, 2, 1, \dots, 1)$	$\geq \left\lfloor 2\sqrt{M_2} - \frac{2}{\sqrt{M_2}} \right\rfloor - 1 \geq 4$
(2, 2)	$\in \{12, \dots, 15\}$	$(2, 2, 2, 1, \dots, 1)$	$\geq \left\lfloor 2\sqrt{M_2} - \frac{3}{\sqrt{M_2}} \right\rfloor - 1 > 4$
(3, 1)	$\in \{10, \dots, 15\}$	$(3, 1, \dots, 1)$	$\geq \left\lfloor 2\sqrt{M_2} - \frac{3}{\sqrt{M_2}} \right\rfloor - 1 \geq 4$
(2, 1)	$\in \{5, \dots, 8\}$	$(2, 1, \dots, 1)$	$\geq \left\lfloor 2\sqrt{M_2} - \frac{1}{\sqrt{M_2}} \right\rfloor - 1 \geq 3$
(1, 1)	2	(1, 1)	$\to$ exceptional case
(1, 1)	3	(1, 1, 1)	$\geq \left\lfloor 2\sqrt{3} - \frac{2}{\sqrt{3}} \right\rfloor - 1 = 2$

Note that if $Y \neq Y_{\mathrm{cl}}$ in the first exceptional case then the needed estimate holds true:

$$d + 1 \geq \left\lfloor 2\sqrt{8} - \tfrac{1}{\sqrt{8}} \right\rfloor - 1 \geq 4\,.$$

Thus, the only 3 exceptions are given by the schemes $Z(1, 1)$, $Z(2, 2)$, $Z^s(A_3)$.

To obtain (b), we may apply the induction hypothesis to the residue scheme $Y : L$. Recall that $\deg(Y : L) = \deg Y - m_1 - m_2$. Hence, either $Y : L$ is one of the three exceptional cases ($Z(1, 1)$, $Z(2, 2)$, $Z^s(A_3)$), or

$$\begin{aligned} \mathrm{reg}(Y : L) &\leq \sqrt{\alpha M_2(Y : L)} + \frac{\deg(Y : L)}{\sqrt{\alpha M_2(Y : L)}} - 2 \\ &\leq \sqrt{\alpha M_2(Y)} + \frac{\deg Y - m_1 - m_2}{\sqrt{\alpha M_2(Y)}} - 2\,, \end{aligned}$$

which does not exceed $d - 1$, due to (3.6.0.14).

It remains to analyse those Y which lead to one of the exceptional cases. Since $d + 1 \geq \deg(Y \cap L)$, the wanted inequality $\mathrm{reg}(Y : L) \leq d - 1$ is satisfied if $\deg(Y \cap L) - 2 \geq 1 = \mathrm{reg}^{\mathrm{an}}(Z(1, 1))$, respectively $\deg(Y \cap L) - 2 \geq 3$ in the other two cases. It follows that it is only left to consider the following two types of zero-dimensional schemes Y:

- if $Y = Z(1, 1, 1, 1)$ then $d - 1 \geq \left\lfloor 2\sqrt{4} \right\rfloor - 3 = 1$;
- if $Y = Z^s(A_2) \cup Z^s(A_2)$ then $d - 1 \geq \left\lfloor 2\sqrt{12} - \frac{2}{\sqrt{12}} \right\rfloor - 3 = 3$. □

Proposition 3.6.5 *Let Z be a zero-dimensional scheme and Y a generic element of* Iso(Z). *Moreover, assume that a generic curve $C \in \left|H^0\big(\mathcal{J}_{Y/\mathbb{P}^2}(d)\big)\right|$ decomposes into reduced, irreducible components $C_1, \dots, C_s$ of multiplicities $\ell_1, \dots, \ell_s$, respectively. Then*

$$\deg(C_i)^2 \geq \sum_{q\in\mathcal{T}(Y)} \frac{\mathrm{mt}(C_i, q)^2}{2}\,,$$

where $\mathscr{T}(Y) = \mathscr{T}(Y_{\mathrm{cl}})$ denotes the constellation of the cluster defined by Y.

Moreover, if C_i is not a straight line containing exactly two (infinitely near) points of $\mathscr{T}(Y)$ then

$$\deg(C_i)^2 \geq \sum_{q \in \mathscr{T}(Y)} \frac{3\,\mathrm{mt}(C_i, q)^2}{4}.$$

Proof Let $1 \leq i \leq s$ be fixed, and set $d_i := \deg(C_i)$, $\mathscr{T}_i := \mathscr{T}(Y) \cap C_i$.

Step 1. Assume that $Y \cap C_i$ is *irreducible*, that is, $\mathrm{supp}(Y \cap C_i) = \{z\}$ for some $z \in \mathbb{P}^2$.

Choose affine coordinates x, y around $z = (0,0)$ such that at z the axes are transverse to C_i as well as to a generic curve $D \in \left|H^0\big(\mathcal{J}_{Y/\mathbb{P}^2}(d')\big)\right|$, $d' \gg d$. We consider the sequence of algebraic automorphisms $\psi_m \in \mathrm{Aut}(\mathbb{C}^2)$,

$$\psi_m : (x, y) \longmapsto (x, y + tx^m), \quad m \geq 0, \quad t \in \mathbb{C}^2 \setminus \{0\} \text{ generic},$$

and the zero-dimensional schemes $\psi_m(Y) \in \mathrm{Iso}(Y) = \mathrm{Iso}(Z)$.

Note that the constellations $\mathscr{T}(Y)$ and $\mathscr{T}(\psi_m(Y))$ coincide up to points of level m, while the free points of level $m+1$ move with t. In particular, for $m \gg 0$ we obtain $\mathscr{T}(Y) = \mathscr{T}(\psi_m(Y))$. We introduce

$$k := \max\left\{m \geq 0 \,\middle|\, \mathscr{T}_i \not\subset \mathscr{T}(\psi_m(Y))\right\},$$

which geometrically indicates the highest level of a free point in $\mathscr{T}_i$.

By our (genericity) assumption, a generic curve $C' \in \left|H^0\big(\mathcal{J}_{\psi_k(Y)/\mathbb{P}^2}(d)\big)\right|$ decomposes as $C' = C_1' \cup \cdots \cup C_s'$, each reduced, irreducible component C_j' satisfying $\deg(C_j') = \deg(C_j)$ and appearing with multiplicity ℓ_j, $j = 1, \ldots, s$. Moreover, $C_i' \cap \mathscr{T}(\psi_k(Y)) = \psi_k(\mathscr{T}_i)$, in particular, $C_i' \neq C_i$, and we can conclude

$$d_i^2 \geq \sum_{q \in \mathscr{T}_i \cap \mathscr{T}(\psi_k(Y))} \mathrm{mt}(C_i, q)^2. \tag{3.6.0.15}$$

On the other hand, if

$$h(q) := \min\left\{m \geq 0 \,\middle|\, q \in \mathscr{T}(\psi_m(Y))\right\},$$

then we shall show below (in Step 3)

$$\mathrm{mt}(C_i, z)^2 \geq \sum_{\substack{q \in \mathscr{T}_i \\ h(q)=2}} \mathrm{mt}(C_i, q)^2 \geq \cdots \geq \sum_{\substack{q \in \mathscr{T}_i \\ h(q)=k}} \mathrm{mt}(C_i, q)^2 \geq \sum_{\substack{q \in \mathscr{T}_i \\ q \notin \mathscr{T}(\psi_k(Y))}} \mathrm{mt}(C_i, q)^2.$$

The latter, together with (3.6.0.15), implies

$$d_i^2 \;\ge\; \frac{k}{k+1} \cdot \sum_{q \in \mathscr{T}_i} \mathrm{mt}(C_i,q)^2\,,$$

hence, the statement holds if $k \ge 3$. It remains to analyse those cases with $k \le 2$.

Case 1A. $\mathscr{T}_i = \{z\}$.

In this case, we obviously have even $d_i^2 \ge \mathrm{mt}(C_i,z)^2 = \sum_{q\in\mathscr{T}_i} \mathrm{mt}(C_i,q)^2$.

Case 1B. C_i contains exactly two (infinitely near) points z, q of $\mathscr{T}(Y)$.

If C_i is non-singular at z, the claimed inequality is satisfied unless C_i is a straight line ($d_i = 1$). Hence, assume that $\mathrm{mt}(C_i,z) \ge 2$. Then Gudkov-Shustin theorem [GuS] (see a little bit more precise statement in [GLS6, Proposition II.2.19]) implies

$$\begin{aligned} d_i^2 \ge \sum_{w\in U(z)} i_w\big(C_i, \psi_1(C_i)\big) \;&=\; \mathrm{mt}(C_i,z)^2 + \mathrm{mt}(C_i,q)\big(\mathrm{mt}(C_i,q)-1\big) \\ \ge \frac{3}{4}\big(\mathrm{mt}(C_i,z)^2 + \mathrm{mt}(C_i,q)^2\big)\,. & \end{aligned}$$

Case 1C. C_i contains at least three (infinitely near) points of $\mathscr{T}(Y)$.

Then either $\mathscr{T}_i = \{z, q_1, q_2\}$ with all appearing multiplicities equal to 1 (and the genericity assumption implies $d_i \ge 2$), or C_i is singular at z. Assume the latter to be given, and let $f(x,y) = 0$ be an affine equation for C_i. Then $f(x, y + tx^m) = 0$ is an affine equation for $\psi_m(C_i)$, and we obtain

$$\begin{aligned} \sum_{q\in\mathscr{T}_i} \mathrm{mt}(C_i,q)^2 \le i_z\big(C_i, \psi_{k+1}(C_i)\big) \;&\le\; i_0\big(f, x^{k+1}\tfrac{\partial f}{\partial y}\big) \\ &= (k+1)\cdot i_0(f,x) + i_0\big(f, \tfrac{\partial f}{\partial y}\big)\,. \end{aligned}$$

On the other hand, Gudkov-Shustin theorem [GuS] (cf. [GLS6, Proposition II.2.19]) gives

$$d_i^2 \;\ge\; \sum_{w\in U(z)} i_w\big(C_i, \psi_1(C_i)\big) \;\ge\; \mathrm{mt}(f) + i_0\big(f, \tfrac{\partial f}{\partial y}\big)\,.$$

Combining these two inequalities and recalling that, by our choice of coordinates, $i_0(f,x) = \mathrm{mt}(C_i,z)$, we obtain

$$d_i^2 \;\ge\; \sum_{q\in\mathscr{T}_i} \mathrm{mt}(C_i,q)^2 - k\cdot \mathrm{mt}(C_i,z)\,.$$

If $k = 1$, we obtain the wanted inequality unless

$$3\,\mathrm{mt}(C_i,z) \ge \frac{3}{4} \sum_{q\in\mathscr{T}_i} \mathrm{mt}(C_i,q)^2 > d_i^2\,.$$

But, the latter is impossible, since $d_i > \mathrm{mt}(C_i, z) \geq 2$.

Now, assume that $k = 2$. As before, the wanted inequality holds unless

$$6\,\mathrm{mt}(C_i, z) \geq \frac{3}{4} \sum_{q \in \mathscr{T}_i} \mathrm{mt}(C_i, q)^2 > d_i^2\,, \qquad (3.6.0.16)$$

which implies $\mathrm{mt}(C_i, z) \in \{2, 3\}$ and $d_i = \mathrm{mt}(C_i, z) + 1$. Due to Bézout's theorem, the latter is only possible if $\mathrm{mt}(C_i, q) = 1$ for all $q \in \mathscr{T}_i \setminus \{z\}$. Moreover, the assumption $k = 2$ implies that all infinitely near points in $\mathscr{T}_i$ of level ≥ 3 are proximate to z, hence the number of points q with $\mathrm{mt}(C_i, q) = 1$ is bounded by 3, respectively 6. In any case, this contradicts (3.6.0.16).

Step 2. Let $Y \cap C_i$ be *reducible*, that is, $\mathrm{supp}(Y \cap C_i) = \{z_1, \ldots, z_r\}$, $r \geq 2$.

Again, we can apply the reasoning of Step 1, choosing an affine plane $\mathbb{C}^2 \subset \mathbb{P}^2$ containing $\{z_1, \ldots, z_r\}$, and, for each $j = 1, \ldots, r$, affine coordinates x_j, y_j around $z_j = (0, 0)$. Then we consider automorphisms $\psi_{j,m}$ defined as before and the schemes $(Y \setminus Y_j) \cup \psi_{j,m}(Y_j) \in \mathrm{Iso}(Y) = \mathrm{Iso}(Z)$ (Y_j denoting the irreducible component of Y with support $\{z_j\}$). Defining

$$k_j := \max\left\{m \geq 0 \,\middle|\, C_i \cap \mathscr{T}(Y_j) \not\subset \mathscr{T}\big(\psi_{j,m}(Y)\big)\right\},$$

and argueing as before, we obtain the wanted inequality as soon as

$$k + 1 := \sum_{j=1}^{r} (k_j + 1) \geq 4\,.$$

Hence, it is only left to analyse the following cases:

Case 2A. $r = 3$, $k_1 = k_2 = k_3 = 0$.

In this case, all points of $\mathscr{T}_i = C_i \cap \mathscr{T}(Y)$ are of level 0. Let $m_\nu := \mathrm{mt}(C_i, z_\nu)$, $\nu = 1, 2, 3$, and assume that $m_1 \geq m_2 \geq m_3$. Then Lemma 3.2.7 gives

$$d_i^2 \geq m_1^2 + m_2^2 + m_3^2 - m_3\,,$$

which implies the wanted formula unless $m_1 = m_2 = m_3 = 1$. But then the genericity assumption gives $d_i \geq 2$, and we have finished.

Case 2B. $r = 2$, $k_1 = k_2 = 0$.

Here, we can simply copy the reasoning of Step 2A, the only exception being the case of a straight line C_i through z_1, z_2.

Case 2C. $r = 2$, $k_1 = 1$, $k_2 = 0$.

In this case, Gudkov-Shustin theorem [GuS] (cf. [GLS6, Propositions II.2.18 and II.2.19]) gives the estimate (cf. Case 1C)

$$d_i^2 \geq \sum_{\nu=1}^{2} \sum_{w \in U(z_\nu)} i_w\big(C_i, \psi_{1,1}(C_i)\big) \geq \sum_{q \in \mathscr{T}(Y)} \mathrm{mt}(C_i, q)^2 - \sum_{\nu=1}^{2} \underbrace{\mathrm{mt}(C_i, z_\nu)}_{=: m_\nu},$$

and the wanted inequality holds unless

$$3 \cdot (m_1 + m_2) \geq \frac{3}{4} \sum_{q \in \mathscr{T}(Y)} \mathrm{mt}(C_i, q)^2 > d_i^2. \tag{3.6.0.17}$$

By our (genericity) assumption, C_i cannot be a straight line. Hence, Bézout's theorem gives $d_i \geq m_1 + m_2$, and (3.6.0.17) implies $d_i = 2$ and $m_1 = m_2 = 1$. But then all appearing multiplicities would be 1, and there are exactly 3 of them (there are no free points of level $\geq k_\nu$), which contradicts the second inequality in (3.6.0.17).
Step 3. It remains to prove the inequalities

$$\sum_{\substack{q \in \mathscr{T}_i \\ h(q)=m}} \mathrm{mt}(C_i, q)^2 \geq \sum_{\substack{q \in \mathscr{T}_i \\ h(q)=m+1}} \mathrm{mt}(C_i, q)^2, \qquad m \geq 1 \tag{3.6.0.18}$$

claimed in Step 1. To do so, we use a reasoning based on Puiseux expansions. Let $f(x, y) = 0$ be a (local) equation for C_i, and let $g \in I_Y \subset \mathbb{C}\{x, y\}$ be a generic element,

$$f(x, y) = \big(1 + O(x, y)\big) \cdot \prod_{\nu=1}^{\mathrm{mt}(f)} \big(y - \xi_\nu(x)\big), \quad \xi_\nu \in \mathbb{C}\{x^{1/\mathrm{mt}(f)}\},$$

$$g(x, y) = \big(1 + O(x, y)\big) \cdot \prod_{\nu=1}^{\mathrm{mt}(Y)} \big(y - \eta_\nu(x)\big), \quad \eta_\nu \in \mathbb{C}\{x^{1/\mathrm{mt}(Y)}\}.$$

Then, for any $\nu = 1, \ldots, \mathrm{mt}(f)$, there exists some $\rho(\nu)$ satisfying

$$\mathrm{ord}(\xi_\nu - \eta_{\rho(\nu)}) = \max_j \mathrm{ord}(\xi_\nu - \eta_j),$$

and we can compute

$$\begin{aligned} \sum_{q \in \mathscr{T}_i} \mathrm{mt}(C_i, q)^2 &= \sum_{\nu=1}^{\mathrm{mt}(f)} \sum_{\mu=1}^{\mathrm{mt}(f)} \min\big\{\mathrm{ord}(\xi_\mu - \xi_\nu),\ \mathrm{ord}(\xi_\mu - \eta_{\rho(\nu)})\big\} \\ &= \sum_{\nu=1}^{\mathrm{mt}(f)} \sum_{\mu=1}^{\mathrm{mt}(f)} \mathrm{ord}\left(\xi_\mu - \frac{\xi_\nu + t_1 \eta_{\rho(\nu)}}{1 + t_1}\right), \end{aligned}$$

for a generic $t_1 \in \mathbb{C} \setminus \{0\}$. Similarly,

$$\sum_{\substack{q\in\mathcal{T}_i\\ h(q)\le m}} \mathrm{mt}(C_i,q)^2 = \sum_{\nu=1}^{\mathrm{mt}(f)}\sum_{\mu=1}^{\mathrm{mt}(f)} \mathrm{ord}\left(\xi_\mu - \frac{\xi_\nu + t_1\eta_{\rho(\nu)}}{1+t_1} - t\cdot x^m\right),$$

for generic numbers t, t_1. In view of the latter relation, the following series of simple inequalities yields (3.6.0.18):

$$\mathrm{ord}\left(\xi_\mu - \frac{\xi_\nu + t_1\eta_{\rho(\nu)}}{1+t_1} - t\cdot x^{m+1}\right) - \mathrm{ord}\left(\xi_\mu - \frac{\xi_\nu + t_1\eta_{\rho(\nu)}}{1+t_1} - t\cdot x^m\right)$$
$$\le \mathrm{ord}\left(\xi_\mu - \frac{\xi_\nu + t_1\eta_{\rho(\nu)}}{1+t_1} - t\cdot x^m\right) - \mathrm{ord}\left(\xi_\mu - \frac{\xi_\nu + t_1\eta_{\rho(\nu)}}{1+t_1} - t\cdot x^{m-1}\right),$$

$\nu, \mu = 1, \dots, \mathrm{mt}(f)$. □

3.7 Historical Notes and References

Though the H^1-statements discussed in this chapter are properly acknowledged with references, we find it useful to indicate the origin of the ideas behind these statements, especially in relation to the geometry of families of curves.

The Riemann-Roch type H^1-vanishing statements combine the (Serre) duality and verification that the dual sheaf resp. line bundle has a negative degree. In the classical Riemann-Roch theorem [Rie, Roch] applied to a smooth irreducible curve C it reads

$$H^1(C, \mathcal{O}_C(D)) = 0 \quad \text{as long as} \quad \deg D > 2g(C) - 2\,.$$

The idea to use such a statement in the context of geometry of equisingular families of curves is due to Severi [Sev], who, in fact, noticed that, in the case of nodal plane curves, the relevant H^1-vanishing condition can be lifted to the normalization of the given curve. Moreover, Severi applied a version of the Noether's $AF + BG$ theorem (see Sect. 2.1.3.2) in order to split the H^1-vanishing condition for reducible curve to H^1-vanishing for individual components. Segre [Seg1, Seg2] observed that, for plane curves with nodes and ordinary cusps, the H^1-vanishing condition responsible for the smoothness and expected dimension of the equisingular family could also be naturally lifted to the normalization. That property under the name of the "completeness of the characteristic linear series" was discussed from the modern point of view in [Gre1, Tan1, Tan2, Tan3, Wah2, Zar3]. Using the Serre duality for reduced curves with arbitrary planar singularities, Greuel and Karras [GrK] deduced a Riemann-Roch type vanishing in the most general setting by introducing and using the "isomorphism defect". This led to the $3d$-condition for arbitray singularities (discovered by Segre for nodes and cusps). Another generalization to curves with arbitrary singularities was suggested by Shustin in [Shu4], where the original approach by Severi and Segre was supplied with the idea to replace the local ideals at singular points that are more complicated than a cusp, which designate the tangent spaces to local equisingular strata, by certain subideals. In fact, the ideals of the minimal cluster schemes containing the singularity schemes and the conductor scheme, liftable to the normalization or on its irreducible components. At last, the idea to use as the base of the H^1-vanishing condition not the original curve or its normalization but another curve appeared, first, in [Shu1], where the original curve was planar and the base curve was a generic polar, and it led to the so-called $(4d-4)$-condition (in contrast to the $3d$-condition). This was then strengthened and generalized in [GrL1]. Other attempts to use auxiliary base curves different from the original one were undertaken in [Shu6, Shu8].

Concerning the Kodaira and Kawamata-Viewig vanishing theorems [Kod, Kaw, Vie], we only mention that it was Xu [Xu2] who related this stuff with the geometry of families of curves.

The idea to use the Bogomolov-Reider theory of unstable rank two bundles on surfaces for the study of equisingular families of curves is due to Chiantini and Sernesi [ChS], who derived a sufficient smoothness and expected dimension criterion for families of nodal curves on surfaces of general type. In [GLS1] this approach was extended to families of curves with arbitrary singularities.

Hirschowitz [HiA1, HiA2] initiated a new procedure for finding H^1-vanishing criteria for ideal sheaves of zero-dimensional schemes, named after the story *"Horace"* written by Corneille.

> *The three Horace brothers fight against the three Curiace brothers to decide which of their countries will dominate the other. At the very beginning, two of the Horace brothers were killed while the Curiace are wounded, say at 20, 50 and 80 %. Then the surviving Horace runs away. This is the Horace method: he then waits for the Curiace brothers who arrive and are killed one after the other. Of course, you have to wait for them along a suitable divisor.*
> (A. Hirschowitz)

In fact, the works [Shu1, GrL1, Shu6, Shu8] mentioned above in connection with the Riemann-Roch type H^1-vanishing can be regarded as specific incarnations of the Horace method. A direct application of the Horace method to the geometry of equisingular families of projective hypersurfaces was performed in [ShT2] (a $(4d-4)$-condition in higher dimensions). However, the most valuable results are obtained by applying the Horace method and its differential version suggested by Alexander and Hirschowitz [AH1, AH2] to the Harbourne-Hirschowitz problem on the H^1-vanishing of the ideal sheaves of zero-dimensional schemes of fat points in general position (and more complicated schemes also in general position). This is discussed in Sects. 3.4.1 and 3.4.2.

The theory of the Castelnuovo function for zero-dimensional subschemes of the plane elaborated by Davis [Dav] was suggested by Hirschowitz to his student Barkats for a possible application to the geometry of plane curves with nodes and cusps, and Barkats successfully accomplished the task [Bar]. He noticed that the non-vanishing of H^1 in certain degree leads to the existence of "long stairs" in the graph of the Castelnuovo function, which by the Davis lemma yields the existence of a curve of a relatively small degree containing a substantial portion of the considered zero-dimensional scheme. On the other hand, the use of that curve as the base for the Horace method shows that the starting degree cannot be be too large. These observations were strengthened and applied to arbitrary zero-dimensional schemes in [GLS4].

We should like to mention also some other ideas to achieve H^1-vanishing criteria. So, the geometric linear normality was applied by Flamini and Madonna [FlM] for that purpose in order to study families of nodal curves on projective surfaces. The method developed by du Plessis and Wall [DPW, DPW1, DPW2] assumes a direct analysis of the projection of the linear system of hypersurfaces of a given degree onto the direct sum of the local Tjurina algebras for all isolated singularities of a given hypersurface aimed at the surjectivity statement (equivalent to the H^1-vanishing for the ideal sheaf of the corresponding zero-dimensional scheme). We have not included them into the monograph since they either are applied to only nodal singularities (at least in their current version), or provide H^1-vanishing criteria basically coinciding with what we obtain with other tools.

References

[Xu2] Xu, Geng: Ample line bundles on smooth surfaces. J. Reine Angew. Math. **469**, 199–209 (1995)

[HiA1] Hirschowitz, A.: La méthode d'Horace pour l'interpolation à plusieurs variables. Manuscr. Math. **50**, 337–388 (1985)

[HiA2] Hirschowitz, A.: Une conjecture pour la cohomologie des diviseurs sur les surfaces rationelles génériques. J. Reine Angew. Math. **397**, 208–213 (1989)

[Bar] Barkats, D.: *Études des variétés des courbes planes à noeuds et à cusps.* Algebraic geometry (Catania, 1993/Barcelona, 1994), Lect. Notes Pure Appl. Math., **200**, 25–35, Dekker, New York (1998)

[GrK] Greuel, G.-M., Karras, U.: Families of varieties with prescribed singularities. Compos. Math. **69**, 83–110 (1989)

[GrL1] Greuel, G.-M., Lossen, C.: Equianalytic and equisingular families of curves on surfaces. Manuscr. Math. **91**, 323–342 (1996)

[GLS6] Greuel, G.-M., Lossen, C., Shustin, E.: Introduction to Singularities and Deformations. Springer, Berlin (2007)

[HaR2] Hartshorne, R.: Algebraic Geometry. Graduate text in mathematics, vol. 52. Springer, Berlin (1977)

[BHPV] Barth, W., Hulek, K., Peters, C., van de Ven, A.: Compact Complex Surfaces. Springer, Berlin (1984)

[Kod] Kodaira, K.: On a differential-geometric method in the theory of analytic stacks. Proc. Natl. Acad. Sci. U. S. A. **39**, 1268–1273 (1953)

[Kaw] Kawamata, Y.: A generalization of Kodaira-Ramanujam's vanishing theorem. Math. Ann. **261**, 43–46 (1982)

[Vie] Viehweg, E.: Vanishing theorems. J. Reine Angew. Math. **335**, 1–8 (1982)

[ShSo] Shiffman, B., Sommese, A.: Vanishing Theorems on Complex Manifolds. Progress in Math., vol. 56. Birkhäuser (1985)

[Kol1] Kollár, J.: Vanishing theorems for cohomology groups. In: Algebraic Geometry, Bowdoin 1985. Proc. Symp. Pure Math., vol. 46, pp. 233–243 (1987)

[LaR] Lazarsfeld, R.: Lectures on linear series. In: Kollár, J. (ed.) Complex Algebraic Geometry. AMS (1997)

[EiL] Ein, L., Lazarsfeld, R.: *Seshadri constants on smooth surfaces.* In: Journées de Géométrie Algébrique d'Orsay, Juillet 1992. Astérisque, vol. 218, PP. 177–186 (1993)

[Xu1] Xu, Geng: Curves in $\mathbb{P}^2$ and symplectic packings. Math. Ann. **299**, 609–613 (1994)
[GuS] Gudkov, D.A., Shustin, E.: Topology. In: Dold, A., Eckmann, B. (eds.) On the Intersection of the Close Algebraic Curves, vol. 1060, pp. 278–289. Leningrad, SLN (1984)
[KeT] Keilen, T., Tyomkin, I.: Existence of curves with prescribed topological singularities. Trans. Am. Math. Soc. **354**(5), 1837–1860 (2002)
[Mum1] Mumford, D.: Lectures on Curves on an Algebraic Surface. Princeton University Press (1966)
[Bog] Bogomolov, F.A.: Holomorphic tensors and vector bundles on projective varieties. Math. USSR Isvestija **13**, 499–555 (1979)
[Rei] Reid, M.: Bogomolov's theorem $c_1^2 < 4c_2$. In: International Symposium, pp. 623–642. Kyoto (1977)
[Rr] Reider, I.: Vector bundles of rank 2 and linear systems on algebraic surfaces. Ann. Math. **127**, 309–316 (1988)
[ChS] Chiantini, L., Sernesi, E.: Nodal curves on surfaces of general type. Math. Ann. **307**, 41–56 (1997)
[GLS1] Greuel, G.-M., Lossen, C., Shustin, E.: New asymptotics in the geometry of equisingular families of curves. Int. Math. Res. Not. **13**, 595–611 (1997)
[Shu10] Shustin, E.: Smoothness of equisingular families of plane algebraic curves. Int. Math. Res. Not. **2**, 67–82 (1997)
[Schn] Schneider, M.: Halbstetigkeitssätze fr relativ analytische Räume. Invent. Math. **16**, 161–176 (1972)
[HaR1] Hartshorne, R.: Connectedness of the Hilbert scheme. Publ. Math. IHES **29**, 261–304 (1966)
[DaG] Davis, E.D., Geramita, A.V.: The Hilbert function of a special class of 1-dimensional Cohen-Macaulay graded algebras. In: The Curves Seminar at Queen's (Queen's Pap. Pure and Appl. Math.), vol. 67, pp. 1–29 (1984). 67
[ShT2] Shustin, E., Tyomkin, I.: Versal deformations of algebraic hypersurfaces with isolated singularities. Math. Ann. **313**(2), 297–314 (1999)
[Shu12] Shustin, E.: Lower deformations of isolated hypersurface singularities. Algebra i Analiz **10**(5), 221–249 (1999). (English translation in St. Petersburg Math. J. **11**(5), 883–908 (2000))
[Bal] Ballico, E.: Curves of minimal degree with prescribed singularities. Illinois J. Math. **43**(4), 672–676 (1999)
[DPW2] Du Plessis, A.A., Wall, C.T.C.: Singular hypersurfaces, versality and Gorenstein algebras. J. Algebr. Geom. **9**(2), 309–322 (2000)
[AH1] Alexander, J., Hirschowitz, A.: La méthode d'Horace éclatée: application à l'interpolation en degré quatre. Invent. Math. **107**, 585–602 (1992)
[AH2] Alexander, J., Hirschowitz, A.: An asymptotic vanishing theorem for generic unions of multiple points. Invent. Math. **140**(2), 303–325 (2000)
[HaB1] Harbourne, B.: Complete linear systems on rational surfaces. Trans. Am. Math. Soc. **289**, 213–226 (1985)
[HaB2] Harbourne, B.: On Nagata's conjecture. J. Algebra **236**(2), 692–702 (2001)
[Eva1] Evain, L.: *Dimension des systèmes linéares: une approche différentielle et combinatoire* (1997). arXiv:math.AG/9709032
[Eva2] Evain, L.: Computing limit linear systems with infinitesimal methods. Ann. Inst. Fourier (Grenoble) **57**(6), 1947–1974 (2007)
[CiM] Ciliberto, C., Miranda, R.: Nagata's conjecture for a square or nearly-square number of points. Ric. Math. **55**(1), 7178 (2006)
[Roe1] Roé, J.: Maximal rank for schemes of small multiplicity by Évain's differential Horace method. Trans. Am. Math. Soc. **366**, 857–874 (2014)
[Roe] Roé, J.: On the existence of plane curves with imposed multiple points. J. Pure Appl. Algebra **156**(1), 115–126 (2001)
[Mig1] Mignon, T.: Systèmes de courbes planes à singularités imposées: Le cas des multiplicités inférieures ou égales à quatre. J. Pure Appl. Algebra **151**(2), 173–195 (2000)

[Mig2] Mignon, T.: An asymptotic existence theorem for plane curves with prescribed singularities. J. Algebr. Geom. **10**(2), 281–297 (2001)

[Roe2] Roé, J.: Maximal rank for planar singularities of multiplicity 2. J. Algebra **302**, 37–54 (2006)

[Dav] Davis, E.D.: 0-dimensional subschemes of $\mathbb{P}^2$: New applications of Castelnuovo's function. Ann. Univ. Ferrara **32**, 93–107 (1986)

[BGM] Bigatti, A., Geramita, A.V., Migliore, J.C.: Geometric consequences of extremal behavior in a theorem of Macaulay. Trans. Am. Math. Soc. **346**(1), 203–235 (1994)

[GLS4] Greuel, G.-M., Lossen, C., Shustin, E.: Castelnuovo function, zero-dimensional schemes and singular plane curves. J. Algebra Geom. **9**(4), 663–710 (2000)

[Shu13] Shustin, E.: Analytic order of singular and critical points. Trans. Am. Math. Soc. **356**, 953–985 (2004)

[Iar1] Iarrobino, A.: Punctual Hilbert schemes. Mem. Am. Math. Soc. (188) (1977)

[Rie] Riemann, B.: Theorie der Abel'schen Functionen. J. Reine und Angew. Math. **54**, 115–155 (1857)

[Roch] Roch, G.: Ueber die Anzahl der willkurlichen Constanten in algebraischen Functionen. J. Reine Angew. Math. **64**, 372–376 (1865)

[Sev] Severi, F.: *Vorlesungen über algebraische Geometrie.* Teubner (1921), respectively Johnson (1968)

[Seg1] Segre, B.: Dei sistemi lineari tangenti ad un qualunque sistema di forme. Atti Acad. naz. Lincei Rendiconti serie **5**(33), 182–185 (1924)

[Seg2] Segre, B.: Esistenza e dimensione di sistemi continui di curve piane algebriche con dati caraterri. Atti Acad. naz. Lincei Rendiconti serie **6**(10), 31–38 (1929)

[Gre1] Greuel, Gert-Martin: A remark on the paper of A. Tannenbaum. Compos. Math. **51**, 185–187 (1984)

[Tan1] Tannenbaum, A.: Families of algebraic curves with nodes. Compos. Math. **41**, 107–126 (1980)

[Tan2] Tannenbaum, A.: Families of curves with nodes on $K3$-surfaces. Math. Ann. **260**, 239–253 (1982)

[Tan3] Tannenbaum, A.: On the classical characteristic linear series of plane curves with nodes and cuspidal points: two examples of Beniamino Segre. Compos. Math. **51**, 169–183 (1984)

[Wah2] Wahl, J.: Deformations of plane curves with nodes and cusps. Am. J. Math. **96**, 529–577 (1974)

[Zar3] Zariski, O.: Algebraic Surfaces, 2nd edn. Springer, Berlin (1971)

[Shu4] Shustin, E.: On manifolds of singular algebraic curves. Sel. Math. Sov. **10**, 27–37 (1991)

[Shu1] Shustin, E.: Versal deformation in the space of plane curves of fixed degree. Funct. Anal. Appl. **21**, 82–84 (1987)

[Shu6] Shustin, E.: Smoothness and irreducibility of varieties of algebraic curves with nodes and cusps. Bull. SMF **122**, 235–253 (1994)

[Shu8] Shustin, E.: Geometry of equisingular families of plane algebraic curves. J. Algebr. Geom. **5**, 209–234 (1996)

[FlM] Flamini, F., Madonna, C.: Geometric linear normality for nodal curves on some projective surfaces. Bull. Unione Math. Ital. Sez. B Artic. Ric. Math. (8) **4**(1), 269–283 (2001)

[DPW] Du Plessis, A.A., Wall, C.T.C.: Versal deformations in spaces of polynomials of fixed weight. Compos. Math. **114**, 113–124 (1998)

[DPW1] Du Plessis, A.A., Wall, C.T.C.: Discriminants and vector fields. In: Arnold, V.I., et al. (eds.) Singularities the Brieskorn anniversary volume (Progress in Math. vol. 162), pp. 119–140. Birkhäuser (1998)

Chapter 4
Equisingular Families of Curves

We are ready to accomplish our main task, that is to answer the two following questions concerning equisingular families (ESF) of curves:

- whether a family of algebraic curves with a prescribed collection of singularities form a *nonempty*, *T-smooth* (i.e. smooth of expected dimension), *irreducible* stratum in the discriminant in a given linear system $|D|$ on a smooth algebraic surface Σ, and
- what is the local structure of the discriminant in a neighborhood of the above stratum, in particular, when is a family *deformation complete* (i.e. when are any simultaneous local deformations of the curve singularities induced by the given linear system).

Some answers are extended to projective hypersurfaces of arbitrary dimension with isolated singularities.

The main results of this chapter have been achieved by the authors in a series of papers [GrK, GrL2, GLS1, GLS2, GLS3, GLS4, GLS5, Shu10, Shu4, Shu5, Shu13, ShT2, ShW]. They give sufficient conditions for the non-emptiness, T-smoothness, irreducibility, and deformation completeness of and equisingular family of curves which are

- *universal*, that is, they concern curves in any ample linear system with any number of singularities of arbitrary types;
- moreover, with respect to non-emptiness, T-smoothness and deformation completeness our results are *asymptotically proper*, that is, our sufficient conditions have the same asymptotics as the known necessary conditions.

For curves with nodes and cusps our results are even *asymptotically optimal*.

Consider the family of curves $C \in |D| = |dD_0|$ with $D_0 \subset \Sigma$ a fixed divisor, having r isolated singularities of fixed types S_1, ..., S_r. Loosely speaking, we call a condition of the form

$$\sum_{i=1}^{r} a(S_i) \leq f_{\Sigma,D_0}(d) ,$$

G.-M. Greuel et al., *Singular Algebraic Curves*, Springer Monographs in Mathematics, https://doi.org/10.1007/978-3-030-03350-7_4

where $a(S_i)$ is a local numerical invariant of a plane curve singularity S_i, and f_{Σ,D_0} a polynomial in d, whose coefficients depend only on Σ and D_0,

- *asymptotically proper* for the considered property, if the following holds: for r and d going to infinity, the above inequality is, for the property to hold, also necessary up to multiplication of f with a positive constant and up to lower order terms in d.
- If the positive constant is 1 then we call the condition *asymptotically optimal* (cf. Sect. 4.2.1 for precise definitions).

Concerning irreducibility, our results are asymptotically proper if the Tjurina numbers of the singularities are bounded.

We like to emphasize, that previously for none of the above properties any asymptotically proper sufficient bound was known except for the case of plane nodal curves.

Throughout this chapter Σ denotes a non-singular projective algebraic surface over an algebraically closed field of characteristic zero, and $D \subset \Sigma$ an ample divisor such that $\dim |D| > 0$. The variety of singular curves in $|D|$ consisting of curves (resp. irreducible curves) in Σ with only isolated singularities, splits into the union of equisingular families $V_{|D|}(S_1, \cdots, S_r)$ (resp. $V^{irr}_{|D|}(S_1, \ldots, S_r)$). These are the varieties of curves $C \in |D|$ having r isolated singular points of given topological or analytic types $S_1, \ldots, S_r$, respectively (cf. Definition 2.2.35), and may be other singularities, with no condition on their behavior. The main emphasis is on irreducible curves. If $\Sigma = \mathbb{P}^2$, $D = dH$, H a hyperplane divisor, we write simply $V^{irr}_d(S_1, \ldots, S_r)$.

4.1 Overview of New Results and Methods

In this short section, we would like to guide the reader through the extensive list of results, examples, and comments presented in this chapter, by underlining the most important aspects and illustrate them by a few statements on plane singular curves.

First of all, a *complete* answer to the questions formulated in the foreword is known only for the case of plane nodal curves (Severi, Harris, cf. Sect. 4.2.2): the inequality

$$0 \le n \le \frac{(d-1)(d-2)}{2}$$

is necessary and sufficient for the nonemptiness, T-smoothness, and irreducibility of the family of irreducible plane curves of degree d with n nodes as their only singularities, and, additionally, for the independent smoothing of prescribed nodes while keeping the others, induced by the space of plane curves of degree d.

Already for plane curves with ordinary cusps a reasonable complete answer is hardly possible, due to a large gap between the known upper bounds to the number of cusps and known examples of curves with many cusps. Due to the irregular behavior of such examples, it seems unrealistic to expect a sufficient condition for either nonemptiness, or T-smoothness, or irreducibility, which is at the same time necessary (as in the case of plane nodal curves).

This situation has motivated us to pursue the following goal: describe the *regular* part of the discriminant (i.e. the part of the space of singular curves consisting of the nonempty, T-smooth, and irreducible equisingular families of curves) in a possibly precise form which should be

(i) *universal*, i.e. applicable to arbitrary singularities,
(ii) *numerical*, i.e. expressed as relations (inequalities) for numerical invariants of the ambient surfaces, curves and their singularities,
(iii) *asymptotically optimal* or *asymptotically proper*, i.e. having either the same asymptotics or are asymptotically "comparable" with the known examples of irregular equisingular families (empty, non-T-smooth, reducible etc).

Here we illustrate the content of this chapter by featuring selected results about plane curves, referring to the main body of the chapter for more general and stronger statements, which include curves on various classes of surfaces and where detailed proofs are given.

4.1.1 Nonemptiness

It is well-known that the inequality

$$\sum_{i=1}^{r} \mu(S_i) \le (d-1)^2 \qquad (4.1.1.1)$$

is *necessary* for the existence of a plane curve C of degree d with r singular points of (topological or analytic) types $S_1, ..., S_r$, respectively. In fact, this upper bound follows from Bézout's theorem to the intersection of two generic polars of C. Our existence result in Corollary 4.5.15 states that the inequality

$$\sum_{i=1}^{r} \mu(S_i) \le \frac{1}{9}(d^2 - 2d + 3) \qquad (4.1.1.2)$$

is *sufficient* for the existence of such a curve, even of an irreducible one C, such that the germ of the ESF at C is T-smooth. This is a universal, numerical, and asymptotically proper statement (the asymptotics of the right-hand sides is governed by d^2 and differs by a constant factor, here $\frac{1}{9}$).

Furthermore, for the case of curves with nodes and cusps, we prove (Theorem 4.5.4) that the inequality

$$n + 2k \le \frac{d^2 - 4d + 6}{2} \qquad (4.1.1.3)$$

is sufficient for the existence of an irreducible plane curve of degree d with n nodes and k cusps as its only singularities. This condition is even *asymptotically optimal* if we restrict our consideration to curves in T-smooth equisingular families, since the inequality

$$n + 2k \leq \frac{d^2 + 3d}{2} \tag{4.1.1.4}$$

(with the same asymptotics $\frac{1}{2}d^2$ in the right-hand side) is necessary for a curve of degree d with n nodes and k cusps to belong to a T-smooth equisingular family. To see the necessity, note that the projective space of all plane curves of degree d has dimension $(d^2 + 3d)/2$ and that a node imposes one and a cusp two conditions. Hence the *expected dimension* for the space of plane curves of degree d with n nodes and k cusps is

$$\frac{d^2 + 3d}{2} - n - 2k \, ,$$

which must be non-negative for T-smooth curves.

The previously known condition $\mu(S) \leq d + 1$ for existence, which is linear in d, is not asymptotically proper. It can be easily derived for any individual singularity S taking into account the determinacy bound in Lemma 1.2.5. Our general existence results in terms of the sum of the Milnor numbers and d^2 are illustrated in Fig. 4.1.

Notice that a sufficient asymptotically proper existence condition for curves with arbitrary singularities has never been formulated before.

4.1.2 T-Smoothness and Deformation Completeness

The classically known Segre-Zariski sufficient condition

$$k < 3d \tag{4.1.2.1}$$

for the T-smoothness of the family with any given number of nodes and k cusps, and for the independence of deformations of these nodes and cusps induced by the space of curves of degree d, has extensions to arbitrary singularities in the form of a $3d$*-condition* (cf. [GrK, Shu4] and Corollaries 4.3.4 and 4.3.5 and Theorem 4.4.1 below) or a $4d$*-condition* (cf. [GrL1, Shu1] and Theorem 4.3.8 and Proposition 4.4.2 below). All allow only *linearly* many (w.r.t. the degree d) singularities different from nodes, what is far away from the right-hand sides of the existence conditions (4.1.1.1)–(4.1.1.4).

In turn, the main results presented in Sects. 4.3 and 4.4 provide sufficient conditions for the T-smoothness and independence of deformations of singular points with a *quadratic* in d right-hand side (cf. Theorems 4.3.9 and 4.4.1 below):

$$\sum_{i=1}^{r} \gamma'(C, z_i) \leq (d + 3)^2 \tag{4.1.2.2}$$

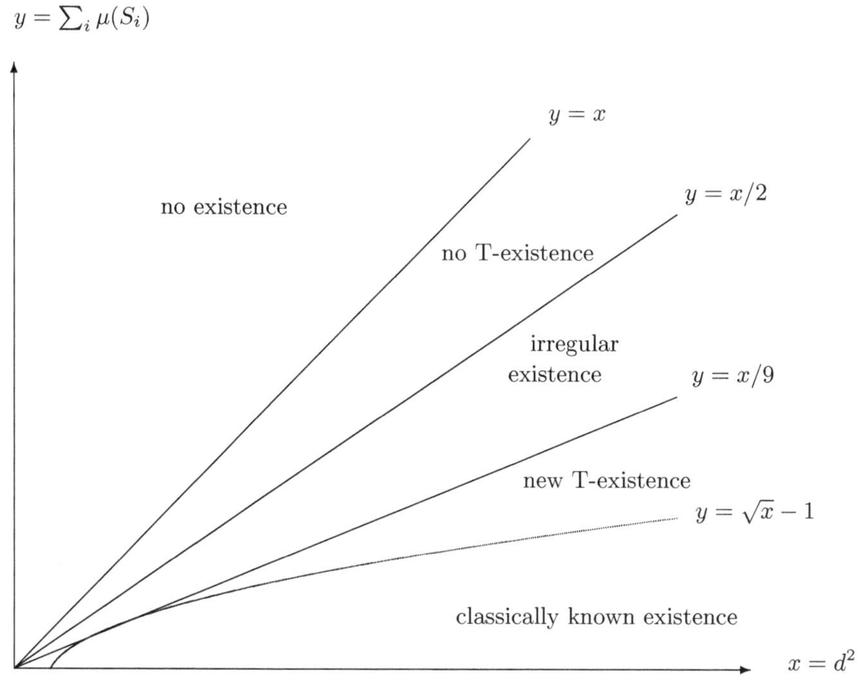

Fig. 4.1 Necessary and sufficient conditions for existence: The statement holds asymptotically for the region between the corresponding boundary curves

with γ-invariants defined in Sect. 1.2.3. Here γ' denotes γ^{ea} for analytic types resp. γ^{es} for topological types. Note that $\gamma' \leq (\tau' + 1)^2$ with τ' being τ resp. τ^{es} (cf. Definition 1.1.63), but that this estimate is rather rough. In fact, we believe that the new γ-invariants are the "right" local singularity invariants with a quadratic upper bound in d.

Particularly, for the case of nodes and cusps we have the sufficient T-smoothness and deformation completeness condition (cf. Corollary 4.3.11 below)

$$4n + 9k \leq (d + 3)^2 . \tag{4.1.2.3}$$

For the case of ordinary singularities of multiplicities $m_1, \ldots, m_r$, we have (cf. Corollary 4.3.12 below)

$$4 \cdot \#(\text{nodes}) + \sum_{m_i \geq 3} 2 \cdot m_i^2 \leq (d + 3)^2 . \tag{4.1.2.4}$$

Furthermore, examples of non-T-smooth equisingular families (Theorem 4.3.23 and Remark 4.3.23.1, Theorems 4.3.24 and 4.3.25) show that condition (4.1.2.2) for simple singularities and condition (4.1.2.4) are *asymptotically proper*, and, moreover, condition (4.1.2.3) is *asymptotically optimal*.

4.1.3 Irreducibility

Our general irreducibility condition for equisingular families of plane curves reads (cf. Corollary 4.6.5 below):

$$\frac{25}{2}\cdot\#(\text{nodes})+18\cdot\#(\text{cusps})+\frac{10}{9}\cdot\sum_{\tau'(S_i)\geq 3}\big(\tau'(S_i)+2\big)^2 < d^2, \qquad (4.1.3.1)$$

$$\text{and}\quad \max_{i=1..r}\tau'(S_i)\leq (2/5)\cdot d-1 \qquad (4.1.3.2)$$

which is *quadratic* in d, and hence is asymptotically proper for any collection of singularities with uniformly bounded Tjurina numbers τ' (i.e. τ for analytic types and τ^{es} for topological types). In general, i.e., for non-bounded Tjurina numbers, the asymptotic properness, resp. optimality of (4.1.3.1), (4.1.3.2) remains an open question.

In Fig. 4.2 we illustrate the new results related to plane curves with nodes and cusps.

4.1.4 Comments on the Methods

In any of the considered questions an important step is to reduce the problem to an H^1-vanishing condition of the ideal sheaf of a certain 0-dimensional scheme.

The idea to apply H^1-vanishing criteria for the treatment of smoothness of equisingular families of curves (with nodes and cusps) and independence of deformations of curve singularities can be traced back to Severi. We go, however, further, by reformulating *all* the considered problems as H^1-vanishing (or non-vanishing) for ideal sheaves of appropriate zero-dimensional subschemes of the ambient surface (or of the ambient variety in the higher-dimensional case), and then we apply the whole repertoire and power of H^1-vanishing criteria developed in Chap. 3. At this point we would like to indicate the role of specific criteria in our approach:

- *T-smoothness and independence of simultaneous singularity deformations.* The classical linear sufficient conditions for plane curves and their generalizations for curves on other surfaces come from the Riemann–Roch theorem applied to the considered curve (or its polar in the planar case). Surprisingly, a relatively simple Castelnuovo function theory provides the best known quadratic sufficient conditions for plane curves. The lack of a similar theory for other surfaces is compensated by the Reider-Bogomolov theory which again provides quadratic sufficient conditions for curves on surfaces of various types. None of this applies in a higher-dimensional situation. Here we use the Horace method of Hirschowitz, resulting in a classical-like "$4d-4$" sufficient condition. The same Horace method appears as a main ingredient in the proof of the non-T-smoothness, resp. obstructedness in our negative examples.

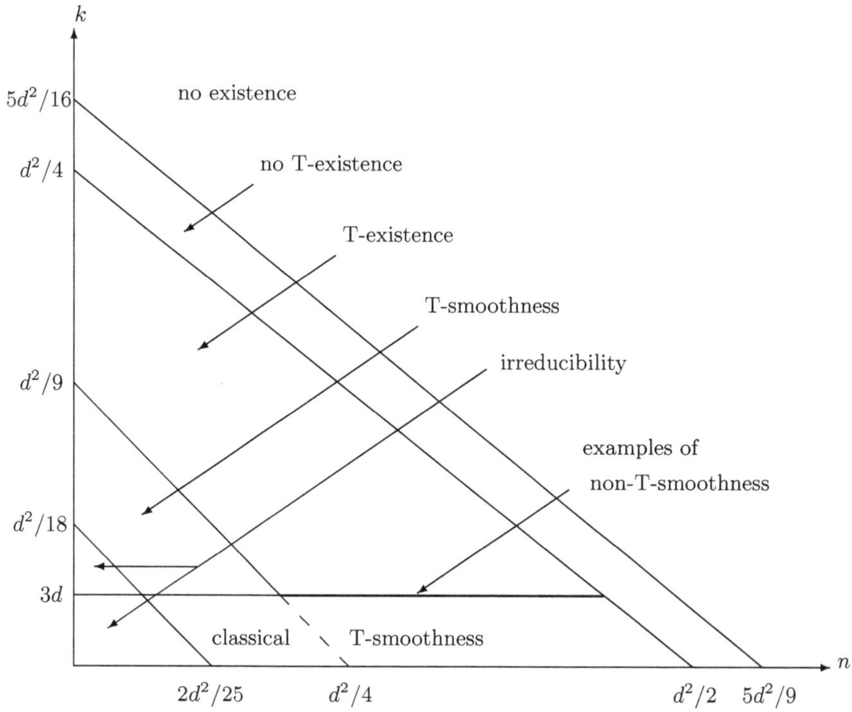

Fig. 4.2 Curves of degree d with n nodes and k cusps: For (n, k) in the region below a line we have the indicated property. The line $k = 3d$ gives a series of non-T-smooth curves of degree d with k cusps

- *Existence*. The background idea of our existence results for singular curves, which cover large ranges of numerical characteristics (and not only sporadic values as in previously known constructions), is the theory of zero-dimensional schemes in general position and the patchworking method. The Castelnuovo function theory again leads to the best results for plane curves with arbitrary singularities, whereas the patchworking method combined with the Riemann–Roch and Horace H^1-vanishing criteria is used for curves on other surfaces or for plane curves with specific singularities (nodes, cusps, ADE-singularities).
- *Irreducibility*. This problem turns to be the hardest one; the quadratic sufficient irreducibility criteria for curves with arbitrary singularities require an ingenious combination of the Kodaira vanishing, the Reider-Bogomolov theory, and, in the planar case, again the Castelnuovo function theory.

4.2 Formulation of Problems, Discussion of Results, Examples

4.2.1 *Statement of the Problem*

After the informal introduction in the last section we formulate, in a rigorous way, the main problems concerning equisingular families (ESF) of curves on algebraic surfaces. The methods developed in the preceding chapters will be then applied to answer the following basic questions on the geometry of ESF or, to be more precise, of the respective equisingular strata.

We keep the notations of the beginning of this chapter. Σ is a non-singular projective algebraic surface, $D \subset \Sigma$ is an ample divisor such that $\dim |D| > 0$ and we assume that the generic member of $|D|$ is irreducible and non-singular. The part of the discriminant (i.e. the variety of singular curves), consisting of irreducible curves in Σ with only isolated singularities, splits into the union of equisingular families $V^{irr}_{|D|}(S_1, \dots, S_r)$ of irreducible curves $C \in |D|$ having r isolated singular points of given topological or analytic types $S_1, \dots, S_r$, respectively (cf. Definition 2.2.35), and may be other singularities with no condition on their behavior. If $\Sigma = \mathbb{P}^2$, $D = dH$, H a hyperplane divisor, we write simply $V^{irr}_d(S_1, \dots, S_r)$. We focus on

(A) *Existence problem*: Is $V^{irr}_{|D|}(S_1, \dots, S_r)$ non-empty, that is, does there exist a curve $C \in |D|$ with the given collection of singularities ? In particular, the question about the minimal degree of a plane curve having a given singularity is of special interest.

(B) *T-smoothness problem*: If $V^{irr}_{|D|}(S_1, \dots, S_r)$ is non-empty, is it smooth and of "expected" dimension or, in other words, is it T-smooth (we are looking for conditions expressible via local invariants of the singularities)? That is, let $C \in V^{irr}_{|D|}(S_1, \dots, S_r)$ have singular points $z_1, \dots, z_r$ of types $S_1, \dots, S_r$, respectively. We say that the family $V^{irr}_{|D|}(S_1, \dots, S_r)$ is *T-smooth* at C if, for any $i = 1, \dots, r$, the germ at C of the family of curves $C' \in |D|$ with a singular point of type S_i in a neighbourhood of z_i is smooth and has the expected dimension (Definition 2.2.53), and, furthermore, all these r germs intersect transversally at C (whence the name "T-smooth").

(C) *Irreducibility problem*: Is $V^{irr}_{|D|}(S_1, \dots, S_r)$ irreducible?

Closely related to problem (B) is

(D) *Deformation problem*: What are the adjacency relations of ESF in the discriminant? In other words, which simultaneous deformations of singular points of $C \in V^{irr}_{|D|}(S_1, \dots, S_r)$ can be realized by variation of C in $|D|$? In fact, the T-smoothness of $V^{irr}_{|D|}(S_1, \dots, S_r)$ for analytic singularities $S_1, \dots, S_r$ means that the linear system $|D|$ induces a joint versal deformation of all singular points of any member $C \in V^{irr}_{|D|}(S_1, \dots, S_r)$, in which case we say that $V^{irr}_{|D|}(S_1, \dots, S_r)$ is *deformation complete*. The T-smoothness of ESF for topological singularities

$S_1, \ldots, S_r$ means the independence of simultaneous "lower" (w.r.t. the Newton diagram) deformations of singular points of $C \in V^{irr}_{|D|}(S_1, \ldots, S_r)$.
This relation of smoothness and deformation problems is described in Theorem 2.3.13, which claims the independence of simultaneous versal, respectively lower, deformations of singular points of a curve under some H^1-vanishing condition, which is actually equivalent to T-smoothness. Moreover, we specify the statements of Theorem 2.3.13 by replacing the H^1-vanishing condition by a numerical condition.

Of course, the same questions can be posed for $V_{|D|}(S_1, \ldots, S_r)$, the variety of reduced (not necessarily irreducible) curves $C \in |D|$ with given singularities $S_1, \ldots, S_r \in \mathcal{S}$.

No complete solution (in the sense of conditions which are necessary and sufficient at the same time) to the above problems is known except basically for the case of plane nodal curves. However, in this chapter we demonstrate that the approach using deformation theory and cohomology vanishing enables us to obtain reasonably proper sufficient conditions for the affirmative answers to the problems stated. More precisely, we intend to give sufficient conditions, which are

- *numerical*, i.e., presented in the form of inequalities relating numerical invariants of the surface, linear system, and singularities,
- *universal*, i.e., applicable to curves in any ample linear system with any number of arbitrary singularities, and
- *asymptotically proper*, i.e., comparable with necessary conditions (cf. the definition below).

We should like to comment on the latter in more detail. Let $D = dD_0$ with a given divisor $D_0 \subset \Sigma$ (e.g., $D_0 = H$ a straight line if $\Sigma = \mathbb{P}^2$). Then the known general restrictions to the singular points of a curve $C \in V^{irr}_{|D|}(S_1, \ldots, S_r)$ read as upper bounds to some total singularity invariants by a quadratic function in d. As an example, consider the bound obtained by the *genus formula* [HiH] for irreducible curves,

$$\sum_{i=1}^{r} \delta(S_i) \leq \frac{1}{2}(d^2 D_0^2 + dK_\Sigma D_0) + 1 .$$

Sufficient conditions for the "regular" properties of ESF appear in a similar form of upper bounds to total singularity invariants by a linear or quadratic function in the parameter d. In this sense we speak of *linear*, or *quadratic* sufficient conditions, the latter revealing the relevant asymptotics.

Furthermore, among the quadratic sufficient conditions we emphasize the *asymptotically proper* ones. We say that the inequality

$$\sum_{i=1}^{r} a(S_i) \leq f_{\Sigma, D_0}(d) ,$$

with $a(S)$ a local invariant of singularities $S \in \mathcal{S} = (S_1, ..., S_r)$ and f_{Σ,D_0} a polynomial in d, is an *asymptotically proper* sufficient condition for a regular property (non-emptiness, smoothness, irreducibility, etc.) of ESF, if

(1) for arbitrary $r, d \geq 1$ and $S_1, \ldots, S_r \in \mathcal{S}$ it provides the required property of the ESF $V^{irr}_{|dD_0|}(S_1, \ldots, S_r)$,
(2) there exists an absolute constant $A \geq 1$ such that, for any singularity S, there is an infinite sequence of growing integers d, r and (maybe, empty) finite collections of singularities $\mathcal{S}_d \subset \mathcal{S}$ such that[1]

$$r \cdot a(S) = A \cdot f_{\Sigma,D_0}(d) + o(f_{\Sigma,D_0}(d)), \quad \sum_{S' \in \mathcal{S}_d} a(S') = o(f_{\Sigma,D_0}(d)),$$

and the ESF $V^{irr}_{|dD_0|}(r \cdot S, \mathcal{S}_d)$ does not have that regular property.

If $A = 1$ in (2), we speak of an *asymptotically optimal* sufficient condition.

In the following Sects. 4.3–4.6 we obtain universal *quadratic* sufficient conditions for non-emptiness, T-smoothness, and irreducibility of ESF of curves in the plane and some special algebraic surfaces, which are asymptotically proper, or even optimal in some cases. We provide series of examples of obstructed, non-smooth or reducible ESF, demonstrating the properness, resp. optimality of the obtained sufficient conditions.

4.2.2 *Curves with Nodes and Cusps: from Severi to Harris*

4.2.2.1 Results on T-Smoothness

Already the Italian geometers Severi and Segre [Sev, Seg1, Seg2] noticed that it is possible to express the T-smoothness of the variety of plane curves with (fixed number of) nodes and cusps infinitesimally. The problem was historically called the *"completeness of the characteristic linear series of complete continuous systems"* (of plane curves with nodes and cusps). Best known is certainly Severi's [Sev] result saying that the variety of nodal curves in $\mathbb{P}^2$ is always T-smooth.

But for more complicated singularities (beginning with ordinary cusps) there are examples of irreducible curves $C \subset \mathbb{P}^2$ of degree d with

$$h^1\big(\mathcal{J}_{Z^{es}(C)/\mathbb{P}^2}(d)\big) > 0, \quad \text{respectively} \quad h^1\big(\mathcal{J}_{Z^{ea}(C)/\mathbb{P}^2}(d)\big) > 0,$$

that is, where T-smoothness fails (cf. Theorem 2.2.40(c)). We discuss such examples in Sect. 4.3.3.

[1] We use the little-o notation, meaning that the correction term $o(f(d))$ includes only lower powers in d than $f(d)$.

On the other hand, various sufficient conditions for T-smoothness were found. The classical result is that the variety $V_d^{irr}(n \cdot A_1, k \cdot A_2)$ of irreducible plane curves of degree d with n nodes and k cusps as only singularities has the T-property if

$$k < 3d\,. \tag{4.2.2.1}$$

For arbitrary singularities and other algebraic surfaces, several generalizations and extensions of (4.2.2.1) were found [GrK, GrL1, Shu1, Shu4, Tan2, Vas]. All of them are of the form that the sum of certain invariants of the singularities is bounded from above by a linear function in d, whereas the number of singularities might be of order d^2, that is, all these results cannot be asymptotically proper.

4.2.2.2 Results on Existence

Various *restrictions* for the existence of plane curves with prescribed singularities $S_1, \ldots, S_r$ have been found. First, one should mention the classical bounds (for irreducible curves)

$$\sum_{i=1}^{r} \delta(S_i) \;\le\; \frac{(d-1)(d-2)}{2}\,, \tag{4.2.2.2}$$

resulting from the genus formula (2.1.4.6), respectively (for reduced, not necessarily irreducible curves)

$$\sum_{i=1}^{r} \mu(S_i) \;\le\; (d-1)^2,$$

resulting from the intersection of two generic polars of C and Bézout's theorem. If C has only nodes and (ordinary) cusps as singularities then the Plücker formulae (2.1.4.4) give (among others) the necessary conditions, see [Lef],

$$2 \cdot \#(\text{nodes}) + 3 \cdot \#(\text{cusps}) \le d^2 - d - 2\,,$$
$$6 \cdot \#(\text{nodes}) + 8 \cdot \#(\text{cusps}) \le 3d^2 - 6d\,.$$

Of course, Bézout's theorem leads to various further necessary conditions for the existence of the curve C such as, for instance by considering a line through 2 points or a conic through 5 points,

$$\max_{i \neq j} \big(\mathrm{mt}(S_i) + \mathrm{mt}(S_j)\big) \;\le\; d\,, \quad \max_{\#(I)=5} \sum_{i \in I} \mathrm{mt}(S_i) \;\le\; 2d\,.$$

By applying the log-Miyaoka inequality, F. Sakai obtained the following necessary condition for existence:

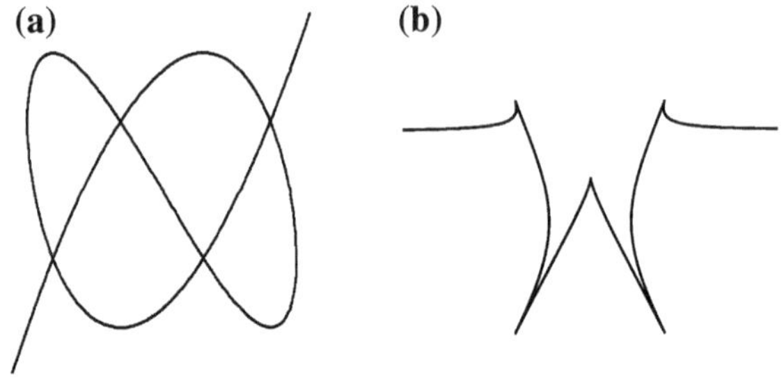

Fig. 4.3 Plane quintics **a** with 4-nodes: $x^5 - \frac{5}{4}x^3 + \frac{5}{16}x - \frac{1}{4}y^3 + \frac{3}{16}y = 0$, **b** with 5 cusps: $\frac{129}{8}x^4y - \frac{85}{8}x^2y^3 + \frac{57}{32}y^5 - 20x^4 - \frac{21}{4}x^2y^2 + \frac{33}{8}y^4 - 12x^2y + \frac{73}{8}y^3 + 32x^2 = 0$

$$\sum_{i=1}^{r} \mu(S_i) < \frac{2\nu}{2\nu+1} \cdot \left(d^2 - \frac{3}{2}d\right),$$

where ν denotes the maximum of the multiplicities mt S_i, $i = 1, \ldots, r$ (cf. [Sak, Thm. A]). If $S_1, \ldots, S_r$ are simple singularities then we have

$$\sum_{i=1}^{r} \mu(S_i) < \begin{cases} \frac{3}{4}d^2 - \frac{3}{2}d + 2 & \text{if } d \text{ is even}, \\ \frac{3}{4}d^2 - d + \frac{1}{4} & \text{if } d \text{ is odd}, \end{cases}$$

(cf. [HiF, Ivi], respectively [Sak]).

Example 4.2.0.1 (*Cuspidal curves*) The maximal number $k_{\max}(d)$ of ordinary cusps on an irreducible cuspidal curve C of degree d is bounded from above by $(5/16)d^2 - (3/8)d \sim 0.3125 \cdot d^2 + O(d)$ [Ivi, Sak] and, moreover, satisfies (cf. [Lan, Formula (11.2.1)])

$$\limsup_{d\to\infty} \frac{k_{\max}(d)}{d^2} \leq \frac{125+\sqrt{73}}{432} \sim 0.3091\,.$$

Again, these bounds are obtained by applying the log-Miyaoka inequality and its extensions. We discuss this in more detail in Sect. 4.5.2.1.

For small degrees d (Fig. 4.3), there are sharper bounds for $k_{\max}(d)$ (as shown in [Zar3, p. 221, 222], [Var2]; respectively resulting from the semicontinuity of the singularity spectrum [Var3]). The following table shows bounds for $k_{\max}(d)$ for small degrees, whose "lower bound" means that cuspidal curves with that many cusps are known (for the results, we refer to [Hi, Sak, Koe, Shu11] and to Example 4.5.4.2 for $k_{\max}(11) \geq 30$):

degree	3	4	5	6	7	8	9	10	11	12
upper bound	1	3	5	9	10	15	21	26	31	40
lower bound	1	3	5	9	10	15	20	26	30	39

Example 4.2.0.2 (*Computing an upper bound for the number of singularities*) The semicontinuity of the singularity spectrum can be used to compute an upper bound for the number of isolated hypersurface singularities of a given type occurring on a hypersurface $V \subset \mathbb{P}^n$ of degree d. More precisely, one can apply the following two facts:

- any hypersurface of degree d with isolated singularities can be obtained as small deformation of $V(f_d)$, where $f_d(x_1, ..., x_n)$ is a nondegenerate d-form; furthermore, the space of nondegenerate d-forms is connected; hence for the bound it is sufficient to compute the spectrum of $f = x_1^d + \cdots + x_n^d$;
- if precisely N of the spectral numbers (counted with multiplicities) of the singularity defined by $f = x_1^d + \cdots + x_n^d$ are in the interval $(a, a+1)$, $a \in \mathbb{R}$, then the singularities (close to 0) of a small deformation of $V(f)$ altogether can have at most N spectral numbers in the interval $(a, a+1)$ (cf. [Var3]).

For instance, we can look for an upper bound for the number of cusps which may appear on a curve of degree 11 by using SINGULAR ([DGPS] and [Schu]):

```
LIB "gmssing.lib";
ring r=0,(x,y),ds;
poly g=x^2-y^3;
list s1=spectrum(g);     // spectral numbers of a cusp (with mult's)
s1;
//-> [1]:
//->        _[1]=-1/6  _[2]=1/6
//-> [2]:
//->        1,1

poly f = x^11+y^11;
list s2 = spectrum(f);  // spectral numbers of f (with mult's)
s2;
//-> [1]:
//->       _[1]=-9/11  _[2]=-8/11  _[3]=-7/11  _[4]=-6/11  _[5]=-5/11
//->       _[6]=-4/11  _[7]=-3/11  _[8]=-2/11  _[9]=-1/11  _[10]=0
//->       _[11]=1/11  _[12]=2/11  _[13]=3/11  _[14]=4/11  _[15]=5/11
//->       _[16]=6/11  _[17]=7/11  _[18]=8/11  _[19]=9/11
//-> [2]:
//->       1,2,3,4,5,6,7,8,9,10,9,8,7,6,5,4,3,2,1
```

Having computed the spectral numbers, we can either look ourselves for an appropriate interval $(a, a+1)$ (here: $a = -\frac{2}{11}$), or use the SINGULAR procedure `spsemicont`:

```
spsemicont(s2,list(s1),1);
// -> [1]: 31
```

to read the sharpest bound for the number of singularities obtainable in the above way (here: at most 31 cusps can appear on a curve of degree 11).

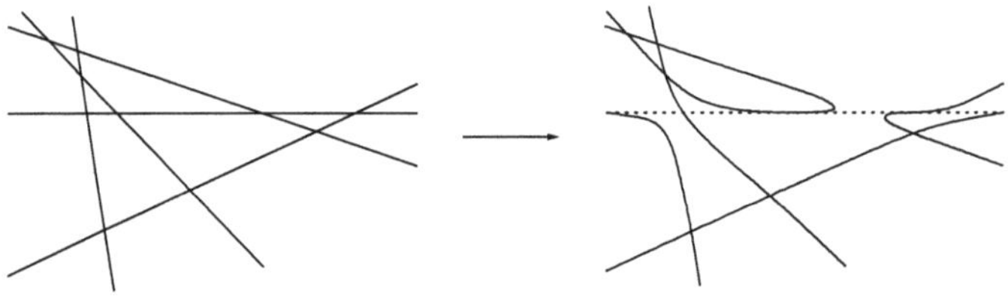

Fig. 4.4 Severi's construction of irreducible nodal curves

4.2.2.3 Methods of Construction

The first method is to construct (somehow) a curve of the given degree, which is degenerate with respect to the required curve, and then to deform it in order to obtain the prescribed singularities.

Example 4.2.0.3 Severi [Sev] showed that singular points of a nodal curve, irreducible or not, can be smoothed, or preserved, independently (cf. Sect. 4.3). The construction is easy: take the union of d generic straight lines in $\mathbb{P}^2$, meeting in $d(d-1)/2$ nodes. To make this curve irreducible, smooth away the $d-1$ nodes at intersection points of one line with the others. The resulting irreducible curve has $(d-1)(d-2)/2$ nodes, the maximum possible. Smoothing away other nodes we get an irreducible curve with any number of nodes, allowed by (4.2.2.2), cf. Fig. 4.4.

Attempts to extend this construction to other singularities give curves with a number of singularities bounded from above by a linear function in the degree d (see, for example, [GrM] for curves with nodes, cusps and ordinary triple points), because of the very restrictive requirement of independence of deformations.

The second method consists of a construction especially adapted to the given degree and given collection of singularities. It may be based on a sequence of birational transformations of the plane applied to a more or less simple initial curve in order to obtain the required curve. Or it may consist of an invention of a polynomial defining the required curve. This is illustrated by constructions of singular curves of small degrees as, for instance, in [Wal1], [Wal2], or by Hirano's construction of cuspidal curves:

Example 4.2.0.4 (*Cuspidal curves*) Hirano [Hi] constructed a series of (irreducible) cuspidal curves of degree $d = 2\cdot 3^k, k \geq 1$ with precisely $\frac{9}{32}(d^2-4)$ ordinary cusps. Since each cusp imposes two conditions, in this case the number of conditions imposed by the cusps is

$$\frac{9}{16}(d^2-4)\,,$$

that is, $d^2/16 + O(d)$ more than the dimension of the space of curves of degree d. Since for these curves the expected dimension is negative, they belong to the irregular existence region in Fig. 4.1. In particular, one can hardly expect that all intermediate quantities of cusps may be realized.

An even higher asymptotics has been achieved by Vik. S. Kulikov [Ku] and A. Calabri et al. [CPS], who constructed series of curves of degrees $d \to \infty$ with

$$\frac{283}{960}d^2 + O(d) = 0.294...d^2 + O(d) \quad \text{and} \quad \frac{2567}{8640}d^2 + O(d) = 0.297...d^2 + O(d)$$

cusps, respectively.

Two main difficulties do not allow to apply this approach to a wide class of degrees and singularities:

- for any new degree or singularity one has to invent a new construction,
- even if one has constructed a curve with many singularities, it is hard to check that these singular points can be smoothed independently and any intermediate number of singularities can be realized.

4.2.2.4 Results on Irreducibility

Severi [Sev] claimed that all non-empty ESF $V_d^{irr}(n \cdot A_1)$ of irreducible plane nodal curves are not only T-smooth but also irreducible. However, as was realized later, his proof of the irreducibility was incomplete.

After some progress, in particular, by Fulton [Ful1] and Deligne [Del], Harris [HaJ] gave a rigorous proof for the irreducibility of the varieties $V_d^{irr}(n \cdot A_1)$, by inventing a new specialization method and by using the irreducibility of the moduli space of curves of a given genus. Since nodal curves form an open dense subset of any *Severi variety* [Alb, Nob2], that is, the variety of irreducible plane curves of a given degree and genus, Harris' theorem extends to all Severi varieties as well. Later Ran [Ran1] and Treger [Tre] gave different proofs. Then Ran [Ran2] generalized Harris' theorem to ESF $V_d^{irr}(S, n \cdot A_1)$ with an ordinary singularity S and any number of nodes $n \le (d-1)(d-2)/2 - \delta(S)$. Kang [Kan1, Kan2] obtained the irreducibility of the families $V_d^{irr}(n \cdot A_1, k \cdot A_2)$ for[2]

$$k \le 3\,, \quad \text{or} \quad \frac{d^2 - 4d + 1}{2} \le n \le \frac{d^2 - 3d + 2}{2}\,.$$

We do not describe these proofs, since they are completely presented in the monograph [HaM]. The main idea consists of using moduli spaces of curves, special degenerations of nodal curves, or degenerations of rational surfaces. In any case, the proofs rely heavily on the independence of simultaneous deformations of nodes for plane curves, which in general does not hold for more complicated singularities, even for ordinary cusps.

[2]The missing cases of $k = 3$ and $d = 5$ or 6 are covered in [AD, Theorem 2.4], [Deg1], [Deg2, Section II.7.3].

4.2.3 Examples of Obstructed and Reducible ESF

4.2.3.1 Cuspidal Curves

Already Segre [Seg2] (see also [Tan3]) constructed a series of irreducible plane curves such that the corresponding germs of ESF are not T-smooth:

Example 4.2.0.5 Let F_{2m}, respectively G_{3m} be two generic homogeneous polynomials of degree $2m$, respectively $3m$. Then the curve $C_{6m} \subset \mathbb{P}^2$ of degree $d = 6m$ given by

$$\big(F_{2m}(x, y, z)\big)^3 + \big(G_{3m}(x, y, z)\big)^2 = 0$$

is irreducible and has precisely $6m^2$ cusps as only singularities (in the intersection points of $\{F_{2m} = 0\}$ and $\{G_{3m} = 0\}$). Moreover, the dimension of the family $V' \subset V_{6m}^{irr}(6m^2 \cdot A_2)$ of such curves is

$$\frac{2m(2m+3)}{2} + \frac{3m(3m+3)}{2} + 1 = \frac{d(d+3)}{2} - 12m^2 + \frac{(m-1)(m-2)}{2}.$$

It follows that for $m \geq 3$ the variety $V_{6m}^{irr}(6m^2 \cdot A_2)$ contains a component which has a bigger dimension than the expected one. In particular, this component is *non-T-smooth*. Note that $h^1\big(\mathcal{J}_{Z^{ea}(C_{6m})/\mathbb{P}^2}(6m)\big) = \frac{(m-1)(m-2)}{2} \neq 0$.

On the other hand, $h^0\big(\mathcal{J}_{Z^{ea}(C_{6m})/\mathbb{P}^2}(6m)\big) = \frac{d(d+3)}{2} - 12m^2 + \frac{(m-1)(m-2)}{2}$, whence we can conclude that the variety $V_{6m}^{irr}(6m^2 \cdot A_2)$ is *smooth* at C_{6m} by Theorem 2.2.40 (a).

By a slight modification of these (classical) examples, we obtain

Example 4.2.0.6 The varieties $V = V_{7m-3}^{irr}(6m^2 \cdot A_2)$, $m \geq 3$, contain a non-T-smooth component. Indeed, the family of curves with homogeneous equations

$$A_{m-3}(x, y, z) \cdot \big(F_{2m}(x, y, z)\big)^3 + B_{m-3}(x, y, z) \cdot \big(G_{3m}(x, y, z)\big)^2 = 0,$$

A_{m-3}, B_{m-3}, F_{2m}, G_{3m} generic homogeneous polynomials of degree $m-3$, respectively $2m$, respectively $3m$, has dimension $\frac{(7m-3)7m}{2} - 12m^2 + 1$, which is 1 more than the expected dimension of V. Note that in this example[3]

$$\#(\text{cusps}) = \frac{6}{49} d^2 + O(d),$$

while, by Corollary 4.3.11 below, a family of cuspidal curves is always T-smooth if

$$\#(\text{cusps}) \leq \frac{1}{9} d^2 + O(d).$$

[3] With the capital O-notation, meaning that the correction term $O(d)$ is linear in d.

4.2.3.2 Curves with Ordinary Triple Points

Example 4.2.0.7 Let $S = D_4$ be an ordinary triple point. The variety $V = V^{irr}_{3m+1}(m^2 \cdot D_4)$, contains the system of curves with homogeneous equations

$$L_1(x, y, z) \cdot \big(F_m(x, y, z)\big)^3 + L_1'(x, y, z) \cdot \big(G_m(x, y, z)\big)^2 = 0,$$

$L_1,\ L_1',\ F_m,\ G_m$ generic homogeneous polynomials of degree 1, respectively m. In particular, V has a component of dimension at least m^2+3m+4, which is greater than the expected dimension $\frac{m^2+15m}{2}+2$ if $m \geq 9$. Note that here

$$\#(\text{ordinary triple points}) = \frac{1}{9}d^2 + O(d).$$

However, it is easy to show that in these examples $V^{irr}_d(S_1, \ldots, S_r)$ is smooth (but of bigger dimension than the expected one).[4] By Corollary 4.3.12, the variety of curves with only D_4-singularities is T-smooth if

$$\#(\text{ordinary triple points}) \leq \frac{1}{12}d^2 + O(d).$$

4.2.3.3 Further Examples and Reducible ESF

In 1987, Luengo [Lue1, Lue2] provided the first examples of curves C such that the corresponding ESF is *of expected dimension, but non-smooth*:

Example 4.2.0.8 The homogeneous polynomial $x^9+z(xz^3+y^4)^2$ defines an irreducible projective plane curve C with exactly one singular point (at $(0:0:1)$), which is of type A_{35}. Moreover, $V^{irr}_9(A_{35})_{\text{red}}$ has the expected dimension 19 but is non-smooth at C.

Concerning reducible ESF, there is mainly one classical example due to Zariski [Zar3]:

Example 4.2.0.9 The variety $V^{irr}_6(6 \cdot A_2)$ is reducible.

More precisely, Zariski shows that there exist exactly two components, both being T-smooth. One component consists of cuspidal sextics C whose singularities lie on a conic. For such curves Zariski computes the fundamental group $\pi_1(\mathbb{P}^2 \setminus C)$ to be the *non-abelian* group $\mathbb{Z}_2 * \mathbb{Z}_3$. On the other hand, he shows that there exist curves C whose singularities do not lie all on a conic and whose complement in $\mathbb{P}^2$ has an

---

[4] This phenomenon has first been observed by Wahl in 1974 [Wahl], where he presented the first series of irreducible plane curves C (with only nodes and cusps) such that the corresponding ESF is non-smooth, but has a smooth reduction: $V^{irr}_{104}(3636 \cdot A_1,\ 900 \cdot A_2)$.

abelian fundamental group. In particular, they cannot be obtained by an *equisingular* deformation from curves in the first component.

Actually, this example belongs to a series [Shu6]:

Example 4.2.0.10 Let $p \geq 2$. Then the variety $V_{6p}^{irr}(6p^2 \cdot A_2)$ is reducible.

By Proposition 4.6.10, for $p \geq 3$ there exist components with different dimensions (cf. Segre's Example 4.2.0.5). For $p = 2$ there exist two different T-smooth components, as in Zariski's example. See also Sect. 4.6.3 for further examples and thorough discussion.

4.3 T-Smoothness

Again let Σ be a smooth projective surface and $D \subset \Sigma$ an effective divisor such that $\dim |D| > 0$. Let $D = dD_0$ with a given divisor D_0, which we assume to be a straight line if $\Sigma = \mathbb{P}^2$. We consider $V_{|dD_0|}(S_1, ..., S_r)$, the equisingular family of curves in $|dD_0|$ with fixed topological or analytic types (and may be other singularities). Sufficient criteria for T-smoothness in terms of H^1-vanishing are formulated in Theorems 2.2.38 and 2.2.40 (and in Theorem 2.2.55 for arbitrary smooth singularity types).

We present numerical sufficient conditions for the T-smoothness of $V_{|dD_0|}(S_1, \ldots, S_r)$, obtained by applying several of the H^1-vanishing theorems presented in Chap. 3. As mentioned before, we distinguish between

- *linear conditions*, that is, upper bounds for total singularity invariants by a linear function in d, and
- *quadratic conditions*, where the bound is given by a quadratic function in d.

Moreover, we put special emphasis on the asymptotical properness of the latter conditions. To do so, we present several series of obstructed ESF (Sect. 4.3.3).

Finally, we address ESF of hypersurfaces in a smooth projective variety of dimension $n > 2$. We present the (to our knowledge) only known general numerical sufficient condition for T-smoothness in higher dimensions (which generalizes the linear condition (3.1.0.8)) and prove its sharpness.

4.3.1 *Linear Conditions*

Throughout the following, let $S_1, \ldots, S_r$ be topological or analytic types of plane curve singularities. To shorten notation, we introduce the invariants

$$\tau'(S_i) := \begin{cases} \tau^{es}(S_i) & \text{if } S_i \text{ is a topological type}\,, \\ \tau(S_i) & \text{if } S_i \text{ is an analytic type}\,, \end{cases}$$

and

$$\operatorname{isod}'(S_i) := \begin{cases} \operatorname{isod}_z\big(Z^{es}(C), C\big) & \text{if } S_i \text{ is a topological type}\,, \\ \operatorname{isod}_z\big(Z^{ea}(C), C\big) & \text{if } S_i \text{ is an analytic type}\,, \end{cases}$$

with (C, z) being any plane curve singularity representing S_i. Recall that $\operatorname{isod}'(S_i)$ is a positive integer (cf. Lemma 1.2.19, respectively Exercise 1.2.20) appearing in the H^1-vanishing condition coming from Riemann–Roch (cf. Sect. 3.1).

4.3.1.1 Curves on Arbitrary Surfaces

Let Σ be a regular surface, that is,

$$H^1(\Sigma, \mathcal{O}_\Sigma) = 0\,, \tag{4.3.1.1}$$

and $D_0 \subset \Sigma$ an effective divisor. By Theorem 2.2.38 (c2) and Remark 2.2.39.1, a non-empty ESF $V = V_{|dD_0|}(S_1, \ldots, S_r)$ is T-smooth at a point $C \in V$, if $H^1\big(\mathcal{J}_{Z'(C)/\Sigma}(dD_0)\big) = 0$. Here $Z'(C) \subset \Sigma$ denotes the disjoint union of the zero-dimensional schemes

$$Z'(C, z_i) := \begin{cases} Z^{es}(C, z_i) & \text{if } S_i \text{ is a topological type}\,, \\ Z^{ea}(C, z_i) & \text{if } S_i \text{ is an analytic type}\,, \end{cases}$$

$i = 1, \ldots, r$. Note that $\deg(Z'(C)) = \sum_i \tau'(S_i)$. The Riemann–Roch theorem (in the form of Corollary 3.1.3), the exact sequence

$$H^1(\Sigma, \mathcal{O}_\Sigma) \longrightarrow H^1\big(\mathcal{J}_{Z'(C)/\Sigma}(dD_0)\big) \longrightarrow H^1\big(\mathcal{J}_{Z'(C)/C}(dD_0)\big)\,,$$

and (4.3.1.1) imply as immediate consequence for irreducible, resp. reducible C:

Theorem 4.3.1 ([GrK], Theorem 6.1, and [GrL1], Theorem 3.6) *Let S_1, ..., S_r be analytic or topological types.*

(a) $V^{irr}_{|dD_0|}(S_1, \ldots, S_r)$ is T-smooth or empty if

$$\sum_{i=1}^{r} \big(\tau'(S_i) - \operatorname{isod}'(S_i)\big) \;<\; (-K_\Sigma \cdot D_0) \cdot d\,. \tag{4.3.1.2}$$

(b) $V_{|dD_0|}(S_1, \ldots, S_r)$ is T-smooth at $C = C_1 \cup \ldots \cup C_s$, C_j irreducible, if, for any $j = 1, \ldots, s$,

$$\deg\big(Z'(C) \cap C_j\big) - \operatorname{isod}_{C_j}\big(Z'(C), C\big) \;<\; -K_\Sigma \cdot C_i\,.$$

For S_i a node resp. a cusp $\tau'(S_i) = 1$ resp. 2, while $\operatorname{isod}'(S_i) = 1$ in both cases (Lemma 1.2.19, Exercise 1.2.20). Hence nodes do not count in the l.h.s. of (4.3.1.2) while cusps count as 1.

If $S_1, \ldots, S_r$ are topological singularity types, then, due to Corollary 1.2.22, we can formulate a T-smoothness criterion via β-invariants (cf. Definition 1.2.13):

Theorem 4.3.2 ([Shu4], Theorem 1) *Let $S_1, \ldots, S_r$ be topological types.*

(a) $V^{irr}_{|dD_0|}(S_1, \ldots, S_r)$ is T-smooth or empty if

$$\sum_{i=1}^{r} \beta(S_i)\big) < (-K_\Sigma \cdot D_0) \cdot d\,.$$

(b) $V_{|dD_0|}(S_1, \ldots, S_r)$ is T-smooth at $C = C_1 \cup \ldots \cup C_s$, C_j irreducible, if, for any $j = 1, \ldots, s$,

$$\sum_{z \in \operatorname{Sing} C} \beta_{C_i}(C, z) < -K_\Sigma \cdot C_i\,.$$

In particular, for ESF of curves with n nodes (A_1) and k ordinary cusps (A_2), we can use Lemma 1.2.15 to obtain

Corollary 4.3.3 *$V^{irr}_{|dD_0|}(n \cdot A_1, k \cdot A_2)$ is T-smooth or empty if*

$$k < (-K_\Sigma \cdot D_0) \cdot d\,. \tag{4.3.1.3}$$

In particular, if $-K_\Sigma$ is ample, then each ESF of nodal curves, $V^{irr}_{|dD_0|}(n \cdot A_1)$, is T-smooth.

Note that in (4.3.1.3) the number of nodes does not appear. For example, $-K_\Sigma$ is ample if $\Sigma = \mathbb{P}^2_r$, the projective plane blown-up at $r \leq 8$ generic points.

4.3.1.2 Plane Curves

In the case of plane curves of degree d, Theorem 4.3.1 reads

Corollary 4.3.4 *(a) $V^{irr}_d(S_1, \ldots, S_r)$ is T-smooth (or empty) if*

$$\sum_{i=1}^{r} \big(\tau'(S_i) - \operatorname{isod}'(S_i)\big) < 3d\,. \tag{4.3.1.4}$$

(b) $V_d(S_1, \ldots, S_r)$ is T-smooth at $C = C_1 \cup \ldots \cup C_s$, C_j irreducible and of degree d_j, if for all $j = 1, \ldots, s$,

$$\deg\big(Z'(C) \cap C_j\big) - \operatorname{isod}_{C_j}\big(Z'(C), C\big) < 3d_j\,.$$

Again note that nodes do not count in the l.h.s. of (4.3.1.4). For topological singularity types $S_1, \ldots, S_r$, Theorem 4.3.2 reads

Corollary 4.3.5 (3d–condition)

(a) $V_d^{irr}(S_1, \ldots, S_r)$ *is T-smooth (or empty) if*

$$\sum_{i=1}^{r} \beta(S_i) < 3d . \tag{4.3.1.5}$$

(b) $V_d(S_1, \ldots, S_r)$ *is T-smooth at $C = C_1 \cup \ldots \cup C_s$, C_j irreducible and of degree d_j, if for all $j = 1, \ldots, s$,*

$$\sum_{z \in \operatorname{Sing} C} \beta_{C_i}(C, z) < 3d_j .$$

Recall from Corollary 1.2.22 the inequality $\beta(S_i) \geq \tau^{es}(S_i) - \operatorname{isod}^{es}(S_i)$ for S_i a topological type ($\operatorname{isod}^{es}(S_i) = \operatorname{isod}_z\big(Z^{es}(C), C\big)$), and that we conjecture equality (Conjecture 1.2.23). On the other hand, the beta invariant is more easy to compute than the isomorphism defect.

Combining Corollary 4.3.5 and Lemma 1.2.15, we obtain a numerical sufficient condition for T-smoothness of families of (reducible) plane curves with nodes and cusps as only singularities.

Corollary 4.3.6 *Let $C \in V = V_d(n \cdot A_1, k \cdot A_2)$ and $C = C_1 \cup \ldots \cup C_s$ its decomposition into irreducible components with* $\deg(C_j) = d_j$*. Then V is T-smooth at C if for $j = 1, \ldots, s$ the number k_j of cusps on the component C_j is bounded by*

$$k_j < 3d_j .$$

In particular, the variety of nodal curves $V_d(n \cdot A_1)$ is either empty or T-smooth.

In the same way, for ESF of curves with ordinary multiple points as only singularities we obtain the criterion of Giacinti-Diebolt [Gia]. Recall that $V_d(m_1, ..., m_r)$ denotes the ESF of degree d curves $C \subset \mathbb{P}^2$ having r ordinary multiple points of multiplicity $m_1, ..., m_r$, respectively.

Corollary 4.3.7 *Let $C \in V = V_d(m_1, \ldots, m_r)$ and $C = C_1 \cup \ldots \cup C_s$ its decomposition into irreducible components with* $\deg(C_j) = d_j$*. Then V is T-smooth at C if for $j = 1, \ldots, s$*

$$\sum_{\substack{x \in C_j \cap \operatorname{Sing} C \\ \operatorname{mt}(C,x) > 2}} \operatorname{mt}(C_j, x) < 3d_j .$$

In particular, $V_d^{irr}(m_1, \ldots, m_r)$ is T-smooth (or empty) if

$$\sum_{m_i > 2} m_i < 3d .$$

For ESF of reducible curves with arbitrary singularities, Corollary 3.1.9(b) leads to a more handy numerical condition:

Theorem 4.3.8 (4d–condition [Shu1], Theorem 2, and [GrL1], Corollary 3.9(d)) *$V_d(S_1, \ldots, S_r)$ is T-smooth (or empty) if*

$$\sum_{i=1}^{r} \tau'(S_i) < 4d - 4. \qquad (4.3.1.6)$$

Note that we don't need to except the case when $V = V_d(S_1, \ldots, S_r)$ contains the union of d lines through the origin (that is, $r = 1$, S_1 an ordinary d fold point), since in this case we know that each element of V corresponds to the union of d lines through the origin and the smoothness of V is easily seen. Actually, it is also implied by Corollary 4.3.7, below.

Usually, one refers to condition (4.3.1.4) as the *3d–condition*, while (4.3.1.6) is called the *4d–condition*. Note that there are T-smooth ESF of irreducible curves where the 3d-condition is satisfied, but not the 4d-condition, and vice versa.

Example 4.3.8.1 (a) Nodes do not contribute to the left-hand side of the 3d-condition, while their contribution to the left-hand side of the 4d-condition is 1. In particular, one can conclude Severi's result that each non-empty ESF of (reducible or irreducible) nodal plane curves is T-smooth from the 3d-condition, but not from the 4d-condition.

(b) $V_6(A_{19})$ is a non-empty T-smooth ESF, where the 3d-condition fails, but not the 4d-condition (cf. [GrL1]).

(c) The examples of du Plessis and Wall (Theorem 4.3.26) shows that the 4d-condition can in general not be improved if we use τ as a local singularity invariant. However, Theorem 4.3.9 provides a quadratic condition in d if we use γ instead of τ.

4.3.2 *Quadratic Conditions*

Two of the approaches presented in Chap. 3, namely the ones being based on the Reider-Bogomolov theory of unstable rank 2 vector bundles, respectively on the Castelnuovo function, lead to quadratic sufficient conditions for T-smoothness of ESF of plane curves.

4.3.2.1 Plane curves

It turns out that the condition obtained by applying Theorem 3.5.7 is slightly stronger than the condition obtained via the Reider-Bogomolov approach (cf. also Sect. 4.3.2.2). We obtain

Theorem 4.3.9 ([GLS4], Theorem 1(a)) *Let $C \subset \mathbb{P}^2$ be an irreducible curve of degree $d > 5$ having r singularities $z_1, \ldots, z_r$ of topological or analytic types $S_1, \ldots, S_r$ as its only singularities. Then $V_d^{irr}(S_1, \ldots, S_r)$ is T-smooth at C if*

$$\sum_{i=1}^{r} \gamma'(C, z_i) \leq (d+3)^2, \tag{4.3.2.1}$$

with $\gamma'(C, z_i) = \gamma^{es}(C, z_i)$ for S_i a topological type, resp. $\gamma'(C, z_i) = \gamma^{ea}(C, z_i)$ for S_i an analytic type (cf. Sects. 1.2.3.1 and 1.2.3.2).

At this point, we should like to recall that, by Proposition 1.2.26, the invariant γ' is bounded by $(\tau'_{ci}+1)^2$, where

$$\tau'_{ci}(C, z) := \max\left\{\deg(Z) \mid Z \subset Z^{es}(C, z) \text{ locally a complete intersection}\right\}$$

for (C, z) representing a topological type S, respectively

$$\tau'_{ci}(C, z) := \max\left\{\deg(Z) \mid Z \subset Z^{ea}(C, z) \text{ locally a complete intersection}\right\}$$

for (C, z) representing an analytic type S (cf. Sect. 1.2). Noting that $\tau'_{ci}(C, z) \leq \tau'(C, z)$ (with $\tau' = \tau^{es}$, resp. τ, for topological, resp. analytic, types) we obtain

Corollary 4.3.10 *Under the assumptions of Theorem 4.3.9, $V_d^{irr}(S_1, \ldots, S_r)$ is T-smooth at C if*

$$\sum_{i=1}^{r} \left(\tau'_{ci}(C, z_i) + 1\right)^2 \leq (d+3)^2 ,$$

which holds, in particular, if

$$\sum_{i=1}^{r} (\tau'(C, z_i) + 1)^2 \leq (d+3)^2 .$$

Remark 4.3.10.1 $(\tau' + 1)^2$ is only a very rough estimate for γ' and we expect γ' to be the "correct" invariant for a quadratic upper bound in d. In fact, as τ' appears quadratic in the inequality of Corollary 4.3.10 we get only a linear upper bound in d if we use τ' instead of γ'. Note that this bound is bigger than that in the $4d$-condition of Theorem 4.3.8 for many singularities but lower for only one singularity (i.e., $r = 1$).

For ESF of irreducible curves with nodes and cusps, Propositions 1.2.28(a) and Theorem 4.3.9 (together with Corollary 4.3.6, which covers the case of small degree $d \leq 5$) give an asymptotically optimal condition (by Theorem 4.3.23).

Corollary 4.3.11 *$V_d^{irr}(n \cdot A_1, k \cdot A_2)$ is T-smooth or empty if*

$$4n + 9k \leq (d+3)^2 . \tag{4.3.2.2}$$

Finally, for ESF of irreducible plane curves with only ordinary singularities, we use Proposition 1.2.28(b) and obtain

Corollary 4.3.12 *$V_d^{irr}(m_1, \dots, m_r)$ is T-smooth or empty if*

$$4 \cdot \#(nodes) + \sum_{m_i \geq 3} 2 \cdot m_i^2 \leq (d+3)^2.$$

In particular, it follows that Theorem 4.3.9 is asymptotically proper for ordinary singularities, since the inequality

$$\sum_{i=1}^{r} m_i(m_i - 1) \leq (d-1)(d-2)$$

is necessary for the existence of an irreducible curve with ordinary singularities of multiplicities $m_1, \dots, m_r$, using the δ-invariant and the genus formula.

4.3.2.2 Curves on Arbitrary Surfaces

Let Σ be a smooth projective surface and D_0 an effective divisor on Σ. In this situation, for lack of a generalization of the Castelnuovo function approach, only the Reider-Bogomolov approach leads to a quadratic sufficient condition for the T-smoothness of (non-empty) ESF $V_{|dD_0|}^{irr}(S_1, \dots, S_r)$. We introduce

$$A(\Sigma, D_0) := \frac{(D_0 \cdot K_\Sigma)^2 - D_0^2 K_\Sigma^2}{4},$$

which is a non-negative number if D_0, or K_Σ, is ample by the Hodge index Theorem 3.2.3. Then the H^1-vanishing statement of Theorem 3.3.4 together with Theorem 2.2.38(c2) implies the following sufficient criterion for T-smoothness:

Theorem 4.3.13 ([GLS1], Theorem 1) *Let $C \subset \Sigma$ be an irreducible curve with r singular points $z_1, \dots, z_r$ of topological (respectively analytic) types $S_1, \dots, S_r$, ordered such that $\tau'_{ci}(C, z_1) \geq \dots \geq \tau'_{ci}(C, z_r)$. Assume that C, $C - K_\Sigma$ are ample and $C^2 \geq \max\{K_\Sigma^2, A(\Sigma, C)\}$. If*

$$\sum_{i=1}^{r} \tau'_{ci}(C, z_i) < \frac{(C - K_\Sigma)^2}{4} \tag{4.3.2.3}$$

and, for any $1 \leq s \leq r$,

$$\left(\sum_{i=1}^{s} \left(\tau'_{ci}(C, z_i) + 1\right)\right)^2 < \sum_{i=1}^{s} \left(C^2 - C \cdot K_\Sigma\left(\tau'_{ci}(C, z_i) + 1\right)\right) - A(\Sigma, C) \tag{4.3.2.4}$$

then $V_{|C|}^{irr}(S_1, \dots, S_r)$ is T-smooth at C.

Note that, in general, the conditions (4.3.2.4) are not *quadratic* sufficient conditions in the above sense, since the local singularity invariant τ'_{C_i} appears on the r.h.s., but in many cases they are. For instance, when one a priori restricts the set of allowed singularity types. So, in the case of only nodal singularities, Flamini [Fla, Theorem] derives a bunch of better (in linear terms) sufficient T-smoothness conditions. Another important case, in which we obtain quadratic sufficient conditions for families of curves with arbitrary singularities is that of $-K_\Sigma$ being nef:

Corollary 4.3.14 *Let Σ be a smooth projective surface with $-K_\Sigma$ nef, D_0 an ample divisor on Σ, and $d > 0$ such that $d^2 \geq \max\{K_\Sigma^2/D_0^2, A(\Sigma, D_0)\}$. If*

$$\sum_{i=1}^{r}\left(\tau'_{ci}(S_i)+1\right)^2 \;<\; D_0^2 \cdot d^2 - 2(D_0 \cdot K_\Sigma)\cdot d - A(\Sigma, D_0)$$

then $V^{irr}_{|dD_0|}(S_1, \ldots, S_r)$ is T-smooth or empty.

In particular, we can apply Corollary 4.3.14 to the case $\Sigma = \mathbb{P}^2$, where we obtain the sufficient smoothness condition

$$\sum_{i=1}^{r}\left(\tau'_{ci}(S_i)+1\right)^2 \;<\; d^2 + 6d\,,$$

which is slightly weaker than Corollary 4.3.10 above.

Here and in the following τ'_{ci} may be replaced by τ', since $\tau'_{ci}(S) \leq \tau'(S)$.

For ESF of curves with nodes and cusps, respectively ESF of curves with ordinary singularities we deduce (cf. Sect. 1.2):

Corollary 4.3.15 *With the assumptions of Corollary 4.3.14, let*

$$4n + 9k \;<\; D_0^2 \cdot d^2 - 2(D_0 \cdot K_\Sigma)\cdot d - A(\Sigma, D_0)$$

then $V^{irr}_{|dD_0|}(n \cdot A_1, k \cdot A_2)$ is T-smooth or empty.

Corollary 4.3.16 *With the assumptions of Corollary 4.3.14, let*

$$4 \cdot \#(nodes) + \sum_{m_i \geq 3}\left\lfloor \frac{m_i^2+2m_i+5}{4}\right\rfloor^2 < D_0^2 \cdot d^2 - 2(D_0 \cdot K_\Sigma)\cdot d - A(\Sigma, D_0)$$

then $V^{irr}_{|dD_0|}(m_1, \ldots, m_r)$ is T-smooth or empty.

Finally, we should like to put special emphasis on important examples, ESF of curves on a smooth hypersurface $\Sigma \subset \mathbb{P}^3$ and on the plane $\mathbb{P}^2_k$ blown-up at k generic points.

Theorem 4.3.17 *Let $\Sigma \subset \mathbb{P}^3$ be a smooth hypersurface of degree $e \geq 5$, $D_0 = (e-4)H$, H a hyperplane section, and $d \in \mathbb{Q}$, $d > 1$. Suppose, moreover, that* $\max_i \tau'_{ci}(S_i) \leq d-2$. *If*

$$\sum_{i=1}^{r} \tau'_{ci}(S_i) < \frac{e(e-4)^2}{4} \cdot (d-1)^2 \tag{4.3.2.5}$$

and

$$\sum_{i=1}^{r} \frac{\left(\tau'_{ci}(S_i)+1\right)^2}{1-\frac{\tau'_{ci}(S_i)+1}{d}} < e(e-4)^2 \cdot d^2 \tag{4.3.2.6}$$

then $V^{irr}_{|dD_0|}(S_1, \ldots, S_r)$ is T-smooth or empty.

Proof We apply Theorem 4.3.13. To do so, assume there exists an irreducible curve $C \in V^{irr}_{|dD_0|}(S_1, \ldots, S_r)$. Note that $K_\Sigma \equiv D_0$, hence, $A(\Sigma, C) = 0$ and $C^2 \geq K_\Sigma^2$. Since, by our assumptions, $(d-1)(e-4) > 0$, both C and $C - K_\Sigma$ are ample. Moreover,

$$(C - K_\Sigma)^2 = (d-1)^2(e-4)^2 H^2 = (d-1)^2(e-4)^2 e\,.$$

Hence, condition (4.3.2.3) is equivalent to (4.3.2.5) and it remains to show that (3.3.0.2) is satisfied. $(C.K_\Sigma)/C^2 = 1/d$, and the Cauchy inequality shows that for any $s \leq r$ the quotient of the left- by the right-hand side of (3.3.0.2) can be estimated as

$$\frac{\left(\sum\limits_{i=1}^{s}(\tau'(S_i)+1)\right)^2}{C^2 \cdot \sum\limits_{i=1}^{s}\left(1-\frac{\tau'(S_i)+1}{d}\right)} \leq \frac{1}{C^2}\sum_{i=1}^{s}\frac{\left(\tau'(S_i)+1\right)^2}{\left(1-\frac{\tau'(S_i)+1}{d}\right)}.$$

Therefore, (4.3.2.6) implies (4.3.2.4), hence the statement. □

For ESF of nodal curves, respectively curves with nodes and cusps, we can easily conclude the following quadratic sufficient conditions for T-smoothness:

Corollary 4.3.18 *Let $\Sigma \subset \mathbb{P}^3$ be a smooth hypersurface of degree $e \geq 5$, $D_0 = (e-4)H$, H a hyperplane section, and $d \in \mathbb{Q}$.*

(a) If $d \geq 3$ and

$$4n < e(e-4)^2 \cdot d(d-2) \tag{4.3.2.7}$$

then $V^{irr}_{|dD_0|}(n \cdot A_1)$ is T-smooth or empty.

(b) If $d \geq 4$ and

$$4n + 9k < e(e-4)^2 \cdot d(d-3)$$

then $V^{irr}_{|dD_0|}(n \cdot A_1, k \cdot A_2)$ is T-smooth or empty.

Remark 4.3.18.1 In the case of a quintic surface $\Sigma \subset \mathbb{P}^3$ ($e = 5$), Chiantini and Sernesi [ChS] provide examples of curves $C \subset \Sigma$, $C \equiv d \cdot H$, $d \geq 6$, having $(5/4) \cdot (d-1)^2$ nodes such that $V^{irr}_{|C|}(n \cdot A_1)$ is non-T-smooth at C. In particular, these examples show that the exponent 2 for d in the right-hand side of (4.3.2.7) is the best possible. Actually, it even shows that for families of nodal curves on a quintic surface the condition (4.3.2.7) is asymptotically optimal.

Note that the case of $\Sigma = \mathbb{P}^2_k$, the plane blown-up at $k \leq 9$ generic points is covered by Corollary 4.3.14. Moreover, Corollary 4.3.14 covers also the case $k > 9$ and $D \cdot K_\Sigma \leq 0$. So, the next statement deals with the case of $\Sigma = \mathbb{P}^2_k$, $k > 9$, and the divisor D on Σ with $D \cdot K_\Sigma > 0$. Denote by $E \subset \mathbb{P}^2_k$ the preimage of a generic straight line in $\mathbb{P}^2$, and by $E_1, \ldots, E_k$ the exceptional divisors of the blowing up.

Theorem 4.3.19 *In the notation of Theorem 4.3.13, let*

$$D = dE - d_1E_1 - \ldots - d_kE_k$$

be a divisor on $\Sigma = \mathbb{P}^2_k$ ($\mathbb{P}^2$ blown up at k points), where $E \subset \mathbb{P}^2_k$ is the preimage of a generic straight line in $\mathbb{P}^2$, and $E_1, \ldots, E_k$ are the exceptional divisors, preimages of the blown-up points. Assume that

$$\max_{1\leq i\leq k} d_i \leq \frac{2}{5}d, \quad \sum_{i=1}^{k} d_i > 3d, \quad \sum_{i=1}^{k}(d_i+1)^2 < \frac{9}{10}(d+3)^2\,, \tag{4.3.2.8}$$

$$\sum_{i=1}^{r} \tau'_{ci}(C, z_i) \; < \; \frac{(d+3)^2 - (d_1+1)^2 - \ldots - (d_k+1)^2}{4}\,,$$

and

$$\sum_{i=1}^{k}(\tau'_{ci}(C, z_i)+1)^2 + \frac{d_1+\cdots+d_k-3d}{r} \cdot \sum_{i=1}^{k}(\tau'_{ci}(C, z_i)+1)$$
$$< d^2 - d_1^2 - \ldots - d_k^2 - \frac{k}{4r}\,.$$

Then $V^{irr}_{|D|}(S_1, \ldots, S_r)$ is T-smooth at C.

Proof Since $K_\Sigma = -3E + E_1 + \cdots + E_k$ and

$$\frac{A(C,\Sigma)}{C^2} = \frac{(k-9)(d^2-d_1^2-\ldots-d_k^2)+(3d-d_1-\ldots-d_k)^2}{4(d^2-d_1^2-\ldots-d_k^2)} \leq \frac{k}{4}\,,$$

the statement immediately follows from Theorem 4.3.13 and the ampleness criterion in Proposition 3.2.6. □

Corollary 4.3.20 *In the notation of Theorem 4.3.19, let (4.3.2.8) hold true as well as*

$$n \le \frac{(d+3)^2 - (d_1+1)^2 - \ldots - (d_k+1)^2 - 2}{4}.$$

Then $V^{irr}_{|D|}(n \cdot A_1)$ is T-smooth.

We note that this T-smoothness criterion for $V^{irr}_{|D|}(n \cdot A_1)$ is asymptotically stronger than that in [GLS2].

Remark 4.3.20.1 Performing more thorough computations in the proof of Theorem 3.3.4, one can improve the result of Theorem 4.3.13 for surfaces with Picard number 1 or 2 (see [Kei2]). For example, the following statement, due to Keilen [Kei2], generalizes the T-smoothness criterion (4.3.2.1) for plane curves:

Theorem 4.3.21 *Let Σ be a surface with Neron-Severi group $L \cdot \mathbb{Z}$, L being ample, $D = d \cdot L$, $S_1, \ldots, S_r$ topological or analytic singularity types, and let $K_\Sigma = k \cdot L$. Suppose that $d \ge \max\{k+1, -k\}$, and*

$$\sum_{i=1}^{r} \gamma'(S_i) < \alpha \cdot (D - K_\Sigma)^2, \quad \alpha = \frac{1}{\max\{1, 1+k\}}.$$

Then either $V^{irr}_{\Sigma,|D|}(S_1, \ldots, S_r)$ is empty or it is T-smooth.

It is interesting that the same invariant $\gamma'(S)$ is obtained here by using the Bogomolov-Reider theory of unstable rank two vector bundles on surfaces instead of the Castelnuovo function theory, as is done in the proof of (4.3.2.1).

4.3.3 Obstructed Equisingular Families

In this section, we present several series of non-T-smooth ESF of plane curves and compare the respective data with the sufficient numerical conditions for T-smoothness obtained in Sects. 4.3.1 and 4.3.2. To be precise, we show that

- for plane curves with nodes and cusps condition (4.3.2.2) is asymptotically optimal (Theorem 4.3.23),
- for plane curves with simple singularities of types $A_{6k-1+s}, D_{6k+1+s}, s = 0, \ldots, 4$, condition (4.3.2.1) is asymptotically proper (Theorems 4.3.24, 4.3.25), and, finally,
- the $4d$-condition (4.3.1.6) is sharp, in the sense that it cannot be improved by adding constant terms on the right-hand side (Theorem 4.3.26).

Before starting with concrete examples, we should like to formulate a useful lemma which enables us to conclude that a given ESF of curves is not only non-T-smooth, but even *non-smooth*.

Lemma 4.3.22 *Using the notation of Sect. 4.3.1, let $V \subset V_d(S_1, \ldots, S_r)$ be an irreducible component with*

(a) $H^1\big(\mathcal{J}_{Z'(C)/\mathbb{P}^2}(d)\big) = 0$ *for generic* $C \in V$, *but*
(b) there exists a $C_0 \in V$ *such that* $H^1\big(\mathcal{J}_{Z'(C_0)/\mathbb{P}^2}(d)\big) \neq 0$.

Then either V is non-smooth, or another component of $V_d(S_1, \ldots, S_r)$ meets V at C_0. In particular, $V_d(S_1, \ldots, S_r)$ is non-smooth.

Proof The above conditions imply that $V_d(S_1, \ldots, S_r)$ has the expected dimension at C, but the Zariski tangent space at C_0 has a larger dimension (Theorem 2.2.38). Hence the statement. □

4.3.3.1 Plane Curves with Nodes and Cusps

Theorem 4.3.23 ([GLS4], Proposition 4.5) *Let $p \geq 6$, $q \geq 9$. Then the variety $V_d^{irr}(n \cdot A_1,\ k \cdot A_2)$ has a component without T-property if*

(a) $d = 3p,\ n = 0,\ k = p^2+3p$, *or*
(b) $d = 2q,\ n = q^2-9q,\ k = 6q$,

or, more generally, if $q \geq 2p - 3$ and

(c) $d = 2q,\ k = pq,\ 4n+8k = d^2+6d$.

Remark 4.3.23.1 The series of examples (a) satisfies $\alpha_0 n+9k = d^2+9d$ (for any $\alpha_0 > 0$), whence the coefficient 9 in the sufficient smoothness condition (4.3.2.2) is sharp.

The series (b) satisfies $4n+\beta_0 k = d^2+O(d)$ (for any $\beta_0 > 0$), that is, the coefficient 4 in (4.3.2.2) cannot be improved. Note that the series (b) is the minimal possible violation (with respect to the number of cusps) of the classical Severi-Segre-Zariski bound $k < 3d$.

Proof of Theorem 4.3.23. Let p, q, r, d be non-negative integers such that

$$a := d-3p-2r \geq 0\,, \quad b := d-2q \geq 0\,. \tag{4.3.3.1}$$

Let A, B, P, Q, R be generic homogeneous polynomials of degrees a, b, p, q, r, respectively, and consider the curve C of degree d given by the vanishing of $AP^3R^2+BQ^2 \in H^0\big(\mathcal{O}_{\mathbb{P}^2}(d)\big)$, which is irreducible.

This curve has a node at each point where R and Q both vanish (and $APB \neq 0$), a cusp (tangent to $\{Q = 0\}$) at each point where P and Q vanish (and $ARB \neq 0$), and no other singularities. So, in the initial notation, $n = rq$, $k = pq$.

To show that $V_d(nA_1, kA_2)$ is not T-smooth at C we have, by Theorem 2.2.40, to look for sufficient conditions on a, b, p, q, r such that $H^1\big(\mathcal{J}_{Z^{ea}(C)/\mathbb{P}^2}(d)\big)$ does not vanish, or, in other words, for

$$h^0\big(\mathcal{J}_{Z^{ea}(C)/\mathbb{P}^2}(d)\big) > \frac{(d+1)(d+2)}{2} - n - 2k\,. \tag{4.3.3.2}$$

Note that each $F \in H^0\big(\mathcal{J}_{Z^{ea}(C)/\mathbb{P}^2}(d)\big)$ vanishes at all points where P^2R and Q vanish. Moreover, at any point $z \in \{Q = 0\} \cap \{P^2R = 0\}$ it intersects the curve $\{Q = 0\}$ with multiplicity greater or equal to the local intersection multiplicity $i_z(Q, P^2R)$. Hence, by Noether's fundamental theorem (Theorem 2.1.25), F can be written as

$$F = MP^2R + NQ, \tag{4.3.3.3}$$

with some $M \in H^0\big(\mathcal{O}_{\mathbb{P}^2}(d-2p-r)\big)$, $N \in H^0\big(\mathcal{O}_{\mathbb{P}^2}(d-q)\big)$. On the other hand, any F of type (4.3.3.3) belongs to $H^0\big(\mathcal{J}_{Z^{ea}(C)/\mathbb{P}^2}(d)\big)$ (which simply means that the curve $F = 0$ passes through each point of $\{R = 0\} \cap \{Q = 0\}$ and is quadratically tangent to $\{Q = 0\}$ at each point of $\{P = 0\} \cap \{Q = 0\}$). Moreover, two homogeneous polynomials $MP^2R + NQ$ and $M'P^2R + N'Q$ coincide if and only if $(M - M')P^2R = (N' - N)Q$. In particular, by the (generic) choice of P, Q, R, this is only possible if Q divides $M - M'$. Hence, if we assume

$$d - 2p - r < q \tag{4.3.3.4}$$

then M, N in (4.3.3.3) are unique. It follows that

$$h^0\big(\mathcal{J}_{Z^{ea}(C)/\mathbb{P}^2}(d)\big) = \frac{(d-2p-r+1)(d-2p-r+2)}{2} + \frac{(d-q+1)(d-q+2)}{2}\,.$$

Hence, assuming (4.3.3.4), the required inequality (4.3.3.2) is equivalent to

$$d < 2p + r + q - 2 \tag{4.3.3.5}$$

We conclude that (4.3.3.1) and (4.3.3.5) imply the failure of T-smoothness. One obtains the first series of examples, setting $r = 0$, $d = 3p$, $q = p+3$, $p \geq 6$, and series (b),(c), by setting $d = 2q$, $r = q - 2p + 3$, $p \geq 6$, $q \geq 2p - 3 \geq 9$. □

4.3.3.2 Plane Curves with A_k and D_k Singularities

The following two theorems are originated in [GLS1, Theorems 5 and 6].

Theorem 4.3.24 *Let $\ell \geq 2$, $0 \leq s \leq \ell - 2$, $k > 2\ell - s$, $q \geq 3/(\ell - s - 1)$ be integers. Then there exist irreducible curves C_d of degree $d = qk + qs$ having precisely q^2 $A_{\ell k+s-1}$-singularities such that $V_d^{irr}(q^2 \cdot A_{\ell k+s-1})$ is non-T-smooth at C_d.*

Moreover, if $k > 4\ell - s$ then $V_d^{irr}(q^2 \cdot A_{\ell k+s-1})$ has a component of the expected dimension but is singular at C_d.

Supplement: *If* $k \geq \max\{\ell^2+2\ell\,,\ 4\ell+4-s\}$ *then* $V_d^{irr}(q^2\cdot A_{\ell k+s-1})$ *is locally a complete intersection, which is normal (hence, analytically irreducible), but non-smooth, at* C_d.

Theorem 4.3.25 *Let* ℓ, s, q *be as in Theorem 4.3.24,* $k > 2\ell - s + 1$. *Then there exist irreducible curves* C_d *of degree* $d = qk + qs$ *having precisely* q^2 $D_{\ell k+s+1}$-*singularities such that* $V_d^{irr}(q^2\cdot D_{\ell k+s+1})$ *is non-*T*-smooth at* C_d.

Moreover, if $k > 4\ell - s$ *then* $V_d^{irr}(q^2\cdot D_{\ell k+s+1})$ *has a component of the expected dimension but is singular at* C_d.

Supplement: *If* $k \geq \max\{\ell^2+2\ell+2\,,\ 4\ell+4-s\}$ *then* $V_d^{irr}(q^2\cdot D_{\ell k+s+1})$ *is locally a complete intersection, which is normal (hence, analytically irreducible), but non-smooth*, at C_d.

Example 4.3.25.1 In particular, we obtain examples of irreducible curves corresponding to non-T-smooth germs of the equisingular stratum with

- 1 A_{35}-singularity in degree 9 (cf. also [Lue1]),
- 4 A_{20}-singularities in degree 14,
- 9 A_9-singularities in degree 15,
- 9 D_{13}-singularities in degree 18.

Proof of Theorem 4.3.24. We proceed in two steps. First, we construct an irreducible curve $C_d \in V_d^{irr}(q^2\cdot A_{\ell k+s-1})$ such that $h^1\big(\mathcal{J}_{Z^{ea}(C_d)/\mathbb{P}^2}(d)\big)$ is positive (Part A). Then, we show that in a neighbourhood of C_d there is a one-parameter family of curves $C \in V_d^{irr}(q^2\cdot A_{\ell k+s-1})$ with $h^1\big(\mathcal{J}_{Z^{ea}(C)/\mathbb{P}^2}(d)\big) = 0$ (Part B). Finally, the statement follows by applying Lemma 4.3.22.

Part A. The curve C_d is constructed as follows:
Step 1. Let C_ℓ be the non-singular plane projective curve with affine equation $y + y^\ell - x^\ell = 0$. The implicit function theorem gives a parametrization

$$y \;=\; x^\ell + (\text{higher terms in } x) \;=:\; \varphi(x)$$

for the local branch of C_ℓ at the origin.
Step 2. We define the curve C_{k+s} to be the projective closure of the affine curve given by the vanishing of

$$F(x,y) = (y+y^\ell-x^\ell)^2(1+\lambda_1 x^{k+s-2\ell}) + \lambda_2 x^s y^k + y^{k+s}\,,$$

$\lambda_1, \lambda_2 \in \mathbb{C}$ generic, $0 \leq s \leq \ell-2$, $k > 2\ell - s$.

The Bertini Theorems 2.1.20, 2.1.24 imply that C_{k+s} is irreducible and smooth outside the origin: C_{k+s} is a generic element of the linear system $\mathcal{L} \subset \big|H^0\big(\mathcal{O}_{\mathbb{P}^2}(k+s)\big)\big|$ spanned by the projective closure of the affine curves given by $F_1(x,y) = (y+y^\ell-x^\ell)^2 + y^{k+s}$, $F_2(x,y) = (y+y^\ell-x^\ell)^2 x^{k+s-2\ell}$ and $F_3(x,y) = x^s y^k$. Hence, it is irreducible and smooth away from the basepoints of $\mathcal{L}$ (in particular, outside of $\{x=0\}$). On the other hand, an easy consideration shows that the curve defined by F_1 has precisely one singularity on $\{x=0\}$, the origin.

Applying the (local) analytic coordinate transformation $\tilde{y} := y - \varphi(x)$ to $F(x, y)$, we obtain

$$\begin{aligned} F(x, y) \underset{r}{\sim} & \; \tilde{y}^2(1+\lambda_1 x^{k+s-2\ell}) + \lambda_2 x^s \big(\tilde{y} + x^\ell + (\text{higher terms in } x)\big)^k \\ & \qquad\qquad + \big(\tilde{y} + x^\ell + (\text{higher terms in } x)\big)^{k+s} \\ = & \; \underbrace{\tilde{y}^2 + \lambda x^{\ell k+s}}_{=: \, f_0(x, \tilde{y})} + f'(x, \tilde{y}), \end{aligned}$$

(where $\lambda \neq 0$ and all monomials in $f'(x, \tilde{y})$ are of $(2, \ell k+s)$-weighted order at least $2\ell k + 2s + 1$). Hence, the singularity at the origin is of type $A_{\ell k+s-1}$. Moreover, the polar curve given by the vanishing of $F_y(x, y)$ intersects C_ℓ at the origin with multiplicity $i(C_\ell, F_y; 0) = \ell k + s - \ell$.

Step 3. After choosing *generic* projective coordinates x_0, x_1, x_2 we consider the curves with defining homogeneous polynomials

$$C_d(x_0, x_1, x_2) := C_{k+s}(x_0^q, x_1^q, x_2^q) \quad \text{and} \quad C_{\ell q}(x_0, x_1, x_2) := C_\ell(x_0^q, x_1^q, x_2^q).$$

C_d is an irreducible curve having q^2 singularities of type $A_{\ell k+s-1}$ as only singularities, while $C_{\ell q}$ is non-singular. Since the polar F_x intersects C_ℓ at the origin with multiplicity (at least) $\ell k + s - 1 \geq i(C_\ell, F_y; 0) = \ell k + s - \ell$, we obtain from the above

$$\deg\big(Z^{ea}(C_d, z) \cap C_{\ell q}\big) \; = \; i(C_\ell, F_y; 0) \; = \; \ell k + s - \ell$$

for any $z \in \operatorname{Sing} C_d$. It follows that the restriction to $C_{\ell q}$ of the ideal sheaf associated to $Z^{ea}(C_d)$ is given by $\mathcal{J}_{Z^{ea}(C_d) \cap C_{\ell q}/C_{\ell q}}(d) = \mathcal{O}_{C_{\ell q}}(D)$, where D is a divisor of degree

$$\deg(D) \; = \; d \cdot \ell q - \deg(Z^{ea}(C_d) \cap C_{\ell q}) \; = \; \big(\ell + s\,(\ell - 1)\big)\, q^2 .$$

By Riemann–Roch, we obtain

$$\begin{aligned} h^1\big(\mathcal{J}_{Z^{ea}(C_d) \cap C_{\ell q}/C_{\ell q}}(d)\big) &= h^0\big(\mathcal{J}_{Z^{ea}(C_d) \cap C_{\ell q}/C_{\ell q}}(d)\big) - \deg(D) - 1 + g(C_{\ell q}) \\ &= h^0\big(\mathcal{J}_{Z^{ea}(C_d) \cap C_{\ell q}/C_{\ell q}}(d)\big) + \left(\frac{\ell^2}{2} - \ell - s\,(\ell - 1)\right) q^2 - \frac{3\ell}{2}\, q\,. \end{aligned}$$

Step 4. We show that we can estimate

$$h^0\big(\mathcal{J}_{Z^{ea}(C_d) \cap C_{\ell q}/C_{\ell q}}(d)\big) \; \geq \; \frac{s^2 + 1}{2}\, q^2 + \frac{3s + 3}{2}\, q + 1\,,$$

which implies the non-vanishing of $H^1\big(\mathcal{J}_{Z^{ea}(C_d) \cap C_{\ell q}/C_{\ell q}}(d)\big)$ as

$$0 \leq (\ell - s - 1)^2 \cdot q^2 - (\ell - s - 1) \cdot 3q = (\ell - s - 1)^2 \cdot q^2 \cdot \left(q - \frac{3}{\ell - s - 1}\right).$$

Let $X_q(x_0,x_1,x_2)$, $Y_q(x_0,x_1,x_2)$ be the homogeneous polynomials of degree q obtained from x, respectively y, by the construction in Step 3. Then, as long as $s \le \ell-2$, the restriction map induces an injection

$$\left\{X_q^s Y_q^{k-1} G_q + Y_q^k H_{sq}\right\} \hookrightarrow H^0\left(\mathcal{J}_{Z^{ea}(C_d)\cap C_{\ell q}/C_{\ell q}}(d)\right)$$

where G_q, H_{sq} are homogeneous polynomials of degree q, respectively sq. Moreover, two polynomials $X_q^s Y_q^{k-1} G_q + Y_q^k H_{sq}$ and $X_q^s Y_q^{k-1} G'_q + Y_q^k H'_{sq}$ coincide iff $X_q^s(G_q - G'_q) = Y_q(H'_{sq} - H_{sq})$, that is, iff $G_q = G'_q + \alpha Y_q$ and $H'_{sq} = H_{sq} + \alpha X_q^s$ ($\alpha \in \mathbb{C}$ arbitrary). It follows

$$\begin{aligned} h^0\left(\mathcal{J}_{Z^{ea}(C_d)\cap C_{\ell q}/C_{\ell q}}(d)\right) &\ge \frac{(q+1)(q+2)}{2} + \frac{(sq+1)(sq+2)}{2} - 1 \\ &= \frac{s^2+1}{2} q^2 + \frac{3s+3}{2} q + 1\,. \end{aligned}$$

Part B. We construct a 1-parameter family of curves $C_d^{(\alpha)} \in V_d^{irr}(q^2 \cdot A_{\ell k+s-1})$ with $C_d = C_d^{(0)}$, such that $H^1\left(\mathcal{J}_{Z^{ea}(C_d^{(\alpha)})/\mathbb{P}^2}(d)\right)$ vanishes for a generic choice of α. The following deformation of F,

$$\begin{aligned} F^{(\alpha)}(x,y) &= F(x,y) + \alpha\cdot(y+y^\ell - x^\ell)\cdot y^{\lfloor (k+s)/2\rfloor+1} \\ &\underset{r}{\sim} \tilde{y}^2 - \lambda x^{\ell k+s} + f'(x,\tilde{y}) + \alpha\tilde{y}\left(\tilde{y} + x^\ell + (\text{higher terms in } x)\right)^{\lfloor (k+s)/2\rfloor+1} \end{aligned}$$

is easily seen to preserve the $A_{\ell k+s-1}$-singularity at the origin. Moreover, for generic α, the polar curve given by $F_y^{(\alpha)}(x,y)$ intersects C_ℓ with multiplicity $\ell\lfloor (k+s)/2\rfloor + \ell$ at the origin. As before, after choosing generic projective coordinates x_0, x_1, x_2, we obtain curves $C_d^{(\alpha)} \in V_d^{irr}(q^2 \cdot A_{\ell k+s-1})$ given by homogeneous polynomials

$$C_d^{(\alpha)}(x_0,x_1,x_2) := C_{k+s}^{(\alpha)}(x_0^q, x_1^q, x_2^q)\,.$$

Note that for A_μ-singularities the Tjurina ideal is spanned by the germs intersecting a generic polar with multiplicity at least μ. This implies that $C_\ell^2 \supset Z^{ea}(C_{k+s}^{(\alpha)}, 0)$ and

$$\begin{aligned} \deg\left(Z^{ea}(C_d^{(\alpha)}) : C_{\ell q}\right) &= \deg(Z^{ea}(C_d^{(\alpha)})) - \deg(Z^{ea}(C_d^{(\alpha)}) \cap C_{\ell q}) \\ &= q^2 \cdot (\ell k + s - 1 - \ell\lfloor (k+s)/2\rfloor - \ell) \le \frac{\ell k + s - (\ell+2)}{2} q^2\,. \end{aligned}$$

To save notation, let $Z^{ea} := Z^{ea}(C_d^{(\alpha)})$. We have the exact (reduction) sequences

$$0 \to \mathcal{J}_{Z^{ea}:C_{\ell q}/\mathbb{P}^2}(d-\ell q) \to \mathcal{J}_{Z^{ea}/\mathbb{P}^2}(d) \to \mathcal{J}_{Z^{ea}\cap C_{\ell q}/C_{\ell q}}(d) \to 0\,,$$

respectively

$$0 \to \mathcal{O}_{\mathbb{P}^2}(d-2\ell q) \to \mathcal{J}_{Z^{ea}:C_{\ell q}/\mathbb{P}^2}(d-\ell q) \to \mathcal{J}_{Z^{ea}:C_{\ell q}/C_{\ell q}}(d-\ell q) \to 0\,.$$

Finally, the vanishing of $H^1\big(\mathcal{J}_{Z^{ea}(C_d^{(\alpha)})/\mathbb{P}^2}(d)\big)$ follows by applying the (reduction) Lemma 3.4.2 and induction. □

Proof of Theorem 4.3.25. We proceed similarly to the proof of Theorem 4.3.24. Hence, we mainly restrict ourselves to those parts of the proof where it differs from the above.

Part A. Construction of the curve C_d:
We define the curve C_{k+s} to be the projective closure of the affine curve given by the vanishing of

$$F(x,y) = x(y+y^\ell-x^\ell)^2(1+\lambda_1 x^{k+s-2\ell-1}) + \lambda_2 x^s y^k + y^{k+s}\,,$$

$\lambda_1, \lambda_2 \in \mathbb{C}$ generic, $0 \le s \le \ell-2$, $k > 2\ell - s + 1$. Again, by Bertini's Theorem, C_{k+s} is irreducible, smooth outside the origin, and we obtain

$$F(x,y) \underset{r}{\sim} \underbrace{x\tilde{y}^2 + \lambda x^{\ell k+s}}_{=:\, f_0(x,\tilde{y})} + f'(x,\tilde{y})\,,$$

(where $\lambda \ne 0$ and all monomials in $f'(x,\tilde{y})$ are of $(2, \ell k+s-1)$-weighted order at least $2\ell k + 2s + 1$). Hence, the singularity at the origin is of type $D_{\ell k+s+1}$. Again, the polar curve defined by $F_y(x,y)$ intersects C_ℓ at the origin with multiplicity $i(C_\ell, F_y; 0) = \ell k + s - \ell$, and we can copy the Steps 3 and 4 verbatim.

Part B. Construction of the 1-parameter family $C_d^{(\alpha)} \in V_d^{irr}(q^2 \cdot D_{\ell k+s+1})$:
The following deformation of F,

$$F^{(\alpha)}(x,y) = F(x,y) + \alpha \cdot x \cdot (y+y^\ell-x^\ell) \cdot y^{\lfloor (k+s)/2 \rfloor + 1}$$

is easily seen to preserve the $D_{\ell k+s-1}$-singularity at the origin. Moreover, for generic α, the polar curve given by $F_y^{(\alpha)}(x,y)$ intersects C_ℓ with multiplicity $\ell\lfloor (k+s)/2 \rfloor + \ell + 1$ at the origin. Proceeding as before, we obtain curves $C_d^{(\alpha)} \in V_d^{irr}(q^2 \cdot D_{\ell k+s+1})$ given by homogeneous polynomials

$$C_d^{(\alpha)}(x_0,x_1,x_2) := C_{k+s}^{(\alpha)}(x_0^q, x_1^q, x_2^q)\,.$$

Finally, we apply the above reduction procedure to deduce the vanishing of $H^1\big(\mathcal{J}_{Z^{ea}(C_d^{(\alpha)})/\mathbb{P}^2}(d)\big)$.

While in the A_μ-case after two reductions we end up with a locally free sheaf, in the D_μ-case we cannot assume that $C_\ell^2 \supset Z^{ea}(C_d^{(\alpha)})$. Hence, we have to reduce three times. Indeed, after the third reduction step, we obtain a locally free sheaf, since $C_\ell^3 \supset Z^{ea}(C_d^{(\alpha)})$. This can be seen by applying Noether's fundamental theorem (Theorem 2.1.25): the local branches $\mathcal{P}_1$ (with tangent x), $\mathcal{P}_2$ of $F_y^{(\alpha)}$ satisfy

$$\begin{aligned} i_0(C_\ell^3,\mathcal{P}_1) &= 3 \;=\; 1+i_0(F_x^{(\alpha)},\mathcal{P}_1)\,,\\ i_0(C_\ell^3,\mathcal{P}_2) &= 3\cdot(\ell\cdot\lfloor(k+s)/2\rfloor+\ell) \;>\; \ell k+s \;=\; 1+i_0(F_x^{(\alpha)},\mathcal{P}_2)\,, \end{aligned}$$

In the same manner, it follows that for each $z \in \operatorname{Sing} C_d^{(\alpha)})$ the ideal $\mathfrak{m}_z\cdot\langle C_{\ell q}^2\rangle$ is contained in $I^{ea}(C_d^{(\alpha)},z)$. In particular, $\deg(Z^{ea}(C_d^{(\alpha)}) : C_{\ell q}^2) \le q^2$, and we can conclude by applying the (reduction) Lemma 3.4.2 and induction. □

Proof of the Supplements to Theorems 4.3.24 *and* 4.3.25. We start with the case of A_μ-singularities ($\mu = \ell k + s - 1$). Assume that $V := V_d^{irr}(q^2\cdot A_{\ell k+s-1})$ decomposes locally at C_d into (at least) two components V_1, V_2. Then, by Theorem 2.2.40, we would have at C_d

$$\begin{aligned} \dim\bigl(\operatorname{Sing} V\bigr) &\ge \dim(V_1\cap V_2) \;\ge\; \dim(V_1)+\dim(V_2)-\dim(V)\\ &\ge \frac{d(d+3)}{2} - q^2\mu - h^1\bigl(\mathcal{J}_{Z^{ea}(C_d)/\mathbb{P}^2}(d)\bigr)\,, \end{aligned} \tag{4.3.3.6}$$

since V_1, V_2 have at least the expected dimension.

Step 1. Since $h^1\bigl(\mathcal{J}_{Z^{ea}(C_d)/\mathbb{P}^2}(d)\bigr) > 0$, there exist

- a (minimal) subscheme $Z_0 \subset Z^{ea}(C_d)$ with $h^1\bigl(\mathcal{J}_{Z_0/\mathbb{P}^2}(d)\bigr) > 0$
- a curve $\Delta_0 \in \bigl|H^0(\mathcal{J}_{Z_0/\mathbb{P}^2}(\underbrace{d+3-m}_{=:\,\delta_0}))\bigr|$, $m > (d+3)/2$,

such that $(d-\delta_0+3)\cdot\delta_0 \le \deg(Z_0) \le q^2\mu$ (see the proof of Theorem 4.3.13). In particular, as $k \ge \ell^2+2\ell$,

$$\begin{aligned} \delta_0 &\le \frac{d+3}{2} - \sqrt{\frac{(d+3)^2}{4} - q^2\mu}\\ &\le \frac{d+3}{2} - \sqrt{\left(\frac{d+3}{2} - (\ell+1)q\right)^2} \;=\; (\ell+1)q\,. \end{aligned} \tag{4.3.3.7}$$

Moreover, by construction of C_d, there exists a smooth curve $C_{\ell q}$ of degree ℓq satisfying $C_{\ell q}^2 \supset Z_0$. The exact sequences appearing in the reduction process imply that, due to the non-vanishing of $h^1\bigl(\mathcal{J}_{Z_0/\mathbb{P}^2}(d)\bigr)$, we must have either $h^1\bigl(\mathcal{J}_{Z_0\cap C_{\ell q}/C_{\ell q}}(d)\bigr) > 0$ or $h^1\bigl(\mathcal{J}_{Z_0:C_{\ell q}/C_{\ell q}}(d-\ell q)\bigr) > 0$. But, we know that

$$\begin{aligned} \deg(Z_0 : C_{\ell q}) \le \deg(Z^{ea}(C_d) : C_{\ell q}) &= q^2(\mu-\ell k-s+\ell) = q^2(\ell-1)\\ &< \ell q(d-2\ell q+3)\,, \end{aligned}$$

whence, by Corollary 3.1.9, $h^1\bigl(\mathcal{J}_{Z_0:C_{\ell q}/C_{\ell q}}(d-\ell q)\bigr) = 0$. In view of the above, this implies $h^1\bigl(\mathcal{J}_{Z_0\cap C_{\ell q}/C_{\ell q}}(d)\bigr) > 0$. In particular, due to the minimality of Z_0, we have $C_{\ell q} \supset Z_0$, and

$$\deg(Z_0) \;=\; \deg(Z_0\cap C_{\ell q}) \;\ge\; \ell q(d-\ell q+3). \tag{4.3.3.8}$$

Step 2. Let $V_{>0} := \{C \in V \mid h^1(\mathcal{J}_{Z^{ea}(C)/\mathbb{P}^2}(d)) > 0\} \supset \operatorname{Sing} V$. As before, for any $C \in V_{>0}$ there exist

- a minimal subscheme $Z_0(C) \subset Z^{ea}(C)$ with $h^1(\mathcal{J}_{Z_0(C)/\mathbb{P}^2}(d)) > 0$, and
- a curve $\Delta(C) \in |H^0(\mathcal{J}_{Z_0/\mathbb{P}^2}(\delta(C)))|$ of minimal degree $\delta(C) < (\ell+1)q$.

Note that the schemes $Z^{ea}(C)$, $C \in V_{>0}$, define a family of curvilinear[5] zero-dimensional schemes. Hence, by semicontinuity of cohomology (see [HaR2, Section III.12]), we can assume that, for $C \neq C_d$ sufficiently close to C_d in $V_{>0}$, the schemes $Z_0(C) \subset Z^{ea}(C)$ define a flat family, with fibre over C_d a zero-dimensional scheme $Z_0(C_d)$ containing Z_0, and that the curves $\Delta(C)$ define a flat family of plane curves of degree δ, with fibre over C_d

$$\Delta_0' := \Delta(C_d) \in |H^0(\mathcal{J}_{Z_0(C_d)/\mathbb{P}^2}(\delta))| .$$

Moreover, we can assume that the degrees of the irreducible components of Δ_0' do not exceed the degrees of the corresponding irreducible components of $\Delta(C)$, $C \in V_{>0}$ sufficiently close to C_d.

Step 2a. We show that for $C \in V_{>0}$ sufficiently close to C_d the curve $\Delta(C)$ is reduced, irreducible and of degree $\delta \in [\ell q,\ (\ell+1)q]$.

By the above, we know already that $\delta \leq (\ell+1)q$. On the other hand, by Bézout's Theorem and the construction of $C_{\ell q}$, any curve Δ of degree $\leq (\ell+1)q$ not containing $C_{\ell q}$ as a component satisfies

$$\begin{aligned}\deg(Z^{ea}(C_d) \cap \Delta) &\leq (\Delta . C_{\ell q}) \cdot \frac{\deg Z^{ea}(C_d)}{\deg(Z^{ea}(C_d) \cap C_{\ell q})} \\ &= (\Delta . C_{\ell q}) \cdot \frac{\mu}{\ell k+s-\ell} \leq (\ell^2+\ell)q^2 \cdot \frac{\ell k+s-1}{\ell k+s-\ell} \leq \frac{3}{2} \cdot (\ell^2+\ell)q^2 \\ &< \ell q(d-\ell q+3) \overset{(4.3.3.8)}{\leq} \deg(Z_0) \leq \deg(Z_0(C_d)) . \end{aligned} \tag{4.3.3.9}$$

It follows that Δ_0' contains $C_{\ell q}$ as a component. In particular, by the above, $\Delta(C)$ has to contain a component $\Delta'(C)$ of degree $\delta'(C) \geq \ell q$, and by applying Bézout's Theorem to the intersection of $C_{\ell q}$ with the remaining components $\Delta''(C)$ of $\Delta(C)$ (which altogether are of degree $\delta - \delta'(C) \leq \delta - \ell q$), we can estimate

$$\begin{aligned}\deg(Z_0(C) \cap \Delta'(C)) &\geq \deg Z_0 - \deg(Z_0 \cap \Delta''(C)) \\ &\geq \deg Z_0 - \ell q(\delta - \ell q) \cdot \frac{\mu}{\ell k+s-\ell} > \deg Z_0 - \frac{\ell k+s-1}{\ell k+s-\ell} \cdot \ell q^2 \\ &\overset{(4.3.3.8)}{\geq} \ell q(d-\ell q+3) - 2\ell q^2 > \frac{q^2\mu}{2} = \frac{\deg Z^{ea}(C)}{2} . \end{aligned} \tag{4.3.3.10}$$

[5] A zero-dimensional scheme $Z \subset \mathbb{P}^n$ is called *curvilinear* if, for each p in the support of Z, there is a smooth curve germ $(C, p) \subset (\mathbb{P}^n, p)$ and some integer $\ell \geq 1$ such that $\mathcal{O}_{Z,p} = \mathcal{O}_{C,p}/\mathfrak{m}_p^\ell$.

Since $Z_0(C) \subset Z^{ea}(C)$ is curvilinear, we immediately get $(\Delta'(C))^2 \supset Z_0(C)$ and, by (4.3.3.10),

$$\begin{aligned}\deg(Z_0(C) : \Delta'(C)) &\le \deg Z^{ea}(C) - \deg\big(Z_0(C) \cap \Delta'(C)\big)\\ &< q^2\mu - \ell q(d - \ell q + 3) + 2\ell q^2 \;=\; (\ell^2+2\ell-\ell s+s-1)q^2-3\ell q\\ &\le \ell q \cdot (d-2\ell q+3) \;\le\; \delta'(C)\cdot\big(d-2\delta'(C)+3\big).\end{aligned}$$

Recall that $h^1\big(\mathcal{J}_{Z_0(C)/\mathbb{P}^2}(d)\big) > 0$, which, by Lemma 3.4.2 and the above, implies $h^1\big(\mathcal{J}_{Z_0(C)\cap\Delta'(C)/\Delta'(C)}(d)\big) > 0$. Due to the minimality of $Z_0(C)$ (with respect to inclusion), we obtain $Z_0(C) \subset \Delta'(C)$. Hence, $\Delta(C) = \Delta'(C)$.

Step 2b. The non-vanishing of $h^1\big(\mathcal{J}_{Z_0(C)/\mathbb{P}^2}(d)\big) = h^1\big(\mathcal{J}_{Z_0(C)/\Delta(C)}(d)\big)$ implies

$$\deg Z_0(C) \;\ge\; \delta\,(d - \delta + 3)\,. \tag{4.3.3.11}$$

Moreover, by semicontinuity, the geometric genus satisfies

$$g(\Delta(C)) \;\ge\; g(\Delta_0') \;\ge\; g(C_{\ell q}) \;=\; \frac{(\ell q-1)(\ell q-2)}{2}\,,$$

which gives, in particular,

$$\begin{aligned}\sum_{z\in\Delta(C)} \Big(\mathrm{mt}\big(\Delta(C),z\big)-1\Big) &\le \sum_{z\in\Delta(C)} \frac{\mathrm{mt}\big(\Delta(C),z\big)\big(\mathrm{mt}\big(\Delta(C),z\big)-1\big)}{2}\\ &\le \frac{(\delta-1)(\delta-2)-(\ell q-1)(\ell q-2)}{2} = \frac{(\delta-\ell q)(\delta+\ell q-3)}{2}\,.\end{aligned} \tag{4.3.3.12}$$

Step 3. We can estimate the dimension of the germ $\big(V_{>0}, C_d\big)$ by

- N_1, the degree of freedom in the choice of $Z_0(C)$, plus
- N_2, the degree of freedom (for fixed $Z_0(C)$) in the choice of the constellation $\mathcal{T}^{top}(C)$ (cf. Proposition 1.2.11), plus
- N_3, the dimension of the space of curves $C \in V$ with fixed $\mathcal{T}^{top}(C)$.

Step 3a. For an A_μ-singularity $(D, z) \subset (\mathbb{P}^2, z)$ the scheme $Z^{ea}(D, z)$ is given by the intersection of two smooth plane curve germs. Hence, the degree of freedom in the choice of a subscheme $Z_0(D, z) \subset Z^{ea}(D, z)$ of degree δ *on a fixed curve* Δ is bounded by $\mathrm{mt}\big(\Delta, z\big) - 1$. Equation (4.3.3.12) implies that

$$N_1 \;\le\; \frac{\delta(\delta+3)}{2} + q^2 + \sum_{z\in\Delta(C)} \Big(\mathrm{mt}\big(\Delta(C),z\big)-1\Big) \;\le\; \delta^2 - \frac{7q^2-9q}{2}\,.$$

Given a curve germ (D, z) with A_μ-singularity, $\mathcal{T}^{top}(D, z)$ consists of at most $(\mu/2) + 1$ (free) infinitely near points in the resolution tree, and all of them are fixed when fixing $Z^{ea}(D, z)$. Hence, by (4.3.3.11),

$$N_2 \le \deg Z^{ea}(C) - \deg Z_0(C) \le q^2\mu - \delta(d-\delta+3) .$$

Finally, we define the cluster scheme $Z'(C) := Z(C, \mathcal{T}^{top})$ and obtain

$$\begin{aligned} N_3 \le h^0\big(\mathcal{J}_{Z'(C)/\mathbb{P}^2}(d)\big) - 1 &= \frac{d(d+3)}{2} - \deg(Z'(C)) + h^1\big(\mathcal{J}_{Z'(C)/\mathbb{P}^2}(d)\big) \\ &\le \frac{d(d+3)}{2} - \left(\frac{3\mu}{2}+1\right)\cdot q^2 + h^1\big(\mathcal{J}_{Z'(C_d)/\mathbb{P}^2}(d)\big) . \end{aligned}$$

Combining the estimates for N_1, N_2, N_3 with (4.3.3.6), one obtains

$$\begin{aligned} \frac{d(d+3)}{2} - q^2\mu - h^1\big(\mathcal{J}_{Z^{ea}(C_d)/\mathbb{P}^2}(d)\big) &\le N_1 + N_2 + N_3 \\ \le \delta^2 - \frac{7q^2-9q}{2} + q^2\mu - \delta(d-\delta+3) &+ \frac{d(d+3)}{2} - \left(\frac{3\mu}{2}+1\right)\cdot q^2 \\ &+ h^1\big(\mathcal{J}_{Z'(C_d)/\mathbb{P}^2}(d)\big) . \end{aligned}$$

The right-hand side contains the expression $2\delta^2 - d\delta$, which takes its maximum in the interval $[\ell q, (\ell+1)q]$ at $\delta = \ell q$, as $k \ge 4\ell + 4 - s$. Therefore, the latter inequality implies that

$$\begin{aligned} & h^1\big(\mathcal{J}_{Z^{ea}/\mathbb{P}^2}(d)\big) + h^1\big(\mathcal{J}_{Z'/\mathbb{P}^2}(d)\big) \\ & \qquad \ge \left(\frac{k-4\ell+2s}{2}\cdot\ell + \frac{10-s}{2}\right)\cdot q^2 + \left(3\ell - \frac{9}{2}\right)\cdot q , \end{aligned} \tag{4.3.3.13}$$

with the shortened notation $Z^{ea} = Z^{ea}(C_d)$, $Z' = Z'(C_d)$.

Step 3b. On the other hand, we can again reduce by $C_{\ell q}$ and consider the corresponding exact sequences. Since $C_{\ell q}^2 \supset Z' \supset Z^{ea}$, we obtain

$$\begin{aligned} h^1\big(\mathcal{J}_{Z^{ea}/\mathbb{P}^2}(d)\big) &\le h^1\big(\mathcal{J}_{Z^{ea}\cap C_{\ell q}/C_{\ell q}}(d)\big) + h^1\big(\mathcal{J}_{Z^{ea}:C_{\ell q}/C_{\ell q}}(d-\ell q)\big), \\ h^1\big(\mathcal{J}_{Z'/\mathbb{P}^2}(d)\big) &\le h^1\big(\mathcal{J}_{Z'\cap C_{\ell q}/C_{\ell q}}(d)\big) + h^1\big(\mathcal{J}_{Z':C_{\ell q}/C_{\ell q}}(d-\ell q)\big). \end{aligned}$$

We know that $\mathcal{J}_{Z^{ea}\cap C_{\ell q}/C_{\ell q}}(d) = \mathcal{O}_{C_{\ell q}}(D)$, $\mathcal{J}_{Z'\cap C_{\ell q}/C_{\ell q}}(d) = \mathcal{O}_{C_{\ell q}}(D')$, for some effective divisors D, D', and

$$\begin{aligned} \deg(Z^{ea} : C_{\ell q}) &= q^2(\ell-1) \le \ell q(d-2\ell q+3) , \\ \deg(Z' : C_{\ell q}) \le \deg Z' - \deg(Z^{ea}\cap C_{\ell q}) &\le q^2\left(\frac{3\mu-1}{2} - \ell k - s + \ell\right) \\ &= q^2\left(\frac{\ell k+s}{2} + \ell\right) \le \ell q(d-2\ell q+3) . \end{aligned}$$

Hence, $h^1\big(\mathcal{J}_{Z^{ea}:C_{\ell q}/C_{\ell q}}(d-\ell q)\big)$ and $h^1\big(\mathcal{J}_{Z':C_{\ell q}/C_{\ell q}}(d-\ell q)\big)$ vanish, again, by Corollary 3.1.9. Using the Riemann–Roch formula, we can estimate

$$\begin{aligned} h^1\big(\mathcal{J}_{Z^{ea}/\mathbb{P}^2}(d)\big)+h^1\big(\mathcal{J}_{Z'/\mathbb{P}^2}(d)\big) &= h^1\big(\mathcal{O}_{C_{\ell q}}(D)\big)+h^1\big(\mathcal{O}_{C_{\ell q}}(D')\big) \\ &\le 2g(C_{\ell q}) \;=\; (\ell q-1)(\ell q-2), \end{aligned}$$

which contradicts (4.3.3.13) and the choice of k, ℓ, q.

Step 4. We conclude that the germ $\big(V, C_d\big)$ is irreducible and, moreover,

$$\dim\big(\operatorname{Sing} V, C_d\big) \;\le\; N_1+N_2+N_3 \;<\; \dim\big(V, C_d\big)-2,$$

showing that V, being locally a complete intersection, is normal at C_d.

In the case of D_μ-singularities (with $\mu=\ell k+s+1$) we proceed similarly. The inequality (4.3.3.6) remains the same, and (4.3.3.7) holds as $k\ge \ell^2+2\ell+2$. Further, by construction of $C_d, C_{\ell q}$, we have $C_{\ell q}^3 \supset Z^{ea}(C_d)\supset Z_0$. Equation (4.3.3.8) holds true, since

$$\begin{aligned} &\deg(Z_0:C_{\ell q}) \le \deg(Z^{ea}:C_{\ell q}) \;=\; q^2(\ell+1) \;<\; \ell q(d-2\ell q+3), \\ &\deg(Z_0:C_{\ell q}^2) \le q^2 \;<\; \ell q(d-3\ell q+3)\,. \end{aligned}$$

Again, we deduce that, for all curves $C\in\operatorname{Sing} V$ close to C_d, there exists an irreducible curve $\Delta(C)\supset Z_0(C)$ of degree $\delta\in[\ell q,\ (\ell+1)q]$, which in a similar manner implies that

$$N_1 \;\le\; \delta^2-\frac{7q^2-9q}{2}\,, \qquad N_2 \;\le\; q^2\mu-\delta(d-\delta+3)$$

and

$$\begin{aligned} N_3 \;&\le\; \frac{d(d+3)}{2}-\deg(Z'(C))+h^1\big(\mathcal{J}_{Z'(C)/\mathbb{P}^2}(d)\big) \\ &\le\; \frac{d(d+3)}{2}-\frac{3\mu-1}{2}\cdot q^2+h^1\big(\mathcal{J}_{Z'(C_d)/\mathbb{P}^2}(d)\big)\,. \end{aligned}$$

Thus, we come to an inequality similar to (4.3.3.13) and the same contradiction, saying that V is locally irreducible and normal at C_d. □

4.3.3.3 Further Obstructed Examples

The following example is due to du Plessis and Wall [DPW2].

Theorem 4.3.26 *(a) For any $d\ge 5$ the curve $C\subset\mathbb{P}^2$ given by the equation $(x_1^d+x_2^5x_0^{d-5}+x_2^d=0)$ has the unique singular point $z=(1:0:0)$ with Tjurina number $\tau(C,z)=4d-4$, and satisfies*

$$h^1\big(\mathcal{J}_{Z^{ea}(C)/\mathbb{P}^2}(d)\big) > 0\,. \tag{4.3.3.14}$$

(b) Denote by S the analytic type of the plane curve singularity (C, z). If $d \geq 10$ then the family $V_d(S)$ has a component of expected dimension but is singular at C.

Proof (a) Let $F(x_0, x_1, x_2) := x_1^d + x_2^5 x_0^{d-5} + x_2^d$. One easily verifies that z is the only solution to the system $F_{x_0} = F_{x_1} = F_{x_2} = 0$. In the affine coordinates $x = x_1/x_0$, $y = x_2/x_0$, one has $I^{ea}(C, z) = \langle x^{d-1},\ y^4\rangle$. Hence $\tau(C, z) = 4d - 4$ and

$$H^0\big(\mathcal{J}_{Z^{ea}(C)/\mathbb{P}^2}(d)\big) = \left\{ \sum_{i=d-1}^{d}\sum_{j=0}^{d-i} a_{ij}x^i y^j + \sum_{j=4}^{d}\sum_{i=0}^{d-j} a_{ij}x^i y^j \;\middle|\; a_{ij} \in \mathbb{C} \right\},$$

that means

$$h^0\big(\mathcal{J}_{Z^{ea}(C)/\mathbb{P}^2}(d)\big) = \frac{d^2 - 5d + 12}{2} = h^0\big(\mathcal{O}_{\mathbb{P}^2}(d)\big) - \deg Z^{ea}(C) + 1,$$

which implies (4.3.3.14), in virtue of the exact sequence

$$0 \to H^0\big(\mathcal{J}_{Z^{ea}(C)/\mathbb{P}^2}(d)\big) \to H^0\big(\mathcal{O}_{\mathbb{P}^2}(d)\big) \to H^0\big(\mathcal{O}_{Z^{ea}(C)}\big) \to H^1\big(\mathcal{J}_{Z^{ea}(C)/\mathbb{P}^2}(d)\big).$$

(b) Let $d = 5q + p$, $q \geq 2$, $0 \leq p \leq 4$. Consider the family of curves C_λ of degree d, given by the equation $x_1^d + (x_2 x_0^{q-1} - \lambda x_1^q)^5 x_0^p + x_2^d = 0$, or

$$F_\lambda(x, y) := x^d + (y - \lambda x^q)^5 + y^d = 0$$

in the affine coordinates $x = x_1/x_0$, $y = x_2/x_0$.

We claim that $C_\lambda \in V_d(S)$ for generic $\lambda \neq 0$. Indeed, after the coordinate change $(x, y) \mapsto (x,\ y + \lambda x^q)$ the local equation of C_λ at the origin becomes $x^d + y^5 + (y + \lambda x^q)^d = 0$. Since, for $d \geq 10$,

$$(y + \lambda x^q)^d \in \langle y, x^q\rangle^d \subset \langle y, x^q\rangle^2 \cdot \langle x^{d-1}, y^4\rangle\,,$$

Mather's finite determinacy theorem in the quasi-homogeneous form of [GLS6, Section I.2.2, Lemma 2.25], implies that the singular point of C_λ is contact equivalent to $x^d + y^5$ for any λ.

Now we apply the basic Horace method (cf. Sect. 3.4.1) to show that

$$H^1\big(\mathcal{J}_{Z^{ea}(C_\lambda)/\mathbb{P}^2}(d)\big) = 0 \quad \text{as } \lambda \neq 0\,. \tag{4.3.3.15}$$

Let L be the line $\{y = 0\}$, put $Z_0 = Z^{ea}(C_\lambda, z)$, $Z_i = Z_{i-1} : L$, $i \geq 1$, and consider the series of exact (reduction) sequences

$$H^1\big(\mathcal{J}_{Z_{i+1}/\mathbb{P}^2}(d-i-1)\big) \to H^1\big(\mathcal{J}_{Z_i/\mathbb{P}^2}(d-i)\big) \to H^1\big(\mathcal{J}_{Z_i\cap L/L}(d-i)\big),$$

$i = 0, \ldots, 8$.

Observe that $Z_9 = \emptyset$. Indeed, this is equivalent to $y^9 \in I^{ea}(C_\lambda, z)$. After the coordinate change $y \mapsto y + \lambda x^q$, the latter relation turns into

$$(y+\lambda x^q)^9 \in \langle 5y^4 + d(y+\lambda x^q)^{d-1},\ dx^{d-1} - 5q\lambda x^{q-1}y^4\rangle = \langle y^4, x^{d-1}\rangle,$$

where the coincidence of the two ideals comes from the fact that the first one is contained in the second, and both have the same codimension $\tau(C_\lambda, z) = 4d - 4$ in $\mathcal{O}_{\mathbb{P}^2,z}$. At last, we notice that $y^{9-j}x^{qj} \in \langle y^4, x^{d-1}\rangle$, $j = 0, \ldots, 9$.

Since $H^1\big(\mathcal{O}_{\mathbb{P}^2}(d-9)\big) = 0$, we can conclude (4.3.3.15) from the above exact reduction sequences, provided that $H^1\big(\mathcal{J}_{Z_i\cap L/L}(d-i)\big) = 0$ for $i = 0, \ldots, 8$. By Corollary 3.1.9 (a), this is equivalent to

$$\deg(Z_i \cap L) \le d - i + 1, \quad i = 0, \ldots, 8.$$

Note that $\deg(Z_0 \cap L) = 4q = d - q - p \le d - 2$. If $\deg(Z_k \cap L) \ge d - k + 2$ for some $k \in \{1, \ldots, 8\}$ then the inclusions $Z_0 \supset Z_1 \supset \ldots \supset Z_k$ imply

$$\deg(Z_i \cap L) \ge \deg(Z_k \cap L) \ge d - k + 2, \quad i = 0, \ldots, k.$$

Hence $4d - 4 = \deg Z_0 \ge 4q + k(d - k + 2)$, in particular, $k \le 3$. But this would mean $\deg(Z_0 \cap L) \ge d - k + 2 \ge d - 1$, hence a contradiction.

Now (b) implies that $V_d(S)$ is of expected dimension at C_λ, hence at C, and from (a) and Theorem 2.2.40 we get that $V_d(S)$ is singular at C. □

4.3.4 ESF of Hypersurfaces in $\mathbb{P}^n$

We consider now equisingular families of higher dimensional hypersurfaces with isolated singularities. Let $S_1, \ldots S_r$ be smooth singularity types of $(n-1)$-dimensional hypersurface singularities.

4.3.4.1 Sufficient Conditions for T-Smoothness

Denote by $V_d^n(S_1, \ldots, S_r)$ the set of hypersurfaces of degree d in $\mathbb{P}^n$, $n \ge 3$, whose singular locus contains r points of types $S_1, \ldots, S_r$, respectively (cf. Definition 2.2.53). Then Theorem 3.4.13, (see also Remark 3.4.13.1) together with Theorem 2.2.35 immediately implies the main result of [DPW2, ShT2]:

Theorem 4.3.27 *Let $S_1, \ldots, S_r$ be analytic singularity types satisfying*

$$\sum_{i=1}^{r} \tau(S_i) < \begin{cases} 4d-4 & \text{if } d \geq 5, \\ 18 & \text{if } d = 4, \\ 16 & \text{if } d = 3\,. \end{cases} \tag{4.3.4.16}$$

Then the variety $V_d^n(S_1, \ldots, S_r)$ is T-smooth, that is, smooth of the expected codimension $\sum_{i=1}^{r} \tau(S_i)$ in $\left|H^0\big(\mathcal{O}_{\mathbb{P}^n}(d)\big)\right|$.

A similar statement for topological types of singularities of dimension ≥ 2 cannot be true in general, because, for some singularities, the $\mu =$ const stratum (parameterizing a local topological equisingular family) in the semiuniversal deformation base is not smooth [Lue1]. In other words, topological types of embedded hypersurface singularities of dimension ≥ 2 are not smooth singularity types in general.

However, for semiquasi-homogeneous singular points the $\mu =$ const stratum in a versal deformation base is smooth [Var1], implying that they define smooth singularity types in the sense of Definition 2.2.42 resp. Definition 2.2.44. Hence, the above reasoning gives:

Theorem 4.3.28 *Let $S_1, \ldots, S_q$, $q < r$, be analytic singularity types, and $S_{q+1}, \ldots, S_r$ be (topological) types of semiquasi-homogeneous singular points. If*

$$\sum_{i=1}^{q} \tau(S_i) + \sum_{i=q+1}^{r} \deg Z^{sqh}(S_i) < \begin{cases} 4d-4 & \text{if } d \geq 5, \\ 18 & \text{if } d = 4, \\ 16 & \text{if } d = 3\,, \end{cases}$$

then the variety $V_d^n(S_1, \ldots, S_r)$ is T-smooth, that is, smooth of the expected codimension $\sum_{i=1}^{q} \tau(S_i) + \sum_{i=q+1}^{r} \deg Z^{sqh}(S_i)$ in $\left|H^0\big(\mathcal{O}_{\mathbb{P}^n}(d)\big)\right|$.

Remark 4.3.28.1 Du Plessis and Wall [DPW2] consider also linear systems of hypersurfaces having a fixed intersection with a hyperplane, which does not pass through the singular points, and obtain the statement of Theorem 4.3.27 for such linear systems under the condition

$$\sum_{i=1}^{r} \tau(S_i) < \begin{cases} 3d-3 & \text{if } d \geq 4, \\ 8 & \text{if } d = 3\,. \end{cases}$$

Similarly, one can formulate an analogue of Theorem 4.3.28.

4.3.4.2 Non-T-Smooth ESF of Hypersurfaces

The condition (4.3.4.16) in Theorem 4.3.27 is not necessary, as already seen in the case $n = 2$. In the following, we discuss which kind of sufficient conditions one might expect.

It would be natural to extend or generalize the corresponding results for plane curves to higher dimensions. Theorem 4.3.27 is a generalization of the $4d$-condition for plane curves. However, the $3d$-condition cannot be extended to higher dimensions in the same form, because it allows any number of nodes for plane curves, without affecting T-smoothness, while there are surfaces of degree $d \to \infty$ in $\mathbb{P}^3$ with $5d^3/12 + O(d^2)$ nodes [Chm], and these nodes are dependent, since $\dim \left|H^0(\mathcal{O}_{\mathbb{P}^3}(d))\right| = d^3/6 + O(d^2)$.

We also point out another important difference between the case of curves and the case of higher dimensional hypersurfaces. The quadratic numerical sufficient conditions for T-smoothness of ESF of plane curves are close to necessary existence conditions, which are quadratic in the degree d as well. In higher dimensions the situation is different. Namely, necessary conditions for the non-emptiness of $V_d^n(S_1, \ldots, S_r)$, such as

$$\sum_{i=1}^{r} \mu(S_i) \le (d-1)^n ,$$

are of order d^n in the right-hand side, and we shall show below that, for a fixed n, there are hypersurfaces with arbitrary singularities and the total Milnor number of order d^n. However, any possible sufficient condition for T-smoothness, in the form of an upper bound to the sum of certain positive singularity invariants, can have at most a quadratic function in d on the right-hand side. This follows since any example of non-T-smooth ESF of curves in the plane can be lifted to a non-T-smooth ESF of hypersurfaces in $\mathbb{P}^n$ by the following Lemma 4.3.29.

The following lemma generalizes examples of (analytic) ESF of plane curves which are non-T-smooth to higher dimensions such that the degree of the hypersurface and the total Tjurina number are not touched. In particular, together with Theorem 4.3.26 this proves that the inequality (4.3.4.16) is sharp, that is, cannot be improved by adding a constant (Theorem 4.3.30).

Lemma 4.3.29 *Let C be a reduced plane curve of degree $d > 2$. Then, for any $n > 2$, there exists a hypersurface $W \subset \mathbb{P}^n$ of degree d having only isolated singular points, such that $\tau(W) = \tau(C)$ and*

$$h^1\big(\mathcal{J}_{Z^{ea}(C)/\mathbb{P}^2}(d)\big) = h^1\big(\mathcal{J}_{Z^{ea}(W)/\mathbb{P}^n}(d)\big).$$

Proof Let $F(x_0, x_1, x_2) = 0$ be an equation of C in some homogeneous coordinates for $\mathbb{P}^2$ such that the line $\{x_0 = 0\}$ does not pass through singular points of C. Introduce a family of surfaces $\Sigma_\lambda \subset \mathbb{P}^3$, $\lambda \in \mathbb{C}$, with homogeneous equation $F^{(\lambda)}(x_0, x_1, x_2, x_3) := F(x_0, x_1, x_2) + \lambda x_3^2 x_0^{d-2} + x_3^d = 0$.

Clearly, the singular points of Σ_0 are just the singular points of C in the plane $\mathbb{P}^2 = \{x_3 = 0\}$. Hence, for a generic $\lambda \neq 0$, $\operatorname{Sing} \Sigma_\lambda$ coincides with the singular points of $C \subset \mathbb{P}^2$.

Let $z \in \operatorname{Sing} C \subset \mathbb{P}^2 = \{x_3 = 0\}$. Taking some coordinates (x, y, x_3) in a neighbourhood of z in $\mathbb{P}^3$, one obtains that $F^{(\lambda)}_{x_3} = x_3 \varphi \in \mathcal{O}_{\mathbb{P}^3,z}$, where $\varphi \in \mathcal{O}_{\mathbb{P}^3,z}$ is invertible. Hence

$$\mathcal{O}_{\mathbb{P}^3,z}/I^{ea}(\Sigma_\lambda, z) = \mathcal{O}_{\mathbb{P}^3,z}/\langle F^{(\lambda)}, F_x^{(\lambda)}, F_y^{(\lambda)}, F_{x_3}^{(\lambda)}\rangle = \mathcal{O}_{\mathbb{P}^2,z}/\langle F, F_x, F_y\rangle .$$

This implies that $\tau(\Sigma_\lambda, z) = \tau(C, z)$, and $Z^{ea}(\Sigma_\lambda, z) = Z^{ea}(C, z) \subset \mathbb{P}^2$. The exact sequence

$$0 = H^1\big(\mathcal{O}_{\mathbb{P}^3}(d-1)\big) \to H^1\big(\mathcal{J}_{Z^{ea}(\Sigma_\lambda)/\mathbb{P}^3}(d)\big) \to H^1\big(\mathcal{J}_{Z^{ea}(C)/\mathbb{P}^2}(d)\big) \to 0$$

shows that $h^1\big(\mathcal{J}_{Z^{ea}(\Sigma_\lambda)/\mathbb{P}^3}(d)\big) = h^1\big(\mathcal{J}_{Z^{ea}(C)/\mathbb{P}^2}(d)\big) > 0$. Inductively one proceeds to any dimension $n \geq 3$. □

Note that this lemma allows to generalize non-T-smooth ESF of plane curves to non-T-smooth ESF of hypersurfaces in $\mathbb{P}^n$. Moreover, under the construction in the proof, a 1-parameter family of plane curves lifts to a 1-parameter family of hypersurfaces, having the same behaviour of h^1. Hence, in view of Theorem 4.3.26 and Theorem 2.2.35, we obtain:

Theorem 4.3.30 *Let $n \geq 2$, $d \geq 5$. Then there exists a reduced hypersurface $W \subset \mathbb{P}^n$ of degree d having an isolated singular point z with Tjurina number $\tau(W, z) = 4d - 4$ and being smooth outside z such that the corresponding germ at W of the analytic equisingular stratum is non-T-smooth, (respectively non-smooth, if $d \geq 10$).*

4.4 Independence of Simultaneous Deformations

We give explicit numerical conditions which suffice for the independent realization of versal or lower deformations of singular points of a reduced curve in a smooth projective surface (or, more generally, of a hypersurface in a smooth projective algebraic variety) when varying in the given linear system $|D|$. We present three kinds of results:

- for C a member of a family of curves, or hypersurfaces, we give sufficient conditions which imply that the family is deformation complete, i.e. induces a joint versal deformation of all singular points of C,
- we give sufficient conditions for the existence of 1-parametric deformations, whose members have given singularities or given geometric properties,
- we apply the above criteria to concrete examples.

These results are mostly contained in [GLS1, GLS4, GrK, GrL1, DPW2, ShT2, Shu1, Shu2, Shu4, Shu12]. For the definition and computation of the appearing singularity invariants we refer the reader to Sect. 1.2.

4.4.1 *Joint Versal Deformations*

4.4.1.1 Plane Curves

Let $C \subset \mathbb{P}^2$ be a projective curve of degree d with isolated singularities. We give numerical conditions guaranteeing deformation completeness for the given singularities in the space of curves of degree d or in a certain subspace, where the topological type of the remaining singularities is fixed. The following statement combines the results of [GrK, GrL1, GLS4]

Theorem 4.4.1 *Let C be an irreducible plane curve of degree d with isolated singularities $z_1, \ldots, z_r$.*

(a) The germ at C of the space $\left|H^0\big(\mathcal{O}_{\mathbb{P}^2}(d)\big)\right|$ of curves of degree d induces a joint versal deformation of the singularities $z_1, \ldots, z_r$ of C if

$$\sum_{i=1}^{r}\big(\tau(C, z_i) - \operatorname{isod}\big(Z^{ea}(C, z_i), C\big)\big) < 3d\,, \tag{4.4.1.1}$$

or

$$\sum_{i=1}^{r}\gamma^{ea}(C, z_i) \leq (d+3)^2\,. \tag{4.4.1.2}$$

(b) Furthermore, let $1 \leq p < r$, and let S_i denote the topological type of (C, z_i) for $i = p+1, \ldots, r$. Then the germ of

$$V_d(S_{p+1}, \ldots, S_r) \subset |H^0(\mathcal{O}_{\mathbb{P}^2}(d))|$$

at C is smooth and it induces a joint versal deformation of the singularities $z_1, \ldots, z_p$ of C if

$$\sum_{i=1}^{p}(\tau(C, z_i) - \operatorname{isod}(Z^{ea}(C, z_i), C)) + \sum_{i=p+1}^{r}(\tau^{es}(C, z_i) - \operatorname{isod}(Z^{es}(C, z_i), C)) < 3d\,, \tag{4.4.1.3}$$

or

$$\sum_{i=1}^{p}\gamma^{ea}(C, z_i) + \sum_{i=p+1}^{r}\gamma^{es}(C, z_i) \leq (d+3)^2\,. \tag{4.4.1.4}$$

Remark 4.4.1.1 Intuitively, the germ of $V_d(S_{p+1}, \ldots, S_r)$ at C is the set of curves $C' \subset \mathbb{P}^2$ of degree d, close to C, with singularities z_i' close to z_i for $i = p+1, \ldots, r$, such that (C, z_i) and (C', z_i') are topologically equivalent.

The versality statement of Theorem 4.4.1(b) says that any local deformation of the singularities (C, z_i), $i = 1, ..., p$, can be realized in the family of curves of degree d with the topological type of the remaining singularities (C, z_i), $i = p+1, ..., r$, being fixed.

Furthermore, for reducible curves we get the following sufficient versality criterion (cf. [GrK, GrL1, Shu1]):

Proposition 4.4.2 *Let C be a reduced plane curve of degree d consisting of irreducible components $C_1, \ldots, C_s$ and having singular points $z_1, \ldots, z_r$. Then statement (a) of Theorem 4.4.1 holds true if, for any $j = 1, \ldots, s$,*

$$\sum_{i=1}^{r}\Big(\deg(Z^{ea}(C, z_i) \cap C_j) - \mathrm{isod}_{C_j}(Z^{ea}(C, z_i), C)\Big) < 3 \deg C_j, \qquad (4.4.1.5)$$

or if

$$\sum_{i=1}^{r} \tau(C, z_i) < 4d - 4. \qquad (4.4.1.6)$$

Statement (b) of Theorem 4.4.1 holds true if, for any $j = 1, \ldots, s$,

$$\sum_{i=1}^{p}\Big(\deg(Z^{ea}(C, z_i) \cap C_j) - \mathrm{isod}_{C_j}(Z^{ea}(C, z_i), C)) \qquad (4.4.1.7)$$

$$+ \sum_{i=p+1}^{r} (\deg(Z^{es}(C, z_i) \cap C_j) - \mathrm{isod}_{C_j}(Z^{es}(C, z_i), C)\Big) < 3 \deg C_j,$$

or if

$$\sum_{i=1}^{p} \tau(C, z_i) + \sum_{i=p+1}^{r} \tau^{es}(C, z_i) < 4d - 4. \qquad (4.4.1.8)$$

Proof of Theorem 4.4.1 *and Proposition* 4.4.2. The required versality of the considered deformations in (a) follows from Theorem 2.2.40(d) and from the vanishing of $H^1\big(\mathcal{J}_{Z/\mathbb{P}^2}(d)\big)$, where Z denotes the zero-dimensional scheme

$$Z = \bigcup_{i=1}^{r} Z^{ea}(C, z_i), \quad \text{respectively} \quad Z = \bigcup_{i=1}^{p} Z^{ea}(C, z_i) \cup \bigcup_{i=p+1}^{r} Z^{es}(C, z_i),$$

which in turn follows from the assumptions (4.4.1.1), (4.4.1.3), (4.4.1.5), (4.4.1.7) applying Corollary 3.1.4, or from (4.4.1.6), (4.4.1.8) applying Corollary 3.1.9(b), or from (4.4.1.2), (4.4.1.4) applying Theorem 3.5.7.

Statement (b) of Theorem 4.4.1 and the second statement of Proposition 4.4.2 follow in the same way from Supplement 2.2.39. □

Remark 4.4.2.1 (i) Conditions (4.4.1.6), (4.4.1.8) apply to irreducible curves as well, and are not covered by Theorem 4.4.1.

(ii) In view of Lemma 1.2.19 and Corollary 1.2.22, the inequalities (4.4.1.1), (4.4.1.3) can be replaced by

$$\sum_{i=1}^{r}\big(\tau(C,z_i)-1\big) \;<\; 3d\,,$$

respectively by

$$\sum_{i=1}^{p}\big(\tau(C,z_i)-1\big)+\sum_{i=p+1}^{r}\beta(C,z_i) \;<\; 3d.$$

If (C,z_i) is an ADE-singularity then $Z^{ea}(C,z_i)=Z^{es}(C,z_i)$, and we can replace $\deg\big(Z^{ea}(C,z_i)\cap C_j\big)-\operatorname{isod}_{C_j}\big(Z^{ea}(C,z_i),C\big)$ by $\beta_{C_j}(C,z_i)$ in (4.4.1.5), (4.4.1.7).

(iii) In the second statements of Theorem 4.4.1 and Proposition 4.4.2 one can impose extra requirements to the behavior of the singular points $z_{p+1},\dots,z_r$. For instance, to have a fixed multiplicity, or a fixed Newton diagram, or a tangency of a certain order with a given non-singular curve, etc. That these conditions are allowed follows from the fact, that they define smooth singularity types (Definition 2.2.42) and fixing them results in strictly ideal versal deformations (cf. Proposition 2.2.10 and following remark). Furthermore, we may fix the position of some of the singularities.

(iv) In general, the conditions mentioned in (iii) can be formulated as follows. Let G be a fixed non-singular curve in $\mathbb{P}^2$ and consider curves C' of degree d, close to C, having a singular point z_i' in a neighborhood of z_i, $p<i\le r$, such that (C',z_i') is topologically equivalent to (C,z_i) and the intersection multiplicities $i_{z_i'}(C',G)$ and $i_{z_i}(C,G)$ coincide.

However, to make this precise, we need to fix sections through z_i, $p<i\le r$, of the corresponding families C (see Example 2.2.10.2). We formalize this by considering the incidence variety $\mathfrak{N}\subset\Sigma\times|dH|$ of the linear system $|dH|=|H^0(\mathcal{O}_{\mathbb{P}^2}(d))|$ over $V_d(S_{p+1},\dots,S_r)$, where S_i are the topological types of (C,z_i), $p<i\le r$. Let σ_i denote sections of morphisms of germs

$$(\mathfrak{N},\,(C,z_i))\to(V_d(S_{p+1},\dots,S_r),C)$$

composed with the projection $(\mathfrak{N},\,(C,z_i))\to(\Sigma,z_i)$. We denote by $V_d'(S_{p+1},\dots,S_r)$ the subspace of curves C' in $V_d(S_{p+1},\dots,S_r)$ such that

$$i_{\sigma_i(C')}(C',G)=i_{z_i}(C,G),\quad p<i\le r\,.$$

Note that fixing the intersection multiplicity $i_{z_i}(C,G)$ and the topological type of (C,z_i) means fixing the topological type of $(C\cup G,z_i)$.

Under these requirements the sufficient condition for versality of the family $V_d'(S_1,\dots,S_r)$ in Theorem 4.4.1 will be (4.4.1.3) with

$$\tau^{es}(C, z_i) - \text{isod}_{z_i}(Z^{es}(C, z_i), C)$$

being replaced by

$$\deg\big(Z^{es}(C \cup G, z_i) \cap C\big) - \text{isod}_C\big(Z^{es}(C \cup G, z_i), C \cup G\big)$$

for all $i = p + 1, \ldots, r$. Similarly, in Proposition 4.4.2 one obtains (4.4.1.7) with

$$\deg\big(Z^{es}(C, z_i) \cap C_j\big) - \text{isod}_{C_j}\big(Z^{es}(C, z_i), C\big),$$

$i = p + 1, \ldots, r$, being replaced by

$$\deg\big(Z^{es}(C \cup G, z_i) \cap C_j\big) - \text{isod}_{C_j}\big(Z^{es}(C \cup G, z_i), C \cup G\big).$$

4.4.1.2 Curves on Surfaces

For curves in ample linear systems on smooth projective surfaces we have the following sufficient versality criteria (cf. [GrK, GrL1]).

Proposition 4.4.3 *Let D be an ample divisor on a smooth projective surface Σ, and $C \in |D|$ an irreducible curve with isolated singularities $z_1, \ldots, z_r$.*

(a) The germ at C of the linear system $|D|$ induces a joint versal deformation of the singular points $z_1, \ldots, z_r$ of C if

$$\sum_{i=1}^{r}\Big(\tau(C, z_i) - \text{isod}\big(Z^{ea}(C, z_i), C\big)\Big) < -K_\Sigma \cdot D. \qquad (4.4.1.9)$$

(b) Furthermore, let $1 \le p < r$, and let S_i denote the topological type of (C, z_i). Then $(V_d(S_{p+1}, ..., S_r), C)$ is smooth and induces a joint versal deformation of the singularities $z_1, \ldots, z_p$ of C if

$$\sum_{i=1}^{p}\Big(\tau(C, z_i) - \text{isod}(Z^{ea}(C, z_i), C)\Big) \qquad (4.4.1.10)$$

$$+ \sum_{i=p+1}^{r}\Big(\tau^{es}(C, z_i) - \text{isod}(Z^{es}(C, z_i), C)\Big) < -K_\Sigma \cdot D.$$

Proposition 4.4.4 *Let D be an ample divisor on a smooth projective surface Σ, and $C \in |D|$ a reduced curve consisting of irreducible components $C_1, \ldots, C_s$ and having r singular points $z_1, \ldots, z_r$. Then the statement (a) of Proposition 4.4.3 holds true if, for $j = 1, \ldots, s$,*

$$\sum_{i=1}^{r}\big(\deg(Z^{ea}(C,z_i)\cap C_j)-\mathrm{isod}_{C_j}(Z^{ea}(C,z_i),C)\big) < -K_\Sigma\cdot C_j\,.$$

Statement (b) of Proposition 4.4.4 holds true if, for any $j=1,\dots,s$,

$$\sum_{i=1}^{p}\big(\deg(Z^{ea}(C,z_i)\cap C_j)-\mathrm{isod}_{C_j}(Z^{ea}(C,z_i),C)\big) \tag{4.4.1.11}$$

$$+\sum_{i=p+1}^{r}(\deg(Z^{es}(C,z_i)\cap C_j)-\mathrm{isod}_{C_j}(Z^{es}(C,z_i),C)) < -K_\Sigma\cdot C_j\,.$$

Proof of Propositions 4.4.3, 4.4.4. As before, the required versality follows from Theorem 2.2.38 and the vanishing of $H^1\big(\mathcal{J}_{Z/\Sigma}(D)\big)$, where Z denotes the zero-dimensional scheme

$$Z=\bigcup_{i=1}^{r}Z^{ea}(C,z_i)\,,\quad \text{respectively}\quad Z=\bigcup_{i=1}^{p}Z^{ea}(C,z_i)\cup\bigcup_{i=p+1}^{r}Z^{es}(C,z_i)\,.$$

The latter follows from (4.4.1.9)–(4.4.1.11) according to Corollary 3.1.3. □

Remark 4.4.4.1 (i) In both statements of Proposition 4.4.3 one can formulate numerical sufficient conditions for versality analogous to (3.3.0.2), (3.3.0.2) in Theorem 3.3.4, applied to the corresponding scheme Z as in the above proof.

(ii) Remarks 4.4.2.1(ii), (iii) apply to Propositions 4.4.3, 4.4.4, as well.
(iii) It is worth to emphasize that in the conditions (4.4.1.1), (4.4.1.3), (4.4.1.5), (4.4.1.7), (4.4.1.9), (4.4.1.10) nodes do not count at all, cusps count with coefficient 1, singularities of type A_{2k} count with coefficient $2k-1$, each branch of a singular point of type A_{2k+1} counts with coefficient k.

4.4.1.3 Hypersurfaces in $\mathbb{P}^n$

At last, we present a versality criterion for deformations of hypersurfaces in $\mathbb{P}^n$, $n\ge 3$, as in [DPW1, ShT2].

Proposition 4.4.5 *(1) Let W be a hypersurface of degree d in $\mathbb{P}^n$, $n\ge 3$, having isolated singular points $z_1,\dots,z_r$. Then the germ at W of the space $\big|H^0\big(\mathcal{O}_{\mathbb{P}^n}(d)\big)\big|$ of hypersurfaces of degree d induces a joint versal deformation of the singular points $z_1,\dots,z_r$ if*

$$\sum_{i=1}^{r}\tau(W,z_i) < \begin{cases}4d-4 & \text{if } d\ge 5\,,\\ 18 & \text{if } d=4\,,\\ 16 & \text{if } d=3\,.\end{cases} \tag{4.4.1.12}$$

(2) *Furthermore, let $z_{p+1}, \ldots, z_r$ be semiquasi-homogeneous singular points of W, and let S_i denote the semiquasi-homogeneous type of (W, z_i). Then the germ at W of $V_d(S_{p+1}, ..., S_r) \subset |H^0(\mathcal{O}_{\mathbb{P}^n}(d))|$ is smooth and induces a joint versal deformation of the singular points $z_1, \ldots, z_p$ if*

$$\sum_{i=1}^{p} \tau(W, z_i) + \sum_{i=p+1}^{r} \deg Z^{sqh}(W, z_i) < \begin{cases} 4d-4 & \text{if } d \geq 5\,, \\ 18 & \text{if } d = 4\,, \\ 16 & \text{if } d = 3\,. \end{cases} \qquad (4.4.1.13)$$

The proof coincides with the proof of Theorems 4.3.27, 4.3.28.

Remark 4.4.5.1 Imposing the extra condition that the intersection of the hypersurfaces with a given hyperplane L, which does not pass through any of the singular points $z_1, \ldots, z_r$, is fixed, one obtains sufficient conditions for versality by replacing the right-hand side in (4.4.1.12) and (4.4.1.13) by $3d-3$ if $d \geq 4$, and by 8 if $d = 3$, as shown in [DPW2].

4.4.2 1-Parametric Deformations

4.4.2.1 Deformations of SQH Singularities

Theorem 2.3.13 asserts the existence of 1-parametric deformations of SQH singular points matching certain deformation patterns (cf. Definition 2.3.10). We shall show that sometimes any ($\mathcal{S}$-adjacent) 1-parametric deformation of an SQH singular point matches a deformation pattern defined by a polynomial whose Newton polytope lies under the Newton diagram of the singular point. Clearly, this is the case if such polynomials induce a versal deformation of the singular point, for instance, for all simple singularities. However, even if the space of polynomials defined above does not induce a full versal deformation, it can induce patterns matching all 1-parametric deformations which are not equisingular, and we shall demonstrate this for ordinary 5-fold curve singularities.

Introduce the notation $\mathbb{Z}_+ = \{n \in \mathbb{Z} \, : \, n \geq 0\}$.

Proposition 4.4.6 *Let*

$$F(\boldsymbol{x}) = \sum_{\ell(\boldsymbol{i}) \geq 1,\, \boldsymbol{i} \in \mathbb{Z}_+^n} A_{\boldsymbol{i}} \boldsymbol{x}^{\boldsymbol{i}}\,,$$

be a polynomial with the only singularity at the origin, where $A_{\boldsymbol{i}} \in \mathbb{C}$ for $\boldsymbol{i} \in \mathbb{Z}_+^n$ and $\ell : \mathbb{Z}^n \to \mathbb{Q}$ a linear function, $\ell(\boldsymbol{i}) = w_1 i_1 + \ldots + w_n i_n$, with $w_1 \geq \ldots \geq w_n > 0$. Assume that the lower quasi-homogeneous form of F

$$f(\boldsymbol{x}) = \sum_{\ell(\boldsymbol{i}) = 1,\, \boldsymbol{i} \in \mathbb{Z}_+^n} A_{\boldsymbol{i}} \boldsymbol{x}^{\boldsymbol{i}}\,,$$

has no critical points in $\mathbb{C}^n\backslash\{0\}$ *(i.e., F defines an isolated semiquasi-homogeneous singularity at the origin of weight* $\nabla\ell = (w_1, \ldots, w_n)$*). Let*

$$F_t(\boldsymbol{x}) = \sum_{\boldsymbol{i}\in\mathbb{Z}_+^n} A_{\boldsymbol{i}}(t)\boldsymbol{x}^{\boldsymbol{i}}, \quad |t| < \varepsilon ,$$

be an analytic family of polynomials of degree $\le d$ *such that* $F_0 = F$*, and, for* $0 < |t| < \varepsilon$*,* F_t *is an equianalytic family and satisfies* $\mu(F_t) := \sum_{p\in\mathrm{Sing}\,F_t} \mu(F_t, p) < \mu(f, 0)$.

Then there exists a polynomial

$$G(\boldsymbol{x}) = \sum_{\ell(\boldsymbol{i})\le 1,\, \boldsymbol{i}\in\mathbb{Z}_+^n} A_{\boldsymbol{i}}\boldsymbol{x}^{\boldsymbol{i}}$$

such that

$$\sum_{p\in\mathrm{Sing}\,G} \mu(G, p) < \mu(f, 0) , \tag{4.4.2.1}$$

and there exists an analytic family G_t*,* $|t| < \delta$*,* $\delta \le \varepsilon$*,* $G_0 = G$*, of polynomials of degree* $\le d\frac{w_1}{w_n}$ *equianalytic to* F_t *as* $t \ne 0$.

We observe that if the space polynomials $\mathcal{P}_+(\ell)$ spanned by the monomials $\{\boldsymbol{x}^{\boldsymbol{i}},\ \ell(\boldsymbol{i}) \le 1\}$ induces a joint versal deformation of singular points of any hypersurface $\{P = 0\} \subset \mathbb{C}^n$ with total Milnor number less than $\mu(f, 0)$, then the family G_t in Proposition 4.4.6 can be chosen in $\mathcal{P}_+(\ell)$. This was first observed in [Shu1] with the following example.

Example 4.4.6.1 Any 1-parametric deformation of an ordinary 5-fold singular point of a planar algebraic curve matches a deformation pattern chosen among affine reduced curves of degree 5. Indeed, for any affine curve C of degree 5 with total Milnor number less than 16 the space of curves of degree 5 induces a joint versal deformation of the singular points of C according to Proposition 4.4.2.

A similar statement can be proved in the same manner for semiquasi-homogeneous singularities of topological type $x^3 + y^6 = 0$ and for ordinary triple points of surfaces in $\mathbb{C}^3$. A more careful study allows one to establish the same property for semiquasi-homogeneous singularities of topological type $x^4 + y^8 = 0$ [OrS]. For a particular type of 1-parametric deformations, the statement holds also for ordinary 6-fold points of planar real curves [KoS].

Proof of Proposition 4.4.6. Without loss of generality, we can suppose that

$$\sum_{\ell(\boldsymbol{i})\ne 1} |A_{\boldsymbol{i}}(t)| + \sum_{\ell(\boldsymbol{i})=1} |A_{\boldsymbol{i}}(t) - A_{\boldsymbol{i}}| < c_1, \quad |t| < \varepsilon , \tag{4.4.2.2}$$

with an arbitrary prescribed constant $c_1 > 0$, and that $F_0 = f$, i.e.

$$A_i(0) = 0 \text{ as } \ell(\boldsymbol{i}) \neq 1, \quad A_i(0) = A_i \text{ as } \ell(\boldsymbol{i}) = 1\,; \tag{4.4.2.3}$$

indeed, we can achieve the required relations (4.4.2.2), (4.4.2.3) by reducing ε if necessary, observing that $A_i(0) = 0$ as $\ell(\boldsymbol{i}) < 1$, and choosing a suitable $\beta = \beta(t) > 0$, $\beta^{\ell(\boldsymbol{i})-1}(t)A_i(t) \to 0$ for $t \to 0$, in

$$Q_\beta(F_t) := \beta^{-1} F_t(z_1\beta^{w_1}, ..., z_n\beta^{w_n})$$

$$= \sum_{\ell(\boldsymbol{i})<1} \beta^{\ell(\boldsymbol{i})-1} A_i(t)\boldsymbol{x}^{\boldsymbol{i}} + \sum_{\ell(\boldsymbol{i})=1} A_i(t)\boldsymbol{x}^{\boldsymbol{i}} + \sum_{\ell(\boldsymbol{i})>1} \beta^{\ell(\boldsymbol{i})-1} A_i(t)\boldsymbol{x}^{\boldsymbol{i}}\,. \tag{4.4.2.4}$$

Now introduce the ideal

$$I = \left\{ \Phi \in \mathbb{C}[\boldsymbol{x}] \;:\; \Phi = \sum_{\ell(\boldsymbol{i})\geq 1} B_i \boldsymbol{x}^{\boldsymbol{i}} \right\} \subset \mathbb{C}[\boldsymbol{x}]\,,$$

and consider the ring $\mathbb{C}[\boldsymbol{x}]/I$ with the projection $\Phi \in \mathbb{C}[\boldsymbol{x}] \mapsto [\Phi] \in \mathbb{C}[\boldsymbol{x}]/I$ and the norm

$$\big|[\Phi]\big| = \sum_{\ell(\boldsymbol{i})<1} |B_i| \quad \text{as} \quad \Phi = \sum_{\boldsymbol{i}\in\mathbb{Z}_+^n} B_i \boldsymbol{x}^{\boldsymbol{i}}\,.$$

The ideal $J(f) = \langle [f_{z_1}], ..., [f_{z_n}]\rangle \subset \mathbb{C}[\boldsymbol{x}]/I$ is a finite-dimensional $\mathbb{C}$-linear space with basis

$$\mathcal{B} = \left\{ [f_{z_k}\boldsymbol{x}^{\boldsymbol{i}}], \quad \ell(\boldsymbol{i}) < w_k, \quad k = 1, ..., n \right\}\,,$$

(we leave this claim to the reader as an elementary exercise). Selecting a suitable set $\mathcal{B}'$ of monomials, we extend the set $\mathcal{B}$ to a basis $\hat{\mathcal{B}} = \mathcal{B} \cup \mathcal{B}'$ of $\mathbb{C}[\boldsymbol{x}]/I$. Now we prescribe the constant c_1 in (4.4.2.2) so that, first, $\mathcal{B}_t \cup \mathcal{B}'$ form a basis of $\mathbb{C}[\boldsymbol{x}]/I$ for each $|t| < \varepsilon$, where

$$\mathcal{B}_t := \left\{ [(F_t)_{z_k}\boldsymbol{x}^{\boldsymbol{i}}], \quad \ell(\boldsymbol{i}) < w_k, \quad k = 1, ..., n \right\}\,,$$

and, second, there is a (unique) transformation $T_t : \mathbb{C}^n \to \mathbb{C}^n$ given by

$$z_k \mapsto z_k + \sum_{\ell(\boldsymbol{i})<w_k} \gamma_i^{(k)} \boldsymbol{x}^{\boldsymbol{i}}, \quad k = 1, ..., n, \quad \sup|\gamma_i^{(k)}| < c_2 c_1\,,$$

with some absolute constant $c_2 > 0$, such that $\pi[F_t(T_t(\boldsymbol{x}))] = 0$, where $\pi : \mathbb{C}[\boldsymbol{x}]/I \to J(f)$ is the projection onto the space $J(f)$ along $\mathrm{Span}(\mathcal{B}')$. The existence of the required T_t straightforwardly follows from (4.4.2.2), (4.4.2.3), from $\pi[f] = 0$, and

$$F_t(T_t(\boldsymbol{x})) = F_t(\boldsymbol{x}) + \sum_{k=1}^{n} \sum_{\ell(\boldsymbol{i})<w_k} \gamma_i^{(k)} f_{z_k}\boldsymbol{x}^{\boldsymbol{i}} + O((c_1 + \max|\gamma_i^{(k)}|) \cdot \max|\gamma_i^{(k)}|)\,.$$

Notice that $\deg(F_t(T_t(\boldsymbol{x})) \le d\frac{w_1}{w_n}$ and that

- the spaces $J(f)$ and $\mathrm{Span}(\mathcal{B}')$, and the projections π are invariant with respect to the transformations Q_β,
- $[F_t(T_t(\boldsymbol{x}))] \neq 0 \in \mathbb{C}[\boldsymbol{x}]/I$ as $t \neq 0$, since otherwise F_t would have a singular point in a neighborhood of the origin with Milnor number $\mu(f, 0)$, contrary to the assumption $\sum_{p \in \mathrm{Sing}\, F_t} \mu(F_t, p) < \mu(f, 0)$.

The latter observation implies that there is a uniquely defined positive function $\beta(t)$, $0 < |t| < \varepsilon$, such that $\left|[Q_{\beta(t)}F_t(T_t(\boldsymbol{x}))]\right| = c_1$ (cf. formula (4.4.2.4)). We also immediately derive that (i) $\beta(t) \to 0$ as $t \to 0$, (ii) there exists a limit

$$G(\boldsymbol{x}) = \lim_{t \to 0} Q_{\beta(t)}F_t(T_t(\boldsymbol{x})) = f(\boldsymbol{x}) + \sum_{\ell(\boldsymbol{i})<1} \widetilde{A}_{\boldsymbol{i}}\boldsymbol{x}^{\boldsymbol{i}} \ .$$

From the relations $\left|[G]\right| = c_1 > 0$ and $\pi[G] = 0$ we derive that G satisfies (4.4.2.1) and that $G_t = Q_{\beta(t)}F_t(T_t(\boldsymbol{x})))$ is the family asserted in Proposition 4.4.6.

Remark 4.4.6.2 With the notations introduced in Sect. 2.3.4, the family F_t is an $\mathcal{S}$-adjacent deformation of the SQH polynomial $F = F_0$, where $\mathcal{S}$ denotes the analytic types of the singularities of F_t, $t \neq 0$. Moreover, we have $\mu(F_t) < \mu(F) = \mu(f, 0)$ and we are looking for possible $\mathcal{S}$-adjacent patterns (in the space of lower polynomials, i.e. with monomials on and below the Newton diagram) of the family F_t. By an affine transformation of $\mathbb{C}^n$ and a renormalization, the hypersurface $\{F_t = 0\}$ is isomorphic to $\{G_t = 0\}$, $t \neq 0$, with a controlled increase of the degree. As t goes to 0, all strictly upper monomials in G_t vanish and G_t degenerates to a polynomial G in the space of lower polynomials with $\mu(G)$ strictly smaller than $\mu(F)$. In general G is not an $\mathcal{S}$-adjacent pattern of F_t as some singularities in G_t may collide or degenerate in the limit. However, for relatively simple SQH singularities, the condition $\mu(G) < \mu(F)$ is sufficient for the fact that the lower deformations are a joint versal deformation of all singularities of G. Then G may be deformed into an $\mathcal{S}$-adjacent pattern, matching the original family F_t.

4.4.2.2 Joint Deformations of Hypersurface Singularities

In the cases where the versality criteria given in the preceding section fail, one still can control some deformations of singular curves, namely, lower deformations of semiquasi-homogeneous (SQH) and Newton nondegenerate (NND) singular points (in the sense of Sect. 2.3), because the independence of lower deformations imposes less restrictive conditions than those for versality. The answer, in fact, is given in Theorem 2.3.13. Therefore, in the following we just give numerical sufficient conditions for the validity of the hypothesis (2.3.5.1) of Theorem 2.3.13. In particular, the numerical conditions in Propositions 4.4.7–4.4.9 and Corollary 4.4.10 imply the conclusions of Theorem 2.3.13 (existence and transversality of certain given singularity types).

Proposition 4.4.7 *In the notation of Theorem 2.3.13, let* $V = \mathbb{P}^n$ *and* $W \subset V$ *be a hypersurface of degree* d. *Assume that* $\mathcal{S}$ *is a set of singularity types, and that all the singular points of* W *belong to smooth singularity types (cf. Sect. 2.3.6). Then the* H^1*-vanishing condition (2.3.5.1) holds if*

$$\deg \widehat{Z} < \begin{cases} 4d-4 & \text{if } d \neq 3,4\,, \\ 18 & \text{if } d = 4\,, \\ 16 & \text{if } d = 3\,, \end{cases}$$

where $\widehat{Z}$ *is the zero-dimensional scheme*

$$\widehat{Z} = \bigcup_{z \in S_0} Z_z^{\mathcal{S}} \cup \bigcup_{z \in S_1} Z_z^{ea} \cup \bigcup_{z \in S_2} Z_z^{sqh} \cup \bigcup_{z \in S_3} \widetilde{Z}_z^{sqh} \cup \bigcup_{z \in S_4} Z_z^{nnd} \cup \bigcup_{z \in S_5} \widetilde{Z}_z^{nnd}\,.$$

The proof coincides with the proof of Theorems 4.3.27, 4.3.28.

4.4.2.3 Curves in $\mathbb{P}^2$ and on Arbitrary Surfaces

In the case $n = 2$ we give a series of numerical criteria covering a wide range of cases. Moreover, we give an application to real algebraic geometry by constructing real nonsingular plane curves of degree 8 with the maximal numbers of ovals, arranged in two different ways.

Proposition 4.4.8 *In the notation of Theorem 2.3.13, let* W *be a reduced plane curve of degree* d, *and assume that* $\mathcal{S}$ *is a set of smooth singularity types. Then (2.3.5.1) holds if one of the following conditions is satisfied:*

(1) for Π *a generic polar of* W,

$$\sum_{z \in \operatorname{Sing} W} \left(\deg \widehat{Z}_z - \operatorname{isod}_z(\mathcal{J}_{\widehat{Z}/\Pi}, \mathcal{O}_\Pi)\right) < 4(d-1)\,,$$

(2) W *is irreducible and*

$$\sum_{z \in \operatorname{Sing} W} \gamma_0(W, \widehat{Z}_z) \le (d+3)^2\,,$$

(3) W *is irreducible and*

$$\sum_{z\in S^a\cup S_1}\big(\tau(W,z)-\operatorname{isod}(Z^{ea}(W,z),W)\big)$$
$$+\sum_{z\in S^t}\big(\tau^{es}(W,z)-\operatorname{isod}(Z^{es}(W,z),W)\big)$$
$$+\sum_{z\in S_2\cup S_4}\beta(W,z)+\sum_{z\in S_3}\beta(\widetilde{Z}^{sqh};W,z)+\sum_{z\in S_5}\beta(\widetilde{Z}^{nnd};W,z)\ <\ 3d\,, \tag{4.4.2.1}$$

where S^a resp. S^t denote the parts of S, where the singular points are considered up to analytical, respectively topological equivalence,

(4) *W decomposes into irreducible components $W_1,\dots,W_s$, and, for each $i=1,\dots,s$,*

$$\sum_{z\in S^t\cap W_i}\big(\deg(Z^{es}(W,z)\cap W_i)-\operatorname{isod}_{W_i}(Z^{es}(W,z),W)\big)$$
$$+\sum_{z\in S^a\cup S_1\cap W_i}\big(\deg(Z^{ea}(W,z)\cap W_i)-\operatorname{isod}_{W_i}(Z^{ea}(W,z),W)\big)$$
$$+\sum_{z\in(S_2\cup S_4)\cap W_i}\beta_{W_i}(W,z)+\sum_{z\in S_3\cap W_i}\beta_{W_i}(\widetilde{Z}^{sqh};W,z) \tag{4.4.2.2}$$
$$+\sum_{z\in S_5\cap W_i}\beta_{W_i}(\widetilde{Z}^{nnd};W,z)\ <\ 3\cdot\deg W_i\,.$$

Proposition 4.4.9 *In the notation of Theorem 2.3.13, let W be a reduced curve in an ample linear system $|D|$ on a smooth projective algebraic surface Σ, and let $\mathcal{S}$ be a set of smooth singularity types. Then (2.3.5.1) holds if one of the following conditions is satisfied:*

(1) *W is irreducible and*

$$\sum_{z\in S^a\cup S_1}\big(\tau(W,z)-\operatorname{isod}(Z^{ea}(W,z),W)\big)$$
$$+\sum_{z\in S^t}\big(\tau^{es}(W,z)-\operatorname{isod}(Z^{es}(W,z),W)\big)+\sum_{z\in S_2\cup S_4}\beta(W,z) \tag{4.4.2.3}$$
$$+\sum_{z\in S_3}\beta(\widetilde{Z}^{sqh};W,z)+\sum_{z\in S_5}\beta(\widetilde{Z}^{nnd};W,z)\ <\ -K_\Sigma.D\,,$$

where S^a resp. S^t denote the parts of S, where the singular points are considered up to analytical, respectively topological equivalence, or

(2) *W decomposes into irreducible components $W_1,\dots,W_s$, and, for each $i=1,\dots,s$,*

$$\begin{aligned}\sum_{z\in S^t\cap W_i} & \big(\deg(Z^{es}(W,z)\cap W_i) - \mathrm{isod}_{W_i}(Z^{es}(W,z),W)\big) \\ & + \sum_{z\in S^a\cup S_1\cap W_i} \big(\deg(Z^{ea}(W,z)\cap W_i) - \mathrm{isod}_{W_i}(Z^{ea}(W,z),W)\big) \\ & + \sum_{z\in(S_2\cup S_4)\cap W_i} \beta_{W_i}(W,z) + \sum_{z\in S_3\cap W_i} \beta_{W_i}(\widetilde{Z}^{sqh};W,z) \\ & + \sum_{z\in S_5\cap W_i} \beta_{W_i}(\widetilde{Z}^{nnd};W,z) \;<\; -K_\Sigma.W_i\,. \end{aligned} \tag{4.4.2.4}$$

Basically, the proof coincides with the proof of Theorem 4.4.1 and Propositions 4.4.2, 4.4.3 and 4.4.4, taking into account additional terms corresponding to SQH and NND singular points. The latter come into play with the cluster schemes Z^{sqh}, $\widetilde{Z}^{sqh}$, Z^{nnd}, $\widetilde{Z}^{nnd}$.

Once again we remind that the isomorphism defect is always non-negative, and, according to (1.2.2.5) in Corollary 1.2.22, the summands in the above sums over S^t can be estimated as

- $\tau^{es}(W,z) - \mathrm{isod}(Z^{es}(W,z),W) \;\le\; \beta(W,z)$
- $\deg(Z^{es}(W,z)\cap W_i) - \mathrm{isod}_{W_i}(Z^{es}(W,z),W) \;\le\; \beta_{W_i}(W,z)$,

and those summands over S^a and S_1, which correspond to simple (ADE) singularities can be estimated as

- $\tau(W,z) - \mathrm{isod}(Z^{ea}(W,z),W) \;\le\; \beta(W,z)$
- $\deg(Z^{ea}(W,z)\cap W_i) - \mathrm{isod}_{W_i}(Z^{ea}(W,z),W) \;\le\; \beta_{W_i}(W,z)$.

Finally, we should like to point out that Remark 4.4.2.1(iii) applies in this situation. Namely, sufficient conditions for the existence of the corresponding 1-parametric deformation can be obtained from (4.4.2.1), (4.4.2.2), (4.4.2.3), (4.4.2.4) by the modification explained in that remark.

An important particular case is the following statement (stronger than [GLS4, Corollary 4.7]):

Corollary 4.4.10 *Let $W \subset \mathbb{P}^2$ be an irreducible curve of degree d, and* $\mathrm{Sing}\,W = S_2$ *consist of ordinary singular points of multiplicities $m_1,\dots,m_r$. Then (2.3.5.1) holds if*

$$2\cdot\sum_{m_i=2} m_i + 2\cdot\sum_{m_i>3} m_i^2 \;\le\; (d+3)^2\,,$$

or if

$$\sum_{i=1}^{r} m_i \;<\; 3d\,.$$

This follows from Proposition 4.4.8 and the upper bound for $\gamma_0(W, Z_z^{sqh})$ in Sect. 1.2.

Example 4.4.10.1 We illustrate Proposition 4.4.8 by two applications to real algebraic geometry, namely, the construction of real non-singular plane curves of degree 8 with 22 ovals (the maximal possible number of ovals for degree 8 curves) arranged in the real plane as

$$\langle 4\rangle + 3 * \langle 1\langle 5\rangle\rangle \quad \text{or} \quad \langle k\rangle + 1\langle 2 + 1\langle 18 - k\rangle\rangle,\ k = 16, 13, 2, 5$$

(the symbol $\langle n\rangle$ means n ovals outside each other, the symbol $n * \langle A\rangle$ means n disjoint collections of ovals, each arranged along the scheme A and such that no oval of one collection embraces an oval of another collection, finally the symbol $\langle 1\langle A\rangle\rangle$ means that one oval embraces a collection of ovals arranged along the scheme A).

The first curve is obtained by a deformation of a real irreducible singular curve W of degree 8 with three SQH singular points, topologically equivalent to $\{y^4 + x^5 = 0\}$ (see details in [Shu2]). We notice that the tangents of this singular curve at its singular points cannot be coordinate lines by Bézout's theorem, so the original Viro method does not apply. However, one can realize lower deformations of the three singular points with prescribed non-singular patterns, staying in the space of curves of degree 8. Here the singular points belong to S_2, and the sufficient conditions for prescribed deformation of singular points reads

$$\sum_{z \in S_2} \beta(W, z) \leq 3 \deg W - 1\,,$$

which holds since $\beta(W, z) = 7$ and $3 \deg W - 1 = 23$.

The other four curves are constructed in [Ch2] by deforming the union of 4 conics which have one common point where they intersect with multiplicity 4. This singular point is topologically equivalent to $\{y^4 + x^{16} = 0\}$ and cannot have a semiquasihomogeneous representative of degree 8. According to Proposition 4.4.8(4), a sufficient condition for the existence of the given singular point deformation of degree 8 with prescribed non-singular pattern is $\beta_{W_i}(W, z) \leq 3 \deg W_i - 1$, $i = 1, \ldots, 4$, which holds in view of $\beta_{W_i}(W, z) = 4$, $3 \deg W_i - 1 = 5$.

4.5 Existence

4.5.1 Conditions for the Patchworking Construction

We intend to give numerical sufficient conditions for $\mathcal{S}$-, strong $\mathcal{S}$- (also called $\mathcal{S}^s$-), and weak $\mathcal{S}$-transversality of deformation patterns P for semiquasihomogeneous (SQH) (cf. Definition 2.3.10) and Newton nondegenerate (NND) (cf. Definition 2.3.11) singularities.

Recall that, in an informal way, $\mathcal{S}$-transversality means that the conditions imposed by the singular points of type $S \in \mathcal{S}$ on the space of polynomials of given degree d and given Newton polytope are independent, providing a smooth subspace of the expected dimension. The singularity conditions are encoded in the zero-dimensional scheme Z. For the precise definitions, see Sects. 2.3.2 and 2.3.4.

4.5.1.1 Polynomials in Arbitrary Many Variables

In any dimension $n \geq 2$ one has

Proposition 4.5.1 *Let $P \in \mathbb{C}[x_1, \ldots, x_n]$ be a deformation pattern of an ordinary singular point of multiplicity m. Then P is $\mathcal{S}$-transversal if*

$$\deg Z < \begin{cases} 4m - 4 & \text{if } m \neq 3, 4, \\ 18 & \text{if } m = 4, \\ 16 & \text{if } m = 3\,, \end{cases}$$

and it is $\mathcal{S}^s$-transversal if

$$\deg Z < \begin{cases} 3m - 3 & \text{if } m \neq 3, \\ 8 & \text{if } m = 3\,, \end{cases}$$

where $Z = \bigcup_{w \in \mathrm{Sing}(P=0)} Z_w^{sqh}(P)$.

Proof The $\mathcal{S}$-transversality statement follows immediately from Proposition 2.3.14, Corollary 3.1.9 (b), and Theorem 3.4.13 (see also Remark 3.4.13.1).

For the second statement, denote $Y := \{x'_0 = 0\}$, where the projective coordinates $x'_0, \ldots, x'_n$ are introduced by $x_1 = x'_1/x'_0, \ldots, x_n = x'_n/x'_0$. In the notation of Proposition 2.3.14, put $X = Y \cup Z$. Then, in the exact sequence

$$H^1\big(\mathcal{J}_{Z/\mathbb{P}^n}(m-1)\big) \longrightarrow H^1\big(\mathcal{J}_{X/\mathbb{P}^n}(m)\big) \longrightarrow 0\,,$$

the left term vanishes by Theorem 3.4.13 and Remark 3.4.13.1; hence we obtain the $\mathcal{S}^s$-transversality by Proposition 2.3.14. □

In the case $n = 2$ we can formulate general sufficient criteria for the $\mathcal{S}$-transversality of triples.

Introduce the following notation. Let $F \in \mathbb{C}[x, y]$ be a *peripherally nondegenerate* (PND) polynomial with a nondegenerate Newton polygon $\Delta(F)$, F' a proper divisor of F, and $\sigma \subset \Delta(F)$ an edge.

The quasi-homogeneous polynomial $F^\sigma(x, y)$ splits into irreducible factors each of them giving rise to a local branch of the curve $\{F=0\}$. We introduce the number $\sigma \circ F'$ which counts those local branches in this set which belong to the curve $\{F'=0\} \subset \{F=0\}$. In other words, the truncation $F^\sigma(x, y)$ is the leading form of $F(x, y)$ with respect to some quasi-homogeneous order. Then $\sigma \circ F'$ counts the number of irreducible factors in which the leading form $(F')^{\sigma'}(x, y)$ (with respect to the same order) decomposes.

4.5.1.2 Polynomials in Two Variables

Recall the notion of (Newton) triple and (weak) $\mathcal{S}$-transversality from Definition 2.3.3.

Proposition 4.5.2 *In the notation of Sect. 2.3.6, let $n = 2$ and assume that F is not divisible by any of the variables. The triple (Δ, Δ_+, F) is weakly transversal with respect to topological equivalence of singularities if*

- *F is irreducible and*

$$\sum_{z\in \mathrm{Sing}(F=0)\cap(\mathbb{C}^*)^2} \left(\tau^{es}(F, z) - \mathrm{isod}_z(Z^{es}(F, z), F)\right) < \sum_{\sigma\subset\partial\Delta,\,\sigma\not\subset\Delta_+} (\sigma \circ F)\,, \tag{4.5.1.1}$$

- *the polynomial F factors into the irreducible polynomials $F_1, \dots, F_s$, and, for $i = 1, \dots, s$,*

$$\sum_{z\in \mathrm{Sing}(F=0)\cap(F_i=0)\cap(\mathbb{C}^*)^2} \left(\deg(Z^{es}(F, z)\cap F_i) - \mathrm{isod}_{F_i,z}(Z^{es}(F, z), F)\right) < \sum_{\sigma\subset\partial\Delta,\,s\sigma\not\subset\Delta_+} (\sigma \circ F_i)\,. \tag{4.5.1.2}$$

The triple (Δ, Δ_+, F) is weakly transversal with respect to analytic equivalence of singularities if

- *F is irreducible and*

$$\sum_{z\in \mathrm{Sing}(F=0)\cap(\mathbb{C}^*)^2} \left(\tau(F, z) - isod_z(\mathcal{J}_{Z^{ea}(F)/F}, \mathcal{O}_F)\right) < \sum_{\sigma\subset\partial\Delta,\sigma\not\subset\Delta_+} (\sigma \circ F)\,, \tag{4.5.1.3}$$

- *the polynomial F factors into the product of irreducible polynomials $F_1, \dots, F_s$, and, for $i = 1, \dots, s$,*

$$\sum_{z\in \mathrm{Sing}(F=0)\cap(\mathbb{C}^*)^2} \left(\tau(F_i, z) - isod_{F_i,z}(\mathcal{J}_{Z^{ea}(F)\cap F_i/F_i}, \mathcal{O}_{F_i}) + i_z(F_i, \widehat{F_i})\right) < \sum_{\sigma\subset\partial\Delta,\sigma\not\subset\Delta_+} (\sigma \circ F_i)\,, \tag{4.5.1.4}$$

where $\left\{\widehat{F_i} = 0\right\}$ is the union of all components of $\{F = 0\}$ except $\{F_i = 0\}$.

Proof This again follows from Proposition 2.3.14 and the H^1-vanishing criteria of Corollary 3.1.4. For instance, inequalities (4.5.1.2) can be rewritten as

$$\sum_{z\in \mathrm{Sing}(F=0)\cap(F_i=0)\cap(\mathbb{C}^*)^2} \big(\deg(Z^{es}(F,z)\cap F_i) - \mathrm{isod}_{F_i,z}(Z^{es}(F,z),F)\big)$$
$$< 3\deg F_i - \sum_{z\in\mathrm{supp}(\widetilde{Y})} \beta_{C_i}(\widetilde{Y};C,z) \quad i=1,\ldots,s\,, \tag{4.5.1.5}$$

where $C=\{F=0\}$, $C_i=\{F_i=0\}$, and the scheme $\widetilde{Y}$ is defined in Proposition 2.3.14 (a routine computation, leading to (4.5.1.5), is left as an exercise; details can be found in Example 1.2.17.1 and in [Shu11]). Then (4.5.1.5) imply inequalities (3.1.2) in the formulation of Corollary 3.1.4 with $d'=d=\deg F$, since

$$\deg(\widetilde{Y}_z\cap C_i) - \mathrm{isod}_{C_i}(\widetilde{Y}_z, C) \le \beta_{C_i}(\widetilde{Y}:C,z), \quad i=1,...,s\,,$$

for each point $z\in\mathrm{supp}(\widetilde{Y})$ (cf. (1.2.2.5) in Corollary 1.2.22).

In the same manner one treats inequalities (4.5.1.1), (4.5.1.3), and (4.5.1.4). □

We also should like to point out that the isomorphism defect is always non-negative, and the summands in the above sums for topological types can be estimated as

- $\tau^{es}(W,z) - \mathrm{isod}_z(\mathcal{J}_{Z^{es}(W)/W}, \mathcal{O}_W) \le \beta(W,z)$
- $\deg(Z^{es}(W,z)\cap W_i) - \mathrm{isod}_{W_i,z}(\mathcal{J}_{Z^{es}(W)/W}\otimes\mathcal{O}_{W_i}, \mathcal{O}_W) \le \beta_{W_i}(W,z)$

according to (1.2.2.5) in Corollary 1.2.22.

Proposition 4.5.2 allows to give a complete answer in the case of curves and deformation patterns with only nodal singular points.

Corollary 4.5.3 *If $n=2$ then any deformation pattern of an SQH or NND singular point which has only nodes as singularities, is $\mathcal{S}^s$-transversal, $\mathcal{S}$ being a topological or analytic classification.*

Proof Given a deformation pattern with nodes as only singularities, orient the graph G as described in the proof of Proposition 2.3.16. Introduce projective coordinates in $\mathbb{P}^2$ by $x=x_1/x_0$, $y=x_2/x_0$. If a triple $(\Delta,\Delta_+(G),F)$ is an element of this pattern, then the edges $\sigma\subset\Delta$, $\sigma\not\subset\Delta_+(G)$, describe all the local branches of the projective closure of the curve $\{F=0\}$, which are centered on $\{x_1=0\}\setminus\{x_0=0\}$, or on $\{x_2=0\}\setminus\{x_0=0\}$, or at $(0:1:0)$, at which point they are transversal to the line $\{x_2=0\}$. The only polynomials which define a curve without such branches are x_0^r, $r\ge 0$, which cannot be the case for F or its proper divisors. Hence in the inequalities (4.5.1.1), (4.5.1.2) we always have a positive right-hand side with vanishing left-hand side. □

Remark 4.5.3.1 We should like to comment on Corollary 4.5.3 and its proof. The singularity invariants in the right-hand side of (4.5.1.1) and (4.5.1.2) vanish for ordinary nodes. So, by Proposition 4.5.2 the family of plane curves of a given degree, which have a given number of nodes in the torus $(\mathbb{C}^*)^2$ and are irreducible in $(\mathbb{C}^*)^2$, is smooth and intersects transversally with the space of curves, defined by polynomials with given Newton polygon and with fixed coefficients on some but not all edges of the Newton polygon.

In the case of curves which are reducible in $(\mathbb{C}^*)^2$ the requirement is that the components of the curves in the given family are defined by polynomials with certain Newton polygons and fixed coefficients on some but not all edges of their Newton polygons. In the proof of Corollary 4.5.3 these requirements read as $\Delta_+(G) \ne \partial\Delta$, and, in fact, we show that any curve used in the deformation pattern definition, or its irreducible component, has local branches centered on the coordinate lines and not induced by polynomial truncations at edges $\sigma \subset \Delta_+(G)$.

4.5.2 *Plane Curves with Nodes and Cusps*

Questions concerning the number of nodes and cusps on a plane curve of a given degree are classical and not trivial, in particular over the reals.

In the whole section we consider complex or real curves which are irreducible over the complex numbers.

4.5.2.1 Known Results for the Maximal Number

The complete answer for *nodes* is due to Severi [Sev]: a *complex* irreducible plane curve of degree $d \ge 3$ with n nodes as only singularities exists iff

$$n \le \frac{d^2 - 3d + 2}{2},$$

the upper bound coming from the genus formula (4.2.2.2). The construction consists in smoothing any prescribed number of nodes of the union of d generic straight lines, which is possible by the results of Sect. 4.2 (cf. Example 4.2.0.3). Nodes of *real* curves can be of three types:

(a) a non-isolated real node, locally equivalent to $\{x^2 - y^2 = 0\}$,
(b) an isolated real node, locally equivalent to $\{x^2 + y^2 = 0\}$,
(c) an imaginary node.

The complete answer for real nodal curves reads as follows: for any $d \ge 3$ and any non-negative n_1, n_2, n_3 satisfying

$$n_1 + n_2 + 2n_3 \le \frac{d^2 - 3d + 2}{2}$$

there exists a real plane irreducible curve with n_1 non-isolated real nodes, n_2 isolated real nodes and $2n_3$ imaginary nodes. This was proven in [Shu5] by means of the patchworking construction, and in a different way in [Pec1, Pec2].

The case of curves with *nodes and cusps* is much more difficult, no complete answer is known in general, neither in the real case, nor in the complex one. For an integer $d \ge 3$, denote by $k_{max}(d)$ (respectively, $k_{real}(d)$) the maximal possible number of cusps on a complex (respectively, real) curve of degree d.

d	3	4	5	6	7	8	9	10
$k_{max}(d)$	1	3	5	9	10	15	20 or 21	26
$k_{real}(d)$	1	3	5	7	10	≥ 14		

For upper bounds to the number of cusps, we refer to Example 4.2.0.1, respectively the references therein. For constructions of curves with the above (maximal) number of cusps, we refer to [Gud4] ($d = 5$), [ItS2, Shu11] ($d = 6, 7, 8, 9$), [Koe] ($d \le 7$), respectively [Hi] ($d = 10$).

Using consequences of the Bogomolov–Miyaoka–Yau inequality (see [Miy, Yau]), Ivinskis [HiF, Ivi] derived an upper bound for the number n of nodes and number k of cusps on a curve of an even degree $d \ge 10$:

$$\frac{9}{8}n + 2k \le \frac{5}{8}d^2 - \frac{3}{4}d < 0.625d^2 .$$

Langer [Lan, Theorem 1] generalized the Bogomolov–Miyaoka–Yau inequality to orbifold Euler numbers and obtained an upper bound to the number of simple singularities of curves on surfaces. In particular (see [Lan, Theorem 9.4.2 and formula (11.1.1)]), for any reduced curves of degree $d \ge 10$ with n nodes and k cusps it yields the bound

$$(2-\alpha)n + \left(\frac{7}{2} - \frac{3}{2}\alpha - \frac{1}{24\alpha}\right)k \le \left(1 - \frac{\alpha}{3}\right)d^2 - d \qquad (4.5.2.6)$$

with an arbitrary $\frac{3}{10} \le \alpha \le \frac{5}{6}$, which is always better than Ivinskis' bound. Substituting $\alpha = \frac{\sqrt{73}-1}{24}$, one obtains the maximal coefficient of k in (4.5.2.6), and hence

$$\frac{6059 + 7\sqrt{73}}{10512}n + k \le \frac{125 + \sqrt{73}}{432}d^2 - \frac{511 + 11\sqrt{73}}{1752}d .$$

As a consequence, the maximal number $k_{\max}(d)$ satisfies

$$\limsup_{d\to\infty} \frac{k_{\max}(d)}{d^2} \le \frac{125+\sqrt{73}}{432} = 0.3091\ldots.$$

On the other hand, a series of cuspidal curves constructed by A. Calabri et al. [CPS, Theorem 6] shows that

$$\limsup_{d\to\infty} \frac{k_{max}(d)}{d^2} \ge \frac{2567}{8640} = 0.2971\ldots.$$

Note that the expected dimension of equisingular families in these examples is negative, that is, the cusps are dependent. Hence, one cannot expect that there exist curves of given degrees with any intermediate number of cusps. We also mention other constructions of curves with many cusps,

(i) by Zariski [Zar3]: a curve $\{F^2+G^3=0\}$, where $F,G \in \mathbb{C}[x_0,x_1,x_2]$ are generic homogeneous polynomials of degrees $3r$, $2r$, respectively, has degree $d=6r$ and $k=6r^2=d^2/6$ cusps (which may be all real);
(ii) by Ivinskis [Ivi]: let $\{F(x_0,x_1,x_2)=0\}$ be a projective sextic, dual to a non-singular cubic, then the curve $\{F(x_0^r,x_1^r,x_2^r)=0\}$ $(r\ge 1)$, has degree $d=6r$ and $k=9r^2=d^2/4$ cusps.

Note, however, that the cusps in both constructions are dependent, hence again one cannot expect that all intermediate numbers of cusps can be realized by curves of the given degree. The constructions by Ivinskis, Hirano, Kulikov, and Calabri et al. give a rather small number of real cusps.

4.5.2.2 Asymptotically Proper Results

We present here results, obtained by means of the patchworking construction, which give an asymptotically proper answer to the problem stated in the real and complex case. Moreover, the answer becomes asymptotically sharp under the T-smoothness assumption.

Theorem 4.5.4 ([Shu5], Theorems 2.2, 3.3, and 4.1)

(i) For any non-negative integers $d\ge 3$ and k such that

$$k \le \frac{d^2-3d+4}{4}$$

there exists a real plane irreducible curve of degree d having k real cusps as only singularities.

(ii) For any non-negative integers d, n, k such that

$$n+2k \le \frac{d^2-4d+6}{2}, \tag{4.5.2.7}$$

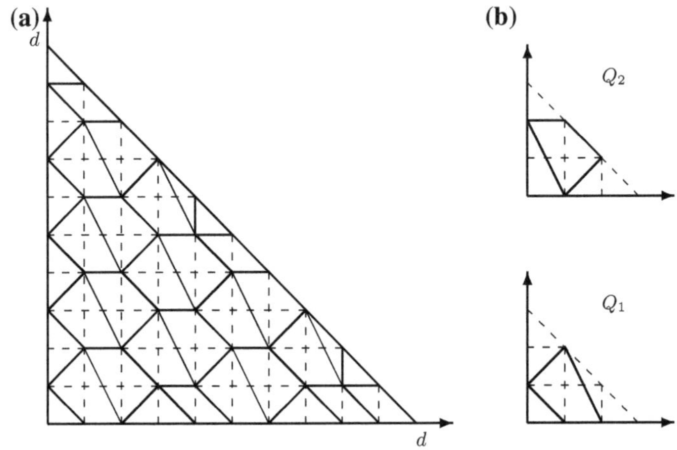

Fig. 4.5 Construction of curves with cusps

there exists a (complex) plane irreducible curve with n nodes and k cusps as only singularities.

Moreover, these curves correspond to T-smooth germs of the respective equisingular families of curves with k cusps in (i) resp. with n nodes and k cusps in (ii).

Remark 4.5.4.1 In fact, the argument presented below allows to prove that, for any non-negative integers n_1, n_2, n_3, k_2, k_2 and $d > 0$ such that

$$n_1 + n_2 + 2n_3 + 2k_1 + 4k_2 \le \frac{d^2}{2} + O(d)\ ,$$

there exists a real irreducible curve of degree d with n_1 non-isolated real nodes, n_2 isolated real nodes, $2n_3$ imaginary nodes, k_1 real cusps and $2k_2$ imaginary cusps as its only singularities [Shu5].

Proof (i) We shall define a convex subdivision of the Newton triangle

$$T_d^2 := \text{conv}\{(0,0),\ (0,d),\ (d,0)\}$$

of a generic polynomial $F \in \mathbb{R}[x, y]$ of degree d and choose suitable coefficients $A_{\boldsymbol{i}}$, $\boldsymbol{i} \in T_d^2$, such that the block polynomials will have real cusps in $(\mathbb{R}^*)^2$. Then we show that altogether they glue up into a real curve with the required number of cusps.

Step 1. Consider the tiling of the plane by hexagons as shown in Fig. 4.5. We divide each hexagon into two quadrangles by a diagonal (see Fig. 4.5(a)), which results in tiling the plane by the translates of the quadrangles Q_1 and Q_2 (Fig. 4.5(b)).

The subdivision of $\mathbb{R}^2$ into these quadrangles is convex. Indeed, it can be defined by a piece-wise linear convex function $\nu : \mathbb{R}^2 \to \mathbb{R}$, where $\nu(i, j) = i^2 + 2j^2$ at any vertex (i, j) of the subdivision. We leave it as an exercise for the reader to verify that this function ν extends linearly on any quadrangle and determines a convex subdivision around any vertex. The restriction of ν on $T_d^2 \cap \mathbb{Z}^2$ extends to a convex piece-wise linear function, which divides T_d^2 into $[(d^2 - 3d + 4)/4]$ translates of Q_1, Q_2 and some other polygons.

Step 2. We show that,

for any given non-zero $a, b, c \in \mathbb{R}$, there exist real PND polynomials

$$\begin{aligned} F_1 &= ay + bx + cx^2 + a_{11}xy + a_{12}xy^2 , \\ F_2 &= ay^2 + bx + cx^2y + b_{11}xy + b_{12}xy^2 , \end{aligned} \tag{4.5.2.8}$$

with Newton polygons Q_1, Q_2, respectively, and such that the curve $\{F_i = 0\}$, $i = 1, 2$, has a real cusp in $(\mathbb{R}^*)^2$.

Namely, take a real cuspidal cubic C and two projective coordinate systems such that

- the line $\{x_1 = 0\}$ intersects C at $(1:0:0)$ and is quadratically tangent to C at $(0:0:1)$, the line $\{x_0 = 0\}$ intersects C at $(0:0:1)$ and is quadratically tangent to C at $(0:1:0)$;
- the line $\{x_1' = 0\}$ intersects C at $(0:0:1)$ and is quadratically tangent to C at $(1:0:0)$, the line $\{x_2' = 0\}$ intersects C at $(1:0:0)$ and is quadratically tangent to C at $(0:1:0)$.

The equations $\widetilde{F}_1(x, y) = 0$ and $\widetilde{F}_2(x, y) = 0$ of C in the affine coordinates $x = x_1/x_0$, $y = x_2/x_0$, or $x = x_1'/x_0'$, $y = x_2'/x_0'$ have Newton polygons Q_1, Q_2 respectively. Now one can choose the coefficients of these polynomials to coincide for the respective terms with those indicated in (4.5.2.8), by means of a suitable transformation

$$F(x, y) \longmapsto \alpha F(\beta x, \gamma y), \quad \alpha, \beta, \gamma \in \mathbb{R} \setminus \{0\} .$$

Step 3. We construct a sequence $A_{\boldsymbol{i}}$, $\boldsymbol{i} \in T_d^2$, inductively. To do so, we introduce the following order for the quadrangles parallel to Q_1, Q_2 in the subdivision of T_d^2. In each quadrangle Q take its interior integral point $\big(i(Q), j(Q)\big)$ and put

$$Q' \succ Q'' \;:\Longleftrightarrow\; j(Q') > j(Q'') \text{ or } j(Q') = j(Q''),\ i(Q') > i(Q'') . \tag{4.5.2.9}$$

Define $A_{\boldsymbol{i}}$, $\boldsymbol{i} \in Q_1$, as the coefficients of F_1. Now, assume that the $A_{\boldsymbol{i}}$ with $\boldsymbol{i}$ in the union of the m first quadrangles are defined. Observe that this fixes at most three coefficients at the boundary points of the $(m+1)$th quadrangle, and these coefficients a, b, c correspond (up to translation) to the coefficients indicated in (4.5.2.8). Then we choose a suitable polynomial F_i, $i = 1, 2$, and take its coefficients as $A_{\boldsymbol{i}}$, $\boldsymbol{i}$ in the $(m+1)$th quadrangle.

Finally, $A_{\boldsymbol{i}}$ with $\boldsymbol{i}$ outside the considered quadrangles should be taken as generic real numbers.

Step 4. Orient the dual graph G of the subdivision of T_d^2 in accordance with the order (4.5.2.9). Then any such quadrangle Q has at least two edges in $\partial Q \setminus Q_+(\widetilde{G})$, that means in all inequalities (4.5.1.1) in Proposition 4.5.2 we have 1 on the left-hand side and ≥ 2 on the right-hand side, which by Theorem 2.3.5 implies the existence of a real irreducible curve $\{F=0\}$ of degree d with $[(d^2-3d+4)/4]$ real cusps as its only singularities. The $\mathcal{S}$-transversality of the triple $(T_d^2, \emptyset, F)$ means that F belongs to a T-smooth component of the equisingular family. Hence one can smooth out any prescribed cusps of the curve $\{F=0\}$, hereby proving the first part of the theorem.

(ii) First, we construct a sequence of irreducible curves $C^{(3)}, C^{(4)}, \dots, C^{(d)}$ of degree d such that $C^{(s)}$ has n_s nodes and k_s cusps, where

$$n_s = \frac{d^2 - 3d - s^2 + 3s}{2}, \quad k_s = \left\lfloor \frac{s^2 - 3s + 4}{4} \right\rfloor .$$

Namely, $C^{(d)}$ is the curve constructed in part (i). If $3 \le s < d$, take a cuspidal curve $\{F(x, y) = 0\}$ of degree s with k_s cusps, constructed in part (i). By Theorem 2.3.5 the leading form $F^{(s)}(x, y)$ splits into the product of s distinct linear factors $L_1, \dots, L_s$. Let $\{G(x, y) = 0\}$ be a generic rational nodal curve of degree $d - s + 1$, which is non-singular and tangent to $\{L_s = 0\}$ at the origin. The curve $\{L_1 \cdots L_{s-1} G = 0\}$ has

$$\frac{(d-s)(d-s-1)}{2} + (s-1)(d-s) = \frac{d^2 - 3d - s^2 + 3s}{2} = n_s$$

nodes, its Newton polygon Δ is the closure of $T_d^2 \setminus T_s^2$ (see Fig. 4.6(a)), and its lowest form is nothing but $F^{(s)}$.

In the notation of Theorem 2.3.5, the only arc of $\widetilde{G}$ is oriented from T_s^2 to Δ, the triple $(T_s^2, \emptyset, F)$ is $\mathcal{S}$-transversal by construction, the triple $(\Delta, \sigma, L_1 \cdots L_{s-1} G)$,

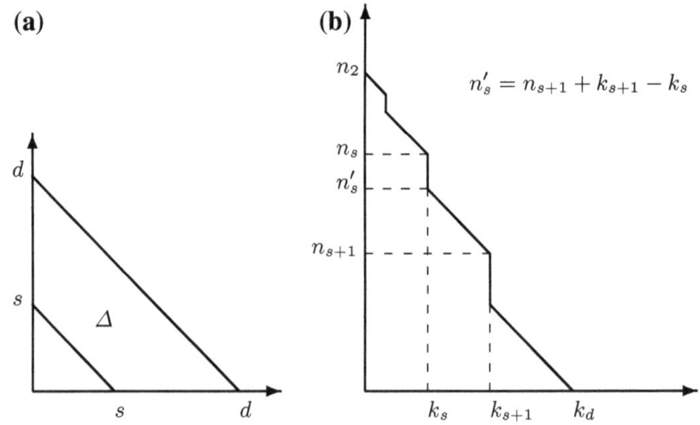

Fig. 4.6 Construction of curves with nodes and cusps

where σ is the edge $[(0, s), (s, 0)]$, is $\mathcal{S}$-transversal, since the inequalities (4.5.1.1) in Proposition 4.5.2 hold true as explained in the proof of Corollary 4.5.3. Hence, by Theorem 2.3.5 we obtain the desired sequence of curves.

Since these curves belong to T-smooth components of the equisingular strata, one can smooth out prescribed nodes and cusps, or deform some cusps into nodes. That means, there exist irreducible curves of degree d with n nodes and k cusps, where (k, n) does not lie above the broken line in Fig. 4.6(b), which implies the second statement of the theorem, because the right-hand side in (4.5.2.7) is the minimal value of $n + 2k$ on the given broken line. □

4.5.2.3 A Plane Curve of Degree 11 with 30 Cusps

We provide also one more application of the patchworking construction which completes the data presented in Example 4.2.0.1.

Example 4.5.4.2 Following [Shu11] we construct a plane curve of degree 11 with 30 cusps, showing that $k_{\max}(11) \geq 30$, the lower bound in the table of Example 4.2.0.1 which was missing in the literature.

Consider the (convex) subdivision S of the triangle T_{11}^2 depicted in Fig. 4.7(a). Up to $SL(2, \mathbb{Z})$-action there are four types of polygons in the subdivision which are shown in Fig. 4.7(b) and which will serve as Newton polygons of suitable peripherally nondegenerate polynomials (cf. Sect. 2.3.2). For the upper left polygon Δ_6, we take a polynomial $F_6(x, y)$ defining a curve of degree 6 with 8 cusps in $(\mathbb{C}^*)^2$ and one cusp at infinity with the x-axis as the tangent. For the upper right polygon Δ_5', we take a polynomial $F_5'(x, y)$ defining a curve of degree 5 with 5 cusps in $(\mathbb{C}^*)^2$ and two flexes at infinity with the x- and y-axes as tangents (cf. Fig. 4.3(b)). For the lower

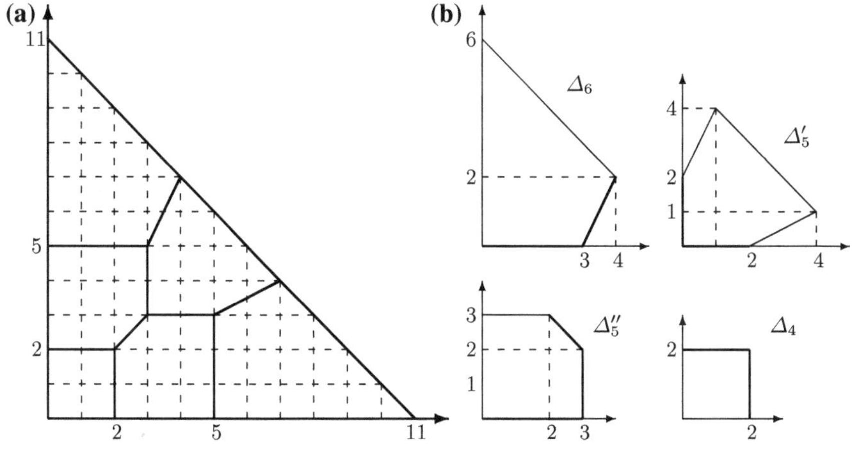

Fig. 4.7 Construction of a curve of degree 11 with 30 cusps

left polygon Δ_5'', we take a polynomial $F_5''(x, y)$ defining a curve of degree 6 with 4 cusps in $(\mathbb{C}^*)^2$ and two more nodes at infinity.[6] At last, for the lower right polygon Δ_4, we take a polynomial $F_4(x, y)$ defining a quartic curve with one cusp in $(\mathbb{C}^*)^2$ and two nodes at infinity.

To complete the construction, we have to find compatible polynomials as above (i.e. with respectively coinciding coefficients along common sides of their Newton polygons), and then to verify the suitable transversality condition. We orient the adjacency graph of the polygons in the subdivision S by letting $\Delta' \prec \Delta''$ if $\Delta' \cap \Delta''$ is a common side and the vector $(-1, -1.1)$ crosses this side pointing from Δ' to Δ''. Then the transversality conditions required by Theorem 2.3.5 amount in the transversality of the triples (cf. Definition 2.3.3)

$$(\Delta_6, (\Delta_6)_+, F_6),\ (\Delta_5', (\Delta_5')_+, F_5'),\ (\Delta_5'', (\Delta_5'')_+, F_5''),\ (\Delta_4, (\Delta_4)_+, F_4)\ ,$$

where Δ_+ in each case denotes the union of the sides shows by fat lines in Fig. 4.7(b). In each case the transversality holds by Proposition 4.5.2, since its hypotheses in our case reduce to the true inequalities

$$\#(\text{cusps in } (\mathbb{C}^*)^2)\ \leq\ \#(\text{integral points in } \partial\Delta\backslash\Delta_+)\ .$$

Furthermore, the same transversality of the above triples means that the map defined on the family of peripherally nondegenerate polynomials with Newton polygon Δ which determine curves with the given number of cusps in $(\mathbb{C}^*)^2$,

$$F\ \mapsto\ (\text{the set of the coefficients of } F \text{ along } \Delta_+)\ \in \mathbb{C}^{|\Delta_+\cap\mathbb{Z}^2|}$$

is dominant in each of the considered situations. Moreover, for the triple $(\Delta_4, (\Delta_4)_+, F_4)$ this map is onto the set of coefficients defining nondegenerate polynomials along the edges of $(\Delta_4)_+$. This immediately implies that, moving through the set of polygons of the subdivision S by the aforementioned order, in each stage we we can choose a polynomial with a given number of cusps which is compatible with the previously defined polynomials.

Thus, Theorem 2.3.5 yields that there exists a curve of degree 11 (i.e. with Newton polygon T_{11}^2) having 30 cusps:

$$\begin{aligned}&2(\#(\text{Sing } F_6 \cap (\mathbb{C}^*)^2)) + \#(\text{Sing } F_5' \cap (\mathbb{C}^*)^2)\\ &+2(\#(\text{Sing } F_5'' \cap (\mathbb{C}^*)^2)) + \#(\text{Sing } F_4 \cap (\mathbb{C}^*)^2)\\ &= 2\cdot 8 + 5 + 2\cdot 4 + 1 = 30.\end{aligned}$$

[6] Such a curve can easily be constructed by applying the Cremona transformation to the 3-cuspidal quartic curve so that one of the axes is quadratically tangent to the quartic, and the fundamental points of the coordinate system are non-singular points of the quartic (we leave details to the reader).

4.5.3 Curves and Hypersurfaces with Simple Singularities

The asymptotically optimal sufficient existence conditions for plane curves with nodes and cusps given in Theorem 4.5.4 can be extended to plane curves and, more generally, projective hypersurfaces of any dimension with arbitrary simple singularities. Following [ShW], we state

Theorem 4.5.5 *(1) For any simple singularity S (i.e. A_m, $m \geq 1$, or D_m, $m \geq 4$, or E_m, $m = 6, 7, 8$) and any integer $n \geq 2$, there exists a polynomial $\varphi_{S,n}(d)$ of degree $n-1$ such that the inequality*

$$k \cdot \mu(S) \leq \frac{d^n}{n!} + \varphi_{S,n}(d) \qquad (4.5.3.1)$$

is sufficient for the existence of a hypersurface of degree d in $\mathbb{P}^n$ having k isolated singular points of type S as its only singularities and belonging to a T-smooth ESF.

(2) Furthermore, for any integers $m \geq 1$ and $n \geq 2$, there exists a polynomial $\psi_{m,n}(d)$ of degree $n-1$ such that the inequality

$$\sum_{i=1}^{r} \mu(S_i) \leq \frac{d^n}{n!} + \psi_{m,n}(d) \qquad (4.5.3.2)$$

with arbitrary r and simple singularities $S_1, ..., S_r$ with Milnor numbers $\leq m$ is sufficient for the existence of a hypersurface of degree d in $\mathbb{P}^n$ having r isolated singular points of types $S_1, ..., S_r$, respectively, as its only singularities and belonging to a T-smooth ESF.

Remark 4.5.5.1 This is a generalization of Theorem 4.5.4 when we put

$$\varphi_{A_2,2}(d) = \frac{-3d+4}{2} \quad \text{and} \quad \psi_{2,2} = -2d+3\,.$$

Inequalities (4.5.3.1) and (4.5.3.2) are indeed asymptotically optimal for T-existence since their left-hand sides are the expected codimensions and the right-hand sides are asymptotically equivalent to $h^0(\mathbb{P}^n, \mathcal{O}_{\mathbb{P}^n}(d))$ as $d \to \infty$. It still remains a question whether it is possible to remove the dependence on m in the second assertion of theorem.

Proof of Theorem 4.5.5. We shall use the patchworking construction, and hence the resulting singular hypersurface will always belong to a T-smooth family (the $\mathcal{S}$-transversality in Theorems 2.3.4 and 2.3.5 reduces in our case to the T-smoothness).

4.5.3.1 Proof for Plane Curves

We start with the case of plane curves presenting, first, suitable patches (Step 1 below), i.e. polynomials with one prescribed simple singularity S in $(\mathbb{C}^*)^2$ and Newton polygon of area $\mu(S)$. Tiling the plane by these patches and checking the required compatibility and transversality hypotheses (Step 2 below), similarly to Step 1 and 4 in the proof of Theorem 4.5.4, we get the desired statement for plane curves. Next (in Step 3 below) we extend the result to higher dimensions again by a suitable patchworking.

Step 1. The patches we use are as follows:

Lemma 4.5.6 *(1) Let Δ be one of the polygons shown in Fig. 4.8(a) and Δ_+ the union of those sides of Δ, whose exterior normals project positively on the vector $(-1, -1/m)$ (marked by fat lines in Fig. 4.8(a)). Then there exists a peripherally nondegenerate polynomial f with Newton polygon Δ such that*

- *f has a singular point of type A_m as its only singularity in $(\mathbb{C}^*)^2$,*
- *the coefficients of f along the part Δ_+ are prescribed nonzero numbers,*
- *the triple (Δ, Δ_+, f) is transversal.*

(2) Let $S = D_m$, $m \geq 4$, or E_m, $m = 6, 7, 8$, Δ_S the polygon shown in Fig. 4.8(b), $(\Delta_S)_+$ the union of those sides of Δ_S whose exterior normals project positively on the vector $(-1, -1)$ (marked there by fat lines in Fig. 4.8(b)) and $(\Delta_S)_-$ the closure of $\partial\Delta_S \backslash (\Delta_S)_+$. Then there exists a polynomial f with Newton polygon Δ_S such that

- *f has two singular points of type S as its only singularity in $(\mathbb{C}^*)^2$,*
- *the triples $(\Delta_S, (\Delta_S)_+, f)$ and $(\Delta_S, (\Delta_S)_-, f)$ are transversal.*

Proof (1) Let $\Delta = \Delta_{A_m}$. We look for the desired polynomial in the form

$$f(x, y) = ay^2 - 2y \cdot Q(x) + cx, \quad ac \neq 0, \quad \deg Q = m, \quad Q(0) \neq 0 .$$

A singular point $(\alpha, \beta) \in (\mathbb{C}^*)^2$ of f must be isolated and of multiplicity 2, that is, of type A_k. The condition that (α, β) is of type A_m means that

$$f(\alpha, \beta) = 0, \quad f_x(\alpha, \beta) = 0, \quad f_y(\alpha, \beta) = 0 ,$$

or, equivalently

$$\beta = \frac{Q(\alpha)}{a}, \quad (Q^2(x) - acx)\big|_{x=\alpha} = 0, \quad (2Q'(x)Q(x) - ac)\big|_{x=\alpha} = 0 ,$$

where $x = \alpha$ is a root of multiplicity m, respectively $m - 1$, of the second, respectively third, equation. To satisfy the requirement, we choose a branch of $\sqrt{acx}$ in a neighborhood of $x = \alpha$, and demand that

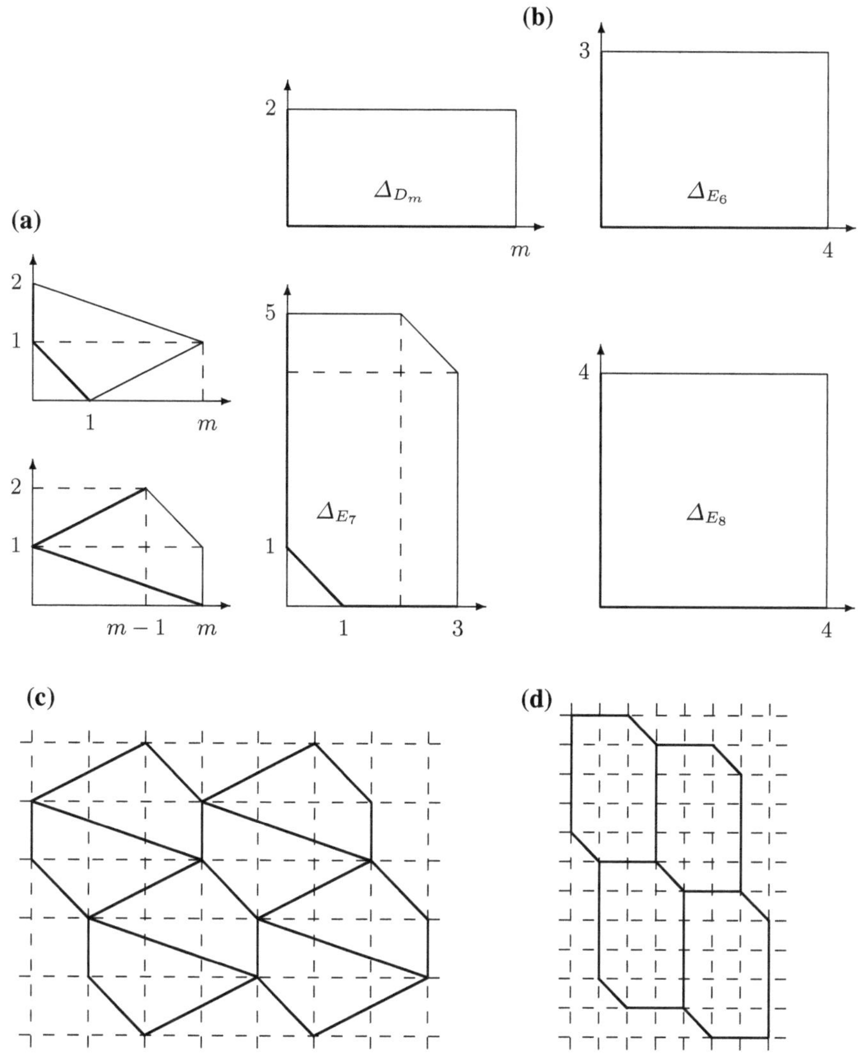

Fig. 4.8 Patchworking curves with simple singularities

$$Q(x) - \sqrt{acx} = (x - \alpha)^{\mu+1}\lambda(x)$$

with a function $\lambda(x)$ holomorphic in a neighborhood of $x = \alpha$. This yields that $Q(x)$ is the m-jet of the Taylor series of $\sqrt{acx}$ at $x = \alpha$. It is an easy exercise to verify that, chosen any three of the coefficients a, $b = 2Q(0)$, c, or $d = \frac{2}{m!}Q^{(m)}(0)$ of f at the vertices of Δ, the rest of the coefficients of f depend (locally) smoothly on the chosen parameters, and this means the transversality of the triple (Δ, Δ_+, f). Furthermore, by acting as $f(x, y) \mapsto \kappa_1 f(\kappa_2 x, \kappa_3 y)$, $\kappa_1, \kappa_2, \kappa_3 \neq 0$, we can prescribe the values of the chosen coefficients.

It remains to notice that the required polynomial with the Newton polygon Δ'_{A_m} can be chosen in the form $x^m y^2 f(x^{-1}, y^{-1})$, with f being constructed above. The required conditions are established in the same manner.

(2) Let $\Delta = \Delta_{D_m}$. Assume that

$$f(x, y) = (x - \alpha_1)(x - \alpha_2)\Big(y^2 \cdot P(x) - 2y \cdot Q(x) + R(x)\Big) = (x - \alpha_1)(x - \alpha_2) \cdot g(x, y) .$$

Let $\alpha = \alpha_i$, $i = 1, 2$. The curve $\{f = 0\}$ has a D_m singularity at $(\alpha, \beta) \in (\mathbb{C}^*)^2$ if and only if the curve $\{g = 0\}$ has an A_{m-3} singularity at (α, β). The latter property is equivalent to

$$\beta = \frac{Q(\alpha)}{P(\alpha)}, \quad R(\alpha) = \frac{Q^2(\alpha)}{P(\alpha)}, \quad R'(\alpha) = \left(\frac{Q^2}{P}\right)'(\alpha) .$$

Since the polar curves $\{g_x = 0\}$, $\{g_y = 0\}$ of g intersect with multiplicity $m - 3$ at (α, β), we obtain that $PR - Q^2 = c \cdot (x - \alpha_1)^{m-2}(x - \alpha_2)^{m-2}$. The latter equation can be solved for P and Q by interpolation: the degree of Q is $m - 2$ and we know the value of Q^2 at the $m - 1$ points $0, x_1, \ldots, x_{\mu-2}$ where x_i are the (distinct) roots of $R(x)$. Note that in particular the value of c is uniquely determined by $P(0)$, $Q(0)$, $R(0)$ and α_1, α_2. Hence we obtain the existence of f and the transversality of (Δ, Δ_+, f), since the coefficients of P, Q depend smoothly on the fixed data $R(x)$, $P(0)$, $Q(0)$, α_1, α_2. In a similar way we can establish the transversality of (Δ, Δ_-, f).

(3) In the case $S = E_m$, $m = 6, 7, 8$, we present concrete polynomials f with the required properties:

$$\begin{aligned} f_{E_6} = &-80x^4y^3 - 792x^4y^2 + 1320x^3y^3 - 2616x^4y + 13056x^3y^2 \\ &- 8160x^2y^3 - 2883x^4 + 11086x^3y + 15384x^2y^2 - 9592xy^3 + 15450x^3 \\ &- 9812x^2y - 2852xy^2 + 8921y^3 - 4488x^2 - 7624xy - 3469y^2 \\ &+ 700x - 12281y + 10328 , \end{aligned}$$

$$\begin{aligned} f_{E_7} = &-3999x^3y^4 - 4000x^2y^5 - 15997x^3y^3 + 2676x^2y^4 - 13334xy^5 \\ &+ 7958x^3y^2 - 8033x^2y^3 + 5628xy^4 - 12742y^5 - 216x^3y - 711x^2y^2 \\ &- 11032xy^3 - 8010y^4 - 243x^3 - 2385x^2y + 7772xy^2 - 9129y^3 - 2160x^2 \\ &- 7248xy + 6613y^2 - 4800x + 548y , \end{aligned}$$

$$\begin{aligned} f_{E_8} = &-67x^4y^4 - 907x^4y^3 + 1119x^3y^4 - 4608x^4y^2 + 15162x^3y^3 \\ &- 6915x^2y^4 - 10413x^4y + 13106x^3y^2 + 2247x^2y^3 - 13187xy^4 \\ &- 8831x^4 + 14451x^3y + 2757x^2y^2 - 788xy^3 + 12925y^4 - 11854x^3 \\ &+ 7076x^2y - 12043xy^2 - 2780y^3 + 8902x^2 + 2816xy - 6533y^2 + 12302x \\ &- 14581y + 12406 . \end{aligned}$$

Since they are defined over $\mathbb{Z}$, one can use the computer algebra system SINGULAR to compute the singular points and the tangent spaces to the ESF, and to verify the required transversality. We leave this to the reader. □

Step 2. First, we cover the plane by the polygons defined above. In the case $S = A_m$ it looks as in Fig. 4.8(c). The convexity of this tiling can be checked as in Step 1 of the proof of Theorem 4.5.4: define the values of the function $\mathbb{R}^2 \to \mathbb{R}$ at the vertices of the tiling by $2i^2 + (m-1)j^2$ and extend linearly on each tile. In the case $S = D_m$, or E_6, or E_8, the tiling and its convexity are evident. In the case $S = E_7$, we naturally cover the plane by the translates of the hexagon Δ_{E_7}. The convexity of the tiling can be checked as above: define the function $\mathbb{R}^2 \to \mathbb{R}$ at the vertices of the tiling by $5i^2 + 2ij + 3j^2$ and verify that it extends linearly on each tile.

Assuming that we have found compatible (i.e. with respectively coinciding coefficients along the intersection of neighboring Newton polygons) polynomials f for each tile, we satisfy the transversality hypotheses in Theorem 2.3.5 when ordering the tiles in the direction of the vector $(1, 1/m)$ (i.e. letting $\Delta' \prec \Delta''$ as $\Delta' \cap \Delta''$ is a common side, and the vector $(1, 1/m)$ crosses this side pointing from Δ' to Δ''). The first statement of Theorem 4.5.5 would then follow, since the area of Δ_{A_m} and Δ'_{A_m} equals m, the area of Δ_S, $S = D_m$ or E_6, E_7, E_8, equals $2\mu(S)$, and we can pack into T_d^2 at least $d^2/(\text{Area}(\Delta_S) + O(d)$ tiles. For the second statement of Theorem 4.5.5, notice that there are only finitely many simple singularities with Milnor number $\le m$. So, we divide T_d^2 into parts with the area proportional to the total Milnor numbers of singularities of each kind among $S_1, \ldots, S_r$, and then cover each part by suitable tiles as described above, we leave details to the reader (cf. Step 4 in the proof of Theorem 4.5.4).

In the case $S = A_m$ one can construct a set of compatible polynomials f using Lemma 4.5.6(1) and the order of tiles defined above.

In the case $S = D_m$, $m \ge 4$, or E_m, $m = 6, 8$, the tiling by rectangles can be obtained from one polygon by successive reflections with respect to sides, and the same reflections can be performed with polynomials in which way we obtain a set of compatible polynomials out of the given f from Lemma 4.5.6(2).

In the case $S = E_7$, we prove the existence of a compatible set of polynomials by induction on the number N of polygons in the tiling. To formulate our claim precisely, introduce the irreducible component M of the equisingular family of peripherally nondegenerate polynomials with Newton polygon $\Delta = \Delta_{E_7}$ having two singular points of type E_7 in $(\mathbb{C}^*)^2$ which contains the polynomial f from Lemma 4.5.6(2). Let us be given the tiling of the plane by the translates of the hexagon $\Delta = \Delta_{E_7}$ as shown in Fig. 4.8(d). Consider the part of the tiling lying in the nonnegative quadrant, partially order it in the direction of the vector $(1, 1)$, and extend this order up to a linear one somehow: $\Delta_1, \Delta_2, \ldots$ For each $N \ge 1$, put

$$\sigma_N = \bigcup_{i=1}^{N} (\Delta_i)_- \cap \partial \left(\bigcup_{i=1}^{N} \Delta_i \right)$$

(shown by fat lines in Fig. 4.8(d)).[7] If $\Delta_i = \Delta + (p, q)$, we define the family of polynomials $M_i = x^p y^q \cdot M$ with Newton polygon Δ_i, $i \geq 1$. Define the family $\widehat{M}_N \subset M_1 \times ... \times M_N$ consisting of the sequences of compatible polynomials $(f_1, ..., f_N)$ (i.e. those coefficients respectively coincide along the common sides of their Newton polygons), and introduce the map $\pi_N : \widehat{M}_N \to \mathbb{C}^{k_N}, k_N = \#(\sigma_N \cap \mathbb{Z}^2)$, sending each sequence $(f_1, ..., f_N)$ to the set of its coefficients along σ_N. We prove by induction on N that $\dim \widehat{M}_N = k_N$ and the map π_N is dominant. For $N = 1$ this follows from Lemma 4.5.6(2), i.e. from the non-emptiness of M and from the transversality of the triple (Δ, Δ_-, f). Now we make step from N to $N + 1$. Notice that

$$\Delta_{N+1} \cap \bigcup_{i=1}^{N} \Delta_i \subset \sigma_N \quad \text{and} \quad \Delta_{N+1} \cap \sigma_N \subset (\Delta_{N+1})_+ ,$$

and that the transversality of the triple (Δ, Δ_+, f) yields that the map $\pi : M \to \mathbb{C}^8$ sending a polynomial to the sequence of its coefficients along Δ_+ is dominant. Together with the dominance of π_N this implies the non-emptiness of $\widehat{M}_{N+1}$ and, moreover, that

$$\dim \widehat{M}_{N+1} \geq \dim \widehat{M}_N - \#(\Delta_{N+1} \cap \sigma_N \cap \mathbb{Z}^2) + 8 = k_{N+1} .$$

On the other hand, if we fix the coefficients along σ_{N+1}, then applying the transversality of the triads $(\Delta_i, (\Delta_i)_-, f_i)$ in the reverse order, we conclude that $\dim \pi_{N+1}(\mathrm{pt}) = 0$ which proves both the relation $\dim \widehat{M}_{N+1} = k_{N+1}$ and the dominance of π_{N+1}.

This completes the proof of the case $n = 2$ in Theorem 4.5.5.

4.5.3.2 Proof for Higher-Dimensional Hypersurfaces

Assume that $n \geq 2$ and make step from n to $n + 1$. Given the simplex

$$T_d^n = \{(\lambda_1, ..., \lambda_n \in \mathbb{R}^n \ : \ \lambda_1, ..., \lambda_n \geq 0, \ \sum_{i=1}^{n} \lambda_i \leq d\} ,$$

introduce the lattice polytopes in $\mathbb{R}^{n+1}$ spanned by the shifted T_d^n and a transverse length 2 segment with the midpoint at a simplex vertex:

$$\begin{aligned}
\Delta'_{k,d} &= \mathrm{conv}\left(\{(\boldsymbol{x}, k) \in \mathbb{R}^{n+1} \ : \ \boldsymbol{x} \in T_d^n\} \cup \sigma'_k\right) , \\
\sigma'_k &= [(0, ..., 0, k - 1), (0, ..., 0, k + 1)] , \\
\Delta''_{k,d} &= \mathrm{conv}\left(\{(\boldsymbol{x}, k) \in \mathbb{R}^{n+1} \ : \ \boldsymbol{x} \in T_d^n\} \cup \sigma''_k\right) , \\
\sigma''_k &= [(0, ..., 0, d - k - 1, k + 1), (0, ..., 0, d - k + 1, k - 1)] .
\end{aligned}$$

[7] For example, $\sigma_1 = (\Delta_1)_-$.

Lemma 4.5.7 *Let $f(x_1, ..., x_n)$ be a peripherally nondegenerate polynomial with Newton simplex T^n_{d-1}, having only isolated singular points in $(\mathbb{C}^*)^n$ and defining a hypersurface in $\mathbb{P}^n$ belonging to a T-smooth analytic ESF. If 2 is not a critical value of f, then the polynomial*

$$F'_{k,f}(x_1, ..., x_n, x_{n+1}) = x_{n+1}^{k-1}\left(x_{n+1}^2 - 2x_{n+1}(f(x_1, ..., x_n) + 1) + 1\right) \quad (4.5.3.3)$$

with the Newton polytope $\Delta'_{k,d}$ is peripherally nondegenerate with only isolated singular points in $(\mathbb{C}^*)^{n+1}$ such that

- $\mathrm{Sing}\, F'_{k,f} \cap (\mathbb{C}^*)^{n+1} = \{(\boldsymbol{x}, 1) \in (\mathbb{C}^*)^{n+1} : \boldsymbol{x} \in \mathrm{Sing}\, f \cap (\mathbb{C}^*)^n\}$,
- *for each* $z \in \mathrm{Sing}\, F'_{k,f} \cap (\mathbb{C}^*)^{n+1}$,

$$I^{ea}(F'_{k,f}, z) = \langle f, f_{x_1}, ..., f_{x_n}, x_{n+1} - 1\rangle\ ,$$

- *the triple $(\Delta'_{k,d}, \emptyset, F'_{k,f})$ is contact analytically transversal.*

Proof All the statements but the last one are straightforward from a simple direct calculation. For the last statement, we notice that, for each point $\boldsymbol{x} \in \mathrm{Sing}\, f \cap (\mathbb{C}^*)^n$, there is a canonical isomorphism $I^{ea}(f, \boldsymbol{x})) \simeq I^{ea}(F'_{k,f}, (\boldsymbol{x}, 1))$, and then consider the diagram

$$\begin{array}{ccc} \mathcal{P}(\Delta'_{k,d}) & \stackrel{x_{n+1}=1}{\longrightarrow} & H^0(\mathbb{P}^n, \mathcal{O}_{\mathbb{P}^n}(d)) \\ \downarrow & & \downarrow \\ \bigoplus_{(\boldsymbol{x},1)\in \mathrm{Sing}\, F'_{k,f}\cap(\mathbb{C}^*)^{n+1}} \frac{\mathcal{O}_{\mathbb{P}^{n+1},(\boldsymbol{x},1)}}{I^{ea}(F'_{k,f},(\boldsymbol{x},1))} & \simeq & \bigoplus_{\boldsymbol{x}\in \mathrm{Sing}\, f\cap(\mathbb{C}^*)^n} \frac{\mathcal{O}_{\mathbb{P}^n,\boldsymbol{x}}}{I^{ea}(f,\boldsymbol{x})} \end{array}$$

Here the upper arrow and the left projection are surjective (the latter one due to the supposed T-smoothness, cf. Theorem 2.2.35(c)). Hence the left arrow is surjective which implies the contact analytic transversality of the triple $(\Delta'_{k,d}, \emptyset, F)$. □

Lemma 4.5.8 *Let, for each $k = 1, ..., d-1$, $f_k \in \mathbb{C}[x_1, ..., x_n]$ be a peripherally nondegenerate polynomial with Newton polygon T^n_k defining a hypersurface $X_k \subset \mathbb{P}^n$ such that the analytic ESF $V_{k,\mathbb{P}^n}(\mathcal{S}_{f_k})$ is T-smooth at X_k.*[8] *Then there exists a peripherally nondegenerate polynomial $F \in \mathbb{C}[x_1, ..., x_{n+1}]$ with Newton polygon T^{n+1}_d such that*

$$\mathcal{S}_F = \sum_{k=1}^{d-1} \mathcal{S}_{f_k}$$

and the analytic ESF $V^{n+1}_d(\mathcal{S}_F)$ is T-smooth at $X \subset \mathbb{P}^{n+1}$, defined by F.

[8] Here $\mathcal{S}_{f_k}$ denotes the collection of contact analytic types of singular points of f_k in $(\mathbb{C}^*)^n$ in the sense of Sect. 2.3.2.

Fig. 4.9 Subpolytopes $\Delta_1, ..., \Delta_{d-1}$ in T_d^{n+1}

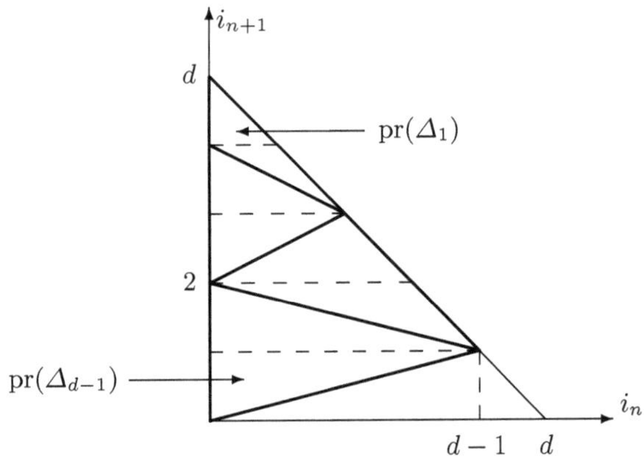

Proof Introduce the subpolytopes

$$\Delta_{d-2k+1} = \Delta'_{2k-1,d-2k+1}, \quad \Delta_{d-2k} = \Delta''_{2k,d-2k}, \quad 1 \le k \le \frac{d}{2},$$

of T_d^{n+1}. It is easy to see that $\Delta_k \cap \Delta_j = \emptyset$ as $|k-j| > 1$, and $\Delta_k \cap \Delta_{k+1}$ is a unit lattice segment (the projections of $\Delta_1, ..., \Delta_{d-1}$ to the plane (i_n, i_{n+1}) are shown in Fig. 4.9). Clearly, there is a convex subdivision of T_d^{n+1} into lattice polytopes, containing $\Delta_1, ..., \Delta_{d-1}$.

Introduce the polynomials

$$F_{d-2k+1} := F'_{2k-1, f_{d-2k+1}}, \quad 1 \le k \le \frac{d}{2},$$

with Newton polytopes Δ_{d-2k+1}, $1 \le k \le d/2$, respectively, defined by (4.5.3.3). For each $1 \le k \le d/2$, take an affine automorphism $\varphi_k : \mathbb{Z}^{n+1} \to \mathbb{Z}^{n+1}$ preserving the hyperplane $i_{n+1} = 2k$ and taking $\Delta'_{2k,d-2k}$ onto $\Delta''_{2k,d-2k}$. Then introduce the polynomials

$$F_{d-2k} := \varphi_* F'_{d-2k, f_{d-2k}}, \quad 1 \le k < \frac{d}{2},$$

with Newton polytopes Δ_{d-2k}, $1 \le k < d/2$, respectively. Observe that

- each polynomial F_{d-2k} is peripherally nondegenerate and has the singularity collection $\mathcal{S}_{f_{d-2k}}$, and, moreover, the triple $(\Delta_{d-2k}, \emptyset, F_{d-2k})$ is contact analytically transverse;
- using transformations $F_r(x_1, ..., x_{n+1}) \mapsto \lambda_r F_r(x_1, ..., x_n, \mu_r x_{n+1})$ with suitable $\lambda_r, \mu_r \in \mathbb{C}^*$, we can make the coefficients of common monomials of $F_1, ..., F_{d-1}$ coinciding;
- the polynomials with the Newton polytopes, completing $\Delta_1 \cup ... \cup \Delta_{d-1}$ up to the given subdivision of T_d^{n+1}, and which are compatible with $F_1, ..., F_{d-1}$, have no singularities in $(\mathbb{C}^*)^{n+1}$.

Thus, the patchworking Theorem 2.3.4 completes the proof. □

Now the higher-dimensional statement of Theorem 4.5.5 follows by induction from Lemma 4.5.8 and the plane curve case settled in Steps 1 and 2.

4.5.4 *Hypersurfaces with Arbitrary Singularities*

For a smooth algebraic variety V of dimension $n \geq 2$ and a linear system $|W|$ on it, we derive here a general sufficient condition for the existence of a divisor in $|W|$ with a collection of isolated singularities, prescribed up to $\mathcal{S}$-equivalence.

4.5.4.1 Order of T-Existence

Definition 4.5.9 Let S be a smooth singularity type (cf. Definition 2.2.42). Then we define $e(S)$ and $e_r(S)$ to be the minimal degrees of polynomials $F_S \in \mathbb{C}[x_1, \dots, x_n]$ and $F_S \in \mathbb{R}[x_1, \dots, x_n]$, respectively, such that

- the leading form of F_S (of degree $e(S)$) has no critical points in $\mathbb{C}^n \setminus \{0\}$,
- $\mathrm{Sing}\{F_S = 0\}$ consists of one point z, and this singular point is of type S,
- $H^1\big(\mathcal{J}_{Z_z^S/\mathbb{P}^n}(e(S) - 1)\big) = 0$.

We call $e(S)$ (resp. $e_r(S)$) the *order* (resp. *real order*) *of T-existence* of the singularity type S.

Lemma 4.5.10 *(1) Let F_S be a polynomial as in Definition 4.5.9. Then the germ M at F_S of the family of polynomials of degree $e(S)$ defining hypersurfaces with a singular point of type S in a neighborhood of z, is smooth of expected dimension, and intersects transversally the space of polynomials with the same leading form as F_S.*

(2) The set of the leading forms of polynomials F of degree $e(S)$, satisfying the conditions of Definition 4.5.9, contains a Zariski open dense subset in the space of homogeneous polynomials of degree $e(S)$ in n variables.

Proof The first statement is reformulated as

$$H^1(\mathcal{J}_{Z_z^S \cup L/\mathbb{P}^n}(e(S)) = 0$$

(cf. Proposition 2.3.14, Sect. 2.3.6) which follows from the exact sequence

$$0 = H^1(\mathcal{J}_{Z_z^S/\mathbb{P}^n}(s(S) - 1)) \to H^1(\mathcal{J}_{Z_z^S \cup L/\mathbb{P}^n}(e(S)) \to H^1(\mathcal{J}_{L/L}(e(S)) = 0$$

($\mathcal{J}_{L/L}$ is the zero sheaf).

For the second statement consider the map of the germ M to the space of homogeneous polynomials of degree $e(S)$ in n variables, which takes a polynomial $F \in M$ into its leading form, and then notice that the above transversality condition means that this map is a submersion. □

4.5.4.2 Construction out of Ordinary Singularities

Now we start by constructing any given singularity types out of ordinary singularities.

Proposition 4.5.11 *Let V be a smooth projective algebraic variety of dimension $n \geq 2$, and $|W|$ a linear system on V. Moreover, let $S_1, \dots, S_r$ be smooth singularity types, and let $W' \in |W|$ be a hypersurface with r singular points of multiplicities at least $e(S_1), \dots, e(S_r)$, respectively, and satisfying*

$$H^1\big(\mathcal{J}_{Z/V} \otimes \mathcal{O}_V(W)\big) = 0\,, \tag{4.5.4.1}$$

where the zero-dimensional scheme Z is supported at the singular points $z_1, \dots, z_r$ of W' and defined by the $(e(S_i)+1)$th power of the respective maximal ideals $\mathfrak{m}_{z_i}$, $i = 1, \dots, r$.

Then there exists a hypersurface $W'' \in |W|$ having r singular points of types $S_1, \dots, S_r$, respectively, as only singularities.

Proof Condition (4.5.4.1) means that, given coordinates in neighborhoods of the points $z_1, \dots, z_r$, any variations of the jets $\mathrm{jet}_{e(S_i)}(W_i')$, $i = 1, \dots, r$, are simultaneously realizable, when varying W' in $|W|$ (here W_i' is the image of the defining equation of W' in $\mathcal{O}_{V,z_i}$). So, first, we deform W' into $\widetilde{W}' \in |W|$ such that $\mathrm{jet}_{e(S_i)}(\widetilde{W}_i')$ is proportional to the leading form of the polynomial $F_{S_i}(\boldsymbol{x})$, $i = 1, \dots, r$. Note that, by Bertini's theorem, $\widetilde{W}'$ is non-singular outside $z_1, \dots, z_r$.

Finally, since $\widetilde{Z}^{sqh}_{z_i}(\widetilde{W}') = Z$, we can apply Theorem 2.3.13 to $\widetilde{W}'$ and the deformation patterns F_{S_i}, $i = 1, \dots, r$, which are $\mathcal{S}$-transversal. □

4.5.5 Plane Curves with Arbitrary Singularities

We shall demonstrate two ways in obtaining numerical sufficient conditions for the existence of plane curves with arbitrary prescribed singularities, both based on the reduction of the problem to H^1-vanishing for the ideal sheaves of certain zero-dimensional schemes associated with topological, respectively analytic, types of plane curve singularities. One way is to associate directly the zero-dimensional scheme corresponding to the given singularities and to produce a sufficient condition using the H^1-vanishing criteria established in Chap. 3. Another way is to construct, first, a curve with ordinary singularities and then to deform it into a curve with the prescribed singularities using the modified Viro method. In turn the H^1-vanishing requirements of the latter construction transform into another numerical sufficient existence condition. We shall see that the sufficient conditions of both types do not cover the other in general.

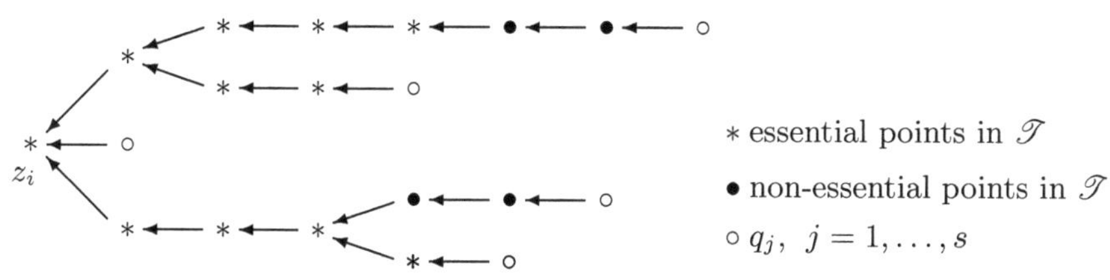

Fig. 4.10 The (graph of the) constellation $\mathscr{T} \subset \mathscr{T}(C, z)$

4.5.5.1 H^1-Vanishing Implies Existence

With a reduced plane curve germ (C, z) we associate the following zero-dimensional schemes

- $Z^s(C, z)$, the singularity scheme (see Sect. 1.1, Definition 1.1.30);
- $Z^s_{st}(C, z) := Z^s(CL, z)$, where (L, z) is a smooth germ, transversal to (C, z);
- $Z^a(C, z)$, the scheme defined by the ideal $I^a(C, z) \subset \mathcal{O}_{\mathbb{P}^2,z}$ (see Sect. 1.1.4, Definition 1.1.66);
- $Z^a_{st}(C, z)$, the scheme defined by the ideal $\mathfrak{m}_z I^a(C, z)$.

Proposition 4.5.12 *(1) Given a zero-dimensional scheme $Z \subset \mathbb{P}^2$, a point $z \in \mathbb{P}^2 \setminus$ supp(Z) and a reduced plane curve germ (C, z) satisfying*

$$H^1\big(\mathcal{J}_{Z \cup Z^s_{st}(C,z)/\mathbb{P}^2}(d)\big) = 0\,.$$

Then there exists a curve $D \in \big|H^0\big(\mathcal{J}_{Z \cup Z^s(C,z)/\mathbb{P}^2}(d)\big)\big|$ such that the germ (D, z) is topologically equivalent to (C, z). Moreover, these curves D form a dense open subset in $\big|H^0\big(\mathcal{J}_{Z \cup Z^s(C,z)/\mathbb{P}^2}(d)\big)\big|$.

(2) In the previous notation, let

$$H^1\big(\mathcal{J}_{Z \cup Z^a_{st}(C,z)/\mathbb{P}^2}(d)\big) = 0\,. \tag{4.5.5.1}$$

Then there exists a curve $D \in \big|H^0\big(\mathcal{J}_{Z \cup Z^a(C,z)/\mathbb{P}^2}(d)\big)\big|$ such that the germ (D, z) is analytically equivalent to (C, z), and these curves D form a dense open subset in $\big|H^0\big(\mathcal{J}_{Z \cup Z^a(C,z)/\mathbb{P}^2}(d)\big)\big|$.

Remark 4.5.12.1 Statement (2) remains true if we replace $Z^a(C, z)$ resp. $Z^a_{st}(C, z)$ by the schemes defined by $\widetilde{I}^a(C, z)$ (Definition 1.1.73) resp. $\mathfrak{m}_z \widetilde{I}^a(C, z)$. Although the statement with Z^a is stronger (since $\widetilde{I}^a \subset I^a$) $\widetilde{I}^a$ has the advantage that it can be computed.

Proof (1) To prove the first statement, we define $\mathscr{T}$ to be the finite subtree of the complete embedded resolution tree $\mathscr{T}(C, z)$ consisting of the essential tree $\mathscr{T}^*(C, z)$ (Definition 1.1.10) and the first infinitely near points q_j outside $\mathscr{T}^*(C, z)$ of all local branches (C_j, z), $j = 1, \ldots, s$, of (C, z), cf. Fig. 4.10.

For each infinitely near point $q \in \mathscr{T}(C, z)$ we consider the zero-dimensional scheme Z_q given by

$$I_{Z_q} := \left\{ g \in I(C, \mathscr{T}) \;\middle|\; \operatorname{mt} \widehat{g}_{(q)} \geq \operatorname{mt} \widehat{C}_{(q)} + 1 \right\}.$$

Note that $C \in I(C, \mathscr{T}) \setminus I_{Z_q}$. Therefore, I_{Z_q} is a *proper* subideal of $I(C, \mathscr{T})$. The exact cohomology sequence induced by

$$0 \longrightarrow \mathcal{J}_{Z \cup Z^s_{st}(C,z)/\mathbb{P}^2} \longrightarrow \mathcal{J}_{Z \cup Z_q/\mathbb{P}^2} \longrightarrow \mathcal{J}_{Z_q/\mathbb{P}^2} \big/ \mathcal{J}_{Z^s_{st}(C,z)/\mathbb{P}^2} \longrightarrow 0$$

shows that $H^1\big(\mathcal{J}_{Z \cup Z^s_{st}(C,z)/\mathbb{P}^2}(d)\big) = 0$ implies $H^1\big(\mathcal{J}_{Z \cup Z_q/\mathbb{P}^2}(d)\big) = 0$.

Moreover, $\mathcal{J}_{Z \cup Z_q/\mathbb{P}^2}$ is a proper subideal of $\mathcal{J}_{Z \cup Z^s(C,z)/\mathbb{P}^2}$, whence there exists a curve

$$G_q \in H^0\big(\mathcal{J}_{Z \cup Z^s(C,z)/\mathbb{P}^2}(d)\big) \setminus H^0\big(\mathcal{J}_{Z_q/\mathbb{P}^2}(d)\big).$$

Obviously, the total transform of a generic curve G in the linear system spanned by the G_q, $q \in \mathscr{T}$, satisfies $\operatorname{mt} \widehat{G}_{(q)} = \operatorname{mt} \widehat{C}_{(q)}$ for any $q \in \mathscr{T}$. Additionally, by construction, $\operatorname{mt} \widehat{G}_{(q_j)} \geq \operatorname{mt} \widehat{C}_{(q_j)}$, and Remark 1.1.11.1 implies that the strict transform of G at q_j has at least multiplicity $\operatorname{mt} C_{(q_j)} = 1$, $j = 1, \ldots, s$. On the other hand, G satisfies the proximity equality (Proposition 1.1.11). Hence, $\operatorname{mt} G_{(q_j)} \leq 1$. It follows that $q_j \in \mathscr{T}(G, z)$ is non-essential. We conclude that $\mathscr{T}^*(G, z) \subset \mathscr{T} \subset \mathscr{T}(G, z)$, $Z(G, \mathscr{T}) = Z(C, \mathscr{T})$, and this proves the first statement in view of Lemma 1.1.33.

(2) By (4.5.5.1), in the exact sequence

$$H^0\big(\mathcal{J}_{Z/\mathbb{P}^2}(d)\big) \longrightarrow \mathcal{O}_{\mathbb{P}^2,z}/I^a_{st}(C, z) \longrightarrow H^1\big(\mathcal{J}_{Z \cup Z^a_{st}(C,z)/\mathbb{P}^2}(d)\big) = 0$$

the first morphism is surjective. Denote by $\varphi \in \mathcal{O}_{\mathbb{P}^2,z}/I^a_{st}(C, z)$ the image of the germ $\psi \in \mathcal{O}_{\mathbb{P}^2,z}$ defined by (C, z). Take $D \in H^0\big(\mathcal{J}_{Z/\mathbb{P}^2}(d)\big)$ which projects to φ. Then $D - \psi \in I^a_{st}(C, z))$ implies that (D, z) is analytically equivalent to (C, z) as well as that such curves D constitute a dense Zariski open subset of $\big|H^0\big(\mathcal{J}_{Z \cup Z^a(C,z)/\mathbb{P}^2}(d)\big)\big|$ with $D \in I^a(C, z)$ by Definition 1.1.73. □

4.5.5.2 Numerical Conditions

The main existence theorems in this section involve quite complicated numerical conditions. Their importance lies in the corollaries with simple numerical bounds.

We shall see that Proposition 4.5.12 and Corollary 3.6.4, imply (cf. [Shu13, Theorem 1])

Theorem 4.5.13 *Let (C_i, z_i), $i = 1, \dots, r$, be reduced plane curve germs.*

(1) If

$$\left\lfloor \sqrt{\frac{4}{3}\sum_{i=1}^{r} M_2(Z^s(C_i, z_i))} + \frac{\sum_{i=1}^{r} \deg Z^s(C_i, z_i)}{\sqrt{4\sum_{i=1}^{r} M_2(Z^s(C_i, z_i))/3}} \right\rfloor \le d + 1\,, \tag{4.5.5.2}$$

then there exists a reduced, irreducible plane curve C of degree d having r singular points topologically equivalent to (C_1, z_1), ..., (C_r, z_r), respectively, as its only singularities.

(2) If

$$\left\lfloor \sqrt{\frac{4}{3}\sum_{i=1}^{r} M_2(Z^a(C_i, z_i))} + \frac{\sum_{i=1}^{r} \deg Z^a(C_i, z_i)}{\sqrt{4\sum_{i=1}^{r} M_2(Z^a(C_i, z_i))/3}} \right\rfloor \le d + 1\,, \tag{4.5.5.3}$$

then there exists a reduced, irreducible plane curve C of degree d having r singular points analytically equivalent to (C_1, z_1), ..., (C_r, z_r), respectively, as its only singularities.

Furthermore, the germ at C of the (topological, respectively analytic) equisingular stratum in the space of curves of degree d is T-smooth.

The hypotheses of Theorem 4.5.13 can be translated (even with improvement in some cases) into more familiar singularity invariants (cf. [Shu13, Theorem 3]).

Theorem 4.5.14 *In the notation of Theorem 4.5.13, let n and k be the number of nodes and cusps, respectively, among the singularities (C_i, z_i), $i = 1, \dots, r$.*

(1) If

$$6n + 10k + \frac{49}{6}t + \frac{625}{48}\sum_{(C_i, z_i) \ne A_1, A_2} \delta(C_i, z_i) \le d^2 - 2d + 3\,, \tag{4.5.5.4}$$

where t is the number of singularities A_{2m}, $m \ge 2$, among the (C_i, z_i), $i = 1, \dots, r$, then there exists a reduced, irreducible curve of degree d having r singular points topologically equivalent to (C_1, z_1), ..., (C_r, z_r), respectively, as its only singularities.

(2) If

$$6n + 10k + \sum_{(C_i, z_i) \ne A_1, A_2} \frac{\left(7\mu(C_i, z_i) + 2\delta(C_i, z_i)\right)^2}{6\mu(C_i, z_i) + 3\delta(C_i, z_i)} \le d^2 - 2d + 3\,, \tag{4.5.5.5}$$

then there exists a reduced, irreducible curve of degree d having r singular points analytically equivalent to (C_1, z_1), ...,(C_r, z_r), respectively, as its only singularities.

Condition (4.5.5.5) can be weakened up to the following simple form (in view of $\delta \le 3\mu/4$ for reduced plane curve singularities different from nodes and cusps):

Corollary 4.5.15 *Let (C_i, z_i), $i = 1, \ldots, r$, be reduced plane curve germs. If*

$$\sum_{i=1}^{r} \mu(C_i, z_i) \le \frac{1}{9}(d^2 - 2d + 3)\,,$$

then there exists a reduced, irreducible curve of degree d having r singular points analytically equivalent to (C_1, z_1), ...,(C_r, z_r), respectively, as its only singularities.

Proof of Theorem 4.5.13. Consider first (1), the case of topological equivalence of singular points. By Lemma 1.1.33 and Propositions 1.1.55, 1.1.57, without loss of generality, we can suppose that $Z^s(C_i, z_i)$ is a generic element of the corresponding Hilbert scheme, $i = 1, \ldots, r$.

Let $Z = Z^s(C_1, z_1) \cup \ldots \cup Z^s(C_r, z_r)$. Then, by Corollary 3.6.4, the inequality (4.5.5.2) implies

$$H^1\big(\mathcal{J}_{Z/\mathbb{P}^2}(d-1)\big) = 0\,. \tag{4.5.5.6}$$

We introduce the zero-dimensional schemes

- $Z \cup \{z\}$, where $z \in \mathbb{P}^2$ is any point different from $z_1, \ldots, z_r$;
- $Z^{(i)} = Z \cup Z^s_{st}(C_i, z_i)$, $i = 1, \ldots, r$,

and claim that H^1 vanishes for each of these schemes in degree d, that is,

$$H^1\big(\mathcal{J}_{Z\cup\{z\}/\mathbb{P}^2}(d)\big) = 0, \quad z \in \mathbb{P}^2 \setminus \{z_1, \ldots, z_r\}\,, \tag{4.5.5.7}$$

$$H^1\big(\mathcal{J}_{Z^{(i)}/\mathbb{P}^2}(d)\big) = 0, \quad i = 1, \ldots, r\,. \tag{4.5.5.8}$$

Indeed, we can argue by using reduction by straight lines (cf. Lemma 3.4.2). Let L_0 be a generic straight line through z, and L_i be a generic straight line through z_i. Then $\deg\big(L_0 \cap (Z \cup \{z\})\big) = 1 \le d + 1$ and

$$\deg(L_i \cap Z^s_{st}(C_i, z_i)) \le \deg(L_i \cap Z^s(C_i, z_i)) + 1 \le d + 1\,,$$

the latter inequality due to (4.5.5.6). In any case, we can conclude

$$H^1\big(\mathcal{J}_{(Z\cup\{z\})\cap L_0/L_0}(d)\big) = H^1\big(\mathcal{J}_{Z^{(i)}\cap L_i/L_i}(d)\big) = 0\,. \tag{4.5.5.9}$$

Moreover, $(Z \cup \{z\}) : L_0 = Z$ and $\big(Z \cup Z^s_{st}(C_i, z_i)\big) : L_i = Z$ Hence, in both cases, we have H^1-vanishing for the residue scheme in degree d, which together with (4.5.5.9) allows to conclude (4.5.5.7), (4.5.5.8).

By Proposition 4.5.12 (1), the H^1-vanishing condition (4.5.5.8) implies that, for any $i = 1, \ldots, r$, there exists a curve $D_i \in \big|H^0\big(\mathcal{J}_{Z/\mathbb{P}^2}(d)\big)\big|$ such that (D_i, z_i) is topologically equivalent to (C_i, z_i), $1 \le j \le r$. Again by Proposition 4.5.12 (1), for a generic curve D of the linear system $\lambda_1 D_1 + \ldots + \lambda_r D_r$, $\lambda_1, \ldots, \lambda_r \in \mathbb{C}$, the germs (D, z_i) are topologically equivalent to (C_i, z_i), $i = 1, \ldots, r$, respectively. Furthermore, we can suppose that D is reduced. Let $w_1, \ldots, w_m$ be all singular points of D outside $z_1, \ldots, z_r$. By virtue of (4.5.5.7), there exist curves $D'_1, \ldots, D'_m \in \left|H^0\big(\mathcal{J}_{Z/\mathbb{P}^2}(d)\big)\right|$ such that $w_j \notin D'_j$, $j = 1, \ldots, m$. By Bertini's Theorem 2.1.20, the singular locus of a generic curve D' of the linear system $\lambda D + \lambda'_1 D'_1 + \ldots + \lambda'_m D'_m$, $\lambda, \lambda'_1, \ldots, \lambda'_m \in \mathbb{C}$, is $\{z_1, \ldots, z_r\}$, and (D', z_i) is topologically equivalent to (C_i, z_i), $i = 1, \ldots, r$.

Finally, we show that D' is irreducible. The above argument shows that the linear system $\left|H^0\big(\mathcal{J}_{Z/\mathbb{P}^2}(d)\big)\right|$ has no fixed part. If all the curves in the linear system $\left|H^0\big(\mathcal{J}_{Z/\mathbb{P}^2}(d)\big)\right|$ would be reducible, then by Bertini's Theorem 2.1.24, a generic curve $D' \in \left|H^0\big(\mathcal{J}_{Z/\mathbb{P}^2}(d)\big)\right|$ would split into irreducible components, which all belong to a one-dimensional linear system. In particular, we would have

$$\dim \left|H^0\big(\mathcal{J}_{Z/\mathbb{P}^2}(d)\big)\right| \le d\,.$$

On the other hand, by the H^1-vanishing (4.5.5.6), we compute

$$\dim \left|H^0\big(\mathcal{J}_{Z/\mathbb{P}^2}(d)\big)\right| = \frac{d(d+3)}{2} - \deg Z\,.$$

The inequality (4.5.5.2) and Corollary 1.1.40 yield

$$\begin{aligned} d + 2 &> \sqrt{\frac{4}{3}M_2(Z)} + \frac{\deg Z}{\sqrt{4M_2(Z)/3}} \\ &\ge \sqrt{\frac{4}{3}\deg Z} + \frac{\deg Z}{\sqrt{4\deg Z/3}} = \frac{7}{\sqrt{12}}\sqrt{\deg Z}\,. \end{aligned}$$

Hence,

$$\begin{aligned} \dim \left|H^0\big(\mathcal{J}_{Z/\mathbb{P}^2}(d)\big)\right| - d &= \frac{d(d+1)}{2} - \deg Z \\ &> \frac{25}{24}\deg Z - \frac{21}{2\sqrt{12}}\sqrt{\deg Z} + 1\,, \end{aligned}$$

which is positive whenever $\deg Z \ge 6$. But the latter is satisfied unless in the case of one single node or one single cusp. But both of these exceptional cases are obviously covered by (4.5.5.2).

The case (2) of analytic equivalence of singular points can be treated in the same way, using Proposition 4.5.12 (2) and Lemma 1.1.75, instead of Proposition 4.5.12 (1) and Lemma 1.1.33.

Finally, we note that, by construction, $Z^s(D, z_i) = Z^s(C_i, z_i)$ in Case (1) of the theorem, and $Z^a(D, z_i) = Z^a(C_i, z_i)$ in Case (2), $i = 1, \ldots, r$. In view of

$$I^s(D, z_i) \subset I^{es}(D, z_i), \quad I^a(D, z_i) \subset I^{ea}(D, z_i), \quad i = 1, \ldots, r\,,$$

Equation (4.5.5.6) implies

$$H^1\big(\mathcal{J}_{Z^{es}(D)/\mathbb{P}^2}(d)\big)=0\,,\quad H^1\big(\mathcal{J}_{Z^{ea}(D)/\mathbb{P}^2}(d)\big)=0\,,$$

respectively to the case considered. In turn the last H^1-vanishing means the T-smoothness of the topological, respectively analytic, equisingular stratum at D in the space of curves of degree d (see Sect. 2.2.2.3 for details). □

Remark 4.5.15.1 In the case of ordinary singularities, one can obtain more precise results. Let (C,z) be a reduced curve germ having an ordinary singularity of multiplicity $m\ge 3$. Then (see Examples 1.1.32.1(b) and 1.1.78.1(a))

$$\deg Z^s(C,z)=\frac{m(m+1)}{2},\quad M_2(Z^s(C,z))=m^2\,,$$

$$\deg Z^a(C,z)\le m^2,\quad M_2(Z^a(C,z))\le(2m-2)^2\,.$$

Assume now that (C_i,z_i), $i=1,\dots,r$, are ordinary singularities of multiplicities $m_i, i=1,\dots,r$, respectively. Applying the argument in the proof of Theorem 4.5.14, presented below, we obtain that

(1) the inequality

$$6n+\sum_{m_i\ge 3}\frac{73m_i^2+18m_i+9}{24}\le d^2-2d+3\,,$$

where n is the number of nodes among (C_i,z_i), $i=1,\dots,r$, is sufficient for the existence of an irreducible curve of degree d with r ordinary singularities of multiplicities $m_1,\dots,m_r$ as its only singularities, and

(2) the inequality

$$6n+\sum_{m_i\ge 3}\frac{137m_i^4-256m_i^2+128}{24(m_i-1)^2}\le d^2-2d+3$$

is sufficient for the existence of an irreducible curve of degree d having r singular points analytically equivalent to $(C_1,z_1),\dots,(C_r,z_r)$ respectively, and no other singularities.

Since $\delta(C,z)=\frac{m(m-1}{2}$ for an ordinary singularity (C,z) of multiplicity m, the l.h.s. of both inequalities above are smaller than the l.h.s. of the corresponding inequalities (4.5.5.4) and (4.5.5.5). Therefore they imply better bounds for the existence of ordinary singularities than those of Theorem 4.5.14.

Proof of Theorem 4.5.14(1). Without loss of generality assume that $d\ge 4$.

Case A. If there are no nodes and cusps among the (C_i,z_i), $i=1,\dots,r$, then by Corollary 1.1.40 and Corollary 1.1.41, we have

- for the germs of type A_{2m}, $m \geq 2$,

$$\deg Z^s(C_i, z_i) = 3\delta(C_i, z_i) + 2, \quad M_2(Z^s(C_i, z_i)) = 4\delta(C_i, z_i) + 2,$$

- for the remaining singularities (C_i, z_i)

$$\deg Z^s(C_i, z_i) \leq 3\delta(C_i, z_i), \quad M_2(Z^s(C_i, z_i)) \leq 4\delta(C_i, z_i)\,.$$

Hence, we can estimate

$$\left(\sqrt{\frac{4}{3}\sum_{i=1}^{r} M_2(Z^s(C_i, z_i))} + \frac{\sum_{i=1}^{r} \deg Z^s(C_i, z_i)}{\sqrt{4\sum_{i=1}^{r} M_2(Z^s(C_i, z_i))/3}}\right)^2 \tag{4.5.5.10}$$

$$\leq \frac{(25\sum_{i=1}^{r} \delta(C_i, z_i) + 14t)^2}{48\sum_{i=1}^{r} \delta(C_i, z_i) + 24t} \leq \frac{625}{48}\sum_{i=1}^{r} \delta(C_i, z_i) + \frac{49}{6}t\,.$$

Thus, (4.5.5.4) implies (4.5.5.2), and we are done.

Case B. Assume now that $n + k > 0$. Put

$$s := \max\left\{t \geq 2 \;\middle|\; \sum_{i=1}^{t-2} (d - i) + 1 \leq 3n + 5k\right\}.$$

Observe that by (4.5.5.4), $s \leq d - 1$. The inequalities (4.5.5.4), (4.5.5.10) also yield

$$\sqrt{\frac{4}{3}\sum_{(C_i,z_i)\neq A_1,A_2} M_2(Z^s(C_i, z_i))} + \frac{\sum_{(C_i,z_i)\neq A_1,A_2} \deg Z^s(C_i, z_i)}{\sqrt{4\sum_{(C_i,z_i)\neq A_1,A_2} M_2(Z^s(C_i, z_i))/3}}$$

$$\leq \sqrt{\frac{625}{48}\sum_{(C_i,z_i)\neq A_1,A_2} \delta(C_i, z_i) + \frac{49}{6}t} \leq \sqrt{d^2 - 2d + 3 - 6n - 10k}$$

$$\leq \sqrt{d^2 - 2d + 3 - 2\sum_{i=1}^{s-2}(d - i) - 2} = \sqrt{d^2 - 2(s-1)(d+s-2) + 1}\,,$$

which does not exceed $d - s + 1$. Hence, by Corollary 3.6.4,

$$H^1\big(\mathcal{J}_{Z'/\mathbb{P}^2}(d - s - 1)\big) = 0\,,$$

where Z' is the part of Z without nodes and cusps. Then we derive (4.5.5.2), and thereby the first statement of Theorem 4.5.14, from Lemma 4.5.16, below. □

Lemma 4.5.16 *Let $L \subset \mathbb{P}^2$ be a straight line, $Z, Y \subset \mathbb{P}^2$ be zero-dimensional schemes such that $L \cap Y = \emptyset$, $Z \cap Y = \emptyset$, and $Z = Z^{(1)} \cup Z^{(2)} \cup Z^{(3)} \cup Z^{(4)}$, where*

- $Z^{(1)}$ *is a union of schemes of degree* 1 *(that is, locally defined by the maximal ideal),*
- $Z^{(2)}$ *is a union of schemes of degree* 2 *(that is, locally defined by ideals like* $\langle y, x^2\rangle$*),*
- $Z^{(3)}$ *is a union of schemes from* $\mathrm{Iso}(Z^s(node))$ *(that is, locally defined by the square of the maximal ideal),*
- $Z^{(4)}$ *is a union of schemes from* $\mathrm{Iso}(Z^s(cusp))$ *(that is, locally defined by ideals like* $\langle y^2, yx^2, x^3\rangle$*).*

Assume that

(i) $H^1\big(\mathcal{J}_{Y/\mathbb{P}^2}(d-s-1)\big)=0$, *where*

$$s := \max\left\{t\geq 2\;\middle|\; \sum_{i=1}^{t-2}(d-i)+1\leq \deg Z\right\}\leq d-1\,;$$

(ii) $Z^{(4)}\cap L=\emptyset$, $\deg(Z\cap L)\leq d$*;*

Moreover, assume that those components of Z which do not meet L are placed in generic position in $\mathbb{P}^2\setminus\mathrm{supp}(Y\cup L)$*. Then*

$$H^1\big(\mathcal{J}_{Z\cup Y/\mathbb{P}^2}(d-1)\big)=0\,. \tag{4.5.5.11}$$

Proof We perform induction on s using the "Horace method" in the form of Lemma 3.4.2.

Case A. If $s=2$, then $\deg Z\leq d-1$. We specialize all the components of Z on the line L with maximal possible intersection with L. First, we note that (4.5.5.11) for the specialized scheme Z implies the same relation for the original Z due to semicontinuity of cohomology. Second,

$$\deg((Z:L)\cap L)\leq\deg(Z\cap L)\leq\deg Z\leq d-1\,,$$

and $Z\subset L^2$. Then (4.5.5.11) follows from the two exact sequences

$$\underbrace{H^1\big(\mathcal{J}_{Y/\mathbb{P}^2}(d-3)\big)}_{=0}\to H^1\big(\mathcal{J}_{(Z:L)\cup Y/\mathbb{P}^2}(d-2)\big)\to\underbrace{H^1\big(\mathcal{J}_{(Z:L)\cap L/L}(d-2)\big)}_{=0},$$

$$H^1\big(\mathcal{J}_{(Z:L)\cup Y/\mathbb{P}^2}(d-2)\big)\to H^1\big(\mathcal{J}_{Z\cup Y/\mathbb{P}^2}(d-1)\big)\to\underbrace{H^1\big(\mathcal{J}_{Z\cap L/L}(d-1)\big)}_{=0}.$$

Case B. Let $s\geq 3$ and $d-1\leq\deg(Z\cap L)\leq d$.
Then, again, $H^1\big(\mathcal{J}_{Z\cap L/L}(d-1)\big)$ vanishes, and the latter sequence reduces (4.5.5.11) to $H^1\big(\mathcal{J}_{(Z:L)\cup Y/\mathbb{P}^2}(d-2)\big)=0$, which holds due to the induction assumption. Indeed, one can easily check that $Z:L$ is the union of zero-dimensional schemes of the above four types, the intersection $(Z:L)\cap L$ is the union of schemes of degree 1, and $\deg\big((Z:L)\cap L\big)<\deg(Z\cap L)\leq d$. Finally, since

$$\sum_{i=1}^{s-1}(d-i)+1 > \deg Z \geq \deg(Z:L)+(d-1),$$

we can easily deduce

$$s-1 \geq \max\left\{t \geq 2 \;\middle|\; \sum_{i=1}^{t-2}(d-1-i)+1 \leq \deg Z\right\}.$$

Case C. Let $s \geq 3$ and $\deg(Z \cap L) \leq d-2$.
If all the components of Z are specialized on L then we complete the proof as in the case $s=2$. If there are components of Z out of L, we specialize some of them on L keeping the three rules:

- $\deg(Z \cap L) \leq d$,
- if $\deg(Z \cap L) \leq d-2$ then a component Z' of degree 2 of Z is specialized on L such that $\deg(Z' \cap L) = 2$ (with tangency),
- if $\deg(Z \cap L) \leq d-3$ then a component $Z' \in \mathrm{Iso}(Z^s(\mathrm{cusp}))$ is specialized on L so that $\deg(Z' \cap L) = 3$ (with tangency).

If we end up with $d-1 \leq \deg(Z \cap L) \leq d$, then the components of $Z:L$ meeting L either are of degree at most 2, or belong to $\mathrm{Iso}(Z^s(\mathrm{node}))$, and $\deg((Z:L)\cap L) \leq \deg(Z \cap L)-1 \leq d-1$; this allows us to complete the proof as in the preceding paragraph. If we end up with $\deg(Z \cap L) \leq d-2$, and all components of Z specialized on L so that the components from $\mathrm{Iso}(Z^s(\mathrm{cusp}))$ are tangent to L, then $Z \subset L^2$, and we complete the proof as in the case $s=2$. □

Proof of Theorem 4.5.14(2). If there are no nodes and cusps among the (C_i, z_i), $i = 1, \dots, r$, then by Corollary 1.1.40 and Lemma 1.1.78,

$$\deg Z^a(C_i, z_i) \leq 2\mu(C_i, z_i), \;\; M_2(Z^a(C_i, z_i)) \leq 2\mu(C_i, z_i) + \delta(C_i, z_i).$$

Hence,

$$\left(\sqrt{\frac{4}{3}\sum_{i=1}^{r} M_2(Z^a(C_i, z_i))} + \frac{\sum_{i=1}^{r}\deg Z^a(C_i, z_i)}{\sqrt{4\sum_{i=1}^{r} M_2(Z^a(C_i, z_i))/3}}\right)^2$$
$$\leq \frac{\left(\sum_{i=1}^{r}(7\mu(C_i, z_i)+2\delta(C_i, z_i))\right)^2}{\sum_{i=1}^{r}(6\mu(C_i, z_i)+3\delta(C_i, z_i))} \leq \sum_{i=1}^{r}\frac{(7\mu(C_i)+2\delta(C_i, z_i))^2}{6\mu(C_i, z_i)+3\delta(C_i, z_i)}.$$

Thus, (4.5.5.5) implies (4.5.5.3), and we are done.

If there are some nodes and cusps among the (C_i, z_i), that is, if $n+k>0$, then we can proceed as in the above proof of Theorem 4.5.14(1). □

4.5.5.3 Curves with One Singular Point

We pay a special attention to the case of curves with exactly one singular point, because it is involved in another construction of curves with prescribed singularities, which is based on the modified Viro method.

The existence results of this section are deduced from explicit constructions of series of equations. Interesting concrete examples are computed resp. verified by using SINGULAR.

In what follows we shall use a refined notion of the *order of T-existence* introduced in Definition 4.5.9.

Definition 4.5.17 Given a reduced plane curve germ (C, z), denote by $e^s(C, z)$, respectively $e^a(C, z)$, the minimal degree m of a plane curve F having only one singular point w, which is topologically (respectively analytically) equivalent to (C, z), and satisfies the condition

$$H^1\big(\mathcal{J}_{Z^{es}(F)/\mathbb{P}^2}(m-1)\big) = 0\,, \quad H^1\big(\mathcal{J}_{Z^{ea}(F)/\mathbb{P}^2}(m-1)\big) = 0\,, \qquad (4.5.5.12)$$

respectively.

We call $e^s(C, z)$ (resp. $e^a(C, z)$) the refined *topological* (resp. *analytical*) *order of T-existence* of (C, z).
These parameters are characterized by the following

Lemma 4.5.18 *(1) Let F be a plane curve as in Definition 4.5.17, and let L be a straight line which does not pass through w. Then the germ at F of the family of curves of degree m having in a neighbourhood of w a singular point topologically (respectively analytically) equivalent to (C, z), is smooth of the expected dimension, and intersects transversally the linear system*

$$\Big\{G \in \big|H^0(\mathcal{O}_{\mathbb{P}^2}(d))\big| \;\Big|\; G \cap L = F \cap L\Big\}\,.$$

(2) Let $L \subset \mathbb{P}^2$ be a straight line. Then the set of m-tuples $\overline{z}$ of distinct points on L, such that there is a curve F of degree m as in Definition 4.5.17, satisfying $F \cap L = \overline{z}$, is Zariski open in $\mathrm{Sym}^m(L)$.

Proof The first statement is equivalent to

$$H^1\big(\mathcal{J}_{Z\cup(F\cap L)/\mathbb{P}^2}(m)\big) = 0\,,$$

where $Z = Z^{es}(F)$, respectively $Z = Z^{ea}(F)$. But the latter follows from (4.5.5.12) and the exact sequence

$$0 = H^1\big(\mathcal{J}_{Z/\mathbb{P}^2}(m-1)\big) \to H^1\big(\mathcal{J}_{Z\cup(F\cap L)/\mathbb{P}^2}(m)\big) \to H^1\big(\mathcal{J}_{F\cap L/L}(m)\big) = 0\,.$$

For the second statement take a curve F as in Definition 4.5.17, which meets L transversally, and consider the germ M at F of the family of curves of degree m having in a neighbourhood of w a singular point topologically (respectively analytically) equivalent to (C, z). The first statement of the lemma implies that the map $M \to \mathrm{Sym}^m(L)$, $G \mapsto G \cap L$ is a submersion, and we are done. □

Next we estimate e^s and e^a.

Theorem 4.5.19 *(1) For any reduced plane curve germ (C, z),*

$$e^s(C, z) \le \sqrt{\frac{4}{3}M_2(Z^s(C, z))} + \frac{\deg Z^s(C, z)}{\sqrt{4M_2(Z^s(C, z))/3}} - 1\,, \quad (4.5.5.13)$$

$$e^a(C, z) \le \sqrt{\frac{4}{3}M_2(Z^a(C, z))} + \frac{\deg Z^a(C, z)}{\sqrt{4M_2(Z^a(C, z))/3}} - 1\,. \quad (4.5.5.14)$$

(2) If (C, z) is a simple (ADE) singularity then

$$e^s(C, z) = e^a(C, z) \begin{cases} \le 2\lfloor\sqrt{m+5}\rfloor & \text{if } (C, z) \text{ of type } A_m,\ m \ge 1\,, \\ \le 2\lfloor\sqrt{m+7}\rfloor + 1 & \text{if } (C, z) \text{ of type } D_m,\ m \ge 4\,, \\ = \lfloor m/2\rfloor + 1 & \text{if } (C, z) \text{ of type } E_m,\ m = 6, 7, 8. \end{cases}$$

If (C, z) is not simple, then

$$e^s(C, z) \le \frac{25}{4\sqrt{3}}\sqrt{\delta(C, z)} - 1\,,$$

$$e^a(C, z) \le \frac{7\mu(C, z) + 2\delta(C, z)}{\sqrt{6\mu(C, z) + 3\delta(C, z)}} - 1 \le 3\sqrt{\mu(C, z)} - 1\,.$$

Proof The bounds (4.5.5.13), (4.5.5.14) are particular cases of Theorem 4.5.13, in view of (4.5.5.6). Moreover, the bounds for non-simple singularities in the second part of the theorem follow from (4.5.5.13), (4.5.5.14), Corollary 1.1.40, Lemma 1.1.78 and Corollary 1.1.41. So we turn our attention to simple singularities.

In Lemma 4.5.20 below we construct a series of reduced curves of degree d having a simple singularity S_0 with Milnor number $\mu(S_0) \ge d^2/2$.

Then we proceed as follows. First, we slightly deform these curves and obtain irreducible curves having a singularity of type S', $\mu(S') \ge d^2/4$, as only singularity and belonging to T-smooth germs of the equisingular stratum $V_d(S') \subset \mathbb{P}\big(H^0\big(\mathcal{O}_{\mathbb{P}^2}(d)\big)\big)$. Second, by deformation theoretic arguments, we can deduce the existence of an irreducible curve of degree d having a simple singularity of type S as only singularity (and corresponding to a T-smooth germ of $V_d(S)$) whenever

$$\mu(S) \le \frac{d^2}{4} - 7\,.$$

Finally, Corollary 4.5.22, respectively Remark 4.5.22.1, will provide the needed estimates. □

The following lemma provides a series of reduced curves of degree d having a simple singularity S_0 with Milnor number $\mu(S_0) \geq d^2/2$. It is, in fact, a particular case of a stronger statement of [GuN2, Proposition 2].

Lemma 4.5.20 *Let $k \in \mathbb{N}$, $k \geq 2$. Then the reduced curve $F_k \subset \mathbb{P}^2$ of degree $d = 2k+1$ with affine equation*

$$f_k(x,y) = y^2 - 2y\Phi_k(x,y) + x^{2k} + x^{2k-1}y^2\,,$$
$$\Phi_k(x,y) := x^k + \textstyle\sum_{i=1}^k a_i x^{k-i} y^{2i}\,,$$

where

$$a_i = \binom{1/2}{i} = \frac{(-1)^{i-1}}{2^i} \cdot \prod_{j=1}^{i-1} \frac{2j-1}{j+1} \quad (i \geq 1),$$

has an A_{2k^2+3k-2}-singularity (at the origin).

Proof Observe that the coefficients a_i satisfy $\left(1 + \sum_{i=1}^\infty a_i z^i\right)^2 = 1 + z$. In particular, for any $\nu \geq 2$ we have $\sum_{i+j=\nu} a_i a_j = 0$. It follows that

$$\Phi_k(x,y)^2 - x^{2k} - x^{2k-1}y^2 = \sum_{\nu=1}^{k} \Bigl(\sum_{\substack{i+j=k+\nu \\ i,j\leq k}} a_i a_j \Bigr) \cdot x^{k-\nu} y^{2k+2\nu}.$$

Hence, applying the coordinate transformation

$$\tilde{y} := y - \Phi_k(x,y) = y \cdot \mathrm{unit} - x^k \in \mathbb{C}\{x,y\}\,, \tag{4.5.5.15}$$

the local equation of F_k at the origin becomes

$$\begin{aligned} &\tilde{y}^2 - \sum_{\nu=1}^{k} \Bigl(\sum_{\substack{i+j=k+\nu \\ i,j\leq k}} a_i a_j \Bigr) \cdot \mathrm{unit} \cdot x^{k-\nu} (\tilde{y} + x^k)^{2k+2\nu} \\ &\quad = \tilde{y}^2 \cdot \mathrm{unit} + \mathrm{const} \cdot \tilde{y} \cdot \bigl(x^{(k-1)+(2k+1)k} + (\text{higher terms in } x) \bigr) \\ &\qquad + \Bigl(\sum_{i+j=k+1} a_i a_j \Bigr) \cdot \mathrm{unit} \cdot x^{(k-1)+(2k+2)k} + (\text{higher terms in } x)\,, \end{aligned}$$

which is right equivalent to $\tilde{y}^2 + x^{2k^2+3k-1}$. □

The following examples are computed using SINGULAR. First, we should like to provide two procedures, the first computes the coefficients a_i, the second generates the above polynomials f_k:

```
proc compute_coeffs (int m)
{
   vector a=(1/2)*gen(1);;
   int j;
   poly p;
   for (int i=2; i<=m; i=i+1)
   {
     p=0;
     for (j=1; 2*j<i; j=j+1) { p=p-a[j]*a[i-j]; }
     if (2*j==i) { p=p-((a[j])^2)/2; }
     a=a+p*gen(i);
   }
   return(a);
};

proc compute_poly (int k)
{
   poly f=y^2+x^(2*k)+x^(2*k-1)*y^2;
   poly A=x^k;
   vector a=compute_coeffs(k);
   for (int i=1; i<=k; i=i+1) { A=A+a[i]*x^(k-i)*y^(2*i); }
   f=f-2*y*A;
   return(f);
};
```

Example 4.5.20.1 (a) The reduced curve F_{14} of degree 29 with affine equation

$$\begin{aligned}&y^2 - 2y\big(x^{14} + \tfrac{1}{2}x^{13}y^2 - \tfrac{1}{8}x^{12}y^4 + \tfrac{1}{16}x^{11}y^6 - \tfrac{5}{128}x^{10}y^8 + \tfrac{7}{256}x^9y^{10}\\ &- \tfrac{21}{1024}x^8y^{12} + \tfrac{33}{2048}x^7y^{14} - \tfrac{429}{32768}x^6y^{16} + \tfrac{715}{65536}x^5y^{18}\\ &- \tfrac{2431}{262144}x^4y^{20} + \tfrac{4199}{524288}x^3y^{22} - \tfrac{29393}{4194304}x^2y^{24} + \tfrac{52003}{8388608}xy^{26}\\ &- \tfrac{185725}{33554432}y^{28}\big) + x^{28} + x^{27}y^2\end{aligned}$$

has an A_{432}-singularity at the origin. Moreover, it is smooth outside the origin (the total Tjurina number $\dim \mathbb{C}[x,y]/(f_{14}, \partial f_{14}/\partial x, \partial f_{14}/\partial y)$, as computed with SINGULAR:

```
LIB "sing.lib";
ring r=0,(x,y),dp;
poly f=compute_poly(14);  // generate F_14
tjurina(f);               // total (affine) Tjurina number
// -> 432
```

coincides with the local Tjurina number and F_{14} is obviously smooth at infinity). Since the only singularity of F_{14} is unibranch, we obtain in particular that the curve F_{14} is irreducible.

(b) The reduced curve of degree 20 with (reducible) affine equation

$$\begin{aligned} x \cdot f_9(x,y) = \ & xy^2 - 2xy\big(x^9 + \tfrac{1}{2}x^8y^2 - \tfrac{1}{8}x^7y^4 + \tfrac{1}{16}x^6y^6 - \tfrac{5}{128}x^5y^8 \\ & + \tfrac{7}{256}x^4y^{10} - \tfrac{21}{1024}x^3y^{12} + \tfrac{33}{2048}x^2y^{14} \\ & - \tfrac{429}{32768}xy^{16} + \tfrac{715}{65536}y^{18}\big) + x^{19} + x^{18}y^2 \end{aligned}$$

has a D_{190}-singularity at the origin. Moreover, deforming this curve by adding y^{20} to the affine equation, we obtain an irreducible curve G_9 having a D_{181}-singularity at the origin as only singularity:

```
poly g=x*compute_poly(9)+y20;
tjurina(g);                    // total (affine) Tjurina number
// -> 181
```

```
ring r1=0,(x,y),ds;
poly g=imap(r,g);
tjurina(g);                    // local Tjurina number
// -> 181
```

The latter computation of the local Tjurina number turns out to be very time (and memory) consuming. The reason is the amazing growth of coefficients during the standard basis computation. The latter can be avoided when working over a field of a sufficiently high, finite characteristic (e.g., 32003), where usually one gets the same result (but, of course, there is no guarantee for that).

Remark 4.5.20.2 (a) If one is interested in the maximal A_k-singularity, which may occur on a curve of degree d, Lemma 4.5.20 gives a positive answer for odd degrees:

$$k_{\max}(d) \geq \frac{d^2}{2} + \frac{d}{2} - 3\,.$$

In [GuN1], Gusein-Zade and Nekhoroshev give an example of a curve of degree 22 with one A_{257}-singularity at the origin and an additional non-simple singularity (of type W_{18}) at infinity. Moreover, in [GuN2] they exhibit a series of plane curves of degree $28s + 9$, $s \geq 0$, possessing a singularity $A_{\mu(s)}$ with Milnor number

$$\mu(s) = 420s^3 + 269s + 42 = \frac{15}{28}d^2 + O(d)\,.$$

Later Orevkov [Ore, Section 4] found a series of curves of degree $d \to \infty$ yielding the bound

$$\limsup_{d\to\infty} \frac{k_{\max}(d)}{d^2} \geq \frac{7}{12} = 0.58333\ldots.$$

(b) One can easily generalize the construction of curves F_k in Lemma 4.5.20 to obtain curves of a possibly small degree having singularities of type $y^q - x^p$. For instance, the curve of degree $d = 6k + 1$ with affine equation

$$y^3 - 3y^2\Psi_k(x, y) + 3y\Psi_k(x, y)^2 - x^{6k} - x^{6k-2}y^3,$$
$$\Psi_k(x, y) := x^{2k} + \textstyle\sum_{i=1}^{k} \binom{1/3}{i} \cdot x^{2(k-i)}y^{3i},$$

has a singularity of type $y^3 - x^p$, $p = 6k^2 + 10k - 2$.

Proposition 4.5.21 *If $k \geq 3$ then there exist reduced, irreducible curves plane C of degree $d = 2k+1$ (respectively $d = 2k+2$) having a singularity of type $S = A_\mu$, $\mu = k^2+4k-1$, (respectively $S = D_\mu$, $\mu = k^2+5k-1$) as its only singularity such that the equisingular stratum $V_d(S)$ is T-smooth at C.*

Proof (A_μ) Consider the curve C with affine equation

$$f(x, y) := f_k(x, y) + t \cdot y^{k+4}, \quad t \neq 0 \text{ generic},$$

(obtained by a small deformation of F_k). Applying the coordinate transformation (4.5.5.15), it is not difficult to see that locally at the origin

$$f(x, y) \overset{r}{\sim} \tilde{y}^2 + \text{unit} \cdot x^{2k^2+3k-1} + x^{(k+4)k} \overset{r}{\sim} \tilde{y}^2 + x^{k^2+4k}.$$

By Bertini's Theorem 2.1.24, C is irreducible and has no singularities outside the base points of the linear system $\{\alpha F_k + \beta y^{k+4}z^{k-3} \mid (\alpha:\beta) \in \mathbb{P}^1\}$, that is, outside $\{y = 0\}$ and the infinite line $\{z = 0\}$. On the other hand, F_k itself is smooth at infinity and has no singularity along $\{y = 0\}$ other than the origin.

It remains to prove the T-smoothness of the germ $(V_d(S), C)$. Since C has a simple singularity at the origin $0 = (0:0:1)$, the equisingularity ideal $I^{es}(C, 0)$ equals the Jacobian ideal, which in the case of an A_μ-singularity can be described as

$$I^{es}(C, 0) = \langle f_x, f_y \rangle = \{g \in \mathcal{O}_{\mathbb{P}^2,0} \mid i(g, \alpha f_x + \beta f_y) \geq \mu = k^2 + 4k - 1\},$$

where $(\alpha:\beta) \in \mathbb{P}^1$ generic and $f_x = \frac{\partial f}{\partial x}$, $f_y = \frac{\partial f}{\partial y}$ denote the partial derivatives of f. Since the generic polar $\alpha f_x + \beta f_y$ is smooth at 0 and intersects the line $\{y = 0\}$ at the origin with multiplicity $i(y, \alpha f_x + \beta f_y) = k$, we obtain

$$y^m \in I^{es}(C, 0) \iff m \geq k + 4. \tag{4.5.5.16}$$

Moreover, let $Z^{es} := Z^{es}(C, 0)$, then

$$\begin{aligned}\deg\big((Z^{es} : y^j) \cap y\big) &= \dim_{\mathbb{C}} \mathbb{C}\{x, y\}/\big(\langle y\rangle + (I^{es}(C, 0) : y^j)\big)\\ &= \dim_{\mathbb{C}} \mathbb{C}\{x, y\}/\big(\langle y\rangle + \langle\, g \in \mathcal{O}_{\mathbb{P}^2,0} \mid i(g, \alpha f_x+\beta f_y) \geq \mu - jk\,\rangle\big),\end{aligned}$$

which is just $i(y, g)$ for a generic g satisfying $i(g, \alpha f_x+\beta f_y) \geq \mu - jk$. In particular, if $k < \mu - jk$, that is, if $0 \leq j \leq k+2$, we obtain

$$\deg\big((Z^{es} : y^j) \cap y\big) = i(y, g) = i(y, \alpha f_x+\beta f_y) = k \leq d - j + 1,$$

which implies the vanishing of $H^1\big(\mathcal{J}_{(Z^{es}:y^j)\cap y/y}(d-j)\big)$. The reduction sequence (3.4.1.3) induce surjective morphisms

$$H^1\big(\mathcal{J}_{Z^{es}:y^{j+1}/\mathbb{P}^2}(d-j-1)\big) \twoheadrightarrow H^1\big(\mathcal{J}_{Z^{es}:y^j/\mathbb{P}^2}(d-j)\big), \tag{4.5.5.17}$$

$0 \leq j \leq k+2$. Finally, by (4.5.5.16) and

$$\deg\big((Z^{es} : y^{k+3}) \cap y\big) \leq \deg Z^{es} - (k+3)\cdot k = k - 1 = d - k - 2,$$

we obtain a surjective morphism

$$0 = H^1\big(\mathcal{O}_{\mathbb{P}^2}(d-k-4)\big) \twoheadrightarrow H^1\big(\mathcal{J}_{Z^{es}:y^{k+3}/\mathbb{P}^2}(d-k-3)\big),$$

and (4.5.5.17) gives the vanishing of $H^1\big(\mathcal{J}_{Z^{es}/\mathbb{P}^2}(d)\big)$.

(D_μ) We proceed as in the A_μ-case, considering the curve C with affine equation $f(x, y) := xf_k(x, y) + s\cdot x^{k-2}y^{k+4} + t\cdot y^{2k+2}$, with $s, t \neq 0$ generic. Again, the main point is to prove the T-smoothness of the germ $\big(V_d(S), C\big)$.

The equisingularity ideal $I^{es}(C, 0)$ is generated by the partial derivatives

$$\begin{aligned}f_x(x, y) &= f_k(x, y) + x\cdot \tfrac{\partial}{\partial x} f_k(x, y) + s\cdot (k-2)\cdot x^{k-3}y^{k+4},\\ f_y(x, y) &= x\cdot \tfrac{\partial}{\partial y} f_k(x, y) + t\cdot (2k+2)\cdot y^{2k+1} + s\cdot (k+4)\cdot x^{k-2}y^{k+3}.\end{aligned}$$

In particular, $\deg(Z^{es} \cap y) = i(y, f_y) = k+1$. Moreover, the (smooth) local branches $\mathcal{P}_1$ (tangent to x), $\mathcal{P}_2$ of f_y satisfy

$$\begin{aligned}i(y^{k+5}, \mathcal{P}_1; 0) &= k+5 > 3 = 1 + i(f_x, \mathcal{P}_1; 0),\\ i(y^{k+5}, \mathcal{P}_2; 0) &= (k+5)k > k^2 + 5k - 2 = 1 + i(f_x, \mathcal{P}_2; 0),\end{aligned}$$

whence the local analogue Noether's fundamental ("$AF + BG$") Theorem (2.1.26) implies that $y^{k+5} \in I^{es}(C, 0)$. In addition, we know that

$$\begin{aligned}\deg\big((Z^{es} : y^{k+3}) \cap y\big) &\leq \frac{(\deg Z^{es} - \deg(Z^{es} \cap y))}{k+3} < k+1,\\ \deg\big((Z^{es} : y^{k+4}) \cap y\big) &\leq \frac{(\deg Z^{es} - \deg(Z^{es} \cap y))}{k+4} < k.\end{aligned}$$

Repeating the above reasoning, we deduce that $H^1\big(\mathcal{J}_{Z^{es}/\mathbb{P}^2}(d)\big) = 0$. □

Corollary 4.5.22 *(a) For any $\mu \geq 1$, $e^s(A_\mu) = e^a(A_\mu) \leq 2\left\lfloor\sqrt{\mu+5}\right\rfloor$.*
(b) For any $\mu \geq 4$, $e(D_\mu) \leq 2\left\lfloor\sqrt{\mu+7}\right\rfloor + 1$.

Proof Due to Theorem 2.2.40(d), the T-smoothness of $V_d(S)$ at C implies that each local deformation of the singularity can be realized by a (global) deformation of the curve C. Hence, it follows from Proposition 4.5.21 that $e^s(A_\mu)$ does not exceed $2k+2$, where k is minimal with $k^2+4k-1 \geq \mu$, and that $e^s(D_\mu) \leq 2k+3$, where k is minimal with $k^2+5k-1 \geq \mu$. This covers the case of A_μ, $\mu \geq 11$, and D_μ, $\mu \geq 14$. The case of smaller values of μ is left to the reader. □

Remark 4.5.22.1 Actually, we can even give concrete equations for curves of low degree with a prescribed simple singularity, since the proof of Proposition 4.5.21 shows that deforming f_k by $x^\alpha y^\beta$, $\alpha + k\beta \leq 2k^2+3k-1$ produces an $A_{\alpha+k\beta-1}$-singularity at the origin.

Remark 4.5.22.1 For ADE-singularities with small Milnor number, one has of course much better estimates, for instance,

$$e^s(A_\mu) = \begin{cases} 4 \text{ if } 3 \leq \mu \leq 7\,, \\ 5 \text{ if } 8 \leq \mu \leq 13\,, \\ 6 \text{ if } 14 \leq \mu \leq 19\,, \end{cases} \qquad e^s(D_\mu) = \begin{cases} 3 \text{ if } \mu = 4\,, \\ 4 \text{ if } \mu = 5\,, \\ 5 \text{ if } 6 \leq \mu \leq 10\,, \\ 6 \text{ if } 11 \leq \mu \leq 13\,, \end{cases}$$

Moreover, $e^s(E_6) = e^s(E_7) = 4$, $e^s(E_8) = 5$. Indeed, the existence of cubic and quartic curves with singularities indicated above is well-known. The existence of quintic and sextic curves with singularities specified above can be found in [AD, Deg1, Deg2]. The T-smoothness follows from the $4d$-criterion in Theorem 4.3.8.

4.5.5.4 Yet Another Numerical Criterion

Proposition 4.5.23 *Let (C_i, z_i), $i = 1, \ldots, r$, be reduced plane curve germs. Let $m_i := e^s(C_i, z_i)$ (respectively $m_i := e^a(C_i, z_i)$), and assume that*

$$H^1\big(\mathcal{J}_{Z(\boldsymbol{m})/\mathbb{P}^2}(d-1)\big) = 0\,,$$

where Z is the (generic) fat point scheme supported at some distinct points $w_1, \ldots, w_r \in \mathbb{P}^2$ and defined by the ideals $\mathfrak{m}_{w_i}^{m_i}$. Moreover, let

$$d > \max_{1\leq i\leq r} e^s(C_i, z_i)\,, \quad \big(\textit{respectively } d > \max_{1\leq i\leq r} e^a(C_i, z_i)\big)\,.$$

Then there exists a reduced, irreducible plane curve $D \subset \mathbb{P}^2$ of degree d with $\operatorname{Sing} D = \{w_1, \ldots, w_r\}$*, and each germ (D, w_i) is topologically (respectively analytically) equivalent to (C_i, z_i), $i = 1, \ldots, r$.*

This is a particular case of Proposition 4.5.26 below, which covers curves on an arbitrary algebraic surface.

Applying the H^1-vanishing criterion in Proposition 3.2.8, we immediately derive

Corollary 4.5.24 *Let (C_i, z_i), $i = 1, \dots, r$, be reduced plane curve germs such that $e^s(C_1, z_1) \geq \dots \geq e^s(C_r, z_r)$. If*

$$\begin{aligned} e^s(C_1, z_1) + e^s(C_2, z_2) &\leq d - 1\,, \quad \text{as } r \geq 2\,, \\ e^s(C_1, z_1) + \dots + e^s(C_5, z_5) &\leq 2d - 2\,, \quad \text{as } r \geq 5\,, \\ \sum_{i=1}^{r} (e^s(C_i, z_i) + 1)^2 &< \frac{9}{10}(d + 2)^2\,, \end{aligned}$$

then there exists a reduced, irreducible plane curve D of degree d with exactly r singular points $w_1, \dots, w_r$, such that each germ (D, w_i) is topologically equivalent to (C_i, z_i), $i = 1, \dots, r$. The same statement holds true when replacing e^s by e^a and the topological equivalence by the analytic one.

Furthermore, taking into account the estimates of Theorem 4.5.19 (2) and the equalities $e^s(A_1) = e^a(A_1) = 2$, $e^s(A_2) = e^a(A_2) = 3$, one can derive

Corollary 4.5.25 *Let (C_i, z_i), $i = 1, \dots, r$, be reduced plane curve germs, among them n nodes and k cusps.*

(1) Assume that $\delta(C_1, z_1) \geq \dots \geq \delta(C_r, z_r)$. If

$$\begin{aligned} \delta(C_1, z_1) + \delta(C_2, z_2) &\leq \frac{4\sqrt{3}}{25}(d + 1)\,, \quad \text{as } r \geq 2\,, \\ \delta(C_1, z_1) + \dots + \delta(C_5, z_5) &\leq \frac{4\sqrt{3}}{25}(2d + 3)\,, \quad \text{as } r \geq 5\,, \end{aligned}$$

and

$$\begin{aligned} 9n + 16k &+ \sum_{\substack{(C_i,z_i)=A_m \\ m\geq 3}} \left(2\left\lfloor\sqrt{m+5}\right\rfloor + 1\right)^2 + \sum_{\substack{(C_i,z_i)=D_m \\ m\geq 4}} \left(2\left\lfloor\sqrt{m+7}\right\rfloor + 2\right)^2 \\ &+ \sum_{\substack{(C_i,z_i)=E_m \\ m=6,7,8}} \left\lfloor\frac{m+5}{2}\right\rfloor^2 + \frac{625}{48} \sum_{(C_i,z_i)\notin ADE} \delta(C_i, z_i) \;<\; \frac{9}{10}(d + 2)^2\,, \end{aligned}$$

then there exists a reduced, irreducible plane curve D of degree d such that $\operatorname{Sing} D = \{w_1, \dots, w_r\}$ *and each germ (D, w_i) is topologically equivalent to (C_i, z_i), $i = 1, \dots, r$.*

(2) Assume that $\mu(C_1,z_1) \geq \cdots \geq \mu(C_r,z_r)$. *If*

$$\mu(C_1,z_1)+\mu(C_2,z_2) \leq \frac{d+1}{3}\,, \quad as\ r \geq 2\,,$$
$$\mu(C_1,z_1)+\cdots+\mu(C_5,z_5) \leq \frac{2d+3}{3}\,, \quad as\ r \geq 5\,,$$

and

$$9n+16k+\sum_{\substack{(C_i,z_i)=A_m\\ m\geq 3}}\Big(2\big\lfloor\sqrt{m+5}\big\rfloor+1\Big)^2+\sum_{\substack{(C_i,z_i)=D_m\\ m\geq 4}}\Big(2\big\lfloor\sqrt{m+7}\big\rfloor+2\Big)^2$$
$$+\sum_{\substack{(C_i,z_i)=E_m\\ m=6,7,8}}\Big\lfloor\frac{m+5}{2}\Big\rfloor^2+9\sum_{(C_i,z_i)\notin ADE}\mu(C_i,z_i) \;<\; \frac{9}{10}(d+2)^2\,,$$

then there exists a reduced, irreducible plane curve D of degree d such that $\operatorname{Sing} D=\{w_1,\ldots,w_r\}$ *and each germ* (D,w_i) *is analytically equivalent to* (C_i,z_i), $i=1,\ldots,r$.

Remark 4.5.25.1 We should like to point out that the existence criterion of Theorem 4.5.14 is better when considering non-simple singularities, whereas the criterion of Corollary 4.5.25 is better for simple singularities.

4.5.6 Curves on Smooth Algebraic Surfaces

We obtain a sufficient condition for the existence of an irreducible curve with prescribed singularities in the given linear system on a smooth algebraic surface. In fact, the following statement is parallel to Proposition 4.5.11, but does not presume the existence of a curve with multiple points.

4.5.6.1 H^1-Vanishing Implies Existence

Proposition 4.5.26 *Let* Σ *be a smooth projective algebraic surface, D a divisor on* Σ, *and* $L\subset\Sigma$ *a very ample divisor. Let* (C_1,z_1), ..., (C_r,z_r) *be reduced singular germs of plane curves. Let*

$$H^1\big(\mathcal{J}_{Z(\boldsymbol{w},\boldsymbol{m})/\Sigma}(D-L)\big)=0\,, \tag{4.5.6.1}$$
$$\max\nolimits_{1\leq i\leq r} m_i < L(D-L-K_\Sigma)-1\,, \tag{4.5.6.2}$$

where $Z(\boldsymbol{w},\boldsymbol{m})$ *is the fat point scheme supported at some distinct points* $w_1,\ldots,w_r\in\Sigma$ *and defined by the ideals* $\mathfrak{m}_{w_i}^{m_i}$, *with* $m_i=e^s(C_i,z_i)$, *respectively* $m_i=e^a(C_i,z_i)$,

$i = 1, \ldots, r$. Then there exists an irreducible curve $C \in |D|$ such that $\operatorname{Sing} C = \{w_1, \ldots, w_r\}$ *and each germ (C, w_i) is topologically, respectively analytically, equivalent to (C_i, z_i), $i = 1, \ldots, r$.*

Proof. Step 1. For any $i = 1, \ldots, r$, define a zero-dimensional scheme Z_i which coincides with $Z = Z(\boldsymbol{w}, \boldsymbol{m})$ at w_j, $1 \le j \le r$, $j \ne i$, and is given by the ideal $(\mathfrak{m}_{w_i})^{m_i+1}$ at w_i. We claim that

$$H^1\big(\mathcal{J}_{Z_i/\Sigma}(D)\big) = 0\,. \tag{4.5.6.3}$$

Indeed, there exists a non-singular curve in $|L|$ (which we again denote by L for the sake of notation) which passes through w_i and does not contain any point w_j, $j \ne i$. Then in the exact sequence

$$0 = H^1\big(\mathcal{J}_{Z/\Sigma}(D-L)\big) \longrightarrow H^1\big(\mathcal{J}_{Z_i/\Sigma}(D)\big) \longrightarrow H^1\big(\mathcal{J}_{Z_i \cap L/L}(D)\big)\,,$$

the last term vanishes since by (4.5.6.2),

$$DL - \deg(Z_i \cap L) \ge DL - m_i - 1 > L^2 + LK_\Sigma = 2g(L) - 2\,.$$

Using (4.5.6.3), for any $i = 1, \ldots, r$, we can find a curve $D_i \in \big|H^0\big(\mathcal{J}_{Z/\Sigma}(D)\big)\big|$ which has an ordinary singular point of multiplicity $e^s(C_i, z_i)$ at w_i, and, in addition, in some fixed local coordinates in a neighbourhood of w_i, the $e^s(C_i, z_i)$-jet at w_i of a local equation of D_i is a generic[9] $e^s(C_i, z_i)$-form.

Step 2. Let w_0 be a generic point in $\Sigma \setminus \{w_1, \ldots, w_r\}$, and let w be any point in $\Sigma \setminus \{w_1, \ldots, w_r, w_0\}$. Since L is very ample, there is a non-singular connected curve in $|L|$ (which we further denote by L as well), which passes through w, w_0 and, may be, through one of $w_1, \ldots, w_r$. In the exact sequence

$$\underbrace{H^1\big(\mathcal{J}_{Z/\Sigma}(D-L)\big)}_{=0} \to H^1\big(\mathcal{J}_{Z \cup \{w, w_0\}/\Sigma}(D)\big) \to H^1\big(\mathcal{J}_{(Z\cap L)\cup\{w,w_0\}/L}(D)\big),$$

the latter term vanishes, since by (4.5.6.2),

$$\begin{aligned}\deg \mathcal{J}_{(Z\cap L)\cup\{w,w_0\}/L}(D) &\ge DL - 2 - \max_{1\le i\le r} e^s(C_i, z_i)\\ &> L^2 + LK_\Sigma = 2g(L) - 2\,.\end{aligned}$$

Hence,

$$H^1\big(\mathcal{J}_{Z\cup\{w,w_0\}/\Sigma}(D)\big) = 0\,.$$

[9] Here "generic" means that the object considered can be chosen arbitrarily in a Zariski open subset of the whole space of objects.

In particular, there exists a curve in $\left|H^0\big(\mathcal{J}_{Z\cup\{w_0\}}(D)\big)\right|$ which does not pass through w. Since w_0 is generic, and w is any point outside $w_0, \ldots, w_r$, by Bertini's Theorem 2.1.24, a generic curve $D_0 \in \left|H^0\big(\mathcal{J}_{Z\cup\{w_0\}}(D)\big)\right|$ is non-singular outside $w_1, \ldots, w_r$, and this linear system has no fixed part. In addition, D_0 is irreducible. Indeed, otherwise, by Bertini's Theorem 2.1.24, D_0 and all close curves in $\left|H^0\big(\mathcal{J}_{Z\cup\{w_0\}}(D)\big)\right|$ would split into variable components which belong to the same one-dimensional algebraic family, but this contradicts the fact that D_0 is non-singular at the fixed point w_0.

Then, a generic curve $\widetilde{D}$ in the linear system spanned by $D_0, D_1, \ldots, D_r$, is irreducible, non-singular outside $w_1, \ldots, w_r$, has an ordinary singular point of multiplicity $e^s(C_i, z_i)$ at w_i, $i = 1, \ldots, r$, and, finally, the $e^s(C_i, z_i)$-jet at w_i of a local equation of $\widetilde{D}$ in some fixed coordinates in a neighbourhood of w_i is a generic $e^s(C_i, z_i)$-form, $i = 1, \ldots, r$.

Step 3. By Lemma 4.5.18 (2), for any $i = 1, \ldots, r$, there is an affine curve F_i of degree $e^s(C_i, z_i)$ with its only singular point topologically equivalent to (C_i, z_i), and such that the leading form of the defining polynomial coincides with the $e^s(C_i, z_i)$-jet at w_i of a local equation of $\widetilde{D}$. Now we apply Theorem 2.3.13 to deform $\widetilde{D}$ into the required curve $C \in |D|$ with prescribed singularities.

Namely, in the assertion of Theorem 2.3.13, $S = S_1 = S_3 = S_4 = S_5 = \emptyset$, $S_2 = \{w_1, \ldots, w_r\}$, the affine curves $F_1, \ldots, F_r$ serve as deformation patterns for the singular points of $\widetilde{D}$, so that these patterns are strongly transversal with respect to the topological equivalence of singular points in view of (4.5.5.12). Furthermore, the H^1-vanishing condition in Theorem 2.3.13 reads as

$$H^1\big(\mathcal{J}_{\widetilde{Z}/\Sigma}(D)\big) = 0\,,$$

where $\widetilde{Z} = \bigcup_i Z^{es}(\widetilde{D}, w_i)$, which immediately follows, say, from (4.5.6.3), since $Z_i \supset \widetilde{Z}$ for any $i = 1, \ldots, r$. □

4.5.6.2 Numerical Conditions

Combining this with the H^1-vanishing criterion in Proposition 3.2.9 and the estimates for e^s, e^a in Theorem 4.5.19 (2), one can derive the following explicit numerical existence criterion:

Theorem 4.5.27 *Let Σ be a smooth algebraic projective surface, D a divisor on Σ with $D - K_\Sigma$ nef, and $L \subset \Sigma$ a very ample divisor. Let (C_1, z_1), ..., (C_r, z_r) be reduced singular germs of plane curves, among them n nodes and k cusps.*

(1) If

$$18n + 32k + \frac{625}{24} \sum_{\delta(C_i, z_i) > 1} \delta(C_i, z_i) \le (D - K_\Sigma - L)^2, \quad (4.5.6.4)$$

$$\frac{25}{4\sqrt{3}} \max_{1 \le i \le r} \sqrt{\delta(C_i, z_i)} + 1 < (D - L - K_\Sigma)L, \quad (4.5.6.5)$$

and, for any irreducible curve B with $B^2 = 0$ and $\dim |B|_a > 0$,

$$\frac{25}{4\sqrt{3}} \max_{1 \le i \le r} \sqrt{\delta(C_i, z_i)} < (D - K_\Sigma - L)B + 1, \quad (4.5.6.6)$$

then there exists a reduced, irreducible curve $C \in |D|$ with r singular points topologically equivalent to (C_1, z_1), ..., (C_r, z_r), respectively, as its only singularities.

(2) If

$$18n + 32k + 18 \sum_{\mu(C_i, z_i) > 2} \mu(C_i, z_i) \le (D - K_\Sigma - L)^2, \quad (4.5.6.7)$$

$$3 \max_{1 \le i \le r} \sqrt{\mu(C_i, z_i)} + 1 < (D - L - K_\Sigma)L, \quad (4.5.6.8)$$

and, for any irreducible curve B with $B^2 = 0$ and $\dim |B|_a > 0$,

$$3 \max_{1 \le i \le r} \sqrt{\mu(C_i, z_i)} < (D - K_\Sigma - L)B + 1, \quad (4.5.6.9)$$

then there exists a reduced, irreducible curve $C \in |D|$ with r singular points analytically equivalent to (C_1, z_1), ..., (C_r, z_r), respectively, as its only singularities.

Here $|B|_a$ means the family of curves algebraically equivalent to B. A discussion of the hypotheses of Theorem 4.5.27 for specific classes of surfaces, as well as concrete examples can be found in [KeT].

Proof of Theorem 4.5.27. Since $e^s(A_1) = e^a(A_1) = 2$, $e^s(A_2) = e^a(A_2) = 3$, and since for $\delta(C_i, z_i) > 1$ Theorem 4.5.19 (2) gives

$$e^s(C_i, z_i) \le \frac{25}{4\sqrt{3}} \sqrt{\delta(C_i, z_i)} - 1,$$
$$e^a(C_i, z_i) \le 3\sqrt{\mu(C_i, z_i)} - 1,$$

the conditions (4.5.6.4)–(4.5.6.6) imply

$$\begin{aligned} 2\sum_{i=1}^{r}(e^s(C_i, z_i) + 1)^2 &< (D - K_\Sigma - L)^2 , \\ \max_{1\le i\le r} e^s(C_i, z_i) + 2 &< (D - L - K_\Sigma)L , \\ \max_{1\le i\le r} e^s(C_i, z_i) &< (D - K_\Sigma - L)B , \end{aligned} \tag{4.5.6.10}$$

respectively the conditions (4.5.6.7)–(4.5.6.9) imply

$$\begin{aligned} 2\sum_{i=1}^{r}(e^a(C_i, z_i) + 1)^2 &< (D - K_\Sigma - L)^2 , \\ \max_{1\le i\le r} e^a(C_i, z_i) + 2 &< (D - L - K_\Sigma)L , \\ \max_{1\le i\le r} e^a(C_i, z_i) &< (D - K_\Sigma - L)B , \end{aligned} \tag{4.5.6.11}$$

where the curves B are understood as in (4.5.6.6), (4.5.6.9). The middle inequalities in (4.5.6.10) and (4.5.6.11) yield (4.5.6.2). By Proposition 3.2.9, this yields (4.5.6.1), where $Z = Z(\boldsymbol{m}) \subset \Sigma$ is a generic fat point scheme with multiplicities $m_i = e^s(C_i, z_i)$, respectively $m_i = e^a(C_i, z_i)$, $i = 1, \dots, r$. Thus, Theorem 4.5.27 is reduced to Proposition 4.5.26. □

4.5.6.3 Nodal Curves on Smooth Algebraic Surfaces

In the case of families of nodal curves, the specific nonemptiness (as well as smoothness and irreducibility) criteria appear to be much better than consequences of general results like Theorem 4.5.27. We illustrate this by formulating without proof sufficient existence conditions for nodal curves on the blown-up projective plane, obtained in [GLS2], on arbitrary toric surfaces, on a generic K3 surface, found by Xi Chen [Che], and on a surface of general type in $\mathbb{P}^3$, which are due to Chiantini and Ciliberto [ChC]. The detailed proofs can be found in these sources, and we only mention here the main ideas. In all cases, one constructs nodal rational curves, whose singular points are independently smoothable, thus, obtaining nodal curves of arbitrary genus. For the blown-up plane, the construction of rational nodal curves in [GLS2] is reduced to the case of plane curves with nodes and fixed multiple points. For toric surfaces, one can write down an explicit parametrization of a nodal rational curve in the tautological linear system. In [Che, ChC] the construction is more involved and it includes a degeneration of the considered surfaces, for example, in [Che] a quartic (K3) surface is degenerated into the union of two smooth quadrics, while in [ChC] a surface in $\mathbb{P}^3$ of degree $d \geq 5$ degenerates into a surface of degree $d - 1$ and its generic tangent plane. Then one takes a pair of suitable rational curves in each component of the reducible surface and deforms the pair (reducible surface, reducible curve) into a generic smooth surface of a given degree and a nodal rational curve in it. It is worth to notice that the construction methods used in [Che, ChC] share a lot with the patchworking construction as presented in Sect. 2.3.

Remark 4.5.27.1 (1) Nodal curves traditionally are of special interest, since they are tightly related to enumerative geometry, notably, to the Gromov-Witten theory of complex algebraic surfaces. We like to comment on this relation in more detail, especially pointing to the related places of the results discussed in the present book.

In case the moduli space of stable maps $\overline{\mathcal{M}}_{g,n}(X, \beta)$ (X a complex algebraic surface, $\beta \in H_2(X)$) is of pure expected dimension (or there exists an appropriate virtual fundamental class having an enumerative meaning), the Gromov-Witten invariant $GW_g(X, \beta)$ counts curves of genus g, pseudo-holomorphic with respect to a generic almost complex structure tamed by the given symplectic structure on X, that pass through an appropriate number of fixed points in X. In turn, if the complex structure on X is sufficiently generic among the tamed almost complex structures, then the counted curves are algebraic. At last, if the nodal curves form a dense Zariski open subset of the families of curves of genus g in a given homology class, then the Gromov-Witten invariant indeed counts nodal algebraic curves, and its positivity yields the existence of nodal curves of a given genus and in a given homology class. On the other hand, the existence statements presented below and proved by a direct construction, imply that the corresponding Gromov-Witten invariants do not vanish.

The conditions mentioned in the preceding paragraph are in general highly nontrivial to establish. They require, in particular, to compute the dimension of equisingular families for both, the nodal and non-nodal case, and show that the latter dimension is less than the "nodal" one. The results of Sects. 4.3 and 4.4 are directly concerned with problems of this sort.

The rational surfaces and K3 surfaces have attracted a special attention in the Gromov-Witten theory, see, for instance, [GoP], which treats rational curves on the plane blown up at points in generic position. Particularly, in this case, the complex structure is sufficiently generic, and the non-vanishing of the genus zero Gromov-Witten invariants indeed yields the existence of rational curves, which appear to be nodal. However, other rational surfaces, like Hirzebruch surfaces (a particular case of toric surfaces), do not fit well with the Gromov-Witten theory, which, for example, reduces the case of Hirzebruch surfaces to the quadric and the plane blown up at one point, while the count of algebraic curves leads to different results for different Hirzebruch surfaces (cf. [AB] and [Wel]). Notice that the existence of nodal curves on Hirzebruch surfaces (and on all other toric surfaces) can be derived from the Mikhalkin's correspondence theorem and the existence of suitable tropical curves of a given degree and genus [Mik1, Mik2]. The K3 surfaces (i.e., two-dimensional Calabi-Yau) turn out to be even more interesting. The celebrated Yau-Zaslow formula [YauZ] (proved in [KMPS]) implies the positivity of the genus zero Gromov-Witten invariants. However, the first rigorous proof of the existence of rational nodal curves on a general K3 surface has been done by Xi Chen [Che] via an appropriate construction (see also [ChL] for the statement that all rational curves on a generic K3 surface are nodal).

According to [GLS2], the problem of the existence of nodal curves in $\mathbb{P}^2_r$, the projective plane blown-up at r points in generic position, is approached for $r \leq 9$ and $r \geq 10$ separately. If $r \leq 9$, Theorems 4.5.28 and 4.5.29 give the complete answer, while for $r \geq 10$, we obtain an asymptotically nearly optimal sufficient criterion (Theorem 4.5.30).

In the following the r blown up points are always in generic position. We denote by $V^{irr}(d;\ d_1, \ldots, d_r;\ k)$ the variety of all irreducible curves C in $|dE_0 - \sum_{i=1}^r d_i E_i|$ with k nodes as only singularities, where E_0 resp. $E_1, \ldots, E_r \subset \mathbb{P}^2_r$ are the strict transform of a generic line in $\mathbb{P}^2$ and the exceptional divisors, respectively.

Theorem 4.5.28 ([GLS2], Theorem 3) *Let* $r = 1$ *then* $V^{irr}(d; d_1; k) \neq \emptyset$ *iff*

$$0 \leq k \leq \frac{(d-1)(d-2) - d_1(d_1-1)}{2}.$$

Let $r = 2$ *and* d, d_1, d_2 *be positive integers. Then* $V^{irr}(d; d_1, d_2; k) \neq \emptyset$ *iff*

$$0 \leq k \leq \frac{(d-1)(d-2) - d_1(d_1-1) - d_2(d_2-1)}{2}$$

and either $(d_1 + d_2 \leq d)$ *or* $(d = d_1 = d_2 = 1)$.

Let $3 \leq r \leq 9$, then we introduce the following equivalence relation on $\mathbb{Z}^{r+1}_{\geq 0}$:

$$\big(d; d_1, \ldots, d_r\big) \sim \big(\widetilde{d}; \widetilde{d}_1, \ldots, \widetilde{d}_r\big)$$

iff there is a finite sequence of Cremona maps and a permutation σ transforming $(d; d_1, \ldots, d_r)$ to $\big(\widetilde{d}; \widetilde{d}_{\sigma(1)}, \ldots, \widetilde{d}_{\sigma(r)}\big)$. Here, by a *Cremona map* we denote a mapping

$$\begin{array}{rccc} \Sigma_{j,\ell,n}: & \mathbb{Z}^{r+1} & \longrightarrow & \mathbb{Z}^{r+1} \\ & (d; d_1, \ldots, d_r) & \mapsto & (d'; d'_1, \ldots, d'_r) \end{array}$$

with $d' = 2d - d_j - d_\ell - d_n$, $d'_i = d_i$ for each $i \notin \{j, \ell, n\}$, and $d'_j = d - d_\ell - d_n$, $d'_\ell = d - d_j - d_n$, $d'_n = d - d_j - d_\ell$. Such a Cremona map corresponds to the standard Cremona transformation in $\mathbb{P}^2$ inducing the base change in $\mathrm{Pic}(\mathbb{P}^2_r)$:

$$\begin{cases} E'_0 = 2E_0 - E_j - E_\ell - E_n \\ E'_j = E_0 - E_\ell - E_n \\ E'_\ell = E_0 - E_j - E_n \\ E'_n = E_0 - E_j - E_\ell \\ E'_i = E_i \ \text{ for each } i \notin \{j, \ell, n\}. \end{cases} \tag{4.5.6.12}$$

Since the Cremona transformation preserves the generality of the blown-up points, such a transformation maps elements in $V^{irr}(d;\ d_1, \ldots, d_r;\ k)$ to elements in $V^{irr}(d'; d'_1, \ldots, d'_r; k)$ supposing d, d', d_i, d'_i $(1 \leq i \leq r)$ to be non-negative. We deduce that the non-emptiness of $V^{irr}(d;\ d_1, \ldots, d_r;\ k)$ is equivalent to the existence of a curve in $V^{irr}(d'; d'_1, \ldots, d'_r; k)$. An (ordered) tuple $(d; d_1, \ldots, d_r) \in \mathbb{Z}^{r+1}_{\geq 0}$, $d_1 \geq d_2 \geq \ldots \geq d_r$, is called *minimal*, if it satisfies the (minimality) condition

$$\max_{\#\{j,\ell,n\}=3} (d_j+d_\ell+d_n) \;=\; d_1+d_2+d_3 \;\le\; d\,. \tag{4.5.6.13}$$

Theorem 4.5.29 ([GLS2], Theorem 4) *Let* $3 \le r \le 9$ *and positive integers* $d \ge d_1 \ge \ldots \ge d_r$ *satisfy the condition*

$$\sum_{i=1}^{r} d_i \;\le\; 3d-1\,. \tag{4.5.6.14}$$

Then $V^{irr}(d;\, d_1,\ldots,d_r;\, k) \ne \emptyset$ *iff*

$$0 \;\le\; k \;\le\; \frac{(d-1)(d-2)}{2} - \sum_{i=1}^{r} \frac{d_i(d_i-1)}{2}$$

and $(d;d_1,\ldots,d_r)$ *is equivalent to a minimal tuple* $\big(\widetilde{d};\widetilde{d}_1,\ldots,\widetilde{d}_r\big)$ *of non-negative integers or to the tuple* $(1;1,1,0,\ldots,0)$.

Remark 4.5.29.1 (a) Condition (4.5.6.14) is necessary in the following sense. By Bézout's Theorem and the generic position of the blown-up points, the only type of an irreducible curve not satisfying (4.5.6.14) is the smooth cubic through the 9 generic points.

(b) For $3 \le r \le 8$ the tuple $(d;d_1,\ldots,d_r)$ is equivalent to a minimal one exactly if the following conditions are satisfied

$$\begin{aligned}
d &\ge d_1+d_2\\
2d &\ge d_1+d_2+d_3+d_4+d_5\\
3d &\ge 2d_1+d_2+d_3+d_4+d_5+d_6+d_7\\
4d &\ge 2d_1+2d_2+2d_3+d_4+d_5+d_6+d_7+d_8\\
5d &\ge 2d_1+2d_2+2d_3+2d_4+2d_5+2d_6+d_7+d_8\\
6d &\ge 3d_1+2d_2+2d_3+2d_4+2d_5+2d_6+2d_7+2d_8
\end{aligned}$$

(c) The exceptional case $(d;d_1,\ldots,d_r) \sim (1;1,1,0,\ldots,0)$ corresponds exactly to the exceptional curves with data

$(2;1,1,1,1,1)$	the conic through 5 of the generic points
$(3;2,1,1,1,1,1,1)$	the cubic through 7 of the generic points having a node at one of them
$(4;2,2,2,1,1,1,1,1)$	the quartic through 8 generic points having nodes at three of them
$(5;2,2,2,2,2,2,1,1)$	the quintic through all 8 generic points having nodes at 6 of them
$(6;3,2,2,2,2,2,2,2)$	the sextic having nodes at 7 of the generic points and a triple point at the remaining one

Theorem 4.5.30 ([GLS2], Theorem 5) *Let $r \geq 10$, and let the positive integers $d, d'; d_1, \dots, d_r$ satisfy $d \geq d'$ and*

$$\sum_{i=1}^{r} \frac{d_i(d_i+1)}{2} < \frac{d'^2 + 6d' - 1}{4} - \left\lfloor \frac{d'}{2} \right\rfloor . \qquad (4.5.6.15)$$

Then for any integer k such that

$$0 \leq k \leq \frac{(d-1)(d-2)}{2} - \frac{(d'-1)(d'-2)}{2} , \qquad (4.5.6.16)$$

there exists a reduced, irreducible curve $C \in |dE_0 - \sum_{i=1}^r d_i E_i|$ on $\mathbb{P}^2_r$, having k nodes as its only singularities, that is, $V^{irr}(d;\, d_1, \dots, d_r;\, k) \neq \emptyset$.

Corollary 4.5.31 *Let $r \geq 10$ and $d;\, d_1, \dots, d_r$ be positive integers such that*

$$2\sum_{i=1}^{r} d_i(d_i+1) \leq d^2 .$$

Then $V^{irr}(d;\, d_1, \dots, d_r;\, k) \neq \emptyset$ if

$$0 \leq k \leq \frac{(d-1)(d-2)}{2} - \sum_{i=1}^{r} d_i(d_i+1) .$$

The latter follows from Theorem 4.5.30, since $d' := \sqrt{2}\sqrt{\sum_{i=1}^r d_i(d_i+1)}$ satisfies

$$\sum_{i=1}^{r} \frac{d_i(d_i+1)}{2} < \frac{d'^2 + 6d' - 1}{4} - \frac{d'}{2} .$$

Theorem 4.5.32 *Let $\Delta \subset \mathbb{R}^2$ be a nondegenerate lattice polygon, $\Sigma = \mathrm{Tor}(\Delta)$ a toric surface associated with Δ, $\mathcal{L}_\Delta$ the tautological invertible sheaf on Σ, and $D = c_1(\mathcal{L}_\Delta) \in \mathrm{Pic}(\Sigma)$. Denote by s the number of edges of Δ and put $p_a(\Delta) = |\,\mathrm{Int}(\Delta) \cap \mathbb{Z}^2|$. Then, for any integer $0 \leq g \leq p_a(\Delta)$, there exists an irreducible nodal curve $C_g \in |D|$ of genus g.*

Proof. We shall present an explicit parametrization of a rational curve $C_0 \in |D|$. Choose distinct points $t_1, \dots, t_m \in \mathbb{P}^1$, $m = -DK_\Sigma$ and split this set into disjoint subsets labeled by the toric divisors of Σ (i.e., by the edges of Δ) assigning to each toric divisor D_i, a set of DD_i points. Denote $\overline{n}_0 = (-1, -1), \overline{n}_1 = (1, 0), \overline{n}_2 = (0, 1)$ and represent $\mathbb{Z}^2$ as the union of the semigroups $\mathbb{Z}^2_{ij} = \langle \overline{n}_i, \overline{n}_j \rangle$, $0 \leq i < j \leq 2$. For any t_j $(1 \leq j \leq m)$ labeled by the toric divisor D_k, set $\overline{n}(t_j)$ to be the primitive integral normal vector to the corresponding side S_k of Δ oriented inwards. Then, for all $i = 0, 1, 2$ and $j = 1, \dots, m$, define a_{ij} by the relation $\overline{n}(t_j) = a_{ij}\overline{n}_i + a_{ik}\overline{n}_k$

if $\overline{n}(t_j) \in \mathbb{Z}^2_{ik}$, and define $a_{ij} = 0$ if $\overline{n}(t_j) \notin \mathbb{Z}^2_{ik}$ for $0 \le k \le 2, k \ne i$.. The required parametrization is

$$x_i = \prod_{j=1}^{m} (t - t_j)^{a_{ij}}, \quad i = 0, 1, 2 .$$

It is an easy exercise to show that it defines a nodal curve for a generic choice of $t_1, ..., t_m$. The independence of deformations of the nodes of the constructed curve follows, for instance, from Proposition 4.4.3(a), which applies, since the constructed curve does not hit possible singular points of Σ, which then can be resolved. The sufficient condition (4.4.1.9) holds, since the nodes do not contribute to the left-hand side, while in the right-hand side one has $-DK_\Sigma = m > 0$. □

Nodal curves on surfaces of degree ≤ 3 in $\mathbb{P}^3$ are, in fact, treated in Theorems 4.5.28 and 4.5.29. A generic surface of degree 4 in $\mathbb{P}^3$ is K3, and this case is covered in the following statement.

Theorem 4.5.33 ([Che], Theorem 1.2, and [ChC], Theorem 1.2 and its consequences) *For any $g \ge 3$, given a general K3 surface Σ of the principal series in $\mathbb{P}^g$, any $m > 0$ and n satisfying*

$$0 \le n \le \dim \left|H^0\big(\mathcal{O}_\Sigma(m)\big)\right| ,$$

there exists an irreducible curve in the linear system $\left|H^0\big(\mathcal{O}_\Sigma(m)\big)\right|$ with n nodes as its only singularities.

Next we notice that a generic surface of degree ≥ 5 in $\mathbb{P}^3$ is of general type. Nodal curves on such surface are subject of

Theorem 4.5.34 ([ChC], Theorem 3.1) *Let Σ be a generic surface of degree $d \ge 5$ in $\mathbb{P}^3$. Then, for all $m \ge d$ and $0 \le n \le \dim \left|H^0\big(\mathcal{O}_\Sigma(m)\big)\right|$ there exists an irreducible curve in the linear system $\left|H^0\big(\mathcal{O}_\Sigma(m)\big)\right|$ having n nodes as its only singularities.*

4.6 Irreducibility

The question about the irreducibility of equisingular strata $V^{irr}_{|D|}(S_1, \dots, S_r)$ is more delicate than the existence and smoothness problem, in particular, if one tries to find sufficient conditions for the irreducibility. Rather difficult from the algebraic-geometric point of view, it is of special topological interest, being connected with the problem on the fundamental group of the complement of a plane algebraic curve (cf. the discussion of Zariski's example in Sect. 4.6.3).

Even the case of plane *nodal* curves appeared to be very hard (cf. Sect. 4.2.2): already Severi [Sev] claimed to have the complete answer, but unfortunately his proof was incomplete (see details in [ArC, HaJ]). It took more than half a century until

Arbarello and Cornalba [ArC] succeeded in giving at least a partial answer. Finally, it was Harris [HaJ] who proved that each non-empty ESF of plane nodal curves is irreducible.

In Sect. 4.6.3, we list examples, which show that for more complicated singularities, beginning with ordinary cusps, possible numerical sufficient conditions for the irreducibility should be different from the necessary existence conditions (as discussed in Sect. 4.5).

The results in this direction concern basically plane curves, and we shall pay most attention to this case.

4.6.1 Approaches to the Irreducibility Problem

(1) One possible approach (for ESF of plane curves) consists of building for any two curves in the equisingular stratum a connecting path, using explicit equations of the curves, respectively of projective transformations. This method works *for small degrees only.* Besides the classical case of conics and cubics, it is proven in this way that all equisingular strata of quartic curves and quintic curves are irreducible (cf. [BrG, Wal2]). But for degrees $d > 5$, this is no longer true and the method is no more efficient (except for some very special cases).

(2) Arbarello and Cornalba [ArC] suggested another approach. It consists of relating the equisingular stratum to the moduli space of plane curves of a given genus, which is known to be irreducible (cf. [DeM]). This gave some particular results on families of plane nodal curves and plane curves with nodes and cusps. Namely, Kang [Kan1] proved that the variety $V_d^{irr}(n \cdot A_1, k \cdot A_2)$ is irreducible whenever
$$\frac{d^2-4d+1}{2} \leq n \leq \frac{(d-1)(d-2)}{2}, \qquad k \leq \frac{d+1}{2}.$$

(3) Harris introduced a new idea to the irreducibility problem, which completed the case of plane nodal curves (cf. [HaJ]). This new idea was to proceed inductively from rational plane nodal curves (whose family is classically known to be irreducible) to any family of plane nodal curves of a given genus. Further development of this idea lead to new results by Ran [Ran2]: let S be an ordinary singularity, then any variety $V_d^{irr}(n \cdot A_1, 1 \cdot S)$ is irreducible (or empty), and by Kang [Kan2]: let $k \leq 3$, then $V_d^{irr}(n \cdot A_1, k \cdot A_2)$ is irreducible (or empty). Note that the requirement to study all possible deformations of the considered curves does not allow to extend such an approach to more complicated singularities, or to a large number of any singularities different from nodes.

(4) Up to now, there is mainly one approach which is applicable to equisingular families of curves of any degree with any quantity of arbitrary singularities (and even to projective hypersurfaces of any dimension). The basic idea is to find an irreducible analytic space $\mathcal{M}(S_1, \ldots, S_r)$ and a dominant morphism

$$V^{irr}_{|D|}(S_1,\ldots,S_r) \longrightarrow \mathcal{M}(S_1,\ldots,S_r)$$

with equidimensional and irreducible fibres (cf. Proposition 1.1.56). It turns out, that in such a situation proving the irreducibility of $V^{irr}_{|D|}(S_1,\ldots,S_r)$ can be reduced to an H^1-vanishing problem: let $Z(C) = Z^s(C)$ (respectively $Z(C) = Z^a(C)$) be the zero-dimensional schemes encoding the topological (respectively analytic) type of the singularities as introduced in Chap. 1.

Then $V^{irr}_{|D|}(S_1,\ldots,S_r)$ is irreducible if $H^1\big(\mathcal{J}_{Z(C)/\Sigma}(D)\big) = 0$ for any $C \in V^{irr}_{|D|}(S_1,\ldots,S_r)$.

For a detailed discussion of the latter approach, we refer to the following section.

4.6.2 Plane Curves with Arbitrary Singularities

Throughout the following, let $S_1,\ldots,S_r$ be topological, respectively analytic types. To be more precise, we allow that some of the S_i are topological types, some are analytic types, but it any case, none of the analytic types S_i is contained in one of the topological types.

Our goal are *universal* numerical sufficient conditions for the irreducibility of equisingular strata $V^{irr}_d(S_1,\ldots,S_r)$. Hence, we may restrict ourselves to the fourth of the above approaches.

Let's assume that the topological, respectively analytic, types $S_1,\ldots,S_{r'}$, $r' \le r$, are pairwise distinct and that, for any $i = 1,\ldots,r'$, the type S_i occurs precisely r_i times in $S_1,\ldots,S_r$. Then we define

$$\mathcal{M}(S_1,\ldots,S_r) := \prod_{i=1}^{r'} \operatorname{Sym}^{r_i}\big(\mathbb{P}^2 \times \mathcal{H}_0(S_i)\big),$$

where $\mathcal{H}_0(S_i)$ denotes the *irreducible* variety (by Theorem 1.1.57 resp. Lemma 1.1.71) parameterizing the zero-dimensional schemes $Z^s(C,z)$, respectively $Z^a(C,z)$, associated to plane curve singularities of topological, respectively analytic, type S_i at a fixed point z (cf. Definition 1.1.58 and Remark 1.1.60.1, respectively Definition 1.1.69.1 and Remark 1.1.70.1). In Chap. 1, we proved the existence of a morphism Ψ_d (cf. (1.1.3.2) resp. (1.1.4.1)),

$$V^{irr}_d(S_1,\ldots,S_r) \underset{\text{dense}}{\supset} V \xrightarrow{\Psi_d} \mathcal{M}(S_1,\ldots,S_r),$$

which associates to a curve $C \in V$ the r-tuple of pairs $\big(z_i, \tau_{zz_i}(Z^s(C,z_i))\big)$, respectively $\big(z_i, \tau_{zz_i}(Z^a(C,z_i))\big)$, $i = 1,\ldots,r$, where $z_1,\ldots,z_r$ are the singular points of C, and τ_{zz_i} denotes the translation mapping z_i to the fixed point z (cf. (1.1.3.2) in Remark 1.1.60.1, respectively (1.1.4.1) in Remark 1.1.70.1).

Note that, for any $C \in V$, the fibre $\Psi_d^{-1}(\Psi_d(C))$ is the open dense subset of $\left|H^0\big(\mathcal{J}_{Z(C)/\mathbb{P}^2}(d)\big)\right|$ consisting of curves $C' \in V$ with $Z(C') = Z(C)$. Here, we denote by $Z(C)$ the disjoint union of the zero-dimensional schemes

$$Z(C, z_i) := \begin{cases} Z^s(C, z_i) & \text{if } S_i \text{ is a topological type}\,, \\ Z^a(C, z_i) & \text{if } S_i \text{ is an analytic type}\,, \end{cases}$$

$i = 1, \ldots, r$.

We claim:

$$H^1\big(\mathcal{J}_{Z(C)/\mathbb{P}^2}(d)\big) = 0 \text{ for each } C \in V\,, \tag{4.6.2.1}$$

implies the irreducibility of $V_d^{irr}(S_1, \ldots, S_r)$.

Namely, (4.6.2.1) implies that all fibres of Ψ_d are smooth and equidimensional. Moreover, it is a stronger assumption than the H^1-vanishing condition for T-smoothness (cf. Theorem 2.2.40). Hence, any irreducible component V^* of V is smooth of the expected dimension

$$\begin{aligned} \dim V &= \frac{d(d+3)}{2} - \sum_{S_i \text{ top. type}} \tau^{es}(S_i) - \sum_{S_i \text{ anal. type}} \tau(S_i) \\ &= \frac{d(d+3)}{2} - \deg Z(C) + \dim \mathcal{M}(S_1, \ldots, S_r) \\ &= \underbrace{h^0\big(\mathcal{J}_{Z(C)/\mathbb{P}^2}(d)\big)}_{= \dim \Psi_d^{-1}(\Psi_d(C))} + \dim \mathcal{M}(S_1, \ldots, S_r) \end{aligned}$$

(cf. Proposition 1.1.62, respectively Lemma 1.1.71). Since $\mathcal{M}(S_1, ..., S_r)$ is irreducible, we can conclude that the restriction of Ψ_d to V^* is dominant. Since the fibres of Ψ_d are irreducible (as open subsets of projective spaces) and equidimensional, this is only possible if V and hence $V_d^{irr}(S_1, \ldots, S_r)$ is irreducible (by Proposition 1.1.56).

Remark 4.6.0.1 The condition (4.6.2.1) has one main disadvantage: one has to prove H^1-vanishing for *each* curve $C \in V$. As already discussed in Chap. 3, in many cases a very special position of $Z(C)$ with respect to a small degree curve prevents $H^1\big(\mathcal{J}_{Z(C)/\mathbb{P}^2}(d)\big)$ from vanishing. But, we can obviously weaken the requirement of (4.6.2.1) to the condition that the (open) subvariety

$$V_{reg}^{(2)} := \left\{C \in V \;\middle|\; h^1\big(\mathcal{J}_{Z(C)/\mathbb{P}^2}(d)\big) = 0\right\} \subset V$$

is dense (using the map $\Psi_d | V_{reg}^{(2)}$). In particular, V has the expected dimension (which is, indeed, the minimal possible dimension) and it suffices to show that

$$\dim\big(V \setminus V_{reg}^{(2)}\big) < \dim V\,.$$

This approach was applied first in [Shu6, Shu7, Shu8]. Moreover, by Theorem 2.2.40 (e),

$$V_{gen} := \{C \in V \mid \operatorname{Sing} C \text{ consists of points in generic position}\}$$

is a dense subset of

$$V_{reg}^{(1)} := \{C \in V \mid h^1(\mathcal{J}_{Z'_{\mathrm{fix}}(C)/\mathbb{P}^2}(d)) = 0\},$$

where $Z'_{\mathrm{fix}}(C)$ denotes the disjoint union of the zero-dimensional schemes

$$Z'_{\mathrm{fix}}(C, z_i) := \begin{cases} Z^{es}_{fix}(C, z_i) & \text{if } S_i \text{ is a topological type},\\ Z^{ea}_{fix}(C, z_i) & \text{if } S_i \text{ is an analytic type}, \end{cases}$$

$i = 1, \ldots, r$. Hence, it suffices to show that V_{gen} is a subset of $V_{reg}^{(2)}$ and that $V_{reg}^{(1)}$ is dense in V. Since, by definition, $V_{reg}^{(1)}$ is either empty or smooth of the expected dimension, instead of the latter condition, it is sufficient to show

$$\dim(V \setminus V_{reg}^{(1)}) < \dim V .$$

As we shall see, the key point is to estimate the dimension of the subspace of curves $C \in V$ whose singularities lie on a curve of relatively small degree.

4.6.2.1 Linear Conditions

The following condition, which is linear in d, implies H^1-vanishing for every $C \in V_d^{irr}(S_1, \ldots, S_r)$. For the definition of the 0-dimensional schemes Z^s resp. Z^a see Definitions 1.1.31 resp. 1.1.66.

Theorem 4.6.1 *Let $S_1, \ldots, S_r$ be topological, respectively analytic types, and d a positive integer. Then $V_d^{irr}(S_1, \ldots, S_r)$ is irreducible (or empty) if*

$$\begin{aligned} &\sum_{S_i \text{ top. type}} \left(\deg Z^s(C, z_i) - \operatorname{isod}(Z^s(C, z_i), C)\right) \\ &\quad + \sum_{S_i \text{ anal. type}} \left(\deg Z^a(C, z_i) - \operatorname{isod}(Z^a(C, z_i), C)\right) < 3d . \end{aligned} \tag{4.6.2.2}$$

for any irreducible curve $C \in V_d^{irr}(S_1, \ldots, S_r)$ with singular points $z_1, \ldots, z_r$ of types $S_1, \ldots, S_r$, respectively.

Proof By Corollary 3.1.4, the condition (4.6.2.2) implies (4.6.2.1), hence the irreducibility of $V_d^{irr}(S_1, \ldots, S_r)$. □

For simple singularities (C, z) we have (see Example 1.1.41.1):

	A_{2k-1}	A_{2k}	D_{2k-1}	D_{2k}	E_6	E_7	E_8
$\deg Z^s(S)$	$3k$	$3k+2$	$3k-1$	$3k$	9	10	11
$\mathrm{isod}(Z^s(S), C)$	$k+1$	k	k	$k+1$	3	4	4
$\deg Z^s(S) - \mathrm{isod}(Z^s(S), C)$	$2k-1$	$2k+2$	$2k-1$	$2k$	6	6	7

Example 4.6.1.1 The variety $V_6^{irr}(2 \cdot E_8, 1 \cdot A_1)$ of irreducible plane sextics having 2 E_8-singularities and 1 node as only singularities is *non-empty* (such a sextic can be obtained from an elliptic quartic with two cusps by a suitable Cremona transformation), T-*smooth* (Corollary 4.3.4) and *irreducible* (Theorem 4.6.1).

Corollary 4.6.2 *The variety $V_d^{irr}(n \cdot A_1, k \cdot A_2)$ of plane irreducible curves of degree d having n nodes and k cusps as only singularities is irreducible or empty if*

$$2n + 4k < 3d\,. \tag{4.6.2.3}$$

Note that Theorem 4.5.4 implies that the varieties $V_d^{irr}(n \cdot A_1, k \cdot A_2)$, $d \geq 6$, satisfying (4.6.2.3), are indeed non-empty, hence irreducible.

Corollary 4.6.3 *The variety $V_d^{irr}(m_1, \ldots, m_r)$ of plane irreducible curves of degree d having r ordinary multiple points of multiplicities $m_1, \ldots, m_r$ as only singularities is irreducible or empty if*

$$\sum_{i=1}^{r} m_i < 3d\,.$$

Proof The statement follows from Theorem 4.6.1, since for an ordinary plane curve singularity (C, z) of multiplicity m, we obtain (by Example 1.2.21.1)

$$\deg Z^s(C, z) - \mathrm{isod}(Z^s(C, z), C) = \frac{m(m+1)}{2} - \delta(C, z) = m\,.$$

□

We should like to recall that the condition (4.6.2.2) is sufficient but not at all necessary for the irreducibility of $V_d^{irr}(S_1, \ldots, S_r)$. For high degrees this will be obvious when we discuss quadratic sufficient conditions for the irreducibility. But even for small degrees applying the approach presented in Remark 4.6.0.1 can lead to particular improvements of the sufficient condition (4.6.2.2).

Example 4.6.3.1 The variety $V := V_6^{irr}(3 \cdot A_1, 3 \cdot A_2)$ is irreducible.

Proof Using the notation of Remark 4.6.0.1, we show that $V_{gen} \subset V_{reg}^{(2)}$, and that each $C \in V$ satisfies $H^1\big(\mathcal{J}_{Z^{es}_{fix}(C)/\mathbb{P}^2}(6)\big) = 0$. Note that the latter is guaranteed by the $3d$-criterion, since

$$\deg Z^{es}_{fix}(C) - \mathrm{isod}(Z^{es}_{fix}(C), C) = 2n + 3k = 15 < 18 = 3d\,.$$

On the other hand, if the 6 singular points are in generic position, then $Z^s(C)$ is contained in a generic fat point scheme $Z := Z(3,3,3,2,2,2)$. Since $\deg Z = 27 \le h^0\big(\mathcal{O}_{\mathbb{P}^2}(6)\big)$, we know that $H^1\big(\mathcal{J}_{Z/\mathbb{P}^2}(6)\big)$ vanishes, which allows to conclude that V_{gen} is, indeed, a subset of $V_{reg}^{(2)}$. □

4.6.2.2 Quadratic Conditions

We derive now sufficient conditions for the irreducibility of $V_d^{irr}(S_1,\dots,S_r)$, where certain numerical invariants of the S_i are bounded by a quadratic function in d which are better than the linear conditions for high degrees.

Let $f \in \mathcal{O}_{\Sigma,z}$ define the reduced plane curve singularity $(C,z) \subset (\Sigma,z)$ and let S be the analytic, resp. topological, type of (C,z). Then

$$\nu^a(S) = \min\{\nu \,:\, \mathfrak{m}_z^{\nu+1} \subset I^a(f)\},$$
$$\nu^s(S) = \min\{\nu \,:\, \mathfrak{m}_z^{\nu+1} \subset I^s(f)\}\,,$$

with $\mathfrak{m}_z \subset \mathcal{O}_{\Sigma,z}$ the maximal ideal, denotes the analytic, resp. topological, singularity level of S. Moreover, $\nu'(S)$ stands for $\nu^a(S)$ is S is an analytic type and for $\nu^s(S)$ if S is a topological type.

Similarly, $\tau'(S)$ denotes the Tjurina number $\tau(C,z)$ if S is an analytic type and $\tau^{es}(C,z)$ if S is a topological type (cf. Definition 1.1.63). Moreover, $Z(C,z)$ denotes $Z^a(C,z)$ resp. $Z^s(C,z)$ if S is analytic resp. topological type. The following statements are originated in [GLS4].

Theorem 4.6.4 *Let $S_1,\dots,S_r$ be topological, respectively analytic, types of plane curve singularities, and d an integer. If*

$$\max_{i=1..r} \nu'(S_i) \le \frac{2}{5}d - 1 \tag{4.6.2.4}$$

and

$$\sum_{i=1}^{r} \big(\nu'(S_i)+2\big)^2 < \frac{9}{10}d^2\,, \tag{4.6.2.5}$$

$$\frac{25}{2}\cdot \#(nodes) + 18\cdot\#(cusps) + \sum_{\tau'(S_i)\ge 3}\big(\tau'(S_i)+2\big)^2 < d^2 \tag{4.6.2.6}$$

then $V_d^{irr}(S_1,\dots,S_r)$ is non-empty and irreducible.

In particular, by Lemma 1.2.9, respectively Proposition 1.2.5, we obtain the following, slightly weaker statement:

Corollary 4.6.5 *Let $S_1,\dots,S_r$ be topological, respectively analytic, types of plane curve singularities, and d an integer. If* $\max\limits_{i=1..r} \tau'(S_i) \le (2/5)\cdot d - 1$ *and*

$$\frac{25}{2}\cdot\#(nodes)+18\cdot\#(cusps)+\frac{10}{9}\cdot\sum_{\tau'(S_i)\geq 3}\big(\tau'(S_i)+2\big)^2 \;<\; d^2,$$

then $V_d^{irr}(S_1,\ldots,S_r)$ *is non-empty and irreducible.*

Corollary 4.6.6 *Let* $d\geq 8$. *Then* $V_d^{irr}(n\cdot A_1, k\cdot A_2)$ *is irreducible if*

$$\frac{25}{2}n+18k \;<\; d^2.$$

Corollary 4.6.7 *Let* S_i *be the topological type of an ordinary* m_i*-fold point and let* $\max m_i\leq(2/5)d$. *Then* $V_d^{irr}(S_1,\ldots,S_r)$ *is non-empty and irreducible if*

$$\frac{25}{2}\cdot\#(nodes)+\sum_{m_i\geq 3}\frac{m_i^2(m_i+1)^2}{4} \;<\; d^2.$$

Proof This follows from Theorem 4.6.4, since for an ordinary m_i-tuple point (C,z_i) we have computed $\tau^{es}(C,z_i)+2=\deg Z^{es}_{fix}(C,z_i)=m_i(m_i+1)/2$, and $\nu^s(C,z_i)=m_i-1$. □

Proof of Theorem 4.6.4. First, note that the non-emptiness of $V_d^{irr}(S_1,\ldots,S_r)$ follows immediately from the inequalities (4.6.2.4), (4.6.2.5) and Corollary 4.5.24 in view of an evident inequality $\nu^s(S)\leq e(S)$ (cf. Definition 4.5.9).

Using the approach of Remark 4.6.0.1, it suffices to derive sufficient numerical conditions for the following two facts

(1) $V_{gen}=\big\{C\in V_d^{irr}(S_1,\ldots,S_r)\,\big|\,\text{Sing}\,C\text{ in generic position}\big\}$ is a subset of $V_{reg}^{(2)}=\big\{C\in V_d^{irr}(S_1,\ldots,S_r)\,\big|\,H^1\big(\mathcal{J}_{Z(C)/\mathbb{P}^2}(d)\big)=0\big\}$.
(2) $V_{reg}^{(1)}=\big\{C\in V_d^{irr}(S_1,\ldots,S_r)\,\big|\,H^1\big(\mathcal{J}_{Z'_{\text{fix}}(C)/\mathbb{P}^2}(d)\big)=0\big\}$ is a dense subset of $V_d^{irr}(S_1,\ldots,S_r)$.

For the first fact (Lemma 4.6.8) we apply a vanishing theorem for generic fat point schemes, while for the second fact (Lemma 4.6.9) we use an involved argumentation being based on the Castelnuovo function associated to $Z'_{\text{fix}}(C)$.

Lemma 4.6.8 *Let* $C\in V_d^{irr}(S_1,\ldots,S_r)$ *be a curve that has its singularities in generic position* $z_1,\ldots,z_r$. *If* $2d>5\cdot\max_{i=1..r}\nu'(C,z_i)+4$ *and*

$$\frac{9}{10}\cdot(d+3)^2 \;>\; \sum_{i=1}^{r}\big(\nu'(C,z_i)+2\big)^2$$

then $h^1\big(\mathcal{J}_{Z(C)/\mathbb{P}^2}(d)\big)$ *vanishes. In particular,* $V_{gen}\subset V_{reg}^{(2)}$.

Proof Let $i\in\{1,\ldots,r\}$ and $\nu_i=\nu'(C,z_i)$. By definition of ν_i, the scheme $Z(C,z_i)$ is contained in the ordinary fat point scheme given by the ideal $\mathfrak{m}_{z_i}^{\nu_i+1}$. Hence, it suffices to show that

$$h^1\big(\mathcal{J}_{Z(\nu_1+1,\dots,\nu_r+1)/\mathbb{P}^2}(d)\big) = 0\,,$$

where $Z(\nu_1+1, \dots, \nu_r+1)$ is the zero-dimensional scheme of r ordinary fat points of multiplicities $\nu_1+1, \dots, \nu_r+1$ in generic position. Now, the statement follows from Proposition 3.2.8. □

Lemma 4.6.9 *Let $d \geq 6$ be an integer and $C \in V_d^{irr}(S_1, \dots, S_r)$ such that*

$$d^2 > \sum_{i=1}^{r} \big(\deg Z'_{fix}(C, z_i)\big)^2, \tag{4.6.2.7}$$

$$2\,(d+3)^2 > \sum_{i=1}^{r} \big(\deg Z'_{fix}(C, z_i) + 2\big)^2, \tag{4.6.2.8}$$

$$\frac{9}{10}\, d^2 > \sum_{i=1}^{r} \max\big\{ \big(\deg D \cap Z'_{fix}(C, z_i)\big)^2 \;\big|\; D \text{ smooth} \big\}, \tag{4.6.2.9}$$

$$(d-1)^2 > \sum_{i=1}^{r} \max\left(\begin{array}{c} \big\{ \big(\deg D \cap Z'_{fix}(C, z_i)\big)^2 \;\big|\; D \text{ smooth} \big\} \\ \cup \big\{ \tfrac{1}{2} \cdot \big(\deg Z'_{fix}(C, z_i)\big)^2 \big\} \end{array} \right), \tag{4.6.2.10}$$

$$\frac{16}{15}\,(d+3)^2 > \sum_{i=1}^{r} \max\left(\begin{array}{c} \big\{ \big(\deg D \cap Z'_{fix}(C, z_i) + \tfrac{16}{15}\big)^2 \;\big|\; D \text{ smooth} \big\} \\ \cup \big\{ \tfrac{1}{2} \cdot \big(\deg Z'_{fix}(C, z_i) + \tfrac{32}{15}\big)^2 \big\} \end{array} \right). \tag{4.6.2.11}$$

Then $V_{reg}^{(1)}$ is dense in $V_d^{irr}(S_1, \dots, S_r)$, that is, $h^1\big(\mathcal{J}_{Z'_{fix}(C)/\mathbb{P}^2}(d)\big) = 0$ for generic $C \in V_d^{irr}(S_1, \dots, S_r)$.

Remark 4.6.9.1 Note that for any reduced plane curve singularity (C, z) and any smooth curve germ D at z we have

$$\begin{aligned} \deg Z'_{\text{fix}}(C, z) &= \tau'(C, z) + 2 \;\geq\; \nu'(C, z) + 2\,, \\ \deg\big(D \cap Z'_{\text{fix}}(C, z)\big) &\geq \nu'(C, z) + 1\,; \end{aligned}$$

the latter follows, since, by definition of the determinacy bounds,

$$\mathfrak{m}^{\nu^s(C,z)+1} \subset I^s(C, z) \subset I^{es}_{fix}(C, z)\,, \qquad \mathfrak{m}^{\nu^a(C,z)+1} \subset I^a(C, z) \subset I^{ea}_{fix}(C, z)\,.$$

For instance, in the case of nodes and cusps, we have

$$\deg Z'_{\text{fix}}(C, z) = \begin{cases} 3 & \text{for a node,} \\ 4 & \text{for a cusp,} \end{cases} \qquad \nu'(C, z) = \begin{cases} 1 & \text{for a node,} \\ 2 & \text{for a cusp,} \end{cases}$$

$$\max\big\{\deg(D \cap Z'_{\text{fix}}(C, z)) \;\big|\; D \text{ smooth}\big\} = \begin{cases} 2 & \text{for a node,} \\ 3 & \text{for a cusp.} \end{cases}$$

Hence, it is not difficult to see that (4.6.2.5) and (4.6.2.6) imply (4.6.2.7)–(4.6.2.11).

Proof of Lemma 4.6.9. Assume that $V_d^{irr}(S_1, \ldots, S_r)$ has an irreducible component $V^* \subset V_d^{irr}(S_1, \ldots, S_r) \setminus V_{reg}^{(1)}$, that is, a generic element C of V^* satisfies $h^1\big(\mathcal{J}_{Z'_{\text{fix}}(C)/\mathbb{P}^2}(d)\big) > 0$.

We consider the morphism $\Phi_d : V_d^{irr}(S_1, \ldots, S_r) \to \text{Sym}^r\mathbb{P}^2$, mapping a curve $C \in V_d^{irr}(S_1, \ldots, S_r)$ to the non-ordered tuple of its singularities. Let $\Pi^* \subset \text{Sym}^r\mathbb{P}^2$ denote the closure of $\Phi_d(V^*)$ then we obtain a commutative diagram

$$\begin{array}{ccccc} V^* & \hookrightarrow & V_d^{irr}(S_1, \ldots, S_r) \setminus V_{reg}^{(1)} & \hookrightarrow & V_d^{irr}(S_1, \ldots, S_r) \\ {\scriptstyle\Phi_d}\downarrow & & & & \downarrow{\scriptstyle\Phi_d} \\ \Phi_d(V^*) & \underset{\text{dense}}{\hookrightarrow} & \Pi^* & \underset{\text{closed}}{\hookrightarrow} & \text{Sym}^r\mathbb{P}^2 =: \Pi\,. \end{array}$$

Recall that the dimension of the fibre $\Phi_d^{-1}\big(\Phi_d(C)\big)$ at C is just the dimension of $V_{d,\text{fix}}(S_1, \ldots, S_r)$ at C, that is, by Theorem 2.2.40 (b),

$$\dim\big(\Phi_d^{-1}(\Phi_d(C)), C\big) \le h^0\big(\mathcal{J}_{Z'_{\text{fix}}(C)/\mathbb{P}^2}(d)\big) - 1\,.$$

To obtain the statement of Lemma 4.6.9, it suffices to show that under the given (numerical) conditions we have

$$h^1\big(\mathcal{J}_{Z'_{\text{fix}}(C)/\mathbb{P}^2}(d)\big) < \text{codim}_\Pi \Pi^*, \tag{4.6.2.12}$$

because this would imply that

$$\begin{aligned} \dim V^* &\le \dim \Pi^* + h^0\big(\mathcal{J}_{Z'_{\text{fix}}(C)/\mathbb{P}^2}(d)\big) - 1 \\ &< \dim \Pi + h^0\big(\mathcal{J}_{Z'_{\text{fix}}(C)/\mathbb{P}^2}(d)\big) - h^1\big(\mathcal{J}_{Z'_{\text{fix}}(C)/\mathbb{P}^2}(d)\big) - 1 \\ &= 2r + h^0\big(\mathcal{O}_{\mathbb{P}^2}(d)\big) - 1 - \deg Z'_{\text{fix}}(C)\,, \end{aligned}$$

which is the expected dimension of $V_d^{irr}(S_1, \ldots, S_r)$, contradicting the fact that each component of $V_d^{irr}(S_1, \ldots, S_r)$ has at least the expected dimension.

Step 1. For $d \ge 6$ the condition (4.6.2.8) implies, in particular, that

$$\deg Z'_{\text{fix}}(C) \le \sum_{i=1}^{r} \frac{(\deg Z'_{\text{fix}}(C, z_i) + 2)^2}{8} < \frac{(d+3)^2}{4} \le \frac{d(d+1)}{2}\,,$$

whence $d > \min\big\{i \mid h^0\big(\mathcal{J}_{Z'_{\text{fix}}(C)/\mathbb{P}^2}(i)\big) > 0\big\}$. By Lemma 3.5.6, we obtain the existence of a curve C_k of degree $k \ge 3$ such that the subscheme

$$Y = C_k \cap Z'_{\text{fix}}(C) \subset C_k \cap C$$

is non-decomposable with

$$h^1\big(\mathcal{J}_{Y/\mathbb{P}^2}(d)\big) \;=\; h^1\big(\mathcal{J}_{Z'_{\text{fix}}(C)/\mathbb{P}^2}(d)\big) \;\leq\; \frac{r_0(r_0+1)}{2}\,, \tag{4.6.2.13}$$

where $1 \leq r_0 := \mathcal{C}_{Z'_{\text{fix}}(C)}(d+1) \leq k-2$ (cf. Remark 3.5.6.1).

Additionally, by (4.6.2.8), we suppose that $\deg Y \leq \deg Z'_{\text{fix}}(C) < \frac{(d+3)^2}{4}$. Hence, (cf. Lemma 3.5.6 and Remark 3.5.6.1)

$$\deg Y \;\geq\; \max\left\{k(d+3-k),\;\; k(d+2+r_0-k) + h^1\big(\mathcal{J}_{Y/\mathbb{P}^2}(d)\big) - \frac{r_0(r_0+1)}{2}\right\}. \tag{4.6.2.14}$$

We can estimate the codimension of Π^* in Π: given the curve C_k, the number of conditions on $Z'_{\text{fix}}(C)$ imposed by fixing the support of the subscheme Y on C_k, respectively its singular locus, is at least $\#Y + \#(Y|_{\operatorname{Sing} C_k})$ if C_k is a reduced curve and at least $\#Y$ if C_k is non-reduced. On the other hand, the dimension of the variety of reduced (respectively non-reduced) curves C_k of degree k is given by $h^0\big(\mathcal{O}_{\mathbb{P}^2}(k-1)\big) - 1$ (respectively $h^0\big(\mathcal{O}_{\mathbb{P}^2}(k-2)\big) + 2$). Thus, in place of (4.6.2.12), it suffices to show that

$$h^1\big(\mathcal{J}_{Y/\mathbb{P}^2}(d)\big) \;<\; \min\left\{\#Y + \#(Y|_{\operatorname{Sing} C_k}) - \frac{k(k+3)}{2}\,,\;\; \#Y - \frac{k^2-k}{2} - 2\right\}. \tag{4.6.2.15}$$

Step 2. Recall that $k \geq 3$ and, by (4.6.2.13),

$$h \;:=\; h^1\big(\mathcal{J}_{Y/\mathbb{P}^2}(d)\big) \;\leq\; \frac{(k-2)(k-1)}{2}\,.$$

Step 2a. Assume $h = h^1\big(\mathcal{J}_{Y/\mathbb{P}^2}(d)\big) = (k-2)(k-1)/2$.
Note that this implies that the Castelnuovo functions (cf. Sect. 3.5) of Y and $C_k \cap C$ coincide, in particular, $\deg Y = kd$, that is, $Y = C_k \cap C$. In this case the condition (4.6.2.15) is satisfied whenever

$$0 \;<\; \min\left\{\#Y + \#(Y|_{\operatorname{Sing} C_k}) - k^2 - 1\,,\;\; \#Y - k^2 + 2k - 3\right\}. \tag{4.6.2.16}$$

We have to consider two cases

Case 1: $\#(Y|_{\operatorname{Sing} C_k}) \geq 1$. Then the right-hand side is bounded from below by

$$\#Y - k^2 = \#Y - \frac{(\deg Y)^2}{d^2}\,.$$

Hence, due to the Cauchy inequality, it suffices to have

$$d^2 \;>\; \sum_{i=1}^{r} \deg\big(\underbrace{Z'_{\text{fix}}(C, z_i) \cap C_k}_{=:\,Y_i}\big)^2,$$

which is implied by (4.6.2.7).

Case 2: $\#(Y|_{\operatorname{Sing} C_k}) = 0$. Then, as $k \geq 3$, the right-hand side is bounded from below by $\#Y - k^2 - 1 \geq \#Y - (10/9)\,k^2$, whence (4.6.2.16) holds whenever

$$\sum_{i=1}^{r} (\deg Y_i)^2 \;<\; \frac{9}{10}\,d^2$$

with $\deg Y_i \;\leq\; \max\,\big\{\deg D \cap Z'_{\text{fix}}(C, z_i) \mid D \text{ smooth}\big\}$, which is a consequence of (4.6.2.9).

Step 2b. Assume $h = h^1\big(\mathcal{J}_{Y/\mathbb{P}^2}(d)\big) < (k-2)(k-1)/2$, in particular, $k \geq 4$. As we have seen in (4.6.2.15), it suffices to show that

$$\max_{k,h}\Big\{h + \underbrace{\frac{k^2-k}{2}+2}_{=:p_1(k)}\Big\} < \#Y\,, \quad \max_{k,h}\Big\{h + \underbrace{\frac{k(k+3)}{2}}_{=:p_2(k)}\Big\} \;<\; \#Y + \#(Y\big|_{\operatorname{Sing} C_k})\,.$$

Introducing

$$\rho_j \;:=\; \min\left\{\frac{(\deg Y)^2}{p_j(k)+h} \;\middle|\; \begin{array}{l} 1 \leq h \leq \min\big\{\frac{r_0(r_0+1)}{2},\, \frac{k(k-3)}{2}\big\} \\ 4 \leq k\,,\ 1 \leq r_0 \leq k-2\,. \end{array}\right\}, \quad j = 1, 2\,.$$

and applying the Cauchy inequality, this holds whenever

$$\rho_1 > \sum_{i=1}^{r} (\deg Y_i)^2\,, \quad \rho_2 > \sum_{z_i \notin \operatorname{Sing} C_k} (\deg Y_i)^2 + \sum_{z_i \in \operatorname{Sing} C_k} \frac{(\deg Y_i)^2}{2}\,. \tag{4.6.2.17}$$

It remains to estimate ρ_1 and ρ_2 as functions in d. By (4.6.2.14), we have for any $j = 1, 2$

$$\frac{(\deg Y)^2}{p_j(k)+h} \;\geq\; \frac{\big(2k(d+2-k+r_0) + 2h \,-\, r_0(r_0+1)\big)^2}{4\cdot(p_j(k)+h)} \;=:\; f_j(k,h,r_0)\,,$$

that is,

$$\rho_j \;\geq\; \min\left\{ f_j(k,h,r_0) \;\middle|\; \begin{array}{l} 1 \leq h \leq \min\big\{\frac{r_0(r_0+1)}{2},\, \frac{k(k-3)}{2}\big\} \\ 4 \leq k\,,\ 1 \leq r_0 \leq k-2\,. \end{array}\right\}, \quad j = 1, 2\,.$$

Note that for fixed k, h the functions $f_j(k, h, \ )$ are increasing in r_0 (on $\big[0, k - \frac{1}{2}\big]$). Hence, they take their minima for the minimal possible value, that is, for r_0 satisfying

$$\left(\frac{r_0(r_0+1)}{2} = h\,,\ \ r_0 \leq k-3\right) \ \text{ or } \ \left(r_0 = k-2\,,\ \ h \geq \frac{(k-3)(k-2)}{2} + 1\right).$$

Case 1: $r_0 = k-2, k(k-3) \geq 2h \geq (k-3)(k-2)+2$. In this case we can estimate

$$f_j(k,h,r_0) \geq \frac{\big(2kd + (k-3)(k-2) + 2 - (k-2)(k-1)\big)^2}{4p_j(k) + 2k(k-3)} = \frac{2\,(kd-k+3)^2}{2p_j(k) + k(k-3)}.$$

Hence, due to $k < (d+3)/2$,

$$f_1(k,h,r_0) \geq \frac{(kd-k+3)^2}{k^2-2k+2} \geq \frac{k^2(d-1)^2}{k^2-2k+2} \geq d^2\,,$$

$$f_2(k,h,r_0) \geq \left(d-1+\frac{3}{k}\right)^2 \geq (d-1)^2\,.$$

Thus, (4.6.2.17) is a consequence of (4.6.2.7) and (4.6.2.10).

Case 2: $2h = r_0(r_0+1),\ r_0 \leq k-3$. It follows that

$$f_j(k,h,r_0) \geq \frac{2k^2(d+2-k+r_0)^2}{2p_j(k)+r_0(r_0+1)} =: g_j(k,r_0)\,, \quad j=1,2\,.$$

We fix $k \geq 4$, and look for the minimum of $g_j(k,r_0)$. Since the derivative

$$\frac{\partial}{\partial r_0}\,g_j(k,r_0) = \underbrace{\frac{2k^2(d+2-k+r_0)}{\big(2p_j(k)+r_0(r_0+1)\big)^2}}_{>0} \cdot \big(r_0\underbrace{(2k-2d-3)}_{<0} + 4p_j(k) + k - d - 2\big)$$

changes sign at most once (from positive to negative), the minimum is taken at one of the endpoints, that is, $\rho_j \geq \min\,\{g_j(k,1),\ g_j(k,k-3)\}$, $j=1,2$. By definition, we have

$$g_1(k,1) = \frac{2k^2}{k^2-k+6}\cdot(d+3-k)^2\,, \qquad g_2(k,1) = \frac{2k^2}{k^2+3k+2}\cdot(d+3-k)^2.$$

Recall that due to (4.6.2.14) we can estimate

$$d+3-k \geq \frac{d+3}{2} + \sqrt{\frac{(d+3)^2}{4} - \deg Y}\,,$$

whence we obtain

$$g_1(k,1) \geq \begin{cases} \min\,\left\{\frac{16}{9}(d-1)^2,\ \frac{25}{13}(d-2)^2\right\} \geq d^2 & \text{if } k \in \{4,5\}\,,\\ \frac{1}{2}\big(d+3+\sqrt{(d+3)^2-4\deg Y}\,\big)^2 & \text{if } k \geq 6\,,\end{cases}$$

$$g_2(k,1) \geq \tfrac{4}{15}\big(d+3+\sqrt{(d+3)^2-4\deg Y}\,\big)^2.$$

On the other hand, we know that $k < (d+3)/2$, which implies

$$g_1(k,k-3) = \frac{k^2(d-1)^2}{k^2-3k+5} > d^2\,,$$
$$g_2(k,k-3) = \frac{k^2(d-1)^2}{k^2-k+3} > (d-1)^2.$$

Thus, if (4.6.2.7) and (4.6.2.10) are satisfied then the condition (4.6.2.17) holds whenever

$$\sum_{i=1}^{r}(\deg Y_i)^2 < \frac{\left(d+3+\sqrt{(d+3)^2-4\deg Y}\,\right)^2}{2}\,, \qquad (4.6.2.18)$$

and

$$\sum_{z_i\notin \mathrm{Sing}\, C_k}(\deg Y_i)^2 + \sum_{z_i\in \mathrm{Sing}\, C_k}\frac{(\deg Y_i)^2}{2} < \frac{4\left(d+3+\sqrt{(d+3)^2-4\deg Y}\,\right)^2}{15}\,. \qquad (4.6.2.19)$$

Step 3. We analyse the conditions (4.6.2.18) and (4.6.2.19). We write $\sum\frac{(\deg Y_i)^2}{\varepsilon_i}$ to denote the left-hand side of (4.6.2.18), respectively (4.6.2.19). As above, we introduce the numbers

$$\alpha_{Y,\varepsilon} := \frac{\sum_{i=1}^{r}\frac{(\deg Y_i)^2}{\varepsilon_i}}{(d+3)^2}\,, \quad \beta_{Y,\varepsilon} := \frac{\sum_{i=1}^{r}\frac{(\deg Y_i)^2}{\varepsilon_i}}{\deg Y}$$

and look for the possible values of $\alpha_{Y,\varepsilon}$ such that (4.6.2.18), respectively (4.6.2.19), holds. This is the case whenever

$$\alpha_{Y,\varepsilon} < \frac{K}{4}\cdot\left(1+\sqrt{1-\frac{4\alpha_{Y,\varepsilon}}{\beta_{Y,\varepsilon}}}\right)^2,$$

where $K=2$, respectively $K=16/15$, that is, if

$$\sum_{i=1}^{r}\frac{(\deg Y_i)^2}{\varepsilon_i} = \alpha_{Y,\varepsilon}\cdot(d+3)^2 < \frac{K\cdot\beta_{Y,\varepsilon}^2}{(\beta_{Y,\varepsilon}+K)^2}\cdot(d+3)^2\,.$$

Note that this restriction can be reformulated as

$$\left(\sum_{i=1}^{r}\frac{(\deg Y_i)^2}{\varepsilon_i}\right)\cdot\left(1+\frac{K}{\beta_{Y,\varepsilon}}\right)^2 < K(d+3)^2,$$

where, by the Cauchy inequality, the left-hand side can be estimated as

$$\left(\sum_{i=1}^{r}\frac{(\deg Y_i)^2}{\varepsilon_i}\right)\cdot\left(1+\frac{K}{\beta_{Y,\varepsilon}}\right)^2=\frac{\left(\sum_{i=1}^{r}\frac{(\deg Y_i)^2}{\varepsilon_i}+K\cdot\sum_{i=1}^{r}\deg Y_i\right)^2}{\sum_{i=1}^{r}\frac{(\deg Y_i)^2}{\varepsilon_i}}$$
$$\leq\sum_{i=1}^{r}\frac{(\deg Y_i+K\cdot\varepsilon_i)^2}{\varepsilon_i}\,.$$

Finally, since $Y_i\subset Z'_{\text{fix}}(C,z_i)$, the conditions (4.6.2.18) and (4.6.2.19) are satisfied if we suppose (4.6.2.8) and (4.6.2.11). □

4.6.3 Reducible Equisingular Families

The following example is probably the first series of reducible ESF's of irreducible plane cuspidal curves, where the different components V_i cannot be distinguished by the fundamental group of the complement of the curves $C\in V_i$.

Let C_1 and C_2 be two complex irreducible algebraic curves in $\mathbb{P}^2$ of the same degree d and the same number and type S_1, ..., S_r of singularities such that $\pi_1(\mathbb{P}^2\setminus C_1)\not\cong\pi_1(\mathbb{P}^2\setminus C_2)$. Then C_1 and C_2 must belong to two different irreducible components V_1 and V_2 of the ESF $V_d^{irr}(S_1,\ldots,S_r)$, and we call (V_1,V_2) (or (C_1,C_2)) a *Zariski pair* (cf. Example 4.2.0.9). Otherwise it is called an *anti-Zariski pair.*

4.6.3.1 ESF of Plane Curves with Cusps

The following proposition provides a series of reducible ESF's of cuspidal curves such that any two components are an anti-Zariski pair.

Proposition 4.6.10 ([GLS4], Proposition 5.4) *Let p,d be integers satisfying*

$$p\geq 3,\quad 6p\leq d\leq 12p-\tfrac{3}{2}-\sqrt{35p^2-15p+\tfrac{1}{4}}\,. \tag{4.6.3.1}$$

Then the variety $V_d^{irr}(6p^2\cdot A_2)$ of irreducible curves of degree d with $6p^2$ ordinary cusps has components of different dimensions.

Moreover, if $d>6p$, for all curves $C\in V_d^{irr}(6p^2\cdot A_2)$ the fundamental group of the complement is abelian, that is, $\pi_1(\mathbb{P}^2\setminus C)=\mathbb{Z}/d\mathbb{Z}$.

Example 4.6.10.1 The variety $V_{91}^{irr}(1350\cdot A_2)$ is reducible, with any two components not a Zariski pair.

Proof Note that $d>6p$ implies $d^2>36p^2=6\cdot 6p^2$. Hence, due to Nori's theorem [Nor, Proposition 3.27], $\pi_1(\mathbb{P}^2\setminus C)=\mathbb{Z}/d\mathbb{Z}$ for all curves $C\in V_d^{irr}(6p^2\cdot A_2)$.

We show that there are (at least) two different components of $V_d^{irr}(6p^2\cdot A_2)$: by (4.6.3.1),

$$6p^2 < \frac{(6p-1)(6p-2)+2}{4} \leq \frac{(d-1)(d-2)+2}{4},$$

and Theorem 4.5.4 gives the existence of a component V' of $V_d^{irr}(6p^2 \cdot A_2)$ having the expected dimension

$$\dim V' = \frac{d(d+3)}{2} - 12p^2 .$$

On the other hand, we construct a family V'' of bigger dimension: let C_{2p}, C_{3p}, C'_{d-6p}, C''_{d-6p} be generic curves of degrees $2p, 3p, d-6p, d-6p$, respectively. The curve $C_d = C_{2p}^3 C'_{d-6p} + C_{3p}^2 C''_{d-6p}$ has degree d and $6p^2$ ordinary cusps as its only singularities, one at each intersection point in $C_{2p} \cap C_{3p}$. Varying the coefficients of $C_{2p}, C_{3p}, C'_{d-6p}, C''_{d-6p}$ in the spaces of curves of degrees $2p,\ 3p,\ d-6p,\ d-6p$, respectively, we obtain a subfamily V'' in $V_d^{irr}(6p^2 \cdot A_2)$. Note that the equality

$$C_d \ = \ C_{2p}^3 C'_{d-6p} + C_{3p}^2 C''_{d-6p} \ = \ \widehat{C}_{2p}^3 \widehat{C}'_{d-6p} + \widehat{C}_{3p}^2 \widehat{C}''_{d-6p} \ = \ \widehat{C}_d$$

with slightly deformed curves $\widehat{C}_{2p}, \widehat{C}_{3p}, \widehat{C}'_{d-6p}, \widehat{C}''_{d-6p}$ implies

$$C_{2p} = \widehat{C}_{2p}, \quad C_{3p} = \widehat{C}_{3p}, \quad C'_{d-6p} = \widehat{C}'_{d-6p}, \quad C''_{d-6p} = \widehat{C}''_{d-6p} .$$

Indeed, if $C_d = \widehat{C}_d$ then they have $6p^2$ common cuspidal points belonging to C_{2p} and $\widehat{C}_{2p}$. Hence, by Bézout's theorem, $C_{2p} = \widehat{C}_{2p}$. The tangent line to $C_d = \widehat{C}_d$ at each cusp is tangent to both, C_{3p} and $\widehat{C}_{3p}$. This means that the intersection number of C_{3p} and $\widehat{C}_{3p}$ is at least $12p^2$, whence $C_{3p} = \widehat{C}_{3p}$. Finally, $C_{2p}^3(C'_{d-6p} - \widehat{C}'_{d-6p}) = C_{3p}^2(\widehat{C}''_{d-6p} - C''_{d-6p})$ and $d - 6p < 2p$, imply that $C'_{d-6p} = \widehat{C}'_{d-6p}, C''_{d-6p} = \widehat{C}''_{d-6p}$. Therefore, we can determine $\dim V''$ by counting the parameters for $C_{2p}, C_{3p}, C'_{d-6p}$, C''_{d-6p} and get

$$\begin{aligned}
\dim V'' &= \frac{2p(2p+3)}{2} + \frac{3p(3p+3)}{2} + 2 \cdot \frac{(d-6p)(d-6p+3)}{2} + 1 \\
&= \frac{d(d+3)}{2} - 12p^2 + \left(\frac{d^2}{2} - d\left(12p - \frac{3}{2}\right) + \frac{109p^2 - 21p + 2}{2}\right)
\end{aligned}$$

which is larger than the expected dimension if (4.6.3.1) holds true. □

4.6.3.2 ESF of Curves with Ordinary Singularities

We deduce now a similar result for families of irreducible curves in $\mathbb{P}^2$, having only ordinary singularities.

Proposition 4.6.11 ([GLS5], Proposition 1.1) *Let $m \geq 9$. Then there exists an integer $k_0 = k_0(m)$ such that for any $k \geq \max\{k_0, m\}$, and any s satisfying*

$$\frac{k-1}{2} \leq s \leq k\left(1-\sqrt{\frac{2}{m}}\right)-\frac{3}{2}$$

the variety $V_{km+s}^{irr}(m,\ldots,m)$ *of plane irreducible curves of degree* $d = km+s$, *having* k^2 *ordinary* m*-fold points as only singularities, is reducible.*

More precisely, $V_d^{irr}(m,\ldots,m)$ *has at least two components, one regular component (of the expected dimension) and one component of higher dimension. Moreover,* $\pi_1(\mathbb{P}^2 \setminus C) \simeq \mathbb{Z}/d\mathbb{Z}$ *for any* $C \in V_d^{irr}(m,\ldots,m)$.

Proof Let $F_k^{(0)}, G_k^{(0)} \in H^0\big(\mathcal{O}_{\mathbb{P}^2}(k)\big)$ define two distinct, reduced, smooth, irreducible plane curves of degree k (also denoted by $F_k^{(0)}, G_k^{(0)}$), intersecting each other transversally in k^2 points.

Moreover, let $Q_1 \in F_k^{(0)}$, $Q_2 \in G_k^{(0)}$ be generic, and denote by $\mathcal{J}_{Q_i} \subset \mathcal{O}_{\mathbb{P}^2}$ the ideal sheaf defining the (reduced) point Q_i, $i=1,2$. Finally, let $1 \leq s < k$, and $R_s^{(0)},\ldots,R_s^{(m)} \in H^0\big(\mathcal{O}_{\mathbb{P}^2}(s)\big)$ be generic. We consider the subfamily

$$\mathcal{F} := \left\{ \sum_{i=0}^{m} \widetilde{R}_s^{(i)} F_k^i G_k^{m-i} \;\middle|\; \begin{array}{l} \widetilde{R}_s^{(i)} \in H^0\big(\mathcal{O}_{\mathbb{P}^2}(s)\big) \text{ close to } R_s^{(i)} \\ F_k \in H^0\big(\mathcal{J}_{Q_1}(k)\big) \text{ close to } F_k^{(0)} \\ G_k \in H^0\big(\mathcal{J}_{Q_2}(k)\big) \text{ close to } G_k^{(0)} \end{array} \right\}$$

of $H^0\big(\mathcal{O}_{\mathbb{P}^2}(s+mk)\big)$. The respective conditions being open, we can assume

- the curves $\widetilde{R}_s^{(i)}, F_k, G_k$, appearing in the definition of $\mathcal{F}$, are reduced, smooth, irreducible, as $R_s^{(i)}, F_k^{(0)}, G_k^{(0)}$ are,
- each element of $\mathcal{F}$ defines a plane curve of degree $km+s$ having precisely k^2 ordinary m-fold points (the k^2 intersection points of F_k and G_k) as only singularities,
- all curves in $\mathcal{F}$ are irreducible (applying Bertini's theorem).

Step 1. We show (by counting parameters) that the family $\mathcal{F}$ has dimension

$$\dim \mathcal{F} = 2\left(\frac{(k+1)(k+2)}{2}-1\right)+(m+1)\cdot\frac{(s+1)(s+2)}{2}-2\,.$$

Suppose we can write an element $F \in \mathcal{F}$ in two ways as

$$\sum_{i=0}^{m} \widetilde{R}_s^{(i)} F_k^i G_k^{m-i} = F = \sum_{i=0}^{m} \widetilde{T}_s^{(i)} F_k'^{\,i} G_k'^{\,m-i}\,.$$

Then the intersection of the irreducible curves defined by F_k, G_k, F_k', G_k' contains the k^2 singular points of the curve defined by F. Additionally, F_k and F_k' both vanish at Q_1, G_k and G_k' both vanish at Q_2. Hence, Bézout's theorem implies $F_k' = \alpha F_k$, $G_k' = \beta G_k$, $\alpha, \beta \in \mathbb{C}$. It follows that

$$\left(\widetilde{R}_s^{(0)} - \beta^m \widetilde{T}_s^{(0)}\right) \cdot G_k^m = \sum_{i=1}^{m} \left(\alpha^i \beta^{m-i} \widetilde{T}_s^{(i)} - \widetilde{R}_s^{(i)}\right) F_k^i G_k^{m-i}$$
$$= F_k \cdot \sum_{i=1}^{m} \left(\alpha^i \beta^{m-i} \widetilde{T}_s^{(i)} - \widetilde{R}_s^{(i)}\right) F_k^{i-1} G_k^{m-i} .$$

Since F_k does not divide G_k, and $s < k$, we obtain $\widetilde{R}_s^{(0)} = \beta^m \widetilde{T}_s^{(0)}$. Proceeding by induction, the same reasoning gives $\widetilde{R}_s^{(i)} = \alpha^i \beta^{m-i} \widetilde{T}_s^{(i)}$, $i = 1, \ldots, m$. The above statement follows, since

$$\dim H^0\big(\mathcal{J}_{Q_i}(k)\big) = \frac{(k+1)(k+2)}{2} - 1, \quad \dim H^0\big(\mathcal{O}_{\mathbb{P}^2}(s)\big) = \frac{(s+1)(s+2)}{2} .$$

Step 2. There exists a regular component $V_{reg} \subset V$ having the expected dimension

$$\dim V_{reg} = \frac{(s+mk)(s+mk+3)}{2} - k^2 \left(\frac{m(m+1)}{2} - 2\right) .$$

Let $W = W(k^2, m)$ denote a generic union of k^2 fat points of multiplicity m, that is, the zero-dimensional scheme concentrated in a (generic) set of points $z_1, \ldots, z_{k^2} \subset \mathbb{P}^2$ and defined by the ideals $I_{W,z_i} := \mathfrak{m}_{z_i}^m \subset \mathcal{O}_{\mathbb{P}^2, z_i}$, $i = 1, \ldots, k^2$. The vanishing theorem of Alexander and Hirschowitz (Theorem 3.4.22) gives the existence of $k_0 = k_0(m)$ such that $H^1\big(\mathcal{J}_W(km+s-1)\big)$ vanishes for any $k \geq k_0$ and any $s \geq 0$ satisfying $H^0\big(\mathcal{J}_W(km+s)\big) > 0$, that is,

$$\begin{aligned} 0 &\leq (km+s-1)(km+s+2) - k^2 m(m+1) \\ &= s^2 + s(2km+1) - k^2 m + km - 2 , \end{aligned}$$

or, equivalently,

$$km \cdot \left(\sqrt{1 + \frac{1}{m} + \frac{9}{4k^2 m^2}} - 1\right) - \frac{1}{2} \leq s < k . \tag{4.6.3.2}$$

On the other hand, the vanishing of $H^1\big(\mathcal{J}_W(km+s-1)\big)$ implies the existence of an irreducible curve $C \subset \mathbb{P}^2$ belonging to a regular component of V. For instance, this can be seen by applying the construction in the proof of Theorem 4.5.13.

Step 3. Comparison of dimensions: let $\overline{\mathcal{F}} \subset V$ denote the image of $\mathcal{F}$. Then

$$\dim \overline{\mathcal{F}} - \dim V_{reg} = \frac{m}{2} \cdot \left(s^2 + s \cdot (3-2k) + (k^2 - 3k + 2)\left(1 - \frac{2}{m}\right)\right)$$

is positive as

$$0 \le s < k - \frac{3}{2} - \sqrt{\frac{2}{m}} \cdot \sqrt{k^2 - 3k + 2 + \frac{m}{8}} . \qquad (4.6.3.3)$$

We conclude that there exist at least 2 components of V if $k \ge k_0$ and (4.6.3.2), (4.6.3.3) hold true. In particular, this is given for any $k \ge \max\{k_0, m\}$ and

$$\frac{k-1}{2} \le s \le k\left(1 - \sqrt{\frac{2}{m}}\right) - \frac{3}{2} .$$

□

Remark 4.6.11.1 Due to Theorem 2.2.38 (e), the regular component of $V_{km+s}^{irr}(m, \dots, m)$ (obtained in Step 2 of the proof) dominates $\mathrm{Sym}^r(\mathbb{P}^2)$ under the morphism $V_{km+s}^{irr}(m, \dots, m) \to \mathrm{Sym}^r(\mathbb{P}^2)$, $r = k^2$, mapping a curve to its singular locus.

Finally, we claim that $\pi_1(\mathbb{P}^2 \setminus C) = \mathbb{Z}/d\mathbb{Z}$ for any irreducible curve C of degree d with only ordinary singularities $S_1, \dots S_r$ satisfying

$$2 \cdot \#(nodes) + \sum_{\mathrm{mt}\, S_i \ge 3} (\mathrm{mt}\, S_i)^2 < d^2 .$$

Indeed, this follows from [Nor, Proposition 3.27] applied to the plane blown up at the ordinary singular points of C of multiplicity ≥ 3. In particular, this holds for any $C \in V_{km+s}^{irr}(m, \dots, m)$.

4.6.4 *Irreducibility of ESF of Nodal Curves on Blown-Up $\mathbb{P}^2$*

Let $\mathbb{P}^2_r$ be the blow up of $\mathbb{P}^2$ at r generic points and let E_0 resp. E_1, ..., E_r denote the strict transform of the generic line in $\mathbb{P}^2$ and the exceptional divisors, respectively. Let $V^{irr}(d; d_1, ..., d_r; k)$ denote the variety of irreducible curves C in $|D| = |dE_0 - d_1E_1 - ... - d_rE_r|$ with k nodes as only singularities.

For $\mathbb{P}^2_1$, the projective plane blown up at one point p_1, Ran shows in [Ran2] that the variety of all irreducible nodal curves $C \in |dE_0 - d_1E_1|$ having exactly k nodes, none of them lying on the exceptional divisor E_1, is irreducible, which yields the irreducibility of $V^{irr}(d; d_1; k)$, while for $r \ge 2$ we proved (cf. [GLS2]) the following irreducibility criterion for $V^{irr}(d;\ d_1, \dots, d_r;\ k)$:

Theorem 4.6.12 ([GLS2], Theorem 2) *Let $r \ge 2$, and let the positive integers $d;\ d_1, \dots, d_r$ satisfy the two (irreducibility) conditions*

$$\left[\sqrt{2k}\,\right] < \frac{d}{4} + 1 - \frac{1}{4}\sqrt{\sum_{i=1}^{r} d_i^2} \tag{4.6.4.1}$$

$$\left[\sqrt{2k}\,\right] < \frac{d}{2} + 1 - \frac{\sqrt{2}}{2}\sqrt{\sum_{i=1}^{r} (d_i+2)^2}\,. \tag{4.6.4.2}$$

Then $V^{irr}(d;\, d_1, \ldots, d_r;\, k)$ *is (smooth and) irreducible.*

Let $C \subset \mathbb{P}^2_r$ *be a reduced nodal curve,* $C \in |D|$*. If for each irreducible component* $C_\nu \in |d^{(\nu)}E_0 - \sum_{i=1}^{r} d_i^{(\nu)} E_i|$, $1 \le \nu \le n$, *of* C *(having precisely* $k^{(\nu)}$ *nodes) the variety* $V_{irr}(d^{(\nu)};\, d_1^{(\nu)}, \ldots, d_r^{(\nu)};\, k^{(\nu)})$ *is smooth and irreducible, then* $V(|D|;\, C)$ *is irreducible.*

The main idea of our proof is as follows. We show that for an irreducible curve $C \in |dE_0 - \sum_{i=1}^{r} d_i E_i|$ in an open dense subset $\tilde{V} \subset V^{irr}(d;\, d_1, \ldots, d_r;\, k)$ the cohomology group $H^1(C, \mathcal{N}^{\operatorname{Sing} C}_{C/\mathbb{P}^2_r})$, $\mathcal{N}^{\operatorname{Sing} C}_{C/\mathbb{P}^2_r} = \operatorname{Ker}\Big(\mathcal{O}_C(C) \to \mathcal{T}^1_{C,\operatorname{Sing} C} = \mathcal{O}_{Z^{es}(C)}\Big)$, vanishes, especially, that the conditions imposed by fixing the k singular points are independent. It follows that the restricted morphism

$$\begin{aligned} \pi_{\tilde{V}} : \tilde{V} &\longrightarrow Sym^k(\mathbb{P}^2_r) \\ C &\longmapsto \operatorname{Sing} C \end{aligned}$$

is dominant, its fibres are all equidimensional and irreducible as open subsets of the linear system $\big|H^0\big(\mathbb{P}^2_r, \operatorname{Ker}(\mathcal{O}(dE_0 - \sum_i d_i E_i) \to \mathcal{T}^1_{C,\operatorname{Sing} C})\big)\big|$. Hence $\tilde{V}$ is irreducible, which implies the irreducibility of $V^{irr}(d;\, d_1, \ldots, d_r;\, k)$ by Proposition 1.1.56. The second statement is a consequence of the fact that for a fixed reduced nodal curve C as above, each generic member $\tilde{C}$ of a component of $V(|D|;\, C)$ decomposes into components $\tilde{C}_\nu$, $1 \le \nu \le n$, which are generic elements of $V_{irr}(d^{(\nu)};\, d_1^{(\nu)}, \ldots, d_r^{(\nu)};\, k^{(\nu)})$. Hence there is a well–defined dominant morphism

$$\prod_{\nu=1}^{n} U_\nu \longrightarrow V(|D|;\, C)$$

where U_ν is open dense in $V_{irr}(d^{(\nu)};\, d_1^{(\nu)}, \ldots, d_r^{(\nu)};\, k^{(\nu)})$.

4.6.5 Irreducibility of ESF of Curves on Other Surfaces

Let Σ be a smooth projective algebraic surface, D a divisor on Σ such that the complete linear system $|D|$ has a positive dimension. We ask about the irreducibility of $V^{irr}_{|D|}(S_1, \ldots, S_r)$, the variety of irreducible curves in $|D|$ with precisely r isolated singular points of the given topological or analytic types $S_1, \ldots, S_r$, respectively. The main sufficient irreducibility condition we get looks like

$$\sum_{i=1}^{r} (\tau'(S_i) + 2)^2 < \kappa \cdot (D - K_\Sigma)^2 ,$$

where τ' stands for τ^{es} resp. $\tau^{ea} = \tau$ (the degrees of $Z^{es}(S)$ resp. $Z^{ea}(S)$) in the notation of Sect. 4.6.2, and κ is a positive number depending on the surface Σ and sometimes on the divisor D.

Our treatment of the irreducibility problem for equisingular families of curves with arbitrary singularities on algebraic surfaces different from the plane follows partly [Kei1], which approaches the problem similarly to Sect. 4.6.2, replacing the Castelnuovo function argument with the Bogomolov instability one (see also Sect. 4.3.2.2).

4.6.5.1 Formulation of Results

We use again the notations of Sect. 4.6.2. In particular, $\mathrm{NS}(\Sigma)$ denotes the Néron-Severi group of Σ and $\sim_a$ algebraic equivalence.

Theorem 4.6.13 *Let Σ be a smooth projective surface with $\mathrm{NS}(\Sigma) = L \cdot \mathbb{Z}$, L ample, and $H^1(\mathcal{O}_\Sigma(C)) = 0$ whenever C is effective. Let D be a divisor on Σ, and let $S_1, \ldots, S_r$ be topological or analytic singularity types. Suppose that $D - K_\Sigma$ is big and nef, $D + K_\Sigma$ is nef, and that*

$$\sum_{i=1}^{r} (\tau'(S_i) + 2) < \lambda \cdot (D - K_\Sigma)^2 ,$$

$$\sum_{i=1}^{r} (\tau'(S_i) + 2)^2 < \alpha \cdot (D - K_\Sigma)^2 ,$$

where

$$\lambda \leq \frac{1}{4}, \quad \alpha = \min\left\{\frac{1}{2}, \frac{(1+\sqrt{1-4\lambda})^2 \cdot L^2}{4\chi(\mathcal{O}_\Sigma) + \max\{0, 2L.K_\Sigma\} + 6L^2}\right\} .$$

Then either $V^{irr}_{|D|}(S_1, \ldots, S_r)$ is empty or it is irreducible and T-smooth.

Corollary 4.6.14 *Let $\Sigma \subset \mathbb{P}^3$ be a smooth surface of degree $n \geq 4$, let H be a plane section, and suppose that the Picard number of Σ is one. Then the statement of Theorem 4.6.13 holds true for $D = dH$, $d > n - 4$, provided that*

$$\sum_{i=1}^{r} (\tau'(S_i) + 2)^2 < \min\left\{\frac{1}{2}, \frac{6n^2(n^3 - 3n^2 + 8n - 6)}{(n^3 - 3n^2 + 10n - 6)^2}\right\} \cdot (d + 4 - n)^2 .$$

Corollary 4.6.15 *Let Σ be a smooth K3-surface with* $\mathrm{NS}(\Sigma) = L \cdot \mathbb{Z}$ *with L ample, and set $n = L^2$. Then the statement of Theorem 4.6.13 holds true for $D = dH, d > 0$, provided that*

$$\sum_{i=1}^{r} (\tau'(S_i) + 2)^2 < \frac{n^2(54n + 72)}{(11n + 12)^2} d^2 .$$

Both corollaries follow immediately from Theorem 4.6.13, since in the situations considered $H^1\big(\mathcal{O}_\Sigma(C)\big) = 0$ holds true for any effective divisor C due to the Kodaira vanishing Theorem 3.2.1.

Beside the situation considered in Theorem 4.6.13 ($\mathrm{NS}(\Sigma) = L \cdot \mathbb{Z}$), we can also give numerical irreducibility criteria for product, respectively ruled, surfaces (both satisfying $\mathrm{NS}(\Sigma) \cong \mathbb{Z}^2$):

Theorem 4.6.16 *Let C_1 and C_2 be two smooth projective curves of genera g_1 and g_2, respectively, with $g_1 \geq g_2 \geq 0$, such that, for $\Sigma = C_1 \times C_2$, the Néron-Severi group is* $\mathrm{NS}(\Sigma) = C_1\mathbb{Z} \oplus C_2\mathbb{Z}$.

Let D be a divisor on Σ such that $D \sim_a aC_1 + bC_2$ with $a > |2g_2 - 2|$, $b > |2g_1 - 2|$, and let $S_1, \ldots, S_r$ be topological or analytic singularity types satisfying

$$\sum_{i=1}^{r} (\tau'(S_i) + 2)^2 < \alpha(a - 2g_2 + 2)(b - 2g_1 + 2) ,$$

$$\max_{1 \leq i \leq r} \tau'(S_i) < \min\{(a - 2g_2,\ b - 2g_1\} ,$$

where the positive constant $\alpha = \alpha(\Sigma, D)$ is given by the following formulas (with $\lambda := (a - 2g_2 + 2)/(b - 2g_1 + 2) > 0$):

- *if $g_1 = g_2 = 0$ then $\alpha = \dfrac{1}{12}$;*
- *if $g_1 = 1,\ g_2 = 0$ then $\alpha = \min\left\{\dfrac{1}{16}, \dfrac{1}{\lambda}\right\}$;*
- *if $g_1 \geq 2,\ g_2 = 0$ then $\alpha = \min\left\{\dfrac{1}{12 + 8g_1}, \dfrac{1}{2g_1\lambda}\right\}$;*
- *if $g_1 = g_2 = 1$ then $\alpha = \min\left\{\dfrac{1}{16}, \dfrac{1}{\lambda}, \lambda\right\}$;*
- *if $g_1 \geq 2,\ g_2 \geq 1$ then $\alpha = \min\left\{\dfrac{1}{12 + 8g_1 + 8g_2}, \dfrac{1}{2g_1\lambda}, \dfrac{\lambda}{2g_2}\right\}$.*

Then either $V^{irr}_{|D|}(S_1, \ldots, S_r)$ is empty or it is irreducible and T-smooth.

Note that for C_1, C_2 both being elliptic curves the condition on $\mathrm{NS}(\Sigma)$ means just that the two curves are non-isogenous.

Let $\pi : \Sigma = \mathbb{P}(\mathcal{E}) \to C$ be a geometrically ruled surface with normalized bundle $\mathcal{E}$ (in the sense of [HaR2, V.2.8.1]). Then $\mathrm{NS}(\Sigma) = C_0\mathbb{Z} \oplus F\mathbb{Z}$ with the intersection matrix

$$\begin{pmatrix} -e & 1 \\ 1 & 0 \end{pmatrix},$$

where $F \cong \mathbb{P}^1$ is a fibre of π, C_0 is a section of π with $\mathcal{O}_\Sigma(C_0) \cong \mathcal{O}_{\mathbb{P}(\mathcal{E})}(1)$, $g = g(C)$ the genus of C, and $e = -\deg(\Lambda^2\mathcal{E}) \geq -g$. For the canonical divisor we have $K_\Sigma \sim_a -2C_0 + (2g-2-e)F$.

Theorem 4.6.17 *Let $\pi : \Sigma \to C$ be a geometrically ruled surface with $e \leq 0$. Let $D = aC_0 + bF$ with $a \geq 2$, $b > 2g - 2 + ae/2$, and if $g = 0$ then $b \geq 2$. Let $S_1, \dots, S_r$ be topological, respectively analytic, singularity types satisfying*

$$\sum_{i=1}^{r} (\tau'(S_i) + 2)^2 < \alpha \cdot (a+2)(2b + 4 + 2e - 4g - ea),$$

$$\max_{1 \leq i \leq r} \tau'(S_i) < a,$$

where the positive constant $\alpha = \alpha(\Sigma, D)$ is given by the following formulas (with $\lambda := (2a+4)/(2b+4-4g-ae)$):

- *if $g = e = 0$ then $\alpha = \frac{1}{24}$;*
- *if $g = 1,\ e = 0$ then $\alpha = \min\left\{\frac{1}{24}, \frac{1}{2\lambda}\right\}$;*
- *if $g = 1,\ e = -1$ then $\alpha = \min\left\{\frac{\lambda}{4\lambda^2 + 30\lambda + 16}, \frac{2}{13\lambda}\right\}$;*
- *if $g \geq 2,\ e = 0$ then $\alpha = \min\left\{\frac{1}{24 + 16g}, \frac{1}{4ga}\right\}$;*
- *if $g \geq 2,\ e < 0$ then α is the minimum of $\frac{1}{4ga - 9ea}$ and*

$$\max\left\{\frac{1}{24 + 16g - 9ea}, \frac{ea}{18ea + 16eag - 9e^2a^2 - 16}\right\}.$$

Then either $V^{irr}_{|D|}(S_1, \dots, S_r)$ is empty or it is irreducible and T-smooth.

4.6.5.2 Plan of the Proof

As in Sect. 4.6.2, we shall show that, under the conditions of Theorems 4.6.13–4.6.17, the following holds true:

- $H^1(\mathcal{J}_{Z/\Sigma}(D)) = 0$ for $Z = Z(\boldsymbol{z}, \boldsymbol{m}) \subset \Sigma$ a generic fat point scheme concentrated in $\boldsymbol{z} = z_1, \dots, z_r$, with multiplicities $m_i = \nu'(S_i) + 1$, $i = 1, \dots, r$;
- The dimension of the space of all curves $C \in V^{irr}_{|D|}(S_1, \dots, S_r)$ satisfying $h^1\big(\mathcal{J}_{Z'_{fix}(C)/\Sigma}(D)\big) > 0$ is strictly smaller than $\dim\big(V^{irr}_{|D|}(S_1, \dots, S_r)\big)$. Here,

$Z'_{fix}(C) = \bigcup_{i=1}^{r} Z'_{fix}(C, w_i)$ denotes the zero-dimensional scheme concentrated at the singular points $w_1, \ldots, w_r$ of C.

The first condition follows from the hypotheses of Theorems 4.6.13–4.6.17, since by Proposition 3.2.9, the requirements are

$$\sum_{i=1}^{r} (\nu'(S_i) + 2)^2 \leq \frac{1}{2}(D - K_\Sigma)^2$$

and

$$\max_{1 \leq i \leq r} \nu'(S_i) < \begin{cases} \min\{(D - K_\Sigma).A,\ (D - K_\Sigma).B\} & \text{in Theorem 4.6.16,} \\ (D - K_\Sigma).F & \text{in Theorem 4.6.17.} \end{cases}$$

One can easily verify these inequalities, noticing that $\nu'(S_i) \leq \tau'(S_i)$ (Lemma 1.2.9 and Proposition 1.2.5).

The second condition is reduced to the following: if we consider, as in Sect. 4.6.2, the projection $V^{irr}_{|D|}(S_1, \ldots, S_r) \to \mathrm{Sym}^r(\Sigma)$, then the image of $\{C \in V^{irr}_{|D|}(S_1, \ldots, S_r) \mid h^1(\mathcal{J}_{Z'_{fix}(C)/\Sigma}(D)) = 0\}$ is dense in $\mathrm{Sym}^r(\Sigma)$, with fibres of codimension $\deg Z'_{fix}(C)$. On the other hand, the fibres of the projection of $\{C \in V^{irr}_{|D|}(S_1, \ldots, S_r) \mid h^1(\mathcal{J}_{Z'_{fix}(C)/\Sigma}(D)) = p\}$, $p > 0$, to $\mathrm{Sym}^r(\Sigma)$ have dimension at most $\deg Z'_{fix}(C) + p$. We intend to show that in the latter case some parts Z^0_i of $Z'_{fix}(C)$ lie on certain curves $\Delta_i \subset \Sigma$, $i = 1, \ldots, m$. The latter imposes at least

$$\sum_{i=1}^{m} \left(\#Z^0_i - \dim |\Delta_i|\right)$$

conditions on the image of $\{C \in V^{irr}_{|D|}(S_1, \ldots, S_r) \mid h^1(\mathcal{J}_{Z'_{fix}(C)/\Sigma}(D)) = p\}$. So, finally, we prove the irreducibility of $V^{irr}_{|D|}(S_1, \ldots, S_r)$ by establishing that

$$\sum_{i=1}^{m} \left(\#Z^0_i - \dim |\Delta_i|\right) > h^1\left(\mathcal{J}_{Z'_{fix}(C)/\Sigma}(D)\right).$$

We prove this only in the case of surfaces with $\mathrm{NS}(\Sigma) = \mathbb{Z}$, referring to [Kei1] for an essentially similar but rather tedious treatment of surfaces with $\mathrm{NS}(\Sigma) = \mathbb{Z}^2$.

4.6.5.3 Estimation of H^1

We prove the existence of certain curves $\Delta_1, \ldots, \Delta_m$ passing through the singular locus of any curve $C \in V^{irr}_{|D|}(S_1, \ldots, S_r)$ satisfying $h^1\left(\mathcal{J}_{Z'_{fix}(C)/\Sigma}(D)\right) > 0$:

Lemma 4.6.18 *Let a surface Σ and a divisor $D \subset \Sigma$ satisfy the hypotheses of Theorem 4.6.13. Let $Z_0 \subset \Sigma$ be a zero-dimensional scheme satisfying*

(1) there exists $C_0 \in |D|$ irreducible with $Z_0 \subset C_0$,
(2) $h^1\big(\mathcal{J}_{Z_0/\Sigma}(D)\big) > 0$, *and*
(3) $\deg(Z_0) < \alpha_1 \cdot (D - K_\Sigma)^2$ *for some* $0 < \alpha_1 \leq \frac{1}{4}$.

Then there are curves $\Delta_1, \ldots, \Delta_m \subset \Sigma$ and locally complete intersection schemes $Z_i^0 \subseteq Z_{i-1} \cap \Delta_i$, $i = 1, \ldots, m$, with $Z_i = Z_{i-1} : \Delta_i$, such that

(a) $h^1\big(\mathcal{J}_{Z_m/\Sigma}(D - \Delta_1 - \ldots - \Delta_m)\big) = 0$,

and for $i = 1, \ldots, m$,

(b) $h^1\big(\mathcal{J}_{Z_i^0/\Sigma}(D - \Delta_1 - \ldots - \Delta_{i-1})\big) = 1$,
(c) $D.\Delta_i \geq \deg(Z_{i-1} \cap \Delta_i) \geq \deg\big(Z_i^0\big)$
$\geq \big(D - K_\Sigma - \Delta_1 - \ldots - \Delta_i\big).\Delta_i \geq \Delta_i^2 \geq 0$,
(d) $\big(D - K_\Sigma - \Delta_1 - \ldots - \Delta_{i-1} - 2\Delta_i\big)^2 > 0$,
(e) $\big(D - K_\Sigma - \Delta_1 - \ldots - \Delta_{i-1} - 2\Delta_i\big).H > 0$ *for $H \in \mathrm{Div}(\Sigma)$ ample, and*
(f) $D - K_\Sigma - \Delta_1 - \ldots - \Delta_i$ *is big and nef.*

Moreover, it follows

$$0 \leq \frac{1}{4}(D - K_\Sigma)^2 - \sum_{i=1}^{m} \deg\big(Z_i^0\big) \leq \left(\frac{1}{2}(D - K_\Sigma) - \sum_{i=1}^{m} \Delta_i\right)^2 . \qquad (4.6.5.1)$$

Proof We are going to find the curves Δ_i and the zero-dimensional schemes Z_i^0 recursively. Therefore, let us suppose that we have already found the curves $\Delta_1, \ldots, \Delta_{i-1}$ and the schemes $Z_1^0, \ldots, Z_{i-1}^0$ satisfying (b)–(f), and suppose that $h^1\big(\mathcal{J}_{Z_{i-1}/\Sigma}\big(D - \sum_{k=1}^{i-1} \Delta_k\big)\big) > 0$. Then we choose $Z_i^0 \subseteq Z_{i-1}$ as a minimal scheme, still satisfying $h^1\big(\mathcal{J}_{Z_i^0/\Sigma}\big(D - \sum_{k=1}^{i-1} \Delta_k\big)\big) > 0$.

Step 1. We show that (b) is satisfied.

By the hypotheses of Theorem 4.6.13, $D - K_\Sigma$ is big and nef and $D + K_\Sigma$ is nef, and hence for $i = 1$, respectively by assumption (f) for $i \geq 2$, and the Kawamata–Viehweg vanishing Theorem 3.2.2, we get $h^1\big(\mathcal{O}_\Sigma\big(D - \sum_{k=1}^{i-1} \Delta_k\big)\big) = 0$. Thus Z_i^0 cannot be empty, that is, $\deg\big(Z_i^0\big) \geq 1$, and we may choose a subscheme $Y \subset Z_i^0$ of degree $\deg(Y) = \deg\big(Z_i^0\big) - 1$.

The inclusion $\mathcal{J}_{Z_i^0} \hookrightarrow \mathcal{J}_Y$, respectively the corresponding long exact cohomology sequence implies

$$h^1\Big(\mathcal{J}_{Y/\Sigma}\Big(D - \sum_{k=1}^{i-1} \Delta_k\Big)\Big) \geq h^1\Big(\mathcal{J}_{Z_i^0/\Sigma}\Big(D - \sum_{k=1}^{i-1} \Delta_k\Big)\Big) - 1 ,$$

and the minimality of Z_i^0 implies $h^1\big(\mathcal{J}_{Z_i^0/\Sigma}\big(D - \sum_{k=1}^{i-1} \Delta_k\big)\big) = 1$.

Step 2. We prove the inequality $\deg\left(Z_i^0\right) \le \deg(Z_0) - \sum_{k=1}^{i-1} \deg(Z_{k-1} \cap \Delta_k)$.

The case $i = 1$ follows immediately from the inclusion $Z_1^0 \subseteq Z_0$, while for $i \ge 2$ the inclusion $Z_i^0 \subseteq Z_{i-1} = Z_{i-2} : \Delta_{i-1}$ implies

$$\deg\left(Z_i^0\right) \le \deg(Z_{i-2} : \Delta_{i-1}) = \deg(Z_{i-2}) - \deg(Z_{i-2} \cap \Delta_{i-1})\,.$$

Thus, it suffices to show, that

$$\deg(Z_{i-2}) - \deg(Z_{i-2} \cap \Delta_{i-1}) = \deg(Z_0) - \sum_{k=1}^{i-1} \deg(Z_{k-1} \cap \Delta_k).$$

If $i = 2$, there is nothing to show. Otherwise $Z_{i-2} = Z_{i-3} : \Delta_{i-2}$ implies

$$\begin{aligned}\deg(Z_{i-2}) - \deg(Z_{i-2} \cap \Delta_{i-1}) &= \deg(Z_{i-3} : \Delta_{i-2}) - \deg(Z_{i-2} \cap \Delta_{i-1}) \\ &= \deg(Z_{i-3}) - \deg(Z_{i-3} \cap \Delta_{i-2}) - \deg(Z_{i-2} \cap \Delta_{i-1})\,,\end{aligned}$$

and we are done by induction.

Step 3. We prove that there exists a "suitable" locally free rank-two vector bundle E_i.

By the Serre-Grothendieck duality (see [HaR2, Section III.7], we see that the non-vanishing $H^1\left(\mathcal{J}_{Z_i^0/\Sigma}\left(D - \sum_{k=1}^{i-1} \Delta_k\right)\right) \neq 0$ implies that $\operatorname{Ext}^1\left(\mathcal{J}_{Z_i^0/\Sigma}\left(D - K_\Sigma - \sum_{k=1}^{i-1} \Delta_k\right)\right) \neq 0$. Hence, there exists an extension

$$0 \longrightarrow \mathcal{O}_\Sigma \longrightarrow E_i \longrightarrow \mathcal{J}_{Z_i^0/\Sigma}\left(D - K_\Sigma - \sum_{k=1}^{i-1} \Delta_k\right) \longrightarrow 0\,. \qquad (4.6.5.2)$$

The minimality of Z_i^0 implies that E_i is locally free (cf. [LaR, Prop. 3.9]) and, hence, that Z_i^0 is locally a complete intersection (cf. [LaR, p. 175]). Moreover, we have (cf. [LaR, Exe. 4.3])

$$c_1(E_i) = D - K_\Sigma - \sum_{k=1}^{i-1} \Delta_k\,, \qquad c_2(E_i) = \deg\left(Z_i^0\right). \qquad (4.6.5.3)$$

Step 4. We show that E_i is Bogomolov unstable.

According to Theorem 3.3.2, we have to show $c_1(E_i)^2 > 4c_2(E_i)$. Since $(4\lambda - 1) \cdot (D - K_\Sigma)^2 \le 0$ due to (3), and since $\Delta_k^2 \ge 0$, we deduce:

$$\begin{aligned}
4c_2(E_i) &= 4\deg\left(Z_i^0\right) \overset{\text{Step 2}}{\leq} 4\deg(Z_0) - 4\sum_{k=1}^{i-1}\deg(Z_{k-1}\cap\Delta_k)\\
&\overset{(3),(c)}{<} 4\lambda(D-K_\Sigma)^2 - 2\sum_{k=1}^{i-1}\Delta_k.\big(D-K_\Sigma-\sum_{j=1}^{k}\Delta_j\big) - 2\sum_{k=1}^{i-1}\Delta_k^2\\
&= \left(D-K_\Sigma-\sum_{k=1}^{i-1}\Delta_k\right)^2 + (4\lambda-1)\cdot(D-K_\Sigma)^2 - \sum_{k=1}^{i-1}\Delta_k^2\\
&\leq \left(D-K_\Sigma-\sum_{k=1}^{i-1}\Delta_k\right)^2 = c_1(E_i)^2.
\end{aligned}$$

Step 5. Find the curve Δ_i.

Since E_i is Bogomolov unstable, there exists a zero-dimensional scheme Z_i and a divisor $\Delta_i^0 \in \mathrm{Div}(\Sigma)$ such that the sequence

$$0 \to \mathcal{O}_\Sigma\left(\Delta_i^0\right) \to E_i \to \mathcal{J}_{Z_i/\Sigma}\left(D-K_\Sigma-\sum_{k=1}^{i-1}\Delta_k-\Delta_i^0\right) \to 0 \qquad (4.6.5.4)$$

is exact (cf. [LaR, Thm. 4.2]) and

(d') $\left(2\Delta_i^0 - D + K_\Sigma + \sum_{k=1}^{i-1}\Delta_k\right)^2 \geq c_1(E_i)^2 - 4\cdot c_2(E_i) > 0,$

(e') $\left(2\Delta_i^0 - D + K_\Sigma + \sum_{k=1}^{i-1}\Delta_k\right).H > 0$ for $H \in \mathrm{Div}(\Sigma)$ ample.

Tensoring (4.6.5.4) with $\mathcal{O}_\Sigma\left(-\Delta_i^0\right)$ leads to the following exact sequence

$$0 \to \mathcal{O}_\Sigma \to E_i\left(-\Delta_i^0\right) \to \mathcal{J}_{Z_i/\Sigma}\left(D-K_\Sigma-\sum_{k=1}^{i-1}\Delta_k-2\Delta_i^0\right) \to 0,$$

and we deduce that $h^0\left(\Sigma, E_i\left(-\Delta_i^0\right)\right) > 0$. On the other hand, tensoring (4.6.5.2) with $\mathcal{O}_\Sigma\left(-\Delta_i^0\right)$ leads to the exact sequence

$$0 \to \mathcal{O}_\Sigma\left(-\Delta_i^0\right) \to E_i\left(-\Delta_i^0\right) \to \mathcal{J}_{Z_i^0/\Sigma}\left(D-K_\Sigma-\sum_{k=1}^{i-1}\Delta_k-\Delta_i^0\right) \to 0.$$

We consider the corresponding long exact cohomology sequence. By (e) and the assumptions that $D-K_\Sigma$ is big and nef and $D+K_\Sigma$ is nef (cf. Theorem 4.6.13) for $i=1$, respectively (f) for $i \geq 2$, we get the inequality

$$-\Delta_i^0.H < -\frac{1}{2}\left(D - K_\Sigma - \sum_{k=1}^{i-1} \Delta_k\right).H \leq 0$$

for each ample divisor H, hence $-\Delta_i^0$ cannot be effective. In particular, the first term in the long exact sequence, $H^0(\Sigma, -\Delta_i^0)$, vanishes. By the above, the second term, $H^0(\Sigma, E_i(-\Delta_i^0))$ is non-vanishing. Hence, we can choose

$$\Delta_i \in \left| H^0\left(\mathcal{J}_{Z_i^0/\Sigma}\left(D - K_\Sigma - \sum_{k=1}^{i-1} \Delta_k - \Delta_i^0\right)\right)\right| .$$

Observe that the curve Δ_i satisfies (d)–(f). Indeed, this immediately follows from (d) and (e), and from the hypotheses of Theorem 4.6.13.

Step 6. We prove that Δ_i satisfies (c).

We would like to apply Bézout's theorem to C_0 and Δ_i. Thus suppose that the irreducible curve C_0 is a component of Δ_i and let H be any ample divisor. Applying (d) and the fact that $D + K_\Sigma$ is nef, we derive the contradiction

$$0 \leq (\Delta_i - C_0).H < -\frac{1}{2} \cdot \left(D + K_\Sigma + \sum_{k=1}^{i-1} \Delta_k\right).H \leq -\frac{1}{2} \cdot (D + K_\Sigma).H \leq 0.$$

Since $Z_{i-1} \subseteq Z_0 \subset C_0$, Bézout's theorem implies

$$D.\Delta_i = C_0.\Delta_i \geq \deg(Z_{i-1} \cap \Delta_i).$$

By definition, $Z_i^0 \subseteq Z_{i-1}$ and $Z_i^0 \subset \Delta_i$, thus $\deg(Z_{i-1} \cap \Delta_i) \geq \deg\left(Z_i^0\right)$. By assumption, each curve (in particular, Δ_i) is nef. Hence,

$$\left(D - K_\Sigma - \sum_{k=1}^{i} \Delta_k\right).\Delta_i = \Delta_i^0.\Delta_i \geq \Delta_i^2 \geq 0.$$

Finally from (d') and by (4.6.5.3) it follows that

$$\left(\Delta_i^0 - \Delta_i\right)^2 \geq c_1(E_i)^2 - 4 \cdot c_2(E_i) = \left(\Delta_i^0 + \Delta_i\right)^2 - 4 \cdot \deg\left(Z_i^0\right),$$

and, thus, $\deg\left(Z_i^0\right) \geq \Delta_i^0.\Delta_i$.

Step 7. After finitely many steps $h^1\left(\mathcal{J}_{Z_m/\Sigma}\left(D - \Delta_1 - \ldots - \Delta_m\right)\right) = 0$.

As we have mentioned in Step 1, $\deg\left(Z_i^0\right) > 0$. This ensures that

$$\begin{aligned}\deg(Z_i) = \deg(Z_{i-1}) - \deg(Z_{i-1} \cap \Delta_i) &\leq \deg(Z_{i-1}) - \deg\left(Z_i^0\right) \\ &< \deg(Z_{i-1}),\end{aligned}$$

that is, the degree of Z_i is strictly diminished each time. Thus the procedure must stop after a finite number m of steps, which is equivalent to the fact that $h^1\big(\mathcal{J}_{Z_m/\Sigma}\big(D - \sum_{i=1}^m \Delta_i\big)\big) = 0$.

Step 8. It remains to show (4.6.5.1).

By assumption, the curves Δ_i are nef, in particular $\Delta_i.\Delta_j \geq 0$ for all i, j. Thus, (c) implies

$$\sum_{i=1}^{m} \deg\left(Z_i^0\right) \geq \sum_{i=1}^{m} \left(D - K_\Sigma - \sum_{k=1}^{i} \Delta_k\right).\Delta_i$$
$$= (D - K_\Sigma).\sum_{i=1}^{m} \Delta_i - \frac{1}{2}\left(\left(\sum_{i=1}^{m} \Delta_i\right)^2 + \sum_{i=1}^{m} \Delta_i^2\right)$$
$$\geq (D - K_\Sigma).\sum_{i=1}^{m} \Delta_i - \left(\sum_{i=1}^{m} \Delta_i\right)^2 .$$

But then, taking condition (3) into account,

$$0 \leq \frac{(D - K_\Sigma)^2}{4} - \deg(Z_0) \leq \frac{(D - K_\Sigma)^2}{4} - \sum_{i=1}^{m} \deg\left(Z_i^0\right)$$
$$\leq \frac{(D - K_\Sigma)^2}{4} - (D - K_\Sigma).\left(\sum_{i=1}^{m} \Delta_i\right) + \left(\sum_{i=1}^{m} \Delta_i\right)^2$$
$$= \left(\frac{(D - K_\Sigma)}{2} - \sum_{i=1}^{m} \Delta_i\right)^2 .$$

□

It is our overall aim to compare the dimension of a cohomology group of the form $H^1\big(\mathcal{J}_{Z_0/\Sigma}(D)\big)$ with some invariants of the Z_i^0 and Δ_i. The following lemma will be vital for the necessary estimations.

Lemma 4.6.19 *Let a surface Σ and a divisor $D \subset \Sigma$ satisfy the hypotheses of Theorem 4.6.13. Moreover, let a zero-dimensional scheme $Z_0 \subset \Sigma$, the schemes $Z_i = Z_{i-1} : \Delta_i$, $i = 1, \ldots, m$, curves $\Delta_1, \ldots, \Delta_m \subset \Sigma$, and zero-dimensional schemes $Z_i^0 \subseteq Z_{i-1}$, $i = 1, \ldots, m$ satisfy the conditions (a)–(f) in Lemma 4.6.18. Then*

$$
\begin{aligned}
h^1\big(\mathcal{J}_{Z_0/\Sigma}(D)\big) &\leq \sum_{i=1}^{m} h^1\big(\mathcal{J}_{Z_{i-1}\cap\Delta_i/\Delta_i}\big(D-\Delta_1-\cdots\Delta_{i-1}\big)\big)\\
&\leq \sum_{i=1}^{m}\Big(1+\deg(Z_{i-1}\cap\Delta_i)-\deg\big(Z_i^0\big)\Big)\\
&\leq \sum_{i=1}^{m}\Big(\Delta_i\cdot\big(K_\Sigma+\Delta_1+\cdots+\Delta_i\big)+1\Big).
\end{aligned}
$$

Proof Throughout the proof we use the following notation

- $\mathcal{G}_i := \mathcal{J}_{Z_{i-1}\cap\Delta_i/\Delta_i}\big(D-\Delta_1-\cdots-\Delta_{i-1}\big),\ i=1,\ldots,m,$
- $\mathcal{G}_i^0 := \mathcal{J}_{Z_i^0/\Delta_i}\big(D-\Delta_1-\cdots-\Delta_{i-1}\big),\ i=1,\cdots,m,$
- $\mathcal{F}_i := \mathcal{J}_{Z_i/\Sigma}\big(D-\Delta_1-\cdots-\Delta_i\big),\ i=0,\ldots,m.$

Since $Z_{i+1}=Z_i:\Delta_{i+1}$ we have the short exact sequences

$$
0\longrightarrow\mathcal{F}_{i+1}\xrightarrow{\cdot\Delta_{i+1}}\mathcal{F}_i\longrightarrow\mathcal{G}_{i+1}\longrightarrow 0
$$

$i=0,\ldots,m-1$, and the corresponding long exact cohomology sequences

$$
\begin{aligned}
0\to H^0(\mathcal{F}_{i+1})\to H^0(\mathcal{F}_i)\to H^0(\mathcal{G}_{i+1})\to H^1(\mathcal{F}_{i+1})\to H^1(\mathcal{F}_i)\to\\
\to H^1(\mathcal{G}_{i+1})\to H^2(\mathcal{F}_{i+1})\to H^2(\mathcal{F}_i)\to H^2(\mathcal{G}_{i+1})=0
\end{aligned}
$$

Step 1. We prove that $h^1(\mathcal{F}_i)\leq h^1(\mathcal{G}_{i+1})+\cdots+h^1(\mathcal{G}_m)$ for $i=0,\ldots,m-1$.

We prove the claim by descending induction on i. From the above long exact cohomology sequence and (a) we deduce $h^1(\mathcal{F}_{m-1})\leq h^1(\mathcal{G}_m)$, which proves the case $i=m-1$.

Therefore, we may assume that $1\leq i\leq m-2$. Once more, from the long exact cohomology sequence we deduce

$$
\begin{aligned}
a &:= h^0(\mathcal{F}_{i+1})-h^0(\mathcal{F}_i)+h^0(\mathcal{G}_{i+1})\geq 0,\\
b &:= h^2(\mathcal{F}_{i+1})-h^2(\mathcal{F}_i)\geq 0,
\end{aligned}
$$

and, finally, applying the induction hypothesis,

$$
\begin{aligned}
h^1(\mathcal{F}_i) &= h^1(\mathcal{G}_{i+1})+h^1(\mathcal{F}_{i+1})-a-b \leq h^1(\mathcal{G}_{i+1})+h^1(\mathcal{F}_{i+1})\\
&\leq h^1(\mathcal{G}_{i+1})+\sum_{j=i+2}^{m}h^1(\mathcal{G}_j) = \sum_{j=i+1}^{m}h^1(\mathcal{G}_j).
\end{aligned}
$$

Step 2. $h^1(\mathcal{G}_i)=h^0(\mathcal{G}_i)-\chi\big(\mathcal{O}_{\Delta_i}\big(D-\sum_{k=1}^{i-1}\Delta_k\big)\big)+\deg(Z_{i-1}\cap\Delta_i).$

The result follows from the long exact cohomology sequence associated to $0 \to \mathcal{G}_i \to \mathcal{O}_{\Delta_i}\big(D - \sum_{k=1}^{i-1} \Delta_k\big) \to \mathcal{O}_{Z_{i-1}\cap\Delta_i/\Delta_i}\big(D - \sum_{k=1}^{i-1} \Delta_k\big) \to 0$.

Step 3. $h^0(\mathcal{G}_i^0) - \chi\big(\mathcal{O}_{\Delta_i}\big(D - \sum_{k=1}^{i-1} \Delta_k\big)\big) = h^1(\mathcal{G}_i^0) - \deg(Z_i^0)$.

This follows analogously, replacing Z_{i-1} by Z_i^0, since $Z_i^0 = Z_i^0 \cap \Delta_i$.

Step 4. $h^1(\mathcal{G}_i^0) \le h^1\big(\mathcal{J}_{Z_i^0/\Sigma}\big(D - \sum_{k=1}^{i-1} \Delta_k\big)\big) = 1$.

Note that $Z_i^0 : \Delta_i = \emptyset$, and, hence, $\mathcal{J}_{Z_i^0:\Delta_i/\Sigma} = \mathcal{O}_\Sigma$. We thus have the following short exact sequence

$$0 \to \mathcal{O}_\Sigma\left(D - \sum_{k=1}^{i} \Delta_k\right) \xrightarrow{\cdot\Delta_i} \mathcal{J}_{Z_i^0/\Sigma}\left(D - \sum_{k=1}^{i-1} \Delta_k\right) \to \mathcal{G}_i^0 \to 0\,. \qquad (4.6.5.5)$$

By assumption (f), the divisor $D - K_\Sigma - \sum_{k=1}^{i} \Delta_k$ is big and nef and, hence,

$$0 = h^0\left(\mathcal{O}_\Sigma\left(-D + K_\Sigma + \sum_{k=1}^{i} \Delta_k\right)\right) = h^2\left(\mathcal{O}_\Sigma\left(D - \sum_{k=1}^{i} \Delta_k\right)\right).$$

Thus, the long exact cohomology sequence of (4.6.5.5) gives the inequality $h^1(\mathcal{G}_i^0) \le h^1\big(\mathcal{J}_{Z_i^0/\Sigma}\big(D - \sum_{k=1}^{i-1} \Delta_k\big)\big)$. However, by assumption (b), the latter is just 1.

Step 5. $h^1(\mathcal{G}_i) \le 1 + \deg(Z_{i-1} \cap \Delta_i) - \deg\big(Z_i^0\big)$.

We note that $\mathcal{G}_i \hookrightarrow \mathcal{G}_i^0$, and, thus, $h^0(\mathcal{G}_i) \le h^0(\mathcal{G}_i^0)$. But then

$$\begin{aligned} h^1(\mathcal{G}_i) &\overset{\text{Step 2,3}}{\le} h^1(\mathcal{G}_i^0) - \deg\big(Z_i^0\big) + \deg(Z_{i-1} \cap \Delta_i) \\ &\overset{\text{Step 4}}{\le} 1 - \deg\big(Z_i^0\big) + \deg(Z_{i-1} \cap \Delta_i). \end{aligned}$$

Step 6. Completion of proof.

The first inequality follows from Step 1, while the second is a consequence of Step 5, and the last inequality follows from assumption (c). □

Lemma 4.6.20 *Let a surface Σ, a divisor $D \subset \Sigma$, and a zero-dimensional scheme $Z_0 \subset \Sigma$ satisfy the hypotheses of Theorem 4.6.13 and the conditions (1)–(3) from Lemma 4.6.18, and, additionally,*

$$(4)\ \sum_{z\in\Sigma} \big(\deg(Z_{0,z})\big)^2 < \frac{\big(1 + \sqrt{1 - 4\lambda}\big)^2 \cdot L^2}{4\chi(\mathcal{O}_\Sigma) + \max\{0, 2K_\Sigma.L\} + 6 \cdot L^2} \cdot (D - K_\Sigma)^2$$

Then, using the notation of Lemma 4.6.18 and setting $Z_S := \bigcup_{i=1}^{m} Z_i^0$,

$$h^1\big(\mathcal{J}_{Z_0/\Sigma}(D)\big) + \sum_{i=1}^{m} \Big(h^0\big(\mathcal{O}_\Sigma(\Delta_i)\big) - 1\Big) < \#Z_S\,,$$

where $\#Z_S$ is the number of points in the support of Z_S.

Proof We introduce the integers $d, \kappa, \delta_1, \ldots, \delta_m, \ell$ by setting

$$D \sim_a d \cdot L\,, \quad K_\Sigma \sim_a \kappa \cdot L\,, \quad \Delta_i \sim_a \delta_i \cdot L\,, \quad \ell := \sqrt{L^2} > 0.$$

Furthermore, the coefficient in the right-hand side of (4) can be written as $\alpha := \big(1 + \sqrt{1-4\lambda}\big)^2/4u$, where

$$u := \frac{4\chi(\mathcal{O}_\Sigma) + \max\{0, 2K_\Sigma.L\} + 6L^2}{4 \cdot L^2} = \begin{cases} \dfrac{\chi(\mathcal{O}_\Sigma)}{\ell^2} + \dfrac{\kappa+3}{2}, & \text{if } \kappa \geq 0, \\[2mm] \dfrac{\chi(\mathcal{O}_\Sigma)}{\ell^2} + \dfrac{3}{2}, & \text{if } \kappa < 0, \end{cases}$$

Now, (4.6.5.1) gives

$$\sum_{i=1}^{m} \delta_i \cdot \ell \leq \frac{(d-\kappa)\cdot \ell}{2} - \sqrt{\frac{(d-\kappa)^2 \cdot \ell^2}{4} - \deg(Z_S)}\,. \tag{4.6.5.6}$$

Moreover, by Lemma 4.6.19 we know:

$$h^1\big(\mathcal{J}_{Z_0/\Sigma}(D)\big) \leq \kappa \cdot \sum_{i=1}^{m} \delta_i \cdot \ell^2 + \frac{1}{2}\left(\Big(\sum_{i=1}^{m}\delta_i\Big)^2 + \sum_{i=1}^{m}\delta_i^2\right)\cdot \ell^2 + m.$$

Until now, we have not yet used the second assumption on the surface: $h^1(\mathcal{O}_\Sigma(C)) = 0$ for each effective divisor C. In particular, this holds for Δ_i, and Riemann–Roch gives

$$\begin{aligned} \sum_{i=1}^{m} \Big(h^0\big(\mathcal{O}_\Sigma(\Delta_i)\big) - 1\Big) &\leq -m + m\cdot\chi(\mathcal{O}_\Sigma) + \frac{1}{2}\sum_{i=1}^{m}\big(\Delta_i^2 - K_\Sigma.\Delta_i\big) \\ &= m \cdot \big(\chi(\mathcal{O}_\Sigma) - 1\big) + \frac{\ell^2}{2}\cdot\sum_{i=1}^{m}\delta_i^2 - \frac{\kappa\ell^2}{2}\cdot\sum_{i=1}^{m}\delta_i. \end{aligned} \tag{4.6.5.7}$$

Finally, using (4.6.5.6)–(4.6.5.7) and $\deg(Z_S) \leq \deg(Z_0) \leq \lambda \cdot (d-\kappa)^2 \cdot \ell^2$, we can conclude:

$$
\begin{aligned}
& h^1\big(\mathcal{J}_{Z_0/\Sigma}(D)\big) + \sum_{i=1}^{m} \Big(h^0\big(\mathcal{O}_\Sigma(\Delta_i)\big) - 1\Big) \\
& \leq m \cdot \chi(\mathcal{O}_\Sigma) + \ell^2 \cdot \sum_{i=1}^{m} \delta_i^2 + \frac{\kappa \ell^2}{2} \cdot \sum_{i=1}^{m} \delta_i + \frac{\ell^2}{2} \cdot \left(\sum_{i=1}^{m} \delta_i\right)^2 \\
& \leq u \cdot \left(\ell \cdot \sum_{i=1}^{m} \delta_i\right)^2 \leq u \cdot \left(\frac{(d-\kappa)\cdot\ell}{2} - \sqrt{\frac{(d-\kappa)^2\cdot\ell^2}{4} - \deg(Z_S)}\right)^2 \\
& \leq u \cdot \left(\frac{2\cdot\deg(Z_S))}{(d-\kappa)\cdot\ell + \sqrt{(d-\kappa)^2\cdot\ell^2 - 4\cdot\deg(Z_S)}}\right)^2 \\
& \leq \frac{4u}{\big(1+\sqrt{1-4\lambda}\big)^2 \cdot (d-\kappa)^2\cdot\ell^2} \cdot \big(\deg(Z_S)\big)^2 \\
& \leq \frac{\#Z_S}{\alpha\cdot(D-K_\Sigma)^2} \cdot \sum_{z\in\Sigma} \deg(Z_{S,z})^2 \overset{(4)}{<} \#Z_S.
\end{aligned}
$$

□

4.7 Open Problems and Conjectures

Though some results of the last chapter are sharp, others seem to be far from a final form, and here we start with a discussion and conjectures about the expected progress in the geometry of families of singular curves. Further discussion concerns possible generalizations of the methods and questions, stressing on links between the methods used in this monograph and further methods, or questions.

4.7.1 *Equisingular Families of Curves*

Quasi-projectivity of Equisingular Families of Curves. Theorem 2.2.32 states that the reduction of any equisingular family of hypersurfaces in a smooth projective variety is a quasi-projective subvariety of the corresponding Hilbert scheme. As pointed in Remark 2.2.32.1, we do not know whether the equisingular families themselves are quasi-projective, except for the case of families of plane curves with nodes and cusps [Wah2, Theorem 3.3.5]. We, however, expect that the quasi-projectivity always holds:

Conjecture 4.7.1 *Let Σ be a smooth projective variety and Hilb^h_Σ the Hilbert scheme parameterizing proper families of hypersurfaces on Σ with fixed Hilbert polynomial h. Moreover, let $\boldsymbol{S} = (S_1, ..., S_r)$ be a sequence of analytic or topological types (assuming that S_i are all analytic types if $\dim \Sigma > 2$). Then the* *locally closed analytic subspace* $V_h(\mathcal{S}) = V_h(S_1, ..., S_r) \subset \mathrm{Hilb}^h_\Sigma$ *representing* $\mathcal{S}$-*equisingular families is a quasi-projective algebraic subvariety of* Hilb^h_Σ.

Existence of Curves with Prescribed Singularities. A natural question about the existence results for algebraic curves, given in Sect. 4.3, is:

How to improve the constant coefficients in the general sufficient existence conditions (which seem to be far from the necessary ones)?

Concerning our method based on H^1-vanishing for the ideal sheaves of generic zero-dimensional schemes, a desired improvement would come from strengthening lower bounds for $\mathrm{reg}^{an}(Z)$, or proving the Harbourne-Hirschowitz Conjecture 3.4.21, the best possible H^1-vanishing criterion for ideal sheaves of schemes of generic fat points. We discuss this below in detail.

We formulate a few more questions about curves with specific singularities. The known sufficient and necessary conditions for the existence of singular plane curves look as bounds to sums of singularity invariants, which implies that existence is guaranteed for all possible values of singularity invariants if their sum is below the given bound. But this seems to be not true in general. The simplest question of such kind is:

Are there $k' < k$ *and* d *such that a curve of degree* d *with* k *cusps does exist, but with* k' *cusps does not?*

A candidate could be series of cuspidal curves in [Hi, Ku, CPS] (see details in Example 4.2.0.4).

A real structure imposes extra restrictions to singular algebraic curves, for example, the number of real cusps on a plane sextic does not exceed 7, whereas the number of complex cusps can be 9 [ItS2]. We conjecture that Theorem 4.5.4 (i) gives an asymptotically sharp lower bound to the maximal number of real cusps, that is,

Conjecture 4.7.2 *The number of real cusps on a real plane curve does not exceed* $d^2/4 + O(d)$.

T-Smoothness and Versality of Deformations. The following conjecture about the asymptotic properness of the sufficient conditions for the T-smoothness of topological ESF of plane curves established in Sect. 4.3, seems to be quite realistic.

Conjecture 4.7.3 *There exists an absolute constant* $A > 0$ *such that for any topological singularity* S *there are infinitely many pairs* $(r, d) \in \mathbb{N}^2$ *such that* $V^{irr}_d(r \cdot S)$ *is empty or non-smooth or has dimension greater than the expected one and* $r \cdot \gamma(S) \le A \cdot d^2$.

In fact, the examples of Sects. 4.3.3 and 4.6.3 show that the sufficient bounds for T-smoothness in Sect. 4.3 are asymptotically proper for simple and ordinary singularities, i.e. the conjecture holds for S being one of these singularities.

We propose a similar conjecture for analytic ESF of plane curves, though it is confirmed only for simple singularities (in which case it coincides with the conjecture for topological ESF).

A closely related question, belonging to local singularity theory, concerns the γ-invariant:

Problem 4.7.4 Find an explicit formula, or algorithm to compute $\gamma(Z^{es})$, $\gamma(Z^{ea})$. Find (asymptotically) proper lower and upper bounds for these invariants. Is $\gamma(Z^{es})$ a topological invariant?

Irreducibility Problem. Our sufficient irreducibility conditions seem to be far from optimal. We state the problem:

Problem 4.7.5 Find asymptotically proper sufficient conditions for the irreducibility of ESF of plane curves, or show that the conditions in Sect. 4.6 are asymptotically proper.
A similar question concerns the results of Sect. 4.6.5.

We also raise the following important question:

Question 4.7.6 *Does there exist a pair (C, D) of plane irreducible algebraic curves of the same degree with the same collection of singularities such that C and D belong to different components of an equisingular stratum but are topologically isotopic in $\mathbb{P}^2$ (anti-Zariski pair)?*

The examples in Sect. 4.6 provide candidates for this. They are reducible ESF, whose members have the same (abelian) fundamental group of the complement.

Positive Characteristic. The questions about existence, T-smoothness and expected dimension, and irreducibility make perfectly sense for equisingular and equianalytic families of reduced plane curves resp. hypersurfaces with isolated singularities defined over an algebraically closed field of positive characteristic. However, very little is known in this case. We just mention here several results that can be regarded as first steps towards a systematic theory of equisingular families of curves and hypersurfaces with isolated singularities in positive characteristic.

The local theory of equisingular deformations of an algebriod plane curve singularity in arbitrary characteristic has been fully developed by Campillo, Greuel and Lossen in [CGL1, CGL2]. In good characteristic, that is, when the characteristic is 0 or does not divide the multiplicity of any branch of the singularity, the results are analogous to the complex case. However, in bad characteristic one has to distinguish between (strongly) equisingular deformations, as in our treatment, and weakly equisingular deformations, which are those that become equisingular after a finite, dominant base change. Similarly one has to distinguish equianalytic and weakly equianalytic deformations. Strongly es-deformations are unobstructed and have a smooth semiuniversal base space in any characteristic, but for an arbitray family an es-stratum (like the μ-constant stratum over $\mathbb{C}$) may not exist in bad characteristic. On the other hand, in bad characteristic a weak es-stratum always exists but it is in general not smooth (it becomes smooth after a finite base change).

Problem 4.7.7 Study equisingular and equianalytic families of plane projective curves and hypersurfaces in positive characteristic. Find asymptotically proper bounds for existence, T-smoothness and expected dimension, and irreducibility for such families.

We expect that, in good characteristic, such families can be constructed as quasiprojective subvarieties of some Hilbert scheme of zero-dimensional schemes, similar as in this book. However, due to a lack of some H^1-vanishing theorems in positive characteristic, the numerical bounds may be weaker. We note also that for our existence bounds we used finite determinacy bounds for the singularities. In positive characteristic such bounds exist, but they are in general bigger [BoGM, GrPh]. Concerning ESF the situation in bad characteristic is, however, unclear due to the non-existence of equisingular or equianalytic strata. One has to consider probably weakly equisingular and weakly equianalytic families.

The variety of reduced projective curves in $\mathbb{P}^2$ of given degree and genus is usually called the *Severi variety*. Over the complex numbers it is the closure of the variety of nodal curves [Zar2], which is irreducible by [HaJ]. The same holds for Severi varieties for Hirzebruch surfaces [Nob1, Tyo1]. Furthermore, Tyomkin [Tyo3, Tyo4] has extended the Zariski characterization of Severi varieties on toric surfaces to the case of arbitrary characteristic, but he also exhibited examples of Severi varieties parameterizing non-nodal curves and of reducible Severi varieties whose components parameterize curves with different singularities. A general picture of Severi varieties in positive characteristic is still open.

4.7.2 Related Problems

Hypersurfaces in Higher-Dimensional Varieties. One can formulate the existence, T-smoothness, and irreducibility problems for families of hypersurfaces with isolated singularities, belonging to (very) ample linear systems of projective algebraic varieties of arbitrary dimension. To find a relevant approach to these problems is the most important question. Constructions of curves with prescribed singularities as presented in Sect. 4.5 can, in principle, be generalized to higher dimensions. An expected analogue of the results for curves could be

Conjecture 4.7.8 *Given a very ample linear system $|W|$ on a projective algebraic variety X of dimension n. There exists a constant $A = A(X, W) > 0$ such that, for any collection $S_1, \dots, S_r$ of singularity types and positive integer d satisfying $\sum_{i=1}^r \mu(S_i) < Ad^n$, there is a hypersurface $W_d \in |dW|$ with exactly r isolated singular points of types $S_1, \dots, S_r$, respectively.*

In view of Proposition 4.5.11, to prove the conjecture it is enough to consider the case of ordinary multiple points and to answer the following analogue of one of the above questions in the affirmative:

Question 4.7.9 *Does there exists an $A = A(n) > 0$ such that for any analytic type S of isolated singular points of hypersurfaces in $\mathbb{P}^n$, there exists a hypersurface of degree $d \le A(\mu(S))^{1/n}$, which has a singular point of type S and no other singularities?*

This is known only for simple singularities [Wes] (see also [Shu13]).

Hypersurfaces with specific singularities, for example nodes, attracted attention of many researches, mainly looking for the maximal possible number of singularities on hypersurfaces of a given degree (see, for instance, [Chm]). We would like to raise the question of an analogue of the Chiantini-Ciliberto theorem for nodal curves on surfaces as a natural counterpart, concerning the domain with *regular* behavior of ESF:

Problem 4.7.10 Given a smooth projective algebraic variety X and a very ample linear system $|W|$ on it with a non-singular generic member. Prove that, for any $n \le \dim |W|$ there exists $W_n \in |W|$ with n nodes as its only singularities such that the germ of the ESF $V_{X,W}(n \cdot A_1)$ is T-smooth at W_n.

Zero-Dimensional Schemes. It is interesting to find the right analogue of the Harbourne-Hirschowitz Conjecture 3.4.21 for zero-dimensional cluster schemes which naturally generalize schemes of "fat" points. We suggest the following version of the Harbourne-Hirschowitz conjecture:

Conjecture 4.7.11 *Given clusters $(\mathcal{K}_1, \boldsymbol{m}_1), \ldots, (\mathcal{K}_r, \boldsymbol{m}_r)$. Prove that*

$$h^1\big(\mathcal{J}_{Z/\mathbb{P}^2}(d)\big) = 0\,, \qquad (4.7.2.1)$$

where $Z = Z_1 \cup \ldots \cup Z_r$ is a zero-dimensional scheme supported at r generic points of $\mathbb{P}^2$ with Z_i a generic member of the Hilbert scheme $\mathrm{Hilb}(\mathcal{K}_i, \boldsymbol{m}_i)$, $i = 1, \ldots, r$, and

$$\deg Z \le \binom{d+2}{2}, \qquad \max_{z_1,z_2,z_3} \left(\mathrm{mt}(Z, z_1) + \mathrm{mt}(Z, z_2) + \mathrm{mt}(Z, z_3)\right) \le d\,,$$

where z_1, z_2, z_3 run over all triples of distinct closed and infinitely near points determined by the scheme Z.

For an arbitrary zero-dimensional scheme $Z \subset \mathbb{P}^2$, the condition $\deg Z \le (d+1)(d+2)/2 = h^0(\mathcal{O}_{\mathbb{P}^2}(d))$ is clearly necessary for (4.7.2.1), but not sufficient even under a generality assumption. We propose the following conjectural statement, which, for schemes of fat points is known as the Alexander-Hirschowitz theorem [AH2]:

Conjecture 4.7.12 *For any m there exists $N(m)$ such that (4.7.2.1) holds for any zero-dimensional scheme Z, which is generic in its isomorphism class and satisfies*

$$\deg Z \leq \binom{d+2}{2}, \qquad \max_i \deg Z_i \leq m, \quad d \geq N(m)\,,$$

where Z_i runs over all irreducible components of Z.

For many purposes, like the existence problem for singular curves, the upper bound to the degree of irreducible components is too restrictive, and we suggest the following conjectural assertions (in the notation of Sect. 3.6).

Conjecture 4.7.13 *(1) There is an absolute constant $k \geq 0$ such that any zero-dimensional scheme $Z \subset \mathbb{P}^2$, generic in its isomorphism class and of degree $\deg Z \geq k$, satisfies*

$$\mathrm{reg}^{an}(Z) \leq \sqrt{M_2(Z)} + \frac{\deg Z}{\sqrt{M_2(Z)}} - 2. \tag{4.7.2.2}$$

(2) Any zero-dimensional scheme $Z \subset \mathbb{P}^2$, generic in its isomorphism class, satisfies

$$\mathrm{ord}^{an}(Z) \leq 2\sqrt{\deg Z} + O(1)\,. \tag{4.7.2.3}$$

In view of $M_2(Z) < 2\deg Z$, (4.7.2.2) yields $\mathrm{reg}^{an}(Z) \leq (3/\sqrt{2})\sqrt{\deg Z} - 2$, which is worse than (4.7.2.3). The statement of Corollary 3.6.4 yields

$$\mathrm{reg}^{an}(Z) \leq \frac{11}{2\sqrt{6}}\sqrt{\deg Z} - 2 = 2.245...\sqrt{\deg Z} - 2\,,$$

which is not far from (4.7.2.3). On the other hand, the coefficient 2 in (4.7.2.3) cannot be diminished, what one can see in an example of Z consisting of two fat points of multiplicity $m \gg 0$, which is always generic and satisfies $\deg Z = m(m+1)$, $\mathrm{reg}^{an}(Z) = 2m - 1 > 2\sqrt{\deg Z} - 2$.

An evidence for this conjecture in the case of A-D-E singularity schemes is provided by [Roe1, Theorem 3].

It would be interesting to find an analogue of the above conjectures for zero-dimensional schemes in projective spaces of arbitrary dimension, especially for Conjecture 4.7.13, which potentially may lead to asymptotically proper existence results for hypersurfaces with isolated singularities in higher-dimensional projective spaces. The ideas and results of [BGM] about zero-dimensional schemes in projective spaces of any dimension can be used for this purpose.

The Harbourne-Hirschowitz conjecture is directly related to the Nagata conjecture:

Conjecture 4.7.14 (Nagata) *If a plane curve of degree d has (non-negative) multiplicities $m_1, \ldots, m_k$ at $k > 9$ generic points of $\mathbb{P}^2$ then*

$$\frac{m_1 + \cdots + m_k}{d} < \sqrt{k}\,. \tag{4.7.2.4}$$

Nagata proved this for $k = n^2, n \geq 4$, [Nag1, Nag2], which was the key point of his construction of counterexamples to Hilbert's 14th problem. For other values of k it is sufficient to establish a non-strict inequality in (4.7.2.4).

We shall explain the well-known fact that Nagata's conjecture follows from the Harbourne-Hirschowitz one. Denote by Z the zero-dimensional scheme of k generic fat points of multiplicities $m_1, \dots, m_k$, respectively. Assume that there exists an irreducible curve of degree d passing through Z, and such that

$$\frac{m_1 + \cdots + m_k}{d} > \sqrt{k}\,.$$

Clearly, $d > 3$. Thanks to the generality of the position of fat points, for any permutation of multiplicities we again have an irreducible curve of degree d with interchanged multiplicities at the given points. Hence, we can take $m_1 = \cdots = m_k = m$ and suppose $km^2 > d^2$. By Bézout's theorem applied to the intersection of this curve C with a cubic through 9 of the given points, we have $3m \leq d$, which is one of the conditions of the Harbourne-Hirschowitz conjecture. Choose any natural number s, take the s-multiple of C (i.e. given by F^s if C is defined by F) and consider the minimal d' such that

$$\frac{(sd + d' + 1)(sd + d' + 2)}{2} \geq k\frac{sm(sm + 1)}{2}\,.$$

One can easily compute that $d' = s(\sqrt{k}m - d) + O(1)$ and

$$\frac{(sd + d' + 1)(sd + d' + 2)}{2} - k\frac{sm(sm + 1)}{2} \leq s\sqrt{k}m + O(1)\,.$$

By the Harbourne-Hirschowitz conjecture, the left-hand side in the latter inequality must be $h^0\big(\mathcal{J}_{Z/\mathbb{P}^2}(sd + d')\big)$ for the scheme of k generic fat points of multiplicity sm. On the other hand,

$$H^0\big(\mathcal{J}_{Z/\mathbb{P}^2}(sd + d')\big) \supset F^s \cdot H^0\big(\mathcal{O}_{\mathbb{P}^2}(d')\big)\,,$$

that is,

$$h^0\big(\mathcal{J}_{Z/\mathbb{P}^2}(sd + d')\big) \geq \frac{(d' + 1)(d' + 2)}{2} = s^2(\sqrt{k}m - d)^2 + O(s)\,,$$

contradicting the above upper bound for sufficiently large s.

A natural analogue of Nagata's conjecture can be proposed for cluster schemes:

Conjecture 4.7.15 *If a curve of degree d contains a cluster scheme Z, which has $k > 9$ closed and infinitely near points and is generic in its Hilbert scheme, then $\sum_i m_i < d\sqrt{k}$, where m_i runs over the multiplicities of all closed and infinitely near points of Z.*

Some other extensions of the Nagata conjecture can be found in [DHKRS] and [CHMR].

Non-Isolated Singularities. None of the problems discussed in this book is even well-stated for non-reduced curves, or hypersurfaces with non-isolated singularities. We simply indicate this case as a direction for further study.

4.7.3 Enumeration of Singular Curves

The global geometry of equisingular families of curves besides the non-emptiness, smoothness and irreducibility problems, discussed in this book, naturally includes the following question.

Given a smooth algebraic surface Σ, a (very) ample divisor $D \subset \Sigma$, and a collection of topological or analytic singularity types $S_1, ..., S_r$. What is $\deg V^{irr}_{|D|}(S_1, \ldots, S_r)$ *(provided that $V^{irr}_{|D|}(S_1, \ldots, S_r)$ is non-empty, irreducible, of expected dimension)? In the enumerative setting, the question can be formulated as: Given a configuration w of* $\dim V^{irr}_{|D|}(S_1, \ldots, S_r)$ *distinct points of Σ in generic position, what is the number of curves $C \in V^{irr}_{|D|}(S_1, \ldots, S_r)$ passing through w?*

The question makes sense if $V^{irr}_{|D|}(S_1, \ldots, S_r)$ is not irreducible, but is of the pure expected dimension. Even if $V^{irr}_{|D|}(S_1, \ldots, S_r)$ fails to be like that, a reasonable question is: *what is the degree of the union of the irreducible components of $V^{irr}_{|D|}(S_1, \ldots, S_r)$ having expected dimension?*

Here we shortly review the main ideas of modern enumerative geometry and we indicate some problems, in which the geometry of equisingular families of curves can be applied.

(1) Enumeration of rational curves in a given second homology class on an algebraic variety appears to be of great importance in the theory of Gromov-Witten invariants, quantum cohomology, mirror symmetry, and string theory (see, for example, [FuP, Get, GoP, KoM, RuT, Vak]). This problem has been solved in several cases, including rational surfaces using the moduli spaces of stable maps of marked rational curves, more precisely, by means of certain relations in the Picard group of these moduli spaces. The Gromov-Witten invariants (when they have an enumerative meaning) are related to enumeration of smooth or nodal curves (see a more detailed discussion in Sect. 4.5.6.3). A careful study of the Picard group of the moduli spaces of stable maps has led to the enumeration of plane rational curves of a given degree having nodes and one extra singularity, either an ordinary cusp, or a tacnode, or an ordinary triple point, see [Pan, Zin]. Possible extensions to positive genera are limited by the fact that the moduli spaces of stable maps of curves of a positive genus reveal pathological geometric behavior, for instance, have components of unexpected dimension.

So, consider for simplicity plane rational curves. Fix $d \geq 1$ and $n \geq 0$. Then (see, [FuP, Theorem 2]) the moduli space $\overline{\mathcal{M}}_{0,n}(\mathbb{P}^2, d)$ is a projective variety of expected dimension $3d - 1 + n$. Let $S_1, ..., S_r$ be topological singularity types different from nodes and satisfying relation (4.3.1.4) in Corollary 4.3.4 or (4.3.1.5) in Corollary 4.3.5. Then the subspace

$$\overline{\mathcal{M}}_{0,n}(\mathbb{P}^2, d, S_1, ..., S_r) \subset \overline{\mathcal{M}}_{0,n}(\mathbb{P}^2, d)$$

defined as the closure of the set of elements $[\rho : \mathbb{P}^1 \to \mathbb{P}^2, (p_1, ..., p_n)] \in \mathcal{M}_{0,n}(\mathbb{P}^2, d)$ such that $\rho(\mathbb{P}^1) \in V_d^{irr}(S_1, ..., S_r, A_1, ..., A_1)$ (where A_1 is repeated $(d-1)(d-2)/2 - \sum_{i=1}^r \delta(S_i)$ times), is either empty or a (possibly reducible) subvariety of expected dimension $3d - 1 + n - \sum_{i=1}^r \tau^{es}(S_i)$. For example, if $S_1 = ... = S_r = A_2$ (an ordinary cusp), then the required restriction is $r < 3d$ and the expected dimension $\dim \overline{\mathcal{M}}_{0,n}(\mathbb{P}^2, d, r \times A_2) = 3d - 1 + n - r$. Thus, we pose

Question 4.7.16 *Under the above conditions, is $\overline{\mathcal{M}}_{0,n}(\mathbb{P}^2, d, S_1, ..., S_r)$ expressible via intersections of divisors as an element of the Chow ring $A^*\overline{\mathcal{M}}_{0,n}(\mathbb{P}^2, d) \otimes \mathbb{Q}$? If so, can such an expression be effectively computed?*

(2) Another approach assumes enumeration of singular curves in a "sufficiently ample" linear system. So, the works related to the Göttsche conjecture (see [Goet, GoS1, GoS2]) provide a series of universal polynomials in $\delta \geq 0$ counting irreducible curves with δ nodes belonging to the linear system $|D|$ on a smooth algebraic surface Σ, and the universality means that the coefficients of these polynomials depend only on D^2, DK_Σ, and K_Σ^2. The polynomials give correct enumerative answers under the condition that D is sufficiently ample with respect to δ. The approach heavily relies on the geometry of the Hilbert scheme $\mathrm{Hilb}_\Sigma^\delta$, and for the validity of the enumerative formulas it is necessary that the natural map $V_{|D|,\Sigma}^{irr}(\delta \times A_1) \to \mathrm{Hilb}_\Sigma^\delta$ is dominant.

Universal formulas for curves with more complicated singularities have been suggested in [FeR, Kaz, Liu] (see also [BM1, BM2, Ker, KlP, Vai]). The application of this method is strictly limited by the necessity to know all possible deformations of the considered singularities, and explicit enumerative results are restricted to singularities with total Milnor number ≤ 8. Furthermore, the validity of these enumerative answers is conditioned by the requirement that the linear system must be sufficiently ample with respect to the considered collection of singularities. In particular, good geometric properties (expected dimension, irreducibility) of the equisingular family are not sufficient for getting a correct answer via universal formulas (see some examples and discussion in [Ker]).

So, we pose the problem to describe the applicability range of the above universal enumerative formulas:

Problem 4.7.17 Given a smooth algebraic surface Σ. Under what conditions on a (very) ample divisor $D \subset \Sigma$ and topological singularity types $S_1, ..., S_r$, $\deg V_{|D|}^{irr}(S_1, \ldots, S_r)$ is given by the corresponding universal formula? More specifically, given a smooth algebraic surface Σ, a (very) ample divisor $D \subset \Sigma$, and topological singularity types $S_1, ..., S_r$, find $d_0 \geq 0$ such that $\deg V_{|dD|}^{irr}(S_1, ..., S_r)$ coincides with some polynomial in d as far as $d \geq d_0$.

(3) The Caporaso-Harris formula [CaH2] allows one to enumerate nodal plane curves of any degree and genus. Similar formulas were obtained by Ran [Ran1, Ran4]. Extensions to arbitrary del Pezzo surfaces were elaborated in [Vak, ShSh, Br]. All these results are related only to nodal curves, though they are free of any genus or degree restrictions. The geometric background of this approach is that the counted curves are specialized so that they split off a fixed line (or, more generally, a smooth divisor). A recursive formula arising in this way requires a careful analysis of degenerations as well as deformations of degenerate curves. A crucial step in this analysis is the computation of the dimension of relevant equisingular families of curves. This seems to be hardly possible in presence of complicated singularities. We, however, think that the following problem is accessible with the techniques developed in this book:

Problem 4.7.18 Derive a Caporaso-Harris type recursive formula enumerating plane curves of any degree and genus having nodes and one or several ordinary cusps.

The problem of enumeration of plane curves of a positive genus with one cusp and any number of nodes still remains open.

(4) Tropical enumerative geometry has led to a number of impressive results in counting curves in toric varieties (see, for example, [Mik2, Mik3, NiS]). Up to now the curves counted in this way have been smooth or nodal. An attempt to count non-nodal curves in a toric surface via tropical geometry has been committed in [GaS] (curves with an arbitrary number of nodes and one cusp). A further advance in enumeration of non-nodal curves in the plane or in another toric surface requires that certain specific equisingular families of curves on toric surfaces have the expected dimension. This is important both, in the analysis of tropical limits, and in the patchworking part of the correspondence between algebraic and tropical curves (see details in Sect. 2.4.5). Namely, Let $\Delta \subset \mathbb{R}^2$ be a nondegenerate lattice polygon, $\Sigma = \mathrm{Tor}(\Delta)$ a toric surface associated with Δ, $\mathcal{L}_\Delta$ the tautological invertible sheaf on Σ, and $D = c_1(\mathcal{L}_\Delta) \in \mathrm{Pic}(\Sigma)$. Denote by s the number of edges of Δ and put $p_a(\Delta) = |\,\mathrm{Int}(\Delta) \cap \mathbb{Z}^2|$. Given an integer $0 \le g \le p_a(\Delta)$ and topological singularity types $S_1, \dots, S_r$, consider the family V of irreducible curves $C \in |D|$ of genus g having r singular points of types $S_1, \dots, S_r$, respectively, and possibly some ordinary nodes, and intersecting each toric divisor of Σ in one point.

Question 4.7.19 *In the above notations, suppose that the expected dimension $s - 1 + g - \sum_{i=1}^{r}(\tau^{es}(S_i) - \delta(S_i))$ of the family V is non-negative. Is the family V T-smooth?*

One should not expect that this is so in general. However, one can easily derive from Proposition 4.5.2 that the answer is affirmative if either $r = 1$ and $S_1 = A_2$ or A_3, or $r = 2$ and $S_1 = S_2 = A_2$. This means, in particular, that the current techniques of tropical geometry allows one to enumerate plane curves of any degree and genus with one or two cusps, or with one tacnode (and any number of nodes). We think that one can give an affirmative answer to Question 4.7.19 for some other collections of singularities (cf. examples in [KuS, Theorem 4.1]).

4.8 Historical Notes and References

Before commenting on specific problems and results we remark that the keynote of this monograph, the concept of asymptotic optimality and asymptotic properness as a relevant measure for results in the geometry of equisingular families of curves has been stated as a research program in [GrS, GLS3] and consistently pursued by the authors in a series of joint papers.

It has already be noticed in Sect. 3.7 that most of the specific H^1-vanishing criteria presented in Chap. 3 were invented for the problem of T-smoothness of equisingular families of curves on an algebraic surfaces and the problem of independence of simultaneous deformations of isolated singularities of a given curve that varies in a certain linear system on the surface. Resembling somehow the story presented in Sect. 3.7, we stress here on other details that are relevant to the contents of this chapter.

The foundational work by Severi [Sev] settled the case of plane nodal curves and indicated the main ideas of deformation theory related to the geometry of families of singular curves. Later, Segre [Seg1, Seg2] and Zariski [Zar3] extended this up to the "$3d$-condition" for plane curves with nodes and cusps. Brusotti [Bru] extended Severi's theorem to real plane nodal curves[10] Gudkov rediscovered the $3d$-condition for plane curves with nodes and cusps and described all deformations of real nodal-cuspidal curves, and also extended the $3d$-condition to the case of curves with nodes, cusps and ordinary multiple points [GuU] (first, published in Russian in 1962). Linear sufficient conditions (in particular the $3d$-condition) for the T-smoothness of families of curves of arbitrary analytic types and any algebraic surface together with the independence of versal deformations of singularities were found in [GrK]. Independently, linear sufficient conditions for families of plane curves with arbitrary topological singularities were suggested in [Shu4]. The "$(4d-4)$-condition" was discovered in [Shu1]. All these results were unified and strengthened in [GrL1]. The fact that the T-smoothness of families of curves with given analytic singularities and the independence of simultaneous versal deformations of singular points are equivalent was understood from the very beginning (cf. [GrK]). On the other hand, in [Shu12] the T-smoothness of topological equisingular families was associated with the independence of "lower deformations", that is, deformations locally induced by monomials under the Newton diagram. The first quadratic sufficient condition for the T-smoothness was announced in [Shu3], later published with proofs [Shu6, Shu8]. About the same time, independently and with a rather different techniques, quadratic conditions for the T-smoothness (and irreducibility) of families of plane nodal-cuspidal curves were found by Barkats [Bar].

Sporadic examples of non-T-smooth families of plane singular curves were discovered by Zariski [Zar3], Wahl [Wah2], and Luengo [Lue1]. Later, series of new examples were found as consequence of the failure of the corresponding H^1-vanishing condition. In this connection we mention the authors' examples from

[10] Note that, over the reals, the deformation problem is much more delicate than over the complex field.

[GLS4], which proved the sharpness of the classical sufficient $3d$-condition for T-smoothness of families of plane nodal-cuspidal curves, and the examples by du Plessis and Wall [DPW2] that proved the sharpness of the $(4d-4)$-condition (both for plane curves and for higher-dimensional projective hypersurfaces). We mention also the work [GG] showing that, in the above examples, the linear system of hypersurfaces of a given degree induces not a versal but a complete joint deformation of all singular points.

The question whether an equisingular family of curves is irreducible is usually harder than that on the T-smoothness. So, the irreducibility of Severi varieties for the plane was proved by Harris [HaJ] in 1985, after more than 60 years after it was announced by Severi [Sev]. This approach requiring a very restrictive independence of deformations of all singular points. It was only extended to some families of plane nodal-cuspidal curves by Kang [Kan1, Kan2] and to Severi varieties for Hirzebruch surfaces by Tyomkin [Tyo1]. More complicated singularities seem to be hardly accessible with this technique. Another idea to prove irreducibility of an equisingular family is to represent it as a fibration over an irreducible base with equidimensional fibres being Zariski open subsets of projective spaces. This is applied to curves with arbitrary singularities and is much in the spirit of the approach to the preceding problems, since the equidimensionality of the fibres again reduces to certain H^1-vanishing conditions. For families of plane curves with arbitrary topological singularities is was, first, realized in [Shu4] with an outcome in the form of $3d$-type sufficient conditions for the irreducibility. It was first independently realized by Barkats [Bar] and Shustin [Shu6] for families of plane nodal-cuspidal curves with quadratic sufficient irreducibility conditions, obtained with different techniques. The authors elaborated this idea in a series of works [GLS2, GLS4, GLS5] finally coming up with asymptotically proper sufficient irreducibility conditions for curves with arbitrary singularities as well as with series of examples of reducible equisingular families. Besides application and refinement of various H^1-vanishing conditions, an important task was to define appropriate zero-dimensional singularity schemes and to study the geometry of the corresponding Hilbert schemes as bases of fibrations for equisingular families [GLS3, GLS4].

The problem of existence of curves with prescribed singularities admits an extensive list of restrictions as well as numerous specific constructions. Severi's theorem [Sev] on the existence of plane curves of any degree with any number of nodes respecting the genus bound appeared to be the first general existence result and, in fact, is the only known *complete* answer. Shustin [Shu5] exhibited the first existence result for plane nodal-cuspidal curves covering an asymptotically proper range of the total Milnor number, and the idea was to combine the Viro patchworking construction with a suitable deformation theory for singular curves on toric surfaces. Developing and refining this idea, the authors proved existence results for plane curves with arbitrary topological [GLS3] and analytic [Shu13] singularities, covering an asymptotically proper range for the total Milnor number.

References

[GrK] Greuel, G.-M., Karras, U.: Families of varieties with prescribed singularities. Compos. Math. **69**, 83–110 (1989)

[GrL2] Greuel, G.-M., Lossen, C.: The geometry of families of singular curves. In: Siersma, D., et al. (eds.) New Developments in Singularity Theory. Proceedings of the NATO Advanced Study Institute on New Developments in Singularity Theory. Cambridge University, Cambridge (2000) (NATO Science Series II, Mathematics, Physics and Chemistry, vol. 21, pp. 159–192. Kluwer, Dordrecht, 2001)

[GLS1] Greuel, G.-M., Lossen, C., Shustin, E.: New asymptotics in the geometry of equisingular families of curves. Int. Math. Res. Not. **13**, 595–611 (1997)

[GLS2] Greuel, G.-M., Lossen, C., Shustin, E.: Geometry of families of nodal curves on the blown-up projective plane. Trans. Am. Math. Soc. **350**, 251–274 (1998)

[GLS3] Greuel, G.-M., Lossen, C., Shustin, E.: Plane curves of minimal degree with prescribed singularities. Invent. Math. **133**, 539–580 (1998)

[GLS4] Greuel, G.-M., Lossen, C., Shustin, E.: Castelnuovo function, zero-dimensional schemes and singular plane curves. J. Algebr. Geom. **9**(4), 663–710 (2000)

[GLS5] Greuel, G.-M., Lossen, C., Shustin, E.: The variety of plane curves with ordinary singularities is not irreducible. Int. Math. Res. Not. **11**, 543–550 (2001)

[Shu10] Shustin, E.: Smoothness of equisingular families of plane algebraic curves. Int. Math. Res. Not. **2**, 67–82 (1997)

[Shu4] Shustin, E.: On manifolds of singular algebraic curves. Selecta Math. Sov. **10**, 27–37 (1991)

[Shu5] Shustin, E.: Real plane algebraic curves with prescribed singularities. Topology **32**, 845–856 (1993)

[Shu13] Shustin, E.: Analytic order of singular and critical points. Trans. Am. Math. Soc. **356**, 953–985 (2004)

[ShT2] Shustin, E., Tyomkin, I.: Versal deformations of algebraic hypersurfaces with isolated singularities. Math. Annalen **313**(2), 297–314 (1999)

[ShW] Shustin, E., Westenberger, E.: Projective hypersurfaces with many singularities of prescribed types. J. Lond. Math. Soc. (2) **70**(3), 609–624 (2004)

[GrL1] Greuel, G.-M., Lossen, C.: Equianalytic and equisingular families of curves on surfaces. Manuscr. math. **91**, 323–342 (1996)

[Shu1] Shustin, E.: Versal deformation in the space of plane curves of fixed degree. Funct. Anal. Appl. **21**, 82–84 (1987)

[HiH] Hironaka, H.: On the arithmetic genera and the effective genera of algebraic curves. Mem. Coll. Sci. Univ. Kyoto **30**, 177–195 (1957)

[Sev] Severi, F.: Vorlesungen über algebraische Geometrie. Teubner (1921), respectively Johnson (1968)

[Seg1] Segre, B.: Dei sistemi lineari tangenti ad un qualunque sistema di forme. Atti Acad. naz. Lincei Rendiconti serie **5**(33), 182–185 (1924)

[Seg2] Segre, B.: Esistenza e dimensione di sistemi continui di curve piane algebriche con dati caraterri. Atti Acad. naz. Lincei Rendiconti serie **6**(10), 31–38 (1929)

[Tan2] Tannenbaum, A.: Families of curves with nodes on $K3$- surfaces. Math. Ann. **260**, 239–253 (1982)

[Vas] Vassiliev, V.A.: Stable cohomology of complements to the discriminants of deformations of singularities of smooth functions. J. Sov. Math. **52**, 3217–3230 (1990)

[Lef] Lefschetz, S.: On the existence of loci with given singularities. Trans. AMS **14**, 23–41 (1913)

[Sak] Sakai, F.: Singularities of plane curves. Geometry of Complex Projective Varieties, Seminars and Conferences, vol. 9, pp. 257–273. Mediterranean Press, Rende (1993)

[HiF] Hirzebruch, F.: Singularities of algebraic surfaces and characteristic numbers. Contemp. Math. **58**, 141–155 (1986)

[Ivi] Ivinskis, K.: Normale Flächen und die Miyaoka-Kobayashi-Ungleichung. University of Bonn, Diplomarbeit (1985)

[Lan] Langer, A.: Logarithmic orbifold Euler numbers of surfaces with applications. Proc. Lond. Math. Soc. **3**(86), 358–396 (2003)

[Zar3] Zariski, O.: Algebraic Surfaces, 2nd edn. Springer, Berlin (1971)

[Var2] Varchenko, A.N.: Asymptotics of integrals and Hodge structures. Mod. Probl. Math. **22** (Itogi nauki i tekhniki VINITI), 130–166 (Russian) (1983)

[Var3] Varchenko, A.N.: On semicontinuity of the spectrum and an upper estimate for the number of singular points of a projective hypersurface. Sov. Math. Dokl. **27**, 735–739 (1983); translation from Dokl. Akad. Nauk SSSR **270**, 1294–1297 (1983)

[Hi] Hirano, A.: Constructions of plane curves with cusps. Saitama Math. J. **10**, 21–24 (1992)

[Koe] Koelman, R.J.: Over de cusp. University of Leiden, Diplomarbeit (1986)

[Shu11] Shustin, E.: Gluing of singular and critical points. Topology **37**(1), 195–217 (1998)

[DGPS] Decker, W., Greuel, G.-M., Pfister, G., Schönemann, H.: Singular 3-1-6 — A computer algebra system for polynomial computations (2012). http://www.singular.uni-kl.de

[Schu] Schulze, M.: A singular 3-1-6 library for computing invariants related to the Gauss-Manin system of an isolated hypersurface singularity (2012) (gmssing.lib.)

[GrM] Gradolato, M.A., Mezzetti, E.: Families of curves with ordinary singular points on regular surfaces. Ann. mat. pura et appl. **150**, 281–298 (1988)

[Wal1] Wall, C.T.C.: Geometry of quartic curves. Math. Proc. Camb. Philos. Soc. **117**, 415–423 (1995)

[Wal2] Wall, C.T.C.: Highly singular quintic curves. Math. Proc. Camb. Philos. Soc. **119**, 257–277 (1996)

[Ku] Vik, K.S.: The generalized Chisini conjecture. Proc. Steklov Inst. Math. **241**(2), 110–119 (2003)

[CPS] Calabri, A., Paccagnan, D., Stagnaro, E.: Plane algebraic curves with many cusps with an appendix by Eugenii Shustin. Annali di Matematica Pura ed Applicata (4) **193**(3), 909–921 (2014)

[Ful1] Fulton, W.: On the fundamental group of the complement of a node curve. Ann. Math. (2) **111**(2), 407–409 (1980)

[Del] Deligne, P.: Le groupe fondamental du complement d'une courbe plane n'ayant que des points doubles ordinaires est abelien (d'apres W. Fulton). Bourbaki Seminar, vol. 1979/80. Lecture Notes in Mathematics, vol. 842, pp. 1–10. Springer, Berlin (1981)

[HaJ] Harris, J.: On the severi problem. Invent. Math. **84**, 445–461 (1985)

[Alb] Albanese, G.: Sulle condizioni perchè una curva algebraica riducible si possa considerare come limite di una curva irreducibile. Rend. Circ. Mat. Palermo **2**(52), 105–150 (1928)

[Nob2] Nobile, A.: On specialization of curves I. Trans. Am. Math. Soc. **282**(2), 739–748 (1984)

[Ran1] Ran, Z.: On nodal plane curves. Invent. Math. **86**, 529–534 (1986)

[Tre] Treger, R.: On the local Severi problem. Bull. Am. Math. Soc. **19**(1), 325–327 (1988)

[Ran2] Ran, Z.: Families of plane curves and their limits: enriques' conjecture and beyond. Ann. Math. **130**(1), 121–157 (1989)

[Kan1] Kang, P.-L.: On the variety of plane curves of degree d with δ nodes and k cusps. Trans. Am. Math. Soc. **316**(1), 165–192 (1989)

[Kan2] Kang, P.-L.: A note on the variety of plane curves with nodes and cusps. Proc. Am. Math. Soc. **106**(2), 309–312 (1989)

[AD] Akyol, A., Degtyarev, A.: Geography of irreducible plane sextics. Proc. Lond. Math. Soc. (3) **111**(6), 1307–1337 (2015)

[Deg1] Degtyarev, A.I.: Isotopic classification of complex plane projective curves of degree 5. Algebra i Analiz **1**(4), 78–101 (1989, Russian); English translation Leningrad Math. J. **1**(4), 881–904 (1990)

[Deg2] Degtyarev, A.: Topology of Algebraic Curves. An Approach via Dessins D'enfants. De Gruyter Studies in Mathematics. Walter de Gruyter, Berlin (2012)

[HaM] Harris, J., Morrison, I.: Moduli of Curves. Graduate Texts in Mathematics, vol. 187. Springer, Berlin (1998)

[Tan3] Tannenbaum, A.: On the classical characteristic linear series of plane curves with nodes and cuspidal points: two examples of Beniamino Segre. Compos. Math. **51**, 169–183 (1984)

[Wah1] Wahl, J.: Equisingular deformations of plane algebroid curves. Trans. Am. Math. Soc. **193**, 143–170 (1974)

[Lue1] Luengo, I.: The μ- constant stratum is not smooth. Invent. Math. **90**, 139–152 (1987)

[Lue2] Luengo, I.: On the existence of complete families of projective plane curves, which are obstructed. J. Lond. Math. Soc. **2**(36), 33–43 (1987)

[Shu6] Shustin, E.: Smoothness and irreducibility of varieties of algebraic curves with nodes and cusps. Bull. SMF **122**, 235–253 (1994)

[Gia] Giacinti-Diebolt, C.: Variétés des courbes projectives planes de degré et lieu singulier donnés. Math. Ann. **266**, 321–350 (1984)

[Fla] Flamini, F.: Some results of regularity for severi varieties of projective surfaces. Commun. Algebra **29**(6), 2297–2311 (2001)

[ChS] Chiantini, L., Sernesi, E.: Nodal curves on surfaces of general type. Math. Ann. **307**, 41–56 (1997)

[Kei2] Keilen, T.: Smoothness of equisingular families of curves. Trans. Am. Math. Soc. **357**(6), 2467–2481 (2005)

[HaR2] Hartshorne, R.: Algebraic Geometry. Graduate Text in Mathematics, vol. 52. Springer, Berlin (1977)

[DPW2] Du Plessis, A.A., Wall, C.T.C.: Singular hypersurfaces, versality and Gorenstein algebras. J. Algebraic Geom. **9**(2), 309–322 (2000)

[GLS6] Greuel, G.-M., Lossen, C., Shustin, E.: Introduction to Singularities and Deformations. Springer, Berlin (2007)

[Var1] Varchenko, A.N.: A lower bound for the codimension of the stratum $\mu =$ const in terms of the mixed Hodge structure. Moscow Univ. Math. Bull. **37**(2), 28–31 (1982)

[Chm] Chmutov, S.V.: Examples of projective surfaces with many singularities. J. Algebraic Geom. **1**(2), 191–196 (1992)

[Shu2] Shustin, E.: New M-curve of the 8th degree. Math. Notes Acad. Sci. USSR **42**, 606–610 (1987)

[Shu12] Shustin, E.: Lower deformations of isolated hypersurface singularities. Algebra i Analiz **10**(5), 221–249 (1999) (English translation in St. Petersburg Math. J. **11**(5), 883–908 (2000)

[DPW1] Du Plessis, A.A., Wall, C.T.C.: Discriminants and vector fields. In: Arnold, V.I., et al. (eds.) Singularities, the Brieskorn Anniversary Volume. Progress in Mathematics, vol. 162, pp. 119–140. Birkhäuser, Basel (1998)

[OrS] Orevkov, S., Shustin, E.I.: Real algebraic and pseudoholomorphic curves on the quadratic cone and smoothings of singularity X_{21}. St. Petersburg Math. J. **28**, 225–257 (2017)

[KoS] Korchagin, A.B., Shustin, E.I.: Affine curves of degree 6 and smoothings of the ordinary 6th order singular point. Math. USSR Izvestia **33**(3), 501–520 (1989)

[Ch2] Chevallier, B.: Four M-curves of degree 8. Funktsional. Anal. i Prilozhen. **36**(1), 90–93 (2002) (English translation in Funct. Anal. Appl. **36**(1), 76–78 (2002)

[Pec1] Pecker, D.: Note sur la réalité des points doubles des courbes gauches. C. R. Acad. Sci. Paris, Sér. I **324**, 807–812 (1997)

[Pec2] Pecker, D.: Sur la réalité des points doubles des courbes gauches. Ann. Inst. Fourier (Grenoble) **49**, 1439–1452 (1999)

[Gud4] Gudkov, D.A.: On the curve of 5th order with 5 cusps. Funct. Anal. Appl. **16**, 201–202 (1982)

[ItS2] Itenberg, I., Shustin, E.: Real algebraic curves with real cusps. Am. Math. Soc. Transl. **2**(173), 97–109 (1996)

[Miy] Miyaoka, Y.: On the Chern numbers of surfaces of general type. Invent. Math. **42**, 225–237 (1977)

[Yau] Yau, S.-T.: Calabi's conjecture and some new results in algebraic geometry. Proc. Natl. Acad. Sci. USA **74**, 1798–1799 (1977)

[GuN2] Gusein-Zade, S.M., Nekhoroshev, N.N.: Singularities of type A_k on plane curves of a chosen degree. Funct. Anal. Appl. **34**(3), 214–215 (2000); translation from Funkts. Anal. Prilozh. **34**(3), 69–70 (2000)

[GuN1] Gusein-Zade, S.M., Nekhoroshev, N.N.: Contiguity of A_k- singularities at points of the stratum $\mu = $ const of a singularity. Funct. Anal. Appl. **17**, 312–313 (1983)

[Ore] Orevkov, S.Y.: Some examples of real algebraic and real pseudoholomorphic curves. Perspectives in Analysis, Geometry, and Topology. Progress in Mathematics, vol. 296, pp. 355–387. Birkhäuser, Basel (2012)

[KeT] Keilen, T., Tyomkin, I.: Existence of curves with prescribed topological singularities. Trans. Am. Math. Soc. **354**(5), 1837–1860 (2002)

[Che] Chen, X.: Rational curves on K3 surfaces. J. Algebraic Geom. **8**, 245–278 (1999). Thesis, Harvard University (1997)

[ChC] Chiantini, L., Ciliberto, C.: On the severi varieties of surfaces in $\mathbb{P}^3$. J. Algebraic Geom. **8**(1), 67–83 (1999)

[GoP] Göttsche, L., Pandharipande, R.: The quantum cohomology of blow-ups of $\mathbb{P}^2$ and enumerative geometry. J. Diff. Geom. **48**(1), 61–90 (1998)

[AB] Abramovich, D., Bertram, A.: The formula $12 = 10 + 2 \times 1$ and its generalizations: counting rational curves on $\boldsymbol{F}_2$. Advances in Algebraic Geometry Motivated by Physics (Lowell, MA, 2000). Contemporary Mathematics, vol. 276, pp. 83–88. American Mathematical Society, Providence (2001)

[Wel] Welschinger, J.-Y.: Courbes algébriques réelles et courbes flexibles sur les surfaces réglées de base $\mathbb{C}P^1$. Proc. Lond. Math. Soc. **85**(2), 367–392 (2002)

[Mik1] Mikhalkin, G.: Counting curves via the lattice paths in polygons. Comptes Rendus Math. **336**(8), 629–634 (2003)

[Mik2] Mikhalkin, G.: Enumerative tropical algebraic geometry in $\mathbb{R}^2$. J. Am. Math. Soc. **18**, 313–377 (2005)

[YauZ] Yau, S.-T., Zaslow, E.: BPS states, string duality, and nodal curves on K3. Nucl. Phys. B **471**, 503–512 (1996)

[KMPS] Klemm, A., Maulik, D., Pandharipande, R., Scheidegger, E.: Noether-Lefschetz theory and the Yau-Zaslow conjecture. J. Am. Math. Soc. **23**(4), 1013–1040 (2010)

[ChL] Chen, X., Lewis, J.D.: Density of rational curves on K3 surfaces. Math. Ann. **356**(1), 331–354 (2013)

[ArC] Arbarello, E., Cornalba, M.: A few remarks about the variety of irreducible plane curves of given degree and genus. Ann. Sci. École Norm. Sup. **16**(3), 467–488 (1983)

[BrG] Bruce, J.W., Gibblin, P.J.: A stratification of the space of plane quartic curves. Proc. Lond. Math. Soc. **442**, 270–298 (1981)

[DeM] Deligne, P., Mumford, D.: The irreducibility of the moduli space of curves with given genus. Publ. Math. IHES **36**, 75–100 (1969)

[Shu7] Shustin, E.: Smoothness and irreducibility of families of plane algebraic curves with ordinary singularities. Israel Mathematical Conference Proceedings, vol. 9, pp. 393–416. AMS, Providence (1996)

[Shu8] Shustin, E.: Geometry of equisingular families of plane algebraic curves. J. Algebraic Geom. **5**, 209–234 (1996)

[Nor] Nori, M.: Zariski conjecture and related problems. Ann. Sci. Ec. Norm. Sup. **4**(16), 305–344 (1983)

[Kei1] Keilen, T.: Irreducibility of equisingular families of curves. Trans. Am. Math. Soc. **355**(9), 3485–3512 (2003)

[LaR] Lazarsfeld, R.: Lectures on linear series. In: Kollár, J. (ed.) Complex Algebraic Geometry. AMS, Providence (1997)

[Wah2] Wahl, J.: Deformations of plane curves with nodes and cusps. Am. J. Math. **96**, 529–577 (1974)

[CGL1] Campillo, A., Greuel, G.-M., Lossen, C.: Equisingular deformations of plane curves in arbitrary characteristic. Compos. Math. **143**, 829–882 (2007)

[CGL2] Campillo, A., Greuel, G.-M., Lossen, C.: Equisingular calculations for plane curve singularities. J. Symb. Comput. **42**(1–2), 89–114 (2007)

[BoGM] Boubakri, Y., Greuel, G.-M., Markwig, T.: Invariants of hypersurface singularities in positive characteristic. Rev. Mat. Complut. **25**, 61–85 (2012)

[GrPh] Greuel, G.-M., Pham, T.H.: On finite determinacy for matrices of power series. Math. Z. (2018). https://doi.org/10.1007/s00209-018-2040-2

[Zar2] Zariski, O.: Studies in equisingularity I–III. Am. J. Math. **87**, 507–536 and 972–1006 (1965), respectively Am. J. Math. **90**, 961–1023 (1968)

[Nob1] Nobile, A.: Families of curves on surfaces. Math. Zeitschrift **187**(4), 453–470 (1984)

[Tyo1] Tyomkin, I.: On severi varieties on Hirzebruch surfaces. Int. Math. Res. Not. **23**, Art. ID rnm109, 31 pp (2007)

[Tyo3] Tyomkin, I.: On Zariski's theorem in positive characteristic. J. Eur. Math. Soc. (JEMS) **15**(5), 1783–1803 (2013)

[Tyo4] Tyomkin, I.: An example of a reducible severi variety. In: Proceedings of the Gökova Geometry-Topology Conference 2013. Gökova Geometry/Topology Conference (GGT), Gökova, pp. 33–40 (2014)

[Wes] Westenberger, E.: Existence of hypersurfaces with prescribed simple singularities. Commun. Algebra **31**(1), 335–356 (2003)

[AH2] Alexander, J., Hirschowitz, A.: An asymptotic vanishing theorem for generic unions of multiple points. Invent. Math. **140**(2), 303–325 (2000)

[Roe1] Roé, J.: Maximal rank for schemes of small multiplicity by Évain's differential Horace method. Trans. Am. Math. Soc. **366**, 857–874 (2014)

[BGM] Bigatti, A., Geramita, A.V., Migliore, J.C.: Geometric consequences of extremal behavior in a theorem of Macaulay. Trans. Am. Math. Soc. **346**(1), 203–235 (1994)

[Nag1] Nagata, M.: On the 14-th problem of Hilbert. Am. J. Math. **81**, 766–772 (1959)

[Nag2] Nagata, M.: On rational surfaces II. Mem. Coll. Sci. Univ. Kyoto Ser. A **33**, 271–293 (1960)

[DHKRS] Dumnicki, M., Harbourne, B., Küronya, A., Roé, J., Szemberg, T.: Very general monomial valuations of $\mathbb{P}^2$ and a Nagata type conjecture. Commun. Anal. Geom. **25**(1), 125–161 (2017)

[CHMR] Ciliberto, C., Harbourne, B., Miranda, R., Roé, J.: Variations on Nagata's conjecture. In: Hassett, B., et al. (eds.) A Celebration of Algebraic Geometry. Clay Mathematics Proceedings, vol. 18, pp. 185–203 (2013)

[FuP] Fulton, W., Pandharipande, R.: Notes on stable maps and quantum cohomology. Algebraic geometry—Santa Cruz 1995. Proceedings of Symposia in Pure Mathematics, vol. 62, Part 2, pp. 45–96. American Mathematical Society, Providence (1997)

[Get] Getzler, E.: Intersection theory on $\overline{\mathcal{M}}_{1,4}$ and elliptic Gromov-Witten invariants. J. Am. Math. Soc. **10**(4), 973–998 (1997)

[KoM] Kontsevich, M., Manin, Yu.: Gromov-Witten classes, quantum cohomology and enumerative geometry. Commun. Math. Phys. **164**, 525–562 (1994)

[RuT] Ruan, Y., Tian, G.: A mathematical theory of quantum cohomology. J. Differ. Geom. **42**(2), 259–367 (1995)

[Vak] Vakil, R.: Counting curves on rational surfaces. Manuscr. math. **102**(1), 53–84 (2000)

[Pan] Pandharipande, R.: Intersections of $\mathbb{Q}$- divisors on Kontsevich's moduli space $\overline{\mathcal{M}}_{0,n}(\mathbb{P}^r, d)$ and enumerative geometry. Trans. Am. Math. Soc. **351**(4), 1481–1505 (1999)

[Zin] Zinger, A.: Counting rational curves of arbitrary shape in projective spaces. Geom. Topol. **9**, 571–697 (2005)

[Goet] Göttsche, L.: A conjectural generating function for numbers of curves on surfaces. Commun. Math. Phys. **196**, 523–533 (1998)

[GoS1] Göttsche, L., Shende, V.: The χ_{-y}- genera of relative Hilbert schemes for linear systems on Abelian and K3 surfaces. Algebraic Geom. **2**(4), 405–421 (2015)

[GoS2] Göttsche, L., Shende, V.: Refined curve counting on complex surfaces. Geom. Topol. **18**, 2245–2307 (2014)

[FeR] Fehér, L., Rimányi, R.: On the structure of Thom polynomials of singularities. Bull. Lond. Math. Soc. **39**, 541–549 (2007)

[Kaz] Kazarian, M.E.: Multisingularities, cobordisms, and enumerative geometry. Russ. Math. Surv. **58**(4), 665–724 (2003)

[Liu] Liu, A.K.: Family blowup formula, admissible graphs and the enumeration of singular curves I. J. Diff. Geom. **56**(3), 381–579 (2000)

[BM1] Basu, S., Mukherjee, R.: Enumeration of curves with two singular points. Bull. Sci. Math. **139**(6), 667–735 (2015)

[BM2] Basu, S., Mukherjee, R.: Enumeration of curves with one singular point. J. Geom. Phys. **104**, 175–203 (2016)

[Ker] Kerner, D.: Enumeration of singular algebraic curves. Isr. Math. J. **155**, 1–56 (2006)

[KlP] Kleiman, S., Piene, R.: Enumerating singular curves on surfaces. Algebraic geometry: Hirzebruch 70 (Warsaw, 1998). Contemporary Mathematics, vol. 241, pp. 209–238. American Mathematical Society, Providence (1999)

[Vai] Vainsencher, I.: Counting divisors with prescribed singularities. Trans. Am. Math. Soc. **267**(2), 399–422 (1981)

[CaH2] Caporaso, L., Harris, J.: Counting plane curves of any genus. Invent. Math. **131**(2), 345–392 (1998)

[Ran4] Ran, Z.: On the quantum cohomology of the plane, old and new, and a K3 analogue. Collect. Math. **49**(2–3), 519–526 (1998)

[ShSh] Shoval, M., Shustin, E.: On Gromov-Witten invariants of del pezzo surfaces. Int. J. Math. **24**(7), 44 pp (2013). https://doi.org/10.1142/S0129167X13500547

[Br] Brugallé, E.: Floor diagrams relative to a conic, and GW-W invariants of del pezzo surfaces. Adv. Math. **279**, 438–500 (2015)

[Mik3] Mikhalkin, G.: Tropical geometry and its applications. In: Sanz-Solé, M., et al. (ed.) Proceedings of the ICM, Madrid, Spain, 22–30 August 2006. Volume II: Invited Lectures, pp. 827–852. European Mathematical Society, Zürich (2006)

[NiS] Nishinou, T., Siebert, B.: Toric degenerations of toric varieties and tropical curves. Duke Math. J. **135**(1), 1–51 (2006)

[GaS] Ganor, Y., Shustin, E.: Enumeration of plane unicuspidal curves of any genus via tropical geometry (2018). arXiv:1807.11443

[KuS] Kulikov, V.S., Shustin, E.: On rigid plane curves. Eur. J. Math. **2**(1), 208–226 (2016)

[GrS] Greuel, G.-M., Shustin, E.: Geometry of equisingular families of curves. In: Bruce, B., Mond, D. (eds.) Singularity Theory. Proceedings of the European Singularities Conference, Liverpool, August 1996. Dedicated to C.T.C. Wall in Occasion of His 60th Birthday (London Mathematical Society Lecture Note Series, vol. 263), pp. 79–108. Cambridge University Press, Cambridge (1999)

[Bru] Bruce, J.W.: Singularities on a projective hypersurface. Bull. Lond. Math. Soc. **13**, 47–50 (1981)

[GuU] Gudkov, D.A., Utkin, G.A.: The topology of curves of degree 6 and surfaces of degree 4. Uchen. Zap. Gorkov. Univ. **87** (Russian), English translation Transl. AMS **112** (1978)

[Shu3] Shustin, E.: Geometry of discriminant and topology of algebraic curves. Proceedings of the International Congress of Mathematicians, Kyoto 1990, vol. 1. Springer, Berlin (1991)

[Bar] Barkats, D.: Études des variétés des courbes planes à noeuds et à cusps. Algebraic Geometry (Catania, 1993/Barcelona, 1994). Lecture Notes in Pure and Applied Mathematics, vol. 200, pp. 25–35. Dekker, New York (1998)

[GG] Gourevich, A., Gourevich, D.: Geometry of obstructed equisingular families of projective hypersurfaces. J. Pure Appl. Algebra **213**(9), 1865–1889 (2009)

Appendix
Patchworking Real Algebraic Varieties by Oleg Viro

Introduction

This paper is a translation of the first chapter of my dissertation[1] which was defended in 1983. I do not take here an attempt of updating.

The results of the dissertation were obtained in 1978–80, announced in [Vir79a, Vir79b, Vir80], a short fragment was published in detail in [Vir83a] and a considerable part was published in paper [Vir83b]. The latter publication appeared, however, in almost inaccessible edition and has not been translated into English.

In [Vir89] I presented almost all constructions of plane curves contained in the dissertation, but in a simplified version: without description of the main underlying patchwork construction of algebraic hypersurfaces. Now I regard the latter as the most important result of the dissertation with potential range of application much wider than topology of real algebraic varieties. It was the subject of the first chapter of the dissertation, and it is this chapter that is presented in this paper.

In the dissertation the patchwork construction was applied only in the case of plane curves. It is developed in considerably higher generality. This is motivated not only by a hope on future applications, but mainly internal logic of the subject. In particular, the proof of Main Patchwork Theorem in the two-dimensional situation is based on results related to the three-dimensional situation and analogous to the two-dimensional results which are involved into formulation of the two-dimensional Patchwork Theorem. Thus, it is natural to formulate and prove these results once

[1]This is not a Ph.D., but a dissertation for the degree of Doctor of Physico-Mathematical Sciences. In Russia there are two degrees in mathematics. The lower, degree corresponding approximately to Ph.D., is called Candidate of Physico-Mathematical Sciences. The high degree dissertation is supposed to be devoted to a subject distinct from the subject of the Candidate dissertation. My Candidate dissertation was on interpretation of signature invariants of knots in terms of intersection form of branched covering spaces of the 4-ball. It was defended in 1974.

G.-M. Greuel et al., *Singular Algebraic Curves*, Springer Monographs in Mathematics, https://doi.org/10.1007/978-3-030-03350-7

for all dimensions, but then it is not natural to confine Patchwork Theorem itself to the two-dimensional case. The exposition becomes heavier because of high degree of generality. Therefore the main text is prefaced with a section with visualizable presentation of results. The other sections formally are not based on the first one and contain the most general formulations and complete proofs.

In the last section another, more elementary, approach is expounded. It gives more detailed information about the constructed manifolds, having not only topological but also metric character. There, in particular, Main Patchwork Theorem is proved once again.

I am grateful to Julia Viro who translated this text.

A.1 Patchworking Plane Real Algebraic Curves

This section is introductory. I explain the character of results staying in the framework of plane curves. A real exposition begins in Sect. A.2. It does not depend on Sect. A.1. To a reader who is motivated enough and does not like informal texts without proofs, I would recommend to skip this section.

A.1.1 The Case of Smallest Patches

We start with the special case of the patchworking. In this case the patches are so simple that they do not demand a special care. It purifies the construction and makes it a straight bridge between combinatorial geometry and real algebraic geometry.

Proposition A.1.1 (Initial Data) *Let m be a positive integer number [it is the degree of the curve under construction]. Let Δ be the triangle in $\mathbf{R}^2$ with vertices $(0,0)$, $(m,0)$, $(0,m)$ [it is a would-be Newton diagram of the equation]. Let T be a triangulation of Δ whose vertices have integer coordinates. Let the vertices of T be equipped with signs; the sign (plus or minus) at the vertex with coordinates (i, j) is denoted by $\sigma_{i,j}$.*

See Fig. A.1.

For $\varepsilon, \delta = \pm 1$ denote the reflection $\mathbb{R}^2 \to \mathbb{R}^2 : (x, y) \mapsto (\varepsilon x, \delta y)$ by $S_{\varepsilon,\delta}$. For a set $A \subset \mathbb{R}^2$, denote $S_{\varepsilon,\delta}(A)$ by $A_{\varepsilon,\delta}$ (see Fig. A.2). Denote a quadrant $\{(x, y) \in \mathbb{R}^2 \mid \varepsilon x > 0, \delta y > 0\}$ by $Q_{\varepsilon,\delta}$.

The following construction associates with Initial Data Proposition A.1.1 above a piecewise linear curve in the projective plane.

Proposition A.1.2 (Combinatorial patchworking) *Take the square Δ_* made of Δ and its mirror images Δ_{+-}, Δ_{-+} and Δ_{--}. Extend the triangulation $\mathbb{T}$ of Δ to a triangulation $\mathbb{T}_*$ of Δ_* symmetric with respect to the coordinate axes. Extend the distribution of signs $\sigma_{i,j}$ to a distribution of signs on the vertices of the extended triangulation which satisfies the following condition: $\sigma_{i,j}\sigma_{\varepsilon i,\delta j}\varepsilon^i\delta^j = 1$ for any vertex*

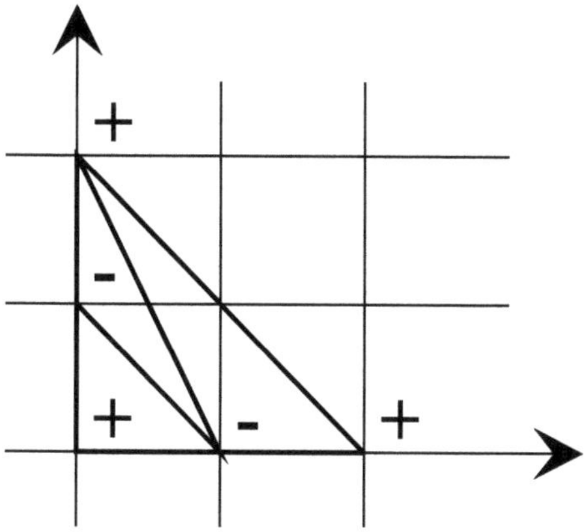

Fig. A.1 Triangulation of the Newton polygon and sign distribution

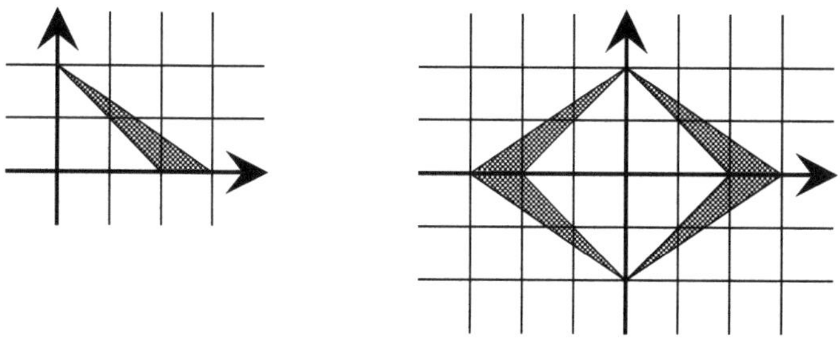

Fig. A.2 Newton polygon and its four copies

(i, j) *of* $\mathbb{T}$ *and* $\varepsilon, \delta = \pm 1$. (In other words, passing from a vertex to its mirror image with respect to an axis we preserve its sign if the distance from the vertex to the axis is even, and change the sign if the distance is odd.)[2]

If a triangle of the triangulation $\mathbb{T}_*$ *has vertices of different signs, draw the midline separating the vertices of different signs. Denote by* L *the union of these midlines. It is a collection of polygonal lines contained in* Δ_*. *Glue by* S_{--} *the opposite sides of* Δ_*. *The resulting space* $\bar{\Delta}$ *is homeomorphic to the projective plane* $\mathbb{R}P^2$. *Denote by* $\bar{L}$ *the image of* L *in* $\bar{\Delta}$.

Let us introduce a supplementary assumption: the triangulation $\mathbb{T}$ of Δ is *convex*. It means that there exists a convex piecewise linear function $\nu : \Delta \to \mathbb{R}$ which is linear on each triangle of $\mathbb{T}$ and not linear on the union of any two triangles of $\mathbb{T}$. A function ν with this property is said to *convexify* $\mathbb{T}$.

In fact, to stay in the frameworks of algebraic geometry we need to accept an additional assumption: a function ν convexifying $\mathbb{T}$ should take integer value on each

[2]More sophisticated description: the new distribution should satisfy the modular property: $g^*(\sigma_{i,j} x^i y^j) = \sigma_{g(i,j)} x^i y^j$ for $g = S_{\varepsilon\delta}$ (in other words, the sign at a vertex is the sign of the corresponding monomial in the quadrant containing the vertex).

vertex of $\mathbb{T}$. Such a function is said to *convexify* $\mathbb{T}$ over $\mathbb{Z}$. However this additional restriction is easy to satisfy. A function $\nu : \Delta \to \mathbb{R}$ convexifying $\mathbb{T}$ is characterized by its values on vertices of $\mathbb{T}$. It is easy to see that this provides a natural identification of the set of functions convexifying $\mathbb{T}$ with an open convex cone of $\mathbb{R}^N$ where N is the number of vertices of $\mathbb{T}$. Therefore if this set is not empty, then it contains a point with rational coordinates, and hence a point with integer coordinates.

Proposition A.1.3 (Polynomial patchworking) *Given Initial Data m, Δ, $\mathbb{T}$ and $\sigma_{i,j}$ as above and a function ν convexifying $\mathbb{T}$ over $\mathbb{Z}$. Take the polynomial*

$$b(x, y, t) = \sum_{\substack{(i,j) \text{ runs over} \\ \text{vertices of } \mathbb{T}}} \sigma_{i,j} x^i y^j t^{\nu(i,j)}.$$

and consider it as a one-parameter family of polynomials: set $b_t(x, y) = b(x, y, t)$. Denote by B_t the corresponding homogeneous polynomials:

$$B_t(x_0, x_1, x_2) = x_0^m b_t(x_1/x_0, x_2/x_0).$$

Proposition A.1.4 (Patchwork Theorem) *Let m, Δ, $\mathbb{T}$ and $\sigma_{i,j}$ be an initial data as above and ν a function convexifying $\mathbb{T}$ over $\mathbb{Z}$. Denote by b_t and B_t the non-homogeneous and homogeneous polynomials obtained by the polynomial patchworking of these initial data and by L and $\bar{L}$ the piecewise linear curves in the square Δ_* and its quotient space $\bar{\Delta}$ respectively obtained from the same initial data by the combinatorial patchworking.*

Then there exists $t_0 > 0$ such that for any $t \in (0, t_0]$ the equation $b_t(x, y) = 0$ defines in the plane $\mathbb{R}^2$ a curve c_t such that the pair $(\mathbb{R}^2, c_t)$ is homeomorphic to the pair (Δ_*, L) and the equation $B_t(x_0, x_1, x_2) = 0$ defines in the real projective plane a curve C_t such that the pair $(\mathbb{R}P^2, C_t)$ is homeomorphic to the pair $(\bar{\Delta}, \bar{L})$.

Example A.1.4.1 Construction of a curve of degree 2 is shown in Fig. A.3. The broken line corresponds to an ellipse. More complicated examples with a curves of degree 6 are shown in Figs. A.4, A.5.

For more general version of the patchworking we have to prepare patches. Shortly speaking, the role of patches was played above by lines. The generalization below is a transition from lines to curves. Therefore we proceed with a preliminary study of curves.

A.1.2 Logarithmic Asymptotes of a Curve

As is known since Newton's works (see [New67]), behavior of a curve $\{(x, y) \in \mathbb{R}^2 \mid a(x, y) = 0\}$ near the coordinate axes and at infinity depends, as a rule, on the coefficients of a corresponding to the boundary points of its Newton polygon $\Delta(a)$.

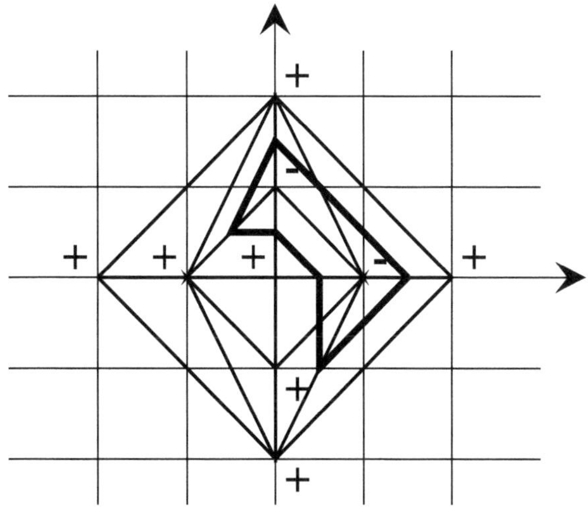

Fig. A.3 Combinatorial patchworking of the initial data shown in Fig. A.1

Fig. A.4 Harnack's curve of degree 6

We need more specific formulations, but prior to that we have to introduce several notations and discuss some notions.

For a set $\Gamma \subset \mathbb{R}^2$ and a polynomial $a(x, y) = \sum_{\omega \in \mathbb{Z}^2} a_\omega x^{\omega_1} y^{\omega_2}$, denote the polynomial $\sum_{\omega \in \Gamma \cap \mathbb{Z}^2} a_\omega x^{\omega_1} y^{\omega_2}$ by a^Γ. It is called the *Γ-truncation* of a.

For a set $U \subset \mathbb{R}^2$ and a real polynomial a in two variables, denote the curve $\{(x, y) \in U \mid a(x, y) = 0\}$ by $V_U(a)$.

Fig. A.5 Gudkov's curve of degree 6

The complement of the coordinate axes in $\mathbb{R}^2$, i.e. a set $\{(x, y) \in \mathbb{R}^2 \mid xy \neq 0\}$, is denoted[3] by $\mathbb{RR}^2$.

Denote by l the map $\mathbb{RR}^2 \to \mathbb{R}^2$ defined by formula $l(x, y) = (\ln|x|, \ln|y|)$. It is clear that the restriction of l to each quadrant is a diffeomorphism.

A polynomial in two variables is said to be *quasi-homogeneous* if its Newton polygon is a segment. The simplest real quasi-homogeneous polynomials are binomials of the form $\alpha x^p + \beta y^q$ where p and q are relatively prime. A curve $V_{\mathbb{RR}^2}(a)$, where a is a binomial, is called *quasiline*. The map l transforms quasilines to lines. In that way any line with rational slope can be obtained. The image $l(V_{\mathbb{RR}^2}(a))$ of the quasiline $V_{\mathbb{RR}^2}(a)$ is orthogonal to the segment $\Delta(a)$.

It is clear that any real quasi-homogeneous polynomial in 2 variables is decomposable into a product of binomials of the type described above and trinomials without zeros in $\mathbb{RR}^2$. Thus if a is a real quasi-homogeneous polynomial then the curve $V_{\mathbb{RR}^2(a)}$ is decomposable into a union of several quasilines which are transformed by l to lines orthogonal to $\Delta(a)$.

A real polynomial a in two variables is said to be *peripherally nondegenerate* if for any side Γ of its Newton polygon the curve $V_{\mathbb{RR}^2}(a^\Gamma)$ is nonsingular (it is a union of quasilines transformed by l to parallel lines, so the condition that it is nonsingular

[3]This notation is motivated in Sect. A.2.3 below.

means absence of multiple components). Being peripherally nondegenerate is typical in the sense that among polynomials with the same Newton polygons the peripherally nondegenerate ones form nonempty set open in the Zarisky topology.

For a side Γ of a polygon Δ, denote by $DC^-_\Delta(\Gamma)$ a ray consisting of vectors orthogonal to Γ and directed outside Δ with respect to Γ (see Fig. A.6 and Sect. A.2.2).

The assertion in the beginning of this section about behavior of a curve nearby the coordinate axes and at infinity can be made now more precise in the following way.

Proposition A.1.5 *Let $\Delta \in \mathbb{RR}^2$ be a convex polygon with nonempty interior and sides $\Gamma_1, ..., \Gamma_n$. Let a be a peripherally nondegenerate real polynomial in 2 variables with $\Delta(a) = \Delta$. Then for any quadrant $U \in \mathbb{RR}^2$ each line contained in $l(V_U(a^{\Gamma_i}))$ with $i = 1,...,n$ is an asymptote of $l(V_U(a))$, and $l(V_U(a))$ goes to infinity only along these asymptotes in the directions defined by rays $DC^-_\Delta(\Gamma_i)$.*

Theorem generalizing this proposition is formulated in Sect. A.6.3 and proved in Sect. A.6.4. Here we restrict ourselves to the following elementary example illustrating Proposition A.1.5.

Example A.1.5.1 Consider the polynomial $a(x, y) = 8x^3 - x^2 + 4y^2$. Its Newton polygon is shown in Fig. A.6. In Fig. A.7 the curve $V_{\mathbb{R}^2}(a)$ is shown. In Fig. A.8 the rays $DC^-_\Delta(\Gamma_i)$ and the images of $V_U(a)$ and $V_U(a^{\Gamma_i})$ under diffeomorphisms $l|_U : U \to \mathbb{R}^2$ are shown, where U runs over the set of components of $\mathbb{RR}^2$ (i.e. quadrants). In Fig. A.9 the images of $DC^-_\Delta(\Gamma_i)$ under l and the curves $V_{\mathbb{R}^2}(a)$ and $V^2_{\mathbb{R}}(a^{\Gamma_i})$ are shown.

A.1.3 Charts of Polynomials

The notion of a chart of a polynomial is fundamental for what follows. It is introduced naturally via the theory of toric varieties (see Sect. A.3). Another definition, which is less natural and applicable to a narrower class of polynomials, but more elementary,

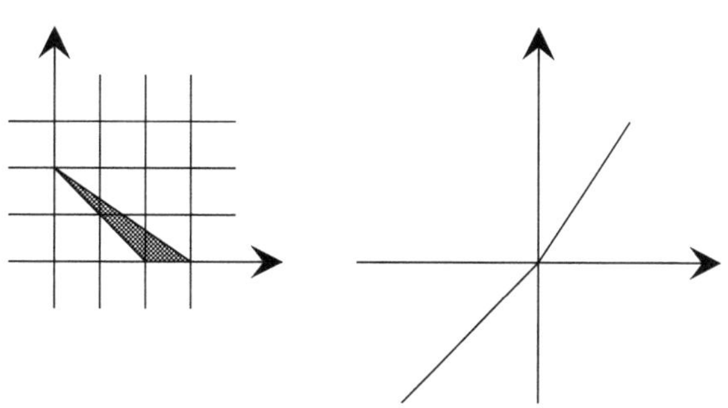

Fig. A.6 Newton polygon and asymptotic rays

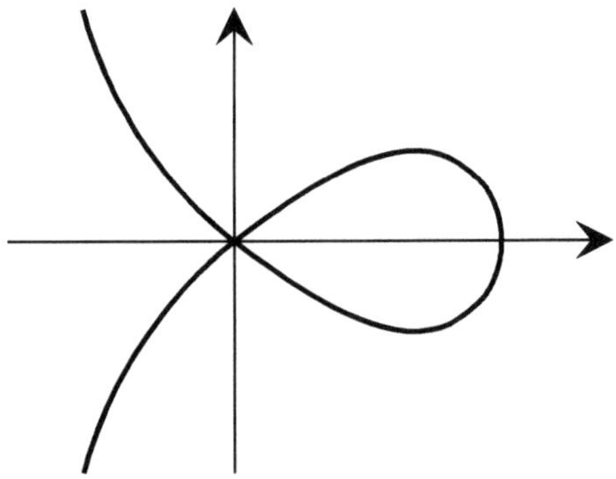

Fig. A.7 The real curve $V_{\mathbb{R}^2}(8x^3 - x^2 + 4y^2)$

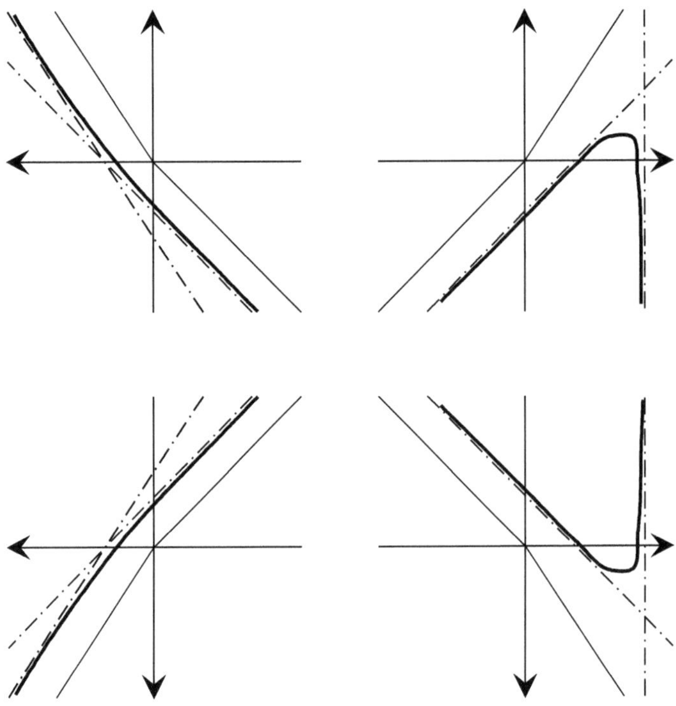

Fig. A.8 The logarithmic image of the curve in each quadrant

can be extracted from the results generalizing Proposition A.1.5 (see Sect. A.6). In this section, first, I try to give a rough idea about the definition related with toric varieties, and then I give the definitions related with Proposition A.1.5 with all details.

To any convex closed polygon $\Delta \subset \mathbb{R}^2$ with vertices whose coordinates are integers, a real algebraic surface $\mathbb{R}\Delta$ is associated. This surface is a completion of $\mathbb{R}\mathbb{R}^2$ $(= (\mathbb{R} \setminus 0)^2)$. The complement $\mathbb{R}\Delta \setminus \mathbb{R}\mathbb{R}^2$ consists of lines corresponding to sides

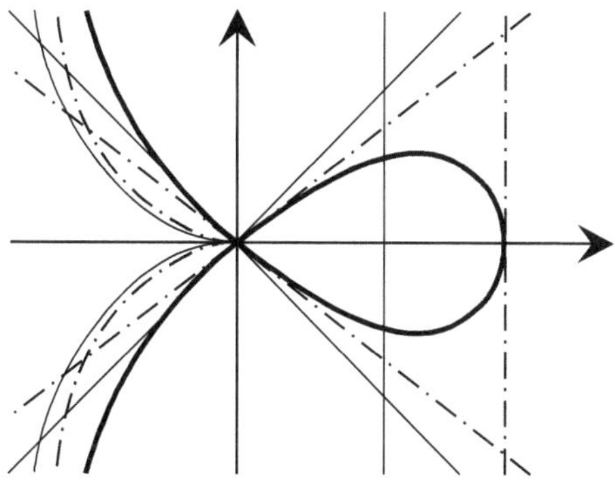

Fig. A.9 The asymptotic rays and the real curve $V_{\mathbb{R}^2}(8x^3 - x^2 + 4y^2)$

of Δ. From the topological viewpoint $\mathbb{R}\Delta$ can be obtained from four copies of Δ by pairwise gluing of their sides. For a real polynomial a in two variables we denote the closure of $V_{\mathbb{R}\mathbb{R}^2}(a)$ in $\mathbb{R}\Delta$ by $V_{\mathbb{R}\Delta}(a)$. Let a be a real polynomial in two variables which is not quasi-homogeneous. (The latter assumption is not necessary, it is made for the sake of simplicity.) Cut the surface $\mathbb{R}\Delta(a)$ along lines of $\mathbb{R}\Delta(a) \setminus \mathbb{R}\mathbb{R}^2$ (i.e. replace each of these lines by two lines). The result is four copies of $\Delta(a)$ and a curve lying in them obtained from $V_{\mathbb{R}\Delta(a)}(a)$. The pair consisting of these four polygons and this curve is a chart of a.

Recall that for $\varepsilon, \delta = \pm 1$ we denote the reflection $\mathbb{R}^2 \to \mathbb{R}^2 : (x, y) \mapsto (\varepsilon x, \delta y)$ by $S_{\varepsilon,\delta}$. For a set $A \subset \mathbb{R}^2$ we denote $S_{\varepsilon,\delta}(A)$ by $A_{\varepsilon,\delta}$ (see Fig. A.2). Denote a quadrant $\{(x, y) \in \mathbb{R}^2 \mid \varepsilon x > 0, \delta y > 0\}$ by $Q_{\varepsilon,\delta}$.

Now define the charts for two classes of real polynomials separately.

First, consider the case of quasi-homogeneous polynomials. Let a be a quasi-homogeneous polynomial defining a nonsingular curve $V_{\mathbb{R}\mathbb{R}^2}(a)$. Let (w_1, w_2) be a vector orthogonal to $\Delta = \Delta(a)$ with integer relatively prime coordinates. It is clear that in this case $V_{\mathbb{R}^2}(a)$ is invariant under $S_{(-1)^{w_1},(-1)^{w_2}}$. A pair (Δ_*, υ) consisting of Δ_* and a finite set $\upsilon \subset \Delta_*$ is called *a chart* of a, if the number of points of $\upsilon \cap \Delta_{\varepsilon,\delta}$ is equal to the number of components of $V_{Q_{\varepsilon,\delta}}(a)$ and υ is invariant under $S_{(-1)^{w_1},(-1)^{w_2}}$ (remind that $V_{\mathbb{R}^2}(a)$ is invariant under the same reflection).

Example A.1.5.2 In Fig. A.10 it is shown a curve $V_{\mathbb{R}^2}(a)$ with $a(x, y) = 2x^6y - x^4y^2 - 2x^2y^3 + y^4 = (x^2 - y)(x^2 + y)(2x^2 - y)y$, and a chart of a.

Now consider the case of peripherally nondegenerate polynomials with Newton polygons having nonempty interiors. Let Δ, $\Gamma_1, \dots, \Gamma_n$ and a be as in Proposition A.1.5. Then, as it follows from Proposition A.1.5, there exist a disk $D \subset \mathbb{R}^2$ with center at the origin and neighborhoods $D_1, \dots, D_n$ of rays $DC^-_{\Delta}(\Gamma_1), \dots, DC^-_{\Delta}(\Gamma_n)$ such that the curve $V_{\mathbb{R}\mathbb{R}^2}(a)$ lies in $l^{-1}(D \cup D_1 \cup \dots \cup D_n)$ and for $i = 1, \dots, n$ the curve $V_{l^{-1}(D_i \setminus D)}(a)$ is approximated by $V_{l^{-1}(D_i \setminus D)}(a^{\Gamma_i})$ and can be contracted (in itself) to $V_{l^{-1}(D_i \cap \partial D)}(a)$.

A pair (Δ_*, υ) consisting of Δ_* and a curve $\upsilon \subset \Delta_*$ is called a *chart* of a if

1. for $i = 1, \dots, n$ the pair $(\Gamma_{i*}, \Gamma_{i*} \cap \upsilon)$ is a chart of a^{Γ_i} and

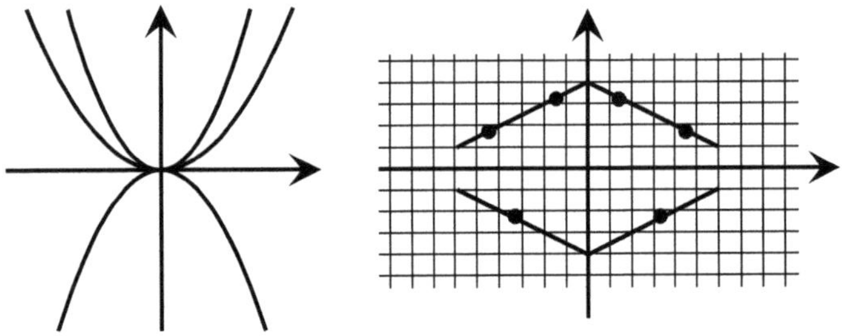

Fig. A.10 The real curve $V_{\mathbb{R}^2}((x^2 - y)(x^2 + y)(2x^2 - y)y)$ and its chart

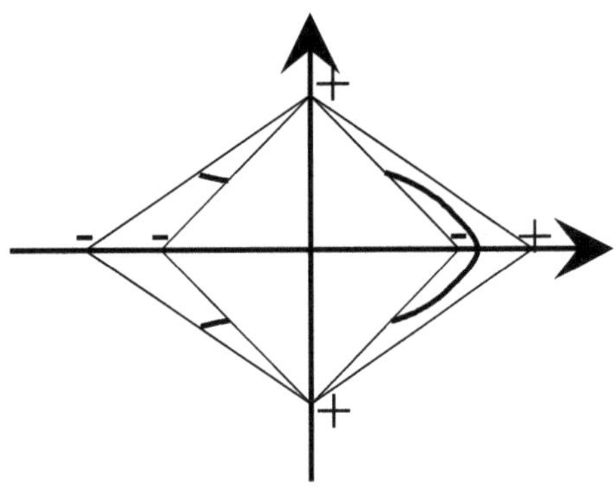

Fig. A.11 The chart of the curve $V_{\mathbb{R}^2}(8x^3 - x^2 + 4y^2)$

2. for $\varepsilon, \delta = \pm 1$ there exists a homeomorphism $h_{\varepsilon,\delta} : D \to \Delta$ such that $\upsilon \cap \Delta_{\varepsilon,\delta} = S_{\varepsilon,\delta} \circ h_{\varepsilon,\delta} \circ l(V_{l^{-1}(D) \cap Q_{\varepsilon,\delta}}(a))$ and $h_{\varepsilon,\delta}(\partial D \cap D_i) \subset \Gamma_i$ for $i = 1, \dots, n$.

It follows from Proposition A.1.5 that any peripherally nondegenerate real polynomial a with $\operatorname{Int} \Delta(a) \neq \varnothing$ has a chart. It is easy to see that the chart is unique up to a homeomorphism $\Delta_* \to \Delta_*$ preserving the polygons $\Delta_{\varepsilon,\delta}$, their sides and their vertices.

Example A.1.5.3 In Fig. A.11 it is shown a chart of $8x^3 - x^2 + 4y^2$ which was considered in Example A.1.5.1.

Proposition A.1.6 (Generalization of Example A.1.5.3) *Let*

$$a(x, y) = a_1 x^{i_1} y^{j_1} + a_2 x^{i_2} y^{j_2} + a_3 x^{i_3} y^{j_3}$$

be a non-quasi-homogeneous real polynomial (i. e., a real trinomial whose the Newton polygon has nonempty interior). For $\varepsilon, \delta = \pm 1$ *set*

$$\sigma_{\varepsilon i_k, \delta j_k} = sign(a_k \varepsilon^{i_k} \delta^{j_k}).$$

Then the pair consisting of Δ_* *and the midlines of* $\Delta_{\varepsilon,\delta}$ *separating the vertices* $(\varepsilon i_k, \delta j_k)$ *with opposite signs* $\sigma_{\varepsilon i_k, \delta j_k}$ *is a chart of* a.

Proof Consider the restriction of a to the quadrant $Q_{\varepsilon,\delta}$. If all signs $\sigma_{\varepsilon i_k,\delta j_k}$ are the same, then $aQ_{\varepsilon,\delta}$ is a sum of three monomials taking values of the same sign on $Q_{\varepsilon,\delta}$. In this case $V_{Q_{\varepsilon,\delta}}(a)$ is empty. Otherwise, consider the side Γ of the triangle Δ on whose end points the signs coincide. Take a vector (w_1, w_2) orthogonal to Γ. Consider the curve defined by parametric equation $t \mapsto (x_0 t^{w_1}, y_0 t^{w_2})$. It is easy to see that the ratio of the monomials corresponding to the end points of Γ does not change along this curve, and hence the sum of them is monotone. The ratio of each of these two monomials with the third one changes from 0 to $-\infty$ monotonically. Therefore the trinomial divided by the monomial which does not sit on Γ changes from $-\infty$ to 1 continuously and monotonically. Therefore it takes the zero value once. Curves $t \mapsto (x_0 t^{w_1}, y_0 t^{w_2})$ are disjoint and fill $Q_{\varepsilon,\delta}$. Therefore, the curve $V_{Q_{\varepsilon,\delta}}(a)$ is isotopic to the preimage under $S_{\varepsilon,\delta} \circ h_{\varepsilon,\delta} \circ l$ of the midline of the triangle $\Delta_{\varepsilon,\delta}$ separating the vertices with opposite signs.

Proposition A.1.7 *If a is a peripherally nondegenerate real polynomial in two variables then the topology of a curve $V_{\mathbb{RR}^2}(a)$ (i.e. the topological type of pair $(\mathbb{RR}^2, V_{\mathbb{RR}^2}(a))$) and the topology of its closure in $\mathbb{R}^2$, $\mathbb{R}P^2$ and other toric extensions of $\mathbb{RR}^2$ can be recovered from a chart of a.*

The part of this proposition concerning to $V_{\mathbb{RR}^2}(a)$ follows from Proposition A.1.5. See below Sects. A.2 and A.3 about toric extensions of $\mathbb{RR}^2$ and closures of $V_{\mathbb{RR}^2}(a)$ in them. In the next subsection algorithms recovering the topology of closures of $V_{\mathbb{RR}^2}(a)$ in $\mathbb{R}^2$ and $\mathbb{R}P^2$ from a chart of a are described.

A.1.4 Recovering the Topology of a Curve from a Chart of the Polynomial

First, I shall describe an auxiliary algorithm which is a block of two main algorithms of this section.

Proposition A.1.8 (Algorithm. Adjoining a side with normal vector (α, β)) Initial data: *a chart (Δ_*, υ) of a polynomial.*

If Δ $(= \Delta_{++})$ has a side Γ with $(\alpha, \beta) \in DC^-_\Delta(\Gamma)$ then the algorithm does not change (Δ_*, υ). Otherwise:

1. Drawn the lines of support of Δ orthogonal to (α, β).

2. Take the point belonging to Δ on each of the two lines of support, and join these points with a segment.

3. Cut the polygon Δ along this segment.

4. Move the pieces obtained aside from each other by parallel translations defined by vectors whose difference is orthogonal to (α, β).

5. Fill the space obtained between the pieces with a parallelogram whose opposite sides are the edges of the cut.

6. Extend the operations applied above to Δ to Δ_* using symmetries $S_{\varepsilon,\delta}$.

7. Connect the points of edges of the cut obtained from points of υ with segments which are parallel to the other pairs of the sides of the parallelograms inserted,

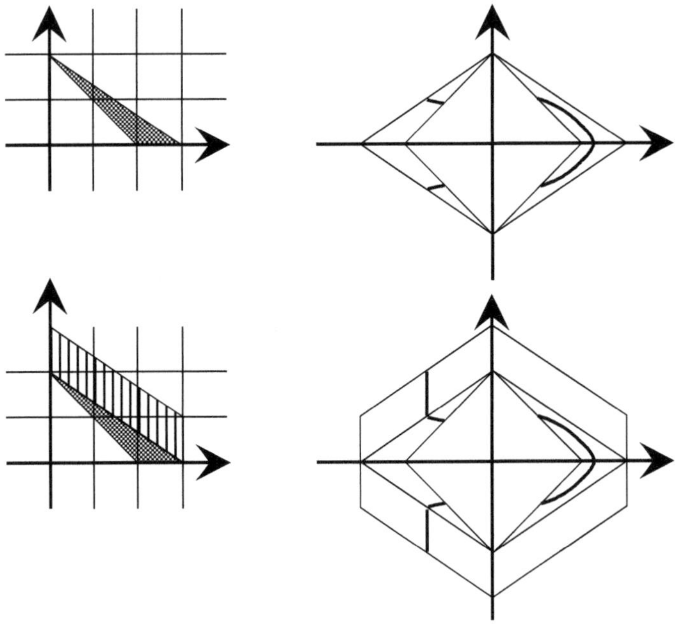

Fig. A.12 Extension of the chart by adjoining a side

and adjoin these segments to what is obtained from υ. The result and the polygon obtained from Δ_* constitute the chart produced by the algorithm.

Example A.1.8.1 In Fig. A.12 the steps of Algorithm A.1.8 are shown. It is applied to $(\alpha, \beta) = (-1, 0)$ and the chart of $8x^3 - x^2 + 4y^2$ shown in Fig. A.11.

Application of Algorithm A.1.8 to a chart of a polynomial a (in the case when it does change the chart) gives rise a chart of polynomial

$$(x^\beta y^{-\alpha} + x^{-\beta} y^\alpha) x^{|\beta|} y^{|\alpha|} a(x, y).$$

If Δ is a segment (i.e. the initial polynomial is quasi-homogeneous) and this segment is not orthogonal to the vector (α, β) then Algorithm A.1.8 gives rise to a chart consisting of four parallelograms, each of which contains as many parallel segments as components of the curve are contained in corresponding quadrant.

Proposition A.1.9 (Algorithm) RECOVERING THE TOPOLOGY OF AN AFFINE CURVE FROM A CHART OF THE POLYNOMIAL. Initial data: *a chart (Δ_*, υ) of a polynomial.*

1. Apply Algorithm A.1.8 with $(\alpha, \beta) = (0, -1)$ to (Δ_*, υ). Assign the former notation (α, β) to the result obtained.

2. Apply Algorithm A.1.8 with $(\alpha, \beta) = (0, -1)$ to (Δ_*, υ). Assign the former notation (α, β) to the result obtained.

3. Glue by $S_{+,-}$ the sides of $\Delta_{+,\delta}$, $\Delta_{-,\delta}$ which are faced to each other and parallel to $(0, 1)$ *(unless the sides coincide).*

4. Glue by $S_{-,+}$ the sides of $\Delta_{\varepsilon,+}$, $\Delta_{\varepsilon,-}$ which are faced to each other and parallel to $(1, 0)$ *(unless the sides coincide).*

5. Contract to a point all sides obtained from the sides of Δ whose normals are directed into quadrant $P_{-,-}$.

6. Remove the sides which are not touched on in blocks 3, 4 and 5.

Algorithm A.1.9 turns the polygon Δ_* to a space Δ' which is homeomorphic to $\mathbb{R}^2$, and the set υ to a set $\upsilon' \subset \Delta'$ such that the pair (Δ', υ') is homeomorphic to $(\mathbb{R}^2, \mathcal{C}\ell\ V_{\mathbb{R}\mathbb{R}^2}(a))$, where $\mathcal{C}\ell$ denotes closure and a is a polynomial whose chart is (Δ_*, υ).

Example A.1.9.1 In Fig. A.13 the steps of Algorithm A.1.9 applying to a chart of polynomial $8x^3y - x^2y + 4y^3$ are shown.

Proposition A.1.10 (Algorithm) RECOVERING THE TOPOLOGY OF A PROJECTIVE CURVE FROM A CHART OF THE POLYNOMIAL. Initial data: *a chart (Δ_*, υ) of a polynomial.*

1. Block 1 of Algorithm A.1.9.

2. Block 2 of Algorithm A.1.9.

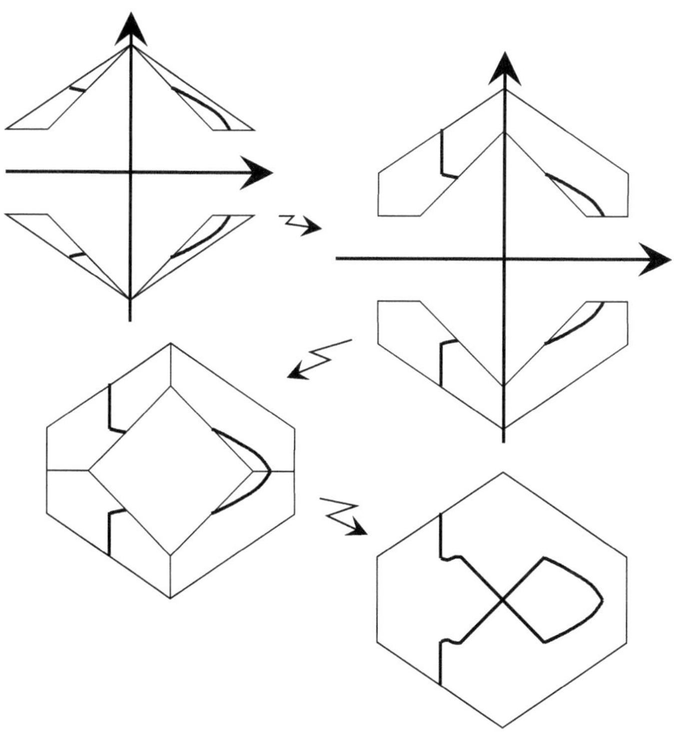

Fig. A.13 Recovering the topology of a projective curve from a chart of the polynomial

3. Apply Algorithm A.1.8 with $(\alpha, \beta) = (1, 1)$ *to* (Δ_*, υ)*. Assign the former notation* (Δ_*, υ) *to the result obtained.*

4. Block 3 of Algorithm A.1.9.

5. Block 4 of Algorithm A.1.9.

6. Glue by $S_{-,-}$ *the sides of* Δ_{++} *and* Δ_{--} *which are faced to each other and orthogonal to* $(1, 1)$.

7. Glue by $S_{-,-}$ *the sides of* Δ_{+-} *and* Δ_{-+} *which are faced to each other and orthogonal to* $(1, -1)$.

8. Block 5 of Algorithm A.1.9.

9. Contract to a point all sides obtained from the sides of Δ *with normals directed into the angle* $\{(x, y) \in \mathbb{R}^2 \mid x < 0,\ y + x > 0\}$.

10. Contract to a point all sides obtained from the sides of Δ *with normals directed into the angle* $\{(x, y) \in \mathbb{R}^2 \mid y < 0,\ y + x > 0\}$.

Algorithm A.1.10 turns polygon Δ_* to a space Δ' which is homeomorphic to projective plane $\mathbb{R}P^2$, and the set υ to a set υ' such that the pair (Δ', υ') is homeomorphic to $(\mathbb{R}P^2, V_{\mathbb{R}\mathbb{R}^2}(a))$, where a is the polynomial whose chart is the initial pair (Δ_*, υ).

A.1.5 Patchworking Charts

Let $a_1, \dots, a_s$ be peripherally nondegenerate real polynomials in two variables with $\operatorname{Int} \Delta(a_i) \cap \operatorname{Int} \Delta(a_j) = \varnothing$ for $i \neq j$. A pair (Δ_*, υ) is said to be obtained by *patchworking* if $\Delta = \bigcup_{i=1}^{s} \Delta(a_i)$ and there exist charts $(\Delta(a_i)_*, \upsilon_i)$ of $a_1, \dots, a_s$ such that $\upsilon = \bigcup_{i=1}^{s} \upsilon_i$.

Example A.1.10.1 In Figs. A.11 and A.14 charts of polynomials $8x^3 - x^2 + 4y^2$ and $4y^2 - x^2 + 1$ are shown. In Fig. A.15 the result of patchworking these charts is shown.

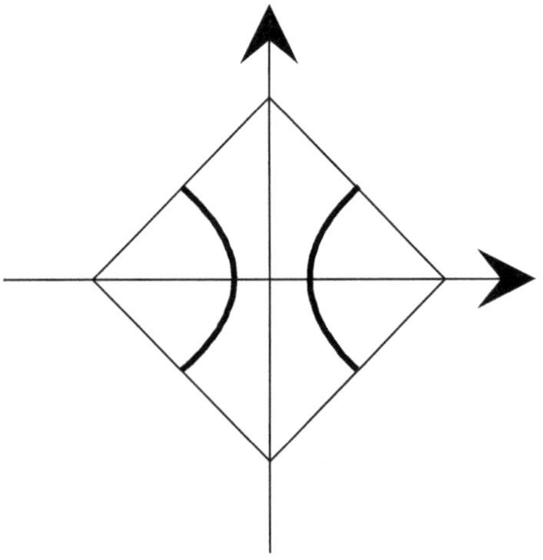

Fig. A.14 The chart of the polynomial $4y^2 - x^2 + 1$

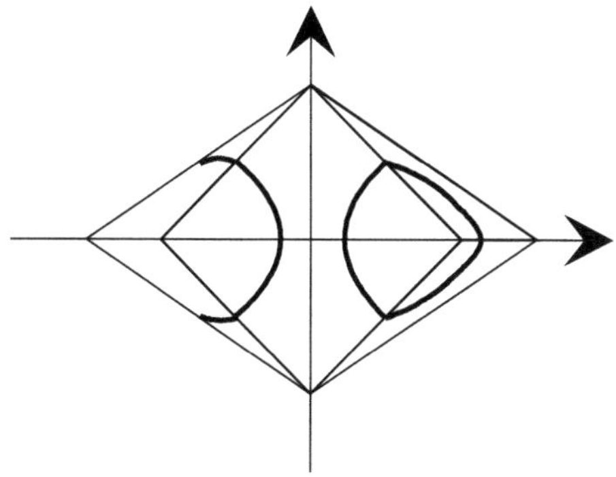

Fig. A.15 Patchworking of the charts of the polynomials $8x^3 - x^2 + 4y^2$ and $4y^2 - x^2 + 1$

A.1.6 Patchworking Polynomials

Let $a_1, \ldots, a_s$ be real polynomials in two variables with $\operatorname{Int} \Delta(a_i) \cap \operatorname{Int} \Delta(a_j) = \varnothing$ for $i \neq j$ and $a_i^{\Delta(a_i)\cap\Delta(a_j)} = a_j^{\Delta(a_i)\cap\Delta(a_j)}$ for any i, j. Suppose the set $\Delta = \bigcup_{i=1}^{s} \Delta(a_i)$ is convex. Then, obviously, there exists the unique polynomial a with $\Delta(a) = \Delta$ and $a^{\Delta(a_i)} = a_i$ for $i = 1, \ldots, s$.

Let $\nu : \Delta \to \mathbb{R}$ be a convex function such that:

1. restrictions $\nu|_{\Delta(a_i)}$ are linear;
2. if the restriction of ν to an open set is linear then the set is contained in one of $\Delta(a_i)$;
3. $\nu(\Delta \cap \mathbb{Z}^2) \subset \mathbb{Z}$.

Then ν is said to *convexify* the partition $\Delta(a_1), \ldots, \Delta(a_s)$ of Δ.

If $a(x, y) = \sum_{\omega\in\mathbb{Z}^2} a_\omega x^{\omega_1} y^{\omega_2}$ then we put

$$b_t(x, y) = \sum_{\omega\in\mathbb{Z}^2} a_\omega x^{\omega_1} y^{\omega_2} t^{\nu(\omega_1,\omega_2)}$$

and say that polynomials b_t are obtained by patchworking $a_1, \ldots, a_s$ by ν.

Example A.1.10.2 Let $a_1(x, y) = 8x^3 - x^2 + 4y^2$, $a_2(x, y) = 4y^2 - x^2 + 1$ and

$$\nu(\omega_1, \omega_2) = \begin{cases} 0, & \text{if} \quad \omega_1 + \omega_2 \geq 2 \\ 2 - \omega_1 - \omega_2, & \text{if} \quad \omega_1 + \omega_2 \leq 2. \end{cases}$$

Then $b_t(x, y) = 8x^3 - x^2 + 4y^2 + t^2$.

A.1.7 The Main Patchwork Theorem

A real polynomial a in two variables is said to be *completely nondegenerate* if it is peripherally nondegenerate (i.e. for any side Γ of its Newton polygon the curve $V_{\mathbb{RR}^2}(a^\Gamma)$ is nonsingular) and the curve $V_{\mathbb{RR}^2}(a)$ is nonsingular.

Proposition A.1.11 *If $a_1, \dots, a_s$ are completely nondegenerate polynomials satisfying all conditions of Sect. A.1.6, and b_t are obtained from them by patchworking by some nonnegative convex function ν convexifying $\Delta(a_1), \dots, \Delta(a_s)$, then there exists $t_0 > 0$ such that for any $t \in (0, t_0]$ the polynomial b_t is completely nondegenerate and its chart is obtained by patchworking charts of $a_1, \dots, a_s$.*

By Propositions A.1.6 and A.1.11 generalizes Proposition A.1.4. Theorem generalizing Proposition A.1.11 is proven in Sect. A.4.3. Here we restrict ourselves to several examples.

Example A.1.11.1 Polynomial $8x^3 - x^2 + 4y^2 + t^2$ with $t > 0$ small enough has the chart shown in Fig. A.15. See Examples A.1.10.1 and A.1.10.2.

In the next section there are a number of considerably more complicated examples demonstrating efficiency of Proposition A.1.11 in the topology of real algebraic curves.

A.1.8 Construction of M-Curves of Degree 6

One of central points of the well known 16th Hilbert's problem [Hil01] is the problem of isotopy classification of curves of degree 6 consisting of 11 components (by the Harnack inequality [Har76] the number of components of a curve of degree 6 is at most 11). Hilbert conjectured that there exist only two isotopy types of such curves. Namely, the types shown in Fig. A.16a and b. His conjecture was disproved by Gudkov [GU69] in 1969. Gudkov constructed a curve of degree 6 with ovals' disposition shown in Fig. A.16c and completed solution of the problem of isotopy classification of nonsingular curves of degree 6. In particular, he proved, that any curve of degree 6 with 11 components is isotopic to one of the curves of Fig. A.16.

Gudkov proposed twice — in [Gud73] and [Gud71] — simplified proofs of realizability of the third isotopy type. His constructions, however, are essentially more complicated than the construction described below, which is based on Proposition A.1.11 and besides gives rise to realization of the other two types, and, after a small modification, realization of almost all isotopy types of nonsingular plane projective real algebraic curves of degree 6 (see [Vir89]).

Construction In Fig. A.17 two curves of degree 6 are shown. Each of them has one singular point at which three nonsingular branches are second order tangent to each other (i.e. this singularity belongs to type J_{10} in the Arnol'd classification [AVGZ82]). The curves of Fig. A.17a and b are easily constructed by the Hilbert method [Hil91], see in [Vir89], Section 4.2.

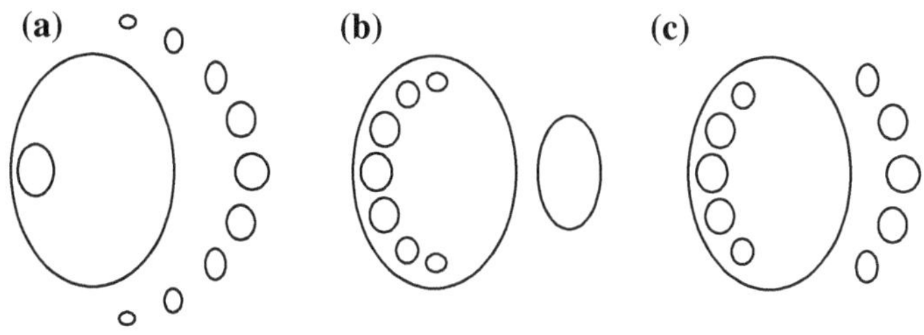

Fig. A.16 Real curves of degree 6 with 11 components

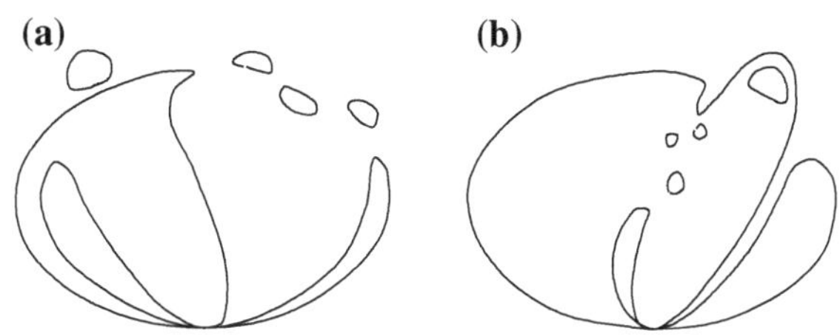

Fig. A.17 Curves of degree 6 with singularity J_{10}

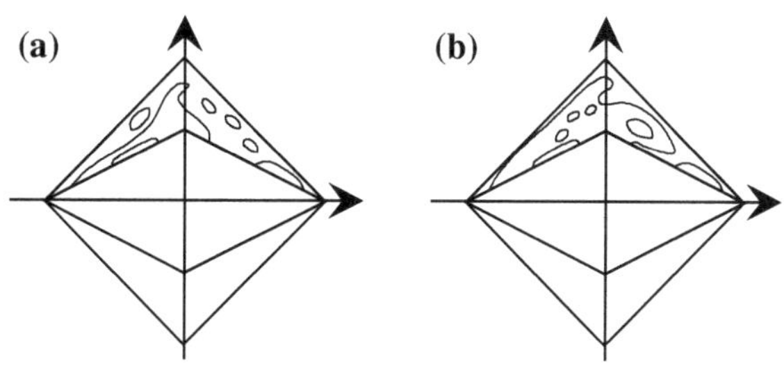

Fig. A.18 Charts of the curves shown in Fig. A.17

Choosing in the projective plane various affine coordinate systems, one obtains various polynomials defining these curves. In Figs. A.18 and A.19 charts of four polynomials appeared in this way are shown. In Fig. A.20 the results of patchworking charts of Figs. A.18 and A.19 are shown. All constructions can be done in such a way that Proposition A.1.11 (see [Vir89], Section 4.2) may be applied to the corresponding polynomials. It ensures existence of polynomials with charts shown in Fig. A.20.

A.1.9 Behavior of Curve $V_{\mathbb{R}\mathbb{R}^2}(b_t)$ as $t \to 0$

Let $a_1, \dots, a_s$, Δ and ν be as in Sect. A.1.6. Suppose that polynomials $a_1, \dots, a_s$ are completely nondegenerate and $\nu|_{\Delta(a_1)} = 0$. According to Proposition A.1.11, the polynomial b_t with sufficiently small $t > 0$ has a chart obtained by patchworking charts of $a_1, \dots, a_s$. Obviously, $b_0 = a_1$ since $\nu|_{\Delta(a_1)} = 0$. Thus when t comes to zero the chart of a_1 stays only, the other charts disappear.

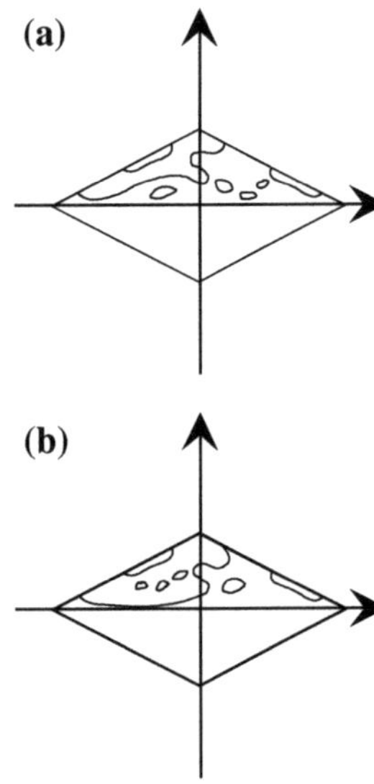

Fig. A.19 Charts of the curves shown in Fig. A.17 after a coordinate change

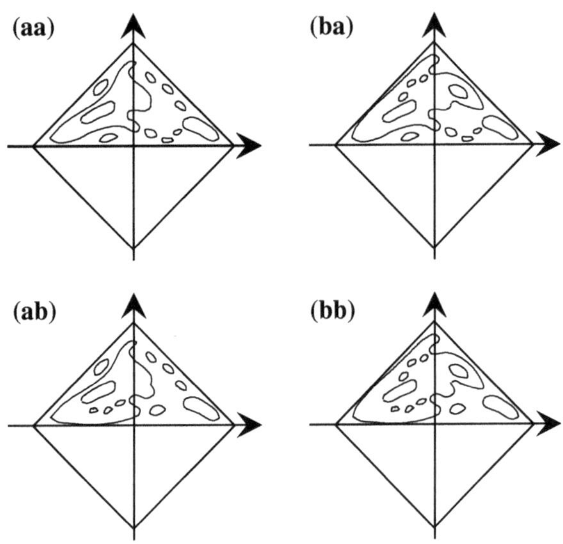

Fig. A.20 Result of the patchworking of the charts shown in Figs. A.18 and A.19

How do the domains containing the pieces of $V_{\mathbb{R}\mathbb{R}^2}(b_t)$ homeomorphic to $V_{\mathbb{R}\mathbb{R}^2}(a_1)$, ..., $V_{\mathbb{R}\mathbb{R}^2}(a_s)$ behave when t approaches zero? They are moving to the coordinate axes and infinity. The closer t to zero, the more place is occupied by the domain, where $V_{\mathbb{R}\mathbb{R}^2}(b_t)$ is organized as $V_{\mathbb{R}\mathbb{R}^2}(a_1)$ and is approximated by it (cf. Sect. A.6.7).

It is curious that the family b_t can be changed by a simple geometric transformation in such a way that the role of a_1 passes to any one of $a_2, \dots, a_s$ or even to a_k^{Γ}, where Γ is a side of $\Delta(a_k)$, $k = 1, \dots, s$. Indeed, let $\lambda : \mathbb{R}^2 \to \mathbb{R}$ be a linear function, $\lambda(x, y) = \alpha x + \beta y + \gamma$. Let $\nu' = \nu - \lambda$. Denote by b'_t the result of patchworking $a_1, \dots, a_s$ by ν'. Denote by $qh_{(a,b),t}$ the linear transformation $\mathbb{RR}^2 \to \mathbb{RR}^2 : (x, y) \mapsto (xt^a, yt^b)$. Then

$$V_{\mathbb{RR}^2}(b'_t) = V_{\mathbb{RR}^2}(b_t \circ qh_{(-\alpha,-\beta),t}) = qh_{(\alpha,\beta),t} V_{\mathbb{RR}^2}(b_t).$$

Indeed,

$$\begin{aligned} b'_t(x, y) &= \sum a_\omega x^{\omega_1} y^{\omega_2} t^{\nu(\omega_1,\omega_2)-\alpha\omega_1-\beta\omega_2-\gamma} \\ &= t^{-\gamma} \sum a_\omega (xt^{-\alpha})^{\omega_1} (yt^{-\beta})^{\omega_2} t^{\nu(\omega_1,\omega_2)} \\ &= t^{-\gamma} b_t(xt^{-\alpha}, yt^{-\beta}) \\ &= t^{-\gamma} b_t \circ qh_{(-\alpha,-\beta),t}(x, y). \end{aligned}$$

Thus the curves $V_{\mathbb{RR}^2}(b'_t)$ and $V_{\mathbb{RR}^2}(b_t)$ are transformed to each other by a linear transformation. However the polynomial b'_t does not tend to a_1 as $t \to 0$. For example, if $\lambda|_{\Delta(a_k)} = \nu|_{\Delta(a_k)}$ then $\nu'|_{\Delta(a_k)} = 0$ and $b'_t \to a_k$. In this case as $t \to 0$, the domains containing parts of $V_{\mathbb{RR}^2}(b'_t)$, which are homeomorphic to $V_{\mathbb{RR}^2}(a_i)$, with $i \neq k$, run away and the domain in which $V_{\mathbb{RR}^2}(b'_t)$ looks like $V_{\mathbb{RR}^2}(a_k)$ occupies more and more place. If the set, where ν coincides with λ (or differs from λ by a constant), is a side Γ of $\Delta(a_k)$, then the curve $V_{\mathbb{RR}^2}(b'_t)$ turns to $V_{\mathbb{RR}^2}(a_k^\Gamma)$ (i.e. collection of quasilines) as $t \to 0$ similarly.

The whole picture of evolution of $V_{\mathbb{RR}^2}(b_t)$ when $t \to 0$ is the following. The fragments which look as $V_{\mathbb{RR}^2}(a_i)$ with $i = 1, \dots, s$ become more and more explicit, but these fragments are not staying. Each of them is moving away from the others. The only fragment that is growing without moving corresponds to the set where ν is constant. The other fragments are moving away from it. From the metric viewpoint some of them (namely, ones going to the origin and axes) are contracting, while the others are growing. But in the logarithmic coordinates, i.e. being transformed by $l : (x, y) \mapsto (\ln|x|, \ln|y|)$, all the fragments are growing (see Sect. A.6.7). Changing ν we are applying linear transformation, which distinguishes one fragment and casts away the others. The transformation turns our attention to a new piece of the curve. It is as if we would transfer a magnifying lens from one fragment of the curve to another. Naturally, under such a magnification the other fragments disappear at the moment $t = 0$.

A.1.10 Patchworking as Smoothing of Singularities

In the projective plane the passage from curves defined by b_t with $t > 0$ to the curve defined by b_0 looks quite differently. Here, the domains, in which the curve defined by b_t looks like curves defined by $a_1, \dots, a_s$ are not running away, but pressing more closely to the points $(1 : 0 : 0)$, $(0 : 1 : 0)$, $(0 : 0 : 1)$ and to the axes joining them. At $t = 0$, they are pressed into the points and axes. It means that under the inverse passage (from $t = 0$ to $t > 0$) the full or partial smoothing of singularities concentrated at the points $(1:0:0)$, $(0:1:0)$, $(0:0:1)$ and along coordinate axes happens.

Example A.1.11.2 Let a_1, a_2 be polynomials of degree 6 with $a_1^{\Delta(a_1)\cap\Delta(a_2)} = a_2^{\Delta(a_1)\cap\Delta(a_2)}$ and charts shown in Figs. A.18a and A.19b. Let ν_1, ν_2 and ν_3 be defined by the following formulas:

$$\nu_1(\omega_1,\omega_2) = \begin{cases} 0, & \text{if } \omega_1 + 2\omega_2 \le 6 \\ 2(\omega_1 + 2\omega_2 - 6), & \text{if } \omega_1 + 2\omega_2 \ge 6 \end{cases}$$

$$\nu_2(\omega_1,\omega_2) = \begin{cases} 6 - \omega_1 - 2\omega_2, & \text{if } \omega_1 + 2\omega_2 \le 6 \\ \omega_1 + 2\omega_2 - 6, & \text{if } \omega_1 + 2\omega_2 \ge 6 \end{cases}$$

$$\nu_3(\omega_1,\omega_2) = \begin{cases} 2(6 - \omega_1 - 2\omega_2), & \text{if } \omega_1 + 2\omega_2 \le 6 \\ 0, & \text{if } \omega_1 + 2\omega_2 \ge 6 \end{cases}$$

(note, that ν_1, ν_2 and ν_3 differ from each other by a linear function). Let b_t^1, b_t^2 and b_t^3 be the results of patchworking a_1, a_2 by ν_1, ν_2 and ν_3. By Proposition A.1.11 for sufficiently small $t > 0$ the polynomials b_t^1, b_t^2 and b_t^3 have the same chart shown in Fig. A.20a, b, but as $t \to 0$ they go to different polynomials, namely, a_1, $a_1^{\Delta(a_1)\cap\Delta(a_2)}$ and a_2.The closure of $V_{\mathbb{RR}^2}(b_t^i)$ with $i = 1, 2, 3$ in the projective plane (they are transformed to one another by projective transformations) are shown in Fig. A.21. The limiting projective curves, i.e. the projective closures of $V_{\mathbb{RR}^2}(a_1)$, $V_{\mathbb{RR}^2}(a_1^{\Delta(a_1)\cap\Delta(a_2)})$, $V_{\mathbb{RR}^2}(a_2)$ are shown in Fig. A.22. The curve shown in Fig. A.22b is the union of three nonsingular conics which are tangent to each other in two points.

Fig. A.21 Sextic curves obtained by the patchworking shown in Fig. A.20(ab) for different functions ν_1, ν_2, ν_3

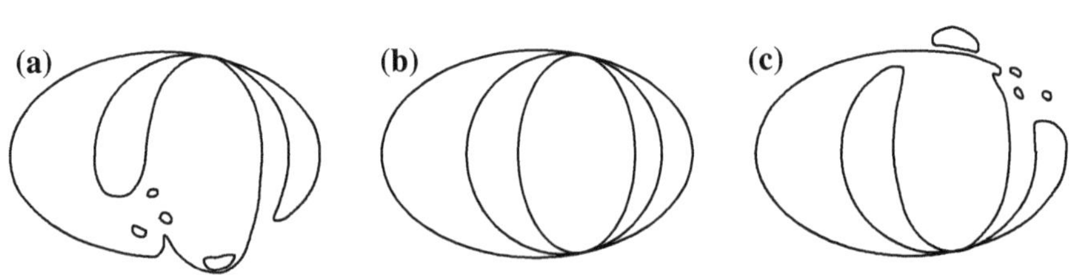

Fig. A.22 Sextic curves with the charts shown in Figs. A.18(a) and A.19(b) and their common limit (the middle figure)

Curves of degree 6 with eleven components of all three isotopy types can be obtained from this curve by small perturbations of the type under consideration (cf. Sect. A.1.8). Moreover, as it is proven in [Vir89], Section 5.1, nonsingular curves of degree 6 of almost all isotopy types can be obtained.

A.1.11 Evolvings of Singularities

Let f be a real polynomial in two variables. (See Section 5, where more general situation with an analytic function playing the role of f is considered.) Suppose its Newton polygon $\Delta(f)$ intersects both coordinate axes (this assumption is equivalent to the assumption that $V_{\mathbb{R}^2}(f)$ is the closure of $V_{\mathbb{R}\mathbb{R}^2}(f)$). Let the distance from the origin to $\Delta(f)$ be more than 1 or, equivalently, the curve $V_{\mathbb{R}^2}(f)$ has a singularity at the origin. Let this singularity be isolated. Denote by B a disk with the center at the origin having sufficiently small radius such that the pair $(B, V_B(f))$ is homeomorphic to the cone over its boundary $(\partial B, V_{\partial B}(f))$ and the curve $V_{\mathbb{R}^2}(f)$ is transversal to ∂B (see [Mil68], Theorem 2.10).

Let f be included into a continuous family f_t of polynomials in two variables: $f = f_0$. Such a family is called a *perturbation* of f. We shall be interested mainly in perturbations for which curves $V_{\mathbb{R}^2}(f_t)$ have no singular points in B when t is in some segment of type $(0, \varepsilon]$. One says about such a perturbation that it *evolves* the singularity of $V_{\mathbb{R}^2}(f_t)$ at zero. If perturbation f_t evolves the singularity of $V_{\mathbb{R}^2}(f)$ at zero then one can find $t_0 > 0$ such that for $t \in (0, t_0]$ the curve $V_{\mathbb{R}^2}(f_t)$ has no singularities in B and, moreover, is transversal to ∂B. Obviously, there exists an isotopy $h_t : B \to B$ with $t_0 \in (0, t_0]$ such that $h_{t_0} = \mathrm{id}$ and $h_t(V_B(f_0)) = V_B(f_t)$, so all pairs $(B, V_B(f_t))$ with $t \in (0, t_0]$ are homeomorphic to each other. A family $(B, V_{\mathbb{R}^2}(f_t))$ of pairs with $t \in (0, t_0]$ is called an *evolving* of singularity of $V_{\mathbb{R}^2}(f)$ at zero, or an *evolving* of germ of $V_{\mathbb{R}^2}(f)$.

Denote by $\Gamma_1, \dots, \Gamma_n$ the sides of Newton polygon $\Delta(f)$ of the polynomial f, faced to the origin. Their union $\Gamma(f) = \bigcup_{i=1}^n \Gamma_i$ is called the *Newton diagram* of f.

Suppose the curves $V_{\mathbb{R}\mathbb{R}^2}(f^{\Gamma_i})$ with $i = 1, \dots, n$ are nonsingular. Then, according to Newton [New67], the curve $V_{\mathbb{R}^2}(f)$ is approximated by the union of $\mathcal{C}\ell\ V_{\mathbb{R}\mathbb{R}^2}(f^{\Gamma_i})$ with $i = 1, \dots, n$ in a sufficiently small neighborhood of the origin. (This is a local version of Proposition A.1.5; it is, as well as Proposition A.1.5, a corollary of Theorem A.6.10.) Disk B can be taken so small that $V_{\partial B}(f)$ is close to $\partial B \cap V_{\mathbb{R}\mathbb{R}^2}(f^{\Gamma_i})$, so the number and disposition of these points are defined by charts $(\Gamma_{i*}, \upsilon_i)$ of f_i^{Γ}. The union $(\Gamma(f)_*, \upsilon) = (\bigcup_{i=1}^n \Gamma_{i*}, \bigcup_{i=1}^n \upsilon_i)$ of these charts is called a *chart of germ* of $V_{\mathbb{R}^2}(f)$ at zero. It is a pair consisting of a simple closed polygon $\Gamma(f_*)$, which is symmetric with respect to the axes and encloses the origin, and finite set υ lying on it. There is a natural bijection of this set to $V_{\partial B}(f)$, which is extendable to a homeomorphism $(\Gamma(f)_*, \upsilon) \to (\partial B, V_{\partial B}(f))$. Denote this homeomorphism by g.

Let f_t be a perturbation of f, which evolves the singularity at the origin. Let B, t_0 and h_t be as above. It is not difficult to choose an isotopy $h_t : B \to B, t \in (0, t_0]$ such that its restriction to ∂B can be extended to an isotopy $h'_t : \partial B \to \partial B$ with

$t \in [0, t_0]$ and $h_0'(V_{\partial B}(f_{t_0})) = V_{\partial B}(f)$. A pair (Π, τ) consisting of the polygon Π bounded by $\Gamma(f)_*$ and an 1-dimensional subvariety τ of Π is called a *chart of evolving* $(B, V_B(f_t)), t \in [0, t_0]$ if there exists a homeomorphism $\Pi \to B$, mapping τ to $V_{\partial B}(f_{t_0})_*$, whose restriction $\partial\Pi \to \partial\Pi$ is the composition $\Gamma(f)_* @ > g >> \partial B @ > h_0' >> \partial B$. It is clear that the boundary $(\partial\Pi, \partial\tau)$ of a chart of germ's evolving is a chart of the germ. Also it is clear that if polynomial f is completely nondegenerate and polygons $\Delta(f_t)$ are obtained from $\Delta(f)$ by adjoining the region restricted by the axes and $\Pi(f)$, then charts of f_t with $t \in (0, t_0]$ can be obtained by patchworking a chart of f and chart of evolving $(B, V_B(f_t)), t \in [0, t_0]$.

The patchworking construction for polynomials gives a wide class of evolvings whose charts can be created by the modification of Proposition A.1.11 formulated below.

Let $a_1, \ldots, a_s$ be completely nondegenerate polynomials in two variables with $\operatorname{Int} \Delta(a_i) \cap \operatorname{Int} \Delta(a_j) = \varnothing$ and $a_i^{\Delta(a_i)\cap\Delta(a_j)} = a_j^{\Delta(a_i)\cap\Delta(a_j)}$ for $i \neq j$. Let $\bigcup_{i=1}^s \Delta(a_i)$ be a polygon bounded by the axes and $\Gamma(f)$. Let $a_i^{\Delta(a_i)\cap\Delta(f)} = f^{\Delta(a_i)\cap\Delta(f)}$ for $i = a, \ldots, s$. Let $\nu : \mathbb{R}^2 \to \mathbb{R}$ be a nonnegative convex function which is equal to zero on $\Delta(f)$ and whose restriction on $\bigcup_{i=1}^s \Delta(a_i)$ satisfies the conditions 1, 2 and 3 of Sect. A.1.6 with respect to $a_1, \ldots, a_s$. Then a result f_t of patchworking $f, a_1, \ldots, a_s$ by ν is a perturbation of f.

Proposition A.1.11 cannot be applied in this situation because the polynomial f is not supposed to be completely nondegenerate. This weakening of assumption implies a weakening of conclusion.

Proposition A.1.12 (Local version of Proposition A.1.11) *Under the conditions above perturbation f_t of f evolves a singularity of $V_{\mathbb{R}^2}(f)$ at the origin. A chart of the evolving can be obtained by patchworking charts of $a_1, \ldots, a_s$.*

An evolving of a germ, constructed along the scheme above, is called a *patchwork evolving*.

If $\Gamma(f)$ consists of one segment and the curve $V_{\mathbb{R}\mathbb{R}^2}(f^{\Gamma(f)})$ is nonsingular then the germ of $V_{\mathbb{R}^2}(f)$ at zero is said to be *semi-quasi-homogeneous*. In this case for construction of evolving of the germ of $V_{\mathbb{R}^2}(f)$ according the scheme above we need only one polynomial; by Proposition A.1.12, its chart is a chart of evolving. In this case geometric structure of $V_B(f_t)$ is especially simple, too: the curve $V_B(f_t)$ is approximated by $qh_{w,t}(V_{\mathbb{R}^2}(a_1))$, where w is a vector orthogonal to $\Gamma(f)$, that is by the curve $V_{\mathbb{R}^2}(a_1)$ contracted by the quasihomothety $qh_{w,t}$. Such evolvings were described in my paper [Vir80]. It is clear that any patchwork evolving of semi-quasi-homogeneous germ can be replaced, without changing its topological models, by a patchwork evolving, in which only one polynomial is involved (i.e. $s = 1$).

A.2 Toric Varieties and Their Hypersurfaces

A.2.1 Algebraic Tori $K\mathbb{R}^n$

In the rest of this chapter K denotes the main field, which is either the real number field $\mathbb{R}$, or the complex number field $\mathbb{C}$.

For $\omega = (\omega_1, \dots, \omega_n) \in \mathbb{Z}^n$ and ordered collection x of variables $x_1, \dots, x_n$ the product $x_1^{\omega_1} \dots x_n^{\omega_n}$ is denoted by x^ω. A linear combination of products of this sort with coefficients from K is called a *Laurent polynomial* or, briefly, *L-polynomial* over K. Laurent polynomials over K in n variables form a ring $K[x_1, x_1^{-1}, \dots, x_n, x_n^{-1}]$ naturally isomorphic to the ring of regular functions of the variety $(K \smallsetminus 0)^n$.

Below this variety, side by side with the affine space K^n and the projective space KP^n, is one of the main places of action. It is an algebraic torus over K. Denote it by $K\mathbb{R}^n$.

Denote by l the map $K\mathbb{R}^n \to \mathbb{R}^n$ defined by formula $l(x_1, \dots, x_n) = (\ln|x_1|, \dots, \ln|x_n|)$.

Put $U_K = \{x \in K \mid |x| = 1\}$, so $U_\mathbb{R} = S^0$ and $U_\mathbb{C} = S^1$. Denote by ar the map $K\mathbb{R}^n \to U_K^n$ $(= U_K \times \dots \times U_K)$ defined by $ar(x_1, \dots, x_n) = (\dfrac{x_1}{|x_1|}, \dots, \dfrac{x_n}{|x_n|})$.

Denote by la the map

$$x \mapsto (l(x), ar(x)) : K\mathbb{R}^n \to \mathbb{R}^n \times U_K^n.$$

It is clear that this is a diffeomorphism.

$K\mathbb{R}^n$ is a group with respect to the coordinate-wise multiplication, and l, ar, la are group homomorphisms; la is an isomorphism of $K\mathbb{R}^n$ to the direct product of (additive) group $\mathbb{R}^n$ and (multiplicative) group U_K^n.

Being Abelian group, $K\mathbb{R}^n$ acts on itself by translations. Let us fix notations for some of the translations involved into this action.

For $w \in \mathbb{R}^n$ and $t > 0$ denote by $qh_{w,t}$ and call a *quasi-homothety* with weights $w = (w_1, \dots, w_n)$ and coefficient t the transformation $K\mathbb{R}^n \to K\mathbb{R}^n$ defined by formula $qh_{w,t}(x_1, \dots, x_n) = (t^{w_1}x_1, \dots, t^{w_n}x_n)$, i.e. the translation by $(t^{w_1}, \dots, t^{w_n})$. If $w = (1, \dots, 1)$ then it is the usual homothety with coefficient t. It is clear that $qh_{w,t} = qh_{\lambda^{-1}w,t}$ for $\lambda > 0$. Denote by qh_w a quasi-homothety $qh_{w,e}$, where e is the base of natural logarithms. It is clear, $qh_{w,t} = qh_{(\ln t)w}$.

For $w = (w_1, \dots, w_n) \in U_K^n$ denote by S_w the translation $K\mathbb{R}^n \to K\mathbb{R}^n$ defined by formula

$$S_w(x_1, \dots, x_n) = (w_1x_1, \dots, w_nx_n),$$

i.e. the translation by w.

For $w \in \mathbb{R}^n$ denote by T_w the translation $x \mapsto x + w : \mathbb{R}^n \to \mathbb{R}^n$ by the vector w.

Proposition A.2.1 *Diffeomorphism $la : K\mathbb{R}^n \to \mathbb{R}^n \times U_K^n$ transforms $qh_{w,t}$ to $T_{(\ln t)w} \times \mathrm{id}_{U_K^n}$, and S_w to $\mathrm{id}_{\mathbb{R}^n} \times (S_w|_{U_K^n})$, i.e.*

$$la \circ qh_{w,t} \circ la^{-1} = T_{(\ln t)w} \times \mathrm{id}_{U_K^n} \quad \textit{and}$$

$$la \circ S_w \circ la^{-1} = \mathrm{id}_{\mathbb{R}^n} \times (S_w|_{U_K^n}).$$

□

In particular, $la \circ qh_w \circ la^{-1} = T_w \times \mathrm{id}$.

A hypersurface of $K\mathbb{R}^n$ defined by $a(x) = 0$, where a is a Laurent polynomial over K in n variables is denoted by $V_{K\mathbb{R}^n}(a)$.

If $a(x) = \sum_{\omega \in \mathbb{Z}^n} a_\omega x^\omega$ is a Laurent polynomial, then by its *Newton polyhedron* $\Delta(a)$ is the convex hull of $\{\omega \in \mathbb{R}^n \mid a_\omega \neq 0\}$.

Proposition A.2.2 *Let a be a Laurent polynomial over K. If $\Delta(a)$ lies in an affine subspace Γ of $\mathbb{R}^n$ then for any vector $w \in \mathbb{R}^n$ orthogonal to Γ, a hypersurface $V_{K\mathbb{R}^n}(a)$ is invariant under $qh_{w,t}$.*

Proof Since $\Delta(a) \subset \Gamma$ and $\Gamma \perp w$, then for $\omega \in \Delta(a)$ the scalar product $w\omega$ does not depend on ω. Hence

$$a(qh_{w,t}^{-1}(x)) = \sum_{\omega \in \Delta(a)} a_\omega (t^{-w}x)^\omega = t^{-w\omega} \sum_{\omega \in \Delta(a)} a_\omega x^\omega = t^{-w\omega} a(x),$$

and therefore

$$qh_{w,t}(V_{K\mathbb{R}^n}(a)) = V_{K\mathbb{R}^n}(a \circ qh_{w,t}^{-1}) = V_{K\mathbb{R}^n}(t^{-w\omega}a) = V_{K\mathbb{R}^n}(a).$$

Proposition A.2.2 is equivalent, as it follows from Proposition A.2.1, to the assertion that under hypothesis of Proposition A.2.2 the set $la(V_{K\mathbb{R}^n}(a))$ contains together with each point $(x, y) \in \mathbb{R}^n \times U_K^n$ all points $(x', y) \in \mathbb{R}^n \times U_K^n$ with $x' - x \perp \Gamma$. In other words, in the case $\Delta(a) \subset \Gamma$ the intersections of $la(V_{K\mathbb{R}^n}(a))$ with fibers $\mathbb{R}^n \times y$ are cylinders, whose generators are affine spaces of dimension $n - \dim \Gamma$ orthogonal to Γ.

The following proposition can be proven similarly to Proposition A.2.2.

Proposition A.2.3 *Under the hypothesis of Proposition A.2.2 a hypersurface $V_{K\mathbb{R}^n}(a)$ is invariant under transformations $S_{(e^{\pi i w_1},\dots,e^{\pi i w_n})}$, where $w \perp \Gamma$,*

$$w \in \begin{cases} \mathbb{Z}^n, \ \textit{if}\ K = \mathbb{R} \\ \mathbb{R}^n, \ \textit{if}\ K = \mathbb{C}. \end{cases}$$

□

In other words, under the hypothesis of Proposition A.2.2 the hypersurface $V_{K\mathbb{R}^n}(a)$ contains together with each its point $(x_1, \dots, x_n)$:

1. points $((-1)^{w_1}x_1, \dots, (-1)^{w_n}x_n)$ with $w \in \mathbb{Z}^n$, $w \perp \Gamma$, if $K = \mathbb{R}$,
2. points $(e^{iw_1}x_1, \dots, e^{iw_n}x_n)$ with $w \in \mathbb{R}^n$, $w \perp \Gamma$, if $K = \mathbb{C}$.

A.2.2 Polyhedra and Cones

Below by a *polyhedron* we mean *closed convex* polyhedron lying in $\mathbb{R}^n$, which are not necessarily bounded, but have a finite number of faces. A polyhedron is said to be *integer* if on each of its faces there are enough points with integer coordinates to define the minimal affine space containing this face. All polyhedra considered below are assumed to be integer, unless the contrary is stated.

The set of faces of a polyhedron Δ is denoted by $\mathcal{G}(\Delta)$, the set of its k-dimensional faces by $\mathcal{G}_k(\Delta)$, the set of all its proper faces by $\mathcal{G}'(\Delta)$.

By a *halfspace* of vector space V we will mean the preimage of the closed halfline $\mathbb{R}_+(=\{x \in \mathbb{R} : x \geq 0\})$ under a non-zero linear functional $V \to \mathbb{R}$ (so the boundary hyperplane of a halfspace passes necessarily through the origin). By a *cone* it is called an intersection of a finite collection of halfspaces of $\mathbb{R}^n$. A cone is a polyhedron (not necessarily integer), hence all notions and notations concerning polyhedra are applicable to cones.

The minimal face of a cone is the maximal vector subspace contained in the cone. It is called a *ridge* of the cone.

For $v_1, \dots, v_k \in \mathbb{R}^n$ denote by $\langle v_1, \dots, v_k \rangle$ the minimal cone containing v_1, ..., v_k; it is called the *cone generated by* $v_1, \dots, v_k$. A cone is said to be *simplicial* if it is generated by a collection of linear independent vectors, and *simple* if it is generated by a collection of integer vectors, which is a basis of the free Abelian group of integer vectors lying in the minimal vector space which contains the cone.

Let $\Delta \subset \mathbb{R}^n$ be a polyhedron and Γ its face. Denote by $C_\Delta(\Gamma)$ the cone $\bigcup_{r \in \mathbb{R}_+} r \cdot (\Delta - y)$, where y is a point of $\Gamma \setminus \partial \Gamma$. The cone $C_\Delta(\Delta)$ is clearly the vector subspace of $\mathbb{R}^n$ which corresponds to the minimal affine subspace containing Δ. The cone $C_\Gamma(\Gamma)$ is the ridge of $C_\Delta(\Gamma)$. If Γ is a face of Δ with $\dim \Gamma = \dim \Delta - 1$, then $C_\Delta(\Gamma)$ is a halfspace of $C_\Delta(\Delta)$ with boundary parallel to Γ.

For cone $C \subset \mathbb{R}^n$ we put

$$D^+C = \{x \in \mathbb{R}^n \mid \forall a \in C \quad ax \geq 0\},$$

$$D^-C = \{x \in \mathbb{R}^n \mid \forall a \in C \quad ax \leq 0\}.$$

These are cones, which are said to be *dual* to C. The cones D^+C and D^-C are symmetric to each other with respect to 0. The cone D^-C permits also the following more geometric description. Each hyperplane of support of C defines a ray consisting of vectors orthogonal to this hyperplane and directed to that of two open halfspaces bounded by it, which does not intersect C. The union of all such rays is D^-C.

It is clear that $D^+D^+C = C = D^-D^-C$. If $v_1, \dots, v_n$ is a basis of $\mathbb{R}^n$, then the cone $D^+\langle v_1, \dots, v_n \rangle$ is generated by dual basis $v_1^*, \dots, v_n^*$ (which is defined by conditions $v_i \cdot v_j * = \Delta_{ij}$).

A.2.3 Affine Toric Variety

Let $\Delta \subset \mathbb{R}^n$ be an (integer) cone. Consider the semigroup K-algebra $K[\Delta \cap \mathbb{Z}^n]$ of the semigroup $\Delta \cap \mathbb{Z}^n$. It consists of Laurent polynomials of the form $\sum_{\omega \in \Delta \cap \mathbb{Z}^n} a_\omega x^\omega$. According to the well known Gordan Lemma (see, for example, [Dan78], 1.3), the semigroup $\Delta \cap \mathbb{Z}^n$ is generated by a finite number of elements and therefore the algebra $K[\Delta \cap \mathbb{Z}^n]$ is generated by a finite number of monomials. If this number is greater than the dimension of Δ, then there are nontrivial relations among the generators; the number of relations of minimal generated collection is equal to the difference between the number of generators and the dimension of Δ.

An affine toric variety $K\Delta$ is the affine scheme Spec $K[\Delta \cap \mathbb{Z}^n]$. Its less invariant, but more elementary definition looks as follows. Let

$$\{\alpha_1, \dots, \alpha_p \mid \sum_{i=1}^{p} u_{1,i}\alpha_i = \sum_{i=1}^{p} v_{i,1}\alpha_i, \dots, \sum_{i=1}^{p} u_{p-n,i}\alpha_i = \sum_{i=1}^{p} v_{p-n,i}\alpha_i\}$$

be a presentation of $\Delta \cap \mathbb{Z}^n$ by generators and relations (here u_{ij} and v_{ij} are non-negative); then the variety $K\Delta$ is isomorphic to the affine subvariety of K^p defined by the system

$$\begin{cases} y_1^{u_{11}} \dots y_p^{u_{1p}} = y_1^{v_{11}} \dots y_p^{v_{1p}} \\ \dots\dots\dots\dots\dots\dots\dots\dots \\ y_1^{u_{p-n,1}} \dots y_p^{u_{p-n,p}} = y_1^{v_{p-n,1}} \dots y_p^{v_{p-n,p}}. \end{cases}$$

For example, if $\Delta = \mathbb{R}^n$, then $K\Delta = \text{Spec}\, K[x_1, x_1^{-1}, \dots, x_n, x_n^{-1}]$ can be presented as the subvariety of K^{2n} defined by the system

$$\begin{cases} y_1 y_{n+1} = 1 \\ \dots\dots\dots \\ y_n y_{2n} = 1 \end{cases}$$

Projection $K^{2n} \to K^n$ induces an isomorphism of this subvariety to $(K \setminus 0)^n = K\mathbb{R}^n$. This explains the notation $K\mathbb{R}^n$ introduced above.

If Δ is the positive orthant $A^n = \{x \in \mathbb{R}^n \mid x_1 \geq 0, \dots, x_n \geq 0\}$, then $K\Delta$ is isomorphic to the affine space K^n. The same takes place for any simple cone. If cone is not simple, then corresponding toric variety is necessarily singular. For example, the angle shown in Fig. A.23 corresponds to the cone defined in K^3 by $xy = z^2$.

Let a cone Δ_1 lie in a cone Δ_2. Then the inclusion in : $\Delta_1 \to \Delta_2$ defines an inclusion $K[\Delta_1 \cap \mathbb{Z}^n] \hookrightarrow K[\Delta_2 \cap \mathbb{Z}^n]$ which, in turn, defines a regular map

$$\text{in}^* : \text{Spec}\, K[\Delta_2 \cap \mathbb{Z}^n] \to \text{Spec}\, K[\Delta_1 \cap \mathbb{Z}^n],$$

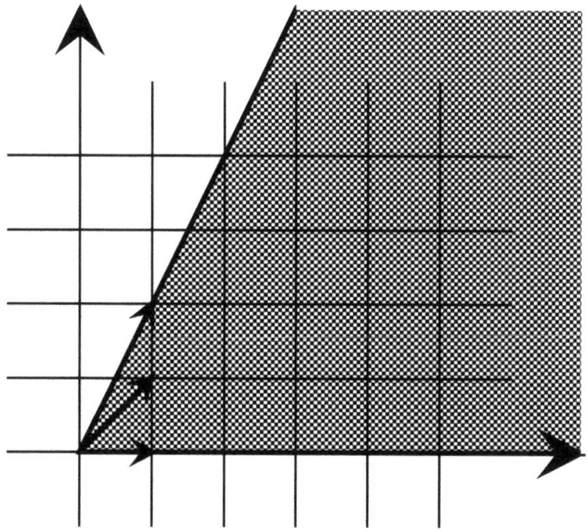

Fig. A.23 The cone defined by the polynomial $xy - z^2$

i.e. a regular map $\text{in}^* : K\Delta_2 \to K\Delta_1$. The latter can be described in terms of subvarieties of affine spaces in the following way. The formulas, defining coordinates of point $\text{in}^*(y)$ as functions of coordinates of y, are the multiplicative versions of formulas, defining generators of semigroup $\Delta_1 \cap \mathbb{Z}^n$ as linear combinations of generators of the ambient semigroup $\Delta_2 \cap \mathbb{Z}^n$.

In particular, for any Δ there is a regular map of $KC_\Delta(\Delta) \cong K\mathbb{R}^{\dim \Delta}$ to $K\Delta$. It is not difficult to prove that it is an open embedding with dense image, thus $K\Delta$ can be considered as a completion of $K\mathbb{R}^{\dim \Delta}$.

An action of algebraic torus $KC_\Delta(\Delta)$ in itself by translations is extended to its action in $K\Delta$. This extension can be obtained, for example, in the following way. Note first, that for defining an action in $K\Delta$ it is sufficient to define an action in the ring $K[\Delta \cap \mathbb{Z}^n]$. Define an action of $K\mathbb{R}^n$ on monomials $x^\omega \in K[\delta \cap \mathbb{Z}^n]$ by formula $(\alpha_1, \dots, \alpha_n)x^\omega = \alpha_1^{\omega_1} \dots \alpha_n^{\omega_n}$ and extend it to the whole ring $K[\Delta \cap \mathbb{Z}^n]$ by linearity. Further, note that if $V \subset \mathbb{R}^n$ is a vector space, then the map $\text{in}^* : K\mathbb{R}^n \to KV$ is a group homomorphism. Elements of kernel of $\text{in}^* : K\mathbb{R}^n \to KC_\Delta(\Delta)$ act identically in $K[\Delta \cap \mathbb{Z}^n]$. It allows to extract from the action of $K\mathbb{R}^n$ in $K\Delta$ an action of $KC^\Delta(\Delta)$ in $K\Delta$, which extends the action of $KC_\Delta(\Delta)$ in itself by translations.

With each face Γ of a cone Δ one associates (as with a smaller cone) a variety $K\Gamma$ and a map $\text{in}^* : K\Delta \to K\Gamma$. On the other hand there exists a map $\text{in}_* : K\Gamma \to K\Delta$ for which $\text{in}^* \circ \text{in}_*$ is the identity map $K\Gamma \to K\Gamma$. Therefore, in_* is an embedding whose image is a retract of $K\Delta$. From the viewpoint of schemes the map in_* should be defined by the homomorphism $K[\Delta \cap \mathbb{Z}^n] \to K[\Gamma \cap \mathbb{Z}^n]$ which maps a Laurent polynomial $\sum_{\omega \in \Delta \cap \mathbb{Z}^n} a_\omega x^\omega$ to its Γ-truncation $\sum_{\omega \in \Gamma \cap \mathbb{Z}^n} a_\omega x^\omega$. In terms of subvarieties of affine space, $K\Gamma$ is the intersection of $K\Delta$ with the subspace $y_{i_1} = y_{i_2} = \dots = y_{i_s} = 0$, where $y_{i_1}, \dots, y_{i_s}$ are the coordinates corresponding to generators of semigroup $\Delta \cap \mathbb{Z}^n$ which do not lie in Γ.

Varieties $\mathrm{in}_*(K\Gamma)$ with $\Gamma \in \mathcal{G}_{\dim \Delta - 1}(\Delta)$ cover $K\Delta \smallsetminus \mathrm{in}^*(KC_\Delta(\Delta))$. Images of algebraic tori $KC_\Gamma(\Gamma)$ with $\Gamma \in \mathcal{G}(\Delta)$ under the composition

$$KC_\Gamma(\Gamma) \xrightarrow{\mathrm{in}^*} K\Gamma \xrightarrow{\mathrm{in}^*} K\Delta$$

of embeddings form a partition of $K\Delta$, which is a smooth stratification of $K\Delta$. Closure of the stratum $\mathrm{in}_* \mathrm{in}^*(KC_\Gamma(\Gamma))$ in $K\Delta$ is $\mathrm{in}_*(K\Gamma)$. Below in the cases when it does not lead to confusion we shall identify $K\Gamma$ with $\mathrm{in}_* K\Gamma$ and $KC_\Gamma(\Gamma)$ with $\mathrm{in}_* \mathrm{in}^* KC_\Gamma(\Gamma)$ (i.e. we shall consider $K\Gamma$ and $KC_\Gamma(\Gamma)$ as lying in $K\Delta$).

A.2.4 *Quasi-projective Toric Variety*

Let $\Delta \subset \mathbb{R}^n$ be a polyhedron. If Γ is its face and Σ is a face of Γ, then $C_\Gamma(\Sigma)$ is a face of $C_\Delta(\Gamma)$ parallel to Γ, and $C_{C_\Delta(\Sigma)}(C_\Gamma(\Sigma)) = C_\Delta(\Gamma)$, see Fig. A.24. In particular, $C_\Delta(\Sigma) \subset C_\Delta(\Gamma)$ and, hence, the map $\mathrm{in}^* : KC_\Delta(\Gamma) \to KC_\Delta(\Sigma)$ is defined. It is easy to see that this is an open embedding. Let us glue all $KC_\Delta(\Gamma)$ with $\Gamma \in \mathcal{G}(\Delta)$ together by these embeddings. The result is denoted by $K\Delta$ and called the *toric variety* associated with Δ. This definition agrees with the corresponding definition from the previous section: if Δ is a cone and Σ is its ridge then $C_\Delta(\Sigma) = \Delta$ and, since the ridge is the minimal face, all $KC_\Delta(\Gamma)$ with $\Gamma \in \mathcal{G}(\Delta)$ are embedded in $KC_\Delta(\Sigma)$ and the gluing gives $KC_\Delta(\Sigma) = K\Delta$.

For any polyhedron Δ the toric variety $K\Delta$ is quasi-projective. If Δ is bounded, it is projective (see [GK73] and [Dan78]).

A polyhedron $\Delta \subset \mathbb{R}^n$ is said to be *permissible* if $\dim \Delta = n$, each face of Δ has a vertex and for any vertex $\Gamma \in \mathcal{G}_0(\Delta)$ the cone $C_\Delta(\Gamma)$ is simple. If polyhedron Δ is permissible then variety $K\Delta$ is nonsingular and it can be obtained by gluing affine spaces $KC_\Delta(\Gamma)$ with $\Gamma \in \mathcal{G}_0(\Delta)$. The gluing allows the following description. Let us associate with each cone $C_\Delta(\Gamma)$ where $\Gamma \in \mathcal{G}_0(\Delta)$ an automorphism $f_\Gamma : K\mathbb{R}^n \to K\mathbb{R}^n$: if $C_\Delta(\Gamma) = \langle v_1, \dots, v_n \rangle$ and $v_i = (v_{i1}, \dots, v_{in})$ for $i = 1, \dots, n$, then we

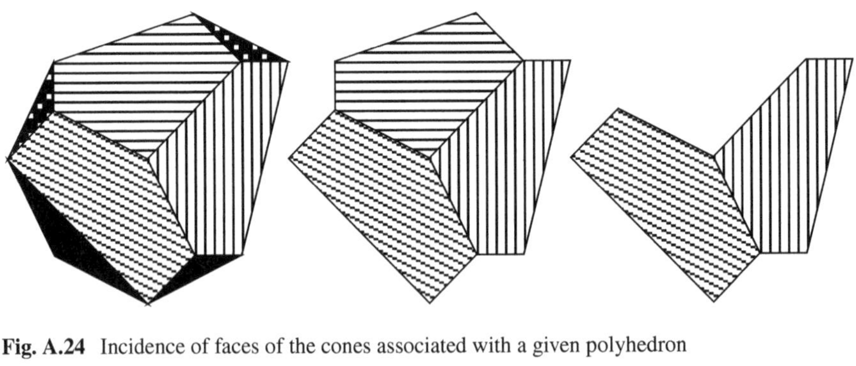

Fig. A.24 Incidence of faces of the cones associated with a given polyhedron

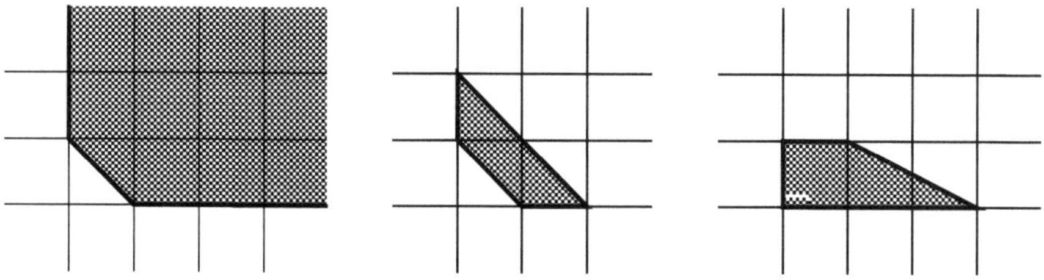

Fig. A.25 Polyhedra defining the affine plane blown up at one point, the projective plane blown up at one point, and a ruled toric surface

put $f_\Gamma(x_1, \dots, x_n) = (x_1^{v_{11}} \dots x_n^{v_{1n}}, \dots, x_1^{v_{n1}} \dots x_n^{v_{nn}})$. The variety $K\Delta$ is obtained by gluing to $K\mathbb{R}^n$ copies of K^n by maps $K\mathbb{R}^n \xrightarrow{f_\Gamma} K\mathbb{R}^n \hookrightarrow K^n$ for all vertices Γ of Δ. (Cf. Khovansky [Kho77].)

The variety $K\Delta$ is defined by Δ, but does not define it. Indeed, if Δ_1 and Δ_2 are polyhedra such that there exists a bijection $\mathcal{G}(\Delta_1) \to \mathcal{G}(\Delta_2)$, preserving dimensions and inclusions and assigning to each face of Δ_1 a parallel face of Δ_2, then $K\Delta_1 = K\Delta_2$.

Denote by P^n the simplex of dimension n with vertices

$$(0, 0, \dots, 0), (1, 0, \dots, 0), (0, 1, 0, \dots, 0), \dots, (0, 0, \dots, 1).$$

It is permissible polyhedron. KP^n is the n-dimensional projective space (this agrees with its usual notation).

Evidently, $K(\Delta_1 \times \Delta_2) = K\Delta_1 \times K\Delta_2$. In particular, if $\Delta \subset \mathbb{R}^2$ is a square with vertices $(0, 0)$, $(1, 0)$, $(0, 1)$ and $(1, 1)$, i.e. if $\Delta = P^1 \times P^1$, then $K\Delta$ is a surface isomorphic to nonsingular projective surface of degree 2 (to hyperboloid in the case of $K = \mathbb{R}^2$).

Polyhedra shown in Fig. A.25 define the following surfaces: $K\Delta_1$ is the affine plane with a point blown up; $K\Delta_2$ is projective plane with a point blown up ($\mathbb{R}\Delta_2$ is the Klein bottle); $K\Delta_3$ is the linear surface over KP^1, defined by sheaf $\mathcal{O} + \mathcal{O}(-2)$ ($\mathbb{R}\Delta_3$ is homeomorphic to torus).

The variety $KC_\Delta(\Delta)$ is isomorphic to $K\mathbb{R}^{\dim \Delta}$, open and dense in $K\Delta$, so $K\Delta$ can be considered as a completion of $K\mathbb{R}^{\dim \Delta}$. Actions of $KC_\Delta(\Delta)$ in affine parts $KC_\Delta(\Gamma)$ of $K\Delta$ correspond to each other and define an action in $K\Delta$ which is an extension of the action of $KC_\Delta(\Delta)$ in itself by translations. Transformations of $K\Delta$ extending $qh_{w,t}$ and S_w are denoted by the same symbols $qh_{w,t}$ and S_w.

The complement $K\Delta \smallsetminus KC_\Delta(\Delta)$ is covered by $K\Sigma$ with $\Sigma \in \mathcal{G}(C_\Delta(\Gamma))$, $\Gamma \in \mathcal{G}'(\Delta)$ or, equivalently, by varieties $KC_\Gamma(\Sigma)$ with $\Sigma \in \mathcal{G}(\Gamma)$, $\Gamma \in \mathcal{G}'(\Delta)$. They comprise varieties $K\Gamma$ with $\Gamma \in \mathcal{G}'(\Delta)$, which also cover $K\Delta \smallsetminus KC_\Delta(\Delta)$. The varieties $K\Gamma$ are situated with respect to each other in the same manner as the corresponding faces in the polyhedron: $K(\Gamma_1 \cap \Gamma_2) = K\Gamma_1 \cap K\Gamma_2$. Algebraic tori $KC_\Gamma(\Gamma) = K\Gamma \smallsetminus \bigcup_{\Sigma \in \mathcal{G}'(\Gamma)} K\Sigma$ form partition of $K\Delta$, which is a smooth stratification; they are orbits of the action of $KC_\Delta(\Delta)$ in $K\Delta$.

We shall say that a polyhedron Δ_2 is *richer* than a polyhedron Δ_1 if for any face $\Gamma_2 \in \mathcal{G}(\Delta_2)$ there exists a face $\Gamma_1 \in \mathcal{G}(\Delta_1)$ such that $C_{\Delta_2}(\Gamma_2) \supset C_{\Delta_1}(\Gamma_1)$ (such a face Γ_1 is automatically unique), and for each face $\Gamma_1 \in \mathcal{G}(\Delta_1)$ the cone $C_{\Delta_1}(\Gamma_1)$ can be presented as the intersection of several cones $C_{\Delta_2}(\Gamma_2)$ with $\Gamma_1 \in \mathcal{G}(\Delta_2)$. This definition allows a convenient reformulation in terms of dual cones: a polyhedron Δ_2 is richer than polyhedron Δ_1 iff the cones $D^+C_{\Delta_2}(\Gamma_2)$ with $\Gamma_2 \in \mathcal{G}(\Delta_2)$ cover the set, which is covered by $D^+C_{\Delta_1}(\Gamma_1)$ with $\Gamma_1 \in \mathcal{G}(\Delta_1)$, and the first covering is a refinement of the second.

Let a polyhedron Δ_2 be richer than Δ_1. Then the inclusions $C_{\Delta_1}(\Gamma_1) \hookrightarrow C_{\Delta_2}(\Gamma_2)$ define for any $\Gamma_2 \in \mathcal{G}(\Delta_2)$ a regular map $KC_{\Delta_2}(\Gamma_2) \xrightarrow{\text{in}^*} KC_{\Delta_1}(\Gamma_1) \hookrightarrow K\Delta_1$. Obviously, these maps commute with the embeddings, by which $K\Delta_2$ and $K\Delta_1$ are glued from affine pieces, thus a regular map $K\Delta_2 \to K\Delta_1$ appears.

One can show (see, for example, [GK73]) that for any polyhedron Δ_1 there exists a richer polyhedron Δ_2, defining a nonsingular toric variety $K\Delta_2$. Such a polyhedron is called a *resolution* of Δ_1 (because it gives a resolution of singularities of $K\Delta_1$). If $\dim \Delta = n$ (= the dimension of the ambient space $\mathbb{R}^n$), then a resolution of Δ can be found among permissible polyhedra.

A.2.5 Hypersurfaces of Toric Varieties

Let $\Delta \subset \mathbb{R}^n$ be a polyhedron and a be a Laurent polynomial over K in n variables. Let $C_{\Delta(a)}(\Delta(a)) \subset C_\Delta(\Delta)$. Then there exists a monomial x^ω such that $\Delta(x^\omega a) \subset C_\Delta(\Delta)$. The hypersurface $V_{KC_\Delta(\Delta)}$ does not depend on the choice of x^ω and is denoted simply by $V_{KC_\Delta(\Delta)}(a)$. Its closure in $K\Delta$ is denoted by $V_{K\Delta}(a)$.[4] Thus, to any Laurent polynomial a over K with $C_{\Delta(a)}(\Delta(a)) \subset C_\Delta(\Delta)$, a hypersurface $V_{K\Delta}(a)$ of $K\Delta$ is related. For Laurent polynomial $a(x) = \sum_{\omega \in \mathbb{Z}^n} a_\omega x^\omega$ and a set $\Gamma \subset \mathbb{R}^n$ a Laurent polynomial $a(x) = \sum_{\omega \in \Gamma \cap \mathbb{Z}^n} a_\omega x^\omega$ is denoted by a^Γ and called the *Γ-truncation* of a.

Proposition A.2.4 *Let $\Delta \subset \mathbb{R}^n$ be a polyhedron and a be a Laurent polynomial over K with $C_{\Delta(a)}(\Delta(a)) \subset C_\Delta(\Delta)$. If $\Gamma_1 \in \mathcal{G}'(\Delta(a))$, $\Gamma_2 \in \mathcal{G}'(\Delta)$ and $C_{\Delta(a)}(\Gamma_1) \subset C_\Delta(\Gamma_2)$ then $K\Gamma_2 \cap V_{K\Delta}(a) = V_{K\Gamma_2}(a^{\Gamma_1})$.*

Proof Consider $KC_\Delta(\Gamma_2)$. It is a dense subset of $K\Gamma_2$. Since $C_{\Delta(a)}(\Gamma_1) \subset C_\Delta(\Gamma_2)$, there exists a monomial x^ω such that $\Delta(x^\omega a)$ lies in $C_\Delta(\Gamma_2)$ and intersects its ridge exactly in the face obtained from Γ_1. Since on $K\Gamma_2 \cap KC_\Delta(\Gamma_2)$ all monomials, whose exponents do not lie on ridge $C_{\Gamma_2}(\Gamma_2)$ of $C_\Delta(\Gamma_2)$, equal zero, it follows that the intersection $\{x \in KC_\Delta(\Gamma_2) \mid x^\omega a(x) = 0\} \cap K\Gamma_2$ coincides with $\{x \in KC_\Delta(\Gamma_2) \mid [x^\omega a]^{C_{\Gamma_2}(\Gamma_2)}(x) = 0\} \cap K\Gamma_2$. Note finally, that the latter coincides with $V_{K\Gamma_2}(a_1)$.

[4]Here it is meant the closure of $K\Delta$ in the Zarisky topology; in the case of $K = \mathbb{C}$ the classic topology gives the same result, but in the case of $K = \mathbb{R}$ the usual closure may be a nonalgebraic set.

Proposition A.2.5 *Let Δ and a be as in Proposition A.2.4 and Γ_2 be a proper face of the polyhedron Δ. If there is no face $\Gamma_1 \in \mathcal{G}'(\Delta(a))$ with $C_{\Delta(a)}(\Gamma_1) \subset C_\Delta(\Gamma_2)$ then $K\Gamma_2 \subset V_{K\Delta}(a)$.*

The proof is analogous to the proof of the previous statement. □

Denote by $SV_{K\mathbb{R}^n}(a)$ the set of singular points of $V_{K\mathbb{R}^n}(a)$, i.e. a set $V_{K\mathbb{R}^n}(a) \cap \bigcap_{i=1}^n V_{K\mathbb{R}^n}(\frac{\partial a}{\partial x_i})$.

A Laurent polynomial a is said to be *completely nondegenerate* (over K) if, for any face Γ of its Newton polyhedron, $SV_{K\mathbb{R}^n}(a^\Gamma)$ is empty and, hence, $V_{K\mathbb{R}^n}(a^\Gamma)$ is a nonsingular hypersurface. A Laurent polynomial a is said to be *peripherally nondegenerate* if for any proper face Γ of its Newton polyhedron $SV_{K\mathbb{R}^n}(a^\Gamma) = \varnothing$.

It is not difficult to prove that completely nondegenerate L-polynomials form Zarisky open subset of the space of L-polynomials over K with a given Newton polyhedron, and the same holds true also for peripherally nondegenerate L-polynomials.

Proposition A.2.6 *If a Laurent polynomial a over K is completely nondegenerate and $\Delta \subset \mathbb{R}^n$ is a resolution of its Newton polyhedron $\Delta(a)$ then the variety $V_{K\Delta}(a)$ is nonsingular and transversal to all $K\Gamma$ with $\Gamma \in \mathcal{G}'(\Delta)$. See, for example, [Kho77].*
□

Proposition A.2.6 allows various generalizations related with possibilities to consider singular $K\Delta$ or only some faces of $\Delta(a)$ (instead of all of them). For example, one can show that if under the hypothesis of Proposition A.2.4 a truncation a^Γ of a is completely nondegenerate then under an appropriate understanding of transversality (in the sense of stratified space theory) $V_{K\Delta}(a)$ is transversal to $K\Gamma_2$. Without going into discussion of transversality in this situation, I formulate a special case of this proposition, generalizing Proposition A.2.6.

Proposition A.2.7 *Let Γ be a face of a polyhedron $\Delta \subset \mathbb{R}^n$ with nonempty $\mathcal{G}_0(\Gamma)$ and with simple cones $C_\Delta(\Sigma)$ for all $\Sigma \in \mathcal{G}_0(\Gamma)$. Let a be a Laurent polynomial over K in n variables and Γ_1 be a face of $\Delta(a)$ with $C_{\Delta(a)}(\Gamma_1) \subset C_\Delta(\Gamma)$. If a^Γ is completely nondegenerate, then the set of singular points of $V_{K\Delta}(a)$ does not intersect $K\Gamma$ and $V_{K\Delta}(a)$ is transversal to $K\Gamma$.*

The proof of this proposition is a fragment of the proof of Proposition A.2.6. □

Proposition A.2.8 (Proposition of A.2.2 and A.2.3) *Let Δ and a be as in Proposition A.2.4. Then for any vector $w \in C_\Delta(\Delta)$ orthogonal to $C_{\Delta(a)}(\Delta(a))$, a hypersurface $V_{K\Delta}(a)$ is invariant under transformations $qh_{w,t} : K\Delta \to K\Delta$ and $S_{(e^{\pi i w_1},\dots,e^{\pi i w_n})} : K\Delta \to K\Delta$ (the latter in the case of $K = \mathbb{R}$ is defined only if $w \in \mathbb{Z}^n$).* □

A.3 Charts

A.3.1 Space $\mathbb{R}_+\Delta$

The aim of this subsection is to distinguish in $K\Delta$ an important subspace which looks like Δ. More precisely, it is defined a stratified real semialgebraic variety $\mathbb{R}_+\Delta$, which is embedded in $K\Delta$ and homeomorphic, as a stratified space, to the polyhedron Δ stratified by its faces. Briefly $\mathbb{R}_+\Delta$ can be described as the set of points with nonnegative real coordinates.

If Δ is a cone then $\mathbb{R}_+\Delta$ is defined as a subset of $K\Delta$ consisting of the points in which values of all monomials x^ω with $\omega \in \Delta \cap \mathbb{Z}^n$ are real and nonnegative. It is clear that for $\Gamma \in \mathcal{G}'(\Delta)$ the set $\mathbb{R}_+\Gamma$ coincides with $\mathbb{R}_+\Delta \cap K\Gamma$ and for cones $\Delta_1 \subset \Delta_2$ a preimage of $\mathbb{R}_+\Delta_1$ under $\text{in}^* : K\Delta_2 \to K\Delta_1$ (see Sect. A.2.3) is $\mathbb{R}_+\Delta_2$.

Now let Δ be an arbitrary polyhedron. Embeddings, by which $K\Delta$ is glued form $KC_\Delta(\Gamma)$ with $\Gamma \in \mathcal{G}(\Delta)$, embed the sets $\mathbb{R}_+C_\Delta(\Gamma)$ in one another; a space obtained by gluing from $\mathbb{R}_+C_\Delta(\Gamma)$ with $\Gamma \in \mathcal{G}(\Delta)$ is $\mathbb{R}_+\Delta$. It is clear that if $\Gamma \in \mathcal{G}'$ then $\mathbb{R}_+\Gamma = \mathbb{R}_+\Delta \cap K\Gamma$.

$\mathbb{R}_+\mathbb{R}^n$ is the open positive orthant $\{x \in \mathbb{R}^n \mid x_1 > 0, \dots, x_n > 0\}$. It can be identified with the subgroup of quasi-homotheties of $K\mathbb{R}^n$: one assigns $qh_{l(x)}$ to a point $x \in \mathbb{R}_+\mathbb{R}^n$.

If $A^n = \{x \in \mathbb{R}^n | x_1 \geq 0, \dots, x_n \geq 0\}$ then $KA^n = K^n$ (cf. Sect. A.2.3) and $\mathbb{R}_+A^n = A^n$.

If P^n is the n-simplex with vertexes $(0, 0, \dots, 0)$, $(1, 0, \dots, 0)$, $(0, 1, 0, \dots, 0)$, ..., $(0, 0, \dots, 1)$, then KP^n is the n-simplex consisting of points of projective space with nonnegative real homogeneous coordinates.

The set $\mathbb{R}_+\Delta$ is invariant under quasi-homotheties. Orbits of action in $\mathbb{R}_+\Delta$ of the group of quasi-homotheties of $\mathbb{R}_+\mathbb{R}^n$ are sets $\mathbb{R}_+C_\Gamma(\Gamma)$ with $\Gamma \in \mathcal{G}(\Delta)$. Orbit $\mathbb{R}_+C_\Gamma(\Gamma)$ is homeomorphic to $\mathbb{R}^{\dim \Gamma}$ or, equivalently, to the interior of Γ. Closures $\mathbb{R}_+\Gamma$ of $\mathbb{R}_+C_\Gamma(\Gamma)$ intersect one another in the same manner as the corresponding faces: $\mathbb{R}_+\Gamma_1 \cap \mathbb{R}_+\Gamma_2 = \mathbb{R}_+(\Gamma_1 \cap \Gamma_2)$. From this and from the fact that $\mathbb{R}_+\Gamma$ is locally conic (see [Loj64]) it follows that $\mathbb{R}_+\Delta$ is homeomorphic, as a stratified space, to Δ. However, there is an explicitly constructed homeomorphism. It is provided by the Atiyah moment map [Ati81] and in the case of bounded Δ can be described in the following way.

Choose a collection of points $\omega_1, \dots, \omega_k$ with integer coordinates, whose convex hull is Δ. Then for $\Gamma \in \mathcal{G}(\Delta)$ and $\omega_0 \in \Gamma \smallsetminus \partial\Gamma$ cone $C_\Delta(\Gamma)$ is $\langle \omega_1 - \omega_0, \dots, \omega_k - \omega_0 \rangle$. For $y \in KC_\Delta(\Gamma)$ denote by y^ω a value of monomial x^ω where $\omega \in C_\Delta(\Gamma) \cap \mathbb{Z}^n$ at this point. Put

$$M(y) = \frac{\sum_{i=1}^k |y^{\omega_i - \omega_0}| \omega_i}{\sum_{i=1}^k |y^{\omega_i - \omega_0}|} \in \mathbb{R}^n.$$

Obviously $M(y)$ lies in Δ, does non depend on the choice of ω_0 and for $y \in KC_\Delta(\Gamma_1) \cap KC_\Delta(\Gamma_2)$ does not depend on what face, Γ_1 or Γ_2, is used for the definition of $M(y)$. Thus a map $M : K\Delta \to \Delta$ is well defined. It is not difficult to show that $M|_{\mathbb{R}_+\Delta} : \mathbb{R}_+\Delta \to \Delta$ is a stratified homeomorphism.

A.3.2 Charts of $K\Delta$

The space $K\mathbb{R}^n$ can be presented as $\mathbb{R}_+\mathbb{R}^n \times U_K^n$. In this section an analogous representation of $K\Delta$ is described.

$\mathbb{R}_+\Delta$ is a fundamental domain for the natural action of U_K^n in $K\Delta$, i.e. its intersection with each orbit of the action consists of one point.

For a point $x \in \mathbb{R}_+C_\Gamma\Gamma$ where $\Gamma \in \mathcal{G}(\Delta)$, the stationary subgroup of action of U_K^n consists of transformations $S_{(e^{\pi i w_1},\dots,e^{\pi i w_n})}$, where vector $(w_1, \dots, w_n)$ is orthogonal to $C_\Gamma(\Gamma)$. In particular, if $\dim \Gamma = n$ then the stationary subgroup is trivial. If $\dim \Gamma = n - r$ then it is isomorphic to U_K^r. Denote by U_Γ a subgroup of U_K^n consisting of elements $(e^{\pi i w_1}, \dots, e^{\pi i w_n})$ with $(w_1, \dots, w_n) \perp C_\Gamma(\Gamma)$.

Define a map $\rho : \mathbb{R}_+\Delta \times U_K^n \to K\Delta$ by formula $(x, y) \mapsto S_y(x)$. It is surjection and we know the partition of $\mathbb{R}_+\Delta \times U_K^n$ into preimages of points. Since ρ is proper and $K\Delta$ is locally compact and Hausdorff, it follows that $K\Delta$ is homeomorphic to the quotientspace of $\mathbb{R}_+\Delta \times U_K^n$ with respect to the partition into sets $x \times yU_\Gamma$ with $x \in \mathbb{R}_+C_\Gamma(\Gamma)$, $y \in U_K^n$.

Consider as an example the case of $K = \mathbb{R}$ and $n = 2$. Let a polyhedron Δ lies in the open positive quadrant. We place $\Delta \times U_\mathbb{R}^2$ in $\mathbb{R}^2$ identifying $(x, y) \in \Delta \times U_\mathbb{R}^2$ with $S_y(x) \in \mathbb{R}^2$. $\mathbb{R}_+\Delta \times U_\mathbb{R}^2$ is homeomorphic $\Delta \times U_\mathbb{R}^2$, so the surface $\mathbb{R}\Delta$ can be obtained by an appropriate gluing (namely, by transformations taken from U_Γ) sides of four polygons consisting $\Delta \times U_\mathbb{R}^2$. Figure A.26 shows what gluings ought to be done in three special cases.

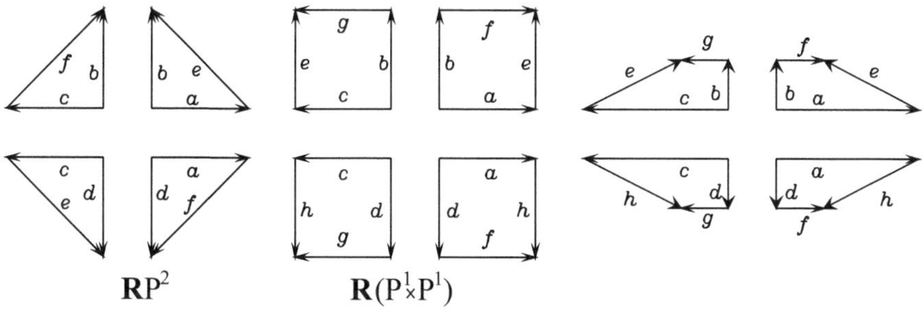

Fig. A.26 Gluing of quadrants of toric surfaces

A.3.3 Charts of L-Polynomials

Let a be a Laurent polynomial over K in n variables and Δ be its Newton polyhedron. Let h be a homeomorphism $\Delta \to \mathbb{R}_+\Delta$, mapping each face to the corresponding subspace, and such that for any $\Gamma \in \mathcal{G}(\Delta)$, $x \in \Gamma$, $y \in U_K^n$, $z \in U_\Gamma$

$$h(x, y, z) = (pr_{\mathbb{R}_+\Gamma} h(x, y),\ zpr_{U_K^n} h(x, y)).$$

For h one can take, for example, $(M|_{\mathbb{R}_+\Delta})^-$.

A pair consisting of $\Delta \times U_K^n$ and its subset υ which is the preimage of $V_{K\Delta}(a)$ under

$$\Delta \times U_K^n \xrightarrow{h\times \mathrm{id}} \mathbb{R}_+\Delta \times U_K^n \xrightarrow{\rho} K\Delta$$

is called a (nonreduced) K-*chart* of L-polynomial a.

It is clear that the set υ is invariant under transformations id $\times S$ with $S \in U_\Delta$ and its intersection with $\Gamma \times U_K^n$, where $\Gamma \in \mathcal{G}'(\Delta)$ is invariant under transformations id $\times S$ with $S \in U_\Gamma$.

As it follows from Proposition A.2.4, if Γ is a face of Δ, and $(\Delta \times U_K^n, \upsilon)$ is a nonreduced K-chart of L-polynomial a, then $(\Gamma \times U_K^n, \upsilon \cap (\Gamma \times U_K^n))$ is a nonreduced K-chart of L-polynomial a^Γ.

A nonreduced K-chart of Laurent polynomial a is unique up to homeomorphism $\Delta \times U_K^n \to \Delta \times U_K^n$, satisfying the following two conditions:

1. it map $\Gamma \times y$ with $\Gamma \in \mathcal{G}(\Delta)$ and $y \in U_K^n$ to itself and
2. its restriction to $\Gamma \times U_K^n$ with $g \in \mathcal{G}(\Delta)$ commutes with transformations id $\times S$: $\Gamma \times U_K^n \to \Gamma \times U_K^n$ where $S \in U_\Gamma$.

In the case when a is a usual polynomial, it is convenient to place its K-chart into K^n. For this, consider a map $A^n \times U_K^n \to K^n : (x, y) \mapsto S_y(x)$. Denote by $\Delta_K(a)$ the image of $\Delta(a) \times U_K^n$ under this map. Call by a (reduced) K-*chart* of a the image of a nonreduced K-chart of a under this map. The charts of peripherally nondegenerate real polynomial in two variables introduced in Sect. A.1.3 are $\mathbb{R}$-charts in the sense of this definition.

Proposition A.3.1 *Let a be a Laurent polynomial over K in n variables, Γ a face of its Newton polyhedron, $\rho : \mathbb{R}_+\Delta(a) \times U_K^n \to K\Delta(a)$ a natural projection. If the truncation a^Γ is completely nondegenerate then the set of singular points of hypersurface $\rho^{-1}V_{K\Delta(a)}(a)$ of $\mathbb{R}_+\Delta(a) \times U_K^n$ does not intersect $\mathbb{R}_+\Gamma \times U_K^n$, and $\rho^{-1}V_{K\Delta(a)}(a)$ is transversal to $\mathbb{R}_+\Gamma \times U_K^n$.*

Proof Let Δ be a resolution of polyhedron $\Delta(a)$. Then a commutative diagram

$$\begin{array}{ccc} (\mathbb{R}_+\Delta \times U_K^n,\ \rho'^{-1}(V_{K\Delta}(a))) & \xrightarrow{\rho'} & (K\Delta,\ V_{K\Delta}(a)) \\ \Big\downarrow{\scriptstyle (\mathbb{R}_+ s\times \mathrm{id})} & & \Big\downarrow{\scriptstyle s} \\ (\mathbb{R}_+\Delta(a) \times U_K^n,\ \rho^{-1}(V_{K\Delta(a)}(a))) & \xrightarrow{\rho} & (K\Delta(a),\ V_{K\Delta(a)}(a)) \end{array}$$

appears. Here s is the natural regular map resolving singularities of $K\Delta(a)$, ρ and ρ' are natural projections and $\mathbb{R}_+ s$ is a map $\mathbb{R}_+\Delta \to \mathbb{R}_+\Delta(a)$ defined by s. The preimage of $K\Gamma$ under ρ is the union of $K\Sigma$ with $\Sigma \in \mathcal{G}'(\Delta)$ and $C_\Delta(\Sigma) \supset C_{\Delta(a)}(\Gamma)$. By A.2.7, the set of singular points of $V_{K\Delta}(a)$ does not intersect $K\Sigma$, and $V_{K\Delta}(a)$ is transversal to $K\Sigma$.

If $\Sigma \in \mathcal{G}'(\Delta)$, $C_\Delta(\Sigma) \supset C_{\Delta(a)}(\Gamma)$ and $\dim \Sigma = \dim \Gamma$, then $\mathbb{R}_+ s$ defines an isomorphism $\mathbb{R}_+ C_\Sigma(\Sigma) \to \mathbb{R}_+ C_\Gamma(\Gamma)$, and if $\Sigma \in \mathcal{G}'(\Delta)$, $C_\Delta(\Sigma) \supset C_{\Delta(a)}(\Gamma)$ and $\dim \Sigma > \dim \Gamma$, then $\mathbb{R}_+ s$ defines a map $\mathbb{R}_+ C_\Sigma(\Sigma) \to \mathbb{R}_+ C_\Gamma(\Gamma)$ which is a factorization by the action of quasi-homotheties $qh_{w,t}$ with $w \in C_\Sigma(\Sigma)$, $w \perp C_\Gamma(\Gamma)$. By A.2.8, in the latter case variety $V_{K\Sigma}(a^\Gamma)$ coinciding, by Proposition A.2.4, with $V_{K\Delta}(a) \cap K\Sigma$ is invariant under the same quasi-homotheties. Hence $V_{K\Delta}(a) = s^{-1} V_{K\Delta(a)}(a)$ and hypersurface $\rho^{-1} V_{K\Delta}(a)$, being the image of $\rho'^{-1} V_{K\Delta}(a)$ under $\mathbb{R}_+ \times \mathrm{id}$, appears to be nonsingular along its intersection with $\mathbb{R}_+\Gamma \times U_K^n$ and transversal to $\mathbb{R}_+\Gamma \times U_K^n$.

A.4 Patchworking

A.4.1 Patchworking L-Polynomials

Let Δ, Δ_1, ..., $\Delta_s \subset \mathbb{R}^n$ be (convex integer) polyhedra with $\Delta = \bigcup_{i=1}^s \Delta_i$ and $\operatorname{Int}\Delta_i \cap \operatorname{Int}\Delta_j = \varnothing$ for $i \neq j$. Let $\nu : \Delta \to \mathbb{R}$ be a nonnegative convex function satisfying to the following conditions:

1. all the restrictions $\nu|_{\Delta_i}$ are linear;
2. if the restriction of ν to an open set is linear then this set is contained in one of Δ_i;
3. $\nu(\Delta \cap \mathbb{Z}^n) \subset \mathbb{Z}$.

Remark A.4.0.1 Existence of such a function ν is a restriction on a collection $\Delta_1, \dots, \Delta_s$. For example, the collection of convex polygons shown in Fig. A.27 does not admit such a function.

Let $a_1, \dots, a_s$ be Laurent polynomials over K in n variables with $\Delta(a_i) = \Delta$. Let $a_i^{\Delta_i \cap \Delta_j} = a_j^{\Delta_i \cap \Delta_j}$ for any i, j. Then, obviously, there exists an unique L-polynomial a

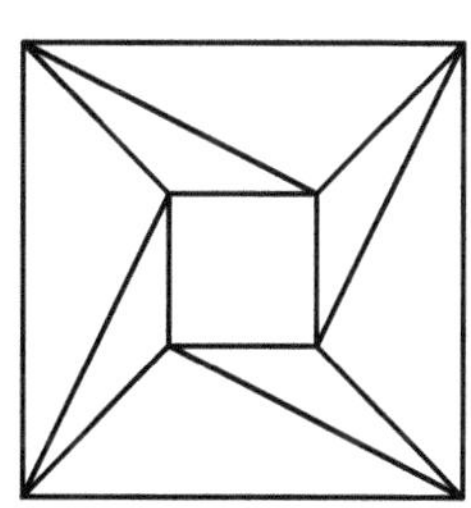

Fig. A.27 Collection of polygons that does not lift to a graph of a convex piecewise linear function

with $\Delta(a) = \Delta$ and $a^{\Delta_i} = a_i$ for $i = 1, \dots, s$. If $a(x_1, \dots, x_n) = \sum_{\omega \in \mathbb{Z}^n} a_\omega x^\omega$, we put $b(x, t) = \sum_{\omega \in \mathbb{Z}^n} a_\omega x^\omega t^{\nu(\omega)}$. This L-polynomial in $n+1$ variables is considered below also as a one-parameter family of L-polynomials in n variables. Therefore let me introduce the corresponding notation: put $b_t(x_1, \dots, x_n) = b(x_1, \dots, x_n, t)$. L-polynomials b_t are said to be obtained by *patchworking* L-polynomials $a_1, \dots, a_s$ by ν or, briefly, b_t is a patchwork of L-polynomials $a_1, \dots, a_s$ by ν.

A.4.2 Patchworking Charts

Let $a_1, \dots, a_s$ be Laurent polynomials over K in n variables with $\operatorname{Int} \Delta(a_i) \cap \operatorname{Int} \Delta(a_j) = \varnothing$ for $i \neq j$. A pair $(\Delta \times U_K^n, v)$ is said to be obtained by *patchworking* K-charts of Laurent polynomials $a_1, \dots, a_s$ and it is a *patchwork* of K-charts of L-polynomials $a_1, \dots, a_s$ if $\Delta = \bigcup_{i=1}^s \Delta(a_i)$ and one can choose K-charts $(\Delta(a_i) \times U_K^n, v_i)$ of Laurent polynomials $a_1, \dots, a_s$ such that $v = \bigcup_{i=1}^s v_i$.

A.4.3 The Main Patchwork Theorem

Let Δ, Δ_1, ..., Δ_s, ν, a_1, ..., a_s, b and b_t be as in Sect. A.4.1 (b_t is a patchwork of L-polynomials a_1, ..., a_s by ν).

Proposition A.4.1 *If L-polynomials $a_1, \dots, a_s$ are completely nondegenerate then there exists $t_0 > 0$ such that for any $t \in (0, t_0]$ a K-chart of L-polynomial b_t is obtained by patchworking K-charts of L-polynomials $a_1, \dots, a_s$.*

Proof Denote by $\mathcal{G}$ the union $\bigcup_{i=1}^s \mathcal{G}(\Delta_i)$. For $\Gamma \in \mathcal{G}$ denote by $\tilde{\Gamma}$ the graph of $\nu|_\Gamma$. It is clear that $\Delta(b)$ is the convex hull of graph of ν, so $\tilde{\Gamma} \in \mathcal{G}(\Delta(b))$ and thus there is an injection $\mathcal{G} \to \mathcal{G}(\Delta(b)) : \Gamma \mapsto \tilde{\Gamma}$. Restrictions $\tilde{\Gamma} \to \Gamma$ of the natural projection $pr : \mathbb{R}^{n+1} \to \mathbb{R}^n$ are homeomorphisms, they are denoted by g.

Let $p : \Delta(b) \times U_K^{n+1} \to K\Delta(b)$ be the composition of the homeomorphism

$$\Delta(b) \times U_K^{n+1} \xrightarrow{h \times \mathrm{id}} \mathbb{R}_+ \Delta(b) \times U_K^{n+1}$$

and the natural projection $\rho : \mathbb{R}_+ \Delta(b) \times U_K^{n+1} \to K\Delta(b)$ (cf. Sect. A.3.3), so the pair $\left(\Delta(b) \times U_K^{n+1},\ p^{-1} V_{K\Delta(b)}(b)\right)$ is a K-chart of b. By Proposition A.2.4, for $i = 1, \dots, s$ the pair

$$\left(\widetilde{\Delta(a_i)} \times U_K^{n+1},\ p^{-1}(V_{K\Delta(b)}(b) \cap \widetilde{\Delta(a_i)} \times U_K^n)\right)$$

is a K-chart of L-polynomial $b^{\widetilde{\Delta(a_i)}}$.

The pair

$$\left(\widetilde{\Delta(a_i)} \times U_K^n,\ p^{-1}(V_{K\Delta(b)}(b) \cap \widetilde{\Delta(a_i)} \times U_K^n)\right)$$

which is cut out by this pair on $\widetilde{\Delta(a_i)} \times U_K^n$ is transformed by $g \times \mathrm{id} : \widetilde{\Delta(a_i)} \times U_K^n \to \Delta(a_i) \times U_K^n$ to a K-chart of a_i. Indeed, $g : \widetilde{\Delta(a_i)} \to \Delta(a_i)$ defines an isomorphism $g^* : K\Delta(a_i) \to K\widetilde{\Delta(a_i)}$ and since $b^{\widetilde{\Delta(a_i)}}(x_1, \dots, x_n, 1) = a_i(x_1, \dots, x_n)$, it follows that $g^* : V_{K\Delta(a_i)}(a_i) = V_{K\widetilde{\Delta(a_i)}}(b^{\widetilde{\Delta(a_i)}})$ and g defines a homeomorphism of the pair $\left(\widetilde{\Delta(a_i)} \times U_K^n,\ p^{-1}(V_{K\Delta(b)}(b) \cap \widetilde{\Delta(a_i)} \times U_K^n)\right)$ to a K-chart of L-polynomial a_i.

Therefore the pair

$$\left(\bigcup_{i=1}^{s} \widetilde{\Delta(a_i)} \times U_K^n,\ p^{-1}(V_{K\Delta(b)}(b) \cap \bigcup_{i=1}^{s} \widetilde{\Delta(a_i)} \times U_K^n)\right)$$

is a result of patchworking K-charts of $a_1, \dots, a_s$.

For $t > 0$ and $\Gamma \in \mathcal{G}'(\Delta)$ let us construct a ring homomorphism

$$K[C_{\Delta(b)}(\Delta(b) \cap pr^{-1}(\Gamma)) \cap \mathbb{Z}^{n+1}] \to K[C_\Delta(\Gamma) \cap \mathbb{Z}^n]$$

which maps a monomial $x_1^{\omega_1} \dots x_n^{\omega_n} x_{n+1}^{\omega_{n+1}}$ to $t^{\omega_{n+1}} x_1^{\omega_1} \dots x_n^{\omega_n}$. This homomorphism corresponds to the embedding

$$KC_\Delta(\Gamma) \to KC_{\Delta(b)}(\Delta(b) \cap pr^{-1}(\Gamma))$$

extending the embedding

$$K\mathbb{R}^n \to K\mathbb{R}^{n+1} : (x_1, \dots, x_n) \mapsto (x_1, \dots, x_n, t).$$

The embeddings constructed in this way agree to each other and define an embedding $K\Delta \to K\Delta(b)$. Denote the latter embedding by i_t. It is clear that $V_{K\Delta}(b_t) = i_t^{-1} V_{K\Delta(b)}(b)$.

The sets $\rho^{-1} i_t K\Delta$ are smooth hypersurfaces of $\Delta(b) \times U_K^{n+1}$, comprising a smooth isotopy. When $t \to 0$, the hypersurface $\rho^{-1} i_t K\Delta$ tends (in C^1-sense) to

$$\bigcup_{i=1}^{s} \widetilde{\Delta(a_i)} \times U_K^n.$$

By Proposition A.3.1, $\rho^{-1} V_{K\Delta(b)}(b)$ is transversal to each of

$$\mathbb{R}_+ \widetilde{\Delta(a_i)} \times U_K^{n+1}$$

and hence, the intersection $\rho^{-1}(i_t K\Delta) \cap \rho^{-1}(V_{K\Delta(b)}(b))$ for sufficiently small t is mapped to

$$V_{K\Delta(b)}(b) \cap \bigcup_{i=1}^{s} \widetilde{\Delta(a_i)} \times U_K^n$$

by some homeomorphism

$$\rho^{-1} i_t K\Delta \to \bigcup_{i=1}^{s} \widetilde{\Delta(a_i)} \times U_K^n.$$

Thus the pair

$$\big(\rho^{-1} i_t K\Delta,\ \rho^{-1} i_t K\Delta \cap \rho^{-1} V_{K\Delta(b)}(b)\big)$$

is a result of patchworking K-charts of L-polynomials $a_1, \dots, a_s$ if t belongs to a segment of the form $(0, t_0]$. On the other hand, since $V_{K\Delta}(b_t) = i_t^{-1} V_{K\tilde{\Delta}}(b)$,

$$\rho^{-1} i_t K\Delta \cap \rho^{-1} V_{K\Delta(b)}(b) = \rho^{-1} i_t V_{K\Delta}(b_t)$$

and, hence, the pair

$$\big(\rho^{-1} i_t K\Delta,\ \rho^{-1} i_t K\Delta \cap \rho^{-1} V_{K\Delta(b)}(b)\big)$$

is homeomorphic to a K-chart of L-polynomial b_t.

A.5 Perturbations Smoothing a Singularity of Hypersurface

The construction of the previous section can be interpreted as a purposeful smoothing of an algebraic hypersurface with singularities, which results in replacing of neighborhoods of singular points by new fragments of hypersurface, having a prescribed topological structure (cf. Sect. A.1.10). According to well known theorems of theory of singularities, all theorems on singularities of algebraic hypersurfaces are extended to singularities of significantly wider class of hypersurfaces. In particular, the construction of perturbation based on patchworking is applicable in more general situation. For singularities of simplest types this construction together with some results of topology of algebraic curves allows to get a topological classification of perturbations which smooth singularities completely.

The aim if this section is to adapt patchworking to needs of singularity theory.

A.5.1 Singularities of Hypersurfaces

Let $G \subset K^n$ be an open set, and let $\varphi : G \to K$ be an analytic function. For $U \subset G$ denote by $V_U(\varphi)$ the set $\{x \in U \mid \varphi(x) = 0\}$.

By *singularity* of a hypersurface $V_G(\varphi)$ at the point $x_0 \in V_G(\varphi)$ we mean the class of germs of hypersurfaces which are diffeomorphic to the germ of $V_G(\varphi)$ at x_0. In other words, hypersurfaces $V_G(\varphi)$ and $V_H(\psi)$ have the same singularity at points x_0 and y_0, if there exist neighborhoods M and N of x_0 and y_0 such that the pairs $(M, V_M(\varphi))$, $(N, V_N(\psi))$ are diffeomorphic. When considering a singularity of hypersurface at a point x_0, to simplify the formulas we shall assume that $x_0 = 0$.

The *multiplicity* or the *Milnor number* of a hypersurface $V_G(\varphi)$ at 0 is the dimension

$$dim_K K[[x_1, \dots, x_n]]/(\partial f/\partial x_1, \dots, \partial f/\partial x_n)$$

of the quotient of the formal power series ring by the ideal generated by partial derivatives $\partial f/\partial x_1, \dots, \partial f/\partial x_n$ of the Taylor series expansion f of the function φ at 0. This number is an invariant of the singularity (see [AVGZ82]). If it is finite, then we say that the singularity is of *finite multiplicity*.

If the singularity of $V_G(\varphi)$ at x_0 is of finite multiplicity, then this singularity is isolated, i.e. there exists a neighborhood $U \subset K^n$ of x_0, which does not contain singular points of $V_G(\varphi)$. If $K = \mathbb{C}$ then the converse is true: each isolated singularity of a hypersurface is of finite multiplicity. In the case of isolated singularity, the boundary of a ball $B \subset K^n$, centered at x_0 and small enough, intersects $V_G(\varphi)$ only at nonsingular points and only transversely, and the pair $(B, V_B(\varphi))$ is homeomorphic to the cone over its boundary $(\partial B, V_{\partial B}(\varphi))$ (see [Mil68], Theorem 2.10). In such a case the pair $(\partial B, V_{\partial B}(\varphi))$ is called the *link* of singularity of $V_G(\varphi)$ at x_0.

The following Theorem shows that the class of singularities of finite multiplicity of analytic hypersurfaces coincides with the class of singularities of finite multiplicity of algebraic hypersurfaces.

Proposition A.5.1 (Tougeron's theorem) *(see, for example, [AVGZ82], Sect. A.6.3). If the singularity at x_0 of a hypersurface $V_G(\varphi)$ has finite Milnor number μ, then there exist a neighborhood U of x_0 in K^n and a diffeomorphism h of this neighborhood onto a neighborhood of x_0 in K^n such that $h(V_U(\varphi)) = V_{h(U)}(f_{(\mu+1)})$, where $f_{(\mu+1)}$ is the Taylor polynomial of φ of degree $\mu + 1$.*

The notion of Newton polyhedron is extended over in a natural way to power series. The Newton polyhedron $\Delta(f)$ of the series $f(x) = \sum_{\omega \in \mathbb{Z}^n} a_\omega x^\omega$ (where $x^\omega = x_1^{\omega_1} x_2^{\omega_2} \dots x_n^{\omega_n}$) is the convex hull of the set $\{\omega \in \mathbb{R}^n \mid a_\omega \neq 0\}$. (Contrary to the case of a polynomial, the Newton polyhedron $\Delta(f)$ of a power series may have infinitely many faces.)

However in the singularity theory the notion of Newton diagram occurred to be more important. The *Newton diagram* $\Gamma(f)$ of a power series f is the union of the proper faces of the Newton polyhedron which face the origin, i.e. the union of the faces $\Gamma \in \mathcal{G}'(\Delta(f))$ for which cones $D^+C_{\Delta(f)}(\Gamma)$ intersect the open positive orthant Int $A^n = \{x \in \mathbb{R}^n \mid x_1 > 0, \dots, x_n > 0\}$.

It follows from the definition of the Milnor number that, if the singularity of $V_G(\varphi)$ at 0 is of finite multiplicity, the Newton diagram of the Taylor series of φ is compact, and its distance from each of the coordinate axes is at most 1.

For a power series $f(x) = \sum_{\omega \in \mathbb{Z}^n} f_\omega x^\omega$ and a set $\Gamma \subset \mathbb{R}^n$ the power series $\sum_{\omega \in \Gamma \cap \mathbb{Z}^n} f_\omega x^\omega$ is called *Γ-truncation* of f and denoted by f^Γ (cf. Sect. A.2.1).

Let the Newton diagram of the Taylor series f of a function φ be compact. Then $f^{\Gamma(f)}$ is a polynomial. The pair $(\Gamma(f) \times U_K^n, \gamma)$ is said to be a *nonreduced chart* of germ of hypersurface $V_G(\varphi)$ at 0 if there exists a K-chart $(\Delta(f^{\Gamma(f)} \times U_K^n, \upsilon)$ of $f^{\Gamma(f)}$ such that $\gamma = \upsilon \cap (\Gamma(f) \times U_K^n)$. It is clear that a nonreduced chart of germ of hypersurface is comprised of K-charts of f^Γ, where Γ runs over the set of all faces of the Newton diagram.

A power series f is said to be *nondegenerate* if its Newton diagram is compact and the distance between it and each of the coordinate axes is at most 1 and for any its face Γ polynomial f^Γ is completely nondegenerate. In this case about the germ of $V_G(\varphi)$ at zero we say that *it is placed nondegenerately*. It is not difficult to prove that nondegenerately placed germ defines a singularity of finite multiplicity. It is convenient to place the charts of germs of hypersurfaces in K^n by a natural map $A^n \times U_K^n \to K^n : (x, y) \mapsto S_y(x)$ (like K-charts of an L-polynomial, cf. Sect. A.3.3). Denote by $\Sigma_K(\varphi)$ the image of $\Gamma(f) \times U_K^n$ under this map; the image of nonreduced chart of germ of hypersurface $V_G(\varphi)$ at zero under this map is called a *(reduced) chart of germ* of $V_G(\varphi)$ at the origin. It follows from Tougeron's theorem that in this case adding a monomial of the form $x_i^{m_i}$ to φ with m_i large enough does not change the singularity. Thus, without changing the singularity, one can make the Newton diagram meeting the coordinate axes.

In the case when this takes place and the Taylor series of φ is nondegenerate there exists a ball $U \subset K^n$ centered at 0 such that the pair $(U, V_U(\varphi))$ is homeomorphic to the cone over a chart of germ of $V_G(\varphi)$. This follows from Proposition A.5.1 and from results of Sect. A.2.5.

Thus if the Newton diagram meets all coordinate axes and the Taylor series of φ is nondegenerate, then the chart of germ of $V_G(\varphi)$ at zero is homeomorphic to the link of the singularity.

A.5.2 *Evolving of a Singularity*

Now let the function $\varphi : G \to K$ be included as φ_0 in a family of analytic functions $\varphi_t : G \to K$ with $t \in [0, t_0]$. Suppose that this is an analytic family in the sense that the function $G \times [0, t_0] \to K : (x, t) \mapsto \varphi_t(x)$ which is determined by it is real analytic. If the hypersurface $V_G(\varphi)$ has an isolated singularity at x_0, and if there exists a neighborhood U of x_0 such that the hypersurfaces $V_G(\varphi_t)$ with $t \in [0, t_0]$ have no singular points in U, then the family of functions φ_t with $t \in [0, t_0]$ is said to *evolve* the singularity of $V_G(\varphi)$ at x_0.

If the family φ_t with $t \in [0, t_0]$ evolves the singularity of the hypersurface $V_G(\varphi_0)$ at x_0, then there exists a ball $B \subset K^n$ centered at x_0 such that

1. for $t \in [0, t_0]$ the sphere ∂B intersects $V_G(\varphi_t)$ only in nonsingular points of the hypersurface and only transversely,
2. for $t \in (0, t_0]$ the ball B contains no singular point of the hypersurface $V_G(\varphi_t)$,
3. the pair $(B, V_B(\varphi_0))$ is homeomorphic to the cone over its boundary $(\partial B, V_{\partial B}(\varphi_0))$.

Then the family of pairs $(B, V_B(\varphi_t))$ with $t \in [0, t_0]$ is called an *evolving* of the germ of $V_G(\varphi_0)$ in x_0. (Following the standard terminology of the singularity theory, it would be more correct to say not a on family of pairs, but rather a family of germs or even germs of a family; however, from the topological viewpoint, which is more natural in the context of the topology of real algebraic varieties, the distinction between a family of pairs satisfying 1 and 2 and the corresponding family of germs is of no importance, and so we shall ignore it.)

Conditions 1 and 2 imply existence of a smooth isotopy $h_t : B \to B$ with $t \in (0, t_0]$, such that $h_{t_0} = \text{id}$ and $h_t(V_B(\varphi_{t_0})) = V_B(\varphi_t)$, so that the pairs $(B, V_B(\varphi_t))$ with $t \in (0, t_0]$ are homeomorphic to each other.

Given germs determining the same singularity, a evolving of one of them obviously corresponds to a diffeomorphic evolving of the other germ. Thus, one may speak not only of evolvings of germs, but also of *evolvings of singularities* of a hypersurface.

The following three topological classification questions on evolvings arise.

Proposition A.5.2 *Up to homeomorphism, what manifolds can appear as* $V_B(\varphi_t)$ *in evolvings of a given singularity?*

Proposition A.5.3 *Up to homeomorphism, what pairs can appear as* $(B, V_B(\varphi_t))$ *in evolvings of a given singularity?*

Smoothings $(B, V_B(\varphi_t))$ with $t \in [0, t_0]$ and $(B', V_{B'}(\varphi'_t))$ with $t \in [0, t'_0]$ are said to be *topologically equivalent* if there exists an isotopy $h_t : B \to B'$ with $t \in [0, \min(t_0, t'_0)]$, such that h_0 is a diffeomorphism and $V_{B'}(\varphi'_t) = h_t V_B(\varphi_t)$ for $t \in [0, \min(t_0, t'_0)]$.

Proposition A.5.4 *Up to topological equivalence, what are the evolvings of a given singularity?*

Obviously, Proposition A.5.3 is a refinement of Proposition A.5.2. In turn, Proposition A.5.4 is more refined than Proposition A.5.3, since in A.5.4 we are interested not only in the type of the pair obtained in result of the evolving, but also the manner in which the pair is attached to the link of the singularity.

In the case $K = \mathbb{R}$ these questions have been answered in literature only for several simplest singularities.

In the case $K = \mathbb{C}$ a evolving of a given singularity is unique from each of the three points of view, and there is an extensive literature (see, for example, [GZ77]) devoted to its topology (i.e., questions Propositions A.5.2 and A.5.3).

By the way, if we want to get questions for $K = \mathbb{C}$ which are truly analogous to questions Propositions A.5.2–A.5.4 for $K = \mathbb{R}$, then we have to replace evolvings by deformations with nonsingular fibers and one-dimensional complex bases, and the variety $V_B(\varphi_t)$ and the pairs $(B, V_B(\varphi_t))$ have to be considered along with the monodromy transformations. It is reasonable to suppose that there are interesting connections between questions Propositions A.5.2–A.5.4 for a real singularity and their counter-parts for the complexification of the singularity.

A.5.3 Charts of Evolving

Let the Taylor series f of function $\varphi : G \to K$ be nondegenerate and its Newton diagram meets all the coordinate axes. Let a family of functions $\varphi_t : G \to K$ with $t \in [0, t_0]$ evolves the singularity of $V_G(\varphi)$ at 0. Let $(B, V_B(\varphi_t))$ be the corresponding evolving of the germ of this hypersurface and $h_t : B \to B$ with $t \in (0, t_0]$ be an isotopy with $h_{t_0} = \text{id}$ and $h_t(V_B(\varphi_{t_0})) = V_B(\varphi_t))$ existing by conditions 1 and 2 of the previous section. Let $(\Sigma_K(\varphi), \gamma)$ be a chart of germ of hypersurface $V_G(\varphi)$ at zero and $g : (\Sigma_K(\varphi), \gamma) \to (\partial B, V_{\partial B}(\varphi))$ be the natural homeomorphism of it to link of the singularity.

Denote by $\Pi_K(\varphi)$ a part of K^n bounded by $\Sigma_K(\varphi)$. It can be presented as a cone over $\Sigma_K(\varphi)$ with vertex at zero.

One can choose the isotopy $h_t : B \to B$, $t \in (0, t_0]$ such that its restriction to ∂B can be extended to an isotopy $h'_t : \partial B \to \partial B$ with $t \in [0, t_0]$ (i.e., extended for $t = 0$).

We shall call the pair $(\Pi_K(\varphi), \tau)$ a *chart of evolving* $(B, V_B(\varphi_t))$, $t \in [0, t_0]$, if there exists a homeomorphism $(\Pi_K(\varphi), \tau) \to (B, V_B(t_0))$, whose restriction $\Sigma_K(\varphi) \to \partial B$ is the composition $\Sigma_K(\varphi) \xrightarrow{g} \partial B \xrightarrow{h'_0} \partial B$. One can see that the boundary $(\partial \Pi_K(\varphi), \partial \tau)$ of a chart of evolving is a chart $(\Sigma_K(\varphi), \gamma)$ of the germ of the hypersurface at zero, and a chart of evolving is a pair obtained by evolving which is glued to $(\Sigma_K(\varphi), \gamma)$ in natural way. Thus that the chart of an evolving describes the evolving up to topological equivalence.

A.5.4 Construction of Evolvings by Patchworking

Let the Taylor series f of function $\varphi : G \to K$ be nondegenerate and its Newton diagram $\Gamma(f)$ meets all the coordinate axes.

Let $a_1, \dots, a_s$ be completely nondegenerate polynomials over K in n variables with $\operatorname{Int} \Delta(a_i) \cap \operatorname{Int} \Delta(a_j) = \varnothing$ and $a_i^{\Delta(a_i) \cap \Delta(a_j)} = a_j^{\Delta(a_i) \cap \Delta(a_j)}$ for $i \neq j$. Let $\bigcup_{i=1}^{s} \Delta(a_i)$ be the polyhedron bounded by the coordinate axes and Newton diagram $\Gamma(f)$. Let $a_i^{\Delta(a_i) \cap \Delta(f)} = f^{\Delta(a_i) \cap \Delta(f)}$ for $i = 1, \dots, s$. Let $\nu : \bigcup_{i=1}^{s} \Delta(a_i) \to \mathbb{R}$ be a nonnegative convex function which is equal to zero on $\Gamma(f)$ and satisfies conditions 1, 2, 3 of Sect. A.4.1 with polyhedra $\Delta(a_1)$, ..., $\Delta(a_s)$. Then polynomials $a_1, \dots, a_s$ can be "glued to φ by ν" in the following way generalizing patchworking L-polynomials of Sect. A.4.1.

Denote by a the polynomial defined by conditions $a^{\Delta(a_i)} = a_i$ for $i = 1, \dots, s$ and $a^{\bigcup_{i=1}^{s} \Delta(a_i)} = a$. If $a(x) = \sum_{\omega \in \mathbb{Z}^n} a_\omega x^\omega$ then we put

$$\varphi_t(x) = \varphi(x) + (\sum_{\omega \in \mathbb{Z}^n} a_\omega x^\omega t^{\nu(\omega)}) - a^{\Gamma(f)} x.$$

Proposition A.5.5 *Under the conditions above there exists $t_0 > 0$ such that the family of functions $\varphi_t : G \to K$ with $t \in [0, t_0]$ evolves the singularity of $V_G(\varphi)$ at zero. The chart of this evolving is patchworked from K-charts of $a_1, \dots, a_s$.*

In the case when φ is a polynomial, Proposition A.5.5 is a slight modification of a special case of Proposition A.4.1. Proof of Proposition A.4.1 is easy to transform to the proof of this version of Proposition A.5.5. The general case can be reduced to it by Tougeron Theorem, or one can prove it directly, following to scheme of proof of Proposition A.4.1. □

We shall call the evolvings obtained by the scheme described in this section *patchwork* evolvings.

A.6 Approximation of Hypersurfaces of $K\mathbb{R}^n$

A.6.1 Sufficient Truncations

Let M be a smooth submanifold of a smooth manifold X. Remind that by a *tubular neighborhood* of M in X one calls a submanifold N of X with $M \subset \operatorname{Int} N$ equipped with a *tubular fibration*, which is a smooth retraction $p : N \to M$ such that for any point $x \in M$ the preimage $p^{-1}(x)$ is a smooth submanifold of X diffeomorphic to $D^{\dim X - \dim M}$. If X is equipped with a metric and each fiber of the tubular fibration $p : N \to M$ is contained in a ball of radius ε centered in the point of intersection of the fiber with M, then N is called a *tubular ε-neighborhood* of M in X.

We need tubular neighborhoods mainly for formalizing a notion of approximation of a submanifold by a submanifold. A manifold presented as the image of a smooth section of the tubular fibration of a tubular ε-neighborhood of M can be considered as sufficiently close to M: it is naturally isotopic to M by an isotopy moving each point at most by ε.

We shall consider the space $\mathbb{R}^n \times U_K^n$ as a flat Riemannian manifold with metric defined by the standard Euclidian metric of $\mathbb{R}^n$ in the case of $K = \mathbb{R}$ and by the standard Euclidian metric of $\mathbb{R}^n$ and the standard flat metric of the torus $U_\mathbb{C}^n = (S^1)^n$ in the case of $K = \mathbb{C}$.

An ε-sufficiency of truncations of Laurent polynomial defined below and the whole theory related with this notion presuppose that it has been chosen a class of tubular neighborhoods of smooth submanifolds of $\mathbb{R}^n \times U_K^n$ invariant under translations $T_\omega \times \mathrm{id}_{U_K^n}$ and that for any two tubular neighborhoods N and N' of the same M, which belong to this class, restrictions of tubular fibrations $p : N \to M$ and $p' : N' \to M$ to $N \cap N'$ coincide. One of such classes is the collection of all *normal* tubular neighborhoods, i.e. tubular neighborhoods with fibers consisting of segments of geodesics which start from the same point of the submanifold in directions orthogonal to the submanifold. Another class, to which we shall turn in Sects. A.6.7 and A.6.8, is the class of tubular neighborhoods whose fibers lie in fibers $\mathbb{R}^{n-1} \times t \times U_K^{n-1} \times s$ of $\mathbb{R}^n \times U_K^n$ and consist of segments of geodesics which are orthogonal to intersections of the corresponding manifolds with these $\mathbb{R}^{n-1} \times t \times U_K^{n-1} \times s$. The intersection of such a tubular neighborhood of M with the fiber $\mathbb{R}^{n-1} \times t \times U_K^{n-1} \times s$ is a normal tubular neighborhood of $M \cap (\mathbb{R}^{n-1} \times t \times U_K^{n-1} \times s)$ in $\mathbb{R}^{n-1} \times t \times U_K^{n-1} \times s$. Of course, only manifolds transversal to $\mathbb{R}^{n-1} \times t \times U_K^{n-1} \times s$ have tubular neighborhoods of this type.

Introduce a norm in vector space of Laurent polynomials over K on n variables:

$$||\sum_{\omega \in \mathbb{Z}^n} a_\omega x^\omega|| = \max\{|a_\omega| \mid \omega \in \mathbb{Z}^n\}.$$

Let Γ be a subset of $\mathbb{R}^n$ and ε a positive number. Let a be a Laurent polynomial over K in n variables and U a subset of $K\mathbb{R}^n$. We shall say that *in U the truncation a^Γ is ε-sufficient for a* (with respect to the chosen class of tubular neighborhoods), if for any Laurent polynomial b over K satisfying the conditions $\Delta(b) \subset \Delta(a)$, $b^\Gamma = a^\Gamma$ and $||b - b^\Gamma|| \le ||a - a^\Gamma||$ (in particular, for $b = a$ and $b = a^\Gamma$) the following condition takes place:

1. $U \cap SV_{K\mathbb{R}^n}(b) = \varnothing$,
2. the set $la(U \cap V_{K\mathbb{R}^n}(b))$ lies in a tubular ε-neighborhood N (from the chosen class) of $la(V_{K\mathbb{R}^n}(a^\Gamma) \smallsetminus SV_{K\mathbb{R}^n}(a^\Gamma))$ and
3. $la(U \cap V_{K\mathbb{R}^n}(b))$ can be extended to the image of a smooth section of the tubular fibration $N \to la(V_{K\mathbb{R}^n}(a^\Gamma) \smallsetminus SV_{K\mathbb{R}^n}(a^\Gamma))$.

The ε-sufficiency of Γ-truncation of Laurent polynomial a in U means, roughly speaking, that monomials which are not in a^Γ have a small influence on $V_{K\mathbb{R}^n}(a) \cap U$.

Proposition A.6.1 *If a^Γ is ε-sufficient for a in open sets U_i with $i \in \mathcal{J}$, then it is ε-sufficient for a in $\bigcup_{i \in \mathcal{J}} U_i$ too.* □

Standard arguments based on Implicit Function Theorem give the following Theorem.

Proposition A.6.2 *If a set $U \subset K\mathbb{R}^n$ is compact and contains no singular points of a hypersurface $V_{K\mathbb{R}^n}(a)$, then for any tubular neighborhood N of $V_{K\mathbb{R}^n}(a) \smallsetminus SV_{K\mathbb{R}^n}(a)$ and any polyhedron $\Delta \supset \Delta(a)$ there exists $\delta > 0$ such that for any Laurent polynomial b with $\Delta(b) \subset \Delta$ and $||b - a|| < \delta$ the hypersurface $V_{K\mathbb{R}^n}(b)$ has no singularities in U, intersection $U \cap V_{K\mathbb{R}^n}(b)$ is contained in N and can be extended to the image of a smooth section of a tubular fibration $N \to V_{K\mathbb{R}^n}(a) \smallsetminus SV_{K\mathbb{R}^n}(a)$.* □

From this the following proposition follows easily.

Proposition A.6.3 *If $U \in K\mathbb{R}^n$ is compact and a^Γ is ε-sufficient truncation of a in U, then for any polyhedron $\Delta \supset \Delta(a)$ there exists $\delta > 0$ such that for any Laurent polynomial b with $\Delta(b) \subset \Delta$, $b^\Gamma = a^\Gamma$ and $||b - a|| < \delta$ the truncation b^Γ is ε-sufficient in U.* □

In the case of $\Gamma = \Delta(a)$ Proposition A.6.3 turns to the following proposition.

Proposition A.6.4 *If a set $U \subset K\mathbb{R}^n$ is compact and contains no singular points of $V_{K\mathbb{R}^n}(a)$ and $la(V_{K\mathbb{R}^n}(a))$ has a tubular neighborhood of the chosen type, then for any $\varepsilon > 0$ and any polyhedron $\Delta \supset \Delta(a)$ there exists $\delta > 0$ such that for any Laurent polynomial b with $\Delta(b) \subset \Delta$, $||b - a|| < \delta$ and $b^{\Delta(a)} = a$ the truncation $b^{\Delta(a)}$ is ε-sufficient in U.* □

The following proposition describes behavior of the ε-sufficiency under quasi-homotheties.

Proposition A.6.5 *Let a be a Laurent polynomial over K in n variables. Let $U \subset K\mathbb{R}^n$, $\Gamma \subset \mathbb{R}^n$, $w \in \mathbb{R}^n$. Let ε and t be positive numbers. Then ε-sufficiency of Γ-truncation a^Γ of a in $qh_{w,t}(U)$ is equivalent to ε-sufficiency of Γ-truncation of $a \circ qh_{w,t}$ in U.*

The proof follows from comparison of the definition of ε-sufficiency and the following two facts. First, it is obvious that

$$qh_{w,t}(U) \cap V_{K\mathbb{R}^n}(b) = qh_{w,t}(U \cap qh_{w,t}^{-1}(V_{K\mathbb{R}^n}(b))) = qh_{w,t}(U \cap V_{K\mathbb{R}^n}(b \circ qh_{w,t})),$$

and second, the transformation $T_{(\ln t)w} \times \mathrm{id}_{U_K^n}$ of $\mathbb{R}^n \times U_K^n$ corresponding, by Proposition A.2.1, to $qh_{w,t}$ preserves the chosen class of tubular ε neighborhoods. □

A.6.2 Domains of ε-Sufficiency of Face-Truncation

For $A \subset \mathbb{R}^n$ and $B \subset K\mathbb{R}^n$ denote by $qh_A(B)$ the union $\bigcup_{\omega \in A} qh_\omega(B)$.

For $A \subset \mathbb{R}^n$ and $\rho > 0$ denote by $\mathfrak{N}_\rho(A)$ the set $\{x \in \mathbb{R}^n | dist(x, A) < \rho\}$.

For $A, B \subset \mathbb{R}^n$ and $\lambda \in \mathbb{R}$ the sets $\{x + y \mid x \in A, y \in B\}$ and $\{\lambda x \mid x \in A\}$ are denoted, as usually, by $A + B$ and λA.

Let a be a Laurent polynomial in n variables, ε a positive number and Γ a face of the Newton polyhedron $\Delta = \Delta(a)$.

Proposition A.6.6 *If in open set $U \subset K\mathbb{R}^n$ the truncation a^Γ is ε-sufficient for a, then it is ε-sufficient for a in $qh_{\mathcal{C}\ell\, DC_\Delta^-(\Gamma)}(U)$.*[5]

[5]Here (as above) $\mathcal{C}\ell$ denotes the closure.

Proof Let $\omega \in \mathcal{C}\ell\ DC^-_\Delta(\Gamma)$ and $\omega w = \delta$ for $w \in \Gamma$. By Proposition A.6.5, ε-sufficiency of truncation a^Γ for a in $qh_\omega(U)$ is equivalent to ε-sufficiency of truncation $(a \circ qh_\omega)^\Gamma$ for $(a \circ qh_\omega)^\Gamma$ in U or, equivalently, to ε-sufficiency of Γ-truncation of Laurent polynomial $b = e^{-\delta} a \circ qh_\omega$ in U. Since

$$e^{-\delta} a \circ qh_\omega(x) = \sum_{w \in \Delta} e^{-\delta} a_w x^w e^{\omega w} = a^\Gamma(x) + \sum_{w \in \Delta \setminus \Gamma} e^{\omega w - \delta} a_w x^w$$

and $\omega w - \delta \leq 0$ when $w \in \Delta \setminus \Gamma$ and $\omega \in \mathcal{C}\ell\ DC^-_\Delta(\Gamma)$, it follows that b satisfies the conditions $\Delta(b) = \Delta, b^\Gamma = a^\Gamma$ and $||b - b^\Gamma|| \leq ||a - a^\Gamma||$. Therefore the truncation b^Γ is ε-sufficient for b in U and, hence, the truncation a^Γ is ε-sufficient for a in $qh_\omega(U)$. From this, by Proposition A.6.1, the proposition follows.

Proposition A.6.7 *If the truncation a^Γ is completely nondegenerate and la $V_{K\mathbb{R}^n}(a^\Gamma)$ has a tubular neighborhood of the chosen type, then for any compact sets $C \subset K\mathbb{R}^n$ and $\Omega \subset DC^-_\Delta(\Gamma)$ there exists δ such that in $qh_{\delta\Omega}(C)$ the truncation a^Γ is ε-sufficient for a.*

Proof For $\omega \in DC^-_\Delta(\Gamma)$ denote by $\omega\Gamma$ a value taken by the scalar product ωw for $w \in \Gamma$. Since

$$t^{-\omega\Gamma} a \circ qh_{\omega,t}(x) = a^\Gamma(x) + \sum_{w \in \Delta \setminus \Gamma} t^{\omega w - \omega\Gamma} a_w x^w$$

for $\omega \in DC^-_\Delta(\Gamma)$ (cf. the previous proof) and $\omega w - \omega\Gamma < 0$ when $w \in \Delta \setminus \Gamma$ and $\omega \in DC^-_\Delta(\Gamma)$ it follows that the Laurent polynomial $b_{\omega,t} = t^{-\omega\Gamma} a \circ qh_{\omega,t}$ with $\omega \in DC^-_\Delta(\Gamma)$ turns to a^Γ as $\Gamma \to +\infty$. It is clear that this convergence is uniform with respect to ω on a compact set $\Omega \subset DC^-_\Delta(\Gamma)$. By Proposition A.6.4 it follows from this that for a compact set $U \subset K\mathbb{R}^n$ there exists η such that for any $\omega \in \Omega$ and $t \geq \eta$ the truncation $b^\Gamma_{\omega,t}$ of $b_{\omega,t}$ is ε-sufficient in U for $b_{\omega,t}$. By Proposition A.6.5, the latter is equivalent to ε-sufficiency of truncation a^Γ for a in $qh_{\omega,t}(U)$.

Thus if U is the closure of a bounded neighborhood W of a set C then there exists η such that for $\omega \in \Omega$ and $t \geq \eta$ the truncation a^Γ is ε-sufficient for a in $qh_{\omega,t}(U)$. Therefore a^Γ is the same in a smaller set $qh_{\omega,t}(W)$ and, hence, (by Proposition A.6.1) in the union $\bigcup_{t \geq \eta, \omega \in \Omega} qh_{\omega,t}(W)$ and, hence, in a smaller set $\bigcup_{t = \eta, \omega \in \Omega}(C)$. Putting $\delta = \ln \eta$ we obtain the required result.

Proposition A.6.8 *Let Γ is a face of another face Σ of the polyhedron Δ. Let Ω is a compact subset of the cone $DC^-_\Delta(\Sigma)$. If Γ-truncation a^Γ is ε-sufficient for a^Σ in a compact set C, then there exists a number δ such that a^Γ is ε-sufficient for a in $qh_{\delta\Omega}(C)$.*

This proposition is proved similarly to Proposition A.6.7, but with the following difference: the reference to Proposition A.6.4 is replaced by a reference to Proposition A.6.3. □

Proposition A.6.9 *Let $C \subset K\mathbb{R}^n$ be a compact set and let Γ be a face of Δ such that for any face Σ of Δ with* $\dim \Sigma = \dim \Delta - 1$ *having a face Γ the truncation a^Γ is ε-sufficient for a^Σ in C. Then there exists a real number δ such that the truncation a^Γ is ε-sufficient for a in* $qh_{\mathcal{C}\ell\, DC^-_\Delta(\Gamma) \smallsetminus \mathfrak{N}_\delta DC^-_\Delta(\Delta)}(\mathrm{Int}\, C)$.

Proof By A.6.8, for any face Σ of Δ with $\dim \Sigma = \dim \Delta - 1$ and $\Gamma \subset \partial\Sigma$ there exists a vector $\omega_\Sigma \in DC^-_\Delta(\Sigma)$ such that the truncation a^Γ is ε-sufficient for a in $qh_{\omega_\Sigma}(C)$, and, hence, by A.6.6, in $qh_{\omega_\Sigma + \mathcal{C}\ell\, DC^-_\Delta(\Gamma)}(\mathrm{Int}\, C)$. Choose such ω_Σ for each Σ with $\dim \Sigma = \dim \Delta - 1$ and $\Gamma \subset \partial\Sigma$. Obviously, the sets $\omega_\Sigma + \mathcal{C}\ell\, DC^-_\Delta(\Gamma)$ cover the whole closure of the cone $DC^-_\Delta(\Gamma)$ besides some neighborhood of its top, i.e. the cone $DC^-_\Delta(\Delta)$; in other words, there exists a number δ such that $\bigcup_\Sigma(\omega_\Sigma + \mathcal{C}\ell\, DC^-_\Delta(\Gamma)) \supset \mathcal{C}\ell\, DC^-_\Delta(\Gamma) \smallsetminus \mathfrak{N}_\delta DC^-_\Delta(\Delta)$. Hence, a^Γ is ε-sufficient for a in $qh_{\mathcal{C}\ell\, DC^-_\Delta(\Gamma) \smallsetminus \mathfrak{N}_\delta DC^-_\Delta(\Delta)}(\mathrm{Int}\, C) \subset \bigcup_\Sigma qh_{\omega_\Sigma + \mathcal{C}\ell\, DC^-_\Delta(\Gamma)}(\mathrm{Int}\, C)$.

A.6.3 *The Main Theorem on Logarithmic Asymptotes of Hypersurface*

Let $\Delta \subset \mathbb{R}^n$ be a convex closed polyhedron and $\varphi : \mathcal{G}(\Delta) \to \mathbb{R}$ be a positive function. Then for $\Gamma \in \mathcal{G}(\Delta)$ denote by $D_{\Delta,\varphi}(\Gamma)$ the set

$$\mathfrak{N}_{\varphi(\Gamma)}(DC^-_\Delta(\Gamma)) \smallsetminus \bigcup_{\Sigma \in \mathcal{G}(\Delta),\ \Gamma \in \mathcal{G}(\Sigma)} \mathfrak{N}_{\varphi(\Sigma)}(DC^-_\Delta(\Sigma)).$$

It is clear that the sets $D_{\Delta,\varphi}(\Gamma)$ with $\Gamma \in \mathcal{G}(\Delta)$ cover $\mathbb{R}^n$. Among these sets only sets corresponding to faces of the same dimension can intersect each other. In some cases (for example, if φ^Γ grows fast enough when $\dim \Gamma$ grows) they do not intersect and then $\{D_{\Delta,\varphi}(\Gamma)\}_{\Gamma \in \mathcal{G}(\Delta)}$ is a partition of $\mathbb{R}^n$.

Let a be a Laurent polynomial over K in n variables and ε be a positive number. A function $\varphi : \mathcal{G}(\Delta(a)) \to \mathbb{R}^n$ is said to be *describing domains of ε-sufficiency for* a (with respect to the chosen class of tubular neighborhoods) if for any proper face $\Gamma \in \mathcal{G}(\Delta(a))$, for which truncation a^Γ is completely non-degenerate and the hypersurface $la(V_{K\mathbb{R}^n}(a^\Gamma))$ has a tubular neighborhood of the chosen class, the truncation a^Γ is ε-sufficient for a in some neighborhood of $l^{-1}(DC_{\Delta(a),\varphi}(\Gamma))$.

Proposition A.6.10 *For any Laurent polynomial a over K in n variables and $\varepsilon > 0$ there exists a function $\mathcal{G}(\Delta(a)) \to \mathbb{R}$ describing domains of ε-sufficiency for a with respect to the chosen class of tubular neighborhoods.*

In particular, if a is peripherally nondegenerate Laurent polynomial over K in n variables and $\dim \Delta(a) = n$ then for any $\varepsilon > 0$ there exists a compact set $C \subset K\mathbb{R}^n$ such that $K\mathbb{R}^n \smallsetminus C$ is covered by regions in which truncations of $a^{\partial\Delta(a)}$ are ε-sufficient for a with respect to class of normal tubular neighborhoods. In other words, under these conditions behavior of $V_{K\mathbb{R}^n}(a)$ outside C is defined by monomials of $a^{\partial\Delta(a)}$.

A.6.4 Proof of Theorem

Proposition A.6.10 is proved by induction on dimension of polyhedron $\Delta(a)$.

If $\dim \Delta(a) = 0$ then a is monomial and $V_{K\mathbb{R}^n}(a) = \varnothing$. Thus for any $\varepsilon > 0$ any function $\varphi : \mathcal{G}(\Delta(a)) \to \mathbb{R}$ describes domains of ε-sufficiency for a.

Induction step follows obviously from the following Theorem.

Proposition A.6.11 *Let a be a Laurent polynomial over K in n variables, Δ be its Newton polyhedron, ε a positive number. If for a function $\varphi : \mathcal{G}(\Delta) \smallsetminus \{\Delta\} \to \mathbb{R}$ and any proper face Γ of Δ the restriction $\varphi|_{\mathcal{G}(\Gamma)}$ describes domains of ε-sufficiency for a^Γ, then φ can be extended to a function $\bar{\varphi} : \mathcal{G}(\Delta) \to \mathbb{R}$ describing regions of ε-sufficiency for a.*

Proof It is sufficient to prove that for any face $\Gamma \in \mathcal{G}(\Delta) \smallsetminus \{\Delta\}$, for which the truncation a^Γ is completely nondegenerate and hypersurface $V_{K\mathbb{R}^n}(a^\Gamma)$ has a tubular neighborhood of the chosen class, there exists an extension φ_Γ of φ such that truncation a^Γ is ε-sufficient for a in a neighborhood of $l^{-1}(D_{\Delta,\varphi_\Gamma}(\Gamma))$, i.e. to prove that for any face $\Gamma \neq \Delta$ there exists a number $\varphi_\Gamma(\Delta)$ such that the truncation a^Γ is ε-sufficient for a in some neighborhood of

$$l^{-1}(\mathfrak{N}_{\varphi(\Gamma)}(DC^-_\Delta(\Gamma)) \smallsetminus [\mathfrak{N}_{\varphi_\Gamma(\Delta)}(DC^-_\Delta(\Delta)) \bigcup_{\Sigma\in\mathcal{G}(\Delta)\smallsetminus\{\Delta\},\ \Gamma\in\mathcal{G}(\Sigma)} \mathfrak{N}_{\varphi(\Sigma)}(DC^-_\Delta(\Sigma))].$$

Indeed, putting

$$\bar{\varphi}(\Delta) = \max_{\Gamma\in\mathcal{G}(\Delta)\smallsetminus\{\Delta\}} \varphi_\Gamma(\Delta)$$

we obtain a required extension of φ.

First, consider the case of a face Γ with $dim\Gamma = dim\Delta - 1$. Apply Proposition A.6.7 to $C = l^{-1}(\mathcal{C}\ell\,\mathfrak{N}_{\varphi(\Gamma)+1}(0)$ and any one-point set $\Omega \subset DC^-_\Delta(\Gamma)$. It implies that a^Γ is ε-sufficient for a in $qh_\omega(C) = l^{-1}(Cl\mathfrak{N}_{\varphi(\Gamma)+1}(\omega))$ for some $\omega \in DC^-_\Delta(\Gamma)$. Now apply Proposition A.6.6 to $U = l^{-1}(\mathfrak{N}_{\varphi(\Gamma)+1}(\omega))$. It gives that a^Γ is ε-sufficient for a in $qh_{DC^-_\Delta(\Gamma)}(l^{-1}(\mathfrak{N}_{\varphi(\Gamma)+1}(\omega)) = l^{-1}(\mathfrak{N}_{\varphi(\Gamma)+1}(\omega + DC^-_\Delta(\Gamma)))$ and, hence, in the smaller set $l^{-1}(\mathfrak{N}_{\varphi(\Gamma)+1}(DC^-_\Delta(\Gamma))) \smallsetminus \mathfrak{N}_{|\omega|}(DC^-_\Delta(\Delta))$. It is remained to put $\varphi_\Gamma(\Delta) = |\omega| + 1$.

Now consider the case of face Γ with $\dim \Gamma < \dim \Delta - 1$. Denote by E the set

$$\mathfrak{N}_{\varphi(\Gamma)}(DC^-_\Delta(\Gamma)) \smallsetminus \bigcup_{\Sigma\in\mathcal{G}(\Delta)\smallsetminus\{\Delta\},\ \Gamma\in\mathcal{G}(\Sigma)} \mathfrak{N}_{\varphi(\Sigma)}(DC^-_\Delta(\Sigma)).$$

It is clear that there exists a ball $B \subset \mathbb{R}^n$ with center at 0 such that $E = (E \cap B) + \mathcal{C}\ell\ DC^-_\Delta(\Gamma)$. Denote the radius of this ball by β.

If $\Sigma \in \mathcal{G}(\Delta)$ is a face of dimension $\dim \Delta - 1$ with $\partial\Sigma \supset \Gamma$ then, by the hypothesis, the truncation a^Γ is ε-sufficient for a^Σ in some neighborhood of

$$l^{-1}(\mathfrak{N}_{\varphi(\Gamma)}(DC^{-}_{\Sigma}(\Gamma)) \smallsetminus \bigcup_{\Theta\in\mathcal{G}(\Sigma),\ \Gamma\in\mathcal{G}(\Theta)} \mathfrak{N}_{\varphi(\Theta)}(DC^{-}_{\Sigma}(\Theta))$$

and, hence, in neighborhood of a smaller set

$$l^{-1}(\mathfrak{N}_{\varphi(\Gamma)}(DC^{-}_{\Delta}(\Gamma)) \smallsetminus \bigcup_{\Theta\in\mathcal{G}(\Sigma),\ \Gamma\in\mathcal{G}(\Theta)} \mathfrak{N}_{\varphi(\Theta)}(DC^{-}_{\Delta}(\Theta)).$$

Therefore for any face Σ with $\dim\Sigma = \dim\Delta - 1$ and $\Gamma \subset \partial\Sigma$ the truncation a^{Γ} is ε-sufficient for a^{Σ} in some neighborhood of $l^{-1}(E)$. Denote by C a compact neighborhood of $l^{-1}(E \cap B)$ contained in this neighborhood. Applying Proposition A.6.6, one obtains that a^{Γ} is ε-sufficient for a in the set

$$qh_{\mathcal{C}\ell\, DC^{-}_{\Delta}(\Gamma)\smallsetminus\mathfrak{N}_{\delta}DC^{-}_{\Delta}(\Delta)}(\operatorname{Int} C) = l^{-1}(\operatorname{Int} l(C) + \mathcal{C}\ell\; DC^{-}_{\Delta}(\Gamma) \smallsetminus \mathfrak{N}_{\delta} DC^{-}_{\Delta}(\Delta))).$$

It is remained to put $\varphi_{\Gamma}(\Delta) = \delta + \beta$

A.6.5 Modification of Theorem 6.3.A

Below in Sect. A.6.8 it will be more convenient to use not Proposition A.6.10 but the following its modification, whose formulation is more cumbrous, and whose proof is obtained by an obvious modification of deduction of Proposition A.6.10 from Proposition A.6.11.

Proposition A.6.12 *For any Laurent polynomial a over K in n variables and any $\varepsilon > 0$ and $c > 1$ there exists a function $\varphi : \mathcal{G}(\Delta(a)) \to \mathbb{R}$ such that for any proper face $\Gamma \in \mathcal{G}(\Delta(a))$, for which a^{Γ} is completely nondegenerate and $la(V_{K\mathbb{R}^n}(a^{\Gamma}))$ has a tubular neighborhood from the chosen class, the truncation a^{Γ} is ε-sufficient for a in some neighborhood of*

$$l^{-1}(\mathfrak{N}_{c\varphi(\Gamma)}(DC^{-}_{\Delta}(\Gamma)) \smallsetminus \bigcup_{\Sigma\in\mathcal{G}(\Delta),\ \Gamma\in\mathcal{G}(\Sigma)} \mathfrak{N}_{\varphi(\Sigma)}(DC^{-}_{\Delta}(\Sigma)).$$

□

A.6.6 Charts of L-Polynomials

Let a be a peripherally nondegenerate Laurent polynomial over K in n variables, Δ be its Newton polyhedron. Let V be a vector subspace of $\mathbb{R}^n$ corresponding to the smallest affine subspace containing Δ (i.e. $V = C_{\Delta}(\Delta)$). Let $\varphi : \mathcal{G}(\Delta) \to \mathbb{R}$

be the function, existing by Proposition A.6.10, describing for some ε regions of ε-sufficiency for a with respect to class of normal tubular neighborhoods.

The pair $(\Delta \times U_K^n,\ \upsilon)$ consisting of the product $\Delta \times U_K^n$ and its subset υ is a K-*chart* of a Laurent polynomial a if:

1. there exists a homeomorphism $h : (\mathcal{C}\ell\ D_{\Delta,\varphi}(\Delta) \cap V) \times U_K^n \to \Delta \times U_K^n$ such that $h((Cl\ \ D_{\Delta,\varphi}(\Delta) \cap V) \times y) = \Delta \times y$ for $y \in U_K^n$,
$$\upsilon = h(la V_{K\mathbb{R}^n}(a) \cap (\mathcal{C}\ell\ D_{\Delta,\varphi}(\Delta) \cap V) \times U_K^n$$
and for each face Γ of Δ the set $h((\mathcal{C}\ell\ D_{\Delta,\varphi}(\Delta) \cap D_{\Delta,\varphi}(\Gamma) \cap V) \times U_K^n)$ lies in the product of the star $\bigcup_{\Gamma \in \mathcal{G}(\Sigma) \Sigma \in \mathcal{G}(\Delta) \smallsetminus \{\Delta\}} \Sigma$ of Γ to U_K^n;
2. for any vector $\omega \in \mathbb{R}^n$, which is orthogonal to V and, in the case of $K = \mathbb{R}$, is integer, the set υ is invariant under transformation $\Delta \times U_K^n \to \Delta \times U_K^n$ defined by formula $(x,\ (y_1, \dots, y_n)) \mapsto (x,\ (e^{\pi i \omega_1} y_1,\ \dots,\ e^{\pi i \omega_n} y_n))$;
3. for each face Γ of Δ the pair $(\Gamma \times U_K^n,\ \upsilon \cap (\Gamma \times U_K^n))$ is a K-chart of Laurent polynomial a^Γ.

The definition of the chart of a Laurent polynomial, which, as I believe, is clearer than the description given here, but based on the notion of toric completion of $K\mathbb{R}^n$, is given above in Sect. A.3.3. I restrict myself to the following commentary of conditions 1–3.

The set $(\mathcal{C}\ell\ D_{\Delta,\varphi}(\Delta) \cap V) \times U_K^n$ contains, by Proposition A.6.10, a deformation retract of $la V_{K\mathbb{R}^n}(a)$. Thus, condition 1 means that υ is homeomorphic to a deformation retract of $V_{K\mathbb{R}^n}(a)$. The position of υ in $\Delta \times U_K^n$ contains, by 1 and 3, a complete topological information about behavior of this hypersurface outside some compact set. The meaning of 2 is in that υ has the same symmetries as, according to Proposition A.2.3, $V_{K\mathbb{R}^n}(a)$ has.

A.6.7 Structure of $V_{K\mathbb{R}^n}(b_t)$ with Small t

Denote by i_t the embedding $K\mathbb{R}^n \to K\mathbb{R}^{n+1}$ defined by $i_t(x_1, \dots, x_n) = (x_1, \dots, x_n,\ t)$. Obviously,

$$V_{K\mathbb{R}^n}(b_t) = i_t^{-1} V_{K\mathbb{R}^{n+1}}(b).$$

This allows to take advantage of results of the previous section for study of $V_{K\mathbb{R}^n}(b_t)$ as $t \to 0$. For sufficiently small t the image of embedding i_t is covered by regions of ε-sufficiency of truncation $b^{\tilde{\Gamma}}$, where $\tilde{\Gamma}$ runs over the set of faces of graph of ν, and therefore the hypersurface $V_{K\mathbb{R}^n}(b_t)$ turns to be composed of pieces obtained from $V_{K\mathbb{R}^n}(a_i)$ by appropriate quasi-homotheties.

I preface the formulation describing in detail the behavior of $V_{K\mathbb{R}^n}(b_t)$ with several notations.

Denote the Newton polyhedron $\Delta(b)$ of Laurent polynomial b by $\tilde{\Delta}$. It is clear that $\tilde{\Delta}$ is the convex hull of the graph of ν. Denote by $\mathcal{G}$ the union $\bigcup_{i=1}^{s} \mathcal{G}(\Delta_i)$. For $\Gamma \in \mathcal{G}$ denote by $\tilde{\Gamma}$ the graph $\nu|_{\Gamma}$. It is clear that $\tilde{\Gamma} \in \mathcal{G}(\tilde{\Delta})$ and hence an injection $\Gamma \to \tilde{\Gamma} : \mathcal{G} \to \mathcal{G}(\tilde{\Delta})$ is defined.

For $t > 0$ denote by j_t the embedding $\mathbb{R}^n \to \mathbb{R}^{n+1}$ defined by the formula $j_t(x_1, \dots, x_n) = (x_1, \dots, x_n, \ln t)$. Let $\psi : \mathcal{G} \to \mathbb{R}$ be a positive function, t be a number from interval $(0, 1)$. For $\Gamma \in \mathcal{G}$ denote by $\mathcal{E}_{t,\psi}(\Gamma)$ the following subset of $\mathbb{R}^n$:

$$\mathfrak{N}_{\psi(\Gamma)} j_t^{-1}(DC_{\tilde{\Delta}}^{-}(\tilde{\Gamma})) \setminus \bigcup_{\Sigma \in \mathcal{G},\ \Gamma \in \mathcal{G}(\Sigma)} \mathfrak{N}_{\varphi(\Sigma)} j_t^{-1}(DC_{\tilde{\Delta}}^{-}(\tilde{\Sigma})).$$

Proposition A.6.13 *If Laurent polynomials $a_1, \dots, a_s$ are completely non-degenerate then for any $\varepsilon > 0$ there exist $t_0 \in (0, 1)$ and function $\psi : \mathcal{G} \to \mathbb{R}$ such that for any $t \in (0, t_0]$ and any face $\Gamma \in \mathcal{G}$ truncation b_t^{Γ} is ε-sufficient for b_t with respect to the class of normal tubular neighborhoods in some neighborhood of $l^{-1}(\mathcal{E}_{t,\psi}(\Gamma))$.*

Denote the gradient of restriction of ν on $\Gamma \in \mathcal{G}$ by $\nabla(\Gamma)$. The truncation b_t^{Γ}, obviously, equals $a^{\Gamma} \circ qh_{\nabla(\Gamma),t}$. In particular, $b_t^{\Delta_i} = a_i \circ qh_{\nabla(\Delta_i),t}$ and, hence,

$$V_{K\mathbb{R}^n}(b_t^{\Delta_i}) = qh_{\nabla(\Delta_i),t^{-1}}(V_{K\mathbb{R}^n}(a_i)).$$

In the domain, where b_t^{Γ} is ε-sufficient for b_t, the hypersurfaces $laV_{K\mathbb{R}^n}(b_t)$ and $laV_{K\mathbb{R}^n}(b_t^{\Delta_i})$ with $\Delta_i \supset \Gamma$ lie in the same normal tubular ε-neighborhood of $laV_{K\mathbb{R}^n}(b_t^{\Gamma})$ and, hence, are isotopic by an isotopy moving points at most on 2ε. Thus, according to Proposition A.6.13, for $t \le t_0$ to the space $K\mathbb{R}^n$ is covered by regions in which $V_{K\mathbb{R}^n}(b_t)$ is approximated by $qh_{\nabla(\Delta_i),t^{-1}}(V_{K\mathbb{R}^n}(a_i))$.

A.6.8 Proof of Theorem

Put $c = \max\{\sqrt{1 + \nabla(\Delta_i)^2},\ i = 1, \dots, s\}$. Apply Proposition A.6.12 to the Laurent polynomial b and numbers ε and c, considering as the class of chosen tubular neighborhoods in $\mathbb{R}^{n+1} \times U_K^{n+1}$ tubular neighborhoods, whose fibers lie in the fibers $\mathbb{R}^n \times t \times U_K^n \times s$ of $\mathbb{R}^{n+1} \times U_K^{n+1}$ and consist of segments of geodesics which are orthogonal to intersections of submanifold with $\mathbb{R}^n \times t \times U_K^n \times s$. (Intersection of such a tubular neighborhood of $M \subset \mathbb{R}^{n+1} \times U_K^{n+1}$ with the fiber $\mathbb{R}^n \times t \times U_K^n \times s$ is a normal tubular neighborhood of $M \cap (\mathbb{R}^n \times t \times U_K^n \times s)$ in $\mathbb{R}^n \times t \times U_K^n \times s$.) Applying Proposition A.6.12 one obtains a function $\varphi : \mathcal{G}(\tilde{\Delta}) \to \mathbb{R}$. Denote by ψ the function $\mathcal{G} \to \mathbb{R}$ which is the composition of embedding $\Gamma \mapsto \tilde{\Gamma} : \mathcal{G} \to \mathcal{G}(\tilde{\Delta})$ (see Sect. A.6.7) and the function $\frac{1}{c}\varphi : \mathcal{G}(\tilde{\Delta}) \to \mathbb{R}$. This function has the required property. Indeed, as it is easy to see, for $0 < t < e^{-\varphi(\tilde{\Delta})}$ $\mathcal{E}_{t,\psi}(\Gamma)$ is contained, in

$$j_t^{-1}(\mathfrak{N}_{c\varphi(\tilde{\Gamma})}(DC^-_{\tilde{\Delta}}(\tilde{\Gamma})) \smallsetminus \bigcup_{\tilde{\Sigma}\in\mathcal{G}(\tilde{\Delta}),\ \tilde{\Gamma}\in\mathcal{G}(\tilde{\Sigma})} \mathfrak{N}_{\varphi(\tilde{\Sigma})}(DC^-_{\tilde{\Delta}}(\tilde{\Sigma}))),$$

and thus from ε-sufficiency of $b^{\tilde{\Gamma}}$ for b in some neighborhood of

$$l^{-1}(\mathfrak{N}_{c\varphi(\tilde{\Gamma})}(DC^-_{\tilde{\Delta}}(\tilde{\Gamma})) \smallsetminus \bigcup_{\tilde{\Sigma}\in\mathcal{G}(\tilde{\Delta}),\ \tilde{\Gamma}\in\mathcal{G}(\tilde{\Sigma})} \mathfrak{N}_{\varphi(\tilde{\Sigma})}(DC^-_{\tilde{\Delta}}(\tilde{\Sigma}))),$$

with respect to the chosen class of tubular neighborhoods in $\mathbb{R}^{n+1} \times U_K^{n+1}$ if follows that for $0 < t < e^{-\varphi(\tilde{\Delta})}$ the truncation b_t^Γ is ε-sufficient for b_t in some neighborhood of $l^{-1}(\mathcal{E}_{t,\psi}(\Gamma))$ with respect to the class of normal tubular neighborhoods. □

A.6.9 An Alternative Proof of Theorem

Let V be a vector subspace of $\mathbb{R}^n$ corresponding to the minimal affine subspace containing Δ. It is divided for each $t \in (0, 1)$ onto the sets $V \cap j_t^{-1}(DC^-_{\tilde{\Delta}}(\tilde{\Gamma}))$ with $\Gamma \in \mathcal{G}$. Let us construct cells Γ_t in V which are dual to the sets of this partition (barycentric stars). For this mark a point in each $V \cap j_t^{-1}(DC^-_{\tilde{\Delta}}(\tilde{\Gamma}))$:

$$b_{t,\Gamma} \in V \cap j_t^{-1}(DC^-_{\tilde{\Delta}}(\tilde{\Gamma})).$$

Then for Γ with $\dim \Gamma = 0$ put $\Gamma_t = b_{t,\Gamma}$ and construct the others Γ_t inductively on dimension $\dim \Gamma$: if Γ_t for Γ with $\dim \Gamma < r$ have been constructed then for Γ with $\dim \Gamma = r$ the cell Γ_t is the (open) cone on $\bigcup_{\Sigma\in\mathcal{G}(\Gamma)\smallsetminus\{\Gamma\}} \Sigma_t$ with the vertex $b_{t,\Gamma}$. (This is the usual construction of dual partition turning in the case of triangulation to partition onto barycentric stars of simplices.)

By Proposition A.6.13 there exist $t_0' \in (0, 1)$ and function $\psi : \mathcal{G} \to \mathbb{R}$ such that for any $t \in (0, t_0']$ and any face $\Gamma \in \mathcal{G}$ the truncation b_t^Γ is ε-sufficient for b_t in some neighborhood of $l^{-1}(\mathcal{E}_{t,\psi}(\Gamma))$. Since cells Γ_t grow unboundedly when t runs to zero (if $\dim \Gamma \neq 0$) it follows that there exists $t_0 \in (0, t_0']$ such that for $t \in (0, t_0]$ for each face $\Gamma \in \mathcal{G}$ the set $\mathfrak{N}_{\psi(\Gamma)} j_t^{-1}(DC^-_{\tilde{\Delta}}(\Gamma))$, and together with it the set $\mathcal{E}_{t,\psi}(\Gamma)$, lie in the star of the cell Γ_t, i.e. in $\bigcup_{\Gamma\in\mathcal{G}(\Sigma)} \Sigma_t$. Let us show that for such t_0 the conclusion of Proposition A.4.1 takes place.

Indeed, it follows from Proposition A.6.13 that there exists a homeomorphism $h : \Gamma_t \times U_K^n \to \Gamma \times U_K^n$ with $h(\Gamma_t \times y) = \Gamma \times y$ for $y \in U_K^n$ such that $(\Gamma \times U_K^n, h(la(V_{K\mathbb{R}^n}(b_t)) \cap \Gamma_t \times U_K^n))$ is K-chart of Laurent polynomial a^Γ. Therefore the pair

$$(\cup_{\Gamma\in\mathcal{G}} \Gamma_t \times U_K^n,\ la V_{K\mathbb{R}^n}(b_t) \cap (\cup_{\Gamma\in\mathcal{G}} \Gamma_t \times U_K^n))$$

is obtained in result of patchworking K-charts of Laurent polynomials $a_1, \dots, a_s$. The function $\varphi : \mathcal{G}(\Delta) \to \mathbb{R}$, existing by Proposition A.6.10 applied to b_t, can be chosen, as it follows from A.6.11, in such a way that it should majorate any given in advance function $\mathcal{G}(\Delta) \to \mathbb{R}$. Choose φ in such a way that $D_{\varphi,\Delta}(\Delta) \supset \cup_{\Gamma\in\mathcal{G}}\Gamma_t$ and $D_{\varphi,\Delta}(\Sigma) \cap \partial D_{\varphi,\Delta}(\Delta) \supset \mathcal{E}_{t,\psi}(\Sigma) \cap \partial D_{\varphi,\Delta}(\Delta)$ for $\Sigma \in \mathcal{G}(\Delta) \smallsetminus \{\Delta\}$. As it follows from Proposition A.6.13, there exists a homeomorphism

$$(\bigcup_{\Gamma\in\mathcal{G}} \Gamma_t \times U_K^n,\ la V_{K\mathbb{R}^n}(b_t) \cap (\bigcup_{\Gamma\in\mathcal{G}} \Gamma_t \times U_K^n)) \to$$

$$(D_{\varphi,\Delta}(\Delta) \times U_K^n,\ la V_{K\mathbb{R}^n}(b_t) \cap (D_{\varphi,\Delta}(\Delta) \times U_K^n)) \qquad \text{(A.1)}$$

turning $\mathcal{E}_{t,\psi}(\Sigma) \cap \partial(\bigcup_{\Gamma\in\mathcal{G}} \Gamma_t \times U_K^n)$ to $\mathcal{E}_{t,\psi}(\Sigma) \cap \partial D_{\varphi,\Delta}(\Delta)$ for $\Sigma \in \mathcal{G}(\Delta) \smallsetminus \{\Delta\}$. Therefore K-chart of Laurent polynomial b_t is obtained by patchworking K-charts of Laurent polynomials $a_1, \dots, a_s$. □

References

[Ati81] Atiyah, M.F.: Convexity and commuting hamiltonians. Bull. Lond. Math. Soc. **14**, 1–15 (1981)

[AVGZ82] Arnold, V.I., Varchenko, A.N., Gusein-Zade, S.M.: Singularities of differentiable maps. I, "Nauka", Moscow, 1982 (Russian), English transl, Birkhaüser, Basel (1985)

[Dan78] Danilov, V.I.: The geometry of toric manifolds. Uspekhi Mat. Nauk **33** (1978), 85–134 (Russian), English transl., Russian Math. Surv. **33:2** (1978)

[GU69] Gudkov, D.A., Utkin, G.A.: The topology of curves of degree 6 and surfaces of degree 4. Uchen. Zap. Gorkov. Univ., vol. 87, 1969 (Russian), English transl., Transl. AMS **112**

[Gud73] Gudkov, D.A.: Construction of a curve of degree 6 of type $\frac{5}{1}5$. Izv. Vyssh. Uchebn. Zaved. Mat. **3**, 28–36 (1973). (Russian)

[Gud71] Gudkov, D.A.: Construction of a new series of m-curves. Dokl. Akad. Nauk SSSR **200**, 1269–1272 (1971). (Russian)

[GZ77] Gusein-Zade, S.M.: Monodromy groups of isolated singularities of hypersurfaces, Uspekhi Mat. Nauk **32** (1977), 23–65 (Russian), English transl., Russian Math. Surv. **32** (1977)

[Har76] Harnack, A.: über vieltheiligkeit der ebenen algebraischen curven. Math. Ann. **10**, 189–199 (1876)

[Hil91] Hilbert, D.: über die reellen züge algebraischen curven. Math. Ann. **38**, 115–138 (1891)

[Hil01] Hilbert, D.: Mathematische Probleme. Arch. Math. Phys. **3**, 213–237 (1901). (German)

[Kho77] Khovanskiĭ, A.G.: Newton polygons and toric manifolds. Funktsional. Anal. i Prilozhen. **11**, 56–67 (1977). (Russian)

[Loj64] Lojasiewicz, S.: Triangulation of semi-analitic sets. Annali Scu. Norm. Sup. Pisa, Sc. Fis. Mat. Ser. 3, 18:4, 449–474(1964) (English)

[Mil68] Milnor, J.: Singular Points of Complex Hypersurfaces. Princeton University Press, Princeton (1968)

[GK73] Mumford, D., Saint-Donat, B., Kempf, G., Knudsen, F.: Toroidal Embeddings i, vol. 339 (1973)

[New67] Newton, I.: The method of fluxions and infinite series with application to the geometry of curves. In: The Mathematical Papers of Isaac Newton. Cambridge University Press, Cambridge (1967)

[Vir79a] Viro, O.Ya.: Constructing M-surfaces. Funktsional. Anal. i Prilozhen. **13** (1979), 71–72 (Russian), English transl., Functional Anal. Appl

[Vir79b] Viro, O.Ya., Constructing multicomponent real algebraic surfaces. Doklady AN SSSR **248** 279–282 (Russian), English transl. in Soviet Math. Dokl. **20**(1979), 991–995

[Vir80] Viro, O.Ya.: Curves of degree 7, curves of degree 8 and the Ragsdale conjecture. Dokl. Akad. Nauk SSSR **254**, 1305–1310 (1980) (Russian)

[Vir83a] Viro, O.Ya.: Plane real algebraic curves of degrees 7 and 8: new prohibitions. Izv. Akad. Nauk, Ser. Mat. **47** (1983), 1135–1150 (Russian), English transl., in Soviet Math. Izvestia

[Vir83b] Viro, O.Ya.: Gluing algebraic hypersurfaces and constructions of curves. Tezisy Leningradskoj Mezhdunarodnoj Topologicheskoj Konferentsii: Nauka. Nauka, 1983, 149–197 (1982) (Russian)

[Vir89] Viro, O.Ya.: Plane real algebraic curves: constructions with controlled topology. Algebra i analiz 1 (1989), 1–73 (Russian). English translation in Leningrad Math. J. **1**(5), 1059–1134 (1990)

Index

G.-M. Greuel et al., *Singular Algebraic Curves*, Springer Monographs in Mathematics, https://doi.org/10.1007/978-3-030-03350-7

MIX
Papier aus verantwortungsvollen Quellen
Paper from responsible sources
FSC® C105338

FSC
www.fsc.org

If you have any concerns about our products,
you can contact us on
ProductSafety@springernature.com

In case Publisher is established outside the EU,
the EU authorized representative is:
Springer Nature Customer Service Center GmbH
Europaplatz 3, 69115 Heidelberg, Germany

Printed by Libri Plureos GmbH
in Hamburg, Germany